英汉·汉英
新能源技术词典

杜振华　主编

English-Chinese & Chinese-English
Dictionary of New Energy Technologies

中国电力出版社
CHINA ELECTRIC POWER PRESS

内 容 提 要

本词典主要收录了风力发电、太阳能光伏发电、太阳能热发电、地热发电、垃圾发电、生物质发电和潮汐发电等方面的英汉、汉英词汇4万余条，内容涵盖了设计、施工、管理、制造、运行等领域。

本词典可供上述专业的工程设计、施工、管理、科研机构的工程技术人员、商务人员和相关专业的大专院校师生参考使用。

图书在版编目（CIP）数据

英汉·汉英新能源技术词典/杜振华主编. —北京：中国电力出版社，2019.3
ISBN 978-7-5198-2289-7

Ⅰ. ①英… Ⅱ. ①杜… Ⅲ. ①新能源–词典–英、汉 Ⅳ. ①TK01-61

中国版本图书馆 CIP 数据核字（2018）第 175081 号

出版发行：	中国电力出版社
地　　址：	北京市东城区北京站西街 19 号（邮政编码 100005）
网　　址：	http://www.cepp.sgcc.com.cn
责任编辑：	杨伟国（010-63412366） 安小丹　孙建英
责任校对：	黄　蓓　朱丽芳　闫秀英
装帧设计：	张俊霞
责任印制：	吴　迪

印　　刷：	三河市万龙印装有限公司
版　　次：	2019 年 5 月第一版
印　　次：	2019 年 5 月北京第一次印刷
开　　本：	850 毫米×1168 毫米　32 开本
印　　张：	24.25
字　　数：	1026 千字
印　　数：	0001—1000 册
定　　价：	120.00 元

版 权 专 有　侵 权 必 究

本书如有印装质量问题，我社营销中心负责退换

《英汉·汉英新能源技术词典》编辑委员会

主　编　杜振华

副主编　谢秋野　贺　莺　李朝渊　侯　希　李晓珊
　　　　　杨兆玺　毛瑞涛　武俊旭

审　定　吴国强　李亚舒　李　立　李广泽　张炳成
　　　　　陈卫国　杜　一　张　旭

编　委（按拼音顺序排序）

　　　　　曹晶晶　陈爱萍　程　静　程凤彩　崔粉绒
　　　　　杜　盼　杜洋洋　樊树斌　胡敏娜　黄姝梦
　　　　　江　晋　李　颜　李慧晓　李向月　李秀井
　　　　　李雪颖　廉　洁　梁博诚　刘帅楠　刘娱婷
　　　　　卢　琳　罗　琪　马　爽　曲　艺　孙　倩
　　　　　王　磊　谢雯霞　谢宇辉　许盼盼　杨璐璐
　　　　　姚　杰　袁修平　岳　强　张　晗　张　静
　　　　　张瑞瑞　张晓田　赵　倩　周　君

前　　言

由于不可再生能源的日趋枯竭，以及化石能源的利用导致环境与全球气候变暖等问题日益严峻，新能源的开发和利用已引起人们的高度关注。很多新技术、新工艺、新理念、新设备、新的研究方法在新能源领域得以广泛地实践和应用。我国在引进新技术的同时，也对外出口我们的新能源技术、设备和产品，甚至整个工程项目，这些都会用到英文资料。为了促进我国新能源行业与世界接轨，帮助新能源从业人员更好地消化、吸收外来技术和输出我们的技术，很有必要出版一本收词全面、准确、实用的新能源词典。经过全体编辑4年多的辛勤工作和努力，我们编撰了《英汉·汉英新能源技术词典》。

本词典主要收录风力发电、太阳能光伏发电、太阳能热发电、地热发电、垃圾发电、生物质发电和潮汐发电等方面的英汉、汉英词汇，在词汇收集过程中，我们主要参考了国外新能源工程设计手册、施工手册、运行手册、产品说明书等以及国内外出版的词典工具书。我们将选取的大量的词汇进行筛录、翻译、校对、确认，对有些有争议的词汇我们也反复讨论、查阅资料考证，最后经专业人员的审校，以期形成一本全面、准确、实用的认可度比较高的词典。

本词典以英汉、汉英双语对照的形式编排，便于使用者双向翻译查用，收录英汉、汉英词汇各2万余条，共计收词4万余条。另外，本词典还收录了电力行业国内外常用单位名称、国内外常用技术标准、电力工程常用英语缩略词、电力工程常用计量单位等。

　　本词典是在西安外国语大学高级翻译学院的领导下，中国电力工程顾问集团有限公司西北电力设计院的支持下完成的，参加本词典编辑的还有来自中国电力工程顾问集团有限公司、西安建筑科技大学、西安工业大学、西安工程大学、西安欧亚学院等的人员。

　　由于本词典收录的词汇多、数量大，给编写工作带来很大的困难，尽管做了最大努力，难免还有疏漏差错，恳请各位专家、学者和广大读者批评指正，以便再版修订。我们衷心欢迎各位读者对我们的工作提出指导和改进意见，请联系 823308853@qq.com。

编　者

2019 年 4 月

编辑体例

一、选词原则和范围

（1）风力发电、太阳能光伏发电、太阳能热发电、地热发电、垃圾发电、生物质发电和潮汐发电等方面的英汉、汉英词汇。

（2）收词以专业名词为主，收入少量日常词汇、形容词和动词。

（3）收集相关专业的新词。

（4）词意以相关专业的意义为主。与其他字典定义所给出的全部内容比较，只是使用了它们的某些特定或部分含义。

（5）词汇主要收集于国外的工程设计手册、施工手册、运行手册、产品说明书，以及参考了国内外相关专业词典。

二、编排和符号说明

（1）英汉词汇按照英文字母顺序编排。

（2）汉英词汇按照汉语拼音字母顺序编排。

（3）同一词目几个译称之间用逗号","隔开。例如：

absorbing wavemaker 吸收造波机，减震造波机

（4）圆括号"（）"的用法。

1）表示可省略的词，例如：

recording anemometer （自动）记录风速计

2）用来替换前边的词，例如：

rotating vane 转叶（旋转叶片）

3）用来解释词义，例如：

reference device 基准器件（是一种以标准太阳光谱分布为依据，用来测量辐照度或校准太阳模拟器辐射度的光伏器件。）

4）用来表示缩略语，例如：

radio inference (RI) 无线电干扰

Public Utility Commission (PUC) 公共事业委员会

（5）分隔号"/"的用法。

按照使用习惯，对某些缩略语或词组进行分割，如：

analog-to-digital（A/D）converter　模拟/数字转换器

as received basis/wet basis　收到基/湿基

（6）关于连字符"-"。

在本词典中，有许多动词词组或名词词组往往以三种形式出现，如 cross flow、cross-flow 和 crossflow。这些词也出现在各种技术文献、书籍、技术图纸或不同的词典中，其意义没有实质性的变化，有时候既可作名词也可作动词。对于连字符符号的使用，英语中也没有严格的限制，主要是作者或公司的一些习惯使然。还有一种解释是，有些新词在开始使用时用连字符，使用时间长了，大家认可了，连字符也就取消了。

目　录

前言
编辑体例
英汉部分……………………………………………………………………… 1
汉英部分……………………………………………………………………… 327
附录1　电力工程常用缩略语 ……………………………………………… 630
附录2　电力工程常用计量单位 …………………………………………… 649
附录3　电力工程常见的国外、国内组织（机构）或公司名称 …………… 656
附录4　电力工程常用的国外和国内标准 ………………………………… 674
附录5　发电工程招标投标文件中技术图纸目录（样例）………………… 738
附录6　国家或地区名称与货币名称、代码 ………………………………… 741
附录7　中国国家、省（市）政府部门名称汉英对照 ……………………… 752
附录8　主要参考文献 ……………………………………………………… 764

英汉部分

英文粹分

3D seismic surveys 三维地震探测
3D seismic tomography 3维地震层析成像
Ⅰ-Ⅴ characteristic curve of a solar cell 太阳电池的伏一安特性曲线
Ⅱ-Ⅵ group solar cell Ⅱ-Ⅵ族太阳电池
Ⅲ-Ⅴ group solar cell Ⅲ-Ⅴ族太阳电池

A

abatement of wind 风力减弱
abatement system 消除有害影响的系统
abatis 通风隔墙,风挡
abatis=abattis 障碍物
abbertite 黑沥青
Abbot silver disk pyrheliometer 艾伯特银盘式日射强度计
Abbot water pyrheliometer 艾伯特水流式日射强度计
ability to penetrate 渗透能力,穿透能力
ability to store heat 储热能力
ablating material 消融物质
ablative insulating quality 隔热性
abnormal temperature rise 不正常温升
abnormal wave 异常波型
abnormally high-pressure formation 异常高压地层
abnormally pressured formation 异常压力地层,反常压力地层
above sea level 海拔高度
above-critical state 超临界状态
above-ground large-scale storage unit 地面大型储存装置
aboveground pumped hydrostorage 地面抽水储能
abrasion 磨损,磨蚀,磨耗,剥蚀
abrasion (abraded) platform 浪蚀台地,浪成台地
abrasion inspection 磨损检查
abrasion resistance 耐磨性,抗磨损性
abrasion-resistant 耐磨的,抗磨的
abrasive disc 磨料盘
abrasive formation 腐蚀性岩层,研磨性岩层
abrasive index 磨损指数
abrupt junction 突变结
abscissa axis 横坐标
absolute (temperature) scale 绝对温标
absolute alcohol 无水酒精,无水乙醇
absolute black body 绝对黑体
absolute cavity radiometer 腔体式绝对辐射表
absolute encoder 绝对式编码器
absolute humidity 绝对湿度
absolute pyrheliometer 绝对日射测量计,绝对日温计
absolute pyrheliometer measurement 绝对日射测量
absolute radiometer 绝对辐射表
absolute spectral response 绝对光谱响应
absolute spectral sensitivity 绝对光谱灵敏度
absolute zero 绝对零值,绝对零度
absolute-spectral-sensitivity characteristic 绝对光谱灵敏度特性
absorb 吸收
absorbability 吸附力

absorbance index 吸收指数
absorbed dose 吸收剂量
absorbed energy 吸收能
absorbed fraction 吸收份额
absorbed power 被吸收功率，吸收能量
absorbed radiation 吸收辐射
absorbed solar radiation 吸收的太阳能辐射
absorbed striking energy 冲击吸收能量
absorbent bed 吸收床
absorbent carbon 活性炭
absorber 吸收器，吸收塔，吸热体
absorber area of concentrating collector 聚光型集热器的吸热体面积
absorber boom 吸热器悬管
absorber panel 吸热板
absorber plate 太阳能吸热片，板状吸收体
absorber solar absorptance 吸收器的太阳能吸收率
absorber solar emittance 吸收器的太阳能发射率
absorber with air flow 空气流动吸热器
absorbing beach 消波岸坡
absorbing wavemaker 吸收造波机，减震造波机
absorptance 吸收系数，吸收比，吸收能力，吸收率
absorption 合并，吸附，吸收（作用），吸水（作用）
absorption by glass cover 玻璃顶盖吸热
absorption chiller 吸收式冷水机，吸收式制冷机
absorption coefficient 吸收系数，吸收强度
absorption constant 吸收常数
absorption cooling 吸收冷却，吸收式制冷

absorption cooling system 吸收式制冷系统
absorption cross section 吸收截面
absorption current 吸收电流
absorption curve 吸收曲线
absorption cycle heat pump solar cooling system 吸收循环热泵太阳能制冷系统
absorption factor 吸收系数，吸收因子
absorption heat 吸热
absorption heat pump 吸热泵
absorption of radiant energy 辐射能吸收
absorption of the photons 光吸收
absorption ratio 吸收比
absorption reaction 吸收反应
absorption reaction rate 吸收反应率
absorption refrigerating machine 吸收式制冷机
absorption refrigeration 吸收式制冷
absorption refrigeration cycle 吸收制冷循环
absorption refrigeration system 吸收式制冷系统
absorption resonance 吸收共振
absorption spectrometry 吸收光谱测定
absorption spectroscopy 吸收光谱法
absorption spectrum 吸收（光）谱
absorption width （波能）吸收宽度
absorption-desorption cooling unit 吸热解吸制冷装置
absorptive character 吸收特性
absorptive interior wall 吸收内墙
absorptivity 吸收率，吸热率
abstracted heat 放出热，散失热
abyssal deposit 深海沉积
abyssal sea 深海
abyssal sediment 深海沉积物
AC (alternating current) load 交流负载

AC breaker panel 交流断路器，交流断流板
AC excitation current 交流励磁电流
AC excitation system 交流励磁系统
AC frequency converter 交流变频器
AC motor 交流电动机
AC rectifier 交流整流器
AC servo motor 交流伺服电机
AC-adjustable speed drive 交流变速传动
accelerated exposure testing 加速寿命测试
accelerated outdoor exposure testing 户外加速寿命试验
accelerated test 加速试验
accelerated wear 加速磨耗
acceleration amplitude 加速度幅值
acceleration sensor 加速度传感器
acceleration transducer 加速度传感器
accelerometer 加速计
access door 检修门，通道门，视口门，入孔门
accessory system 辅助系统
accidental consumption 事故消耗，意外消耗
accompanying fluid 伴随液体
accumulated heat 蓄热
accumulated temperature 积温
accumulation layer 积累层
accumulator 蓄能器，回收槽，收集器
accumulator battery 蓄电池组
accuracy for WTGS 风力发电机组精度
acetic acid 醋酸，乙酸
acetone-butanol-ethanol (ABE) [process used in the production of butanol] 丙醇—丁醇—乙醇工艺
acid battery 酸性蓄电池

acid deposition 酸沉降
acid former 成酸物质，成酸剂
acid forming 成酸
acid precipitation 酸雨，酸性降水
acid proof coating 耐酸保护层
acid rock 酸性岩
acid smut 酸性煤尘
acid sulfate water 硫酸盐水
acid volcanic 酸性火山岩
acidic metal oxide 酸性金属氧化物
acidification potential 酸化潜势
acidification reaction 酸化反应
acidize 用酸处理，酸化
acidogenesis experimental reactor 酸化实验反应器
Ackert-Keller cycle 闭式燃气轮机循环
acknowledgement 确认
acoustic backscattering 声波反向散射
acoustic Doppler current profiler 声学多普勒海流剖面仪
acoustic Doppler sensor system 声学多普勒感应系统
acoustic emission 声频发射
acoustic pulse 声脉冲
acoustic reference wind speed 声的基准风速
AC-pv module 交流光伏组件
across-wind 横向风
actinic absorption 光化吸收
actinic glass 光化玻璃，闪光玻璃
actinic radiation 光化辐射
actinometer 感光计，曝光计，日光能量测定器，日射表
actinometry 日射测量学
actinoscope 光能测定计
activated carbon 活性炭
activated sludge 活性污泥

activated sludge oxidation 活性污泥氧化
activated sludge process 活性污泥法（污水处理）
activated sludge system 活性污泥系统
activated-charcoal absorber 活性炭吸附器
activation 活动，赋活，激活，活化，激励，启用
activation energy 活性能，激活能
activation power (for wind turbine) 临界功率
activation rotational speed 临界转速
active solar heating system 主动式太阳能加热系统
active solar space heating 主动式太阳能空间加热
active (passive) circuit elements 有（无）源电路元件
active accumulated temperature 活动积温
active air-heating collector 主动式空气加热集热器
active area of a solar cell 单体太阳电池的有效光照面积
active bacteria 活性微生物，活性细菌
active carbon 活性炭
active carbon filter 活性炭过滤器
active carbon storage silo 活性炭储器
active cavity radiometer 有源空腔辐射计
active collector 主动式集热器
active component 有功分量
active control 有效控制，主动控制
active cooling 主动式冷却
active cooling surface 有效冷却表面
active current 有功电流
active dimension of mounted area 有效安装面积尺寸
active dimension of mounted space 有效安装空间尺寸
active dust 放射性尘埃
active fault 活动断层，活性断层
active fuel bed 燃料燃烧层
active harmonic filter 有源谐波滤波器
active power 有功功率
active power filter 有源电力滤波器
active power of wind farm 风电场有功功率
active residential solar heating system 主动式太阳能住宅供暖系统
active sensor 主动探测器
active solar cooling 主动式太阳能冷却系统
active solar energy system 主动式太阳能系统，主动式太阳能装置
active solar heating 主动式太阳能供热，主动太阳能加热
active solar heating system 主动式太阳能供热系统，主动式太阳能供暖系统
active solar house 主动式太阳房
active solar system 主动式太阳能系统（一种由集热器、储热装置和将太阳能转换为热能的传热流体构成的系统）
active solar thermal collecting device 主动式太阳热收集装置
active solar thermal design 主动式太阳热设计
active stall 主动失速
active storage system 主动存储系统
active surface area 活性表面积
active system 能动系统
active valve 主动阀
active volcano 活火山
active wall 主动式（蓄热）墙

active yaw 主动偏航
activity coefficient 活度系数
activity of nitrogen fixing bacteria 固氮细菌活动
activity parameter （厌氧发酵系统的）活性参数
actual station cycle 电站实际（工作）循环，电站实际热力系统
actual wind 实际风
actual wind power 实际风能
actual wind speed frequency distribution 实际风速频率分布
actuator disk 促动盘
actuator rod 制动器杆
acyl dehydrogenase 脂酰脱氢酶
adaptive control 适应控制
added heat 附加热
added mass 附加质量
added mass coefficient 附加质量系数
addendum modification on gear 齿轮的变位
adding binder 添加黏结剂
additional load 附加荷载
adhesive 添加黏合剂
adhesive resin 添加黏胶树脂
adhesive tape 胶带
adiabatic 绝热的，隔热的
adiabatic atmosphere 绝热大气
adiabatic calorimeter 绝热式热量计
adiabatic char gasification 绝热炭气化
adiabatic combustion temperature 绝热燃烧温度
adiabatic conditions 绝热条件，绝热工况，绝热状态
adiabatic curve 绝热曲线
adiabatic efficiency 绝热效率
adiabatic energy storage 绝热储能

adiabatic exchanger 绝热交换器
adiabatic expansion 绝热膨胀
adiabatic exponent 绝热指数
adiabatic heating 绝热增温
adiabatic process 绝热过程
adiabatic reaction temperature 绝热反应温度
adiabatic saturation temperature 绝热饱和温度
adiabatic temperature 绝热温度
adiabatic thermodynamic system 绝热热力系统
adiabatic warming 绝热增温
adiabatic wellbore flow 绝热井筒流动
adjustable angle 可调角度
adjustable choke 可调式节流阀
adjustable focusing collector 跟踪式聚焦型集热器，可调聚焦集热器
adjustable pitch 可调螺距
adjustable spanner 活动扳手
adjusting plate 调整板
adjusting screw 调整螺栓，调节螺栓
administration multiple building 综合楼
admission end 进气端, 进汽端
admittance 导纳
admixture 混合，混合物
adnate algae 联生藻类，并生藻类
adret 阳坡，山阳
adretto 阳坡
adrift 漂流的, 漂移的
adsorbed layer 吸附层
adsorbent 吸附剂
adsorption 吸附作用
adsorption filter 吸附过滤器
adumbral 遮阳的
advance in applied solar energy 太阳能应用进展

advanced biodiesel 先进生物柴油
advanced reciprocating internal combustion engine 高级往复式内燃机
advanced turbine system 高级汽轮机系统
advancing blade 前行叶片
adverse wind 逆风
adversely 逆地，反对地
adze 扁斧
aeolian 风成的，风积的
aeolian accumulation 风（堆）积
aeolian deposit 风积物，风成沉积
aeolian energy 风能
aeolian erosion 风蚀（作用），风力侵蚀
aeolian material 风积物
aeolian plain 风成平原
aeolian soil 风积土
aeolian vibration 微风振动
aeolic energy 风能
aeolotropic 各向异性的
aerate mud 曝气泥
aerated drilling fluid 充气钻井液
aerated fluid 混气液
aerating agent 充气剂
aeration tank 曝气池
aerator for biochemical reaction basin 生化反应池曝气器
aerator for equalization basin 调节池曝气头
aerial 天线
aerial radiation thermometer 空气辐射温度计
aero derivative turbine 航空用汽轮机
aero generator 气力发电机
aerobic aeration bin 好氧曝气器
aerobic bacteria 好氧菌，需氧细菌
aerobic composting 好氧堆肥
aerobic digestion 需氧消化
aerobic digestion of sludge 污泥需氧消化
aerobic fermentation 需氧发酵
aerobic sludge digestion 污泥需氧消化
aerobiology 高空生物学
aerobiont 需氧微生物
aero-derivative gas turbine 航空衍生型燃气轮机
aerodynamic 空气动力的
aerodynamic behavior 空气动力特性
aerodynamic brake 空气动力刹车
aerodynamic braking system 空气动力刹车系统
aerodynamic calculation 空气动力计算
aerodynamic characteristics 空气动力特性
aerodynamic characteristics of rotor 风轮空气动力特性
aerodynamic chord of airfoil 翼型气动弦线
aerodynamic damping 空气动力阻尼
aerodynamic design 空气动力设计
aerodynamic downwash 气动力下洗
aerodynamic efficiency 气动效率
aerodynamic force 空气动力
aerodynamic interaction 气动互制
aerodynamic loss 空气动力损失
aerodynamic moment 空气动力力矩
aerodynamic noise 空气动力噪声
aerodynamic performance 空气动力性能
aerodynamic regulation 空气动力调节
aerodynamic stall 气动失速

aerodynamic torque 气动扭矩
aerodynamics 空气动力学，气体动力学
aerodynamics efficiency 空气动力学效率
aeroelastic instability 空气弹性变形的不稳定（性）
aerofoil 翼型，机翼剖面，机翼
aerofoil blade 螺旋桨叶片
aerofoil cross-section 机翼横截面
aerofoil drag 风翼阻力
aerofoil fan （机翼型）轴流风机
aerofoil flow measuring element 翼型测风装置
aerofoil lift 风翼升力
aerofoil zero lift line 翼型零升力线
aerogenerator 风力发电机，航空发电机
aerogenerator propeller 风能发电机螺旋桨
aeromagnetic survey 航磁测量
aeromagnetics 航空地磁学
aeromotor 风力发动机，航空发动机
aerosol 大气微粒，悬浮微粒，气溶胶，喷雾器
aerosol electricity 气溶胶电（学）
aerosol electrostatic filter 气溶胶静电过滤器
aerosol particle 气溶胶粒子
aerostatics 气体静力学
aeroturbine 风轮机
aerovane 风向风速仪
Aero Vironment 航空环境,航空环境公司
affinity laws 相关法律
after burning chamber 补燃室
aftercooler 附加冷却器，末级冷却器

agar roll tube 琼脂卷管
aged-refuse 矿化垃圾
aged-refuse-based bioreactor 矿化垃圾生物反应床
ageing tests 老化试验
agglomerant 凝聚物
aggregate 合计，总计，集合体
aggressively 侵略地，攻势地
agitator 搅拌器
agri-chemicals 农化产品
agricultural by-product 农副产品
agricultural residues 农业废物
agricultural waste 农业废弃物，农业废料，农业废物
agro-ecological zoning 农业生态分区
agrofuel 农业燃料
agro-industrial by-product 农工副产品
Agulhas current 阿古拉斯海流（沿着非洲南部东岸向西南流的印度洋洋流）
Ah 安培小时
aileron 副翼
aileron control 副翼操纵
air (heating) collector 空气集热器
air accumulator 空气蓄压器，空气储气瓶，空气蓄电池
air analysis (test) 风力筛分试验
air at rest 静止空气
air atomizing burner 空气雾化喷燃器
air attemperator 空气调温器
air barriers 气密层，内衬层
air based solar-heating system 太阳能空气加热系统
air battery 空气蓄电池
air bearing 空气轴承
air blanketing 空气夹层
air blast 鼓风
air bleed 排气，通气

air blow floatation unit 空气鼓风悬浮分离装置
air blow system aeration tank 空气鼓风型曝气槽
air boundary layer 空气附面层，空气边界层
air brake 空气制动，气闸
air braking system 空气制动系统
air breather 通风装置，通风孔
air breathing engine 喷气发动机
air capacity 排气量
air cell 空气室，空气电池
air cell alkaline 碱性空气电池
air chamber 空气室
air change 换气
air change rate 换气次数，换气率
air channel 通风道，排气道
air charging system 供气装置
air chimney 通风竖井，通风道，排气道
air chute 通风道，排气道
air circuit 空气管道系统
air circuit breaker 空气断路器
air circulation 风流
air circulation rate 换气次数
air classification 风力分级
air classifier 风力分级机
air cleaner 空气净化器，滤气器
air cleaning 空气洗涤法，空气净化，风选
air cleaning equipment 空气净化装置
air cleaning facility 空气净化设备
air cleaning system 空气净化系统
air cloth ratio 气布比
air coil 气冷蛇管，空气冷却蛇管
air collector 集气罐
air compressed system 空压站

air compressor 空气压缩机，空压机
air compressor station 空气压缩机房
air condenser 风冷冷凝器，空冷式冷凝器
air conditioned area (class A) 空调场所（A级）
air conditioned location 空调区域
air conditioner 空气调节器
air conditioning 空气调节，风量调节
air connection 通风联络巷道
air consumption 耗气量
air contactor 空气接触器
air contaminant 空气污染物，大气污染物
air contamination 空气污染
air contamination monitor 空气污染监测器
air convection 空气对流
air coolant 空气冷却剂
air cooled 风冷的，空冷的
air cooler 空气冷却器
air course 通风巷道，风道
air crossing 风桥
air current 气流，风流
air curtain 空气幕，空气屏障，空气屏蔽
air cushion 气垫
air cushion drilling platform 气垫钻井平台
air cutting 气侵
air cycle efficiency 空气循环效率
air cycle heat pump 空气式热泵（用空气做循环介质）
air cyclone 气体旋流器，旋风器
air damper 空气缓冲器，风挡板
air damping 空气阻尼
air deflector 导风板
air density 空气密度

air density correction factor 空气密度修正因数
air diffuser 空气扩散器
air diffusion aerator 空气扩散曝气仪
air displacement 排气量，换气量
air distribution 风量分配，气流分配，空气分布
air distribution plate 空气分布板
air distributor 空气分布器
air door 风门
air dose 空气剂量
air draft 通风，气流
air drag 空气阻力，气动阻力
air drain 空气流泄，通风管
air drifter (drill) 气动架式凿岩机
air drilling 空气钻井，压缩空气钻井，气动钻井
air dry basis 空气干燥基
air dry weight 风干重
air dryer 干燥器
air drying machine 空气干燥机
air duct 气流通道
air dynamic noise 空气动力噪声
air economizer 空气节能器
air eddy 空气涡旋
air ejector 抽气机
air electrode 空气电极
air emission 空气排放
air emissions regulations 气体排放法规
air end 风井口
air engine 空气发动机
air entrainment 夹杂空气
air entrainment of high velocity flow 高速水流掺气
air exhaust 排气
air exhaust fan 排风机

air exhauster 排风机
air exhausting device 排气装置
air extraction system 抽真空系统
air extractor 抽气机
air film cooling 气膜冷却
air filter 空气过滤器
air filter and absorber unit 空气过滤吸附机组
air filter unit 空气过滤机组
air filtration unit 空气滤清装置，空气过滤设备
air flat-plate collector 空气型平板集热器
air floatation unit 空气浮选装置（污水处理）
air flow 气流，风流，空气质量流量
air flow guide 导流罩，导流板
air flow meter 空气流量计
air flow rate 风量，气量，空气流量
air flow structure 气流结构
air flow test 空气流动试验
air flue 气道，风道
air foil 机翼
air friction 空气摩擦阻力
air fuel ratio 空气燃料比
air gap 气隙
air gap reluctance 气隙磁阻
air gas 发生炉煤气，含空气煤气，风煤气
air gasification 空气气化
air gauge 气压表
air governor 空气调节器，送风调节器
air grill (e) 格栅风口
air hammer 空气锤，风镐
air hammer drill 风动冲击凿岩机，风钻
air handler 空气处理机
air handling 空气处理

air handling units　空气处理机组，组合式空调机组
air hatch　通风口
air header　集气管
air heading　通风巷道，通风坑道
air heat exchanger　空气换热器
air heater　空气加热器
air heating　空气加热，空气供热
air heating system　空气供热系统，空气供暖装置
air horse power　空气马力
air humidifier　空气增湿器
air humidity　空气湿度
air infiltration　空气渗透，漏风
air injection　空气引射
air inlet　通风口，进风口
air inlet system　空气输入系统
air input　进气量
air intake　采风口，进风口
air intake branch　进气支管
air intake duct　进气管道
air intake heater　进气加热器
air intake opening　进气口开度，进风口开度
air jet　空气喷射，空气射流
air jet floatation machine　喷射式浮选机
air leakage　漏风
air leakage factor　漏风系数
air leakage test for boiler furnace　锅炉漏风试验
air leg　风动伸缩式气腿
air load　风载，气动载荷
air lock　气闸，气阻，气锁
air magnetic power circuit breaker　空气磁力断路器
air main　主风道

air mass　大气质量，空气质量
air mass analysis　气团分析
air mass attenuation　空气质量减弱
air mass density　气团密度
air mass flow　空气流量
air measuring station　测风站
air meter　风速计，空气流量计
air metering device　空气测量设备
air micrometer　空气测微计
air mixing chamber　空气混合箱
air monitor　空气监测器
air monitoring　大气监测
air monitoring equipment　大气监控装置
air monitoring instrument　大气监测仪器，空气污染检测仪
air monitoring network　大气监测网
air monitoring station　大气监测站
air blower　空气鼓风机
air oil reservoir　油气罐
air opening　风口
air operated recording regulator　气动记录调节器
air operated test pump　气动试验泵
air outlet　出风口，送风口
air output　送风量，（空气压缩机）风量，（压缩）空气量
air parameter　空气环境参数
air parcel　气块
air particle　空气粒子
air particle monitor　空气粒子监测器
air passage　通风道，排气道
air permeability　透气性
air pick　风镐
air pilot valve　空气导阀
air pipe　空气管道，风管
air plume　排气烟羽
air pollutant analysis　空气污染分析

air pollution concentration 空气污染浓度
air pollution control 空气污染控制，大气污染控制
air pollution control equipment 空气污染控制设备
air pollution index 空气污染指数
air preheater 空气预热器
air preheating 空气预热
air pressure 气压
air pressure differences 气压差异
air pressure probe 气压探头
air pressure regulator 气压调节器
air pressure ring buoy 风压浮标（一种波浪能装置）
air pump 抽气机
air pump room 抽气机房，气泵房
air purging 气洗，空气吹洗
air purifier 空气净化器
air quality 空气品质
air quality and emission standard 空气质量和排放标准
air quality criteria 大气质量标准
air quality index 大气质量指数
air quality monitoring system 大气质量监测系统
air quality standard 大气质量标准
air quality surveillance 大气质量检测
air quality surveillance network 大气质量检测网
air quantity 风量
air quantity balance 空气量平衡
air quantity control 风量调节
air quantity estimation 风量估算
air ram (mer) 风动锤，风镐，气锤，风冲子
air ratio 空气比

air receiver 空气吸热器
air refrigeration cycle 空气造冷循环
air regenerating device 换气设备，空气循环设备
air register 空气调节器
air regulation 风量调节
air regulator 空窗
air release 放气
air renewal rate 空气更新率，通风率
air resistance 空气阻力，气动阻力
air resonance 空气共振
air return system 回气系统
air return way 回风道
air reversing 反风
air sampler 大气采样器
air sampling 大气采样
air sampling measurement 大气采样测量
air sampling rig 大气采样装置
air saturation value 空气饱和值
air scoop 进气喇叭口，进风口
air screen 空气幕
air scrubber 空气洗涤器
air separator 空气分离器，气分机
air set 空气中凝固，常温自硬，自然硬化
air shaft 风井，风筒，排气竖井
air sinker 气动凿岩机
air sinks 气穴
air solutizer-process 空气强化脱硫过程，助溶剂氧化脱硫法
air spaces in walls 壁内空气空间
air speed integrator 风速积分器
air split 风流分支
air spoiler 扰流板
air spring suspension 空气弹簧悬架
air staging 空气分级
air stopper 气动伸缩式凿岩机

air stopping 风墙
air storage system 储气系统
air strainer 空气过滤器
air stream 风流
air supply 送风，进风，气源
air supply line pipe 供气管道，供风管道
air to rock solar heating system 空气—岩石太阳能供暖系统
air transport 风力运输
air trunk 通风总管，通气总管
air turbine wave motor 空气涡轮波力电动机
air valve 风阀
air velocity 气流速度，风临度
air velocity transducer 风速传感器
air vent 通风，排气口，通气孔
air vent pipe 排气管
air ventilation 通风，换气
air volume 风量
air volume calculation method 风量计算法
air volume meter 风量计
air wave 空气波，气浪
air way 风眼，气道，通气井，风巷
air way characteristic curve （井巷）风阻特性曲线
air well 通风（竖）井，通风筒
air wire 架空线
air-bag 气囊
air-bag support 气囊支架
air-based solar energy system 在母机上的太阳能系统（装置）
airblast 鼓风
air-blast transformer 强迫风冷式变压器
air-blown cooler 吹风式冷却器
air-blown gasifier 氧化气化器

air-blown pressure gasification 氧化压力气化
airborne 气载的，机载的，空运的
airborne electronic survey control 机载电子测量控制
airborne emission 大气排放
airborne survey 机载勘测
airborne thermal infrared imagery 机载热红外成像
air heating collector efficiency factor 空气加热型集热器效率因数
air-conditioning apparatus 空调装置
air-conditioning plant 空气调节装置
air-conditioning system 空气调节系统
air-conditioning unit 空气调节机组，空调机组
air-cooled condenser 风冷冷凝器，风冷式冷凝器
air-cooled moving grate furnace 气冷式移动炉排炉
air-cooled reactor 空气冷却（反应）堆
air-cooled system 空冷系统
aircraft auxiliary power unit 飞机辅助电源装置
air-cushion 空气垫
air-exhaust filter 排风过滤
air-feed drifter 气动给进架式风钻
air-feed drill 气动给进钻机，气动给进钻岩机
air-feed nozzle 供气喷嘴
airflow 风量，气流，空气流量
air-flow guide 导流罩，导流板
airflow line 气流管道，空气管路，流线
airflow noise 气流噪声
airflow pattern 流谱
airflow pulsation 气流脉动
air-flow wattmeter 气流瓦特计

airfoil 机翼，螺旋桨，翼型
airfoil drag 翼型阻力
airfoil drag coefficient 翼型阻力系数
airfoil lift 翼型升力
airfoil section 翼型剖面
airfoil section characteristic 翼型剖面特性
airfoil section drag 翼型剖面阻力
airfoil wind machine 翼型板风力机
airfoil-shaped blade 机翼叶片
air-fuel ratio 空气燃料比
air-gap flux 气隙磁通
air-gap flux distribution 气隙磁通分布
air-gap line 气隙磁化线
air-hammer drilling 气震钻井，压气冲击钻井
air-handling unit 空气处理单元，空气处理机组
air-heating flat plate collector 空气加热式平板集热器
air-heating solar collector 空气加热式太阳能集热器
air-heating system 空气加热系统
airiness 通风
air-intake 进风
air-intake filter 进风过滤器
air-jet vortex generator 射流式涡流发生器
airleg rock drill 气腿凿岩机，风动凿岩机
air-lift (floatation) machine 气升式（浮选）机
airlift method 气升法
air-lift type agitator 气升式搅拌机，注气搅拌机
airlock 气闸，气塞

air-measuring station 风流测量站
airmeter 风速表
air-oil separator 油—空气分离器
air-oven 热空气干燥箱
air-penetrability 透气率，透气性
air-producer gas 空气发生炉煤气
air-purification equipment 空气净化装置
air-resistance brake 动力阻力制动器
air-reversing way 反风道
air-sand process 风砂分选法
airscrew slip 螺旋桨滑流
air-sea temperature difference 大气—海洋温差
airseal 气密
air-shooting 压风落煤
airslide 气力输送
airspace 大气层，大气空间
airspeed head 空速管，气压感受器
air-speed tube 空速管
air-storage 空气储备，露天堆放
airstream 气流
air-suction system 空气抽吸装置
air-termination system 接闪器
airtight cast 气密铸件
air-tight seal 气密
airtight storage 密封储藏
airtightness 气密性
air-to-air heat exchanger 空气—空气换热器
air-to-cloth ratio 气布比
air-to-liquid heat exchanger 气液热交换器
air-to-oil cooler 气—油冷却器
air-to-oil heat pump 空气—油热泵
air-to-water heat exchanger 气水热交换器

air-to-water heat pump 气水热泵（一种利用户外空气作热源或冷源，通过第二介质——水，向需要空调的户内空间供热的热泵）
air-track drill 风动履带式凿岩机
air-turbine type wave power absorber 空气涡轮机型波能吸收器，空气涡轮机型波能发电装置
air-turbine wave motor 空气—涡轮波力发电机
air-type collector 气式集热器
air-type evacuated tube collector 空气型真空管集热器
air-type flat plate collector 气式平板集热器
air-type single-layer glass cover flat plate collector 空气型单层玻璃盖板平板集热器
air-type solar system 气式太阳能系统
air-warmed radiant panel 暖空气散热片
air-water capillary pressure 空气—水毛细压力
air-water cooling 空气—水冷却
air-water interaction 空气—水相互作用
air-water interface 空气—水界面
airway 呼吸道，通气孔，通气井，风道
alarm cut out 报警抑制
albedo 反照率
alcohol fermentation 酒精发酵，乙醇发酵
alcohol fuel 酒精燃料
alder birch 西桦
algae 藻类
algae bloom 藻类高发
algae digester pond 藻类消化池

algae energy conversion 藻类能量转换
algae fuel 藻类燃料
algae growth potential 藻类生长潜力
algae oil based biodiesel 藻类生物柴油
algal lipid upgrading 藻脂改质
algal suspension 藻类悬浮物
algebraic manipulation 代数运算
align 对准，校直
alignment 对准，定位调整
alkali biomass fuel 碱性生物质燃料
alkali chloride 碱金属氯化物
alkali compound 碱化合物
alkali content 碱含量
alkali fuel 碱性燃料
alkali metal salt 碱金属盐
alkali sulphate 碱性硫酸盐
alkaline accumulator 碱性蓄电池
alkaline storage battery 碱性蓄电池
alkali-sensitive 碱性感测
alkane 烷烃
all vanadium redox flow battery 全钒液流电池
all weather 全天候
all weather operation 全天候运行
all-circuits law 全电路定律
allen key 六方
allen wrench 六方扳手
all-glass collector 全玻璃集热器
all-glass evacuated collector tube 全玻璃真空集热管
all-glass evacuated tube collector 全玻璃真空管集热器
allobaric wind 变压风
allohypsic wind 变高风
allotropic transformation 同素异晶转变

allowable load 容许载荷
allowable stress 容许压力
allowable voltage 容许电压
allowed energy 允许能量
allowed energy level 允许能级
alloy 合金
alloy steel 合金钢
all-roof collector 全屋顶集热器
all-solar energy home 全太阳能家庭
alluvial bed 冲积床
all-weather access 全天候通道
alpha disintegration energy α（射线的）蜕变能
Altamont pass wind farm 阿尔塔蒙特山口风电场（世界最大风电场）
altazimuth mount 地平式装置
altazimuth tracker 地平式跟踪器
alteration 变更，改造
alteration zone 蚀变带
altered rock 蚀变岩
alternate energy 非传统能源，可替代能源
alternating current (AC) 交流电
alternating current (AC) machine 交流电机
alternating current circuits 交流回路
alternating pressure 交变压力
alternating stress 交变应力
alternating voltage 交变电压，交流电压
alternative daily cover 垃圾场另类覆盖层
alternative energy 替代能源
alternative energy resource 替代能源
alternative energy sources 替代能源
alternative operating mode 替代操作模式，交替工作模式
alternator 交流发电机

altitude 海拔
altitude angle 高度角
altitude-altitude heliostat 高度—高度定日镜
altitude-azimuth heliostat 高度—方位定日镜
alumina silicate refractory 硅酸铝耐火材料
aluminized mirror 铝反射镜
alumino-silicate 铝硅酸盐，硅酸铝
aluminum 铝
aluminum absorber fin 铝制吸收器翅片
aluminum alloy housing 铝质金属外壳
aluminum blade 铝制叶片
aluminum brass 铝黄铜
aluminum bridge 铝桥
aluminum bronze 铝青铜
aluminum cell 铝电池，铝电解槽
aluminum continuous melting and holding furnaces 连续溶解保温炉
aluminum glass support 铝制玻璃支架
aluminum sheet 铝（吸收）片
aluminum-foil solar collector 铝箔太阳能集热器
AM0 condition AM0 条件（标定和测试空间用，大气质量等于0，太阳电池所规定的辐照度和光谱分布）
AM1.5 condition AM1.5 条件（标定和测试空间用，大气质量等于1.5，太阳电池所规定的辐照度和光谱分布）
AM1.5 solar spectrum AM1.5 太阳光谱
amature winding terminal 电枢绕组出线端
ambient air 环境空气
ambient air quality 环境空气质量
ambient air quality standard 环境空气质量标准

ambient air temperature 周围空气温度，环境空气温度
ambient atmosphere 环境大气
ambient energy sources 环境能源
ambient noise 环境噪声
ambient power density 周围功率密度
ambient pressure 环境压力
ambient temperature 环境温度
ambient wind speed 环境风速
ambient wind velocity 环境风速
American Recovery and Reinvestment Act (ARRA) 美国经济恢复和再投资法案
American windmill 美国风车
ammonia 氨
ammonia machine 氨气制冷机
ammonia measuring pump 氨计量泵
ammonia measuring tank 氨计量箱
ammonia support system 氨保障系统，氨维持系统
ammonium acetate solution 乙酸铵盐
amorphous film 非晶膜，非晶体膜，非晶形膜，非晶薄膜
amorphous material 非晶态材料
amorphous silicon 非晶硅
amorphous silicon solar cell 非晶硅太阳能电池
amorphous silicon panel 太阳能非晶硅组件，非晶硅太阳能面板
amorphous silicon PV cell 非晶硅太阳能电池，非晶硅光伏电池
amorphous silicon solar cell 非晶硅太阳能电池
amorphous solid silicon solar cell 非晶硅太阳能电池
amount of primary energy supply 一次能源供应量

Ampere-hours (Ah) 安（培小）时
Ampere's circuital law 安培环路定律
Ampere's law 安培定律
ampere-turns 安匝（数）
amphibolite 闪岩，角闪岩
amphidromic point 无潮点
amphidromos 转潮点，转风点
ample space 充足的空间
amplidyne 微场扩流发电机
amplifier panel 放大器盘
amplitude 振幅，幅度，幅值
amplitude control 振幅控制，幅度控制
amplitude modulation 调幅
amplitude of first harmonic 基波振幅
amplitude of oscillation 摆幅，振幅
amplitude of stress 应力幅度
amplitude of the tide 潮幅
amplitude of wave 波幅
amplitude of wind tide 风壅水幅度，半风潮差
amplitude permeability 振幅磁导率
amplitude ratio 振幅比，幅值比，幅度比率
amplitude-frequency response 幅—频响应
anaerobe 厌氧性生物，厌氧微生物
anaerobic 没有空气而能生活的，厌氧性的
anaerobic acid 厌氧酸
anaerobic bacteria 厌氧菌
anaerobic bacterial 厌氧细菌的
anaerobic biodegradability 厌氧降解力，厌氧降解率，厌氧降解性
anaerobic biological treatment process 厌氧生物处理过程
anaerobic biomass decomposition 厌氧生物分解

anaerobic breakdown 厌氧分解
anaerobic contact 厌氧接触
anaerobic contact process 厌氧接触法，厌氧接触工艺
anaerobic corrosion 厌氧腐蚀
anaerobic culture 厌氧培养
anaerobic decomposition 厌氧发酵，无氧分解，嫌气分解
anaerobic digester 厌氧消化器，厌氧煮解器
anaerobic digestion 厌氧消化（作用），厌氧煮解
anaerobic digestion 厌氧消化，厌氧发酵
anaerobic digestion with gas diffusion 气体扩散厌氧消化
anaerobic fermentation 厌氧发酵
anaerobic fermentation process 厌氧发酵过程
anaerobic filter 厌氧滤器
anaerobic lagoon 厌氧氧化池
anaerobic microbe 厌氧菌
anaerobic microorganism 厌氧微生物
anaerobic petri dish 厌氧培养皿
anaerobic pond 厌氧塘
anaerobic process 厌氧过程
anaerobic sewage treatment 厌氧污水处理
anaerobic sludge digester 厌氧污泥消化池
anaerobic sludge digestion 厌氧污泥消化
anaerobic treatment 厌氧处理
anaerobic treatment process 厌氧处理方法，厌氧处理过程
anaerobic waste treatment 厌氧废物处理

anaerobically digested sludge 厌氧消化污泥
anaerobiont 厌氧微生物
anaerophyte 厌氧植物
anaflow 上升气流
analog input 模拟输入
analog input terminal 模拟量输入端子
analog oscilloscope 模拟示波器
analog output 模拟输出
analog signal 模拟信号
analog type 模拟型
analog-to-digital converter 模拟/数字转换器
analogue board 模拟盘
analogue control 模拟控制
analyser 分析器
analytical equation 解析公式
anaphase leachate 后期渗滤液
anchor bolt 锚定螺栓
anchorage 锚具
ancillary electrical equipment 辅助电气设备
ancillary service 辅助服务
ancillary system 辅助系统
Anderson cycle 安德森系统，安德森周期
andesite 安山石，中性长石
anemobarometer 风速气压计
anemobigraph 风速风压记录器，（自记）风压计
anemocinemograph 电动风速记录器
anemoclinograph 风向风速仪
anemoclinometer 风斜表，铅直风速表
anemo-electric dynamo 风力发电机
anemogram 风速记录图，风力自记曲线
anemograph 自记风速表，风力记录仪

anemography 测风学
anemology 风学，测风学
anemometer 风速计，测风仪
anemometer of wind mill type 风车式风速计
anemometrograph 自动风速表，风力记录仪
anemorumbometer 风向风速表
anemoscope 风向仪，风速计，测风器
aneroid barometer 无液气压表，无液晴雨表
angle grinder 角锉
angle of attack of blade 叶片几何攻角
angle of incidence 入射角
angle of inclination 倾斜角
angle of inclination for collector 集热器倾角
angle of internal friction 内摩擦角
angle of repose 安息角，休止角，静止角
angle of rupture 破裂角
angle of shade 保护角
angle of shearing strength 剪切强度角
angle of true internal friction 真内摩擦角
angle of wind deflection 风偏角
angle plate 角盘
Angstrom pyrheliometer 埃斯特朗直接日射表
angular frequency 角频
angular induction factor 角干扰系数
angular interval 间隔角，角间隔
angular momentum 角动量
angular momentum equation 角动量方程
angular motion 角向运动，角运动
angular position 角位置

angular speed 角速率，角速度
angular velocity 角速度
angular velocity vector 角速度矢量
animal by-product 畜禽副产品
animal dropping 动物粪便
animal dung 动物粪便
animal electricity 动物电
animal fat 动物油脂
animal fibre 动物化纤维
animal husbandry residue 畜牧业残渣
animal husbandry wastes 畜业废弃物
animal matter 动物有机残留物质
animal oil 动物油
animal remains 动物遗体
animal resin 动物树脂
animal waste 动物废物，动物排泄物
animal waste conversion 动物废物转换
anion 阴离子，负离子
anions exchanger 阴离子交换器
anisotropic energy surface 各向异性等能面
anisotropic sample 各向异性样本
anisotropic stress 各向异性应力
anisotropy 各向异性
Annapolis tidal power station 安娜波利斯潮汐电站
anneal 退火
annealing effect 退火效应
annealing temperature 退火温度
annual amplitude 年变幅，年较差
annual available hours 年利用小时数
annual average 年平均
annual average insolation 年平均日照量
annual average load 年平均负荷
annual average power 年平均功率

annual average sunshine hour 年平均日照小时

annual average wind power density 年平均风能密度

annual average wind speed 年平均风速

annual capacity factor 年均利用率，发电利用率

annual efficiency of collector field 集热场年效率

annual efficiency of concentrator field 聚光场年效率

annual efficiency of solar thermal power plant 太阳能热发电厂年效率

annual electric output of solar thermal power plant 太阳能热发电厂年发电量

annual energy delivery 年能量供给量

annual energy displacement 年能量交换量

annual energy output 年发电量

annual energy production 年发电量

annual extreme daily mean of temperature 年最高日平均温度

annual heat pump coefficient of performance 热泵年度性能系数

annual heat pump plant coefficient of performance 热泵装置年度性能系数

annual heat pump plant utilisation period 热泵装置年度利用期

annual heating value 年度产热量，年度热值

annual load fraction 年照射量百分率

annual maximum 年最高

annual mean solar flux 太阳能通量年平均值

annual mean wind speed 年平均风速

annual operation hours 年利用小时

annual service hours of generating unit 发电机组年运行小时数

annual solar heat 年太阳热量

annual solar radiation 年太阳辐射量

annual solar saving 年太阳能节能量（太阳能建筑相对于非太阳能建筑每年所节省的能量）

annual temperature change 年度温度变化

annual theoretical production 年均理论产值

annual useful heat supplied by heat pump plant 热泵装置年度有效供热

annual utility hours 年利用小时

annual variation 年变化（风速或风功率密度）

annual waste flow 年度废物流

annual wind regime 年风况

annual wind speed frequency distribution 年风速频率分布

annual zone 年轮

annular air gap 环形气隙

annular area 环状面积，环形面积

annular flow 环形流体

annulus 环面

annunciator 信号器

anode 阳极，正极

anode battery 屏极电池组，阳极电池组

anode plate 阳极板

anode ray current 阳极射线电流

anode rest current 阳极静电流

anode spot 阳极斑点

anodic oxidation 阳极氧化

anodization 阳极氧化

anodized aluminum selective coating 铝阳极氧化选择性吸收涂层（在铝材上采用电解着色阳极氧化方法制备的选择性吸收涂层，它具有良好的选择吸收性能）

anomalistic tide 不规则潮汐
anomalous density 反常密度
anomalous heat flow 异常性热流量
anomalous high heat flow 异常高热流量
antarctic circumpolar current 绕南极流，南极绕极流
antechamber diesel engine 预燃室式柴油机
anthracite 无烟煤
anthropogenic-induced subsidence 人因诱发沉降
anticlinal structure 背斜构造
anticline 背斜，背斜构造
anticline model 背斜模型
anticline structure 背斜构造
anticyclone 反气旋，高气压
anticyclonic circulation 反气旋环流
anti-fouling coating 防污涂层，抗污涂料
antifouling paint 防污涂料
antifreeze 防冻液
antifreeze solution 防冻溶液，防冻剂
antifriction 减低或防止摩擦之物，润滑剂
antireflection 减反射
anti-reflection coating 抗反射膜
anti-reflective coating 防反射涂层，防反射膜，抗反射涂层，抗反射膜
antisolar glass 吸热玻璃
antisun 防止日光照射的，抗日光的
aperture 采光口
aperture area 开口面积，采光面积
aperture area of a solar collector 太阳能集热器的窗口（进光口）面积
aperture area of solar cooker 太阳灶采光面积
aperture diaphragm 有效光阑，孔径光阑
aperture plane 采光平面
aphelion 远日点
apiculture 养蜂业
apochlorotic micro-organism 无叶绿素微生物
apochlorotic plant 无叶绿素植物
apogean tidal current 远月潮流
apogean tide 远月潮
apparent (solar) time 真太阳时，视太阳时
apparent chargeability 视荷电率
apparent conductivity 视电导率
apparent power 视在功率，表现功率
apparent resistance 视电阻
apparent solar day 视太阳日
apparent sound power level 视在声功率级
apparent viscosity 表现黏度
apparent wind speed 表观风速
appearance 外观
application drawing 操作图，应用图
application of geothermal energy 地热能应用
application of solar energy 太阳能利用
applied electric field 外加电场
applied load 外施荷载
applied magnetic field 外加磁场
applied moment 作用力矩，外加力矩
applied voltage 外加电压，外施电压
approach channel 引水渠
appropriate site characteristics 合适位置特征

approximate 近似，接近，约计
apron slope 迎波面坡
aqua power barge 水养动力驳船
aquaculture 水产养殖
aquastat 水温自动调节仪
aquatic 水生的
aquatic biota 水中生物群
aquatic ecosystem 水生生态系统
aquatic food chain 水生食物链
aquatic pollution 水污染，水质污染，水生污染
aquatic wildlife 鱼类，水中野生动物
aquation 水合作用
aquifer 含水层，含水地带，蓄水层
aquifer storage 蓄水层储水量
arbor 树阴，凉亭，藤架
arc control device 灭弧装置
arc welder 电焊机
arc welding 电弧焊，弧焊
Archimedes wave swing 阿基米德波浪摆
Archimedes wave swing generator 阿基米德波浪摆发电机
architectural aerodynamics 建筑空气动力学
area for protection of hot springs 热泉保护区
area grid 区域电网
area of field 热田面积
area of reservoir 水库面积，库面积
area of wave generation 波浪发生区
area swept 扫掠面积
area utilization of a solar array 方阵的面积利用率
armature 电枢
armature circuit 电枢电路
armature coil 电枢线圈
armature current 电枢电流，电驱电流

armature loop 电枢绕组元件
armature winding 电枢绕组
aromatic compound 芳香族化合物
aromatic polymer 芳族聚合物
aromaticity 芳香性
array 数组，阵列，大量
array diode 二极管阵列
array field 方阵场
array loss 阵列损失
array of concentrating solar cell 聚光太阳能电池阵列
array of wind turbine generators 风力发电机排列布置
array switching system 方阵联结开关系统
artesian pressure 自流压，承压水压力
articulated 接合，链接，有关节的
articulated blade 铰接桨叶
articulated flat mirror 万向平面反射镜，转向平面反射镜
artificial fracture 人工裂缝
artificial geothermal reservoir 人工地热储层
artificial lift 人工升举，人工举升
artificial upwelling 人工上升流
artificially induced blow 人工引喷
as received basis 收到基
asbestos 石棉，防火布
ash 灰分
ash agglomeration 灰熔聚
ash agglomeration gasifier 灰熔聚炉
ash analysis 灰分分析
ash coalescence 灰聚结
ash content 灰分含量，含灰量，灰分
ash conveyor 灰渣输送机
ash deformation temperature 灰的变形温度

ash deposition 积灰，灰沉积
ash discharge hopper 排灰斗
ash discharger 除灰器
ash disposal 除灰
ash disposal system 除灰系统
ash extractor 出灰机
ash flow temperature 灰的流动温度
ash formation 灰渣形成
ash fusibility/ash melting behaviour 灰熔融性
ash fusion temperature 灰熔融温度
ash granulation 灰造粒
ash hemisphere temperature 灰的半球温度
ash hoist 起灰机
ash hopper 灰斗
ash melter 化灰机
ash melting point 灰熔点
ash recycling 灰回收
ash removal 除灰，清灰
ash sintering 灰烧结
ash softening temperature 灰的软化温度
ash transport 细灰输送机
ash storage silo 储灰仓
ash-melting point 灰熔点
ash-melting temperature 灰熔温度
ash-rich biomass fuel 多灰生物质燃料
ash-rich fuel 多灰燃料
Asia Sustainable and Alternative Energy Program 亚洲可持续和可替代能源计划
aspect ratio 纵横比，高宽比
aspirated turbosail 吸气转筒式风帆
assemble 集合，聚集，装配
assembling loss 组合损失
assessment for geothermal resources 地热资源评价
assessment methods of geothermal resources 地热资源评价方法
assessment of exposure 照射评价
assessment of landscape 地形评价
assessment of wind-farm noise 风电场噪声评估
assessment parameter of geothermal resources 地热资源评价参数
assorted files 分类排列，相匹配（文件）
assume 假定，设想，采取，呈现
astronomical tide 天文潮，特大潮汐
astronomical unit 天文单位
asymmetrical cell 不对称电池
asymmetrical concentrator 非对称太阳能聚光器
asymptotic acceleration potential method 渐进加速势方法
asynchronous generator 异步发电机
asynchronous machine 异步电机
athermancy 不透辐射热性
atmosphere pressure 大气压力，大气压强
atmosphere-wave forecast system 大气波预测系统
atmospheric(al) circulation 大气环流
atmospheric(al) composition 大气组成
atmospheric absorption 大气吸收
atmospheric attenuation 大气衰减
atmospheric boundary layer 大气边界层
atmospheric chemistry 大气化学
atmospheric circulation 大气环流
atmospheric circulation pattern 大气环流模式
atmospheric components 大气成分

atmospheric composition 大气组成
atmospheric counter radiation 大气向下辐射
atmospheric gas 大气气体
atmospheric heat balance 大气热平衡
atmospheric heat radiation 大气热辐射
atmospheric heat transport 大气热量输送
atmospheric impact category 大气影响类别
atmospheric lapse rate 环境递减率
atmospheric optical depth 大气光学厚度
atmospheric optical thickness 大气光学厚度
atmospheric particulates 大气颗粒物
atmospheric photochemistry 大气光化学
atmospheric pollution 大气污染，空气污染
atmospheric precipitation 大气降水
atmospheric pressure 大气压，大气压力，气压
atmospheric processes 大气过程
atmospheric radiation 大气辐射
atmospheric reservoir system 大气水库系统
atmospheric stability 大气稳定度
atmospheric stability class 大气稳定度类别
atmospheric transmissivity 大气透射，大气传递
atmospheric turbidity 大气浑浊度
atmospheric wind 大气风
atom 原子
atom absorption spectroscopy 原子吸收谱
atom spectrum 原子光谱
atomic battery 原子电池
atomic binding energy 原子结合能
atomic bond 原子键
atomic crystal 原子晶体
atomic energy 原子能
atomic energy storage battery 原子能蓄存器
atomic mass unit 原子质量单位
atomic number 原子序数
atomic orbit 原子轨道
atomic packing factor 原子堆积因数
atomic planar density 原子面密度
atomic power 原子能（动力）
atomic power station 原子能电站，原子能发电装置
atomic radiation 原子辐射
atomic spectra 原子光谱
atomic structure 原子结构
atomic weight 原子量
atomiser 喷雾器
atomized liquid droplet 雾化液滴
attached flow regime 附着流动区
attached sunspace 附加阳光间式
attached sunspace passive solar house 附加阳光间式被动太阳房
attack angle 迎角，攻角
attainable depth 可达深度
attemperater 减温器
attemperation 减温器，调温装置
attenuation 衰减
attenuation of solar radiation 太阳辐射衰减
attenuator 衰减器，弱化子
attribute value 属性值
audible noise 可听噪声

audio-frequency magnetotelluric method 音频大地电磁法
auger 螺旋钻
auger drill mining 螺旋钻探采矿
auger drive 螺旋推运器
auger method 钻取法
augering 螺旋钻法
augmented turbine 加力式涡轮
austenite 奥氏体（碳丙铁）
auto transformer 自耦变压器
auto correlation function 自相关函数
auto-correlation 自动关联
automatic dispatch system (ADS) 自动调度系统
automatic feed tray 自动给料盘
automatic flue gas suction fan 自动烟气抽风机
automatic generation control (AGC) 自动发电控制
automatic lathes 自动车床
automatic operation 自动操作
automatic solar module laminator 全自动太阳能电池组件层压机
automatic solar tracker 自动太阳能跟踪装置
automatic station 无人值守电站
automatic temperature recorder 温度自动记录器
automatic voltage regulator 自动电压调节器，自动稳压器，自动调压器
automatic watering system 自动补水系统
auto-reclosing cycle 自动接通周期
auto-regression 自动回归
auto-regressive series 自动海退岩系，自回归序列
autotrophic organism 自养生物

auxiliary burner 辅助燃烧器
auxiliary circuit 辅助电路
auxiliary device 辅助装置
auxiliary energy source 辅助能源
auxiliary feed valve 辅助进给阀
auxiliary generator 辅助发电机
auxiliary heat source 辅助热源
auxiliary heater 辅助加热器
auxiliary heating surface 附加受热面，辅助受热面
auxiliary heating system 辅助加热系统
auxiliary item 辅助生产区
auxiliary motor 辅助电动机
auxiliary power quantity 发电厂厂用电量
auxiliary switch 辅助触点
availability 有效性，可利用率
available energy 可用能，现有能量
available heat 可用热，有效热
available power 有效功率
available power generation 可用发电
available power loss 有效能的损失，可用能量的损失
available work 可用功，有用功
average 平均，平均水平，平均数，海损，一般的，通常的
average cell voltage 平均电池电压
average daily efficiency 平均日效率（在有太阳辐照的一天内，太阳热水器储水所获得的热量与照射到太阳集热器采光表面上的太阳辐射能量之比）
average heat loss coefficient 平均热损系数
average irradiance 平均辐（射）照度
average monthly heating load 月平均

加热负荷
average noise level 平均噪声
average organic loading rate 平均有机加载率
average plate temperature 平板平均温度
average steam production 平均蒸汽产量
average velocity 平均速度
average wind 平均风
average wind speed 平均风速
average wind velocity 平均风速
aviation airfoil 航空翼型
aviation light 航空灯
Avogadro's Law 阿伏伽特罗定律
axial clearance 轴向间隙
axial current density （磁流体发电机内）轴向电流密度
axial flow 轴流
axial flow engine 轴流式发动机
axial flow hydroelectric unit 轴流式水轮机组
axial flow steam turbine 轴流式汽轮机
axial flow turbine 轴流式汽轮机，轴流式涡轮机
axial flow water turbine 轴流式水轮机
axial flux induction machine 轴向磁通感应电机
axial flux machine 轴向磁通电机
axial gap induction machine 轴向间隙感应机
axial induction factor 轴向干扰系数
axial interference factor 轴向干涉因子
axial load 轴向载荷，轴向荷重
axial load failure 轴负载性故障

axial pitch 轴向齿距
axial thrust 轴向推力
axial turbine 轴流式涡轮机，轴流式汽轮机，轴流式水轮机
axial-flow propeller machine 轴向流动推进器
axial-flux motor 轴向磁通电动机
axle 轮轴，车轴
azimuth 方位角
azimuth angle 方位角
azimuth-elevation tracking 方位—俯仰跟踪

B

back contact 后触点
back insulation thermal conductivity （集热器）背面绝热传导率
back pressure 背压，回压
back pressure turbine 背压式涡轮机，背压式汽轮机
back reflection solar cell 背反射太阳能电池
back scatter 反向散射，背散射
back surface field solar cell 背场太阳能电池
back surface field effect solar cell 背场效应太阳能电池
back surface reflection and back surface field solar cell 背场背反射太阳能电池
back surface reflection cell 背反射电池
back surface reflection field cell 背场背反射电池
back surface reflection solar cell 背反射太阳能电池
back surface reflector (BSR) 背面反射器

back up heater   备用加热器，辅助加热器
back wash pump   反冲洗泵
back-draft damper   反向通风调节风门，反向通风气闸
backflow preventer   回流防止器，防回流阀
background colour   背景颜色
background irradiance   背景辐射
background response   背景响应，背景反应，本底响应
backlash   齿隙
back-pressure plant   背压装置
back-pressure steam turbine   背压式汽轮机
back-pressure turbine   背压式汽轮机
backrush   回卷，退浪
back-scattered dose   反向散射辐射剂量
back-scattering   反（向）散射，后向散射
back-seal   背封
backsheet   太阳能电池背膜
backsurface field   背面场
backsurface field cell   背面场电池
backup energy system   备用能源系统（常指太阳能供热系统的备用系统，能在太阳能系统停止工作时，提供全部需求的能量）
backup generator   备用发电机组
backup power   备用电源
backup power source   辅助能源
back-up unit   备用装置
backwater effect   回水效应
bacterial action   细菌作用
bacterial cell   细菌电池
bacterial decay   细菌性腐烂
bacterial fermentation   细菌发酵

bacterial fertilizer   细菌肥料
bacterial hydrogenase   细菌氢化酶
bacterial luminescence   细菌发光
bacterial metabolism   细菌代谢
bacterial population   细菌种群
baffle plate   折流挡板，挡板，隔板
baffle scrubber   挡板式洗涤器
bag breaker   拆袋器
bag filter   袋式过滤器
bagasse   甘蔗渣
bag-breaking machine   破袋机
baghouse   袋式除尘器，袋滤室，集尘室
baghouse filter   布袋除尘器
balance of plant   电厂辅助设施
balance of system (BOS)   系统平衡，光伏平衡系统
balancer-transformer   均压变压器
balancing equipment   平衡设备
balancing valve   平衡阀，均衡阀
baled biofuel   打包生物质燃料
baler   打包机
baling press   压缩打包机，填料压机
ball bearing   滚珠轴承
ball diffuser   球形风口，球形散流器
ball float valve   浮动式球阀
ball saddle   滚珠支撑
ball screw   滚珠螺杆，滚珠丝杆，滚珠丝杠，滚珠导螺杆
ball screw actuator   滚珠螺旋致动器
ball screw shaft   滚珠丝杠
ball valve   球阀
ballast water   压载水，压舱水
ball-eye   球头挂环
ball-hook   球头挂钩
ballistic separator   冲击式分离机
ball-peen hammer   圆头锤
ball-race mill   球跑轨磨

band 区，带，波段带子，波段
band edge energy 带边能量
band gap 能隙，能带隙
band saw 环形锯，带锯机
band separator 人工分选台
band shift 能带移动
band structure 能带结构
bandgap structure 带隙构造
bandwidth 频带宽度
bar magnet 磁条
bare cell 裸电池
bare tube 裸管
barge 驳船
barge type wave energy conversion system 船形波浪能发电装置
barium fuel cell 钡燃料电池（使用钡和氧或钡和氯，把化学能转变为电能的燃料电池）
barometer 气压计，晴雨表
barometric damper 气流调节器
barometric response 气压响应
barrage 拦河坝，水闸
barrel of oil equivalent 桶油当量
barrier energy 势垒能
barrier layer capacity 势垒层电容
baryte 重晶石
basaltic layer 玄武岩层
base 基极，底座
base bottom 底座底
base impedance 基础阻抗
base load 基本负荷，基荷
base load plant 基荷电厂
base load power 基荷动力，基荷功率，基本负荷动力，基本负载功率
base load regulation solar thermal power plant 太阳能热发电基本负荷电厂，基本负荷调节太阳能热电厂

base material 基底材料
base plate 底座
base top 底座面
baseboard heater 护壁板式加热器
baseboard radiator 护壁板散热器，护壁板暖气片
base-load capacity 基荷容量
baseload power 基载电力，基本负荷动力
base-load power 基本负荷动力，基荷动力
base-load unit 基本负荷机组，基荷机组
basic density 基础密度
basic error 基准误差
basic insulation level 基本绝缘等级
basic rock 基岩
basic temperature 基础温度
basic wind speed 基本风速
basin type solar still 盆形太阳能蒸馏器
basin-fill sediment 盆地充填沉积
basin-type solar still 池式太阳能蒸馏器
batch combustion 分批燃烧
batch dryer 分批干燥器
batch furnace 间歇式炉
batch heating 间歇供热
batch reactor 间歇式反应器
batch-fed 按批投料
batch-type solar water heater 间歇式太阳能热水器
batholith 岩基，岩盘
bathymetry 水深测量，海洋测深学
battery 蓄电池
battery bank 电池组
battery charge 电池充电器

battery charge converter 蓄电池充电转换器
battery charge regulator 蓄电池充电调节器
battery charging circuit 蓄电池充电电路
battery discharge 电池放电
battery discharge converter 蓄电池放电转换器
battery discharge regulator 蓄电池放电调节器
battery power drill 电池钻
battery size 电池型号
battery sizing 电池容量确定
battery stack 电池组,成套电池
battery storage 蓄电池
batteryless grid-tie system 无电池并联系统
batteryless grid-tied wind electric system 无电池并网风电系统
baud 波特
baud rate 波特率
bayesian directional method 贝叶斯方法
bayonet 卡口
beam angle 波来角
beam characterization system 直射光特性校准系统
beam compass 长臂圆规
beam energy 束能量,射束能量
beam idler gear 惰轮齿
beam insolation (太阳)直接辐射
beam radiation 太阳直射辐射
beam solar radiation 直接日射
beam trammel 骨架
bearer 支架,托架,支座,载体
bearing 轴承

bearing fittings 轴承配件
bearing processing equipment 轴承加工机
bearing seal 轴承密封
bearing wear 轴承磨损
bed coke 底焦
bed material 床料
bedding plane 层面,层理面,地貌面
bedplate 机舱底架
bedrock 基岩
bedrock wells 基岩井
beginning water level for flood control 防洪限制水位
bell crank 曲柄
Bell laboratories 贝尔实验室
bell valve 钟形阀
bellow type 波纹管式
belt 风带
belt conveyor 皮带运输机,带式运输机
belt drive 皮带传动
belt dryer 带式干燥机
belting 制带的材料,带类,调带装置
bending load 弯曲负荷
bending moment 弯矩
bending of beam 弯曲梁
bentonite 斑脱岩
Berl saddle 贝尔鞍形填料,弧鞍形填料
Bernoulli's equation 伯努利方程
Bessemer 酸性转炉炼钢
Bessemer converter 酸性转炉,贝塞麦转炉
bessemer steel 酸性转炉钢,贝塞麦钢,酸性钢
best available control technology 最佳可行控制技术,最佳可用控制技术
beta absorber β吸收体

Betz law 贝茨定律（贝茨指出，理论上机械设备能够提取的风能最大值为总量的59.3%。实际上，这一效率的极限值约为40%）
Betz limit 贝茨极限，贝兹极限
Betz power coefficient 贝茨功率系数
Betz theory 贝茨理论
Betz's efficiency 贝茨效率
bevel gear 斜齿轮
bevel-edge steel 斜缘薄钢板
beveling 磨斜棱，磨斜边
bicarbonate water 重碳酸盐水
biconcave lens 两面凹镜
biconvex lens 两面凸镜
bidding and tendering system 招投标制
bidirectional current 双向电流
bi-directional ducted horizontal turbine 双指向性管道卧式涡轮机
bi-directional wells turbine 双向井用涡轮机
bifacial combined photovoltaic module 双面光伏模块，双面光伏组件
bifacial flat panel 双面（太阳能）电池板
bifacial photovoltaic flat panel 双面光电地板
bifacial solar cell 双面太阳能电池
bifurcated rivet 开口铆钉，分叉的铆钉
bifurcation 分支，分叉
bilateral circuit 双向电路
billboard receiver 平板式吸热器
billet temperature 钢温
bi-logarithmicg raph 双对数图
bimetallic 双金属的
bimetallic actinograph of Robitzsh 罗比茨型双金属日射计

bimetallic pyrheliometer 双金属直接日射计
bimetallic-stem dial thermometer 双金属杆拨温度计
bimotored 双马达的
binary cycle 双循环
binary cycle generation 双循环发电
binary cycle power plant 双循环发电厂
binary-cycle system 双循环系统
binary-vapor cycle 双汽循环
binder matrix 结合混合料
binding bond 结合键
bio-aeration 生物曝气
biobased bulk chemicals 生物基化学品
biobattery 生物电池
bio-butanol 生物丁醇
biocarbon 生物碳
biocatalysis 生物催化作用
biocatalyst 生物催化剂
biochar 生物碳
biochemical 生化的，生化药剂，生化产品
biochemical action 生（物）化（学）作用
biochemical conversion technology 生化转换技术
biochemical deposit 生物化学沉积
biochemical energy cycle 生化能量循环
biochemical fermentation 生物发酵
biochemical fuel cell 生化燃料电池
biochemical indicator 生化指示剂
biochemical methane potential 生物化学甲烷势，生化甲烷势
biochemical oxidation 生化氧化

biochemical oxygen demand 生化耗氧量
biochemical process 生化过程
biochemical reaction basin 生化反应池
biochemistry conversion technology 生化转换技术
biochip 生物芯片
biocide 杀菌剂
biocontrol 生物控制
bioconversion 生物转化
bioconversion of solar energy 太阳能的生物转换
bio-crude 生质原油
biocycle 生物带（包括陆地、海洋和淡水域），生物环
biodegradation 生物降解
biodiesel 生物柴油
biodiesel blend stock 调合用生物柴油
biodiesel fuel 生物柴油燃料
biodiesel plant 生物柴油厂
biodiesel processing facility 生物柴油处理设施
bioelectricity 生物电，生物电流，生物电性
bio-electricity 生物质发电
bioenergy 生物质能
bioenergy crop 生物能源作物
bio-energy electricity 生物质能发电
bioenergy production 生物能源生产
bio-energy with carbon capture and storage 生物能源与碳捕获与储存
bioengineering 生物工程
bioethanol 生物乙醇
bioethanol economy 生物乙醇的经济性
bioethanol fuel 生物乙醇燃料
biofilm fluidized bed reactor 生物膜流化床反应器
biofilm reactor 生物膜反应器
bio-filter 生物滤池
bioflocculation 生物絮凝（作用）
biofouling 生物淤积，生物污垢（指液体流经的管道等物体的表面上黏附细菌或水下船底贝属动物的沉积或黏附）
biofuel 生物燃料
biofuel blend 生物质燃料掺合物
biofuel blend stock 调合用生物燃料
biofuel briquette 生物质燃料块
biofuel intermediate 生物燃料中间体
biofuel mixture 生物质燃料混合物
biofuel pellet 生物质燃料丸
biofuel preprocessing 生物质燃料预处理
bio-gas 生物质气体
biogas 沼气
biogas anaerobic fermentation 沼气厌氧发酵
biogas comprehensive utilization 沼气综合利用
biogas cooker 沼气灶
biogas desulphurizing 沼气脱硫
biogas dewatering 沼气脱水
biogas digester 沼气池
biogas engineering 沼气工程
biogas fermenting 沼气发酵
biogas generation 沼气发电
biogas generator 沼气池
biogas lamp 沼气灯
biogas microorganism 沼气微生物
biogas muck 沼气池污泥
biogas plant 沼气池
biogas power generation 沼气发电
biogas power station 沼气电站
biogas production rate 产气率

biogas supply system 沼气供应系统
biogenic waste 有机垃圾
bio-jet fuel 生物航油
biokinetics 生物动力学,生物运动学
biologic effect of ultrasound 超声生物效应
biological 生物学的
biological accumulation factor 生物积累因子
biological activity 生物活性
biological affinity 生物亲和性
biological breakdown 生物分解
biological concentration factor 生物浓集因子
biological contact oxidation 生物接触氧化
biological corrosion 生物腐蚀
biological damage （辐射）生物损伤
biological decay 生物衰变
biological decay constant 生物衰变常数
biological decomposition 生物分解
biological depollution 生物（学）去污（染）
biological deposit 生物沉积物
biological diversity 生物多样性
biological dose 生物剂量
biological dosimeter 生物剂量计
biological effectiveness 生物效应
biological energy 生物能源
biological filter 生物滤池
biological filtering membrane 生物过滤膜
biological filtration 生物过滤
biological fluidized bed 生物流化床
biological fouling 生物污染
biological fuel cell 生物燃料电池

biological half-life 生物半衰期
biological hole 生物通道,生物洞
biological oxidation 生物氧化
biological oxygen demand 生物需氧量
biological pollution in cooling system 冷却水生物污染
biological pretreatment 生物预处理
biological probe 生物探头
biological process 生化过程,生物过程
biological protection 生物防护
biological purification 生物净化
biological resource 生物资源
biological sampling 生物取样
biological sensor 生物量传感器
biological shield 生物屏蔽
biological slime 生物黏膜
biological stain 生物染色剂
biological system 生物系统
biological transducer 生物量传感器
biological tunnel （反应堆内）生物实验孔道
biological waste treatment 生物（学）法废物处理
biological wastewater treatment 废水生物处理
biological weathering 生物风化
biologically active floc 生物活性絮体
biology 生物学
biology liquid fuel 生物液体燃料
biomass 生物质,生物燃料
biomass ash 生物质灰
biomass briquette 生物质压块
biomass burner 生物质燃烧炉,生物质燃烧器
biomass burning 生物质燃烧
biomass co-firing 生物质共燃,生物质混燃

biomass co-firing technology 生物质共燃技术
biomass combustion 生物质燃烧
biomass combustion technology 生物质燃烧技术
biomass conversion 生物质能转化
biomass conversion technology 生物能转换技术
biomass cost 生物质成本
biomass crop 生物质作物
biomass downdraft gasifier 下吸式生物质气化炉
biomass electric power generation 生物质能发电
biomass energy 生物质能,生物能
biomass energy conversion 生物质能转换
biomass energy conversion system 生物质能转换系统,生物能转换系统
biomass energy crop 生物质能源作物
biomass energy fuel 生物质能燃料
biomass energy industry 生物质能源产业
biomass energy resource 生物质能资源
biomass energy technology 生物质能源技术
biomass energy utilization 生物质资源利用
biomass facility 生物质设备
biomass feedstock 生物质原料
biomass fuel 生物质燃料
biomass fuel cell power plant 生物燃料电池电厂
biomass fuel consumption 生物质燃料消耗
biomass fuel cost 生物质燃料成本

biomass fuel handling system 生物质燃料处理系统
biomass fuel resource 生物质燃料资源
biomass furnace 生物质燃烧炉,生物质反应堆
biomass gasification 生物质气化
biomass gasification plant 生物质气化装置
biomass gasification power generation 生物质气化发电
biomass gasification power plant 生物质气化发电厂
biomass gasification set 生物质气化炉
biomass gasification technology 生物质气化技术
biomass gasifier 生物质气化炉,生物质气化机组
biomass generating capacity 生物质发电量,生物质发电容量
biomass generation 生物质发电
biomass generator 生物质发电机
biomass intermediate 生物质中间体
biomass internal combustion gas turbine 生物质内燃涡轮机
biomass internal gasification combined cycle 生物质整体气化联合循环发电
biomass liquefaction 生物质液化
biomass material 生物质
biomass membrane bioreactor 膜生物反应器
biomass molding fuel 生物质颗粒燃料,生物质成型燃料
biomass plantation 生物质种植园
biomass power generation 生物质发电

biomass power generation industry 生物质发电产业
biomass power generation technology 生物质发电技术
biomass power plant 生物质能发电厂
biomass production 生物质生产
biomass pyrolysis 生物质热分解
biomass reserve 生物质储备
biomass residue 生物质残留物
biomass resource 生物质资源
biomass thermoconversion 生物质热转换过程
biomass total utilization 生物量全利用
biomass utilization 生物质利用
biomass waste 生物质废弃物
biomass-based power plant 生物质电厂
biomass-derived electricity 生物质电
biomass-derived fuel 生物质衍生燃料
biomass-fired power station 生物质火力发电厂，生物质火力发电站
biomass-to-liquids diesel 生物质柴油
biomaterial 生物材料
biomechanics 生物力学
biomedical dosimetry 生物医学剂量学
biomedical radiation-counting system 生物医学辐射计数系统
biomedical radiography 生物医学射线照相
biometeorology 生物气象学
biomethane 生物甲烷
bionic computer 仿生计算机
bionics 仿生学，仿生电子学
bionics material 仿生材料
bio-oil 生物油
biopower 生物电源
bioreactor 生物反应器
biorefinery 生物质提炼，生物炼油厂

biosensor 生物传感器
biosludge 生物质浆
biosphere 生物层，生物圈
biostatics 生物静力学
biostatistics 生物统计学
bio-synthetic natural gas 生物质合成天然气
biota 生物群
biotic 生物的
Biot-Savart law 毕奥—萨伐尔定律
biowaste 生物废弃物
biphase 双相的
bipolar junction transistor 双极结型晶体管
BIPV product 光伏建筑一体化产品
biquartz transducer 双石英变频器
bital control 数字控制
birch plywood 桦木胶合板
bispectral model 双谱模型
bit 位，比特
bit design 钻头设计
bit life 钻头寿命
bit wear 钻头磨损
bituminous coal 烟煤
bivane 双向风向标
black absorber 黑色吸热体，黑色吸热器
black absorber surface 黑色吸热体表面，黑色吸热器表面
black absorbing area 黑色吸热面积
black absorbing material 黑色吸热材料
black absorbing sheet 黑色吸热板
black blade 黑色叶片
black body 黑体
black body emissive power 黑体辐射力，黑体辐射功率
black body flame temperature 黑体火焰温度

black body radiant exchange 黑体（间）辐射换热
black body radiation 黑体辐射
black body radiator 黑体辐射计
black body receiver 黑体吸热器
black cell 黑（色）电池
black chrome selective coating 黑铬太阳能选择性膜
black coating 黑色涂层
black collector plate 集热器黑化吸热板
black gauze 黑色金属网
black gauze solar collector 黑色金属网式太阳能聚热器
black lead sulfide particle 黑色硫化铅颗粒
black liquor 黑液
black mechanical absorber 黑体机械吸收体
black metal base 黑金属基座
black metal receiver 黑金属吸热器
black metal sheet 黑金属片
black polyvinyl plastic 黑色聚乙烯基塑料
black porous absorbing head 黑色多孔吸热顶盖
black porous evaporator 黑色多孔蒸发器
black radiation 黑体辐射
black solar absorber 黑色太阳能吸收器
black solar cell 太阳能黑电池，无反射太阳能电池
black tray of water 黑色水箱
blackbody radiation 黑体辐射，绝对黑体的辐射
blackened aluminum 黑化铝
blackened aluminum sheet 黑色铝板
blackened heat absorber 涂黑吸热器
blackened metal tube 涂黑金属管
blackened metallic receiver 涂黑金属吸热器
blackened pond bottom 涂黑池底
blackened rough-textured heat absorbing surface 涂黑粗糙吸热面
blackened surface 涂黑表面
blackness 黑度
blackout 停电，断电（事故）
blade 叶片
blade anchor point 叶片固定点，叶片死点
blade angle 叶片倾角
blade assembly 叶片组装
blade axis 叶片轴线
blade bearing 叶片轴承
blade bending 叶片弯曲
blade bending load 叶片弯曲载荷
blade calculation 叶片计算
blade carrier 静叶持环，隔板套
blade cascade 叶栅
blade chord 叶（片）弦
blade composition 叶片材料
blade configuration 叶片结构
blade connection 叶片连接
blade construction 叶片结构
blade contour 叶片轮廓
blade deflection 叶片偏差
blade diameter 风轮直径
blade elastic axis 叶片弹性轴
blade element method 叶素方法
blade element momentum (BEM) theory 叶素动量理论
blade element theory 叶片元理论，叶素理论
blade element velocity 叶片元速度

blade fatigue 叶片疲劳
blade feathering 叶片顺桨
blade flutter 叶片颤振
blade flutter parameter 叶片颤振参数
blade for iron saw 锯刃
blade frequency test 叶片频率试验
blade geometry 叶片几何形状
blade geometry optimum 叶片几何形状优化
blade geometry parameter 叶片几何参数
blade hub 叶片轮毂
blade length 叶片长度
blade loss 叶片损失
blade material 叶片材料
blade motion 叶片运动
blade mounting ring 叶片安装环
blade natural frequency 叶片自然频率
blade noise 叶片噪声
blade of constant chord width 等弦宽叶片
blade operating servomotor 转轮叶片操作接力器,转轮叶片伺服器
blade passing frequency 叶片通过频率
blade path 叶片通流
blade pitch 叶片节距,桨距,叶距
blade pitch angle 叶片桨距角
blade pitch change 叶片桨距变化
blade pitch set angle 叶片桨距角设置角度
blade pitching rate 桨叶节距角变化率
blade pressure 叶片压力
blade profile 叶片翼型
blade radius 叶片半径
blade resonance 叶片共振
blade root 叶片根部,叶根

blade root bending moment 叶根弯矩
blade root fixing 叶片根部组件
blade root load 叶根荷载
blade section 叶片剖面,叶片断面,叶片横截面
blade servomotor 叶片接力器,叶片伺服器
blade shape 叶片形状
blade solidity 叶片坚固性
blade strength 叶片强度
blade structural design 叶片结构设计
blade structure 叶片结构
blade tip 桨叶梢部,叶梢,叶尖
blade tip receptor 叶尖接收器
blade tip speed 叶尖速度,叶片圆周速度
blade top 叶片顶部
blade twist 叶片的扭曲,桨叶扭旋
blade twist distribution 叶片扭曲分布
blade type burner 叶片式燃烧器
blade vibration 叶片振动
blade weight 叶片重量
blade wheel 叶轮
blader 叶片装配工
blade-tower clearance 叶片—塔架的净空
blast 强风
blast cleaning 喷抛清理
blast draft 强制通风
blast furnace 鼓风炉
blasthole drilling 爆破孔钻井,炮眼钻井
bleed 放出(液体、气体等),漏出,漏入,泄漏,色料扩散
bleed off valve 泄压阀
bleed valve 放泄阀,放气阀,排放阀
bleeder current 旁漏电流
bleeder valve 溢流阀
bleeder well 减压井

blind 挡板
blind ram 盲板
blinding plate 盲板
block and tackle 滑轮组
block copolymer 嵌段共聚物
block diagram 框图
block mountain area 断块山区
block relaxation time 岩块弛豫时间
blockage correction 堵塞修正, 阻塞修正, 槽壁修正
blockage effect 截面阻塞影响, 阻塞效应, 堵塞效应
blocked rotor 闭锁转子, 制动转子
blocked rotor test 堵转试验
blocking diode 隔离二极管
blocking effect 阻塞效应
blocking factor 遮挡因子, 遮光系数
blocking loss 遮挡损失
blocking (for wind turbines) 锁定（风力机）
blowdown 排污, 放气, 排料
blower 鼓风机, 吹风机
blowing jet 吹气射流
blowing rate 风量
blowing system of ventilation 压入式通风系统
blowlamp 喷灯
blow-out can 防爆罐
blowout preventer 防喷器
blowout preventer 井喷防护
blowpipe 吹风管
blowtorch 吹管, 喷灯
blue energy 蓝色能量
blue-gray dolomite 蓝灰色白云岩, 蓝灰色白云石
body-centered cubic 体心立方结构
body-mounted type solar (cell) array 壳体式太阳电池阵
bogie type annealing furnace 转向架式退火炉
Bohr atomic model 波尔原子模型
boiler 锅炉, 加热器
boiler ash 锅炉灰
boiler capacity 锅炉容量
boiler efficiency 锅炉效率
boiler feed water pump 锅炉给水泵
boiler fuel 锅炉燃料
boiler plant 锅炉房, 锅炉间
boiler pressure 锅炉压力
boiler steam pressure 锅炉蒸汽压
boiler tube 锅炉管
boiler-turbine-generator unit protection 单元机组保护
boiling coefficient 沸腾传热系数
boiling mud pit 沸腾泥浆坑
boiling mud pool 沸泥塘
boiling point thermometer 沸点温度表
boiling point-gravity constant 沸点—比重常数
boiling water reactor 沸水反应堆
boiling-point-for-depth 深度沸点
bolometer 变阻测辐射热表
bolt 螺栓, 螺钉, 支持, 维持
Boltzmann constant 玻耳兹曼常数, 玻耳兹曼恒量
bomb calorimeter 弹式热量计
bonding bar 等电位连接带
bonding conductor 等电位连接导体
bonding energy 键能
bonding wire 接合线, 焊线
bonnet 汽车发动机罩
boost 增压
boost flat plate collector 增强式平板集热器

boost-buck 升压去磁
booster compressor 升压压风机
booster ejector 接力抽出器
booster heater 辅助加热器
booster pump 升压泵，增压泵，接力泵，前置泵
bore 涌潮
borehole 钻孔，井孔，镗孔，炮眼，（为了解地质水文而挖的）小口径小井
borehole collapse 钻孔坍塌，井塌
borehole deviation 井斜，钻井偏差
borehole heat exchanger 埋管换热器
borehole heat storage 埋管热存储
borehole size 井眼尺寸
borehole thermal resistance 钻孔热阻
borehole wall 井壁
boring head 搪孔头
boring machine 镗床
boron diffused BSF cell 硼漫射背面场电池
boron nitride nanotube 氮化硼纳米管
borosilicate glass 3.3 硼硅酸玻璃3.3（又称特硬玻璃）
both sides welding 双面焊接
bottom ash 底灰
bottom cycle 底循环
bottom dead-centre 下死点，下顶点
bottom friction 海底摩擦力
bottom heat loss （集热器）底部热损失
bottom hole pressure 井底压力
boulder 大块石，大圆石，大卵石，巨石
bound water 结合水
boundary current 边界流
boundary element method 边界元法，边界要素法

boundary layer 边界层，界面层，附面层
boundary layer control 边界层控制
boundary nucleation 晶界形核
Bourdon gauge 波登管式压力计
Bourdon tube 波登管，单圈弹簧管
Bourdon tube pressure gauge 波登管式压力计，弹簧管式压力计
bowl mill 球磨机
bow-spring compass 弓形片弹簧圆规
box solar cooker 箱式太阳灶
box spanner 管钳子
box spanner inset 插入式套筒扳手
boxed absorber 箱型吸收器
box-like compartment （整流器）箱状分隔
Boyle Mariotte law 玻意耳—马略特定律
Boyle's law 玻意耳定律
brace 支柱，带子，振作精神
bracket 支架，托架，括弧，支架
brackish water 半盐水，半咸水，苦咸水，微咸水，淡盐水
brad 曲头钉
bradawl 小锥
Bragg's law 布拉格定律
brake fluid 刹车油
brake disc 刹车盘
brake lining 闸衬片
brake mechanism 制动机构
brake pad 闸垫
brake setting 制动器闭合
brake shoe 闸瓦
brake 制动器
braking mechanism 制动机构
braking releasing 制动器释放
braking system 制动系统
braking torque 制动扭矩

branch connection 分支接续
branch of joint 连接分支
brass 黄铜,黄铜制品
Bravais indice 布拉维指数
Brayton cycle 布雷顿循环
Brayton engine 布雷顿发动机
brazing 铜焊
breadbox hatch heater 面包箱式间歇加热器
breadbox solar water heater 面包箱式太阳能热水器
breadbox-type water heater 面包箱式热水器
break contact, b-contact 开断触头
break induced current 断路感应电流
breakaway force 起步阻力
breakaway starting current of an A.C. motor 交流电动机的最初启动电流
breakdown 击穿
breakdown pressure 破裂压力,破碎压力
breakdown slip 极限转差率,临界转差率
breakdown torque 极限转矩
breaker 碎浪,断路器
Breakers Sizing 断路器规格确定
break-even point 保本点,收支平衡点,损益两平点
breaking 分断
breaking seas 巨浪
breaking strength 抗断强度
breaking wave 破碎波,破浪波,碎波
break-up energy 分裂能
breakwater 防浪堤,挡水板,防浪板
breast drill 胸压手摇钻
breccia zone 角砾岩区
breeching seals 筒尾密封

bridging 桥联性
brightness of solar disk 日轮亮度
brightness distribution of solar disk 日轮亮度分布
brightness of solar image 太阳镜象亮度
brine 盐水
brine composition 盐水组分
brine concentration 盐水浓度,脱水和浓缩
briquette press 压煤砖机
briquetting 压块
british thermal unit (BTU) 英国热量单位
brittleness 脆性
broad diameter tube 大直径管
broadband irradiance 宽频带辐射
bronze 青铜
brown limestone 褐色石灰岩
brown sandy clay 棕色砂质泥土,褐砂质泥土
Brownian movement 布朗运动
brushless A.C. generator 无电刷交流发电机
brushless DC machine 无刷直流电机
Brushless doubly-fed 无刷双馈
brushless synchronous generator 无刷同步发电机
bubble controlling spray pump 翻泡控制喷水泵
bubble controlling sprayer 翻泡控制喷头
bubble flow 泡流(地热蒸汽泡沿管道的上部流动,其速度与液体的速度相近)
bubbler 鼓泡器,起泡器
bubbling fluidized bed 鼓泡流化床
bucket 桶,铲斗
bucket elevator 斗式提升机,斗式运输机

bucking 浸渍
buffer solution 缓冲液，缓冲剂
buffer storage 缓冲储热器
buffer tank 缓冲槽
buffing wheel 抛光轮，弹性磨轮
bug battery 细菌电池
building envelope （太阳能）建筑物外体
building heat gain （太阳能）建筑物热（量）增益
building heat load coefficient 建筑物加热负荷系数
building heat loss 建筑物热量损失
building heating load （太阳能）建筑物加热负荷，（太阳能）建筑物供暖负荷
building integrated photovoltaics (BIPV) 光伏建筑一体化
building load coefficient 建筑物加热负荷系数，（太阳能）建筑物供暖负荷系数
building wake 建筑物尾流
build-up pressure 生成压力
built-in redundancy 内建冗余
bulb casing 灯泡形外壳
bulb hanger 发电机吊架
bulb light 球泡灯
bulb turbine 泡式涡轮机，泡式水轮机
bulb-type generating unit （潮汐电站用）灯泡式发电装置
bulge wave 涨波
bulk 大小，体积，大批，大多数，散装
bulk density 容（体）积密度
bulk fiber 散纤维
bulk material 散装材料
bulk modulus 体积弹性模量
bulk volume 容积

bulk waste 散装废料，散装垃圾
bulkhead 隔壁，防水壁
bull gear 大齿轮
bulldozer 推土机
bundled biofuel 捆扎生物质燃料
bundler 捆束机
bunsen burner 本生灯
buoy 浮标
buoyancy 浮力
buoyancy effect 浮力效应
buoyancy force 浮力
Burgers circuit 柏氏回路
Burgers vector 柏氏矢量
Buried-contact solar cell 埋入式接点太阳电池
burn through 烧蚀
burnable absorber 可燃吸收体
burnable gas 可燃气体
burner block 烧嘴砖
burning section 燃烬段
burnout 燃尽，燃料烧尽
burr 芒刺，刺果植物，针球
bursting 突然破裂，爆发，脉冲
bursting disc 爆炸隔膜，防爆膜
bus bar 母线
bus bar separator 母线间隙垫
bus bar 汇流条，汇流排，母线
bus bar expansion joint 母线伸缩节
bus coupler 总线耦合器
bus duct 母线槽
bus voltage 母线电压
bush 矮树丛，（机械）衬套
bushing 轴衬，套管
butt weld 对接焊缝
butterfly valve 蝶形阀
butterfly-type isolating valve 蝴蝶型隔离阀

butylene rubber 丁烯橡胶
bypass capacity 旁路电容
bypass damper 旁通阀
bypass diode 旁路二极管
bypass filter 旁滤器
bypass valve 旁通阀,旁通活门,防通阀
bypass (shunt) diode 旁路二极管
byte 字节

C

C/N ratio 碳氮比
cabinet converter 变频器柜
cabinet door 柜门
cable 电线,电缆
cable armor 电缆铠装
cable bundle 束,光纤束,捆,卷
cable connection system 锚链连接系统,缆索连接系统
cable cutter 电缆剪
cable fitting 电缆配件
cable gland 电缆衬垫
cable installation 电缆敷设,电缆安装
cable making tool 造线机
cable reel 电缆盘
cable routing 电缆路由选择
cable shear 电缆剪
cable sheath 电缆包皮层
cable shoe 电缆靴
cable slip-ring 锚链滑环
cable tie 电缆带
cable tray 电缆盘
cable trunk 电缆管道
cable twist 扭缆
cadmium 镉
cadmium sulfide based thin film solar cell 硫化镉基薄膜太阳能电池
cadmium sulfide ceramic solar cell 陶瓷硫化镉太阳能电池
cadmium sulphide ceramic solar cell 硫化镉陶瓷太阳电池
cadmium sulphide solar cell 硫化镉太阳能电池
cadmium telluride solar cell 碲化镉太阳能电池
cage 保持器
caisson 沉井,沉箱
calandria 排管式加热体(冷却管)
calcareous algae 石灰藻,钙性藻类
calcite 方解石
calcium carbonate 碳酸钙
calcium compound 钙化合物
calcium oxalate 草酸钙
calcium silicate 硅酸钙
calcium silicide 硅化钙
calcium stearate 硬脂酸钙
calibrated feeder 定量供料器
calibrating 标定(获得标准太阳电池的方法或手段称为标定)
calibration 标度,刻度,校准
calibration factor 校准因子
calibration value 标定值
calicum-aluminium-silicon 硅铝钙合金
California Distributed Energy Resources Guide 加利福尼亚州分布式能源指南
caliper 卡钳
caliper log 钻孔直径记录图
calm 平静的,无风浪的,零级风,无风
calm wind 静风
caloricity 热容,热值
calorific balance 热量平衡
calorific heat 热容量,热值,发热能力
calorific intensity 发热强度

calorific power 发热量，发热值
calorific value 热值，发热量，卡值
calorific value sensor 热值传感器
calorimeter 热量计
calorimeter box 量热箱
calorimetric procedure 热测量法
calorstat 恒温器，温度调节器
cam 凸轮
cam shaft 凸轮轴
camber 曲弧度
camber control 拱度控制
camber line 弧线，弧面曲线
cambered airfoil 有弧度翼型
camelina 亚麻荠属
Campbell diagram 坎贝尔图，共振图
canal lock 船闸
canary grass 金丝雀草
candidate site 候选场址
candle filter 烛式过滤器
canopy density 植冠密度
cantilever 伸臂，悬臂，悬臂梁
cantilevered beam 悬臂梁
cap rock 冠岩，盖岩
capacitance 电容
capacitance effect 电容效应
capacitive load 电容性负载
capacitive method 电容法
capacitive reactance 容抗
capacitor 电容器
capacitor bank 电容器组合
capacitor compensation 电容补偿
capacitor for voltage protection 保护电容器
capacity curve 库容曲线
capacity factor 容量系数，利用率，功率，能力系数
capacity limit 容量限度，极限容量

capacity of reservoir 库容
capillary energy 毛细管能量
capillary wave 表面张力波，界面波
capital fund rule 资本金制
capping ends 顶盖末端
capping mass 覆盖，物表土
capping structure 覆盖层结构
capstan lathe 绞盘车床
capturable wind power 可捕获风能
capture area 进气面积
capture chamber 吸收室
capture cross section 捕获截面
capture energy 摄取能量
capture energy system 摄取能量系统
capture width 捕获宽度，（波能）吸收率
carbide cutter 硬质合金刀具，碳化物刀具
carbohydrate 碳水化合物
carbohydrate fermentation 糖类发酵
carbon brush 碳刷
carbon capture and storage 碳捕获与封存
carbon char 活性炭，活性炭黑
carbon content 碳含量
carbon cycle 碳循环过程
carbon electrode 碳电极
carbon fiber 碳纤维
carbon filament 碳丝
carbon filter 活性炭过滤器
carbon footprint 碳排放量，碳足迹
carbon monoxide 一氧化碳
"carbon neutral" emission "碳中性"排放
carbon to nitrogen ratio 碳氮比
carbonaceous material 碳素材料，碳质材料

carbonate-type geothermal water 碳酸盐型地热水
carbon-containing material 含碳物质
carbon-filament lamp 碳丝灯泡
carbon-free renewable energy 清洁可再生能源
carbon-in-ash value 飞灰含碳值
carbothermic reduction 碳热还原
carburetor 汽化器
carburizing 渗碳,渗碳处理,渗碳剂
cardan shaft 万向轴
Carnot cycle 卡诺循环
Carnot cycle efficiency 卡诺循环效率
Carnot efficiency 卡诺效率
carrier 载波
carrier current 载波电流
carrier medium 媒介载体
Cartesian coordinates 笛卡尔坐标系
cartridge 夹头
cascade heater 串联加热器
cascade multijunction cell 级联多结电池
cascade solar cell 叠层太阳电池(级联太阳电池)
cased depth 覆盖层深度
cash flow analysis 现金流分析
casing collapse 套管挤坏,套管损坏
casing collar location 套管接箍位置
casing coupling 套管接箍,套管连接器
casing diameter 壳体直径
casing erosion 套管侵蚀
casing flange 套管法兰盘
casing head 套管头
casing joint 套管接头
casing perforating 套管射孔
casing setting depth 壳体下入深度
casing shoe 套管鞋,套管靴,套管底环

casing string 套管柱,套管串
casing tonnage 壳体吨位
casing type 壳体种类
cassava 木薯
cast-aluminum rotor 铸铝转子
castellated coupling 牙嵌式连接
casting 铸件,铸造
casting-aluminium 铸铝
casting-copper 铸铜
casting-gray iron 铸灰口铁
casting-malleable iron 可锻铸铁
casting-steel 铸钢
castle nut 槽形螺母,城堡螺母
catabolism 分解代谢
catalyst 催化剂
catalyst bed 催化剂床,触媒床
catalyst poisoning 催化剂中毒
catalytic combustion incinerator 催化燃烧焚烧炉
catalytic combustor 催化燃烧室
catalytic converter 催化转化器,催化转换器,触媒转换器
catalytic gasification reactor 催化气化反应器
catalytic intervention of enzymes 酶的催化作用
catalytic pyrolysis 催化热解
catalytic reaction process 催化反应过程
catalytic reactor 催化反应堆
catalytic tar cracking 催化焦油裂解
catalytic upgrading 催化改质
cataphoretic coating 电泳涂层
catastrophic failure 严重故障
catch pot 油气分离罐
catchment areas 集水区,汇水面积
cathode 阴极,负极

cathode plate 阴极板
cathode spot 阴极斑点
cathode-grid voltage 栅—阴极电压
cathode-ray tube 阴极射线管
cathodic protection system 阴极保护系统
cathodic voltage 阴极电压
cation 阳离子，正离子
cation exchanger 阳离子交换器
cattle slurry 牛粪浆
catwalk 桥上人行道，狭小通道
caulking metal 填隙合金（材料）
cause of scale formation 垢的成因
causeway 堤道，长堤
caustic chemical solution 苛性溶液
cavitation 气穴现象，空穴作用，成穴
cavity absorber （太阳能）腔体吸收器
cavity flow 空泡流，气穴流，涡空流
cavity heater 空腔加热器
cavity radiation 空腔辐射
cavity receiver 腔式吸热器
cavity resonator 空腔共振波能转换装置
CdTe thin-film solar cell 碲化镉薄膜太阳能电池
celestial axis 天轴
celestial equator 天赤道
celestial meridian 天球子午圈
celestial movement 天体运动
celestial pole 天极
celestial sphere 天球
cell barrier 电池阻挡层
cell efficiency 电池效率
cell temperature 电池温度
cell voltage 槽电压，电池电压
cellular localization 细胞定位
cellular solar collector 蜂窝状太阳能集热器

cellulose 纤维素
cellulose complex 纤维素复合体，纤维素复合物
cellulose rich residue 富含植物纤维质残渣
cellulose waste 纤维素废弃物
cellulosic biomass 纤维素类生物质
cellulosic ethanol 纤维素乙醇
cellulosic ethanol plant 纤维素乙醇工厂
cement 水泥
cement grout 水泥浆液，水泥浆
cement kiln 水泥窑，水泥回转窑
cement lined piping 水泥衬里
cement plug 水泥塞
cement sheath 水泥护层，水泥壳，水泥环
cement slurry 固井水泥浆，水泥浆
cementing material 水泥材料
cementite 渗碳体，碳化铁
cenozoic volcanism 新生代火山活动
center bit 中心位
center distance 中心距
center equiaxial crystal zone 中心等轴晶区
center gear 中心轮
center of inertia 惯性中心
center of mass 质量中心，质心
center of mass of the earth 地球质心
center of mass of the moon 月亮质心
center puncher 中心冲
central main control cabinet 中央主控制箱
central air conditioner heat pump water chillers 中央空调热泵冷水机组
central air conditioning system 中央空调系统

central pebble rock pile 中心式卵石床
central power station 中心电力站
central receiver collector 中心接收集热器，中央接收器式集热器
central receiver optical system 中心接收器光学系统
central receiver power system 中心接收器电源装置
central receiver solar-thermal power plant 中央收集系统太阳能发电站
central receiver-heliostat 中心接收器定日镜阵列，中心吸热器定日镜阵
central riser pipe 中央立管
centralized control 集中控制
centralized wind energy conversion systems 集中式风能转换系统
central-tower solar thermal electric powerplant 中心塔太阳热电动力站
central-tower solar-electric system 中心塔太阳能—电能转换系统
centre-of-mass energy 质心能量
centrifugal 离心
centrifugal blower 离心式鼓风机
centrifugal force 离心力，地心引力
centrifugal moment 离心力矩
centrifugal pump 离心泵
centrifugal separator 离心式分离器
centrifugal storage pump 离心式蓄能泵
centrifugal unit 离心单元
ceramic 陶瓷制品
ceramic ball 陶瓷球
ceramic bead 陶瓷小珠
ceramic coating 陶瓷涂层
ceramic crucible 陶瓷坩埚
ceramic fiber 陶瓷纤维
ceramic pebble 陶瓷卵石
ceramic pebble bed 陶瓷卵石床

ceramic seal 陶瓷密封
ceramic solar cell 陶瓷太阳能电池
ceramics 陶瓷，陶瓷技术
cereal combustion 谷物燃烧
cereal straw 谷物秸秆
chain conveyor 链条输送机
chain drag 链板式输送机
chain drive 链传动
chain making tools 造链机
chain trough conveyor 环链式输送机
chain vice 链式钳
chain wheel 滑轮
chain-grate stoker 链条炉排加煤机
chalcedony 玉髓，石髓
chamfer 斜面，凹槽
chamfer machines 倒角机
chamotte brick 耐火黏土砖
change of state 物态变换
change of tide 转潮
change over 转换，转接，改变成，对调位
change over switching 转换开关,换接
changeover switch 切换开关,转换开关
changer speed gear 变速箱,变速齿轮
changing drilling location 改变钻井位置
channel base 沟渠基底
channel style fan 管道风机
char combustion 煤焦燃烧
char oxidation 煤焦氧化
char particle 碳粒
character for geothermal resource 地热资源特征
characteristic curve 特性曲线
characteristic of geothermal resource 地热资源特征
characteristic of instantaneous efficiency 瞬时效率特性

characteristic water level 特征水位
characterization test 表征测试
charcoal 木炭
charcoal combustion 木炭燃烧
charcoal gasifier 木炭气化炉
charge carrier 载流子，带电粒子，载荷子，电荷载体
charge cloud 电荷云
charge controller 充电控制器
charge current 充电电流
charge efficiency 充电效率
charge power 充电功率
charge voltage 充电电压
chargeable heat 消耗热
charged particle 带电粒子
charged particle radiation 带电粒子辐射
charging air inlet 充气嘴
charging efficiency 充电效率，充气效率
charging valve 进给阀
check valve 止回阀
cheese 垫砖
cheese-head screw 有槽凸圆柱头螺钉
chemical alteration 化学蚀变
chemical bath deposition 化学池沉积
chemical bond 化学键
chemical compatibility 化学兼容性，化学相容性
chemical corrosion 化学腐蚀
chemical drilling 化学凿岩
chemical electrolyte 化学电解质
chemical energy 化学能
chemical equilibrium 化学平衡，化学均衡
chemical fractionation technique 化工分离技术

chemical heat pump 化学热泵
chemical interaction 化学相互作用
chemical kinetics 反应动力学，化学动力学
chemical lasing energy 化学激光器能量
chemical oxygen demand (COD) 化学需氧量，化学耗氧量
chemical potential 化学势，化学位
chemical reaction heat energy storage 化学反应热储存
chemical transformation of sulfide 硫化物化学变化
chemical vapor deposition 化学气相沉积
chemistry conversion technology 化学转换技术
chill zone 激冷区，表面等轴晶区
chilled water 冷冻水，冷水，冷却水
chilled-water coil 冷却线圈
chiller capacity 冷却能力，冷水机组容量
chiller plant 制冷设备，制冷设备
chiller system 制冷系统，制冷机系统，冷水系统
chiller-heater 冷温水机组
chillers 冷却装置
chipper 削片机
chipping 修琢
chisel 凿子，砍凿
chloride 氯化物，漂白粉
chloride content 氯化物含量
chloride storage battery 氯蓄电池
chlorine 氯
chlorine content 氯含量
chlorophyll 叶绿素
chlorophyll solar cell 叶绿素电池
chloroplast membrane 叶绿体膜

choice of storage medium	储热介质选择
choke	窒息，阻气门
choke valve	阻气阀
chopped straw	轧断的禾草
chopper circuit	斩波电路
chord	翼弦
chord line	弦线，翼弦线
chrome	铬，铬矿石，氧化铬
chromel-alumel thermocouple	镍铬—镍铝热电偶
chromite refractory	铬矿质耐火材料
chromosphere	色球层
chromosphere eruption	色球爆发
chronometer	精密计时表
chuck	用卡盘夹住，卡盘，轴承座
chunk wood	碎木块
chute of water	水流
cinder	炉渣
circle diagram	圆图
circlip	环形，弹性挡圈
circuit board	电路板
circuit branch	支路
circuit breaker	断路器
circuit component	电路元件
circuit diagram	电路图
circuit parameter	电路参数
circuit voltage	电路电压
circuit-commutated recovery time	电路换向恢复时间
circuitry	电路，电路系统，电路学
circular	圆形，环，循环
circular Fresnel lens	圆形菲涅尔透镜
circular platform of concrete	混凝土圆坯
circular platform of water particle	水质点圆周运动
circular saw	圆锯
circular solar cell	环形太阳能电池
circular solar collector	环形太阳能集热器
circular track	环轨
circulate	循环，流通
circulating cooling water	循环冷却水
circulating cooling water system	循环冷却水系统
circulating current	环流
circulating fluid	循环流体
circulating fluidized bed	循环流化床
circulating pump	循环水泵
circulating steam generator	循环蒸汽发生器
circulating system	循环系统
circulating water pump house	循环水泵房
circulation	循环
circulation control rotor	环量控制型风轮
circulation of hot medium	热介质循环
circulation pressure equipment	循环加压设备
circulation pump	循环泵
circulation supply system	循环供给系统
circumference	圆周，周长
circumferential backlash	圆周侧隙
circumferential joint	周圈接缝
circumferential speed	圆周速度，周向速率
circumsolar radiation	环日辐射
cirrus cloud	卷云
CIS thin-film solar cell	铜铟硒薄膜太阳能电池
citrus waste water	柑桔废水
city garbage-burning power station	燃用城市垃圾电厂
city rubbish	城市垃圾

civil and erection cost 建筑安装工程费
civil engineer 土木工程师
civil works 土建工程，建筑工程
civilian over-current protection circuit breaker 民用过电流保护断路器（小型断路器）
cladding 包层，覆层，电镀，喷镀
clam shell 泡壳包装
clamp 夹子，夹具，夹钳
clamp ammeter 钳形表
clamping force 夹紧力，锁模力
clamping/holding system 夹具/支持系统
clapotis 驻波，反射浪
classification of solar collector 太阳能集热器分类
Claude cycle 克劳德循环
claw hammer 拔钉锤
clay liner 黏土衬垫
clay mineral 黏土矿物
clean burning fuel 完全燃烧燃料
clean development mechanism 清洁发展机制
clean energy 清洁能源，洁净能源
clean energy source 无污染能源
clean source of power 清洁能源
clean steam 纯净蒸汽，不含矿物盐蒸汽
clean value 净值
cleaner energy 清洁能源
clear glass 透明玻璃
clear plexiglass 透明（耐热）有机玻璃
clearance 排除故障，清除
clear-day 晴天
clear-day insolation 晴天接受到的太阳辐射，晴天单位水平表面上接受直接太阳能的速率
clear-day sun-pulse curve 晴天太阳脉冲曲线
clearness factor 清晰度系数
clearness index 晴空指数，清晰度
clerestory window 天窗，高侧窗
clevis U形夹
clevis drawbar 牵引环，联结钩
clevis joint 拖钩，脚架接头
climate 气候
climate conditions 气候条件
climate effect 气候影响
climatic atlas 气候图集
climatic conditions 气象条件
climatic extreme 气候极值
climatic limit loading 气候极限载荷
climatize 适应气候
climatography 气候志
climatology 气候学
clinker formation 熟料形成，熔块形成
close grain 结晶粒
close-coupled solar water heater 紧凑式太阳（能）热水器
closed circuit 闭合电路
closed circuit cooler 闭路冷却器
closed circuit evaporative cooler 闭式蒸发式冷却器
closed cycle 闭式循环
closed cycle engine 闭路式循环动力机
closed cycle gas turbine installation 闭式循环燃气轮机装置
closed cycle ocean thermal 闭路循环海洋热能
closed cycle ocean thermal energy conversion 闭式循环海洋热能转换
closed cycle system 闭式循环系统
closed gas turbine 闭式燃气轮机
closed gas turbine plant 闭式燃气轮机装置

closed landfill leachate （垃圾填埋场）封场后渗滤液
closed liquid solar collector 封闭式液体太阳能集热器
closed loop 闭环，闭合回路
closed loop cycle 闭合式循环周期
closed loop earth coupling 闭环大地耦合
closed loop geothermal heat exchanger 闭环地热换热器
closed loop steam system 闭环蒸汽系统
closed loop system 闭环系统
closed loop 闭合环路，闭合回线，闭环
closed oil cycle 封闭油循环
closed power plant of energy from salinity gradient 闭式盐差能动力装置
closed system 封闭系统
closed thermal cycle 封闭热循环
closed valve 闭阀
closed-cycle 封闭式循环
closed-cycle engine 封闭式循环发动机
closed-cycle ocean thermal energy conversion system 闭式循环海水温差发电系统，闭式循环海洋热能发电系统
closed-cycle sea thermal power plant 闭路循环海洋热能电站
closed-cycle system 封闭式循环系统
closed-cycle turbine system 闭路循环透平系统
closed-cycle-tower 闭路循环塔
closed-gas-turbine power station 闭式循环燃气轮机电厂
closed-in pressure 关井压力，闭井压力
closed-loop active solar water heater 主动式闭环太阳能热水器
closed-loop control system 闭环控制系统
closed-loop domestic water heater 民用闭环太阳能热水器
closed-loop solar heating system 闭路太阳能供暖系统，闭路太阳能加热系统
closed-loop system configuration 闭环系统构型，闭环系统结构
closer spacing 近间距
closure 截流，合龙
cloud cavitation 云状空化
cloud water 云水
clout nail 大帽钉
clutch 离合器，联轴器
clutch brake 离合器制动器
CMU system 卡内基—米隆系统（一种利用氨作工作流体的闭环海洋温差发电系统）
coal blend 配煤
coal briquette 煤砖，煤块
coal bunker 煤仓
coal gasification 煤炭气化，煤气化
coal gasification combined cycle power plant 煤气化联合循环发电厂
coal handling 输煤设备
coal-fired furnace 燃煤炉
coal-fired power station 燃煤电站，燃煤电厂
coanda effect 附壁效应
coarse 粗（糙，略），近似
coarse control 粗调节
coarse floc 粗絮状物
coarse fly-ash particle 粗粉煤灰颗粒
coarse gravel 粗碎石，粗砂砾
coarse material 粗粒物质
coarse sand 粗砂，粗沙
coarse synchronizing 粗同步
coastal current 沿岸流，近岸流

Coastal Engineering Manual 美国海岸工程手册
coastal protection 海岸保护
coastal sand dune 海岸沙丘
coastal tide 海潮，海岸潮
coastal wind energy converter 沿海风能转换器
coastal zone alternative energy 沿岸区新能源，沿岸区替代能源
coastal zone management 海岸带管理
coated collector 有涂层的集热器
coated glass 有涂层的玻璃
coating process 裹贴法，涂层工艺
coaxial 共轴的，同轴的
coaxial cable 同轴电缆
coaxial concentrator 同轴聚光器
co-axial contra rotating propeller 共轴反转式螺旋桨
coaxial microwave ion source 同轴微波离子源
coaxial shaft 同轴
cobamide 谷酰胺
coccolithophore 颗石藻
Cockerel raft wave energy conversion device 科克雷尔筏波能转换器
co-combustion 混燃，混烧
coconut shell 椰子壳，椰壳
co-current combustion 顺流燃烧
co-current flow 并向流
coefficient of losses 损耗系数，漏电系数
coefficient of performance 性能系数
coefficient of pore structure 孔隙结构系数
coefficient of thermal diffusion 热扩散系数
coefficient of torsional rigidity 扭转刚度系数

coefficient of transmission 透射系数
coferment 辅酶
co-fermentation 联合发酵
cofferdam 围堰，围堰坝
co-firing 混烧
co-firing of biomass 生物质共燃
co-firing of biomass with coal 生物质与煤共燃
cogeneration 废热发电，热电联产
cogeneration capacity 热电联产能力
cogeneration plant 热电厂，热电站
co-generation power plant 热电联产电厂
cogeneration system 汽电共生系统，热电联产机组
cogenerator 热电联供装置
coherent boundary 共格界面
coherent twin boundary 共格孪晶界，相干孪晶间界
coil 线圈
coil heater 盘管加热器
coil pitch 线圈节距
coil spring 弹圈
coil spring damper 盘簧减震器
coil winding 线圈绕组
coke bed 焦炭床层，底焦
coking stoker 焦化炉排，焦化加煤机
cold bottom water 底部冷水
cold crushing strength 常温耐压强度
cold gas plume 冷气羽流
cold meteoric water 冷的大气水，冷的天落水
cold nutrient sea water 冷营养海水
cold salt storage tank 冷盐储罐，冷盐储槽
cold shortness 冷脆（性）
cold standby reserve 冷备用

cold water discharge　冷水排放
cold water flow rate　冷水流量
cold water intake pipe　冷水进水管
cold water layer　冷水层
cold water outlet　冷水出口
cold water pipe　冷水管
cold water pump　冷水泵
cold water recharge　冷水再灌
colder deep water　底层冷水
cold-water pipe with buoyancy tank　带浮力箱的冷水管
cold-water pond　冷水池
cold-water-mass　冷水团
collapse of hole well　孔壁坍塌
collapse of production casing　生产套管坍塌
collar　凸缘，套环，轴环，卡圈，安装环
collar bolt　凸缘螺栓
collar locator　接箍定位器
collecting aperture area　集热开口面积
collecting efficiency　集热效率
collecting plate　集热板
collecting system　集热系统
collection and transmission system　（地热流体的）集输系统
collection efficiency　收集效率，集尘效率，除尘效率，集热效率
collection length of day　日集热时间
collection medium　集热介质
collection of bottom hole sample　孔底试样采集
collection of wind energy　风能的收集
collection system　集热系统，集热装置
collector　集电器，集电环，集电极
collector absorber plate　集热器吸热板
collector absorber surface　集热器吸收器表面
collector air flow rate　集热器空气流速，集热器空气流率，集热器空气流量
collector area　集热器面积
collector area requirement　集热器面积要求
collector array　集热器阵列
collector array structure　集热器阵结构
collector azimuth angle　集热器方位角
collector casing　集热器外壳
collector configuration　集热器设置方式
collector cover　集热器盖层
collector cover plate　集热器盖板
collector efficiency　集热器效率
collector efficiency equation　集热器效率方程
collector efficiency factor　集热器效率因子
collector electrode　集电极
collector energy production rate　集热器产能速率
collector equation　集热器方程式
collector frame　集热器框架
collector field　集热场
collector flow factor　集热器流动因子
collector fluid　集热器流体
collector generator　集热器发电机
collector glazing　集热器施釉
collector heat exchanger　集热器的热交换器，集热器的换热器
collector heat loss conductance　集热器热损失传导率
collector heat removal efficiency　集热器热交换效率
collector heat removal factor　集热器热转移因子
collector inlet　集热器进口

collector instantaneous efficiency　集热器瞬时效率
collector insolation　集热器接收到的太阳辐射，集热器单位水平表面上接收直接太阳能的速率
collector lens　集热透镜
collector load　集热器负载
collector loop　集热器回路
collector mirror　集热器反射镜
collector mounting procedure　安装集热器程序
collector of plastic　塑料集热器
collector of solar radiation　太阳能辐射集热器
collector of wind energy　风能收集器
collector optical efficiency　集热器光学效率
collector orientation　集热器方位
collector orientation angle　集热器定向角
collector outlet　集热器出口
collector overall energy loss coefficient　集热器总能量损失系数
collector overall heat loss coefficient　集热器总热损系数
collector performance　集热器性能
collector plumbing layout　集热器管件布置
collector pump　集热器泵
collector reserve current　集电极反向电流
collector ring　集电环，集流环
collector size　集热器尺寸
collector slope　集热器倾斜度
collector storage system　集热器储存系统
collector subsystem　集热器子系统，集热器辅助系统
collector surface　集热器表面
collector system　汇集系统
collector testing　集热器试验
collector theory　集热器理论
collector thermal capacity　集热器热容量
collector tilt　集热器倾角
collector tracker　（太阳能）集热器跟踪器
collector with less curvature　小弧度集热器
collector-and-storage-tank solar water heater　集热器—储罐太阳能热水器（太阳能热水器两种基本类型之一）
collector-heat exchanger correction factor　集热器—热交换器修正系数
collector-heat exchanger efficiency　集热器—热交换器效率，集热器—换热器效率
collector-plate flow distribution　集热器集热板流量分布
collector-tank heat exchanger　集热器—水箱热交换器
collision rate　碰撞率
colloid charging system　胶体充电系统
color identification　彩色识别
color rendering　显色指数
color temperature　色温
color-detection system　彩色检测系统
colorimetric thermometer　比色温度计
combination　结合，合并，化合，化合物
combination burner-boiler　组合式燃烧器锅炉
combination passive solar house　组合式被动太阳房

combination pliers 钢丝钳，平嘴钳
combination solar energy system 组合式太阳能系统，组合式太阳能装置
combined circulation boiler 复合循环锅炉
combined cold heat and power 冷热电联产
combined cycle 联合循环
combined cycle combustion turbine 复合循环燃气轮机
combined cycle plant 复合循环厂
combined cycle power plant 联合循环发电厂
combined heat and power station 热电联产电厂
combined loss 组合损失
combined sample 合成样品
combined solar heating-cooling installation 太阳能供暖制冷联合装置
combined solar-wind energy conversion system 太阳能—风能联合转换系统
combined steam and gas turbine cycle 蒸汽—燃气联合循环
combustible gas 可燃气体，可燃性气体
combustible waste 可燃垃圾，可燃废物
combustion 燃烧，氧化
combustion air 助燃空气，燃烧空气
combustion behavior 燃烧性能
combustion chamber 燃烧室
combustion chamber volume heat release rate 炉膛容积热负荷
combustion characteristics of biomass 生物质的燃烧特性
combustion efficiency 燃烧效率
combustion fixed bed 炉算加热
combustion furnace 燃烧炉，焚烧炉

combustion gas 燃烧气体
combustion of biomass fuel 生物质燃料燃烧
combustion performance 燃烧性能
combustion plant 燃烧车间，燃烧设施
combustion process 燃烧过程
combustion process control 燃烧过程控制
combustion process controller 燃烧过程控制器
combustion rate 燃烧率
combustion reaction 燃烧反应
combustion section 燃烧段
combustion system 燃烧系统
combustion technique 燃烧技术
combustion technology 燃烧技术
combustion temperature 燃烧温度
combustion turbine 燃气轮机
combustion turbine generator 燃气轮发电机
combustion zone 燃烧区
combustor 燃烧器，燃烧室
command 命令
command input 指令输入，命令输入
commencement 开始
commercial natural steam field 工业天然蒸汽田
commercial plant 工业装置，工业设备
commercial solar building 商品性太阳能建筑
commercial tidal power plant 商业性潮汐电站
commercial use of solar energy 太阳能的工业应用，太阳能的商用
comminution 粉碎，捣碎
commissioning test 投运试验

common earthing system　共用接地系统
common sample　共用样品
common-mode-voltage　共模电压
communication cable　通信电缆
communication line　通信线，通信线路
communication port　通信端口
communication tower　通信塔
commutating current　整流电流
commutation　换向
commutation condition　换向状况
commutator　换向器，整流器
commutator pitch　换向器节距
commutator segment　换向片
commutator-brush combination　换向器—电刷总线
compact heat exchanger　紧凑式热交换器
compact tidal power station　紧凑型潮汐发电场，紧凑型潮力发电站
comparable comprehensive energy consumption for unit output of product　产品单位产量可比综合耗能
comparative matrix　比较矩阵
comparative tracking index　相对漏电起痕指数
compensated pyrheliometer　补偿式绝对辐射表
compensating stream　补偿流
compensating wind　补偿风
compensation circuit　补偿电路
compensation pyrheliometer　补偿直接日射（强度）计
complementary energy　余能，补充能量
complete circuit　闭合回路
complete combustion　完全燃烧
complete tree　全树

completion test　完井试验
complex amplitude　复幅值
complex conjugate control　复共轭控制
complex impedance　复数阻抗
complex polysaccharide carbohydrate　络合多糖碳水化合物
complex terrain　复杂地形，复杂地形带
complexed heating system　复合式供暖系统
component　组分，组元
component efficiency　组件效率
component of terrestrial heat flow　大地热流量分量
component of flat-plate collector　平板型集热器元件
composite lens　复合透镜
composite materials　复合材料
composite parabolic condenser　复合抛物面聚光器
composition　成分
composition of hydrothermal fluid　热液流体成分
compositional analysis　成分分析，组成分析
compost material　堆肥
compound　混合物，化合物，复合的，混合，配合
compound generator　复励发电机
compound parabolic concentrator　复合抛物面型集光器
compound parabolic concentrator collector　复合抛物面集热器
compound semiconductor solar cell　化合物半导体太阳能电池
compound-curvature collector　复合曲率集热器

compounded 复励
comprehensisve energy consumption for unit output value 产品单位产量综合耗能（单位产品综合耗能）
comprehensive efficiency of pumped storage station 抽水蓄能电站综合效率
comprehensive energy consumption 综合耗能
comprehensive utilization of geothermal energy 地热能综合利用
compressed air 压缩空气
compressed air energy storage 压缩空气储能
compressed air energy storage (CAES) 压缩空气储能系统
compressed air storage 压缩空气存储
compressed air storage tank 压缩空气储罐
compressed biofuel 压缩生物质燃料
compressed liquid 压缩液体
compressibility 体积压缩率
compression ignition 压缩点火
compression ignition engine 压燃式发动机
compression ratio 压缩比
compression refrigerating cycle 压缩制冷循环
compression stroke 压缩冲程
compressional wave 压缩波
compressor 压缩机，压迫器
compressor control module 压缩机控制模块
computational fluid dynamic model 计算流体力学模型
computational fluid dynamics 计算流体动力学

computational fluid dynamics analysis 水力仿真分析
computer processing unit (CPU) 中央处理器
computer simulation 计算机仿真
concave 凹，凹面
concave lens 凹面透镜
concave mirror 凹面镜
concave-convex lens 凹凸透镜
concentrated cell 聚光电池
concentrated solar power 集中式太阳能发电
concentrating (focusing) collector 聚焦集热器
concentrating collector of parabolic cylindrical type 抛物面柱面型聚光集热器
concentrating mirror surface 聚光镜面
concentrating optics 聚光光学系统
concentrating solar cell 聚光太阳能电池
concentrating solar cogeneration system 聚光太阳能热电联供系统
concentrating solar collector 聚光型太阳能集热器
concentrating solar cooker 聚光型太阳灶
concentrating solar power 聚光太阳能发电，聚光型太阳能热发电，聚焦式太阳能发电
concentration gradient 浓度梯度
concentration of heat source 热源富集
concentration profile 浓度分布曲线，浓度剖视图
concentration range 浓度范围
concentration ratio 聚光率
concentration triangle 浓度三角形
concentration unit 浓度单位
concentrator 聚光器

concentrator aperture area　聚光器采光面积
concentrator cell　聚光电池
concentrator field　聚光场
concentrator field aperture area　聚光场采光面积
concentrator solar cell　聚光太阳能电池
concentrator(solar) collector　聚光型（太阳能）集热器
concentrator surface contour error　聚光器表面轮廓误差
concentrator with single axis　单轴跟踪聚光器
concentrator with two axis　双轴跟踪聚光器
concentric annuli　同心环空
concentric cylinder　同轴圆筒
concentric tubes collector　同心管式（太阳能）集热器
conceptual model　概念模型
concrete　混凝土
concrete block solar thermal storage wall　混凝土块太阳能储热壁
concrete caisson　混凝土沉箱
concrete cap　混凝土柱帽，混凝土顶盖
concrete column　混凝土柱
concrete drill　混凝土钻
concrete pad　混凝土基座
concrete solar collector　混凝土制太阳能集热器
concrete wall　混凝土墙
condensate　浓缩物
condensate pump　凝结水泵
condensate return　凝结水循环
condensate tank　冷凝槽
condensate water　冷凝水，凝结水
condensation　冷凝

condensation aerosol　凝结气胶
condensation loss　凝结损失
condensation zone　凝析带
condensed fluid　凝结液
condenser　凝汽器，冷凝器，凝结器，电容器
condenser cooling surface　凝汽器热交换面
condenser pressure　冷凝器压力,凝汽器压力
condenser temperature　冷凝器温度
condenser-discharge anemometer　电容放电风速计
condenser-heat exchanger　冷凝器—热交换机
condensing plant　凝汽装置
condensing vapor-to-liquid heat exchanger　凝结蒸汽—液体热交换器
condensate drum　冷凝液罐
condition of geothermal heat　地热条件
conditioning tower　增湿塔
conductance　电导
conductance of material　材料的热导
conducting bar　导电排，导条
conducting ring　导电环
conduction band　导带
conduction heat loss　传导热损失
conductive adhesive　导电黏合剂
conductivity　导电性
conductor　导体，隔水套管
conductor clamp　卡线钳
conductor guide　套管导向
conductor holder　夹线器
conductor thickness　导线厚度
conduit　导管，线管
conduit box　导管接线盒
conduit entry　导管引入装置

conduit fittings 导管配件
conduit outlet 电线引出口
cone 锥形物,圆锥体,使成锥形
cone angle of rotor 风轮锥角
cone bit 牙轮钻头,锥形钻头
cone valve 锥形阀
confined vortex 约束涡
confining layer 不透水层,隔水层
confirmed installed capacity estimate 确认装机容量评估
confirmed resource estimate 确认资源评估
conical 圆锥的,圆锥形的
conjugate line 共轭线
conjunction 联合,关联,连接词
connate waters 原生水
connect time 接通时间,通电时间
connected heat pumps 连接热泵
connecting rod 连接杆
connection 联结
connection semiconductor 连接半导体
connection to the grid 并网
connectivity 连通度
connector 接线器
consequent low pumping effort 顺向低泵工作
conservation of vorticity 涡量守恒
consistency 致密度
console 控制台
constant angle 定角
constant chord 固定弦
constant chord blade 等截面叶片
constant current wire anemometer 恒流热线风速计
constant energy gap solar cell 恒能带宽度太阳能电池
constant flow 恒流,定量流动

constant flow system 定流量系统
constant flux layer 等通量层
constant frequency generator 恒频发电机
constant frequency output 恒频输出
constant illumination 恒定光照
constant pressure 恒压,恒定压力
constant pressure operation of turbine 汽轮机定压运行
constant rate 恒定流量,恒速
constant speed 恒速
constant speed constant frequency wind turbine generator system 恒速恒频风力发电机组
constant speed region 恒速区域
constant speed rotor 定转速风轮
constant speed-constant frequency system 恒速恒频发电系统
constant tip-speed ratio scheme 恒定叶尖速比方案
constant torque 恒转矩,恒扭矩
constant volume 定容,恒容,恒定体积
constant-rpm operation 恒速运行
constant-volume and variable-temperature system 定风量变温度系统
constituent material 组成材料
constitutional supercooling 成分过冷
construction and demolition debris 拆建废料
construction and demolition waste 拆建废料
construction critical path schedule 施工控制性进度
construction debris 建筑垃圾
construction general layout 施工总布置
construction general schedule 施工总进度

construction machine management 施工机械管理
construction organization design 施工组织设计
construction safety management 施工安全管理
construction specialization 施工专业化
construction work 施工工程
constructive interference 相长干涉,结构干涉,建设性干涉
consumption charge 按消耗量收费(根据实际消耗量收取公用事业费用,与按需用量收费不同)
contact 触点,触头,接触
contact anemometer 接触式风速计
contact breaker 断路器
contact of solar cell 太阳能电池电极
contact time 接触时间
contact voltage 接触式电压
contact-cup anemometer 接触式风杯风速计
contacting heat exchanger 触热交换器
contactor 接触器
container 箱,罐,容器,集装箱,货柜
container for storing heat 储热器,蓄热器
contaminant release rate 污染物排放率
contaminate 污染,弄污
continental climate 大陆性气候
continental lithosphere 大陆岩石圈
continental season wind climate 大陆性季风气候
continental shelf 大陆架
continuous blowdown flash tank 连续排污扩容器
continuous combustion 连续燃烧
continuous emission monitoring 烟气排放连续监测
continuous flow reactor 连续流反应器
continuous furnace 连续式炉,连续作业炉
continuous incineration 连续焚烧方式
continuous operation 持续运行
continuous power generation (潮汐电站)连续发电
continuous radiant-tube furnace 连续式辐射管式炉
continuous solar drier 连续式太阳能干燥机
continuous system 连续系统
continuous thunder and lightning 连续雷电
continuous tracking collector 连续跟踪集热器
continuous tracking concentrator 连续跟踪式聚光器
continuously controllable apparatus 连续可控设备
continuously stirred tank reactor 连续搅拌反应釜
continuously-flowing fluid reservoir 断裂裂隙型地热
continuum 连续介质
contour curve 等值曲线
contour line 轮廓线,等高线
contour map 等高线图
contouring raft 型线浮箱(一种利用波浪能的设备,由一列或一排浮箱组成,能将浮箱运动中的能量转换为流体中的高压),流面筏
contouring raft type wave energy conversion system 波动筏式波浪能发电装置
contracting wave channel 收缩波道式

contraflow system 逆流式系统
contrary-turning propeller 对转推进器，反转螺旋桨
contrast 衬度
control algorithm 控制算法
control apparatus 控制电器
control cabinet 控制柜
control cable 控制电缆，操纵索
control center 控制中心，调度室
control circuit 控制电路
control circuitry 控制电路
control contact 控制触头
control desk 控制台
control device 控制装置
control electrode 控制电极
control electronics 控制电子设备
control gate 节制闸
control gear 控制机构，控制齿轮，控制设备
control input 控制输入
control panel 控制面板
control panel and console 控制盘台
control platform 控制平台
control scheme 控制方案
control shed 控制棚室
control system 控制系统
control tower 控制塔
control valve 控制阀
control valve actuator 阀控传动机构
control voltage 控制电压
control volume 控制体积
control wiring 控制线路
controllable length blade 可控桨叶长度
controllable orifice 控制孔
controlled flash evaporation （海洋温差电站）受控急速蒸发
controlled flash evaporation process 可控闪蒸过程
controlled source audio-frequency magnetotellurics 可控源音频大地电磁法
controller 控制器
controller action 控制器作用
controlling glare 控制强光
controlling recirculation 控制再循环
convection 传送，运流，对流
convection heat loss 对流热损失
convection heating surface 对流受热面
convection pattern 对流形式
convection superheater 对流过热器
convection type desuperheater 面式减温器
convective cell 对流圈
convective circulation model 对流循环模型
convective heat loss 对流热损失，对流热损耗
convective heat transfer 对流传热
convective heat transfer coefficient 对流换热系数
convective heat transfer effect 对流换热效应
convective heating 对流加热，对流采暖
convective instability 对流不稳定
convective mixing 对流混合
convenience receptacle 电源插座
conventional biodiesel 常规生物柴油
conventional boiler 锅炉房，传统形式锅炉
conventional hydrothermal 常规热液
conventional hydrothermal expansion 常规热液扩散
conventional solar cell 常规太阳电池
conventional solid fuel 常规固体燃料

converging wave channel 聚合波道
conversion device 变换装置，转换设备，变换设备
conversion efficiency （光伏电池）转换效率
conversion efficiency of a pumped storage cycle 抽水蓄能循环转换效率，抽水蓄能系数
conversion of salinity gradient energy 盐度差能转换，盐度梯度能转换
conversion of solar energy 太阳能转换
conversion of solar energy into heat 太阳能转换为热能
conversion of solar heat directly into electricity 太阳热直接转换为电能
conversion of wood to liquid fuel 木柴转换为液体燃料
conversion plant 转换装置，燃料转化厂
conversion process 转化程序
conversion technologies for advanced biofuels 先进生物燃料转化技术
conversion technology 转化技术
converter 整流器，变流器，转化器
converter topologies 变换器拓扑结构
converting solar to chemical energy 太阳能转换成化学能
convex lens 凸透镜
convex surface 凸面
conveying chains 输送链
conveyor belt 传送带，运输带
convolution integral 卷积积分，褶合积分
cool air 冷空气
coolant 冷却剂，冷却介质
coolers 冷却机
cooling 冷却，冷却技术
cooling coil 冷却旋管，冷却盘管

cooling configuration 冷却构型，冷却结构
cooling curve 冷却曲线
cooling cycle 冷却循环
cooling dominated vertical closed loop 冷却主导型垂直闭合环路
cooling fin 散热片，冷却片
cooling load 冷负荷，冷却负载
cooling media loss 冷却介质损耗
cooling medium 冷却介质，载热介质，冷却剂
cooling requirement 冷却需求，冷却要求
cooling separator drain heat exchanger 分离器疏水冷却热交换器
cooling tower 冷却塔
cooling tower system 冷却塔系统
cooling water 冷却水
cooling water system 冷却水系统
cooling-power anemometer 冷却功率式风速计
coordinate system 坐标系
coordinated action on ocean energy 海洋能联合行动
coordination number 配位数
coordination polyhedron 配位多面体
copper indium 铜铟化合物
copper indium diselenide material 铜铟硒化合物材料
copper indium diselenide cell 铜铟联硒化物电池
copper indium selenide solar cell 铜铟硒太阳能电池
copper indium selenium 铜铟硒
copper rod 铜棒，铜条
copper sulfate test column 硫酸铜试验柱

copper to aluminium adapter board 铜铝过渡板
core breaker 岩心提取器，除芯机，碎芯机
core drilling 岩心钻进，岩心钻探，取心钻进
core drilling assembly 取芯钻井装置
core of earth 地核
core sand 型芯沙
core stream of vapor 蒸汽核心流
coring bit 取芯钻头
Coriolis 科里奥利，科氏，科氏力
Coriolis correction 科氏修正
Coriolis deflection 科氏偏转
Coriolis force 科里奥利力，科氏力，地球自转偏向力
cork residue 软木残渣
cork waste 软木废料
corkscrew feeder 螺旋送料机，螺旋给料机
corn husk 玉米壳，玉米皮
corn stalk 玉米秆
corn wet-mill 玉米湿磨机
corncob 玉米棒子
corner flow 拐角流，角落流
corona discharge 电晕放电
correction factor 校正系数
correlate-measure-predict method 关联—测量—预测方法
correspond 符合，协调，通信，相当，相应
corridor 通路
corrosion 腐蚀，浸蚀
corrosion control 腐蚀控制
corrosion mechanism 腐蚀机理
corrosion of metal 金属腐蚀
corrosion protective layer 腐蚀防护层
corrosion resistance 防腐，耐蚀性，耐腐蚀性，抗腐蚀性
corrosion resistance test 耐腐试验
corrosion resistant 抗腐蚀
corrosion resistant material 耐腐蚀材料
corrosive brine 腐蚀性卤水
corrosiveness 腐蚀性
corrosivity of hydrothermal fluid 热液流体腐蚀性
corrosometer probe technique 腐蚀性测定计探测技术
corrugated absorber plate 波纹型吸热板
corrugated blackened aluminum sheet 涂黑波形铝板
corrugated iron 波形铁皮
cosine factor 余弦因子
cosine loss 余弦损失
cost of collector 集热器造价
cost of electric power station output 发电成本
cost of electricity 发电成本
cost per kilowatt 单位千瓦投资
cost per kilowatt hour of the electricity generated by WTGS 风力发电机组度电成本
cost per kilowatt-hour 单位电能投资
cost ratio 成本比率
cotter pin 开口销
cotton fiber 棉花纤维
cottonseed hull 棉籽壳
coulombic force 库伦力
counter clockwise 逆时针方向
counter electrode 对电极，反电极
counter input 脉冲量输入
counter weight 重锤
counter current combustion 逆流燃烧

counter current flow 对向流
counter current regeneration 对流再生（逆流再生）
counter current spray chamber 逆流喷雾室
counterflow evaporator 逆流蒸发器
counterflow heat exchanger 逆流热交换器
counterflow of steam 蒸汽反向流
counterflow of water 水的反向流
counterrotating turbine 对转式汽轮机
counterrotating wind turbine 反向旋转风力机
countersink 埋头孔，暗钉眼
coupled plasma-atomic emission spectrometry technique 耦合等离子体原子发射光谱法
coupling 耦合
coupling bolt 联结，接合，耦合，耦合性，耦合技术
coupling capacitor 结合电容
couplings 联轴器
covalent bond 共价键
covariance function 协方差函数，共变量函数，积差函数
coverage 覆盖，敷层，有效区域
crack 裂纹，裂缝
Cramer's rule 克莱姆法则
cramp 钳位（电路），压（夹）板，卡子，夹子
crank 不稳定的，曲柄
crankcase 曲轴箱，曲柄箱，曲柄轴箱
crankshaft 曲轴，机轴
crate 柳条箱
creosote 木馏油，杂芬油，碳酸
creosoted timber 坑木

crest 波峰，冠，山顶，顶饰
crest elevation 波峰高
crest height 波峰高度
crest length 波峰长度
crib 井壁基环，护壁棚架
crimping tools 卷边工具
criteria air contaminant 标准空气污染物
criterion 标准，判据，准则
critical damping 临界阻尼
critical load case for tower base 塔架基础临界荷载工况
critical loads 临界荷载
critical nucleus 临界晶核
critical nucleus radius 临界晶核半径
critical pressure 临界压力，临界压强
critical value 临界值
critically damped 临界阻尼
crooked hole 弯曲井眼，钻弯的井眼
crop by-product 农作物副产品
crop production residue 农作物生产残渣
crop residue 作物残渣，作物残茬，农田残茬
crop waste 农作物废弃物，农作物废料
cropping 种植作物
cross flow 横流，交叉流动错流，错流，正交流动
cross flow angle 横流角
cross flow cooling tower 交叉流式冷却塔
cross flow heat exchanger 交叉流式热交换器，交叉流式热交换机，横流式热交换器
cross flow turbine 双击式水轮机
cross mark 十字标记
cross product 向量积
cross sea 逆浪，交错浪，逆恶浪

cross slotted screw 十字长孔
cross ventilation 对流通风
cross wind axis wind machine 横风轴风力机
cross wind buffeting 横风抖振
cross wind diffusion 横风扩散
cross wind direction 横风向
cross wind dispersion parameter 横风弥散参数
cross wind displacement 横风位移
cross wind galloping 横风驰振
cross wind gust 横向阵风
cross wind installation 横风装置
cross wind loading 横风载荷
cross wind stability 横风稳定性
cross wind test 横向风试验
cross wind vibration 横风振动
cross wind wake force 横风尾流力
cross-cut end 横切头
cross-draught stove 交叉通风炉
cross-field 穿场（法）
cross-flow spray chamber 错流喷雾室
cross-flow wind turbine 横流式风轮机
crosshead 十字头，丁字头
crossing angle 交叉角
cross-peen hammer 横头锤
cross-section 横断面，横切面，截面
cross-sectional area 横截面积
cross-sectional view 剖视图，横断面视图
cross-spectra 频谱
cross-spectral density 互动率谱密度
crosswind 侧风
crosswind paddles wind machine 横风桨板式风力机
crosswise 斜地，成十字状地，交叉地
crowbar 撬棍，铁棍，起货钩

crown wheel 顶圈
crucible 坩埚
cruise speed 巡航速度
crusher 破碎机
crust 地壳
crustal rock 地壳岩石
crustal structure 地壳结构
cryogenic spill 低温溢出
cryptoclimate 室内小气候
crystal 晶体
crystal defect 晶体缺陷
crystal group 晶族
crystal growth 晶体生长
crystal imperfection 晶体缺陷
crystal lattice 晶格
crystal silicon 晶体硅
crystal structure 晶体结构
crystal surface 晶体表面
crystal system 晶系
crystalline bedrock 水晶基岩
crystalline grain 晶粒
crystalline metamorphic rock 结晶变质岩
crystalline rock 结晶岩石，结晶岩
crystalline semi-conductor 晶体半导体
crystalline silicon 晶体硅
crystalline silicon cell 晶硅电池，晶体硅电池
crystalline silicon solar cell 晶体硅太阳电池
crystalline solar cell 晶体硅太阳电池片
crystalline thin-film tandem cell 多晶薄膜串接电池
crystalline-silicon photovoltaics 晶硅太阳能电池
crystallization 结晶
crystallographic orientation 晶体学取

向关系
crystallography 结晶学
Cu (InGa)Se2 solar cell 铜铟镓硒薄膜太阳能电池
cubic meter 立方米
cubicle 室，箱
culmination 中天，顶点
cumulative capacity 累积容量
cumulative damage 累计损伤
cumulatively compounded motor 积复励电动机
cup anemometer 杯形气流计，杯形风力计，旋杯式风速计
cup-cross anemometer 转杯风速表，杯形风力计
cup-generator anemometer 磁感风杯风速表
cupola furnace 圆顶熔炉
cuprous-oxide photovoltage cell 氧化亚铜光电池，氧化亚铜光伏电池
cup-type anemometer （带有太阳能热量计的）杯式风速计
cure time 固化时间
Curie point 居里温度
current anomaly 电流异常
current circuit breaker 电流断路器
current harmonics 电流谐波
current oscillate 电流振荡
current ration 电流定值
current rms-value 电流有效值
current source converter 电流型变流器
current source inverter 电流型逆变器，电流型变流器
current temperature coefficient 电流温度系数（指在1000W/m² 的试验条件下，被测太阳电池温度每变化1℃，太阳电池短路电流的变化值）
current temperature coefficient of a solar cell 太阳能电池的电流温度系数
current-voltage characteristic 电流—电压特性
curvature function of airfoil 机翼弯度函数
curve correction coefficient 曲线修正系数
curve factor 曲线因数
curve of reservoir capacity 库容曲线
curvillnear groove 曲线槽
cusp collector 尖顶形集热器
customer power systems 用户电力系统
cut biofuel 切割生物质燃料
cut in wind speed 切入风速
cut out valve 截止阀
cut-in speed 切入风速，动风速，启动速度（使风力发电机旋转并产生输出功率的风速）
cut-in wind speed 切入风速，接入风速
cut-in wind velocity 接入风速，起点风速
cut-off wind speed 切断风速
cut-off wind velocity 切除风速
cut-out speed 切出风速
cut-out speed selection 断路器速度选择
cut-out wind speed 切出风速
cut-out wind velocity 停车风速
cutter 刀具，切割机
cutter chip 切割木片
cutting disk 切割盘
cutting feed rate 切割进给速度
cutting in 接通，并入线路
cutting opening 切孔

cutting-off machine 切断机
cycle efficiency 循环效率
cycle fluid （热力装置的）循环流体
cycle heat rate 循环热耗率
cycle thermal efficiency 循环热效率
cyclic driving force 循环驱动力
cyclic durability of storage 储存系统循环使用寿命（年限）
cyclic economy 循环经济性
cyclic irregularity 周期不规则性
cyclic load 循环荷载，周期荷载
cyclic pitch 周变桨距
cyclic service 周期性运行
cyclic shear strain amplitude 周期剪应变振幅
cyclic stress limit 周期性应力极限
cyclic variation 周期性变化
cycling fluidized-bed 循环流化床
cycloconverter 循环换流器
cyclogenesis 气旋生成
cyclolysis 气旋消失
cyclone 气旋，旋风
cyclone ash 旋风灰，气旋灰
cyclone burner 旋流燃烧器
cyclone dust collector 旋风除尘器
cyclone filter 旋风过滤器
cyclone fired boiler 旋风燃烧锅炉
cyclone flow 旋流
cyclone gaswasher 旋风水膜式除尘器
cyclone separator 旋风分离器
cyclone spray chamber 旋风喷雾室
cyclone steam separator 旋风汽水分离器
cyclonic bed reactor 气旋床反应器
cyclonic particle 气旋粒子
cyclonic separator 旋风分离器
cyclonic vorticity 气旋涡度，气旋涡量

cyclonic wave 气旋波
cyclonic whirl 气旋型涡流
cyclonic wind 气旋风
cyclotron absorption 回旋吸收
cylinder 缸体
cylinder block 缸体
cylinder head 缸头
cylinder head gasket 缸头垫片
cylindrical 圆柱形，圆柱体，柱面
cylindrical collector 圆柱形集热器
cylindrical concentrator 柱面聚光型集热器
cylindrical focusing collector 圆柱形聚焦集热器
cylindrical focusing mirror 圆柱形聚光镜
cylindrical Fresnel lens 圆柱形菲涅尔透镜
cylindrical gear 圆柱齿轮
cylindrical geometry heat pipe 圆筒形热管
cylindrical mirror 柱面反射镜
cylindrical optical mounting 柱面光学装置
cylindrical optical system 柱面光学系统
cylindrical parabolic focusing collector 抛物柱面聚焦集热器
cylindrical receiver 圆柱形接收器
cylindrical reflector 圆筒形反射镜
cylindrical roller bearing 圆柱滚子轴承
cylindrical rotor 鼓形转子，隐极转子
cylindrical slat mirror 圆柱形金属条反射镜
Czochralski artefaction system 切克劳斯基晶体生长装置
Czochralski method 晶体生长提拉法

D

DC/AC up(down) converter 直流/交流电压上（下）变换器
DC/DC up(down) converter 直流/直流电压上（下）变换器
Dabancheng wind farm 达坂城风电场
daily amplitude 日变幅
daily clearness index 日晴朗系数（地面水平日辐射总量与大气上界对应平面辐射总量之比）
daily efficiency 日效率
daily extraterrestrial radiation 宇宙日辐射量，大气上界日辐射量
daily heat load 日热负荷
daily insolation 日照值
daily load curve 日负荷曲线
daily load factor 日负荷系数
daily load fluctuating operation 日负荷变动运行
daily maximum temperature 日最高温度
daily mean 日平均值
daily mean temperature 日平均温度
daily mean value 日平均值
daily minimum 日最小
daily minimum temperature 日最低温度
daily observation 日常检查
daily oil tank 日用油箱
daily output 日产量，日出力
daily precipitation amount 日降水量
daily radiation ratio 日辐射比
daily range 日变幅
daily range of air temperature 气温日较差
daily range of temperature 温度日较差
daily solar radiation amount 日太阳辐射量
daily surveillance 日常检查
daily temperature variation 温度日变化
daily terrestrial radiation 地面日辐射量
daily total beam radiation （太阳能）直接辐射日总量
daily total diffuse radiation （太阳能）直接散射日总量
daily total extraterrestrial horizontal radiation 大气上界水平辐射日总量
daily total horizontal radiation （太阳能）水平辐射日总量
daily total scattered radiation （太阳能）散射日总量
daily variation 日变动
daily variation graph of heat consumption in one month 每月日热负荷图
daily water consumption 日耗水量
dam 大坝，水闸
dam board 挡板
dam failure 溃坝，坝失事
damage index 损失指数
damage ratio 损失比
dam-atoll "坝礁"的折射聚焦
damp air 湿空气
damp steam 湿蒸汽
damped alternating current 衰减交流，减幅交流
damped oscillation 阻尼振荡
damped structure 阻尼缓冲结构
damped valve 减震阀
damped vibration 阻尼振动
dampen 使潮湿，使沮丧
dampening force 减震力，缓冲力
damper 防振锤
damper control 挡板调节

damper loss 阻尼损失
damper plate 挡板
damping 阻尼
damping capacity 吸震能力
damping coefficient 阻尼系数，衰减系数
damping constant 阻尼常数，减幅常数
damping ratio 阻尼率
Danish wind turbine concept 丹麦风力机概念
dark characteristic curve 暗特性曲线
dark current 暗电源
dark electrode 暗电极
Dark equation 达肯方程
dark shading 浓阴影
Darrieus 达里厄型（水轮机的一种）
Darrieus machine 达里厄型风力机
Darrieus rotor 达里厄风轮
Darrieus turbine 达里厄汽轮机
Darrieus type wind turbine 达里厄型风力机
Darrieus vertical axis wind-energy conversion system 达里厄竖轴风能转换系统
Darrieus wind turbine 达里厄型风力发电机
Darrieus windmill 达里厄型风车
dashpot 减震器
data acquisition system 数据采集系统
data base 数据库
data circuit 数据电路
data logger 数据记录器
data recording system 数据记录系统
data set for power performance measurement 数据组功率特性测试
data terminal equipment 数据终端设备

Davis hydro turbine 戴维斯水力涡轮机
d-axis reactance 直轴电抗
daylighting in solar building 太阳能建筑物中的日光
daylight effect 日光作用，昼光效应
daylight illumination 日光照明
daylight source 日光源
DC (direct current) load 直流负载
DC bus 直流母线，直流总线，直流汇流排
DC current 直流电流
DC generator 直流发电机
DC load flow 直流潮流
DC motor 直流电动机
DC power supply 直流电源
DC rectifier 直流电整流
DC resistivity sounding 直流电阻率测深
DC/AC inverter 直流—交流逆变器
de novo synthesis 从头合成，全程合成
deacidification 脱酸
deactivate 释放，去激励，停用，退出工作，使无效
dead area 死水区
dead belt 无风带
dead dike (dyke) 堤坝
dead joint 固定连接
dead load 静负荷，静载，恒载
dead loss 净损失
dead tide 最低潮，停潮
dead time 空载时间，滞后时间，失效时间
dead well 枯竭井，已开采过的井
dead wind 逆风
deadband 死谱带，静带
deadening 隔音，隔音作用

deaerated water sampling 除氧水取样器
deaerating unit 除气设备
deaeration 除氧，脱氧，除气，除空气，脱气，去气
deaerator reducer 除氧器减温减压阀
deaquation 脱水
deashing 脱灰
debris 碎片，残骸
decanter 油水分离器
decay area （风浪）平息区
decay constant 衰减常数
decay function 衰减函数
decay parameter 衰减参数
decentralized solar power station 分散型太阳能电站
decentralized system 分散型系统
decentralized wind energy 分散式风能系统
decibel (dB) 分贝
decibel adjusted 调整分贝
decimal 十进的，小数的，小数
declination 赤纬
$DeCO_2$ tower 二氧化碳脱气塔
decode 译码
decomposition energy 分解能
decomposition of silane 硅烷热分解
decomposition of sulphide 硫化物分解（量）
decompression chamber 减压室
decrement 衰减量
dedicated energy crop 专用能源作物
dedicated fluid loop 专用液体循环
dedicated system 专用系统，专属系统
dedusting system 除尘系统
deep crust 地壳深处
deep drilling 深钻井

deep electrical exploration 深层电法勘探
deep geothermal energy 深层地热能
deep level 深能级
deep mantle 深部地幔
deep ocean moored buoy 深海锚定浮标
deep ocean water application 深海水应用，深层海水应用
deep spherical mirror 深凹球面反射镜
deep stall 严重失速
deep temperature monitoring 深度温度检测
deep water 深水的，深海的，靠近海洋的
deep water port 深水港
deep water salinity gradient energy converter 深水盐度差能转换装置，深水盐度梯度能转换装置
deep water wave 深水波
deepwater discharge system 深水排放系统
deepwater drilling 深水钻井
deepwater isotopic current analyzer 深海同位素海流分析仪
defect imperfection 缺陷
defective 有缺陷的，欠缺的
deflecting force 偏转力
deflection 偏向，挠曲，偏差
deflection anemometer 偏转风速表
deflector 导风板，导流板，转向装置，偏转板
deflector baffle 折流挡板
deflector wind machine 导风器式风力机
deformable raft 变形浮箱（一种利用波浪能的设备）
deformation 变形，形变，畸变，失真

deformation of reflective mirror 反射镜表面变形
deformation ratio 变形比
degasifier 除氧器
degassing duct 脱气管道，脱气风管
degrease 脱脂，除油污
degree Celsius 摄氏度
degree of blackness 黑度
degree of consistency 均匀度
degree of curvature 弯度
degree of darkening 暗度
degree of dispersion 弥散度
degree of saturation 饱和度
degree of supercooling 过冷度
degree of superheat 过热度
degree of weathering 风化度
degree of wind sensitivity 风敏感度
degree-day during heating period 采暖期度日数
degree-Kelvin 开氏温标
degree Fahrenheit 华氏度，华氏度数
degree of freedom 自由度
dehumidification 除湿，空气减湿
dehumidifier 干燥器，脱水装置，减湿器
dehydration 脱水
dehydrator 脱水器
deicer 防冰器
deincrustant 除垢剂
deionization plant 除离子装置
deionized water 除离子水
delaying separation 延迟分离
delineated installed capacity estimate 划定装机容量评估
delineated resource estimate 划定资源评估
delineation well 定界井
delivery cock 排污阀
delivery end 卸料端，排出口
delivery head 扬程
delivery lift 扬程
delivery pump 输送泵
delivery regulator 供给量调节器
delivery system 传送系统，分配系统，运载系统
delocalization energy 离域能
delta connection 三角形连接
delta wing 三角形机翼
demand charge 按需用量收费（公用事业部门根据可能的需用量，而不是按实际消耗量收取费用）
demand forecast 需求预测，市场预测
Dembet effect 丹倍效应
demineralization 去矿物质
demineralized water station 脱盐水站
demineralizer 除盐装置，软化器
demister 除雾器
demobilization 复员
demodulation current 检波电流
demodulator 解调器
demolition wood 拆除的木材
demonstration plant 示范厂，示范装置
denature fuel ethanol 变性燃料乙醇
dendritic segregation 枝晶偏析
denitration equipment of biomass-fired power plant 生物质电厂烟气脱硝技术装备
densely built-up city 建筑密集城市
densification 致密化
densified biofuel 致密生物质燃料
density 密度
density difference 密度差
density gradient 密度梯度
density of hot fluid 热流体密度
density of ocean current energy 海流

能密度
density profiles 密度剖面
density ratio 密度比
density separation 密度分选
dependable peaking capacity 可靠峰值输出功率
deploy 展开，配置
depolymerization 解聚，解聚作用
deposit 堆积物，沉淀物
deposit of hydrocarbons 碳氢化合物沉积
deposition 沉积
deposition rate 沉积速率
depress 使沮丧，使消沉，压下，压低
depression 洼地
depressurize 使减压，使降压
depth control equipment 深度控制仪
depth gauge 深度计
derrick 井架，转臂起重机，起重架
desalinated water 脱盐水
desalination 脱盐作用，减少盐分，海水淡化
desalination by solar distillation 太阳能蒸馏淡化（法）
desalination plant 淡化装置
desalination system 脱盐系统
desalting plant 淡化工厂
desalting process 淡化方法
desalting reactor 海水淡化反应器
desander 除砂器
descriptive algorithm 描述性运算
desiccant 干燥剂
design basis storm 设计基准风暴
design building heating load 建筑物加热设计负荷
design capacity 设计容量，设计功率
design considerations 设计要素

design cooling load 冷却设计负荷
design heating load 设计供热负荷，设计加热负荷
design limit 设计极限
design of anaerobic process 厌氧过程设计
design parameter 设计参数
design point 设计点
design point power 设计点功率
design point thermal power of receiver 吸热器额定热功率
design pressure 设计压力
design situation 设计工况
design temperature difference 设计温度差
design tip speed ratio 设计叶尖速比
design tradeoff 设计折衷
design wind load 设计风载
design wind speed 设计风速
desilter 除泥器
desired output 期望输出值
desoxy 脱氧
desoxydation 脱氧
destructive distillation 干馏，分解蒸馏
destructive interference 相消干涉，相消干扰，破坏性干扰
destructive metabolism 分解代谢
destructive test 破坏性试验
destructive vibration 破坏性振动
desulfurization 脱硫
desulphurization 脱硫，脱硫作用
desuperheater 减温器，过热蒸汽降温器，减热器，过热降温器
desuperheating coil 减温器盘管
detachable blade 可拆桨叶
detachment of vortices 漩涡脱体
detecting element 传感元件

detection 检测，探测，检波
detector 检测器，探测器
detergent 清洁剂，去垢剂
deterministic fluctuating wind speed 主脉动风速
deterministic gust 主阵风
deterministic method 定值设计法，确定性方法
deterministic variable 决定性变量
development scheme 开发方式
deviating force 偏转力
deviation 偏差，偏移
deviation angle 偏角
device connection 设备连接，器件连接
device for recovering wave energy 回收波能装置
devolatilization 液化作用,脱挥发分作用
dew point 露点温度
dew point hygrometer 露点湿度表
dew point spread 露点差
dew point temperature 露点温度
dewatering 脱水作用，脱水
dewatering unit 脱水机
diagonal 对角
dial gauge 量规
dialyzer 渗析膜，渗析器
diameter 直径
diametral flow 径向流动
diamond crystal structure 金刚石晶体结构
diamond cubic lattice 立方金刚石晶格
diamond cutters 钻石刀具
diamond drilling 金刚石钻井，金刚石钻进
diaphragm 膜片，隔板，光阑
diaphragm cell 隔膜电池
diaphragm manometer 膜片压力表
diaphragm pump 隔膜泵
diaphragm seal 隔膜密封
diatomaceous earth 硅藻土
diatomaceous silica 硅藻土
diatomite 硅藻土
dicing saws 晶圆切割机
die casting dies 压铸冲模
die casting machine 压铸机
dielectric 电介质，绝缘体
dielectric constant 介电常数
dielectric layer 介电层
dielectric loss 介质损耗
dielectric test 介质试验
diesel engine 柴油发动机
diesel fuel 柴油，柴油机燃料
diesel generation 柴油发电机
diesel generator 柴油发电机,柴油发电机组
diesel genset 柴油发电机组
diesel index 柴油指数
diesel locomotive 内燃机车
diesel number 柴油值
diesel oil 柴油
diesel-powered grid 柴油电网
dies-progressive 连续冲模
diestock 螺丝攻
difference equation 差分方程
difference method 差分法
differential actinometer 示差日射表
differential equation 微分方程
differential gear 差速齿轮
differential pressure 压差
differential protection 差动保护
differential temperature controller 温差控制器
differential temperature sensor 微分温度传感器

differential thermal analysis 差热分析
differential thermometer 示差温度表
differentiation 微分
diffraction 衍射
diffraction angle 衍射角
diffraction force 绕射力
diffraction problem 绕射问题
diffuse energy 散射能量
diffuse illumination 漫射光照
diffuse insolation （太阳）散射辐射
diffuse irradiance 散射辐照度
diffuse irradiation 散射辐射
diffuse light 漫射光
diffuse loss 扩散段损失
diffuse radiation 漫辐射，漫射辐射，散射辐射
diffuse radiation form 漫辐射形态
diffuse reflectance 漫射率
diffuse sky radiation 天空漫射辐射
diffuse solar irradiance 散射（日射）辐照度
diffuse solar radiation 太阳散射辐射
diffuse source 散射源
diffuse sunlight 漫射太阳光
diffuse to beam radiation ratio 散射—直射比
diffuse transmission 扩散透射，漫射传输
diffused junction 扩散结
diffused light 漫射光
diffuseness error 扩散误差
diffuser 扩散器
diffuser augmented wind turbine 扩散体增强型风力机
diffuser-augmented rotor 扩风器增力型风托
diffusing material 漫射材料

diffusing media 散射媒质
diffusing power 漫射功率，散射能力
diffusing source 漫射源
diffusion 散射
diffusion capacity 扩散电容
diffusion coefficient 扩散系数，扩散率
diffusion constant 漫射常数
diffusion equation 扩散方程
diffusion flow rate 扩散流量
diffusion flux 扩散通量
diffusion mechanism 扩散机理
diffusion photo-voltage 扩散光电压
diffusion process 扩散工艺
diffusion property 扩散特性
diffusion pump 扩散泵
diffusion rate 扩散率
diffusion region 扩散区
diffusion temperature 扩散温度
diffusion velocity 扩散速度
diffusion well 补给井
digestant 消化剂
digestate （厌氧分解产生的）沼渣沼液
digested sludge 消化污泥
digester 发酵池
digester energy 煮解能
digester performance 消化器性能
digester slurry 发酵液
digester tank 煮解器槽
digestibility 可消化性
digestion of sewage sludge 污水游渣煮解
digestion process 消化处理
digger 挖掘者 挖掘机
digiquartz pressure 双石英压力
digital counter 数字计数器
digital data loggers 数字数据记录器
digital displacement 数字传输，数字位移

digital elevation model 数字高程模式
digital fault recorder 数字故障记录仪
digital indicator 数字指示器
digital input 数字输入
digital input terminal 数字量输入端子
digital line graph 数字划线地图
digital output 数字输出
digital output terminal 数字量输出端子
digital type 数字型
digital-to-analog converter 数字—模拟转换器
digitizing tablet 数字面板
dike (dyke) 堤坝
dilation 膨胀
dilute 冲淡，变淡，变弱，稀释
dilution air 稀释空气
dilution ratio 稀释率
dimension 尺寸
dimensional analysis 量纲分析
dimensional inspection 尺寸检验
dimensional vector space 维矢量空间
dimensionless coefficient 无量纲系数
dimethylether 二甲醚
dinas firebrick 硅石耐火砖
Dines anemometer 达因风速表
Dines pressure anemograph 达因风压机
diode bridge 二极管电桥
diode bridge rectifier 二极管整流桥
diode module 二极管模块
diode rectifier 二极管整流器
diode 二极管
dioxide 二氧化物
dioxin 二恶英
dip of surface 地表倾斜
dipolar nature 两极性

dipole resistivity 偶极电阻率法（用于地热勘探的一种地球物理方法）
direct absorption receiver 直接吸收接收器
direct acting pump 直接驱动泵
direct axis 直轴
direct axis transient time constant 直轴瞬变时间常数
direct co-firing 直接混烧
direct collection 直接集热
direct combustion 直接燃烧
direct contact condenser 混合式凝汽器，直接接触凝结器
direct contact heat exchanger 混合式热交换器，直接接触式换热器
direct cooling system 直接冷却系统
direct coupling well water 直接耦合井水
direct current 直流电，直流
direct current generator 直流发电机
direct current machine 直流电机
direct drive 直接驱动
direct drive multi-pole 直驱多极
direct drive propeller 直接传动螺旋桨
direct drive wind turbine generator system 直驱式风电机组
direct drive WTGS 无齿轮箱式风电机组
direct energy conversion 能量直接转换
direct energy density 直接能量强度
direct expansion 直接膨胀，直接蒸发
direct expansion low temperature solar thermal power 直接膨胀式低温太阳能热发电
direct feed evaporator 直接供液蒸发器
direct fertilizer 直接肥料
direct fired biomass 生物质直燃
direct fired biomass boiler 生物质直燃

锅炉
direct fired evaporator 直烧蒸发器
direct fired heater 直燃式加热炉
direct fired oven 直接燃烧炉
direct firing 直接火焰加热
direct flow solar water heater 直流式太阳能热水器
direct flow vacuum tube 直通式真空集热管
direct gain 直接受益式
direct gain aperture 直接增益采光面积
direct gain building 直接增益建筑
direct gain passive heating system 直接增益被动式加热系统
direct gain passive system 直接增益被动式系统
direct gain roof 直接增益屋顶
direct gain skylight 直接增益天窗
direct gain system 直接增益系统
direct gain wall 直接增益墙
direct gap material 直接能隙材料，带隙材料
direct gap semiconductor 直接能隙半导体
direct generation 正向发电
direct incident solar flux 太阳能直接入射通量
direct injection stratified charge 直喷分层充气（发动机）
direct insolation 直射太阳辐照
direct irradiance 直射辐照度
direct irradiation 直射辐照
direct mains coupling 直接耦合电源
direct normal radiation 太阳直射通量
direct pumping 正向抽水
direct radiation 直接辐射
direct reading instrument 直读式仪表

direct reduction 直接还原
direct return piping 直接回路管道
direct rotary 直接加热回转
direct rotary rig 直接旋转钻机
direct smelting 直接熔炼
direct solar conversion 直接太阳能转换
direct solar energy 直接太阳能
direct solar gain 直接太阳能增益
direct solar irradiance 直接日射辐照度
direct solar radiation 太阳直射辐射，太阳直接辐射
direct solar radiation 直接太阳辐射
direct sun temperature 阳光直射温度
direct sunlight 阳光直射，直接阳光
direct system 直接系统
direct transition solar cell 直接转换太阳能电池
direct use 直接使用
direct use of solar energy 太阳能直接利用
direct utilization of geothermal energy 地热能直接利用
direct voltage 直流电压
direct-axis magnetizing reactance 直轴磁化电抗
direct-beam radiation 直接光束辐射
direct-contact heat exchanger 混合式热交换器
direct-contact type condenser 混合式热凝汽器
direct-drive batteryless wind-electric system 直驱无电池风电系统
direct-drive generator 直驱式发电机
direct-driven wind turbine generator 直驱式风力发电厂
direct-fired biomass plant 直燃式生物质电厂

direct-fired biomass power plant 直燃式生物质发电厂
direct-fired biomass power station 直燃式生物质发电厂
direct-fired natural gas chiller 直燃式天然气冷冻机
direct-gain passive solar house 直接受益型太阳房
direct-gain passive system 直接增益被动式（太阳能）系统
direct-gain solar house 直接增益（系统），直接受益式太阳房
direct-heat rotary dryer 直接加热式回转干燥器
direction of heat flow 热流方向
direction vane 方向舵
directional comparison protection 方向比较保护，非单元纵联保护
directional comparison protection system 方向比较保护系统
directional control 方向控制，定向控制
directional cooling 定向冷却
directional coupler 定向耦合器
directional drilling 定向钻井
directional radiation 定向辐射
directional reflection 定向反射
directional solidification 定向凝固
directional solidification equipment 定向凝固装置
directional solidification system 定向凝固系统法
directional solidification technology 定向凝固工艺
directional source 定向源
directional spectra 定向光谱，指向光谱
directional spectral density 定向谱线密度，定向频谱密度

directional spreading function 定向扩展函数
directional stop 定向限位架
directional wave spectrum 方向波谱
directivity 指向性
directly irradiated receiver 直接式太阳辐射吸热器
direct-magnetic wave energy converter 定向磁性波能转换器
direct-piped hot water 直接管道热水
direct-pneumatic wave-energy converter 直接气动波能转换器
direct-to-load photovoltaic power generation 直接负载光伏发电
direct-type evaporative cooler 直接式蒸发冷却器
Dirmhirn-Sauberer pyranometer 迪尔姆希尔恩—绍贝勒尔总日射计，迪尔姆希尔恩—绍贝勒尔辐射强度计
disc chipper 盘式削片机
disch 放电
discharge 卸下，卸货，流注，放电
discharge area 出口截面积
discharge characteristics 流量特性
discharge current 放电电流
discharge electrode 放电电极，放电极
discharge hole 排出孔，出口
discharge loss 出口损失
discharge path 放电路径
discharge pipe 排放管，出风管，出液管
discharge ration 流量系数
discharge steam 乏汽
discharge tube 放电管
discharge voltage 放电电压
discharging efficiency 放电效率
disconnect switch 隔离开关
disconnection 解列
discontinuous load 断续负荷

discontinuous phase 非连续相
discrete controller 离散控制器
discrete gust model 离散阵风模型
discrete pulse modulation 离散脉冲调制
discrete system 离散系统
discrete time 离散时间
discretization 离散化
disc-tube reverse osmosis 碟管式反渗透
dish collector 盘形聚热器,盘形集光器
dish concentrator 碟式聚光器
dish solar system 碟式太阳能系统
disk brake 盘式制动器
dislocation 位错
dislocation arrangement 位错排列
dislocation array 位错阵列
dislocation atmosphere 位错气团
dislocation axis 位错轴
dislocation climb 位错爬移
dislocation coalescence 位错聚结
dislocation density 位错密度
dislocation line 位错线
dislocation loop 位错环
dislocation nucleation 位错形核
dislocation slip 位错滑移
dismantle 拆除,拆卸
dismount 拆卸,卸下
disordered solid solution 无序固溶体
disordered-order transition 有序无序转变
dispersed air floatation 曝气浮选
dispersed generation 分散式发电
dispersion 扩散,弥散,散布,色散
dispersion aerosol 分散气溶胶
dispersion coefficient 色散系数
dispersion fuel 弥散燃料

dispersion relation 色散关系,分散关系,频散关系
displacement amplitude 位移幅值,位移幅度
displacement current 位移电流
displacer 平衡浮子
display lamp 指示灯
disposable tool holder bits 舍弃式刀头
disposal 处理,处置,布置,安排,配置,支配
disposition notice 处罚通知书
dissimilation of acretate 醋酸盐异化(作用)
dissipation 分散,浪费,损耗,耗散,消耗
dissolution 分解,解散
dissolved constituents 溶解成分
dissolved floatation 溶气浮选
dissolved gas 溶解气体,消融气体
dissolved matter 溶解物
dissolved mineral 溶解矿物质
dissolved organic carbon 溶解有机碳,溶解性有机碳
dissolved oxygen 溶解氧
dissolved oxygen analyzer 溶解氧分析仪
dissolved solid (matter) 溶解固形物
dissolved solids 溶解性固体,溶解质
dissolver 溶解装置
dissolving tank 溶解箱
distance constant 距离常数
distance ring 间隔环
distilled water 蒸馏水
distillate fuel oil 蒸馏燃料油
distillation 蒸馏,净化,蒸馏法,蒸馏物
distillation process 蒸馏过程
distortion 畸变

distributed capacitance 分布电容，分散式电容
distributed collector 分布式集热器，分散型集热器
distributed computing 分布式计算
distributed control system 分布式控制系统，分散控制系统
distributed energy resources 分布式能源
distributed energy resources customer adoption model 分布式电源客户侧模型
distributed generation 分布式发电
distributed linear collector 分布型线集热器系统
distributed power generation facility 分散式发电装置
distributed power supply technology 分布式供电技术
distributed solar power production system 分散式太阳能发电系统
distributed solar thermal power system 太阳热能分散式发电系统
distributed system 分散式系统，分布系统
distributed winding 分布绕组
distributed-parameter model 参数分布模型
distributing apparatus 配电电器
distribution 分布
distribution board 配电盘，配电板，配电箱，配电屏
distribution coefficient 分配系数，分布系数
distribution company 配电公司，配气公司
distribution feeder 配电线路
distribution infrastructure 配置性基础设施
distribution line 配电线路
distribution of geothermal heat 地热分布
distribution panel 电源分配盘
distribution subsystem 输配子系统
distribution system 产品分配系统，配电系统，配水系统
distribution transformer 配电变压器
distributor 导水装置
distributor plate 分布板，配电板，配电盘
district cooling 区域供冷
district exploration 区域开发
district heat exchange 区域热交换器
district heating 区域供暖，局部供热
district heating plant 区域供热站
district heating station 区域供热站
district heating system 区域供热系统，区域供暖系统
district-heating plant 区域供暖站，区域供热站
disturbing acceleration 扰动加速度
disturbing current 扰动电流
disturbing force 扰动力
disturbing voltage 干扰电压
diurnal 昼行性的
diurnal circle 日循环
diurnal flow 昼夜流
diurnal inequality 日潮不等
diurnal range 日潮差
diurnal tidal component 日分潮，日潮汐分量
diurnal tide 全日潮，日周潮
diurnal variation 日变化（风速或风功率密度）
diurnal wind 日变风

divergence windspeed　发散风速
diversity factor　分散率，分散系数
dividers　圆规
divorsed eutectic　离异共晶
dog clutch　爪形（式）离合器，齿式离合器
doldrums　赤道无风带
dolly　洋娃娃，移动车，台车，移动摄影车
dolomite　白云石，白云岩，白云土
dolomite lime　白云质灰岩
domed nut　圆顶螺母
domestic boiler　家用锅炉，民用锅炉
domestic building　居住建筑
domestic dwelling　民用住宅
domestic hot water heater　家庭（太阳能）热水器
domestic solar energy system　家用太阳能系统
domestic solar heating　家用太阳能供暖，民用太阳能供暖
domestic solar water heater　家用太阳（能）热水器
domestic waste　家庭废物
domestic water heating system　家用水加热系统
dominant wind direction　主导风向
Donghai bridge off-shore wind farm　东海大桥海上风电场（我国首座，也是亚洲首座海上风力发电场）
donor atom　施主原子
dopant　掺杂物，掺杂剂
doped silicon　掺杂硅
Doppler acoustic radar　多普勒声雷达
Doppler coefficient　多普勒系数
Doppler constant　多普勒常数

Doppler displacement　多普勒位移
Doppler effect　多普勒效应
Doppler feedback　多普勒反馈
Doppler shift　多普勒频移
dosage　剂量，配料，定量器
dot product　点乘
double basin three way tidal power plant　双库三向潮汐电站
double basin-single effect scheme　双库一单作用方案
double bladed windmill　双叶片风车
double casing　双层外壳
double clamp　双卡头
double cover collector　双层盖板集热器
double cover glass　双层玻璃盖板
double curvature concentrating collector　双曲率聚光集热器
double curvature concentrator　双曲率聚光器，双曲率聚光型集热器
double curvature device　双曲率装置
double direction decomposition　双向分解
double effect chiller　双效制冷机，双效冷却装置
double exposure flat plate collector　双面吸热式平板集热器
double fluid cell　两液电池
double glass　双层玻璃
double glass cover　双层玻璃盖
double glass cover collector　双层玻璃盖板集热器
double glass flat plate collector　双层玻璃盖板平板集热器
double glass system　双层玻璃系统
double glazed cover collector　双层透明盖板集热器

double glazed flat plate collector 双层玻璃盖板平板集热器
double glazing 双层玻璃窗
double glazing clear glass 双层上釉透明玻璃
double glazing heat-absorbing glass 双层上釉吸热玻璃
double glazing reflecting glass 双层上釉反射玻璃
double helical aerator 双螺旋曝气头
double humped curve 双峰曲线
double humped distribution 双峰分布
double layer antireflection coating 双层减反射涂层
double layer winding 双层式线圈
double lens concentrator 双透镜聚光器，双透镜聚光型太阳能集热器
double loop system 双回路系统
double medium theory 双介质理论
double phase 两相
double pipe cooler 套管冷却器
double pipe exchanger 套管换热器
double rotor 双转子
double sloped glass cover 斜置双层玻璃盖板
double sluice gate 双层泄水闸
double squirrel cage rotor 双鼠笼式转子
double stage pump 双级泵
double step-up gear 双增速齿轮
double suction impeller 双吸式的叶轮
double suction pump 双吸式泵
double tide 双潮，复潮
double tracer technique 双元示踪技术
double vacuum tube collector 双真空管集热器
double wall heat exchanger 双壁热交换器
double-basin transconnected scheme 双库贯连方案
double-curvature concentrating device 双曲率聚光装置
double-effect 双重效果，双重影响
double-effect absorption chiller 双效吸收式冷水机，双效吸收式制冷机
double-effect single-basin scheme 双作用—单库方案
double-effect tidal plant 双作用潮汐电站
double-effect unit 双作用装置
double-exposure flat-plate 双重曝光平板
double-fed asynchronous generator 双馈异步发电机
double-fed induction generator 双馈感应发电机
double-glazed window 双层玻璃窗
double-helical gear 人字形齿轮
double-layer plastic film mirror 双层塑料薄膜反射镜
double-layer sluice 双层泄水闸
double-pane glass 双窗格玻璃
double-pool system （潮汐电站）双库系统
doubler 倍压器
double-regulated generating unit 双可调发电装置
double-walled heat exchanger 双层壁换热器
double-way operation 双向发电（该潮汐电站在涨潮、落潮时均能发电）
double-way sluice 双向泄水闸
doubly fed 双馈
doubly fed induction generator 双馈发电机

doubly salient permanent magnet generator 双凸极永磁发电机
doubly-fed induction machine 双馈感应电机
doubt 不确定，疑惑
dowel 木钉，销子，用暗销接合
down conductor 引下线
down draft 下引风，下行通风，倒烟
down time 停机时间，故障时间，停工时间
down wind 下风向
down-conductor 引下线
down-draught kiln 倒焰窑
downhill diffusion 下坡扩散
downhole geothermal measurement 焊井地热测量
downhole heat pump 井下热泵
down-hole heat-exchanger technology 井下热交换技术，井下热换技术
down-hole measurement 井下测量
downhole operation （地热井）井下操作
downhole recording temperature gauge 井底记录温度仪
downhole temperatures 井下温度
down-shot firing 下射式燃烧
downslope wind 下坡风
downstream 下行，顺流，（体）顺水，下游侧顺流的，顺流地，在下游的
downstream valve 下游阀
downstream wind (downwind) 顺风
downtime 故障停机时间
downward facing cavity receiver 下开口（太阳能）腔体接收器
downwelling （海洋在地壳板块压力下的）下降，沉降流
downwind 下风向，顺风的

downwind configuration 下风向布置
downwind propeller 顺风螺旋桨，下风螺旋桨
downwind rotor 下风向风轮
downwind sector 下风向扇形区
downwind side 顺风面
downwind spacing 顺风间距
downwind turbine 顺风风力机
downwind type of WTGS 下风向式风电机组
down-wind wind turbine 下风向风力机
downwind turbine generator system 下风向式风电机组
d-q axis model 直横轴模型
draft capacity 通风能力
draft fan 排风扇
draft loss 通风阻力，烟道阻力
draft resistance 通风阻力
draftmeter 风压表
drag 阻力
drag area 风阻面积
drag coefficient 阻力系数
drag cup anemometer 阻力型风杯风速计
drag device 拖拽装置
drag force 拖曳力，迎面阻力，曳力，阻力
drag friction 摩擦阻力
drag from lift 升致阻力
drag from pressure 压差阻力
drag reducing device 减阻装置
drag reduction 减阻
drag spoiler 阻力板
drag test 风阻试验
drag wind load 风阻荷载
drag-type 阻力型

drag-type rotor　阻力型风轮
drag-type wind machine　阻力型风力机
drain　泄油，排水沟，消耗，排水
drain tap　排气阀
drain water pump　疏水泵
drain water tank　疏水箱
drainage wind　流泄风
draindown system　排放系统
drainback system　回流系统
drawing board　画图板，制图板
drawing machines　拔丝机
drawing point　绘图点
draw-off temperature　取水温度
drawwork　绞车，钻井绞车
drift　漂移，偏差
drift current　漂移电流
drift effect　漂移效应
drift field　漂移场
drift field cell　漂移场电池
drift force　定偏力，漂移力
drift mobility　漂移率
drift type photovoltaic device　漂移型光伏器件
drift velocity　漂移速度
drift voltage　漂移电压
drill　训练，钻孔，钻头，锥子，钻机
drill collar　钻铤
drill gauge　钻规
drill hole　钻孔
drill pipe　钻管，钻杆
drill rig　钻机
drill stem　钻柱，钻杆，钻具
drill string　钻柱
drill team　钻井队
drilled rock core　钻芯

driller　钻孔者，钻孔机
drilling and completion　钻探与完井
drilling capacity　钻井能力，钻孔容量
drilling contingency factor　钻井权变因素，钻井临时因素，钻井意外因素
drilling contractor　钻井承包商
drilling cost　钻井成本
drilling curve　钻井曲线
drilling depth　钻孔深度，钻进深度
drilling experience　钻井经验
drilling fluid　钻井液，钻液，钻孔液体，钻孔泥浆
drilling fluid circulating system　钻井液循环系统
drilling for geothermal steam　地热蒸汽钻井
drilling for hot water　热水钻井
drilling for steam　蒸汽钻井
drilling Information　钻井信息
drilling machine　钻床
drilling machine bench　钻床工作台
drilling mud　钻井泥浆
drilling mud additives　钻井泥浆添加剂
drilling platform　钻井平台
drilling process　钻井过程
drilling rate　钻速
drilling technique　钻井技术，钻探技术
drilling velocity　钻速
drills　钻头
drip cock　泄放阀，排污阀
drip feeder　滴给器
drip pan　油滴盘
drip pipe　泄放管，排出管
drip pump　排水泵
drive assembly　传动装置
drive chain　传动链

drive mechanism 驱动机构
drive shaft torsional flexibility 传动轴扭转柔性
drive topology 驱动拓扑
drive train 传动系统
drive train inertia 传动系惯性
driven gear 从动齿轮
driver 驱动器
driving fluid 工作液体
driving force 驱动力
driving gear 主动齿轮，驱动齿轮
drop chute 跳水槽
drop hammer 落锤
drop out current 开断电流
drum brake 鼓状刹车
drum chipper 鼓式削片机
drum dryer 鼓式干燥机
drum feeder 鼓式加料机
dry adiabatic lapse rate 干绝热递减率
dry anaerobic composting process 干厌氧堆肥过程
dry ash-free basis 干燥无灰基
dry basis 干基
dry battery 干电池
dry bulb temperature 干球温度
dry cell 干电池
dry cooling tower 干式冷却塔
dry deposition 干沉降
dry digestion 干法分解
dry disk photoelectric cell 干片式光电池
dry etching 干法刻蚀
dry excavation technique 干式开挖技术
dry expansion evaporator 干膨胀式蒸发器
dry fermentation 干发酵
dry fuel 固体燃料，干燃料
dry matter 干物质
dry matter content 干物质含量
dry saturated steam 干饱和蒸汽
dry scrubber 干式洗涤器，干式除尘器
dry solid waste 干固体垃圾，干燥固体废物
dry sorbent injection system 干吸收剂喷射系统
dry steam 干蒸汽
dry steam energy system 干蒸汽能源系统
dry steam field 干蒸汽田
dry steam plant 干蒸汽动力厂
dry steam power plant 干蒸汽发电站
dry steam power plant 干蒸汽电厂
dry steam type 干蒸汽型（地热电厂）
dry tower 干燥塔
dry type transformer 干式变压器
dry willow 枯柳
dry-ash-removal Lurgi gasifier 固态排渣鲁奇气化炉
dry-bulb economizer control 干球节能控制
dry-hot-rock geothermal system 干热岩地热系统
drying 干燥，干性
drying bed 干燥床
drying kiln 干燥窑，烘干窑，干燥室
drying machinery 干燥设备
drying medium 干燥剂
drying section 干燥段
dry-out of vapor-dominated reservoir 蒸汽优势储层的干燥
dry-steam sources 干蒸汽资源
dual admission turbine 双进汽涡轮机
dual circulation 两级循环

dual circulation boiler 两级循环锅炉	dump load 甩负荷，备用负载
dual combustion cycle 双燃循环	dump pit 垃圾坑
dual flashed-steam 二次闪蒸汽	dump steam 排汽，废汽
dual fuel 混合燃料	dung collection 粪便收集
dual pressure boiling water reactor 双压沸水堆	duo directional current 双向电流
dual purpose medium 双效介质	duplex transmission 双工传输
dual rotor turbine 双转子涡轮机	durability 耐久性，耐用性，寿命
dual (multi) bore-hole technology 双（多）井眼（钻井）技术	duration 宽度，持续时间
	duration of ebb 落潮时，落潮历时
dual-axis suntracker 双轴太阳追踪器，双轴太阳跟踪器	duration of flood 涨潮时，涨潮历时
	duration of possible sunshine 可照时数
dual-fluid nozzle 双流体喷嘴	duration of the fault 故障持续时间
dual-powered mill 双动力电磨机	dust cake 尘饼，粉尘层
dual-powered tidal current mill 双动力潮汐电磨机	dust collection electrode 收尘极，收尘电极
dual-tank solar hot-water system 双罐太阳能热水系统	dust emission 粉尘排放，烟尘排放，扬尘量
dual-wall pipe 双重壁管	dust filter 滤尘器，尘土过滤器
dual-zone stove 双层炉	dust injection furnace 尘埃注入炉
duck 震荡波形	dust precipitation 煤尘沉降，除尘
duck wave energy extractor 鸭式波浪能抽提装置	dust precipitation technology 除尘技术
	dust protected 防尘
duck wave power generator 鸭式波浪能发电机	Dutch four arm type mill 荷兰四臂型风车
	Dutch windmill 荷兰风车
duct 管道，管子	duty cycle 运行周期
duct loss 管道阻力，管道损失	duty ratio 负载比
ducted propeller 涵道螺旋桨	duty ratio, dutyfactor 占空比，负载比
ducted rotor 导管式风轮	dye penetrant examination 染料渗透试验法
ducted rotor wind machine 导管式风轮风能发电机	dye-sensitized solar cells 染料敏化太阳能电池
ducted wind turbine 外罩式风力机	dyke 堤，岩脉
ductile iron pipe 球墨铸铁管	dynamic anaerobic 动态厌氧
ductility 延性	dynamic bus impedance 动态总线阻抗
dummy load 假负载，仿真（等效）载荷	dynamic control system 动态控制系统

dynamic controller 动力控制
dynamic coupling 齿啮式连接
dynamic dam 动态大坝
dynamic effect 动力效应
dynamic friction 动态摩擦
dynamic inflow 动态入流
dynamic interaction 动态交互作用
dynamic load testing 动载荷试验
dynamic marine component test facility 动力海洋组件测试设备
dynamic pressure 动压力
dynamic pressure anemometer 动压风速计
dynamic project cost estimate 动态工程投资概算
dynamic resistance 动阻力
dynamic response 动态响应
dynamic simulation 动态模拟
dynamic stability derivative 动稳定导数
dynamic stall 动态失速
dynamic stress 动应力
dynamic tidal power 动态潮汐能
dynamic transfer system 动态存储系统，动力传输系统
dynamic voltage restorer 动态电压恢复器（补偿器、调节器）
dynamic wind load 动态风载
dynamic-state operation 动态运行
dynamo 发电机
dynamometer 测力计，功率计，动力计

E

earth conductor 接地线
earth coupling 大地耦合
earth current 大地电流，接地电流
earth electrode 接地体
earth energy system 地能系统
earth radiation 地球辐射
earth termination system 接地装置
earth thermometer 地温表
earth tide （太阳和月球的引力引起的）地球潮汐，固体潮
earth work & levelling 土方及场地平整
earth's rotation 地球自转
earth-contact cooling 大地接触冷却
earth-coupled heat pump 地耦合热泵
earthed circuit 接地电路
earthing 接地
earthing reference point 接地基准点
earthing switch 接地开关
earthquake intensity 地震烈度
earthquake magnitude 地震震级
earth-termination system 接地装置
ebb channel 落潮水道
ebb current 退潮流
ebb gate 退潮闸门
ebb generation 落潮发电
ebb interval 退潮时间间隙
ebb strength 最大退潮（流速）
ebb tide 落潮
ebb tide current 落潮流
ebb-and-floodcurrent 退涨潮流
ebb-and-flow structure 涨落潮流构造
ebb-generation system 退潮发电系统
EC material (electrochromic material) 电致变色材料
eccentric circulation vortex 偏心环流
eccentric rotor engine 偏心转子发动机

ecological assessment 生态评价，生态评估
ecological efficiency 生态效率
ecological interaction 生态相互作用
ecological process 生态过程
economic analysis 经济性分析
economic assessment 经济性评价
economic evaluation 经济性评价
economic feasibility 经济可行性
economic optimal solar system coast 太阳能系统最佳经济成本（费用）
economic potential 经济潜势
economic viability 经济可行性，经济能力
economics of solar cell 太阳能电池经济学
economics of solar energy 太阳能经济学
economics of thermal storage 储热经济学
economics of tidal power 潮汐能经济学
economiser 节热器
economizer 节约装置
economy of energy 能源经济
ecosystem 生态系统
eco-system conservation 生态保护
ecosystem damage 生态系统破坏
ecosystem research 生态系统研究
ecosystem-based management (EBM) 以生态系为本管理机制
eddy 涡流，旋涡
eddy advection 涡动平流
eddy conductivity 涡动传导率
eddy current 涡流，涡电流
eddy current effect 涡流效应
eddy displacement current 位移涡流
eddy energy 涡动能量

eddy flow 涡流，紊流
eddy flux 紊流，紊流通量
eddy resistance 涡流（动）阻力
eddy water 涡流
eddy zone 涡流区
eddying effect 涡流效应
eddying flow 涡流流动，紊流流动
edge chip 崩边
edge dislocation 刃型位错，边缘位移
edge loss （集热器）边缘热损失
edge retaining system 边缘固定系统（使太阳能集热器各层嵌板的边缘固定就位的金属槽道）
edge shedding 涡流散发
edgewave 边缘波，边波，棱波，海边浪
edgewise bending 侧面弯曲
edgings 边角料
edifice 大型建筑
Edison battery 碱蓄电池，爱迪生电池
Edison Electric Institute 爱迪生电气协会
Edwards balance 爱德华天平（一种测量气体密度的仪器）
effect of exploration on field 地热田开发效应
effect of irradiation 辐照效应
effect of non-condensible gas 非冷凝气体效应
effect of soil warming 土壤加温效应
effect of streamline squeezing 流线密集效应
effect on thermal pollution 热污染效应
effect specular radiation 有效镜面辐射
effective pressure 有效压力
effective radiated power 有效辐射功率
effective acceptance angle 有效受光角
effective accumulated temperature 有效集温

effective aerodynamic downwash 有效气动下洗
effective angle of attack 有效迎角
effective aperture 有效孔径,等效孔径
effective area 有效面积
effective blade area 桨叶有效面积
effective buoyancy 有效浮力
effective camber 有效弯度
effective capacity 有效容量,有效功率
effective carrying capacity 有效载能
effective collection area 有效集热面积
effective concentration ratio 有效聚光比
effective contact area 有效接触面积
effective current 有效电流
effective distribution coefficient 有效分布系数
effective downwash 有效下洗
effective gate voltage 有效栅电压
effective gust velocity 有效阵风速度
effective half life 有效半衰期
effective heating surface 有效受热面
effective height 有效高度
effective horsepower 有效功率,有效马力
effective impedance 有效阻抗
effective luminous intensity 有效光强
effective multiplication factor 有效增值系数
effective nocturnal radiation 夜间有效辐射
effective out-put 有效出力
effective permeability 有效渗透率,有效渗透性
effective porosity 有效空隙率,有效孔隙率
effective power 有效功率
effective range 使用范围
effective solar radiation 有效太阳能辐射
effective source area 有效源面积
effective temperature 有效温度
effective terrestrial radiation 有效地球辐射
effective thermal resistance 有效热阻
effective thrust 有效推力
effective tilt factor 有效倾斜因子
effective uniform temperature 有效均匀温度
effective values 有效值
effective wave height 有效波高
effective wind speed 有效风速
effects of saturation 饱和效应
efficiency 效率
efficiency assessment 效能评估
efficiency data 效能数据
efficiency index 效能指数
efficiency of collector 集热器效率
efficiency of collector field 集热场效率
efficiency of concentrator field 聚光场效率
efficiency of solar drying 太阳能干燥效率
efficiency of solar heating 太阳能供暖效率,太阳能加热效率
efficiency of storing solar energy 太阳能储存效率
efficiency of WECS 风能转换系统效率
efficiency of WTGS 风电机组效率
efficiency ratio 效率比
efficient back radiation of sea surface 海面有效回辐射
efficient photovoltaic cell 高效光电池
efficient solar still 有效太阳能蒸馏器

effluent 污水，流出物，废气
effluent pipe 排水管
effluent pump 排水泵
effluent treatment 废水处理
efflux coefficient 流量系数
egg-beater rotor 打蛋器式叶轮
eigenfrequency 本征频率，特征频率
eigenvalue 特征值
eigenvector 特征向量，本征矢量
Einstein energy 爱因斯坦能
Einstein's equation 爱因斯坦方程
Einstein's mass-energy relation 爱因斯坦质能关系式
ejection nozzle 喷嘴
ejector 喷射器
Ekman layer 埃克曼层
Ekman spiral 埃克曼螺旋，埃克曼螺线，埃克曼螺旋风层
Ekman transport 埃克曼输送
elastic collision 弹性碰撞
elastic coupling 弹性连接
elastic deformation 弹性变形
elastic energy 弹性能
elastic force 弹力
elastic force effect 弹力效应
elastic modulus 弹性模数，弹性模量
elastic resistance 弹性阻力
elastic structure 弹性结构
elastic system 弹力系统
elasticity releasable factor 弹性释水系数
elastomeric hose pump 弹性软管泵
elastomeric seal 人造橡胶密封，合成橡胶密封
elbow 弯管接头
elbow draft tube 肘型尾水管，尾水管里衬

electric active 电活性
electric charge 电荷
electric control 电气控制
electric control valve 电动调节阀
electric converter 电复律器，网络变频器
electric coupling 电耦合
electric current 电流
electric discharge 放电
electric discharge machines 电火花机
electric displacement 电位移
electric double girder travelling crane 电动双桥式起重机
electric drift field 电漂流场
electric energy 电能
electric energy transducer 电能转换器
electric field 电场
electric field intensity 电场强度
electric flux density 电通量密度
electric force 电力
electric generating capacity 发电容量
electric generating wave pipe 发电波管
electric generation 发电
electric grid 电网，电栅极
electric heat tracing 电伴随加热
electric heater 电热器，电暖气
electric level 电平
electric line cutter 电气管路切割器
electric loss 电损失
electric machine 电机
electric meter 电表
electric motor 电动机
electric operator 电动执行机构
electric plant 电站，电力装置
electric plant efficiency 电力装置效率

electric potential by concentration 浓淡电位差，浓差电位
electric power 电力，电能，电功率
electric power bus 电源母线
electric power of ocean energy from concentration gradients 海水浓度差发电
electric power system 电力系统，电网
electric power tools 电动刀具
electric power transmission 电能传输
electric precipitation 电沉淀
electric propulsion 电力牵引，电力推动
electric pump 电动泵
electric shock 触电，电击
electric submersible pump 电潜泵
electric supply grid 供电网
electric utility 电气设施
electric wire and cable 电线电缆
electrical anemometer 电传风速度表
electrical braking 电机制动
electrical capacity 电容量
electrical component 电力组件
electrical conductivity 导电性
electrical connector 电插塞，电连接器，电接插件
electrical contact 电触头
electrical conversion 电复律
electrical degree 电角度
electrical device 电气元件，电气设备
electrical discharge 放电
electrical doule-layer capacitor 双电层电容器
electrical efficiency 电效率，发电效率
electrical endurance 电气寿命
electrical energy 电能，电力
electrical energy generation 发电量
electrical energy load 电能负荷
electrical filtering 电子滤波器
electrical generator 发电机
electrical grid 输电网络
electrical insulator 绝缘体
electrical load 电力负荷
electrical load matching 电力负荷匹配模式
electrical loss 电损耗
electrical material 电气材料
electrical network 电网
electrical noise 电气噪声
electrical panel 配电板，配电盘
electrical phase angle 电相角
electrical potentials 电势，电位
electrical power aggregator 电力聚合器
electrical power connection 电力接头
electrical resistance 电阻
electrical resistivity 电阻率
electrical resistivity method 电阻法，电阻探测法
electrical resistivity of sea water 海水电阻率
electrical rotating machine 旋转电机
electrical schematic 电路简图
electrical span adjuster 电量程调节器
electrical switch box and control 电器开关盒
electrical torque 电磁转矩
electrical unit 电单位，电单元
electrical wire and cable 电线电缆
electrical-thermal cogeneration 热电联产，热电合供，热电联供
electricity from geothermal energy 地热能发电
electricity from solar radiation 太阳辐射发电
electricity from wind power 风力发电

electricity generating system of biomass 生物质发电系统
electricity generation 发电
electricity generation capacity 发电容量，发电能力
electricity generation with geothermal steam 地热蒸汽发电
electricity generator 发电机
electricity grid 电网，电力网格
electricity load 电力负荷
electricity output 电流输出
electricity production 电力生产，发电量
electricity utility 发电网
electro welding 电焊
electrochemical battery 电化学电池
electrochemical cell 电化学电池
electrochemical converter 电化学转换器
electrochemical effect 电化学效应
electrochemical energy conversion 电化学能量转换
electrochemical photovoltaic cell 电化学光电池
electrochemical potential 电化学势能
electrochemical power generation 电化学发电
electro-chemical reaction 电化学反应
electrochemical reduction cell 电化学还原电池
electrochemical replacement 电化学置换
electrochemical solar cell 电化学太阳能电池
electrochemical storage battery 电化学储能电池
electrochemical techniques 电化学技术

electrochromic cell 电致变色元件（器件、装置）
electrochromic coating 电色涂层（薄膜），电致变色涂层（薄膜）
electrochromic device 电色器件，电致变色器件
electrochromic element 电色原件，电致变色原件
electrochromic lay 电色层，电致变色层
electrochromic material 电色材料，电致变色材料
electrochromic multilayer structure 电致变色多层结构
electrochromic reaction 电致变色反应
electrochromic switch 电致变色开关
electrode 电极
electrode boiler 电热锅炉，电极锅炉
electrode dark current 电极暗电流
electrode plate 电极片（板）
electrode potential 电极电位，电极电势
electrode stability 电极稳定性
electrodeless discharge lamp 无极灯
electrodeposited coating 电镀层，电解沉积层
electrodeposition 电解沉淀，电附着，电沉积
electrodialysis 电渗析
electrodialyzer 电渗析器
electroextraction 电解萃取
electrofluid 电流体
electrofluid dynamic 动电流体式
electrofluid dynamic wind driven generator 带电流体风力发电机
electrofluid dynamic wind generator 动电流体式风力发电机
electrogaseous dynamic wind driven ge-

nerator 带电气体风力发电机
electrokinetic effect 电动效应
electrokinetic potential 电动电势
electroless plating 化学镀，无电镀
electroluminescent cell 电致发光元件
electroluminescent diodes 电致发光二极管
electrolysis 电解，电解作用，电蚀
electrolyte 电解质，电解液
electrolyte material 电解质材料
electrolytic bath 电解（电）池
electrolytic cell 电解槽
electrolytic corrosion 电解腐蚀
electrolytic hydrogen energy storage 电解制氢储能
electrolytic solution 电解溶液
electrolytic voltameter 电解电量计
electrolyze 电解
electro-magnetic 电磁场，电磁的
electromagnetic ray 电磁射线
electromagnetic brake 电磁闸，电磁制动器
electromagnetic braking system 电磁制动系
electromagnetic continuous pulling 电磁铸锭法
electromagnetic drum system 电磁鼓系统，电磁滚筒系统
electromagnetic eddy current damper 电磁涡流阻尼器
electromagnetic effect 电磁效应
electromagnetic energy 电磁能
electromagnetic energy storage 电磁能储能
electromagnetic gear 电磁传动
electromagnetic induction 电磁感应，电磁效应，感应电流，电磁离合器

electromagnetic induction method 电磁感应法
electromagnetic interference 电磁干扰
electromagnetic parameter 电磁环境参数
electromagnetic radiation 电磁辐射
electromagnetic spectrum 电磁波谱
electromagnetic torque 电磁转矩
electromagnetic valve 电磁阀
electromagnetic wave 电磁波
electromagnetic wave propagation 电磁波传播
electromagnetically induced voltage 电磁感应电压
electromagnetism 电磁
electromechanical device 机电装置
electromechanical energy conversion 机电能量转换
electromechanical power conversion 机电能量变换
electromechanical transducer 机电换能器
electromechanical transient performance 机电暂态性能
electromechanical yaw mechanism 机电偏航机制
electromotive force 电动势
electromotive series 电动次序
electromotor 电机
electron affinity 电子亲和力
electron back-scattered diffraction 电子背散射衍射
electron diffraction 电子衍射
electron hole 电子空穴
electron lens 电子透镜
electron orbital 电子轨道

electron state 电子状态，电子态
electron-beam-induced current 电子束诱生电流
electron-hole pair 电子空穴对
electronic bathythermograph 电子海水深度温度自动记录仪
electronic grate polycrystalline silicon 电子级多晶硅
electronic load 电子负载
electronic soft starter 电子软起动器
electronic switch 电子开关
electronic warning system 电子报警系统
electron-ion pair 电子离子对
electron-pair bond 电子对键，电子偶键，共价键
electron-sink product 电子汇点产物
electroplating 电镀
electrostatic filter 静电过滤器
electrostatic lens 静电透镜
electrostatic precipitator 静电除尘器，静电集尘器
electrostatics 静电学
electrothermal energy conversion 电热能量转换
elektronen-loch-paar 电子空穴对
elemental analysis 元素分析
elementary reaction rate 基本反应速率，基元反应速率
elevated steam conditions 高蒸汽参数
elevation angle of sun 太阳仰角
elevator shaft (well) 升降机井
eliminate 消除，排除，切断
elliptical integrals 椭圆积分
elongated body 可延长体
embedded temperature detector 内置测温器

ember 灰烬，余烬
emergency auxiliary power 事故备用电源
emergency back-up fuel 应急燃料
emergency braking system 紧急制动系统
emergency cell 应急电池
emergency condition 事故工况，事故状态
emergency core cooling system 应急冷却系统
emergency door 安全门
emergency dump steam 事故排汽
emergency electric supply unit 备用供电设备
emergency governor 危急遮断器，危急保安器
emergency holding pond 应急蓄水池
emergency light 事故信号灯
emergency operating plan 事故运行方式
emergency outage 紧急停运
emergency pump 事故备用泵
emergency shutdown 紧急关机
emergency stack outlet 紧急烟道出口
emergency stop push button 紧急停车按钮
emergency valve 危急遮断阀，紧急止流阀
emergency-stop 紧急停止
emery 金刚砂
emery cloth 砂布，金刚砂布
emery wheel 金刚砂旋转磨石，砂轮
emission 发射
emission abatement system 排放减排系统
emission control 排放控制，废气排出

控制，发射控制
emission control equipment 排放控制设备
emission factor 排放因子，排放系数
emission limit 排放限度，排放限值，排放极限
emission limit value 排放值限定，排放限值
emission permit 排污许可证
emission rate 排放速度，排放速率，发射率
emission reduction credits 减排信用额度
emission scrubber system 排放涤气系统
emissive power 辐射力
emissivity 辐射率，放射率，辐射能力，发射率，辐射系数
emittance 辐射，发射，放射，辐射率
emitted current 发射电流
emitter 发射管，放射器，发射极
emitting area 辐射面积
emitting color 发光色彩
emitting medium 散射介质
emitting surface 辐射面
emollient 软化剂
empirical coefficient 经验系数
empirical relationship 经验关系
empirical test 经验测试
emptying 放空
encapsulant material 封装材料
encode 编码
end cell 终端电池，附加电池
end condition 边界条件
end effect 端部效应
end load 末端荷载
end losses 端部损失

end panel column 抗风柱
end plate 端板
end point energy 边界能量，终点能量
end ring 端环
end shield 端罩
end winding 端部绕组
end-device 终端设备
endergonic conversion 吸能转换
endergonic photochemical reaction 蓄能光化学反应，吸能光化学反应
endergonic reaction 蓄能反应，吸能反应
endothermal change 吸热变化
endothermic (endothermal) reaction 吸热反应
endothermic conversion 吸热反转化
endothermic disintegration 吸热转化（吸收能量）
endothermic process 吸热过程
endplate 终板
end-stop 终点挡板，末端挡板，终点止动装置
endurance limit 持久极限，疲劳极限
endurance test 耐久试验
end-use demand 最终需求量，最大需求量
end-use energy 终端能源
end-use load 终端载荷
end-user facility 终端用户设施
endwise 末端朝前或向上的，向前的
energetic wind 强风
energized hour 通电时长
energy conversion device 能量转换设备，换能器
energy absorbed by storage pumping 抽水蓄能电站抽水耗能量
energy absorbing material 吸能材料

energy absorption 能量吸收
energy agriculture 能源农业
energy alternative 能源替代
energy and gases supply for construction 施工力能供应
energy balance 能量平衡，能量均衡
energy balance analysis 能量平衡分析
energy band 能带
energy band gap 能带宽度
energy band structure 能带结构
energy barrier 能量位垒
energy budget 能量收支
energy by ocean current 海流能
energy by salinity gradient （海水）盐（度）差能
energy by thermal gradient （海洋）温差能
energy by wave motion 波浪运动能
energy calculation 能量计算
energy calorific value 能量的当量值
energy capability factor 产能因子
energy capability of a pumped storage station during turbine operation 抽水蓄能电站水轮机运行期间产生能量
energy capacity 能量容量，能量容限，能源总量
energy capture 能量获取
energy carrier 能源载体
energy conservation 能量守恒，节能
energy consumption 能耗
energy consumption unit 用能单位
energy containing eddy 含能涡旋
energy content 能含量
energy conversion 能量转换
energy conversion efficiency 能量转换效率
energy conversion machine 能量转化机
energy conversion system 能量转换系统
energy converter 能量转换器，电能转换器
energy conversion cycle system （温差）发电循环系统
energy crop 能源作物，含能作物
energy cropping 能源种植
energy current 能流
energy deficit 能量不足
energy delivery factor 能源输送因子
energy density 能量密度
energy density peak 能量密度峰值
energy development 能源开发
energy dissipation 能量耗散
energy dissipation device 耗能装置
energy distribution 能量分布，能源分布
energy efficiency 能源效率
energy efficiency equation (EU) （欧盟）能源效率方程，能量效率方程
energy efficiency in ocean current 海流能量利用率
energy efficiency ratio 能源效率比值，能效比
energy efficiency technology 节能技术
energy equivalent 能源当量，能当量
energy equivalent value 能量的等价值
energy exchange 能量交换
energy extraction 获能，能量提取
energy extraction device 能量开采设备
energy farm 能源种植场
energy farming 能源农作物
energy flow 能流，能量通量
energy flux density 能通量密度
energy forest tree 能量森林树

energy from biomass 生物质能
energy from cereal grain 谷类植物能
energy from earth's interior 地球内部能源地热能
energy from ocean 海洋能
energy from ocean surface wave 海面波能
energy from organic waste 有机废物能
energy from salinity gradient 盐差能
energy from the sun 太阳能
energy from the wave 波浪能
energy from waste 转废为能
energy gain 能量收益
energy gap 能隙,禁带宽度
energy grade 能量等级
energy gradient 能量梯度
energy grass 能量草
energy in ecosystem 生态系统中的能量
energy index of reliability 可靠能源指数
energy information administration 能源信息管理局
energy input 能量输入
energy input peak 输入能量峰值
energy input ratio 能量输入比
energy intensity 能量强度
energy level 能级
energy level diagram 能级图
energy liberation 能量释放
energy loss 能量损耗,能耗
energy management and control system 能源管理与控制系统
energy manager 能源管理器
energy meter 能量计量
energy metabolism 能量代谢
energy mobilization 能量动用
energy mode 能量模式
energy of sea current 海流能

energy output 能量输出
energy output of sun 太阳能量输出
energy payback time 能源回收时间,能源回收期,能源偿还时间
energy penalty 能源损耗
energy plant 能源植物,能源种植场
energy plantation (为用作合成燃料或发电的原料而种植的)能源植物丛
energy plantation tree 能量种植树
Energy Policy Act (EPAct) (美国)能源政策法案
energy poor nation 缺能国家,能源不足国家
energy potential of geothermal field 地热田能量潜力,地热田能源潜力
energy pyramid 能量金字塔
energy radiation from surface of sun 太阳表面能量辐射
energy recovery factor 能量回收系数
energy recovery 能量回收
energy release rate 能量释放率
energy research and development 能源研究和开发
energy resource 能量资源
energy return period 能量生产还本周期
energy saving lamp 节能灯
energy saving performance contracting 节能绩效保证合约
energy security 能源安全
energy sharing 能源共享,能源输出
energy source 能源
energy spectrum 能谱,能量分布
energy spectrum from sun 太阳能谱
energy standard 能源标准
energy storage 蓄能,能量储存
energy storage capacity 蓄能容量
energy storage pond 能量储存池

energy storage system of wind power 风力发电储能系统
Energy Tax Act of 1978 1978 年美国能源税法案，美国《1978 年能源税法案》
energy thickness 能量厚度
energy transfer 能量传递
energy transformation 能量转换
energy transmission 能量传递
energy transport system 能量传送系统
energy unit 能量单位
energy utilization 能源利用
energy without pollution 无污染能源
energy wood 能量木材
energy-consumed medium 耗能工质
energy-dispersive X-ray spectroscopy 能量色散 X 射线光谱法
energy-from-waste 废物再生能源，垃圾焚烧发电
engagement mesh 啮合
engine control system 发动机操纵系统
engine cooling airflow 发动机冷却气流
engine cover 发动机罩
engine cowl 发动机罩
engine cycle 热机循环
engine displacement 发动机排量
engine efficiency 发动机效率
engine manifold 发动机歧管
engine power 动力性，发动机功率
engine starting system 发动机启动系统
engine-driven generator 驱动发动机的发电机
engineering analysis 工程分析
engineering committee on oceanic resource 海洋资源工程委员会
engineering information 工程资料
engineering judgement 工程判断

engineering oceanography 工程海洋学
engineer's pocket thermometer 工程师用便携式温度计
engraving machine 雕刻机
enhanced geothermal systems 人造地热能，强化地热系统
enhancement of permeability 渗透性增强
enoyl hydrase 烯酰水化酶
enriched gas 浓缩汽油
enriched material 浓缩物，浓缩材料
ensilage 青储饲料
enthalpy 焓，热函
enthalpy entropy diagram 焓—熵图
enthalpy wheel 焓轮
entire thermal resistance 总热阻
entrained-flow gasification reactor 气流床，气化反应器
entrainment 夹带，夹杂
entrainment separator 雾沫分离器
entrance pressure 进口压力
entropy of transition 转变熵
entropy production 熵产，熵产生
envelope temperature 罩温，盖板温度
environment condition 环境条件
environmental capacity 环境容量
environmental concern 环保意识
environmental control technology 环境控制技术
environmental impact 环境影响，环保冲击
environmental impact assessment 环境影响评估
environmental impact of solar thermal system 太阳能—热能系统的环境影响
environmental impacts associated with geothermal development 地热开发

对环境的影响
Environmental Management Act 环境管理法
environmental monitoring 环境监测
environmental protection of geothermal development 地热开发环境保护
environmental quality index 环境质量指数
environmental quality pattern 环境质量模式
environmental restriction 环境限制
environmental sensor 环境传感器
environmental wind 环境风
enzymatic breakdown 酶催化分解
enzyme 酶,酵素
enzyme activity 酶活性
enzyme cell 酶电池
enzyme electrode 酶电极
enzyme hydrolysis 酶水解
enzyme induction 酶的诱导
enzyme inhibition 酶抑制
eolation 风蚀
eolian anemometer 风琴式风速计
epicenter 震中
epicyclic gear 行星齿轮,周转齿轮
epipelagic zone 海洋光合作用带（区）
epitaxial diffused process 外延扩散过程
epitaxial growth activation energy 外延生长激活能
epitaxial junction 外延结
epitaxial technology 外延工艺
epithermal energy 超热能
epoch angle 初相角
epoxy 环氧树脂
epoxy adhesive 环氧黏合剂
epoxy coating 环氧涂层

epoxy composite 环氧复合材料
epoxy glue 环氧胶
epoxy lined steel 环氧树脂钢
equalization basin 调节池
equalizer 平衡器,补偿器
equation of time 时差
equation of transient coefficient 非稳态效率方程
equatorial mount 赤道装置
equatorial tide 赤道潮
equatorial tracker 赤道式跟踪器
equatorial wave 赤道波
equiaxed crystal zone 等轴晶区
equilateral triangle 等边三角形
equilibrium 平衡,均衡
equilibrium carrier 平衡载流子
equilibrium distribution coefficient 平衡分布系数
equilibrium position 平衡位置
equilibrium pressure 平衡压力,平衡压强
equilibrium solidification 平衡凝固
equinox 昼夜平分点,二分点
equipment failure 设备故障
equipment failure information 设备故障信息
equipotential bonding 等电位连接
equipotential map 等势图
equivalence ratio 当量比
equivalent capacitance 等效电容
equivalent circuit 等效电路
equivalent circuit of a solar cell 太阳能电池的等效电路
equivalent circuit parameter 等效电路参数
equivalent electrical circuit 等效电路

equivalent full load hours 等效负载小时
equivalent inductance 等效电感
equivalent length 当量长度
equivalent T circuit T型等值电路
erection 直立，竖起，架设
errant algae 浮游藻类
error amplifier 误差放大器，误差信号放大器
error codes 故障代码
error detector 误差检测器
error margin 误差容限，误差幅度
error signal 误差信号
esterification 酯化，酯化作用
ester 酯类
estuarine salinity gradient energy converter 海港盐度差能转换装置
estuary 河口
estuary deposit 港湾沉积
etching machines 蚀刻机
ethanol 乙醇
ethanol blend 乙醇混合物
ethanol degrading bacteria 乙醇降解菌
ethanol gasoline for motor vehicles 车用乙醇汽油
ethanol-cane 甘蔗乙醇
ethyl alcohol 乙醇，酒精
ethyl tertiary butyl ether 乙基叔丁基醚
ethylene glycol 乙二醇，甘醇
Ethylen-Venyl-Acetat 封装太阳能电池板的薄膜
European Wave and Tidal Energy Conference 欧洲波浪和潮汐能源会议
eutectic material 低共熔材料
eutectic reaction 共晶反应
eutectic salt 低溶盐（太阳能储热器的一种储热介质）
eutectic structure 共晶组织
eutectic temperature 低共熔温度，共晶温度
eutectoid 共析体，类低共熔体
evacuated collector 真空集热器
evacuated collector tube 真空集热管
evacuated glass tube collector 真空玻璃管集热器
evacuated glass tube cover 真空玻璃管（圆柱面）盖板
evacuated receiver 真空接收器
evacuated tube absorber 真空管吸收器
evacuated tube collector 真空管集热器
evacuated tubular solar collector 真空管（状）太阳能集热器
evacuated-tube solar collector 真空管式太阳能集热器
evaluation of geothermal resources 地热资源估算
evaporated antireflection coating 蒸镀减反射膜
evaporated dish 蒸发器
evaporated oxide film 蒸发氧化薄膜
evaporating circuit 蒸发回路
evaporating heating surface 蒸发受热面
evaporating surface 汽化面，蒸发面
evaporation 蒸发
evaporation capacity 蒸发量
evaporation factor 蒸发系数，汽煤比
evaporation heat 蒸发热
evaporation loss 蒸发损失
evaporation opportunity 蒸发可能率
evaporation pond 蒸发池
evaporation rate 蒸发率

evaporation ratio 蒸发比率
evaporation tank 大型蒸发器，蒸发池
evaporation zone 蒸发区
evaporative capacity 蒸发量
evaporative condenser 蒸发凝汽器
evaporative cooler 蒸发冷却器，蒸发式冷却器
evaporative cooling system 蒸发冷却系统
evaporative cooling tower 蒸发式冷却塔
evaporative rate 蒸发率
evaporative space 蒸发空间
evaporator 蒸发器
evaporator chamber 蒸发器室
evaporator coil 蒸发器蛇形管，蒸发器盘管
evaporograph 蒸发计
event information 事件信息
ex situ 非原位，天然状态外
excavator 挖掘机
excess air （超过理论所需的）过量空气，过剩空气
excess air coefficient 过量空气系数
excess air control 过量空气控制
excess air quantity 过量空气量
excess air ratio 过量空气比，过剩空气系数
excess current 过载电流
excess heat 余热
excess oxygen 富氧
excess sludge 剩余污泥，剩余活性污泥，废活性污泥
excess voltage 过压
exchange energy 交换能，互换能量
exchange field 交换场
exchange rate 交换率，交换速度
exchanger 热交换器，换热器，散热器，交换剂
excitation 励磁
excitation current 励磁电流
excitation energy 激发能
excitation field 激发场
excitation force 激发力，激振力
excitation frequency 励磁频率
excitation level 激发能级
excitation loss 励磁损失
excitation response 励磁响应
excitation spectrum 激发光谱
excitation system 励磁系统
excitation winding 励磁线圈，励磁线组
exciter 励磁机
exciter rectifier 激励整流器
exciting force 激发力
exciting voltage 励磁电压
exergetic efficiency 火用效率
exergonic conversion 释能转换
exergonic reaction 放能反应，释能反应
exfoliation 剥落
exhaust 排气装置
exhaust cycle 排气循环
exhaust fan 排风机
exhaust gas 尾气，废气
exhaust heat 废热
exhaust heat exchanger 排气加热热交换器
exhaust line 排气线，排气管线
exhaust loss 排气损失
exhaust manifold 排气总管
exhaust pipe 排气管
exhaust pressure 排出压力，排气压力
exhaust regulator 排气调节阀
exhaust resistance 出口阻力
exhaust silencer 排气消音器
exhaust system 排气系统，抽风系统

exhaust valve 排气阀
exhausted cell 放完电的电池
exhausting system 抽风系统
exit dose 出射剂量，离体剂量，引出端剂量
exit loss 出口损失，输出端损失
exothermal 发热的，放热的，放能的
exothermal chemical reaction 放热化学反应
exothermic 放热的，发热的，放（出）能（量）的
exothermic absorber 放热吸收体，放热吸收器
exothermic chemical reaction 放热化学反应
exothermic chemical reaction mechanism 放热化学反应机制
exothermic disintegration 放热衰变，放热转化（放出能量）
exothermic reaction 放热反应
expanded metal 膨胀金属
expander 扩张器，扩充器，扩大器
expansion bolt 膨胀螺栓，扩开螺栓，自攻螺丝
expansion cock 调节阀，安全阀
expansion coefficient 膨胀系数
expansion energy 膨胀能量
expansion joint 伸缩接头
expansion stress 膨胀应力，拉应力
expansion tank 膨胀水箱，涨溢箱
expansion valve 膨胀阀
expected wind speed 期望风速
experimental geothermal electric power plant 实验性地热发电站
experimental modelling 实验模型，实验建模
experimental photovoltaic cell 实验用光电池，实验用光伏打电池
experimental tidal power plant 潮汐试验电站
explanatory quad 填充铅块
exploitable reserves 可开采量
exploitable reservoir 可开发（热）储层
exploration of geothermal power 地热能勘探
exploration of geothermal resource 地热资源勘探
exploration of the ocean 海洋开发
exploration strategy 勘探策略
exploratory drilling 钻探，勘探钻井
exponential distribution 指数分布
exposed receiver 裸露接收器，无玻璃盖板接收器
exposure 空晒
exposure temperature 空晒温度
expression 公式
extended maximum entropy principle 扩张最大熵值法
extended maximum likelihood method 扩张极大似然法
extended producer responsibility 生产者责任延伸
extended surface exchanger 表面展扩热交换器
extended surface tube 鳍片管，肋片管
extension tube 伸缩管
external heat 外来热
external absorber 外表面吸收器，外部受光吸收器
external armature circuit 电枢外电路
external calipers 外卡钳
external capacitor 外部电容
external characteristic 外特性
external combustion cycle 外燃循环

external condition 外部条件
external dimension 外观尺寸
external flow water collector 外流水式集热器
external gear 外齿轮
external heat loss 散热损失
external lightning protection system 外部防雷系统
external load 外荷载
external power 外部动力
external power supply 外部动力源
external pressure 外压力
external pressure coefficient 外压力系数
external pressure loss 外压损失
external receiver 外置式吸热器
external suction 外部吸力
external surface 外表面
external water wave 外部水波浪
external wind load 外部风载
external work 外功
extractable energy 可获能量
extractable power 可提取电力
extraction apparatus 提取器
extraction column 提取塔,萃取塔
extraction distillation 提取蒸馏
extraction efficiency 提取效率,萃取效率
extraction of elements from seawater 海水元素提取
extraction plant 抽提装置
extraction rate 萃取率
extraction system 提取系统
extraction tower 提取塔
extraction turbine 抽汽式汽轮机
extractor 提取器
extra-low head tidal power 超低压头潮汐能发电
extraneous ash 外来灰分
extraneous loading 附加荷载
extrapolated power curve 外推功率曲线
extrapolation technique 外推技术
extraterrestrial irradiance 地外（日射）辐照度
extraterrestrial irradiance 地球大气层外的太阳辐照度
extraterrestrial radiation 大气顶层太阳辐射
extraterrestrial solar flux 大气上界太阳能通量，宇宙太阳能通量
extraterrestrial solar radiation 地外太阳辐射
extratropical cyclone 温带气旋
extreme gradient wind 极端梯度风
extreme gust 阵风极值
extreme lifetime wind speed 寿命极端风速
extreme low tide 最低潮位
extreme maximum 极端最高
extreme mile wind speed 极端英里风速
extreme surface wind 地面极端风
extreme wave height 最大波形高度
extreme wind 极大风
extreme wind speed 极大风速
extrinsic absorption 非本征吸收
extrinsic lifetime 非本征寿命
extrinsic photoconductivity 非本征光电导率
extrinsic semiconductor 非本征半导体
extruded aluminum 挤制铝
extruded blade 挤压叶片
extrusion 挤压,挤压成形
eye screw 螺丝眼

eyebolt 吊环螺栓，吊耳

F

fabric filter 纤维过滤器
fabrication 制造，生产，加工
fabrication drawing 制造图纸，制作图
fabrication tolerance 制造容差
face area 迎风面积
face velocity 迎风风速
face voltage 工作面电压
face width 齿宽
faceted collector 多反射平面集热器
facilitate 使容易，使便利，推动，帮助，促进
facility design 设施规划，设施设计
fact sheet 说明书
factor of safety 安全系数
factory building 厂房
facultative anaerobe 兼性厌氧微生物
facultative bacteria 兼性菌
fahrenheit 华氏温度计，华氏温标
fail-safe 自动防故障装置
fail-safe operation 自动防止故障运转
fail-safe system 自动防故障系统
failure predication 故障预测
fall block 动滑轮
fall slag 漏渣（从焚烧炉炉排间隙漏下的固态物质）
falling film receiver 降膜式吸热器
false bottom 活动底板
family photovoltaic power system 户用光伏电源
fan 风机，风箱，冲积扇
fan aspect ratio 风机展弦比
fan belt 风扇皮带
fan blade 风扇叶片，风机叶片

fan casing 风机导流装置
fan coil 风机盘管
fan coil system 风机盘管系统
fan coil unit 风机盘管机组
fan convector 风机对流器
fan cooler 冷风机
fan cover 风扇罩
fan delivery 风机送风量
fan discharge 通风机排量
fan drive assembly 风机驱动部件
fan driven generator 风力发电机
fan exhauster 排气风机
fan heater 风扇加热器
fan hub 风扇轮毂
fan inlet 风机入口
fan motor 风扇电机
fan outlet 风机出口
fan performance 风机特性，风机性能
fan performance curve 风机特性曲线
fan pitch drive control 风机导叶驱动控制
fan propeller 风扇叶轮
fan room 通风机房
fan rotor 风扇转子
fan suction 风机负压
fan-coil unit 风扇螺管装置，冷热风机组
fan-driven generator 风（机驱）动发电机
fanner 风扇，通风机
fanning strip 扇形片
fantail 扇形尾（一种靠风力本身来操纵的自动调向机构）
far field 远场，远源场
far wake 远尾流
Faraday's law 法拉第定理
Faraday's law of electrolysis 法拉第电解定律

far-field pressure-time history 远场压力—时间历程
far-infrared radiation 远红外辐射
farm windmill 电厂风车
farm-type windmill 农场型风车
far-ultraviolet radiation 远紫外辐射
fast geothermal reconnaissance 地热快速勘探
fast pyrolysis 快速裂解
fast response cup anemometer 灵敏转杯风速表
fast response instrument 快速响应仪表
fast tuning 快速调谐
fast-burning fuel 速燃燃料
fastener 紧固件
fastest mile wind 最大英里风
fat-fuelled power station 脂肪燃料发电厂
Fathrenheit temperature scale 华氏温标
fatigue damage 疲劳损伤
fatigue endurance limit 抗疲劳极限
fatigue life curve 疲劳寿命曲线
fatigue load 疲劳荷载
fatigue loading 疲劳加载，疲劳载荷
fatigue resistance 抗疲劳性，抗疲劳强度
fatigue stress 疲劳强度
fatigue test 疲劳试验
fatty acid 脂肪酸
fatty acid methyl ester 脂肪酸甲酯
fatty acid oxidation 脂肪酸氧化
fault 故障
fault condition 故障状态（事故工况）
fault current 故障电流
fault diagnosis 故障诊断
fault earthing 故障接地
fault system 断层系统，断裂系统，断层系
fault zone 断裂带
fault-controlled spring system 控制断层的泉水系统
favorable pressure difference 顺压差
favorable pressure gradient 顺压梯度
feasibility study 可行性研究
feasibility test 可行性试验
feather position 顺桨位置
fcathered position 顺流位置
feathering 顺桨
feathering airscrew 顺桨螺桨
feathering paddles 活桨叶
Federal Energy Management Program （美国）联邦能源管理计划
feed auger 旋转加料器
feed chute 进料槽
feed circuit 馈电电路
feed hopper 进料斗，装料斗
feed in tariff 上网电价
feed mixer 进料混合器，进料搅拌机
feed preparation 备料
feed pump 给水泵
feed ram 加料推杆
feed rate 加料速度，进给速率
feed screw conveyor 来料螺旋输送机
feed system 供料系统
feed tank 给水箱
feed valve 送料阀，进给阀
feed water supply 给水补给
feed water tank 供水槽
feed water treatment 给水处理
feed well 给水井
feedback compensation 反馈补偿
feedback component 反馈元件

feedback control device 反馈控制装置
feedback control system 反馈控制系统
feedback linearization 回馈线性化，反馈线性化
feedback loop 反馈环，反馈回路
feedback path 反馈路径
feedback signal 反馈信号
feedback sun tracker 反馈式太阳能跟踪器
feedback system 反馈系统
feedback voltage 反馈电压
feedback voltmeter 反馈式伏特计
feeder 馈线
feeder conveyor 装料传送机
feeder line 馈线
feedstock 给料，原料
feedwater cycle 回热循环，给水系统
feedwater flow 给水流量
feedwater line 给水管路
feedwater makeup 补给水
feedwater regulator 给水调节器
feedwater softening 给水软化
feedwater treatment 给水处理
feeler gauge 触规，测隙规
feldspar 长石
felling axe 外轮轴
female rotor 凹形转子，阴转子
fender 防护板，挡泥板，缓冲装置
fermentating bacteria 发酵细菌
fermentating organism 发酵生物
fermentation 发酵
fermentation accelerator 发酵加速剂
fermentation alcohol 发酵醇
fermentation engineering 发酵工程
fermentation ethanol 发酵乙醇
fermentation gas 发酵气

fermentation liquid 发酵液
fermentation of glucose 葡萄糖发酵
fermentation of organic substance 有机物质发酵
fermentation raw material 发酵原料
fermentation tube 发酵管
fermentative assimilation 发酵同化作用
fermentative bacteria 发酵性细菌
fermenter 发酵罐
fermenting tank 发酵池
Fermi energy 费密能量
Fermi level 费米能级
ferrite 铁酸盐，铁素体
ferrite magnet 铁氧体磁铁
ferritic 铁素体的
ferromagnetic material 铁磁材料
ferromagnetic swarf 铁磁屑
ferrosilicon 硅铁
ferrous metal 黑色金属，含铁金属
fertile element 燃料原料
fertilizer effect of biogas slurry 沼气浆肥效
fertilizer feedstock 化肥原料
fibreboard residue 纤维板残渣
fiber enforced concrete 纤维增强混凝土
fiber glass reinforced thermoplastics 玻璃纤维增强热塑性塑料
fiber optic 光纤
fiberglass cloth 玻璃纤维布
fiberglass cover 玻璃纤维盖板
fiberglass insulation 玻璃纤维绝缘（材料）
fiberglass reinforced plastic (FRP) 玻璃钢
fibreglass reinforced plastic valve 加

强的玻璃纤维塑料阀
fiberglass reinforcement 玻璃纤维增强材料
fiberglass structure 玻璃纤维结构
fibre rope 纤维缆绳，纤维缆索
fibre sludge 纤维浆
fibrous filter 纤维过滤器
fibrous thermal insulation 纤维热绝缘层
fibrous vegetable waste 纤维蔬菜废弃物
Fick's first law 菲克第一定律
Fick's second law 菲克第二定律
fictitious load 假荷载，模拟荷载
fidelity 保真度
fiducial temperature 基准温度
field 现场的，现场
field boundary condition 场边界条件
field bus 现场总线
field coil 场线圈，励磁线圈
field corn 饲料玉米
field current 励磁电流
field data 现场数据
field development 地热田开发
field effect 场效应
field drainage system 地热田排水系统
field effect transistor 场效应管
field energy 场能
field enhanced photoconductivity 场增强光电导率
field fabrication 工地制造，现场装配
field information 现场信息，矿区信息
field installation 现场安装
field instrument 携带式仪表
field intensity 电场强度
field measurement 现场测量
field mirror 场镜

field of view angle (of pyrheliometer) （直接日射表）视场角
field radiometer 工作辐射表
field reliability test 现场可靠性试验
field survey 现场调查
field test with turbine 外联机试验
field testing 野外试验，现场试验
field testing of wind energy conversion system 风电机组现场试验
field trial 田间试验，实地测试，实地试验
field well 生产井
field winding 励磁绕组，磁场绕组
field work 现场作业，工地工作
filament 灯丝，细丝
filament winding 纤维缠绕
fill factor 填充因数，填充因子，曲线因子
filled insulation 绝缘填料
filled plastics 填充塑料
filler metal 焊料，焊丝
filler rod 焊条
fillet weld 角焊，填角焊
filling 充水
fills 填充物
film circuit 薄膜电路
film coefficient 膜系数
film density 影片密度
film GaAs solar cell 薄膜砷化镓太阳能电池
film heat transfer coefficient 膜传热系数
film morphology 薄膜形貌
film viewer 底片观察用光源
filter 滤光片，滤光器
filter aid 助滤剂，助滤器
filter cake 滤饼
filter cloth 滤布

filter diaphragm cell 滤膜电池
filter layer 滤层，渗滤层
filter press 压滤机，压滤器
filter-press cell 压滤式电池
filtration equipment 过滤设备
fin efficiency 散热片效率
fin heat exchanger 鳍管热交换器
fin tube absorber 肋状管吸收器
fin type radiator 片式散热器
fin type receiver 鳍形接收器
final anode voltage 末级阳极电压
final evaporator 末级蒸发器
final moisture content 蒸汽终湿度，排气湿度
final operating temperature 最终工作温度
final quality control 成品质量检验
final reheater 末级再热器
final stage deaerator 末级除氧器
final superheater 末级过热器
final system design 最终（风能）系统设计
fine grid 密网格
fine grinding mill 精细粉碎机
fine material 细粒材料，精细材料，细颗粒物料
fine powder 细粉
fine structure of atomic spectra line 原子光谱线的精细结构
fine-meshed heat matrix 细孔热矩阵
fine-tuning 微调，细调
finger baffle 导向隔板
finishing machine 修整机
finite depth 有限深度
finite difference 有限差，有限差分法
finite element method 有限元技术
finite size 有限尺度

finned absorber plate 翅片式吸热板
finned pipe 翅片管
finned surface 鳍片受热面
finned-plate absorber 翅片板吸收器
fire barriers 防火间隔
fire bed 火床
fire fighting water pump 消防水泵
fire forest 薪炭林
fire retardant paint 耐火涂料
fire wood 薪柴
firebrick 耐火砖
fireclay refractory 黏土质耐火材料
fireman's axe 消防斧
fire-tube boiler 火管锅炉
firing angle 开火角
firing battery 点火用电电池
firing voltage 点火电压，开启电压
firmer chisel 凿子
first class pyranometer 一级（工作）总日射表
first class pyrheliometer 一级（工作）直接日射表
first class pyrradiometer 一级（工作）全辐射表
first coat 底涂层
first diffusion equation 第一扩散方程
first guess 首次猜想
first phase 初相
first stage survey （地热田）第一调查阶段
first-in first-out 先进先出
first-order response 一阶响应
first-surface mirror 第一表面反射镜
Fischer-Tropsch diesel 费托柴油
Fischer-Tropsch process 费托合成过程
Fischer-Tropsch reactor 费托合成反

应器
fish migration 洄游，鱼类洄游
fish pass 鱼道
fish passage 鱼道
fish way 鱼道
fish-friendly turbine 亲鱼（型）水轮机
fishing ground 渔场
fission calorimeter 热变量热计
fix bed combustion 固定床燃烧
fix pitch airscrew 定距螺桨
fixed axis 定轴
fixed bed biomass gasifier 生物质固定床气化炉
fixed bed combustion 固定床燃烧
fixed bed gasifier 固定床气化炉
fixed blade 固定式桨叶
fixed carbon 固定碳
fixed circular trough concentrator 固定式圆形槽聚光器
fixed collector 固定式集热器
fixed compound parabolic concentrator 固定式复合抛物面聚光器
fixed concentrator 固定聚光器，固定聚光集热器
fixed contact 静触头
fixed dome digester 固定气箱沼气池
fixed frequency 稳频
fixed grate 固定炉排
fixed hub 固定桨毂
fixed mirror line focus system 固定反射镜线聚焦系统
fixed mirror/distributed focus 固定反射镜配焦
fixed mirror/distributed focus solar-to-electrical conversion 固定反射配焦太阳能—电能转换
fixed parabolic mirror 固定抛物线形反射镜
fixed pitch 定桨距
fixed pitch axial-flow fan 定桨距轴流风机
fixed pitch blade 定桨距叶片
fixed receiver rotating-slat reflector collector 固定吸热器式旋转反射板集热器
fixed sloped surface 固定斜面
fixed solar collector 固定式太阳能收集器
fixed spherical concentrator 固定式球面聚光型集热器
fixed structure 固定式结构物
fixed suspended solid 固定性悬浮固体
fixed system for absorbing wave energy 固定式波能吸收装置
fixed target concentrator 定焦聚光器
fixed trough type collector 固定式槽型聚光器
fixed tilt array 固定倾斜阵列
fixed tilt collector 固定斜置集热器
fixed tilt solar collector 固定式倾斜式太阳能集热器
fixed voltage 稳压
fixed-bed system 固定床系统
fixed-mirror moving absorber system 固定反射镜动吸收器系统
fixed-mirror tracking receiver collector 固定反射镜跟踪吸收器式集热器
fixed-mirror tracking receiver trough like collector 固定反射镜跟踪吸收器类槽式集热器
fixed-speed drive 定速驱动
fixed-speed operation 恒速运行
fixing 固定
fixture 夹具

flame arrestor 阻焰器，阻火器
flame atomic absorption spectrometry 火焰原子吸收光谱法
flame impingement 火焰冲击
flame ionization detector 火焰离子化检测器
flame length 火焰长度
flame luminance 火焰亮度
flame port 火焰通口
flame radiation 火焰辐射
flame scanner 火焰监视器
flame zone 火焰带
flaming combustion 有焰燃烧
flammable gas 可燃气体
flange 边缘，轮缘，凸缘，法兰
flange bolt 法兰螺栓
flange connection 凸缘连接
flange coupling 凸缘联轴器
flange gasket 法兰垫片
flange joint 凸缘接头
flanged joint 法兰接头，法兰连接
flanged nut 凸缘螺母
flanged union 凸缘连接
flank 侧面，侧腹
flap 襟翼
flap air brake 襟翼空气制动
flap angle 襟翼偏转角
flap shutter 转动挡板
flap system 拍动板系列
flap valve 挡板阀
flap wavemaker 襟翼造波机
flapping equation 翼动方程式
flaptype damper 转动叶片式挡板
flaptype wave motor 拍动板型波力发动机
flapwise （风轮）翼面方向的
flapwise bending 摆振弯曲

flare line 火炬管线
flash 闪蒸
flash boiler 快热锅炉，闪蒸锅炉
flash evaporation technique 瞬间蒸发技术
flash evaporator 闪蒸蒸发器，闪蒸器
flash facility 闪蒸设施
flash over voltage 火花放电电压,击穿电压
flash plate 闪熔镀层
flash power plant 闪蒸电厂
flash pyrolysis 瞬间热解
flash steam generator 快速蒸汽发生器
flash steam power plant 闪蒸式电厂，闪蒸蒸汽发电站
flash steam system 闪蒸系统
flash tank 闪蒸罐
flash vessel 闪蒸器
flash welding 闪光焊
flash/binary combined cycle 闪蒸/两相联合循环
flash-evaporation 闪蒸发
flashlight 手电筒，闪光灯
flashlight battery 闪光灯电池
flashover 闪络
flashpoint 飞溅点
flat absorber 平板吸收器
flat bed truck 平板卡车
flat collector 平板集热器
flat die pelletizer 平模制粒机
flat nut 平螺母
flat plate collector 平板型集热器，平板式集热器
flat plate collector with planar reflector 增强式平板集热器（配备有平面反射器的平板式集热器）
flat plate collector with reflecting aluminum

铝反射翼平板太阳能集热器
flat plate converter 平板热转换器（装置）
flat plate module 平板式组件
flat plate solar collector 太阳能平板集热器
flat plate solar energy collector 太阳能平板集热器
flat plate solar heat collector 太阳能平板集热器
flat receiver 平板接收器
flat terrain 平坦地区
flat tube heat exchanger 平管热交换器
flat voltage 稳定电压
flat-bottomed vessel 平底容器
flat-glass solar heat collector 玻璃平板太阳能集热器
flat-head rivet 平头铆钉
flat-plate 平板式的
flat-plate collector 平板集热器
flat-plate drag 平板阻力
flat-plate evacuated tube collector 平板型真空管（太阳能）集热器
flat-plate hearth 平板炉
flat-plate heat exchanger 平板式换热器
flaw 裂缝，缺陷，疵瑕
flex 弯曲，伸缩，折曲
flexbeam 柔性梁
flex-fuel vehicle 灵活燃料车辆
flexible AC transmission systems 灵活交流输电系统
flexible bag 弹性袋（一种波能装置）
flexible conduit 软管
flexible coupling 挠性联轴节
flexible gear 柔性齿轮
flexible hinge 挠性轴

flexible pipe 软管，挠性导管，挠性管
flexible rolled-up solar array 挠性卷缩太阳能阵列
flexible rolling bearing 柔性滚动轴承
flexible rotor 柔性转子，挠性转子
flexible ventilation ducting 通风软管
flexible wire rope 柔性钢丝绳
flexural rigidity 弯曲刚度，抗挠刚度
flexure 弯曲，挠曲
flicker 闪变
flicker coefficient for continuous operation 持续运行的闪变系数
flicker step factor 闪变阶跃系数
flight level 飞行高度
float 浮选
float arm 浮子臂
float chamber 浮子室
float device 漂浮装置，浮子
float gauge 浮子式液面计
float switch 浮动开关，浮控开关，浮球开关
float system 浮子系统，浮筒系统
float valve 浮阀，浮控阀，浮球阀
float zone 浮区，浮区熔法
floating 漂浮的，浮动的，移动的，流动的，不固定的
floating battery 浮充电池组
floating buoy 浮标，浮筒
floating cover digester 浮动气罩沼气池
floating holder digester 浮罩式沼气池
floating ocean thermal energy conversion plant 浮式海洋热能转换站
floating plant 浮水植物，浮式机械设备
floating plant with an open cycle 浮式开式循环（海水温差）电站
floating platform 浮动平板

floating solar pond 浮式太阳能箱
floating solar still 浮式太阳能蒸馏器
floating tidal plant 浮式潮汐电站
floating wind turbine 漂浮风轮机
float-zone-procedure 悬浮区熔法
flocculating agent 絮凝剂
flood control storage 防洪库容
flood current 涨潮流
flood generation 涨潮发电
flood tide 涨潮
flood tide current 涨潮流
flooded electrolyte battery 充斥电解液蓄电池
flooded evaporator 全浸式蒸发器
flooded system 充溢系统
flooded waterplane 浸水水线面
floor radioactive space heating 地热辐射采暖
flora and fauna 动植物，动植物群
flow angle 气流角
flow battery 液流电池
flow chart 流程图
flow circuit 流动回路，循环回路
flow coefficient 流量系数
flow condition 流动状态
flow control 流量调节
flow diagram （工艺）流程图
flow distortion 气流畸变
flow ditch 流沟
flow field 流场
flow friction 流动摩阻
flow measurement 流量测量，流量测定
flow natural distribution 风量自然分配
flow parameter 流道参数，流动参数，水流参数，气流参数
flow path 气体流程

flow rate 流量，流速
flow rate requirement 流量需求
flow reactor 流动反应器，连续流动反应器
flow recirculation zone 回流区
flow regime 流态
flow regulator 流量调节器
flow reversal 倒流
flow sensor 流量传感器
flow separation 气流分离
flow sheet 流程图
flow speed 流速
flow straightener 整流器，稳流器
flow temperature 流动温度
flow uniformity 流动均匀性，均匀流动
flow velocity 流速，流出速度
flow visualization 流动显示，流动可视化
flow volume 流量，注出量
flowability 流动性
flowing phase 自喷阶段，流动相
fluctuating external wind loading 外部脉动风载
fluctuating internal wind loading 内部脉动风载
fluctuating pressure 脉动压力
fluctuation wind speed spectrum 脉动风速谱
fluctuation 波动
flue 烟洞，烟道，暖气管，蓬松的东西
flue ash 烟道灰
flue dust collector 除尘器
flue dust removal system 烟道除尘系统
flue gas 烟道气，烟气，排烟，废气
flue gas cleaning process 废气净化处理过程
flue gas cleaning system 烟气净化系统

flue gas compound 烟道气体化合物
flue gas condensation 烟气冷凝
flue gas condensation unit 烟气冷凝装置
flue gas desulfurization 烟气脱硫
flue gas heat exchanger 烟气换热器
flue gas outlet 烟道气出口
flue gas recirculation 烟气循环
flue gas treatment hall 烟气净化区
flue gas velocity 烟气流速
flue/stack gas 烟（道）气
fluid behavior 流体行为，流体性质
fluid chemistry 流体化学
fluid circulation 流体循环
fluid column 液柱
fluid composition 流体构成
fluid coupling 液力耦合器
fluid dynamics 流体动力学
fluid efficiency 液体效率，压裂液效率
fluid flow 流体流动，流体流量
fluid flow energy conversion 流体流动能量转化
fluid friction 流体摩擦
fluid inlet temperature 工质进口温度
fluid loss 流体损失，滤失量
fluid mechanics 流体力学，液体力学
fluid outlet temperature 工质出口温度，流体出口温度
fluid particle 流体质点
fluid pressure 流体压力
fluid property 流体性质
fluid quality 流体质量，流体品质，液体质量，液体品质
fluid quality 流体质量
fluid resistance 流体阻力
fluid solution 流体
fluid spherical concentrator 液体球面聚光集热器
fluid supply temperature 流体供应温度
fluid surface 流体面，液体表面
fluid temperature 流体温度
fluid viscosity 流体黏度，流体的黏滞性
fluidization vessel 流化容器
fluidized bed 流化床，硫化层
fluidized bed boiler 流化床燃烧锅炉
fluidized bed combustion 流化床燃烧法
fluidized bed combustor 流化床燃烧室
fluidized bed furnace 流化床加热炉
fluidized bed incinerator 流化床焚烧炉
fluidized bed reactor 流化床反应器，伶床反应堆
fluidized bed receiver 流化床吸热器
fluidized-bed gasification reactor 流化床气化反应器
fluidized-bed method 流态床反应法
fluorescence detection of leak 荧光粉检漏
fluorescence spectroscopy 荧光光谱
fluorescent lamp 荧光灯
fluorescent leak detection 荧光粉检漏
fluorescent solar concentrator 荧光聚光器
fluorine 氟
fluorine-doped tin oxide 氟掺杂氧化锡
fluorocarbon polymer 聚合物
flush valve 冲洗阀，冲刷阀
flushing 冲洗，填缝
flutter 颤振
flux 磁通，通量，焊剂，流动，熔化，流出
flux concentration ratio 通量聚光比
flux density 通量密度

flux line 液面线
flux linkage 磁链，磁通匝连数
flux linkage calculation 磁链计算
fluxing agent 溶剂，助溶剂，焊剂
fly ash 粉煤灰
fly ash disposal 飞灰处理
fly ash filter 飞灰过滤器
fly ash stability 飞灰稳定化
fly ball governor 飞球式调速器
flywheel 飞轮，惯性轮，调速轮
flywheel energy storage 飞轮储能
foam core 泡沫芯层
foamed cement 泡沫水泥
focal line 焦线
focal mechanism 震源机制，震源机构
focused solar radiation 太阳聚焦辐射
focusing collector 聚焦集热器
focusing collector of high precision 高精度聚焦型集热器
focusing collector of plastics 聚焦型塑料集热器
focusing efficiency 聚焦效率
focusing mirror 聚焦反射镜
focusing optics 聚焦光学器件，聚焦光学系统
focusing solar collector 聚焦型太阳能集热器
focusing solar energy collector 聚焦型太阳能集热器
focusing system 聚焦系统
foehn wind 焚风
follow-up system 跟踪装置，跟踪系统
food chain energy losses 食物链能量损失
food processing industry residue 食品加工业残渣
foot pump 脚泵

forage chopper 饲料切碎机
force feed system 压力进给系统
force feedback 力觉反馈
force perpendicular 垂直力
force production and calculation 产能和计算
forced circulation 强制循环
forced circulation system 强制循环系统
forced circulation water heater system 强制循环水加热系统
forced cycle solar water heater 强制循环太阳热水器
forced draft fan 鼓风机，压力通风机，强压通风扇
forced draught 强制通风
forced vibration 强迫振动
forced-circulation solar water heater 强迫循环太阳能热水器
forcing function 强制函数，强加函数
forebay 前池，进水前池
foreign material 外来物质
forest and plantation wood 森林和人造林木材
forest biomass 森林生物量
forest chip 森林木片
forest fuel 森林燃料
forest productivity 森林生产率
forest residue 森林残留物，林业废弃物
forest resource 森林资源
forestry residue 森林废弃物
forestry waste 森林垃圾，森林废弃物
forge 炼炉，熔炉
forging dies 锻模
fork 派生，分叉，分支
fork-lift truck 叉架式运货车，铲车
form of solar energy collection 太阳能

集热形式
formation 构造，结构，形成，建立，形式
formation factor 地层因素，地层电阻率因素
formation pressure 地层压力
formation temperature 地层温度
forming 印版
form-wound 模绕
forward osmosis 正向渗透
forward transfer function 正向传递函数
fossil-fuel-based energy source 矿物燃料能量源
fouling 污垢
fouling deposit 结垢沉积物
fouling indice 污染指数
foundation 基础，根本，地基，建立，创立
foundation earth electrode 基础接地体
foundry 铸造，翻砂，玻璃厂，铸造厂
foundry equipment 铸造设备
four probe method 四探针法
Fourier analysis 傅里叶分析，谐波分析
Fourier coefficient 傅里叶系数
Fourier series 傅里叶级数，傅里叶序列
Fourier transform infrared spectroscopy 傅里叶红外分光光谱仪
Fourier transform method 傅里叶转换（方法）
four-jaw chuck 四爪卡盘
four-quadrant frequency converter 四象限变频器
four-quadrant operation 四象限运行

four-stroke 四冲程
four-stroke engine 四冲程发动机
fourth-stage survey （地热田）第四调查阶段
frac tank 压裂液罐
fractional energy savings 节能率
fracture aperture 裂缝开度，裂隙开度
fracture density 裂缝密度
fracture face 断裂面，裂缝面，断口面
fracture network 裂隙网络，裂缝网络
fracture pattern 破裂型式，断裂模式
fracture radius 裂缝半径
fracture spacing 裂隙间距，裂缝间距
fracture system 断裂系，破裂系，断裂系统，裂缝系统
fracture zone 破碎岩层，断裂带，裂缝带，断层
fractured formation 裂隙岩层
fractured rock 裂隙岩体
fracturing technique 压裂技术，压裂工艺
fracturing technology 压裂技术，压裂工艺
frame 帧，画面，框架，机架，机柜
Francia type furnace 弗朗萨式太阳能炉
freak wave 畸形波，异常波
free body 自由体
free convection 自由对流
free convection heat transfer 自由对流传热
free electron 自由电子
free energy 自由能
free expansion 自由膨胀
free floating moored buoy 自由浮碇浮标

free floating wave power buoy 自由浮动式波能浮标
free flow 自由流
free flow turbine 自由流动涡轮
free form concentrator 自由曲面聚光器
free oscillation 自由振动,自有震荡
free propeller 无侧板螺旋桨
free rotor 自由风轮,自由转子,无侧板风轮
free stand tower 独立式塔架
free standing type 独立式
free stream static pressure 自由(气)流静压力
free stream velocity 自由流速度
free stream wind 自由风
free stream wind speed 自由流风速
free surface 自由表面,自由面
free surface energy 自由表面能
free water 游离水,自由水分
free wind 自由风
free yaw rotor 定向风轮
free yaw system 自由偏航系统
freestanding tower 独立塔
freestream wind speed 自由流风速
freewheeling 惰性滑行
freeze-thaw cycle 冻融循环
freeze-thaw temperature 凝结—融化温度
freezing plant 制冷装置,冻结装置
freezing rain 冻雨
Freon 氟利昂
frequency 工作频率
frequency converter 变频器
frequency distribution 频数分布
frequency domain 频率范围,频域,频率域
frequency domain analyser 频域分析器

frequency domain analysis 频域分析
frequency domain measure 频域测量
frequency domain method 频域法
frequency domain migration 频域偏移
frequency domain modelling 频域模型
frequency estimation 频数估计
frequency inverter 变频器
frequency modulation 调频
frequency of wind direction 风向频率
frequency of wind speed 风速频率
frequency ratio 频数比
frequency response 频率响应
frequency shift keying 移频键控
frequency signal injection 高频信号注入
frequency spectrum 频谱,频率谱
frequency-meter anemometer 频率表式风速计
frequent wind speed 常现风速
fresh breeze 五级风,劲风
fresh gale 大风,八级风
fresh solid digestion 新鲜固体消化
freshwater bleed 淡水渗透
freshwater duckweed 淡水生浮萍
Fresnel collector 菲涅耳集热器
Fresnel concentrator 菲涅耳聚光器
Fresnel lens 菲涅耳透镜
Fresnel lens collector 菲涅耳透镜集热器
Fresnel mirror collector 菲涅耳反射镜集热器
Fresnel zone 菲涅耳带,半波区,半周期区
Fresnel's equations 菲涅耳方程
fret-saw 线锯
friction 摩擦,摩擦力

friction coefficient 摩擦系数
friction drag 摩擦阻力
friction factor 摩擦系数
friction grip 摩擦盘
friction head 摩擦水头，摩擦损失
friction loss 摩擦损耗
friction pressure 摩擦压力
friction resistance 摩擦阻力
friction seal 摩擦密封
frictional 摩擦的，摩擦力的
frictional convergence 摩擦辐合
frictional dissipation 摩擦损耗，摩擦消散
frictional divergence 摩擦辐散
frictional drag 摩擦阻力
frictional loss 摩擦损失
frictional pressure loss 摩擦压力损失
frictional resistance 摩擦阻力
front boundary cell 前膜光电管（池）
front boundary layer cell 前膜光电池
front contact 前触点
front end processor 前置机
front port resonant efficiency 前港谐振效应
frontal area 迎风面积
frontal drag 迎风阻力
frontwall thin film solar cell 前壁薄膜太阳能电池
frozen battery 不充电电池
fruit biomass 果实生物质
fruit pit 果核
FT (Fischer-Tropsch) synthesis 费托合成
FT (Fischer-Tropsch) synthesis catalyst 费托合成催化剂
fuel assembly 燃料组件
fuel bed 燃料层，燃料堆，燃料床

fuel bed temperature 燃料床温
fuel bioethanol 燃料乙醇，酒精
fuel cell 燃料电池
fuel cell catalyst 燃料电池催化剂
fuel cell electrolyte 燃料电池电解质
fuel cell fuel 燃料电池的燃料
fuel cell powered vehicle 燃料电池汽车
fuel chute 卸料斜槽
fuel composition 燃料组成，燃料成分
fuel consumption rate 燃料消耗率
fuel cost 燃料成本，燃料费，燃料费用
fuel cost saving 燃料成本节约
fuel crop 燃料作物
fuel distribution 燃料分配
fuel efficiency 燃料效率
fuel electrode 燃料电极
fuel ethanol 燃料乙醇
fuel feed 燃料供给，燃油供给
fuel feed hopper 燃料进料斗
fuel feeder 燃料供给器，燃料加料器
fuel feeding 燃料供给
fuel flexibility 燃料适应性
fuel gas 燃料气，燃气
fuel grass 燃料草
fuel handling 燃料装卸，燃料处理
fuel homogeneity 燃料同质性
fuel hopper 煤斗
fuel injection 燃油喷射，燃料喷射
fuel load 可燃物负荷，可燃物负荷量
fuel particle 燃料颗粒
fuel pellet 燃料芯块，燃料球芯块
fuel powder 生物质燃料末
fuel preparation process 燃料制备过程
fuel pretreatment 燃料预处理
fuel pretreatment system 燃料预处理系统

fuel pretreatment technology 燃料预处理技术
fuel property 燃料属性，燃料特性
fuel quality 燃料质量，燃料品质
fuel ratio 燃料比
fuel saving stove 省柴灶
fuel size 燃料尺寸
fuel storage 燃料存储
fuel substitution rate 燃料替代率
fuel supply 燃料供应，燃油供应
fuel type 可燃物类型
fuel value 燃烧值，燃料值
fuel wood 燃料木材
fuel-air mixture 油气混合物
fuel-air ratio control 燃料—空气比例调节
fuel-cell vehicle 燃料电池车
fuel-energy-to-electrical-conversion efficiency 燃料能—电能换能效率
fuel-feeding and handling system 燃料进给和处理系统
fuelwood 薪材，薪柴
fugitive emission 易散性排放
full bridge circuit 全桥驱动电路
full converter 全额电能转换
full emitter 全辐射体
full feathering propeller 全活叶螺旋桨
full gale 大风，狂风
full load 满载
full radiator 全辐射体
full wave 全波
full-load torque 满载转矩
full-scale prototype 完全模型
full-span pitch control 全叶片变桨
full-speed test 全速测试
fully stalled regime 完全失速区
fumarole （火山区的）喷气，气孔，喷气孔

fumarolic area 冒气地面，冒气区域
function group control 功能组级控制
functional component 功能部件，功能元件
functional device 功能器件
funnel 漏斗
furan 呋喃，氧杂茂
furling 折尾
furling windspeed 收叶风速（风力机械在风速比灾害性风速低得多时的停机风速）
furnace 熔炉
furnace arch 炉拱
furnace burner 高炉燃烧器
furnace chamber 炉腔，炉膛
furnace cross-section 炉膛截面
furnace draft 炉膛负压
furnace dust 炉灰
furnace hearth pressure 炉膛压力，炉膛负压
furnace skin loss 壁热损失
furnace slag 炉渣
furnace surface loss 壁热损失
furnace temperature 炉温
furnace volume 炉膛容积
furnace wall 炉墙
furnish 供应，提供，装备，布置
fuse 熔断器
fused disconnect 熔断
fused quartz 熔融石英，熔凝石英
fuses sizing 电容器规格确定
fusion 熔融
fusion/melting point 熔点
future energy 未来能源
future energy concept 未来能源概念
future energy policy 未来能源政策
future energy price 未来能源价格
future energy prospect 未来能源前景

future growth of energy demand 未来能源需求增长
future source of energy 未来能源

G

GaAs solar cell 砷化镓太阳能电池
gabbro 辉长岩
gage glass 液位玻璃管
gale 大风，八级风
gallium arsenide cell 砷化镓电池
gallium arsenide semiconductor 砷化镓半导体
gallium arsenide solar cell 砷化镓太阳能电池
galvanic battery 伽伐尼电池（组），原电池（组）
galvanizing 电镀
galvanometer 电流表
gamma absorber γ射线吸收体
Gamma function 伽马函数
gamma ray emission 射线发射
gap magnet 气隙磁场
garbage bin 垃圾箱，清洁箱
garbage can 垃圾箱
garbage collection 垃圾收集
garbage collection technology 垃圾接收工艺
garbage disposal 垃圾处置
garbage disposal plant 垃圾处理场
garbage dump 垃圾堆
garbage furnace 垃圾焚化炉
garbage grinder 垃圾磨碎机
garbage incineration disposal device 垃圾焚烧处理设备
garbage incineration power station 垃圾焚烧电站
garbage power 垃圾能
garbage power plant 垃圾电厂
garbage truck 垃圾车，清洁车
garbage-fired boiler 垃圾锅炉
gas absorption 气体吸附
gas agent 气化剂
gas analyzer 气体分析器，气体分析仪
gas atmosphere 气氛
gas baffle 烟气挡板
gas battery 气体电池组
gas burner 煤气灶，煤气火焰
gas cap 气顶，气冠
gas chromatograph 气相色谱仪
gas chromatographic detection 气相色谱检测
gas cleaner 燃气净化器
gas cleaning 气体净化，烟气净化，煤气净化
gas cleaning unit 气体净化装置
gas collector 气体收集器，集气器
gas column 气柱
gas compressor 气体压缩机
gas conditioning 气体净化处理，气体调节
gas constant 气体常数
gas cooler 气体冷却器，煤气冷却器
gas cutting 气割
gas distributor 气体分布器,气体分配器
gas filled barrier layer cell 充气电阻挡层光电池
gas film 气膜
gas fired boiler 燃气锅炉
gas flare 废气燃烧器
gas gasification 煤气化
gas gathering system 集气系统
gas generator 气体发生器

gas heating collector 气热式（太阳能）集热器
gas holder 储气器，煤气储柜
gas installation 煤气设备，煤气工程，气体装置
gas ion 气体离子，气态离子
gas jet 气体喷流，煤气喷嘴
gas law 气体定律
gas leakage decting 检漏，漏气测试
gas lift 气流提升
gas main 煤气总管
gas mixer 气体混合器，气体混合仪
gas partial pressure 气体分压
gas particulate cleanup system 气体微粒净化系统
gas permeability 透气性，气相渗透率
gas phase 气相，气态
gas phase combustion 气相燃烧
gas phase kinetics 气相动力学
gas platform 气井钻台
gas receiver （太阳能）气体接收器
gas reformer 气体转化器
gas scrubber 气体洗涤器，废气洗涤器，净气器
gas source 气源，瓦斯源
gas speed 气速
gas storage tank 储气罐
gas tank 气罐
gas thermometer 气体温度表
gas trap 气体捕集器，气体分油器，气体分离器，集气器
gas turbine 燃气轮机，燃气涡轮
gas turbine combined cycle power plant 燃气轮机联合循环发电厂
gas vapor 蒸汽，燃气
gas velocity 气流速度

gas vent 排气孔
gas voltameter 气解电量计
gas well 气井
gas yield 产气量
gaseous biofuel 气体生物制燃料，气体生物燃料
gaseous diffusion 气体扩散
gaseous emission 气体排放
gaseous fluid 气体
gaseous fuel 气体燃料
gaseous mixture 气体混合物
gaseous pollutant 气体污染物
gaseous radiation 气体辐射
gas-fired burner 燃气燃烧器
gas-fired engine 燃气轮机
gas-fired furnace 煤气炉
gas-fired power plant 燃气发电厂
gas-fired waste combustible incinerator 可燃废物燃气焚烧炉
gas-fueled generator set 燃气发电机组
gas-gas heat exchange 气—气换热
gasification 气化
gasification efficiency 气化效率
gasification fuel cell power plant 气化燃料电池发电厂
gasification reaction 气化反应
gasification reactor 气化反应器
gasification zone 气化带，气化区
gasified efficiency 气化效率
gasifier 气化炉
gasifier reactor vessel 气化炉反应堆容器
gasifier set 气化机组
gasket 垫片，垫圈，接合垫
gasoline additive 汽油添加剂
gasoline alkylate 烷基化汽油

gasoline throttle 油门，节气门
gasometer 气量计
gas-side corrosion 烟气侧腐蚀
gassy system 充气系统
gas-to-boiling-liquid heat exchanger 气—沸腾液体热交换器
gas-to-gas heat exchanger 气—气热交换器
gas-to-liquid heat exchanger 气—液热交换器
gate pole 门极
gate turn off thyristor 门极可关断晶闸管
gate valve 闸阀
gatherer 收集器，导入装置
gathering area 储集区
gathering system 收集系统
gauge 标准尺，规格，量规，量表，测量
gauge board 样板，模板，规准尺
Gauss distribution 高斯分布，正态分布
gavel 槌
GC detector 气相色谱仪鉴定器
gear box 齿轮箱
gear box ratio 齿轮箱变比
gear case 齿轮箱
gear cutting machine 齿轮切削机
gear drive 齿轮传动
gear driven pump 齿轮传动泵
gear fan 齿轮箱冷却风扇
gear hub 齿轮毂
gear lever 变速杆
gear loading 齿轮负荷
gear meshing 齿轮啮合
gear motor 齿轮电动机
gear pair 齿轮副
gear pair with parallel axis 平行轴齿轮副
gear pump 齿轮泵
gear ratio 齿轮速比
gear shifter 变速机构
gear tooth 轮齿
gear train 齿轮系
gear transmission 齿轮传动
gear type pump 齿轮泵
gear water pump 齿轮水泵
gear wheel 齿轮
gear with addendum modification 变位齿轮
gearbox 变速箱，齿轮箱
gearbox bearing 齿轮箱轴承
gearbox gear ratio 齿轮箱齿轮比
geared wind turbine 齿轮风机
gearing 齿轮装置，传动装置
gearless wind drive 无齿轮风力驱动
gearless wind power plant 无齿轮风力发电站
gearless wind turbine generator system 直驱式风电机组
gear-up ratio 增速传动比
Gedser wind turbine （丹麦）盖瑟风力机
Geiger counter 盖格计数器
general algebraic modeling system 通用代数建模系统
general analysis sample 一般分析试样
General Electric Company 通用电气公司
general purpose computer 通用计算机
general purpose diode 普通二极管（整流二极管）
generated electric led-in system 发电接入系统

generated rated output 发电机额定容量
generating caisson 发电沉箱
generating capacity 发电容量,发电量
generating plant 发电厂
generating surface 蒸发受热面
generating unit 发电机组
generating unit availability 发电机组可用率
generation cost 发电成本
generation outrage 发电事故
generation phase 生成阶段
generation rate 产生率,生成率
generation set feeding 发电机组馈electric
generator 发电机
generator bearing 发电机轴承
generator breaker 发电机断路器
generator capability 发电机可能出力
generator circuit breaker 发电机断路器
generator cooler 发电机冷却器
generator electrical capacity 发电机容量
generator fan external 发电机外部风扇
generator fan internal 发电机内部风扇
generator gas 发生炉煤气
generator parameters 电机参数
generator pole 发电机电极
generator rated capacity 发电机额定容量
generator speed 发电机转速
generator terminal 机端
generator voltage 发电机电压
generic generator type 通用发电机类型
generation set 发电机组,发电设备
gentle breeze 三级风,微风

geochemical and geophysical survey 地球化学和物理调查
geochemical composition 地球化学成分
geochemical method 地球化学勘探方法
geochemical sample 地球化学取样
geochemical survey 地球化学勘探
geochemistry 地球化学
geoelectric effect 地电效应
geoelectric method 地电法(勘探地热资源的一种地球物理方法)
geoexchange heat pump 土工交换热泵
geo-exchange system 地质交换系统
geofluid 地热流体,地质流体,岩石孔隙中的各种流体
geographic information system 地理信息系统
geoisotherm 地热等温线
geologic condition 地质条件
geologic control 地质控制
geologic formation 地质构造
geologic mapping 地质制图,地质绘图
geologic setting 地质条件,地质环境
geological condition in geothermal area 地热区地质条件
geological data 地质数据
geological factor 地址因素
geological history of field 热田地史
geological information 地质信息
geological setting 地质背景,地质环境
geological similarity 地质相似度
geological strata 地层
geological thermometer 地质温度计
geological unconformity 地质不整合面
geometric chord 几何弦长

geometric concentration ratio 几何聚光比
geometric radius 几何半径
geometric similarity 几何相似
geometrical concentrator ratio 几何聚光率
geometrical position 几何位置
geometrical transformation 几何变换
geometrics 地热学
geophone 地震检波器,地音探测仪,地音探测器
geophysical method 地球物理方法
geophysical sample 地球物理取样
geophysical survey 地球物理勘探,地球物理探测
geophysical technique 地球物理学技术
geopressured application 超压设施
geopressured aquifer 高压含水层
geopressured basin 高压盆地
geopressured deposit 高压型沉积
geopressured gas 高压气体
geopressured geothermal activity 高压地热能
geopressured geothermal resources 地压型地热资源
geopressured geothermal system 地压地热系统
geopressured reservoir 高压型储层
geopressured resource 地压力资源
geopressured sand 地压砂岩层
geopressured system 超压地热系统
geopressured water 地压水,地热水
geopressured well 高压井
geopressured zone 地热增压区
geopressurised system 地热增压装置
geopressurized zone 地压区域
geoscientific tools 地学工具
geostrophic current 地转流,地转风气流
geostrophic drag law 地转拽力定律
geostrophic wind 地转风
geostrophic wind field 地转风场
geostrophic wind height 地转风高度
geostrophic wind level 地转风级
geostrophic wind vector 地转风矢量
geostrophic wind velocity 地转风速
geosyncline 地槽
geotemperature 地热温度
geothermal 地热的
geothermal abnormity 地热异常
geothermal absorption refrigeration 地热吸收制冷
geothermal activity 地热活动
geothermal alteration 地热蚀变
geothermal anomaly 地热异常
geothermal anomaly area 地热异常区
geothermal anomaly phenomenon 地热异常现象
geothermal application 地热设施
geothermal aquaculture 地热水产养殖
geothermal area 地热区
geothermal belt 地热地带
geothermal boiler 地热锅炉
geothermal bore 地热井孔
geothermal bore hole 地热钻孔
geothermal bore water 地热井水
geothermal bore water circuit 地热井水回路
geothermal brine 地热盐水,地热卤水
geothermal brine deposit 地热盐水矿床,地热卤水矿床
geothermal brine utilization plant 地热盐水利用工厂

geothermal capacity 地热能容量
geothermal characteristic 地热特性
geothermal component test facility 地热成分试验设施
geothermal condition 地热条件
geothermal consultant 地热咨询
geothermal conversion technology 地热转换技术
geothermal cooling 地热能冷却
geothermal core drilling 地热岩心钻探
geothermal data 地热资料, 地热数据
geothermal deposit 地热矿藏
geothermal depth 地热深度
geothermal depth gradient 地热深度梯度
geothermal desalination 地热脱盐
geothermal developer 地热开发商
geothermal development 地热开发
geothermal direct space heating system 地热直接供暖系统
geothermal direct use 地热非电利用, 地热直接利用
geothermal distribution system 地热分布网
geothermal district heating 地热区域供热
geothermal district heating system 地热区域供热系统
geothermal disturbance 地热扰动
geothermal disturbed zone 地热扰动带
geothermal domestic heating 地热民用取暖
geothermal drilling 地热钻井, 地热钻探
geothermal dry steam power 地热干蒸汽发电
geothermal dry steam power plant 地热干蒸汽发电站
geothermal drying 地热干燥
geothermal economics 地热经济学
geothermal efficiency 地热利用率
geothermal electric power plant 地热发电站
geothermal electric power station 地热发电站
geothermal electricity 地热发电
geothermal electricity generating capacity 地热发电能力, 地热发电容量
geothermal electricity generation 地热发电
geothermal electricity generation facility 地热发电设施
geothermal energy 地热能
geothermal energy plant 地热能电站
geothermal energy recovery 地热能回收
geothermal energy reserve 地热能储量
geothermal energy resource 地热能资源
geothermal engineering 地热工程
geothermal event 地热活动
geothermal experimental power station 地热试验电站
geothermal exploitation 地热开发
geothermal exploration 地热勘探
geothermal extraction technology 地热抽提技术
geothermal facility 地热设施
geothermal feature 地热特征
geothermal field 地热田
geothermal field model 地热田模型
geothermal field monitoring 地热田监测

geothermal field output 地热田流量
geothermal field life 地热田寿命
geothermal fluid reinjection strategy 地热流体回灌策略
geothermal fluid 地热流体
geothermal flux 地热流量
geothermal flux density 地热流量密度
geothermal gas 地热气
geothermal generating capacity 地热发电能力
geothermal generating station 地热发电站
geothermal generation 地热发电
geothermal girdle 地热带
geothermal gradient 地热梯度,地热增温率
geothermal gradient change 地热梯度变换
geothermal gradient drilling 地热梯度钻井,地热增温钻井
geothermal gradient survey 地热梯度测量
geothermal greenhouse 地热暖房,地热温室
geothermal growth 地热增长
geothermal heat 地热,地温
geothermal heat exchanger 地热换热器
geothermal heat flow 地热流量
geothermal heat plant 地热站
geothermal heat pump 地源热泵,地热泵
geothermal heat pump modules 地热泵模块
geothermal heat pump system 地热热泵系统
geothermal heat station 地热站
geothermal heating 地热供热
geothermal heating cost 地热供热成本,地热加热费用
geothermal hot water field 地热热水田
geothermal hot water reservoir 地热热水储层
geothermal hot water source 地热热水源
geothermal hot-air space heating 地热热风供暖
geothermal incubation 地热孵化
geothermal indicator 地热指标
geothermal indirect space heating system 地热间接空间供暖系统
geothermal industry 地热产业
geothermal innovation 地热创新
geothermal investigation 地热调查
geothermal lease processing 地热租赁处理
geothermal manifestation 地表热显示,地热现象
geothermal map 地热图
geothermal measurement 地热测量
geothermal media 地热介质
geothermal method 地热法
geothermal mixture condenser 地热混合凝气器
geothermal noise 地热噪声
geothermal normal area 地热正常区
geothermal normal region 地热正常区
geothermal option 地热选项
geothermal plant 地热发电站,地热装置
geothermal potential 地热势
geothermal power 地热能
geothermal power facility 地热发电设施

geothermal power generation 地热发电
geothermal power market 地热发电市场
geothermal power plant 地热电厂
geothermal power project 地热能计划
geothermal power station 地热发电厂
geothermal power system using steam flashed from hot brine 闪蒸地热发电系统
geothermal power system using binary cycle 双循环地热发电系统
geothermal pressure 地热田压力
geothermal prime mover 地热原动机
geothermal problem 地热问题
geothermal producing zone 地热生产层
geothermal profile 地热剖面
geothermal project 地热工程
geothermal prospect 地热勘探
geothermal prospecting 地热勘探
geothermal prospecting procedure 地热勘探程序
geothermal reconaissance 地热普查
geothermal reconnaissance survey 地热勘探
geothermal region 地热区
geothermal reinjection 地热回灌
geothermal research 地热研究
geothermal reservoir 储热层,地热储层
geothermal reservoir analysis 地热储层分析
geothermal reservoir engineering 地热储层工程
geothermal reservoir evaluation 地热储层估计
geothermal resource 地热资源

geothermal resource area 地热资源面积
geothermal resources assessment 地热资源评价
geothermal resources exploration 地热资源勘查
geothermal saturated steam 地热饱和蒸汽
geothermal set 地热机组
geothermal source temperature 地热源温度
geothermal space heating 地热供暖
geothermal spring 地热泉
geothermal stair comprehensive utilization 地热梯级综合利用
geothermal steam 地热蒸汽
geothermal steam collector 地热蒸汽收集器
geothermal steam consumption 地热蒸汽消耗量
geothermal steam ejection 地热蒸汽喷射
geothermal steam extraction plant 地热蒸汽抽提装置
geothermal steam field 地热蒸汽田
geothermal steam injection 地热蒸汽喷射
geothermal steam plant 地热蒸汽发电厂
geothermal steam power 地热蒸汽发电
geothermal steam pressure 地热蒸汽压力
geothermal steam reservoir 地热蒸汽储层
geothermal steam turbine 地热汽轮机
geothermal steam well 地热蒸汽汽井

geothermal stream 地热流
geothermal surface manifestation 地热地表显示（区）
geothermal system 地热系统
geothermal tectonic zone 地热构造带
geothermal temperature gradient 地热温度梯度
geothermal temperature increment ratio 地热增温率
geothermal therapy 地热治疗
geothermal total flow system 全流地热系统
geothermal turbine 地热蒸汽轮机
geothermal underground water 地下地热水
geothermal vault 地热库，地热房，地热控制室
geothermal warm water reservoir 地热暖水储层
geothermal waste heat 地热余热
geothermal water 地热水
geothermal water anticorrosion 地热水防腐
geothermal water reinjection 地热水回灌
geothermal water scale prevention 地热水防垢（处理）
geothermal well 地热井
geothermal well and wellhead installation 地热井与井口装置
geothermal well design 地热井设计
geothermal well fluid measurement 地热井流体测量
geothermal well logging 地热井测井
geothermal wellhead generating unit 地热井口发电机组
geothermal-electric power plant 地热电站，地热发电厂
geothermal-heat extraction 地热热抽提
geothermic 地热的
geothermic degree 地热度
geothermic depth 地下增温深度
geothermic efficiency 地热效应
geothermic electropower station 地热发电站
geothermic step 单位深度地温差
geothermics 地热学
geothermograph 地热温度记录仪
geothermometer 地热计，地热表，地温计
geothermometry 地温测量
geothermy 地热
German wind turbines 德国风力涡轮机
Germany's 2004 Renewable Energy Sources Act 2004年德国可再生能源法，（德国）《2004年可再生能源法》
getter 吸气剂，消气剂（吸附真空器件中残留气体的材料，可分为蒸散型和非蒸散型两大类）
gettering 吸气，收气
geyser 间歇泉，喷泉，间歇井喷
Geysers Geothermal Power Plant (U.S.A) 盖瑟斯地热电站（美国）
Gibbs entropy 吉布斯熵
Gibbs free energy 吉布斯自由能
Gibbs function 吉布斯函数
Gibbs phase rule 吉布斯相律
gimbal 万向接头
gimbals 平衡环，平衡架
gimlet 手钻，螺丝锥
gin pole 起重扒杆
girder structure 大梁式结构
girt gear 齿圈

girth weld 环形焊缝
GL rules=Germanischer Lloyd's Regulation for the Certification of Wind Energy Conversion 德国劳埃德规范
glacial deposits 冰川沉积
glacial outwash deposits 冰川沉积矿床
glacially scoured bedrock 冰川地基岩冲刷
gland 密封管
glass base 玻璃底板
glass coating 玻璃涂层
glass cover 玻璃盖板
glass cover plate （太阳能装置的）玻璃盖板
glass cover sheet 玻璃盖片
glass cover temperature 玻璃盖板温度
glass covered water bag 玻璃盖板下面的水袋
glass cutter 玻璃刀
glass cylinder with vacuum 真空玻璃管
glass diffusion source 玻璃扩散源
glass evacuated tube collector 玻璃真空管集热器
glass fiber 玻璃纤维
glass fiber bundle 玻璃纤维束
glass fiber reinforced concrete 玻璃纤维增强混凝土
glass resistance 玻璃热阻
glass roof 玻璃顶盖
glass seal 玻璃封接
glass semiconductor 玻璃半导体
glass spacer 玻璃隔热
glass transmission 玻璃透光率
glass tube 玻璃管
glass-encapsulated tube 玻璃封装管
glass-roofed solar still （带）玻璃顶盖太阳能蒸馏器
glass-roofed still （带）玻璃顶盖蒸馏器
Glauert empirical relation 葛劳渥经验关系
glaze 釉料，釉面，光滑面，上釉，上光
glazed collector 带透明盖板集热器，上釉集热器
glazed window 玻璃窗
glazing 玻璃，上光材料
glazing cover 玻璃盖板，半透明盖板
glazing insulation 施釉保温，施釉绝缘
glitch 短时脉冲波干扰
global bioenergy partnership 全球生物能伙伴关系
global calibrating method 全球标定法
global circulation 全球环流
global geothermal power 全球地热能
global insolation 全球日照率，全球所受到的太阳能辐射
global irradiance 总辐照度
global pressure 全球气压
global radiation 总辐射
global solar radiation 太阳总辐射
global thermal pollution 全球热污染
global warming potential 全球暖化潜势，全球变暖潜值，全球增温潜势
global wind pattern 全球风力格局
global wind-circulation pattern 全球风环流模式
global wind 全球风
globe valve 球形阀
gloss paint 光滑涂料
glossy 平滑的，有光泽的
glow discharge 辉光放电
glue 胶，胶水，胶合，粘贴，粘合
glycerin 甘油

gneiss 片麻岩
goggle （复数）风镜，护目镜
goons 细打包麻布
gouge 弧口凿，半圆凿
gouging 刨削槽
governing device 调节设备
governing equation 控制方程
governing parameter 控制参数
governor 限速器，调速器
grab 抢夺，攫取，夺取
graded absorbing film 梯度吸收膜
graded band-gap 能隙梯度
graded energy gap solar cell 分级能带宽度太阳能电池
graded index absorber 梯度指数吸收器
grader 分类机，分级机
gradient velocity 梯度风速
gradient wind 梯度风
gradienter 水准仪，倾斜度测定仪
grading ring 均压环
grail 填砾，颗粒，晶粒，粒度，纹理
grain boundary 晶界
grain boundary migration 晶界迁移
grain size 晶粒大小
grain-oriented electrical steel 晶粒取向电用硅钢片
granites 花岗岩
granitic layer 花岗岩层
granodiorite 花岗闪长岩
granular aquifers 粒状含水层，颗粒含水层
graph 图表，曲线图
graphite 石墨
graphite electrode 石墨电极
grass clipping 草屑
grate 炉排

grate bar 炉条，炉栅
grate combustion 炉排燃烧
grate firing 层燃
grate firing boiler 层燃锅炉
grate furnace 炉排炉，层燃炉
grate furnace technology 层燃炉技术
grate heat release rate 炉排热负荷
grate sifting 漏渣
grate surface 炉篦
grate system 炉排系统
grating solar cell 栅式太阳电池
grating type 光栅型（光电电池）
gravel filter 砾石过滤器
gravel pack 砾石过滤层，砾石充填
gravimetric heat pipe 重力热管
gravitational acceleration 重力加速度
gravitational energy 重力能
gravitational field 重力场，引力场
gravitational field measurement 重力场测量
gravitational force 地心引力
gravitational mass 引力质量
gravitational potential 重力位势
gravitational potential energy 重力势能
gravitational settling 重力沉降
gravity base structure 重力式基础
gravity battery 比重液电池
gravity casting machine 重力铸造机
gravity circulation 自然循环，重力循环
gravity filter 重力式过滤器
gravity foundation 重力式基础
gravity load 重力荷载
gravity separator 重力分选机
gravity survey 重力测量，重力调查
gravity tank 重力油箱
gravity wave 重力波，引力波，重点波
gray dolomite 灰白色白云岩，灰白色白云石

gray limestone 灰色石灰岩
gray sandstone 灰色砂岩
gray sticky clay with sand 灰色砂质黏土
graywacke 杂砂岩
grazing angle 掠射角
grease 油膏,润滑油
grease gun 注油枪,滑脂枪
grease nipple 油管
green architecture 绿色建筑
green basis 新鲜基
green chip 新鲜木片
green coal-based power 绿色煤电
green energy machine 绿色能源机器
green energy source 绿色能源
green fuel 新添加的燃料
green house gas emission 温室气体排放
green house gas penalty 温室气体排放惩罚
green power 绿色节能,绿色能源,清洁能源
green power generation 绿色发电
green test 试运转,连续试验
green thermal power 绿色火电
green tide 绿潮
greenfield project 绿地项目,新建项目
greenhouse effect 温室效应
greenhouse gas 温室气体
greening factor of plant area 厂区绿化系数
Greenwich mean time (GMT) 格林尼治平时
grid 网格,谱,电网
grid cell 定位格架栅元,栅格单元,栅电池
grid connected photovoltaic power generation 并网光伏电站

grid connection 网并联,栅极接线
grid connection point for wind farm （风电场）电网连接点
grid delivery point 电厂配送点
grid energy storage 电网储能
grid frequency 电网频率
grid interconnection 电网联网
grid operator 网格因子
grid parity 市电同价,电网平价
grid pattern 网格图形
grid system 电力网
grid valve control 旋转挡板调节
grid-connected 并网的
grid-connected photovoltaic system 光伏并网发电系统
grid-connected power generation 并网发电机设备,并网发电
grid-connected solar photovoltaic system 并网太阳能光伏发电系统
grid-connected variable speed wind power system 并网变速风力发电系统
grid-connected wind plant 风电厂并网发电
grid-connected wind power system 风电并网系统
grind off 磨掉
grinder 磨床
grinder bench 磨床工作台
grinder/crusher 碎渣机,研磨机
grinding 磨的,摩擦的,碾的
grinding drum 磨碎机
grinding dust 磨屑
grinding machine 磨床
grinding tool 磨削工具
grinding wheel 砂轮
grit 沙砾

groin 防砂堤
groove 凹槽，惯例，最佳状态
grooved concentrator 槽型太阳能聚光器
gross array efficiency （太阳能）电池阵列的总效率
gross calorific value 发热量，总热量，恒容高位发热量，高位热值
gross capability 总出力
gross capacity 总容量，总功率
gross collector area 集热器总面积
gross collector array area 集热器阵列总面积
gross density 毛密度
gross generation 总发电量
gross head 毛水头
gross power 总功率，净功率
gross thermal efficiency 总热效率
ground battery 电池组
ground capacitance 对地电容
ground clearance 对地距离
ground coil 接地线圈
ground controller 地面控制器
ground coupled heat pump 热泵或地下耦合热泵
ground cover ratio 地面覆盖率
ground elevation 地面高度
ground equipment station 地面设备站
ground heat exchanger 地下热交换器
ground leakage resistance 漏地电阻
ground logic 地面逻辑
ground loop 接地回路，接地环路
ground loop heat exchanger 接地回路热交换器
ground mounted collector array 地面集热器阵列
ground noise 地噪声
ground noise survey 地噪声测量（用于地热勘探的一种地球物理方法）
ground radiation 地面辐射
ground reflected light 地面反射光
ground reflection 地面反射
ground rod 接地棒
ground source heat pump 地源热泵
ground state 基态
ground station 地面接收站
ground surface friction coefficient 地表摩擦系数
ground temperature 地中温度，地温
ground thermometer 地温表
ground track 地面跟踪
ground water heat pump 地下水热泵
ground water level 地下水位，潜水位
ground water table 地下水位表
ground wind 地面风
ground wood 磨木
ground-couple heat pumps 地下耦合热泵系统，埋管式土壤源热泵系统
groundcoupled heat pump 土壤源热泵
grounding 接地
grounding conductor 接地导体
ground-level plastic solar water heater 地面式塑料太阳能热水器
ground-mounted generator 地面发电机
ground-source system 地源系统
groundwater 地下水
groundwater amount 地下水储量
groundwater movement 地下水运动
groundwater pollution 地下水污染
groundwater reservoir 地下水库，地下水储层
groundwater runoff 地下水径流
ground-water source heat pump 地下水源热泵

groundwater well 地下井
group heating system 群体供热系统,群体供暖系统
group of wells 井群
group solar cell 族太阳电池
group velocity 群速
grout 薄泥浆,水泥浆
grouting material 灌浆材料,浆材料
grouting procedure 灌浆工序
grown junction photocell 生长结光电池
growth hormone 生长激素
growth rate 长大速率
growth rings of tree 树木生长年轮
grub screw 自攻螺丝
guard grating 安全栅,保护栅
guard plate 护板
guide bars 导向棍
guide blade 导流叶片
guide block 导向块
guide plate 导向板
guide ring 导向环
guide rod 导向杆
guide roller 导辊
guide rule 准则,导则
guide sheath 导向护套
guide support 导向支架
guide tool 导向工具
guide tube 导向管
guide tube support plate 压紧顶帽
guide vane 导向叶片
guide vane servomotor 导叶接力器
guide vane trunnion 导叶轴颈,导叶轴
guide vane 导叶,导流叶片,活动导叶
guide wall 导流墙,导水墙
guide wheel 导轮
guide wheel bucket 导向叶片
guide wire 尺度(定距)索,准绳

guidelines for wind energy development 风能开发导则
guideway 导向槽
guiding fin 导向板,垂直定向板
guiding shaft 导向轴
guillotine 闸刀,处斩刑,切(纸)
gulf stream (墨西哥)湾流
gulfweed 马尾藻
Gumbel distribution 耿贝尔分布
gumming 树胶分泌
gusset plate 角撑板,加固板
gust 瞬时风,阵风
gust anemometer 阵风风速计
gust factor 阵风影响系数,阵风系数
gust generator 阵风发生器
gust influence 阵风影响
gust load 阵风载荷
gust loading 阵风荷载(工况)
gust peak speed 阵风最大风速,最高阵风风速
gust recorder 阵风记录仪
gust response factor 阵风响应系数
gust slicing 阵风切片
gust slicing effect 阵风切片效应
gust speed 阵风速度
gustiness 阵风性,阵发性,湍流度
gustiness factor 阵风系数
gustsonde 阵风探空仪,阵风探测仪
gusty 阵风的,多阵风的
guy cable 拉缆
guy clip 线卡子
guy wire 牵索,拉绳
guyed lattice 拉索式格栅
guyed pole 拉线杆
guyed tower 拉索式塔架
gyro power take-off 陀螺动力输出装置
gyroscope 陀螺仪,回转仪,回旋装置,纵舵调整器

gyroscopic effect 回转效应
gyroscopic force 回转力
gyroscopic moment 回转力矩
gyroscopic motion 陀螺运动
gyroscopic torque 陀螺力矩

H

habitat loss 栖息地丧失
hacksaw 可锯金属的弓形锯,钢锯
hail 冰雹
hairspring 细弹簧,游丝
half duplex transmission 半双工传输
half life 半衰期
half mirror 半镜(一种防热玻璃,有透光、反射双重功能)
half shaft 半轴
half sine wave 半正弦波
half tide 半潮
half tide level 半潮位
half value period 半寿期,半衰期
half-duplex transmission 半双工传输
half-round file 半圆锉
halfway 在半途
Hall effect 霍尔效应
Hall sensor 霍尔式传感器
halogen lamp 卤素灯
halogenated organic compound 卤化有机化合物
halophyte 盐生植物
hammer mill 锤式粉碎机
hand tools 手工具
handheld anemometer 手持式风速计
handheld instrument 手持工具
handiness 操纵性
hanger 吊架
hanging 顶端对齐,悬挂
hard damping 强阻尼

hard energy source 硬能源
hard formation 硬地层,坚硬岩层
hardgrove index (HGI) 哈氏(可磨度)指数
hard radiation 强辐射
hard rock 硬岩
hard water 硬水
hard/soft and free expansion sheet making plant 硬(软)板(片)材及自由发泡板机组
hardener 固化剂,硬化剂
hardgrove grindability index 可磨性指数
hardware 硬件
hardware logic 硬件逻辑
hardware platform 硬件平台
hardwood 硬木,硬木材,阔叶树
harmonic 谐波的
harmonic analysis 谐波分析,调和分析
harmonic analysis of tides 潮汐调和分析
harmonic component 谐波分量,调和成分,低分量
harmonic distortion 谐波失真
harmonic elimination 谐波抑制
harmonic frequency 谐振频率
harmonic variation 谐波变化
harmonic voltage source 谐波电压源
harmonic wave 谐波
harmonics 谐波
harness 导线,装备,利用
harrow 耙
harsh ambient conditions 恶劣环境条件
harvesting energy 获能
hatch 舱口,舱口盖,开口,孵,孵出,策划,图谋

hatchet 短柄斧
hazard 冒险(性),相关危险,事故,故障
hazardous air pollutant 有害空气污染物
hazardous waste 危险废物
Hazardous Waste Regulation 危险废物监管
Hazen Williams coefficient 海曾—威廉系数
head 水头,落差
head loss 水头损失
head of tide 潮头
head screw 主轴螺杆
head wind 逆风
header 集管
headrace 导水沟
headstock 主轴箱
headwaters 水源
hearth load 炉负荷
heat absorbing 吸热的
heat absorbing pipe 热吸收管
heat absorbing surface 吸收面,受热面
heat absorption 热量吸取
heat absorption efficiency 吸热效率
heat absorption rate 吸热强度(受热面)
heat abstractor 吸热设备
heat account 热平衡计算
heat activation 热激作用
heat capacity 热容,热容量
heat capacity of flat plate collector 平板型太阳能集热器热容量
heat collecting tube 集热管
heat collection 集热(量)
heat collection (storage) wall 集热(蓄热)墙
heat collection device 集热器,集热装置
heat conducting material 导热材料
heat conduction 热传导
heat conductivity 热导率
heat content 热含量,热函
heat convection 对流放热
heat distribution 热分布,热量分布,热分配
heat efficiency 热效率
heat energy 热能
heat exchange 热交换
heat exchange area 换热面积
heat exchange pipe 换热管
heat exchange system 热交换系统
heat exchanger 换热器,热交换器
heat exchanger circuit 热交换器回路
heat exchanger cleaning system 热交换器清洗系统
heat exchanger coil 热交换器盘管
heat exchanger effectiveness 热交换器效率
heat exchanging 热交换,热传导,换热
heat extraction equipment 热量提取设备
heat extraction from rock 岩石萃热
heat extraction system 排热系统
heat extractor 热交换器
heat floor slab 蓄热地板
heat flow 热流
heat flow contour 热流体等高线
heat flow diagram 热流图
heat flow meter 热流计
heat flux 热流,热通量
heat flux density 热流密度
heat gain 热增益
heat input 热量输入,供热

heat insulating property 绝热性能，保温性能
heat insulation 绝热（性）
heat interchanger 热交换器
heat island 热岛
heat leakage 散热损失
heat liberation 放热
heat load 热负荷
heat loss 热损失
heat loss conductance 热损传导率
heat loss dump load 热损失转储负载
heat loss factor 热损失系数
heat mining 热开采
heat mirror 热镜
heat mirror coating 热镜涂层，热镜薄膜
heat of absorption 吸收热，吸附热
heat of combination 化合热
heat of crystallization 结晶热
heat of decomposition 分散热
heat of formation 形成热
heat of fusion 熔化热
heat of liquefaction 熔解热，液化热
heat of mixing 混合热
heat of radiation 辐射热
heat of solidification 凝固热
heat of solution 熔解热
heat of transformation 相变热，转变热
heat outflow 热溢流
heat output 热功率，热量输出
heat pickup 吸热
heat pipe 热管（太阳电池等传热元件）
heat pipe evacuated collector tube 热管式真空集热管
heat pipe evacuated tube collector 热管式真空管集热器
heat pipe receiver 热管式吸热器

heat pipe vacuum tube 热管真空集热管
heat pollution 热污染
heat preserving furnace 保温炉
heat producing 产生热量的，放热的
heat producing capability 热出力，产热能力
heat production 产热，热产生
heat proof 防热的，耐热的，抗热的
heat pump 热泵，蒸汽泵
heat pump evaporators 热泵蒸发器
heat pump modules 热泵模块
heat pump refrigerant 热泵制冷剂
heat quantity 热量
heat rate 热率，热效率，热耗率，发热量
heat rate of geothermal power plant 地热电厂热能利用率
heat rating 热功率
heat rays 热线
heat reactor 供热（反应）堆
heat receiving surface 吸热面，受热面
heat reclaim 热量回收，热量利用
heat recovery 余热回收
heat recovery boiler 余热锅炉
heat recovery steam generator 余热锅炉，废热回收蒸汽发生器
heat recovery unit 热回收装置
heat reflecting glass 热反射玻璃
heat reflective coating 热反射涂层，热反射薄膜
heat rejection 热量废弃，热量放出，排热，热的耗损
heat rejector 热分离器
heat release 放热，释放热量
heat release rate 放热率，释热率
heat releasing fluid 散热剂

heat removal factor （集热器的）热交换因子
heat reservoir 热库，储热器
heat resistance 热阻，耐热性，耐热的，耐热度
heat resistance paint 耐热漆，耐高温漆
heat resistant 耐热的
heat resistant alloy 耐热合金
heat resistant film 耐热薄膜
heat resistant paint 耐热漆，耐高温漆
heat resisting 耐热的，耐热
heat resisting porcelain 耐热瓷
heat retainer 蓄热体
heat retaining 蓄热的，保温的
heat retaining capacity 蓄热能力
heat retaining mass 蓄热体
heat retention 保温性能，保温力
heat return 热量回收
heat sensor 热量传感器
heat shield 热屏
heat shield capability 热防护能力
heat shield material 隔热材料，保温材料
heat sink 冷槽，散热器，吸热部件
heat source 热资源，热源
heat source for geothermal power 地热能热源
heat source in deep mantle 深部地幔热源
heat source in earth's interior 地热源
heat source strength 热源深度
heat stabilizer 热稳定剂
heat steam coil 加热器蒸汽盘管
heat storage 蓄热，热存储
heat storage boiler 蓄热锅炉
heat storage capacity 蓄热能力
heat storage medium 储热介质
heat storage of material 工质储热
heat storage tank 储热水箱
heat storage wall 储热墙
heat store 热库
heat tracing 防冻加热器
heat transfer 热传递
heat transfer analyses 传热分析
heat transfer by hot spring 热泉换热
heat transfer by radiation 辐射传热
heat transfer coefficient 传热系数
heat transfer cycle 传热过程
heat transfer effectiveness 热传输效率
heat transfer efficiency 传热率
heat transfer fluid 传热流体
heat transfer impairment 热损伤
heat transfer loop 传热回路
heat transfer mechanism 传热机理，换热机理
heat transfer media 传热介质
heat transfer medium 传热介质，换热介质
heat transfer model 传热模型
heat transfer property 换热性能，传热性能
heat transfer rate 传热系数
heat transfer resistance 热阻
heat transfer salt 传热盐
heat transfer surface 传热表面
heat transfer system 传热系统
heat transmission 传热，热传递
heat transport 热量输送
heat transport fluid 传热流体
heat trap 吸热，吸热器
heat treating 热处理
heat tube 热管
heat voltage 灯丝电压，加热电压

heat wave 热浪
heat/thermal flux 热通量
heat-actuated heat pump 热激励热泵
heat-driven absorption chiller 高温驱动吸收式冷冻机
heated blade 加热叶片
heated collector 受热集热器
heater 加热器
heater coil 加热盘管，加热线圈
heater plant 加热设备，加热站
heatflow data 热流数据
heating and ventilation engineering （采）暖通（风）工程
heating appliance 电热器
heating boiler 采暖锅炉，暖气锅炉
heating boxes 加热室
heating capacity 热值
heating coil 加热盘管，加热线圈
heating configuration 加热构型，加热结构
heating distribution system 采暖分配系统
heating dominated vertical closed loop 加热主导型垂直闭合环路
heating element 加热元件，发热体
heating load 热负荷，加热负载
heating of building with solar energy 建筑物太阳能供暖
heating oil 取暖油
heating process 加热过程
heating range 加热范围
heating rate 加热速度，升温速率
heating section 加热区段
heating surface area 受热面面积
heating treatment furnaces 熔热处理炉
heating up 加热，升温
heating up period 加热期，升温时间
heating up time 加热时间，升温时间
heating value 发热量
heating voltage 灯丝电压
heating water 热水，供热水，采暖水
heating wire 电热丝
heating, ventilation and air conditioning 暖通空调（暖气、通风及空气调节）
heat-recovery boiler 余热回收锅炉，废热回收锅炉
heat-storing stove 蓄热炉
heat transfer matrix 传热矩阵
heat-transfer property 传热性能
heaving 隆起，隆胀，垂直模式
heaving and pitching body wave energy conversion device 垂荡和纵摇波能转换装置
heaving sand 流沙
heavy duty 大容量，大功率
heavy duty lathe 重型车床
heavy frame turbine 重型汽轮机
heavy fuel 重质燃料
heavy gust 强阵风
heavy hole 重空穴
heavy hole band 重空穴带
heavy mineral 重矿石，重矿物
heavy sea 巨浪
heavy water cooled reactor 重水冷却反应堆
helical cutter 螺齿铣刀
helical gear 斜齿轮
helical gear 斜齿圆柱齿轮
helical gearing 螺旋齿
helical pump 螺旋泵
helical teeth 螺旋齿
helicoidal anemometer 螺旋桨式风速表
heliodon 日影仪

helioelectric index 总辐射平均强度表指数
helioelectrical process 光电过程
heliograph 日照计
helioplant 太阳能利用装置
heliosphere 日光层
heliostat 定日镜,日光反射装置
heliostat array controller 定日镜阵列控制器
heliostat cluster 定日镜簇
heliostat field 定日镜场
heliostat packing density 定日镜填充密度
heliostat power plant 塔式太阳热能发电厂
heliostat tracking 定日镜跟踪
heliostat type solar power plant 定日镜型太阳能装置
heliostat-type furnace 定日镜型太阳炉
heliotechnics 太阳能技术
heliotherapy 日光疗法
heliothermal process 太阳热过程
heliotropic wind 日成风,日转风
helium neon laser 氦氖激光器
helix 螺旋,螺旋状物
hematite 赤铁矿
hemicellulase 半纤维素酶
hemicellulose 半纤维素
hemispherical bowl mirror 半球形碗形反射镜
hemispherical hollow cylinder 半球形中空圆筒
hemispherical radiation 半球向辐射
hemispherical reflectance 半球向反射比
hemispherical solar irradiance 半球向日射辐照度

hemispherical solar radiation 半球向辐射
hemispherical spectral emissive power 半球光谱辐射能力,半球光谱辐射功率
hemispherical spectral emissivity 半球光谱发射率
hemispherical total emissivity 半球总辐射率
herbaceous biomass fuel 草本生物质燃料
herbaceous energy crop 草本能源作物
herbaceous fuel 草本燃料
herbaceous perennial energy crops 多年生草本能源作物
Herringbone gear 人字(齿)轮,双螺旋齿轮
heterogeneous combustion 复相燃烧
heterogeneous nucleation 非均匀形核
heterojunction cell 异质结电池
heterojunction solar cell 异质结太阳能电池
heterophase boundary 异相界面
heterotrophic algae 异养藻类
hexagon screw die 六角板牙,六角螺丝钢板
hexagon spanner 六方
hexagonal close-packed 六方密排结构
hexagonal nut 六角螺母
high altitude 高空
high angle grain boundary 大角度晶界
high basin (潮汐发电)高库
high boiling point 高沸点
high brightness beam 高亮度光束
high brightness concentration factor 高亮度聚光因子,高亮度聚光度
high capacity 大容量

high collection efficiency 高集热率
high concentration collector 高集聚集热器
high cyclical rotor shaft torque stress 高速旋转叶轮轴扭应力
high definition 高清晰度
high density plastic pipe 高密度塑料管道
high dopant concentration 高掺杂浓度
high duty 高功率，大容量，高负荷
high duty boiler 大容量锅炉
high efficiency 高效率
high efficiency solar cell 高效率太阳能电池
high elevation 高地势
high emissivity coating 高发射率涂料，高辐射率涂料
high energy excitation 高能激励源
high energy radiation 高能辐射
high frequency generator 高频发电机
high frequency heating 高频加热
high frequency voltage 高频电压
high grade energy 高级能
high heating value (HHV) 高发热值
high insolation 高日照
high insolation area 高日照区
high level radiation source 高强度辐射源
high lift 高升力
high osmotic pressure 高渗透压
high performance 高效率，高性能
high pool level 高库水面
high power density 大功率密度
high power LED bulb lamp 大功率 LED 球泡灯
high power LED ceiling sport light 大功率 LED 天花射灯
high power LED flood lamp 大功率 LED 隧道灯
high power LED grid lights 大功率 LED 格栅灯
high power LED sport light 大功率 LED 射灯
high power LED street lamp 大功率 LED 路灯
high power LED wall washer lamp 大功率 LED 洗墙灯
high pressure feed water 高压给水
high pressure stage 高压级
high pressure steam 高压蒸汽
high solid digestion system 高固分解系统
high specific speed turbine 高比速汽轮机
high speed continuous rotary motion 高速连续旋转运动
high strength 高强度
high temperature 高温
high temperature alloy 耐热合金
high temperature coating 高温涂层
high temperature collector 高温集热器
high temperature geothermal field 高温地热田
high temperature insulating separator 高温绝热分离器
high temperature property 高温性能，高温特性
high temperature reactor 高温反应堆
high temperature sealant 高温密封剂
high temperature stability 高温稳定性
high temperature storage 高温蓄热
high temperature storage tank 高温蓄热箱
high temperature strength 高温强度

high tide 高潮，满潮
high transmissivity coverglass 高透射率玻璃罩
high turbulence 高速湍流
high vacuum 高度真空
high viscosity 高黏度，高黏性
high voltage alternating current 高压交流电
high voltage coil 高压线圈
high voltage direct current 高压直流电
high voltage direct current transmission 高压直流输电
high voltage power supply 高压电源
high water 高潮
high water equinoctial spring tide 分点大潮高潮位
high water level 高潮位
high water mark 高潮标
high water stand 高潮位
high wind 大风
high wind speed 高风速
high wind zone 高风带
high yield fracture system 高产断裂系统
high-density polyethylene heat exchanger 高密度聚乙烯换热器
high-density polyethylene pipe 高密度聚乙烯管
high-density polyethylene piping 高密度聚乙烯管道
high-efficiency photovoltaic module 高效光电微型组件
high-flux low-temperature solar collector 高通量低温太阳能集热器
higher heating value 高热值，高位热值
higher high water 高高潮位
higher low water 高低潮位

higher temperature storage 高温储热器
high-frequency harmonics 高频谐波
high-gear ratio 高速齿轮比
high-grade energy 高品位能源
high-level of solar radiation 高能级太阳辐射
highly absorbent surface 高吸收剂表面
highly reflective foil 高反射金属薄片
highly reflective surface 高反射表面
high-order method 高位法
high-performance 高性能的
high-pressure flushing pump 高压冲洗泵
high-pressure sodium lamp 高压钠灯
high-rate activated sludge waste treatment 高速活性污泥废水处理
high-rate digestion 高速消化
high-rate sludge digestion 高速污泥消化
high-solidity rotor 高实度风轮
high-solids content 高固体含量
high-speed lathes 高速车床
high-speed shaft 高速轴
high-strength organic waste 高含量有机废物
high-temperature digestion 高温煮解
high-temperature geothermal field 高温地热田
high-temperature gradient 高温梯度
high-temperature solar conversion system 高温太阳能转换系统
high-temperature upflow area 高温上升液流面积
high-temperature well 高温井
hightensile 高强度
high-tide level 高潮位

high-velocity seawater erosion 高速海水侵蚀
high-volatility coal 高挥发性煤
high-voltage induction motor 高压感应电机
high-voltage photovoltaic converter 高伏特光电转换器，高电压光电转换器
hill shading 阴影法
hindcast 追算，后报
hindcast model 追算模型
hindrance 障碍
hinge spring 弹簧铰链
hinge-spring rotor 铰链弹簧转子
HITEC 西德克（太阳能发电系统中的一种易熔混合物工作介质）
hoist 升起，吊起，推起，起重机，支持物，固定器
hold tank 储存罐
hole saw 空穴锯
Holland windmill 荷兰风车
hollow fiber bundle 空心纤维束
hollow pipe 空心管
hollow-core masonry wall 空心砖石墙
homeostasis 体内平衡
homogeneous nucleation 均匀的核化
homogeneous soil 均质土壤
homogeneous source 均匀热源
homogeneous state 均匀态
homogeneous terrain 均匀地形
homogenization 均化，均性化，均化作用
homogenization heat treatment 均匀化热处理
homojunction 单质结
homojunction cell 同质结电池
homojunction solar cell 同质结太阳电池
homophase boundary 同相界面

honeycomb collector 蜂窝状（太阳能）集热器
honeycomb compound material 蜂窝复合材料（置于透明盖板和集热板之间类似蜂窝状的隔热材料）
honing machine 搪磨机
hook bolt 吊耳
hook spanner 弯脚扳手
Hooke's Law 胡克定理
hook's joint 钩结合，钩（连）接
hopper 单足跳者
horizontal 地平线的，水平的
horizontal area 水平面积
horizontal axis wind turbine 水平轴风车
horizontal axis aerogenerator 水平轴风力发电机
horizontal axis configuration 横轴结构
horizontal axis rotor 水平轴风轮
horizontal axis tidal turbine 水平轴潮汐涡轮机
horizontal axis turbine 水平轴风力机，水平轴涡轮机，横卧式涡轮机
horizontal axis wind turbine 水平轴风力发电机
horizontal axis wind turbine blade 卧轴式风力涡轮机叶片
horizontal axis windmill 横轴风车，卧轴式风车
horizontal black tray 卧式黑色水箱
horizontal bulb type turbine generator 卧式灯泡形涡轮发电机
horizontal circle 地平圈
horizontal closed loop 水平闭合回路
horizontal collector 卧式集热器
horizontal combustion chamber 水平燃烧室
horizontal cylinder 卧式气缸

horizontal discharge valve 水平放泄阀
horizontal displacement 水平位移
horizontal double auger feeding 卧式双螺旋喂料
horizontal earth coupling system 水平大地耦合系统
horizontal fuel bed 水平燃料床
horizontal ground heat exchanger 水平地埋管热换器
horizontal ground-coupled heat pump 水平埋管地源热泵系统
horizontal hydraulic conductivity 水平渗透系数
horizontal line 水平线
horizontal motion 水平运动,水平震动
horizontal multijunctions solar cell 水平多结太阳电池
horizontal pipe 水平管,横管
horizontal plane 地平面
horizontal pressure gradient 水平压力梯度
horizontal resolution 水平清晰度
horizontal shaft bulb unit 水平轴灯泡型装置
horizontal still 卧式蒸馏器
horizontal stress 水平应力
horizontal tailrace 水平尾水渠,水平放水渠
horizontal type wind turbine 水平轴风力机
horizontal wheel 水平轮
horizontal wind field 水平风场
horizontal wind shear 水平风切
horizontal wind vector 水平风矢量
horizontal yoke heliostat 水平支架定日镜
horizontal-axis hydraulic machine 卧轴液压机器
horizontal-axis wind turbine 水平轴风力发电机
horizontal-axis-rotor WECS 水平轴风力机
horizontal-shaft-propeller-type windmill 水平轴式螺旋桨型风车
horse latitude 回归线无风带
horsepower 马力
horseshoe magnet 马蹄形磁铁
horticultural residue 园艺残渣
hose 软管,胶皮管,蛇管
hose clamp 软管支架,软管夹
hose clip 管夹
hose pump 软管泵,蠕动泵
hot air outlet 热空气出口
hot air turbine 热空气汽轮机
hot brine 热卤水,热盐
hot brine field 热盐水田
hot brine pipe 热盐水管
hot brine pool 热盐水池
hot cell 热室
hot dry rock 干热岩,高温岩
hot dry rock geothermal power 干热岩地热能
hot emitted electrode 热发射电极
hot exchange method 热交换法
hot film anemometer 热膜风速计
hot flue gas 热烟气
hot jacket 热套
hot magma body 热岩浆体
hot plate 电炉
hot pressurized water 增压热水
hot probe method 热探针法
hot reserve, spinning reserve 热备用
hot rock 热岩
hot rock formation 热岩构造

hot rock reservoir 热岩层
hot rock technology 热岩技术
hot salt storage tank 熔盐储罐，熔盐储槽
hot spot 热点
hot spring 温泉，热泉
hot spring flow 热泉流量
hot spring fluctuation 热泉变化
hot tower 热塔
hot tub 热槽
hot water field 热水田
hot water geothermal system 热水地热系统
hot water heat store tank 热水储热箱
hot water outlet 温水出口
hot water seepage 热水渗出
hot water tank 热水箱
hot water well 热水井
hot wave 热浪
hot well 热水井
hot zone 热场
hot-dip 热沾
hot-molded glass mirror 热成型玻璃反射镜
hot-water supply 热水供给
hot-wire anemometer 热线风速计，热丝式风速计
hot-wire chemical vapor deposition 热丝化学气相沉积
hour angle 时角
hour circle 时圈
hourly average insolation 时平均接收到的太阳辐射
hourly destruction capacity 每小时销毁容量
hourly furnace oil consumption 每小时炉油损耗

hours of insolation 日照时数
hours of useful collection per day 日有效集热时数
house heating 供热，采暖
house hot water system 住宅热水系统
house size 太阳房尺寸
household hazardous waste 家庭有害垃圾，家庭危险废弃物
housing 灯罩
hovercraft 水翼船
HRS rectifier HRS 整流器
hub 轮毂
hub bore 毂孔
hub height 轮毂高度
hub rigidity 轮毂刚度
hubcap 轮毂罩
human toxicity potential 人体毒性潜值
humidity 湿气，潮湿，湿度
humidity load 湿负荷
humus soil 腐殖土
hurricane 飓风，十二级风
hurricane force 飓风，十二级风
HV transformer system 高压变配电系统
HVAC (heating ventilation air conditioning) system 暖通空调系统
hybrid 混合式
hybrid cooling 混合冷却
hybrid cooling system 混合冷却系统
hybrid cycle system 混合式循环系统
hybrid design 混合设计
hybrid electric vehicle 混合动力汽车
hybrid permanent magnet machine 混合永磁电机
hybrid photovoltaic system 混合式光电系统，混合式光伏打系统
hybrid power system 混合动力系统
hybrid solar (energy) system 混合式太阳能系统

hybrid solar thermal power 混合太阳能热发电，混合太阳能热电
hybrid sorghum 杂交高粱
hybrid system 混合动力系统
hybrid use of solar and wind energy 太阳能和风能的综合利用
hydrated lime 氢氧化钙，熟石灰，消石灰
hydration salt heat storage 水化盐物储热
hydraulic 水压的，液压的
hydraulic accumulator 液压储能器
hydraulic arm 液压臂
hydraulic block 液压块
hydraulic braking system 液压制动系统
hydraulic calculation 水力计算
hydraulic circuit 液压系统，油路
hydraulic components 液压元件
hydraulic conductivity 渗透系数，导水率，水力传导率
hydraulic cylinder 液压缸，液缸，油压缸
hydraulic drag 水力阻力，水力拖拽力
hydraulic drive 液压驱动，液动
hydraulic energy 水力能源，水力能量，水力能
hydraulic feed ram 液压加料推杆
hydraulic feeder 液压给料机
hydraulic filter 液压过滤器
hydraulic fluid 液压机液体
hydraulic fracturing （岩石）水力压裂
hydraulic joint 液压接头
hydraulic losses 水力损失，液体损失
hydraulic motor 液压马达，液压发动机
hydraulic parameter 水力参数，水压参数，液压参数
hydraulic permeability 水力渗透性，透水性

hydraulic piston 液压活塞
hydraulic power 水力，液压动力
hydraulic power tools 液压工具
hydraulic power units 液压动力元件
hydraulic pressure 液压，水压
hydraulic property 水力特性
hydraulic pump 液压泵
hydraulic ram 液压油缸
hydraulic research station rectifier （海洋）水力研究站整流装置
hydraulic reservoir 液压油箱，液压储油柜
hydraulic retention time 水力停留时间
hydraulic rotary cylinders 液压回转缸
hydraulic sealing digester 水封式沼气池
hydraulic stimulation 水力压裂
hydraulic system 液压系统
hydraulic technique of rock fracturing 岩石压裂水力技术
hydraulic turbine 水轮机，液力涡轮
hydraulic unit 液压单元
hydraulic valve 液压阀
hydraulic wheel 水车
hydraulics 水力学
hydrazine from sea water 海水抽提联氨
hydride battery 氢化物电池
Hydril annular preventer 海德里环形防喷器
hydrocarbon 碳氢化合物
hydrocarbon deposit 碳氢化合物沉积
hydrocarbon feedstock 烃原料
hydrocarbon fuel 碳氢燃料，烃类燃料
hydrocarbon liquid fuel 烃类液体燃料
hydrochloric acid solution 盐酸溶液
hydrocyclone 水力旋流器，旋液分离器，涡流除尘器

hydrodeoxygenation 加氢脱氧
hydrodynamic coefficients 水动力系数
hydrodynamic damping 流体动力阻尼
hydrodynamic efficiency 水动力效率
hydrodynamic model 水动力学模型
hydrodynamic performance 水动力性能，水动力学性能
hydroelectricity 水力电气
hydrofracture 水力压裂，水压致裂
hydrogen bond 氢键
hydrogen bromine battery 氢溴电池
hydrogen infusion 氢气浸提
hydrogen oxygen fuel cell 氢氧燃料电池
hydrogen power generation 氢能发电
hydrogen production 制氢
hydrogen sulfide (H_2S) 硫化氢
hydrogenated amorphous silicon 含氢的非结晶性硅
hydrogeochemistry 水文地球化学
hydrogeological flow 水文地质流
hydrogeological investigation 水文地质调查
hydrogeology 水文地质学
hydrographic (al) station 海洋水文站，海洋观察站
hydrokinetic 流体动力学的，流体动力的
hydrokinetic power 液体动力（海浪、潮汐、海流）
hydrokinetic technology 流体动力技术，液压技术
hydrokinetic wave farm 潮汐水力发电厂
hydrologic effect 水文效应
hydrologic tension 水文张力
hydrologist 水文学家

hydrolysis 水解作用
hydrolysis reactor 水解反应器
hydrolytic bacteria 水解细菌
hydrolytic enzyme 水解酶
hydrometeorology 水文气象学
hydrometeorology information 水文气象资料
hydronic system 液体循环加热（冷却）系统
hydropneumatic subsurface power converter 液体气动水下电源转换
hydropneumatic wave power device 水气并动波浪发电装置，水气并动波能装置
hydroponics 水栽法，溶液培养学
hydroponics method 溶液栽培法
hyhydropower development 水利发展，水电开发，水力发展
hydropower station 水电站
hydropulper 水利化浆机
hydrostatic force 静水压力，流体静力
hydrostatic model 静水模型
hydrostatic pressure 静水压力
hydrostatic pressure test 静水压试验
hydrostatic test 流体静力学试验，静水压试验
hydrothermal 水热（作用）的，热液的，热水的
hydrothermal action 热液作用
hydrothermal activity 热液活动
hydrothermal alteration 水热蚀变
hydrothermal anomaly 热液异常
hydrothermal capacity 热液容量，热液产能
hydrothermal circulation 热液循环
hydrothermal electricity production 水热发电

hydrothermal energy （地下蒸气产生的）热液能
hydrothermal eruption 水热喷发
hydrothermal explosion 水热爆炸
hydrothermal fluid 热流体
hydrothermal processing 水热处理
hydrothermal reservoir 热储
hydrothermal resources 热液资源
hydrothermal system 热液系统，水热系统
hydrothermal upgrading 热解改质
hydrotreated vegetable oil 氢化植物油
hydrous salt 水合熔盐
hygrographs 自动湿度记录计
hyperspectral data 高光谱数据
hyperspectral instruments 高光谱仪器
hyperspectral surveys 高光谱测量
hypocenters 地面零点
hysteresis 滞后作用，磁滞现象
hysteresis loop 滞后回线
hysteresis loss 磁滞损耗

I

ice sensor 结冰传感器
iceland 冰洲
ideal black surface 理想黑体表面
ideal blackbody 绝对黑体，理想黑体
ideal power coefficient 理想的功率系数
ideal source 理想电源
ideal synchronizing 理想同步
identification mark 识别标志
idle battery 无负荷电池
idle capacity 空载功率
idler gear 惰轮，空转齿轮，换向齿轮
idler pulley 惰轮，空转轮，导轮
idling 空转

idling temperature 无功温度
igneous intrusion 岩浆岩侵入体
igneous rock 火成岩
ignition temperature 着火点，燃点
illuminance 照明度，施照度
illuminante 发光物，照明剂
illuminate 光源
illuminated cell 照明电池
illuminating gas 照明用气
illuminator 发光器，照明装置
ilmenite 钛铁矿
imaginary part 虚部
immunity 不敏感性，抗扰性
impact assessment 影响评估，影响评价
impact damper 冲击阻尼器
impact induced vibration 冲击致振
impact load 冲击负载
impact pressure 冲击压力，动压力
impact resistance 抗冲击
impact strength 冲击强度
impact torque 冲击扭矩
impact toughness 冲击韧性
impedance 阻抗
impedance voltage 阻抗电压
impeller 推进者，叶轮
impinge 撞击，冲击
impingement plate 防冲板
impingement scrubber 自激式洗涤器
implementation potential 实施潜势
imported energy 进口能源
impounding 蓄水
impoundment 蓄水，拦蓄
impregnated diamond bit 孕镶金刚石钻头，孕镶钻头
imprint 印记

imprison 监禁，关押
impulse 推动，刺激，冲动，推动力，脉冲
impulse buoy 脉冲浮标
impulse current 脉冲电流，冲击电流
impulse load test 冲动载荷试验
impulse response 脉冲响应
impulse turbine 冲力式涡轮机，冲击式汽轮机
impulsive input 脉冲输入
impulsive load 冲击载荷
impurity 杂质
impurity atom 杂质原子
impurity concentration 杂质浓度
impurity content 杂质含量
impurity elimination process 杂质的去除过程
in situ 在原地，在原位
incandescent lamp 白炽灯
Incheon Tidal Power Station 仁川潮汐电站
incidence angle 入射角，迎角，冲角
incidence angle modifier 入射角修正系数
incident 入射的
incident angle modifier 入射角修正系数
incident intensity 入射强度，入射光强度
incident light 入射光
incident solar energy 入射太阳能
incident sunlight 入射阳光
incident wave 入射波
incineration 烧尽，焚烧，焚化
incineration line 焚烧线
incineration of municipal solid wastes 城市生活垃圾焚烧法
incineration plant 焚化场
incinerator 焚烧炉，煅烧炉
incinerator bottom ash 底灰
incinerator residue 焚化炉残渣
incinerator stoker 垃圾焚烧炉
inclination 交角
inclination measurement 井斜测量
inclined reciprocating grate 倾斜式往复炉排
inclined rod grid water film scrubber 斜棒栅水膜除尘器
inclined stainless steel chute 斜不锈钢槽
inclined supporting strut 倾斜支柱，倾斜桩
inclined vibrating screen 倾斜式振动筛
inclusion 杂质
incombustible ash 不燃灰
incoming quality control 来料控制
incoming solar radiation 入射太阳辐射
incomplete combustion 不完全燃烧
incomplete combustion loss 不完全燃烧损耗
increment 增量，增值
incremental effect 递增效应
incremental encoder 增量式编码器
incremental output 增量式输出
incremental technology 渐进性技术
incrustation 水锈，硬壳，结壳，镶嵌物
independent solar photovoltaic system 独立太阳能光伏发电系统
indirect co-firing 间接混烧
indirect expansion low temperature solar thermal power 间接膨胀式低温太阳能热发电
indirect fired oven 间接燃烧炉

indirect form of solar energy 太阳能间接形态
indirect gain 间接增益（系统）
indirect gasification 间接气化
indirect land-use change 间接土地用途变更
indirect solar conversion system 间接太阳能转换系统
indirect solar energy 间接太阳能
indirect system 间接系统
indirectly irradiated solar receiver 间接式太阳辐射吸热器
indirect-pneumatic wave energy converter 间接气动波能转换器
indium-zinc-oxid 铟锌氧化物
individual guide vane servomotor 单导叶接力器
individual waste fuel 个人垃圾燃料，个人废弃物燃料
indoor air quality 室内空气质量
indoor climate 室内气候
indoor design condition 室内设计条件
induced current 感应电流
induced draft fan 引风机，抽风机，抽风式风扇
induced draught 诱导通风
induced force 感应力
induced junction photovoltaic cell 感应结光电池，感应结光伏电池
induced polarization 激发极化
induced velocity at the rotor 叶轮干扰速度
induced velocity field 感应速度场
induced voltage 感应电压
inductance 感应系数，电感
inductance coil 电感线圈
induction generator 感应发电机
induction machine 感应式电机
induction motor 感应电动机，感应电机
induction sorting system 感应分类系统
inductive 诱导的，感应的
inductive circuit 电感电路
inductive load 电感负载
inductive reactance 感抗
inductive sensor 电感式传感器
inductively coupled plasma atomic emission spectrometer 电感耦合等离子体原子发射光谱法
inductively coupled plasma mass spectrometer 电感耦合等离子体质谱法
inductor 感应器
inductor core 电感器铁芯
industrial biogas 工业沼气
industrial computer 工业计算机
industrial organic waste 工业有机废物
industrial silicon 工业硅
industrial timer 工业计时器
industrial waste 工业废料
inert carrier gas 惰性载气
inert material 惰性物质，惰性材料
inert nitrogen 惰性氮
inert working fluid 无效工作流体，惰性工作流体
inertia 惯性
inertia force 惯性力
inertia switch 惯性开关
inertial centrifugal force 惯性离心力
inertial dust separator 惯性除尘器
inertial subrange 湍流惯性负区
infiltration air 渗滤空气
infinite bus 无限大母线
infinite voltage gain 无穷大电压增益

inflammable 易燃的，易怒的
inflated plastic still 充气式塑料蒸馏器
inflated still 充气式蒸馏器
inflow of solar energy 太阳能吸入量
inflow wind 进风
influence by the tower shadow 塔影效应
influence by the wind shear 风切变影响
in-forest residue 森林内残留物
infragravity wave 长重力波
infrared absorption spectrum 红外吸收光谱
infrared gas analyser 红外气体分析仪
infrared heating 红外加热，红外线热处理
infrared imagery 远红外成像（确定热显示区的一种方法）
infrared imagery survey 红外成像探测
infrared pyrometer 红外高温计
infrared pyrometry 红外高温测量
infrared radiant heater 红外辐射加热量
infrared radiation 红外线辐射
infrared radiation detector 红外辐射探测器
infrared radiation thermometer 红外辐射温度计
infrared region （太阳光谱的）红外线区
infrared spectrometry 红外光谱法
infrequently-led-sludge 特殊污泥（指腐败的粪便）
infusion-induced delayed fracture 浸提致滞后断裂
ingot 铸锭，铸坯
inherent ash 固有灰分
inherent damping 固有阻尼

inhomogeneity 不均匀性
inhomogeneous air supply 不均匀送风
initial 最初的，初始的
initial chemical product 原始化学产物
initial cleaning 初清洗
initial cost 最初成本，初建费用
initial flash 首次闪蒸
initial leachate 初期渗滤液
initial pressure 初始压力，起裂压力
initial pressure profile 初始压力分布曲线
Initial setting time 初凝时间
initialize 初始化
injection pump 喷射泵，注射泵，注浆泵
injection well 回灌井，注水井
injector manifold 喷油器歧管，喷油器集箱（母管）
injure 损害，伤害
inlet air 入口空气
inlet air temperature 进气温度
inlet bucket 导风轮叶片，进口导叶
inlet chute 加料斜槽
inlet head 入口压头
inlet pipe 入口管，引入管，输入管
inlet valve 入口阀
inner packing 内包装
inner portion of blade 叶片内部
inner ring 内环
inner shaft 内半轴，内轴
innovative wind energy conversion system 新概念型风能转换系统
inoperative 不起作用的，无效的
in-phase 同相位
in-phase current 同相电流
input 输入
input power 输入功率

input pressure 输入压力
input process quality control 制程控制
input shaft 输入轴
input torque 输入扭矩
input voltage 电压
inrush current 励磁涌流,合闸电流,冲击电流
insert bit 镶齿钻头,镶式钻头,镶嵌钻头
in-service condition 工作状态
in-shore 陆上的
inside blow-out preventer 内防喷器
insolation 曝晒,中暑,日射(率),辐照量
insolation area 日照区
insolation duration 日照时间
insolation rate 日射率
insolation value 日照值
insoluble 不溶的
inspection earthing 检修接地
installation 安装
installation charge 安装费
installation drawing 安装图
installation height 安装高度
installed capacity 装机容量,装置容量
installed wind capacity 风电装机容量
instance 事实,范例,样品,情况,场合
instaneous voltage 瞬时电压
instaneous wind speed 瞬时风速
instantaneous 瞬间的,即刻的,即时的,瞬时,立即
instantaneous efficiency 瞬时效率
instantaneous electric power 瞬时电功率
instantaneous measurement 瞬时测量
instantaneous mechanical power 瞬时机械功率
instantaneous power 瞬时功率
instantaneous power output 瞬时功率输出
instantaneous value 瞬时值
instantaneous variation 瞬时变化
instantaneous wind speed 瞬时风速
instrument air 仪表气源
instrument rack 计测器支架,计测器框架
instrumentation tower 气象观测塔
insulant 绝缘物
insulated gate bipolar transistor 绝缘栅双极晶体管
insulated sleeve 绝缘套管
insulated steel lamination 绝缘硅钢片
insulating board 绝缘板
insulating boots 绝缘靴
insulating brick 隔热砖,绝缘砖
insulating bushing 绝缘套管
insulating castable 保温浇注料,隔热浇注料
insulating glove 绝缘手套
insulating material 隔热材料,绝缘材料
insulating plate 绝热板,绝缘板
insulating separator 绝缘隔板
insulating shutter 隔热百叶窗
insulating sleeving 保温套,绝缘套
insulating varnish 绝缘漆
insulation bed 绝热底座
insulation level 绝缘等级
insulation power factor 绝缘功率因数
insulation resistance 绝缘电阻
insulator 绝缘子
insulator ball guide 球头定位器
insulator 隔热体,绝缘体

intake manifold 进气歧管
intake pipe 进气管，进水道
intake pit 进气槽
intake valve 进气阀
intalox saddle 矩鞍形填料
integer 整数
integral collector/storage system 组合集热器/存储系统
integral collector-storage solar water heater 闷晒式太阳（能）热水器
integral diode solar cell 整体二极管太阳电池
integral solar water heater 闷晒式太阳热水器
integrated a-Si solar cell 集成型非晶硅太阳电池
integrated coal gasification 整体煤气化
integrated coupling 固定连接
integrated food and energy system 粮食能源一体化系统
integrated gasification combined cycle 集成气化组合循环技术，整体煤气化联合循环发电系统
integrated gate commutated thyristor 集成门极换流晶闸管
integrated planning model 综合规划模型
integrated pollution prevention and control directive 《综合污染防治指令》（欧盟）
integrated refrigeration load 集中式制冷负荷
integrated resource planning 综合资源规划
integrated solar combined cycle system 太阳能—燃气联合循环槽式热发电系统

integrated value 累计值
integrated waste management 综合废物管理
integrated α-Si solar cell 集成型非晶硅太阳电池
integrating mariculture and energy production 综合海洋养殖与能源生产
integrating sphere 积分球（内壁镀有漫反射涂层的空腔圆球。球壁上任一面受照后，其反射辐照度在整个空腔圆球内将是均匀一致的，利用积分球的这一特性可测定材料的反射率）
integration 积分下限
intensifying screen 增光屏，光增强屏
intensity of black-body radiation 黑体辐射强度
intensity of combustion 燃烧强度
intensity of incident light 入射光强度
intensity of light 光强
intensity of radiation 辐射强度
intensity of reflected light 反射光强度
intensity of solar radiation 太阳辐射强度
inter-annual variation 年际变化
intercept factor (spillage factor) of receiver 吸热器截断因子（或溢出因子）
interchangeable temperature 可交变温度
interconnected operation 并联运行
inter-connected raft 相互连接的浮箱
interconnected star winding 曲折连接绕组
interconnected system 互联系统，互连系统
interconnection 互连
interconnection of aerogenerators 风力发电机互联

inter-esterification 酯交换（是用于改变油和脂肪理化特性的处理过程之一，是一种甘油分子上的酰基进行重排的反应）
interface 接口
interface circuit 接口电路
interface defect 界面缺陷
interface energy 界面能
interference between wells 井之间的干扰
interference filter 干涉滤光片
interference test 干扰试验
intergranular porosity 粒间孔隙度
intergranular vaporization 颗粒间蒸发（作用）
interlocker 连锁装置
intermediate band solar cell 居间带太阳能电池
intermediate casing 技术套管，中间套管
intermediate ceiling 假顶，假平顶
intermediate depth waves 中深波
intermediate pumping station 中间泵站
intermediate fluid OTEC plant 中间流体海洋温差电站
intermediate load generating unit 中间负荷发电机组
intermetallic 金属间化合物
intermittency of solar energy 太阳能的间歇性，太阳能的间歇现象
intermittent 间歇
intermittent direct sunlight 间歇性阳光直射
intermittent flow 间歇流
intermittent furnace 间歇式炉
intermittent generation （潮汐电站）间断发电
intermittent heater with solar energy 太阳能间歇式加热器
intermittent heating 间歇加热
intermittent load 间歇负载
intermittent operation 间歇运行
intermittent operation control 间歇运行控制
intermittent power generation 间歇发电
internal absorptance 内吸收比
internal calipers 内卡钳
internal circulation 内循环
internal combustion 内燃
internal combustion cycle 内燃循环
internal combustion engine 内燃机
internal concentrative vacuum tube 内聚光式真空集热管
internal energy 内能
internal friction 内摩擦，内耗
internal gains 内部收益
internal gear 内齿轮
internal gear pair 内齿轮副
internal impedance 内电阻，内阻抗
internal lighting protection system 内部防雷系统
internal loss 内损
internal rate of return 内部盈利率，内部收益率，内部报酬率
internal resistance 内阻
internal tide 内潮
internal transmittance 内透射比
internal voltage 闪电压
internal water motion 内部水运动
internal wave of ocean 海洋内波
internal wear 内部摩擦
internal-combustion engine 内燃机
internally condensing still 内冷凝蒸馏器

International Sustainability and Carbon Certification (ISCC) 国际可持续发展和碳认证
International Cogeneration Society (ICS) 国际热电联产学会
International Conference on Large High Voltage Electric System (CIGRE) 国际大电网会议
International Sustainability and Carbon Certification System (ISCC) 国际可持续性和碳认证体系
interoffice 局间的
interseasonal storage 跨季节（能量）储存
interstation impulse response method 站间脉冲响应法
interstitial atom 间隙原子
interstitial diffusion 间隙扩散，填隙式扩散
interstitial site 间隙位置
interstitial solid solution 间隙固溶体，填隙式固溶体
intertidal area 潮间带
intertidal zone 潮间带，潮间地区，潮间区域
interval 间隔，时间间隔，区间
intervene 干涉，干预，插入，介入
intracontinental melting anomalies 陆内熔融异象
intraplate hot spot 板块内热点
intrinsic conversion efficiency 本征转换效率
intrinsic diffusion coefficient 本征扩散系数
intrinsic fill factor (theoretical fill factor) 本征填充因数（理论填充因数）
intrinsic silicon 本征硅
intrinsic spin 固有自旋
intrinsic thermal defect 本征热缺陷
intrusive body 侵入体
inundation model 淹溢模式
inverse 倒数
inverse cosine function 反余弦函数
inverse time-delay operation 反时限动作
inversion height 逆温层高度
inversion point 反向点
inverted air current 风流反向
inverter 变流器，反相器，直交流转换器
inverter control 逆变器控制
inverter efficiency 变换效率
iodine absorber 碘吸收器
ion-conductivity 离子电导率
ionic bonding 离子键
ionic conduction 离子传导
ionic mobility 离子迁移率
ionisation 电离作用
ionosphere 电离层
IR reflection coating 红外线反射涂层
iron cutter 截铁器，切铁刀
iron loss for generator 发电机铁损
ironbark 铁皮木
iron-carbon alloy 铁碳合金
iron-loss 铁损
irradiance 辐照度
irradiance meter 辐照度计
irradiance ratio 辐照率
irradiate energy 送射能量
irradiated fuel 辐射燃料，光渗燃料
irradiated fuel inspection 辐照燃料检验
irradiated fuel reprocessing 辐照燃料后处理

irradiated uranium dioxide fuel element 辐照二氧化铀燃料元
irradiation 辐照量
irradiation creep 辐照蠕变
irradiation damage 辐照损伤
irradiation distortion 辐照畸变
irradiation distribution 光照分布
irradiation dose 辐照剂量
irradiation effect 辐照影响
irradiation embrittlement 辐照脆化
irradiation hazard 辐照危险性
irradiation induced growth 辐照感生生长
irradiation processing of fruit 水果辐照处理
irradiation rig 辐照试验台
irradiation surveillance capsule 辐照监督管
irradiation-induced creep 诱发辐照蠕变
irregular galaxy 不规则星系
irregular wave 不规则波
irregularity 不规则,无规律,不规则性,不正则性
isentropic 等熵的,等熵线
isentropic analysis 等熵分析
isentropic change 等熵变化
isentropic chart 等熵图
isentropic compression 等熵压缩
isentropic condition 等熵条件
isentropic efficiency 等熵效率
isentropic mixing 等熵混合
isentropic motion 等熵运动
isentropic process 等熵过程
isentropic surface 等熵面
isentropy 等熵线
island operation 岛式运行,孤岛运行
Island system 孤岛系统
islanding 电网解列
isobar 等压线
isogeotherms 等地温线
isogradient 等梯度线
isohaline 等盐度线
isohel 等日照线
isohume 等温度线
isokinetic probe 等动能采样头
isokinetic sampling 等动力采样
isolated operation 孤立运行
isolated-gain system 独立增益系统
isolating valve 隔离阀,隔断阀
isolation valve 隔离阀
isolator 隔离开关（刀闸）
isometric projection 等边体
isopach data 等厚图
isorad 等曝辐量线
isosceles triangle 等边三角形
isothermal compressibility 等温压缩率,等温压缩性
isothermal layer 等温层
isothermal line 等温线
isothermic 等温的
isotope 同位素
isotope geothermometers 同位素地温计
isotropic body 各向同性体,均质体
isovent map 风速线图
iterated maximum likelihood method 迭代最大似然法
iterative approach 迭代法,辗转法
iterative earth 重复接地
iterative solution 迭代解
I-V characteristic curve of a solar cell 太阳电池的伏—安特性曲线

J

jack 插孔,插口,插座,千斤顶

jatropha 麻风树
jet 喷嘴，射流
jet stream 急流，射流，喷射气流
jet-fuel 航油，喷气燃料
jet-turbine 喷射式涡轮
jib crane 挺杆臂，转臂起重机
jig 钻模，夹具，定位，模具
jockey pump 管道补压泵
joint 节理
joint probability 联合几率
jointing compound 复合填料
joist 托梁，搁栅
Joule heat 焦耳热
Joule heating 焦耳加热，电阻加热
Joule law 焦耳定律
jubilee clip 连接螺旋夹
jumbo windmill 巨型风车
jumper 跨接线，跨接（片），跳接器
jumper clamp 跳线线夹
junction box 接线盒，套管，连轴器
juxtaposition 邻近位

K

Kaimei experiment （日本）海明波浪发电装置试验
Kaimei wave power facility 海明波浪发电装置
kampometer 热辐射计
kaolinite 高岭石
Kaplan turbine 卡普兰水轮机，转桨式水轮机
keeping house clean 室内消毒（法）
Kelvin 开尔文，开氏度
Kelvin wave 开尔文波
kerf loss 截口损失

kerosene 煤油
keyhole saw 铨孔锯，键孔锯
key-operate 键盘操作
keyway 键槽
K-fuels 卡普曼燃料
kieselguhr 硅藻土
kiln 窑，炉
kilopascal 千帕
kilotonne (kton) 千吨
kilovolt-ampere (kVA) 千伏安
kilowatt (kW) 千瓦
kilowatt electric 千瓦电力
kilowatt thermal 千瓦热能
kilowatt-hour 千瓦时
kilowatt-hour meter 千瓦时计，电度计，电表
kinds of earth heat 地热类型
kinematic wave 运动波
kinetic and gravitational potential energy 动力及重力势能
kinetic hydro energy conversion system 动力水能转换系统
kinetic power hydro system 动态功率水电系统
kiosk substation 配电小间
Kirchhoff's law 基尔霍夫定律
Kirchhoff's voltage law 基尔霍夫电压定律
Kirkendall effect 柯肯达尔效应
kit 成套工具，用具包，工具箱
kite anemometer 风筝式风速计
knee point 曲线弯曲点，拐点
knob 按钮，调节器
knuckle joint 万向接头
knurled nut 凸螺母
knurling 多瘤的，多节的

L

laboratory condition 实验条件
laboratory modeling 室内模型试验
laboratory sample 实验室样品
labyrinth 迷路，迷宫，难解的事物
ladder logic 梯形逻辑
lag-bolt collector 方头螺栓集热器
lagging power factor 滞后功率因数
lagging reactive current 滞后无功电流
lagging reactive power 感性无功
lagoon 泻湖，环礁湖，滨海湖，小片淡水域，（人造）污水储留池
Laguarnes filed 劳加奈斯地热田
Lambda control probe 拉姆达控制探针
lambda sensor 氧传感器
laminar flow 层流
laminar low 层流
laminate 层压（制品），叠片，叠层板，绝缘板，碾压
laminated core 叠片铁芯
laminated panel collector 层压板式集热器
laminated polymer-aluminium-steel reflector material 层压铝钢复合反射层材料
lamination 层压，压块
lamination process 层压
lamp holder 灯座
land 纹间表面
land and sea breeze 海陆风
land breeze 陆风
land configuration 地貌，地形，陆地形状
land drainage 地面排水，土地排水
land mass 大陆块
land slide surge 滑坡涌浪

land subsidence 地面沉降
land-acquisition issues 征地问题
land-based plant with a closed cycle 陆地闭（式循）环（海水温差）电站
land-based power plant 陆上动力装置，固定式动力装置
land-breeze 陆风（自陆地吹向海上的微风）
landfill 垃圾填埋地，垃圾填埋场，垃圾堆
landfill density 填埋密度
Landfill Directive （欧盟）《垃圾掩埋法令》
landfill gas 填埋气体，堆填区沼气，垃圾填埋气
landfill gas recovery 垃圾填埋气回收
landfill methane gas 垃圾填埋场沼气
landfill site 垃圾填埋场，垃圾堆填区
landscape 景观，地貌
landscape architecture 园林建筑学
landscape assessment 景观评价（估）
landscape engineering 环境绿化工程，景观工程
landscape management residue 园林管理残渣
landscaping 环境美化，绿化
land-use change 土地利用变化，土地用途变更（用来反映影响土地资源可持续利用的人类活动，过程和格局的指标）
Langmuir circulation 朗缪尔环流
lap 搭接，重叠，研磨，抛光
lapjoint 搭接
Laplace transform 拉普拉斯变换
lapping machine 精研机
lapse rate 递减率
Larderello geothermal power plant (Italy) 拉德瑞罗地热电站（意大利）

Large Combustion Plant Directive 大型电站能源消费法令
large floating power plant 大型浮式电站
large heat-storage unit 大型储热装置
large power drill 大型钻
large scale energy use 大规模能源利用
large south-facing glazed area 大面积向阳玻璃面
large wind generator array 大型风力发电机阵列
large-area flow counter 大面积流计数器
large-area polycrystalline 大面积多晶薄膜
large-area thin-film photovoltaic cell 大面积薄膜光伏电池，大面积薄膜光电电池
large-scale storage of heat 大规模储热
large-scale weather situation 大范围天气形势
laser anemometer 激光风速计
laser beam 激光束
laser cutting 激光切割
laser Doppler anemometer 激光多普勒风速计
laser drilling 激光凿岩，激光钻孔
laser energy conversion 激光能量转换
laser etching 激光刻蚀
laser profilometry 激光轮廓测量技术
laser scribing 激光划片，激光划线
latch 锁存器
latching 封锁
latent fault dormant failure 潜伏故障
latent heat 潜热
latent heat calorimeter 潜热量热器

latent heat energy storage 潜热储存
latent heat of vaporization 汽化潜热
latent heat storage 潜热储存，相变储热
lateral density variation 横向密度变化
lateral diffusion 横向扩散
lateral dispersion 横向弥散
lateral displacement 侧向移位
lateral force 侧向力
lateral gust 横向阵风
lateral translation 侧向平移
lathe bench 车床工作台
lathe tool 车床工具
lattice 格栅，格子
lattice mast 格构式杆架
lattice parameter 晶格参数
lattice tower 格子形杆架
lava 火山岩
law of thermodynamics 热力学定律
layer level class 层
layout 规划，设计，配线，设计图案，布局图，版面设计
layout drawing 布置图
leach bed process 滤床工艺
leachate collection pipe 渗滤液收集管
leachate disposal 渗滤液处理
leachate recycling 滤液回收
leaching 沥滤，浸出
leaching process 浸出法
lead acid battery 铅酸蓄电池
lead foil 铅箔
lead pipe 铅管
lead-acid battery 铅酸蓄电池
leading edge （机翼的）前缘，前缘缝翼，（螺旋桨叶片的）导边
leading edge of blade 叶片前缘
leading edge radius 前缘半径

leading phase 超前相，领先相
leading power factor 超前功率因数
leading reactive power 超前无功
lead-lag （风轮）摆振方向
lead-lag motion 桨叶摆动
lead-time 提前期
leaf spring 叠簧
leaflet 小叶，传单
leak detection system 泄漏检测系统
leak test 漏泄试验
leakage circuit breaker 漏电断路保护器
leakage current 泄漏电流
leakage field 漏磁场，漏泄场
leakage flux 漏磁通
leakage inductance 漏电感
leakage reactance 漏磁电抗
leaking pipeline 泄漏管道
lease 释放
lee side 背风面，下风弦
leeward 下风（的），背风处
left-hand rule 左手定则
left-hand thread 左旋螺纹
length of blade 叶片长度
length of free pipe 自由管长度
length scale 长度规
lengthwise 纵长的，纵向长的
lens focusing 透镜聚焦
lens solar furnace 透镜式太阳能
Lenz law 楞次定律
less perfect collector 不太理想的集热器
level instrument 位面计，水平仪
level switch 液位开关
levelized cost 平准价格
lever rule 杠杆定律
life cycle 生命周期
life cycle assessment 生命周期评估

life cycle cost 生命周期成本
life cycle impact assessment 生命周期影响评价
life cycle inventory 生命周期清单
life of well 井孔寿命
life span 寿命
life time 寿命试验
life-cycle energy 生命周期能耗
lift coefficient 升力系数
lift cycle system 外压循环系统
lift device 提升装置
lift force 升力，举力
lift-drag ratio 升阻比
lift-drag-type 升阻型
lifting flight 扬料板
lift-type 升降型
lift-type machine 升举式机械，升力型风机
lift-type rotor 升力型叶轮
light 光，日光
light absorption 光吸收
light air 一级风，软风
light breeze 二级风，轻风
light control system 光控系统
light current density 光流密度
light detection and ranging 激光雷达，光探测和测距
light diesel oil 轻柴油
light emitting diode 发光二极管
light energy 光能
light gray sandstone 浅灰色砂岩
light metal spirit level 光金属水平仪
light scatter 光散射
light sensitive detector 光敏探测器
light shading 淡阴影
light source 光源
light trapping 光的增透

light wind 小风
light-emitting diode 发光二极管
light-generated current 光生电流
lighting conductor 光导体
lighting fixture 照明器材
lighting intensity 光强
lightning protection system 防雷系统
lightning current 闪电流，雷电流
lightning flash density 闪电密度
lightning protection 避雷（装置）
lightning protection zone 防雷区
lightning rod 避雷针
lightweight aggregate 轻骨料，轻砂石
lightweight cement 轻质水泥
lignin 木质素
lignin complex 木质素复合体，木质素复合物
lignin residue 木质素残渣
lignite 褐煤
lignocellulose 木质纤维素
lignocellulosic anthracite 木质纤维无烟煤
lignocellulosic bioethanol 木质纤维生物乙醇
lignocellulosic energy crop 木质纤维素能源作物
lignocellulosic feedstock 木质纤维素原料
liming agent 浸灰剂
limit gauge 极限量规，限制量计
limit of solar distillation 太阳能（蒸馏器的）蒸馏量
limit speed switch 限速开关
limit state 极限状态
limit switch 限位开关
limitation 限度
limited current circuit 限流电路

limiter 限幅器
limiting value 极限数
line current 线路电流
line defect 线缺陷
line focus collector 线性聚焦集热器
line focus concentrator 线聚焦聚光器
line focus configuration 线性聚焦型式
line focus system 线聚焦系统
line frequency 行频
line number 行数
line to line voltage 行对行电压
line trap 限波器
line voltage 线电压
linear actuator 线性驱动器，线性传动装置
linear electrical generator 线性发电机
linear focused collector 线性聚焦集热器
linear Fresnel reflector solar power 线性菲涅耳式太阳能热发电
linear generator 直线发电机，线性发电机
linear heat source 线性热源
linear hydraulic pump 直线液压泵，直线水力泵，线性液压泵，线性水力泵
linear kinematic effect 线性运动效应
linear momentum 线性动量，线动量
linear programming 线性规划
linear solar collector 线性太阳能集热器
linear speed 线速度
linear synchronous generator 线性同步发电机
linear system 线性系统
linear theory 线性理论，线性定理
linear velocity 线速度
linear vibration 线性振动

linear wave theory 线性波动理论，直线波动理论
linear zone 线性区
linearized method 线性化法
line-focus collector 线聚焦集热器
line-focus concentrator 线聚焦聚光器
line-to-line voltage 全线电压
line-to-neutral 线与中性点间的
lingo-cellulose 木质纤维素
lining 内衬，衬里
link 链环，连结，联合
linkage bearing 联动装置轴承
linked-basin scheme 连通库方案
lintel 楣，过梁
lintel beam 水平横楣梁
lipid 脂质
liquefaction 液化，流化，溶解，液态
liquefied petroleum gas 液化石油气
liquefy 液化
liquid (heating) collector 液体集热器
liquid biofuel 液体生物质燃料，液体生物燃料
liquid combined solar cell 液体组合式太阳能电池
liquid core 液态地核
liquid deposition 液相沉淀法
liquid fertilizer 液体肥料，液态肥料
liquid flat plate collector 液体平板型集热器
liquid fuel 液体燃料
liquid heating flat plate collector 液体加热平板集热器
liquid heating system 液体供暖系统，液体加热系统
liquid injection incinerator 液体喷射焚烧炉
liquid penetrant examination 液体渗透探伤

liquid phase 液相
liquid piston 液体活塞
liquid pool 液体池，液态池
liquid velocity 液相流速，液体速度
liquid-based solar heating system 液基太阳能加热系统
liquid-dominated field 液体占优势的地热田
liquid-dominated hydrothermal resource 以液体为主的热液资源
liquid-dominated reservoir 液体占优势的储层
liquid-heating collector 液体加热集热器
liquid-in-glass thermometer 液体温度计，充液玻璃温度计
liquid-propane recirculation 液态丙烷再循环
liquid-solid ratio 液固比
liquid-type collector with tube and plate 管板液式集热器
liquid-type solar collector 液式太阳能集热器
liters of diesel equivalent 升柴油当量（能量含量为36.1兆焦/升）
liters of gasoline equivalent 升汽油当量（能量含量为33.5兆焦/升）
lithium stearate 硬脂酸锂
lithium-doped solar cell 掺锂太阳电池
lithium-ion (Li-ion) battery 锂电池
lithium-polymer (Li-poly)battery 锂聚合物电池
lithologic mapping 岩体绘图
lithology 岩性
littoral 沿海地区，沿海的，海滨的
livestock residues 牲畜残余物，牲畜残留物

Lloyd energy system 罗伊德能源系统
load 负载，负荷
load angle 负载角，负荷角
load case 荷载工况
load characteristic 负载特性，负荷特性
load control 负载调节，负荷调节
load current 负载电流，负荷电流
load duration 负载历时，负荷历时
load factor 负载因数，负载系数
load flow 电力潮流
load growth 负载递增，负载成长，负荷增长
load power 负载功率，负荷功率
load profile 负载简档，负载配置文件，负荷曲线
load sharing strategy 负载分配策略
load sharing, load distribution 负载共享，负载分配，均分负载
load shedding 卸载，减载
load transient 负荷暂态
load variation 负载变化，负荷变化，负荷波动
load voltage 负载电压，负荷电压
load waterplane 载重水线面
load-communicated inverter 负载换相变频器
load-duration curve 负荷历时曲线
load-saturation curve 负载饱和曲线
local circulation 局部环流
local control 就地控制
local effect 局部效应
local geology 当地地质
local heat treatment 局部热处理
local panel 现场配电盘
local resistance 局部阻力
local solar intensity 当地太阳能强度
local solidity 当地实度

local speed ratio 当地速比
local temperature disturbance 局部温度扰动
local utility grid 地方公用电网
local winds 局地风
locality 地方性
localized equilibrium 局部平衡
locate 定位，位置
location 单元，位置
location map of thermal seepage 热渗出位置图，地渗出位置图
locational marginal price 节点电价
lock 加锁，锁（紧，定），封闭
Lock number 洛克数
lock washer 锁紧（止动），（防松）垫圈
locked rotor protection 锁定转子保护
locked-rotor 锁定转子
locked-rotor torque 锁定转子转矩
locking device 锁定装置
locking nut 自锁螺母
locking set 锁定装置
locknut 防松螺母
locus 轨迹
Lof house 洛夫太阳房（一种主动式太阳能房）
log law 对数律
log-log plot 双对数曲线图
logarithmic wind shear law 对数风切变律
logging residue 伐木残渣
logic chip 逻辑片，逻辑集成电路
logic diagram 逻辑图
long collection efficiency 低集热效率
long distance heating 长距离供热
long distance line 长距离管道
long stator linear synchronous motor 长定子直线同步发动机

long wave infrared scanner 长波红外线扫描仪
long wavelength radiation 长波辐射
longevity 寿命
long-focus collector 长聚焦集热器
long-focus collector with less curvature 小曲率长聚焦集热器
longitudinal 纵向的
longitudinal flux permanent magnet 横流永磁
longitudinal lamination 纵向层压（薄板）
long-life power source 长寿命能源（如太阳能）
long-life solar array 长寿期太阳能阵列
longshore drift 沿岸漂移，沿岸漂砂，沿岸物质流
long-term energy substitution programme 长期能源替代计划
long-term performance of collector 集热器长期性能
long-term storage 长期储热（系统），长期储存
longwall mining 长壁开采
long-wave emittance 长波辐射源
longwave radiation 长波辐射
long-wavelength infrared radiation 长波红外辐射
long-wavelength radiation 长波辐射
loom 织布机，织机，隐现，迫近
loop 回路
loop current 环流
loop cycle 周期，循环周期
loop flow rate 循环流率
loop operating temperatures 循环操作温度
loop antinode 波腹

loose volume 散体积
loss of circulation zone 井漏带
loss of ignition 炉渣热灼减率
loss of load 甩负荷
loss of phase 失相
loss of wind 风损
loss on ignition 烧失量
lost circulation material 堵漏材料
lost tool 落在井内的工具
lost-circulation 井漏带
lost-circulation zone 井漏区，循环液漏失区
lost-head nail 断头钉
lot 批
loudspeaker 扩音器，喇叭
low angle grain boundary 小角度晶界
low boiling point 低沸点
low concentration 低浓度，低集中度
low concentration collector 低聚光集热器
low emission glass 低辐射玻璃
low emissivity coating 低发射率涂料，低辐射率涂料
low enthalpy 低热焓
low friction loss 低摩擦损失，低摩擦损耗
low heat value 低位热值
low heat value for design 设计低位热值
low impact technology solution 低效技术法（合理利用太阳能的一种方法）
low insolation （接受到的太阳能）低辐射，低日照率
low level recovery 低级热能回收
low melting eutectic mixture 低共熔混合物
low molecular weight gas 低分子量气体
low osmotic pressure 低渗透压

low pool 低库
low pool level 低库水面
low potential thermal energy 低位势热能
low pressure flash evaporator chamber 低压闪蒸室
low pressure flash vessel 低压闪蒸容器
low pressure stage 低压级，低压段
low pressure steam 低压蒸汽
low reservoir 低库
low temperature fuel cell 低温燃料电池
low temperature gasifier 低温气化炉
low temperature heat 低温热
low temperature heat energy 低温热能
low temperature solar thermal power with Kalina cycle 卡琳娜低温太阳能热发电
low temperature flat-plate solar collector 低温平板型太阳能集热器
low tide 低潮
low tide level 低潮位
low voltage 低压电器
low voltage coil 低压线圈
low voltage ride through of wind turbines 风电机组低电压穿越
low water level, low water stand 低潮位
low/medium solid digestion system 低/中固分解系统
low-grade thermal energy 低位热能
low-concentrate solvent 低浓度的
low-cost solar array 低成本太阳能阵列
low-energy solar plasma 太阳低能量等离子体
lower atmosphere 低层大气
lower limit 下限
lower limit LHV of waste for incinerator 焚烧炉下限垃圾低位热值
lower limit waste treatment capacity for incinerator 焚烧炉下限垃圾处理量
lower source temperature 低热源温度，低温热源
lower temperature conductivity 低热导率
lower temperature limit 低温限
lower temperature storage 低温储热
lower-pressure escape route 低压排出回路
low-gear ratio 低速齿轮比
low-grade energy 低位能，低品位能
low-grade geothermal reservoir 低等级地热储集层
low-grade geothermal resources 低等级地热资源
low-grade heat energy 低位热能
low-grade waste heat 低品位余热
low-head 低水头（数米至十余米的发电水头）
low-head bi-directional turbine 低头双向涡轮机
low-head hydro-electric unit 低压头水电装置
low-head wave energy water turbine 低水头波能水轮机
low-heat water-power plant 低压头水电站
low-high junction theory 低—高结理论
low-leakage piston ring 低耗电活塞环
low-level heat recovery 低位热回收
low-level nighttime jet 低级夜间喷嘴

low-level nocturnal jet 低级夜间喷嘴
low-level reservoir 低水位水库
low-level reservoir basin 低库区
low-level reservoir capacity 低库库容
low-level reservoir site 低库库址
low-lying flat 低洼地，海流
low-moderate temperature solar thermal power plant 低中温太阳能热电站
low-moderate-temperature solar collector 低中温太阳能集热器
low-order method 低位法
low-power solar cell 低功率太阳能电池
low-pressure air 低压空气
low-pressure condensing turbine 低压凝汽式汽轮机
low-pressure steam 低压蒸汽，低压水蒸气
low-pressure steam chest 低压蒸汽室
low-solidity rotor 低实度风轮
low-solids content 低固体含量
low-speed multi-blade wind turbine 低速多叶片风力涡轮机
low-speed shaft 低速轴
low-tar fuel 低焦油燃料
low-temperature area 低温区
low-temperature carbonization gas 低温干馏气，低温碳化气
low-temperature collector 低温集热器
low-temperature collector device 低温集热装置
low-temperature collector system 低温集热系统
low-temperature denitration equipment system 低温脱硝装置系统
low-temperature emissivity 低温发射率
low-temperature geothermal fluid 低温地热流体
low-temperature reservoir 低温蓄热库
low-temperature solar conversion system 低温太阳能转换系统
low-velocity layer 低速软弱夹层
lubricant 润滑油
lubricant pump 润滑泵
lubricate 润滑
lubrication system 润滑系统
lubricator 注油机
lug 接线柱，柄，把手，突起
lumber scrap 木材废料
luminescence 发（荧）光
luminescent material 发光材料
luminous efficiency 发光效率
luminous flux 亮度，光通量
luminous intensity 光强，发光强度，照度
lunar diurnal tide 太阴全日潮，太阴日周潮
lunar evectional tide 太阴出差潮
lunar retardation 月球延滞
lunar tide 月潮，太阴潮
lunar-raising force 月球上升力
luni-solar 日月（之间）的
lunitidal interval 月潮间隙，满潮时距
lunnar day 太阴日
Lurgi gasifier 鲁奇气化炉
Lynmouth turbine 林茂次涡轮机

M

mach number 马赫数
machinability 切削性
machine efficiency 机械效率
machine element 机械零件
machine tools 工作母机
machining 机械加工

maelstrom 大漩涡，异形漩流
mafic rock 铁镁岩质
magma 岩浆
magma chamber 岩浆房，岩浆库
magma liquid 液态岩浆
magma solid 固态岩浆，岩浆流体
magma trap 岩浆收集器
magma-chamber 岩浆室
magmatic-hydrothermal system 岩浆热液系统
magnesite 菱镁矿
magnet 磁体，磁石
magnetic 磁的，有磁性的，有吸引力的
magnetic amplifier 磁放大器
magnetic circuit 磁路
magnetic core 磁芯
magnetic core loss 磁芯损耗
magnetic coupling 磁性联轴
magnetic field 磁场
magnetic field distortion 磁场畸变
magnetic flux 磁通量
magnetic flux density 磁通密度
magnetic iron 磁铁
magnetic lens 磁透镜
magnetic method 磁法（用于地热勘探的一种地球物理方法）
magnetic particle examination 磁粉检验
magnetic pole 磁极
magnetic pole pair 磁极对
magnetic saturation 磁路饱和
magnetic separation 磁选，磁力分离
magnetic separator 磁力分离器
magnetic susceptibility 磁化率
magnetic tool 磁性工具
magnetic torque 电磁转矩
magnetisation 磁化
magnetize 吸引，使磁化
magnetizing current 磁化电流
magnetizing reactance 磁化电抗
magneto anemometer 磁电风速计
magneto resistor encoder 磁敏电阻式编码器
magnetohydrodynamic power generation 磁流体发电
magnetomotive force 磁动势
magnetotellurics 大地电磁学
magnetron sputtering coating 磁控溅射镀膜
Magnus effect 马格努斯效应
Magnus effect type wind turbine 马格努斯效应式风力机
main block valve 总闸门
main circuit 主电路
main circuit breaker 主断路器
main contact 主触头
main contactor 主接触器
main DC disconnector 主直流断路器
main drive shaft 主动轴
main engine 主发动机
main exciter 主励磁机
main function 主要功能
main pipe 主管道
main shaft 主轴，总轴
main spar 主梁
main substation 总变电站
maintenance 维护
maintenance interval 维修间隔期，维修周期
maintenance test 维护试验
major construction technical scheme 主要施工技术方案
majority charge carrier 多数载流子

make a bridge 短接
make contact 关合触头
male rotor 凸形轮，凸形转子，阳转子
mandrel 半导体阴极金属心，心轴
manhole 人孔，检修孔
manifold 集管，母管
manometer 压力计
mantle 环梁壳
manual control 手动控制
marginal cost payment time 成本收回时间
marginal sea 领海
mariculture 海洋生物养殖
marine air 海洋气团，海洋空气
marine atmosphere 海洋大气
marine bed 海相层
marine biological pollution 海洋生物污染
marine biology 海洋生物学
marine biomass 海洋生物量，海洋生物质
marine biomass conversion 海洋生物质能转化
marine biota 海洋生物
marine boiler 船用锅炉
marine buoy 海上浮标
marine cable 船用电缆，海底电缆，海底电线
marine chemical resource 海洋化学资源
marine climate 海洋性气候
marine complex 海洋综合设施
marine component testing facility 海洋组件测试设备
marine corrosion 海水腐蚀
marine culture 海洋养殖
marine current 海洋气流，洋流

marine current power 海流能
marine current turbine 海流涡轮机
marine current velocity profiler 海流速度剖面仪
marine deposit 海相沉积
marine design 船舶设计
marine disposal 海洋处置（废物）
marine ecology 海洋生态学
marine ecosystem 海洋生态系统
marine electrochemistry 海洋电化学
marine energy application 海洋能应用
marine energy conversion 海洋能转换
marine energy source 海洋能源
marine engine 海用引擎
marine engineering geology investigation 海洋工程地质调查
marine environment 海洋环境
marine environment quality 海洋环境质量
marine erosion 海蚀，浪蚀
marine exploration equipment 海洋勘探设备
marine facies 海相
marine fouling 海洋污垢
marine gas cooling reactor 船用气冷（反应）堆
marine geological survey 海洋地质调查
marine geomorphy 海洋地貌
marine growth 海洋生物，海生植物丛，船底附生物
marine hydrographic(al) survey 海洋水文测量，海道测量
marine insurance 海上保险，水险
marine life 海洋生物
marine microbial pollution 海洋微生物污染

marine mineral resource 海洋矿物资源
marine oil spill 海上漏油
marine organism 海洋生物，海洋有机物
marine organism corrosion 海洋生物腐蚀
marine petroleum exploitation 海洋石油开发
marine pipeline 海洋管道
marine pipeline trenching 海上管道开沟
marine plain 海蚀平原
marine plant and animal 海洋动植物
marine pollutant 海洋污染物
marine pollution 海洋污染
marine pollution monitoring 海洋污染监测
marine pollution prevention 防止海洋污染
marine pollution prevention law 海洋防污法
marine positioning 海上定位
marine power 海洋能
marine propeller 船用螺旋桨
marine propulsion 船舶推进装置
marine propulsion reactor 船舶推进（反应）堆
marine protected area 海洋保护区
marine protection 海洋保护
marine resource 海洋资源
marine sanctuary 海上禁捕区，海上自然保护区
marine science 海洋科学
marine sediment 海洋沉积物
marine soil mechanics 海洋土力学
marine thermodynamics 海洋热力学
marine wave energy conversion system 海洋波能转换系统

marine wind regime 海洋风况
marine map 海图
marinetime engineering 海洋工程，海岸工程
maritime 海的，海事的，沿海的
maritime air 海洋性空气
maritime climate 海洋性气候
maritime spatial planning 海洋空间规划
Martensite 马氏体
Martensite phase transformation 马氏体相变
masonry drill 石钻
masonry wall 砖石墙
mass balance analysis 物料平衡分析（法）
mass burn system 集中式焚烧，全量焚烧系统
mass flow 质量流，质量流量
mass flow rate 质量流率
mass load of grate 炉排机械负荷
mass moment 质量矩
mass-reduction 缩分
master blade 主叶
master control system 主控制系统
master cylinder 主液压缸
master fuel trip 总燃料跳闸
master schedule 主要图表，综合图表，主要作业表
master valve 主阀，总阀
mat 字模，垫块
material 材料
material certificate 材料合格证
material fatigue 材料疲劳
material flux 物质通量
material output 材料产出
material property 材料特性

material recovery facility 物料回收设备
material substitution 材料置换
material supply plan for construction 施工物资供应计划
material for film resistance 薄膜电阻材料
material for regulation resistance 调节电阻材料
material for thermocouple 热电偶材料
materials optics 材料光学
material recovery facility 材料回收设备
mathanization 甲烷化
mathanogen 产甲烷生物
mathematical programming 数学规划
mating gear 配对齿轮
matrix converter 矩阵式变换器
matrix crack 基体开裂
matrix material 复合材料
matrix operation 矩阵运算
matrix solution 矩阵解
matted crystal 子晶，雏晶
maximum achievable control technology 最大可实现控制技术，最大可完成控制技术
maximum average temperature 最高平均温度
maximum average wind speed for 10 minutes 10 分钟最大平均风速
maximum average wind velocity 平均最大风速
maximum bare table acceleration 空载最大加速度
maximum bending moment 最大弯曲力矩
maximum brightness concentration 最大亮度集聚
maximum concentration ratio 最大聚光比

maximum continuous rating 最大连续蒸发量（功率/出力）
maximum daily output 最大日供应量
maximum designed wind speed 最大设计风速
maximum energy 最大能量
maximum energy density 最大能量密度
maximum energy level 最高能量级
maximum lift coefficient 最大升力系数
maximum measured power 最大测量功率
maximum mooring force 最大系泊力
maximum operating plan 最大运行方式
maximum operating pool 最大运行库
maximum operating temperature 最高工作温度
maximum permitted power 最大允许功率
maximum power 最大功率
maximum power current 最大功率电流
maximum power of wind turbine 风力机最大功率
maximum power point 最大功率点
maximum power point tracker 最大功率点跟踪器
maximum power point tracking 最大功率点跟踪
maximum power tracking 最大功率跟踪（法）
maximum power voltage 最大功率电压
maximum reflector collector 最大反射集热器
maximum rotational speed 最大转速

maximum service temperature 最高工作温度
maximum solar radiation 太阳辐射最大值
maximum summer insolation 夏季最大日照率
maximum system voltage 最大系统电压
maximum temperature 最高温度
maximum thermometer 最高温度表
maximum turning speed of rotor 风轮最高转速
maximum value 极大值，最大值
maximum value of wind speed 风速最大值
maximum wave height 最大波高
maximum wave period 最大波周期
maximum wind speed 最大风速
maximum wind velocity 最大风速
maximum wind velocity of definite period observation 定时的最大风速
maximum power output 最大输出功率
Maxwell's equations 麦斯威尔方程
mean 平均的
mean annual thunderstorm day 年平均雷暴日
mean annual wind power potential 年均风能潜势
mean annual wind speed 年均风速
mean camber line 中弧线
mean current energy density in position 位置平均（海流）能密度
mean current energy density of time 时间平均（海流）能密度
mean discharge 平均流量
mean earth-sun distance 日地平均距离
mean flow 平均流量
mean fluid temperature 工质平均温度
mean geometric chord length 平均几何弦长
mean high water 平均高潮位
mean high water lunitidal interval 平均月潮高潮间隙
mean high water neaps 小潮平均高潮位
mean high water spring 大潮平均高潮位
mean higher high water 平均高高潮面
mean hourly speed 时平均风速
mean life 平均寿命
mean line 中弧线
mean load 平均荷载，平均负荷
mean low water 平均低潮面
mean low water lunitidal interval 平均月潮低潮间隙
mean low water neaps 平均小潮低潮面
mean low water springs 平均大潮低潮面
mean lower low water 平均低低潮面
mean monthly temperature 月平均温度
mean noon 平正午
mean power 平均功率
mean power output 平均能量输出
mean range, mean range of tide 平均潮差
mean rise interval 平均涨潮间隙
mean river level 平均河水位
mean sea level 平均海平面，平均海面
mean solar day 平太阳日
mean solar time 平（太阳）时
mean speed 平均速度
mean spring rise 平均大潮升

mean stay 平均停留
mean sun 平太阳
mean tide level 平均潮面
mean tide range 平均潮差
mean time 平均时间
mean value 平均值
mean wave period 平均波浪周期
mean wind direction 平均风向
mean wind speed 平均风速
measured speed 测量速度
measurement, correlation and prediction technique 度量、关联和预测技术
measurement of terrestrial heat flow 地热流测量
measurement parameter 测量参数
measurement period 测量周期
measurement power curve 测量功率曲线
measurement seat 测量位置
measurement sector 测量扇区
measurement while drilling 随钻测量
measuring circuit 测量电路
mechanic 技工，机械师
mechanical anemometer 机械风速计，机械风速仪
mechanical biological treatment 机械生物处理
mechanical brake 机械制动器
mechanical braking system 机械制动系
mechanical collector 机械除尘器，机械集尘器
mechanical commutator 机械换向器，机械转换器
mechanical control 机械控制
mechanical control system 机械控制系统
mechanical draft 机械通风，强制通风

mechanical endurance 机械寿命
mechanical energy 机械能
mechanical failure 机械故障
mechanical force 机械力
mechanical gear 机械齿轮
mechanical load 机械负载
mechanical loss 机械损失
mechanical mechanism 力学机制
mechanical motion 机械运动
mechanical polishing 机械抛光
mechanical power 机械功率
mechanical processing 机械加工
mechanical radian 机械弧度
mechanical resilience 机械韧性
mechanical resonance 机械共振
mechanical room layout 机械房布局
mechanical rotating salinity gradient energy conversion 机械转动式浓度差能发电
mechanical shake cleaning 机械振动布袋清灰
mechanical strength 机械强度
mechanical stress 机械应力
mechanical system 机械系统
mechanical turbine torque 机械转矩
mechanical turbulence 机械湍流
mechanical instability 机械性不稳定
mechanically sorted municipal solid waste 机械分类城市生活垃圾
mechanism of heat transfer 换热机理，地热机制
medical waste 医疗垃圾
medium density fibreboard 中密度纤维板
medium-term leachate 中期渗滤液
medium-voltage 中压
megajoule (MJ) （功或能的单位）兆焦耳

megawatt (MW)　兆瓦
megawatt hour (MWh)　兆瓦时
megawatt wind turbine　兆瓦级风电机组
melting curve　溶解曲线，熔融曲线
melting furnace　熔化炉，熔窑，熔解炉
melting latent heat　熔化潜热
melting temperature　熔化温度，熔融温度
melting zone　熔化带，熔化层
membrane　膜，薄膜，羊皮纸
membrane biochemical reactor　膜生化反应器
membrane module　膜组件
memory function　记忆函数
mercaptan　硫醇
mercury removal system　脱汞系统
mesh type porous absorber　网型多孔吸热器
meshing interference　啮合干涉
mesophilic anaerobic digestion　适温厌氧消化
mesophilic anaerobic sludge digestion　适温厌氧污泥消化
mesophilic digester　中温发酵池，中温消化池
mesophilic organism　适温性生物
mesophilic temperature optimum　最佳嗜热温度
metal box duct absorber plate　金属盒通道吸热板
metal fatigue　金属疲劳
metal fibre　金属纤维
metal halogen lamp　金属卤化物灯
metal hydride　金属氢化物
metal organic chemical vapor deposition　有机金属化学气相沉积法

metal oxide semiconductor field effect transistor　金属氧化物半导体场效应晶体管
metal shaft　金属轴
metal-inert gas welding　金属惰性气体焊接
metal-insulator-N/P silicon (MINP)　金属—绝缘体—N型/P型硅
metal-insulator-N/P junction solar cell　金属—绝缘体—N型/P型硅太阳电池
metal-insulator-semiconductor solar cell　金属—绝缘体—半导体太阳电池
metal-insulator-silicon　金属绝缘硅
metallic bond　金属键
metallic component　金属部件
metallic lithium　金属锂
metallic salt-hydrate　金属盐水合物
metallic scrap　金属废料，废金属
metallurgical coke　冶金焦
metallurgical grade silicon　冶金级硅
metallurgy　冶金，冶金术
metamorphic rock　变质岩，烧变岩
metamorphic water　变质水
metastable phase　介稳态过渡相
meteoric water　大气降水
meteorological　气象的
meteorological condition　气象条件
meteorological element　气象要素
meteorological elements over the years　历年气象要素
meteorological essential　气象要素
meteorological forecast　气象预报
meteorological institution　气象站
meteorological mast　气象塔
meteorological observatory　气象站，气象观测台
meteorological optical range　气象视距

meteorological potential 气象潜势
meteorological station 气象站
meteorological system 气象系统
meteorological tide 气象潮
meteorology 气象学，气象条件
meteorology conditions for design 设计气象条件
methane 甲烷，沼气
methane bacterium 甲烷菌
methane cooker 沼气灶
methane fermentation 甲烷发酵，沼气发酵
methane gas 甲烷气
methane gas reservoir 甲烷气柜
methane hydrate 甲烷水合物
methane indicator 甲烷指示器
methane ionization chamber 甲烷离子化室
methane layering 甲烷分层
methane liberation 甲烷放出，沼气释放
methane monitoring system 甲烷监控系统
methane outburst 甲烷喷出，沼气喷出
methane potential 甲烷势
methane recorder 甲烷记录器
methane rich gas 富含甲烷气
methane tanker 甲烷运输船
methane tester 甲烷测定器
methane-oxidizing bacteria 甲烷厌氧菌
methane-producing bacteria 甲烷产菌
methane-producing digester 甲烷煮解器
methane-rich gas 富含甲烷气
methanide 甲烷化物
methanobacillus omelianskii 奥氏甲烷芽孢杆菌
methanogenesis 甲烷生成
methanogenic bacteria 产甲烷菌
methanol 甲醇
methanol-air battery 甲醇—空气电池
method of bins 比恩法，分组方法
method of chemical explosive stimulation 化学爆炸激励法
method of Ellis and Ritchie 埃利斯—里奇法（进行碱度测定的一种方法）
methods for modelling and analysis 建模和分析方法
methyl tertiary butyl ether 甲基叔丁基醚
metres per second (m/s) 米/秒
metric ton 公吨
mica 云母
Michelson bimetallic pyrheliometer 迈克尔森双金属直接日射计
micro grid 微型电网，微网
micro wind power system 微型风力发电系统
micro-ammeter 微型安培计
microbial decomposition 微生物分解
microbial ecology 微生物生态学
microbial formation 微生物形成
microbial fuel cell 微生物燃料电池
microbiocell 微生物电池
microbiofouling rate 微生物腐烂速率，微生物结垢速率
microbiological cell 微生物电池
micro-CHP system 微热电联产系统
microcomputer control 微机控制
microcosm 微型生态系统
micro-crystalline silicon 微多晶硅
microearthquake 微震
microelectronic circuit 微电子电路
microfiltration 微滤
micro-fuel cell 微燃料电池

microgrid (μGrid)　微电网
micro-hydro plant　微型水电站
micromorph　非微晶叠层
micron　微米
micro-organism　微生物
microorganism consortium　微生物组合
microplankton　小型浮游生物
microseismicity　微震活动性
micrositing　微观选址
microstructure　显微结构，微观结构
microtidal estuary　弱潮河口
microturbine　微型燃气轮机
microwave current　微波电流
microwave energy　微波能
microwave photo conductive decay　微波光电导衰减仪
microwave power transmission　微波能（量）传送
microwave radiation　微波辐射
microwave radio　微波无线电，微波通信，数位微波传输设备
microwave radiometry　微波辐射测量
microwave survey　微波测量（用于地热勘探的一种地球物理方法）
microwave transmission of solar energy　太阳能微波传输
mid-ocean　海洋中部的，海中央
midshot　水流冲击（潮轮）中部而转动的
migratory fish　洄游鱼类
mild-steel welding　低碳钢焊接
Million Solar Roof Initiative　美国"百万太阳能屋顶计划"
million tons of oil equivalent　百万公吨当量
millisecond　毫秒
mimic photosynthesis　模拟光合（作用）
mineral assemblage　矿物组合，矿物共生

mineral composition　矿物成分
mineral deposition　矿物质沉积
mineral dissolution　矿物溶解
mineral extraction　采矿，矿物提取
mineral impurity　矿物杂质
mineral particle　矿物颗粒
mineral precipitation　矿物质析出
mineral recovery　矿产回收
mineral-free dry steam　无矿物干蒸汽
mineralogical analysis　矿物分析
mineralogical composition　矿物组成
Minerals Management Service (MMS)　美国矿产管理局
minicomputer program control　微机程序控制
minimum angle of incidence　最小入射角
minimum ebb　最小落潮流
minimum excess air　最低过量空气
minimum flood　最小涨潮流
minimum load　最小负荷
minimum loading of power plant　发电厂最小出力
minimum operating plan　最小运行方式
minimum operating pool　最小运行库
minimum value of wind direction　风向最小值
minimum value of wind speed　风速最小值
minimum α control　最小触发角控制
minimum γ control　最小关断角控制
mini-OTEC plant　小型海洋热能转换电站
mini-tidal power station　小型潮汐电站
minority carrier　少数载流子
minority carrier lifetime　少子寿命

minority charge carrier 少数电荷载流子
minus sign 减号
mirror 镜，镜子，反向镜
mirror cleanliness 镜面清洁度
mirror collector 反射集热器
mirror concentrator 反射镜聚光器
mirror Fresnel collector 菲涅尔反射镜集热器
miscanthus 芒草，芒属，芒属植物
misdirected solar radiation 错瞄太阳能辐射，不直接的太阳能辐射
misplaced atom 误位原子
mist eliminator 除雾器
mist lift cycle 雾滴提升循环
mitigation technique 减震技术
mixed current 混合潮流
mixed cycle system 混合循环系统
mixed dislocation 混合位错
mixed fluid 混合流体
mixed integer linear programming 混合整数线性规划
mixed liquor suspended solid 混合液悬浮固体
mixed liquor volatile suspended solid 混合液挥发性悬浮固体
mixed tide 混合潮
mixed-flow storage pump 混流式蓄能泵
mixing 混合
mixing chamber 混合室，预燃室
mixing layer 混合层
mixing tank 混合槽，配料仓
mixing valve 混合阀
mobile desalination plant 移动式海水淡化装置
mobile electro-generating unit 移动式发电装置
mobile emergency power supply 移动式应急电源
mobile generator set 移动式电机组
mobile rig 轻便钻机
mobilization 活化作用
modal vibration analysis 振动模态分析
mode controller 模式控制器
mode of frequency regulation 调频方式
mode of operation 运行工况，运行方式
mode shape 振型
mode speed 模式速度
model development 模型开发
model linearization 模型线性化
model predictive control 模型预测控制
modem 调制解调器
moderate breeze 四级风，和风
moderate gale 七级风，疾风
modern wind turbine 现代风机
modified siemens process 改良西门子法
modified yeast 改性发酵剂
modular 组装式的
modular two-stage combustion 模块化两级燃烧
modularity 模块化
module 模数
module efficiency 组件效率
module surface temperature 组件表面温度
modulus of elasticity 弹性模量
moisture absorption 吸湿，吸湿性
moisture analysis sample 水分分析样品

moisture capacity 湿度，含湿量
moisture carrying capacity 携带水分能力
moisture separator 水分分离器
moisture content 含水量
moisture factor 湿润因子
moisture free 无水的，干的
moisture in the air dry basis 空气干燥基水分
moisture in the general analysis sample 一般分析试样水分
moisture index 湿润指数
moisture indicator 湿度指示器
moisture loss 水分损失
moisture meter 湿度计，湿度表
moisture permeability 透湿度
moisture removal 除湿
moisture resistance 防潮性，耐湿性
moisture resistant cell 抗（耐）湿电池
moisture ton 湿吨
moisture-laden gas 含水煤气
moistureproof 防湿，防潮，耐湿
mol concentration 摩尔浓度
molar free energy 摩尔自由能
molar ratio 摩尔比率
molar refraction 摩尔折射（度）
molar rotation 摩尔旋光（度）
molar surface energy 摩尔表面能
molecular beam mass spectrometry 分子束质谱技术
molecular crystal 分子晶体
molecular energy 分子能
molecular heat 分子热
molecular pump 分子泵
molecular scattering 分子散射
molecular sieve 分子筛
molecular weight 分子重量
molecule 分子
molecule weight 分子量
Moll-Gorchinski pyrheliometer 莫尔—戈尔钦斯基直接日射强度计，莫尔—戈尔钦斯基总日射表
molten 熔化的，熔融的
molten bath 熔池
molten carbonate fuel cell 熔融碳酸盐燃料电池
molten salt 熔融盐
molten salt capacity 熔盐容量
molten salt convertor reactor 熔盐转换堆
molten salt heat storage tank 熔盐蓄热箱
molten salt receiver 熔融盐吸热器，熔盐接收器，熔盐吸热器
molten salt receiver inlet vessel 进口冷盐罐
molten salt receiver outlet vessel 出口热盐罐
molten silicon 熔融硅
molybdenum 钼
molybdenum chloride 二氯化钼
molybdenum oxide 氧化钼
molybdenum tetrabromide 四溴化钼
molybdenum trioxide 三氧化钼
moment equation 力矩方程
moment of couple 力偶矩
moment of inertia 惯性矩
moment of momentum 角动量，动量矩
moment of rotation 转矩
moment of stability 稳性矩
moment of torsion 扭矩
moment (torque) coefficient 力（扭）矩系数

momentary duty 瞬时负荷	monocrystalline wafer 单晶硅片
momentary load 瞬时负荷，短时间负荷	monolithic refractory 不定形耐火材料
momentary maximum wind velocity 瞬时最大风速	monorail system 单轨系统，单轨运输系统
momentary power 瞬时功率	monsoon 季节风
momentum 动量	monsoon circulation 季风环流
momentum boundary layer 动量边界层	monsoon climate 季风气候
	monsoon precipitation 季风降水（量）
momentum conservation 动量守恒	monsoon weather pattern 季风气候
momentum deficit 动量亏损	montmorillonite 蒙脱土
momentum space 动量空间	moored float 系泊浮体
momentum transfer 动量传递	moored vessel 系泊船舶，锚泊浮体
momentum-blade element theory 动量—叶素理论	mooring 系泊用具，停泊处，系泊区
	mooring cable 系缆，系留索，系泊链，系泊缆绳
monitor 监视器，监控器	mooring force 系泊力，船舶系缆力，系船牵引力
monitor photoelectric cell 监控光电管	
monitored information 监视信息	mooring plant 系泊设备
monitoring node 监控点	mooring system 锚泊系统
monitoring well 监测井，检测井	mortar 灰泥，迫击炮
monochlorobenzene 一氯代苯（太阳热间接转换为电能的一种工质）	motor drive 电动机
	motor protection 电机保护
monochromatic absorptivity 单色吸收率	motor synchronizing 自同步，电动机同步
monochromatic emissive power 单色发射能力	motor terminal 电机接线盒
	motor valve 电动阀
monochromatic emissivity 单色发射率	motor winding 电机绕组
monochromatic radiation 单色光辐射	mountain pass 山口，峪
monochromatic response 单色光响应	mountain wind 山风
monochromatic wave 单色波	mountain-valley breeze 山谷风
monocrystal silicon 单晶硅	movable focusing collector 跟踪式聚焦集热器，移动式聚焦集热器
monocrystalline block 单晶硅方棒	
monocrystalline cell 单晶电池片	movable solar collector 移动式太阳能集热器
monocrystalline ingot 单晶硅棒	
monocrystalline silicon 单晶硅	movable solar medication bath equipment 移动式太阳能药浴设备
mono crystalline solar cell 单晶硅太阳能电池	
	moving air 气流，流动空气

moving average 移动平均数
moving charge 移动电荷
moving contact 动触头
moving grate 活动炉栅
moving grate incinerator 移动炉排焚烧炉
moving slider 滑块
moving surface water 移动表面水
moving target concentrator 动焦聚光器
moving-bed gasification reactor 移动床气化反应器
moving-cable 循环索系
mud agitator 泥浆搅拌器
mud circulation 泥浆循环
mud drilling 泥浆钻井
mud-gas separator 泥气分离器
mud line 泥浆管线
mud logger 泥浆测井仪
mud motor 井下动力钻具
mud pump 泥浆泵
mud tank 泥浆罐
muffling technique 消声技术,降噪技术
multi-antireflection coating 多层减反射膜
multiband photography 多波段摄影
multi-bandgap a-si solar cell 多带隙非晶硅太阳电池
multiblade windmill 多叶风车
multi-blade rotor 多锋面转子
multibrid technology wind turbine generator system 半直驱式风电机组
multicrystal silicon 多晶硅
multi-crystalline 多晶的
multicrystalline silicon 多晶硅
multicrystalline silicon ingot 多晶硅锭

multicrystalline wafer 多晶硅硅片
multi-cyclone 复式旋风器,多管旋风分离器
multi-drive clipper 多级旋风分离器
multi-effect distillation 多效蒸馏,多效蒸馏法
multi-function sensor 多功能传感器
multi-function transducer 多功能传感器
multifunctional bioenergy system (MBS) 多功能生物能源系统
multi-jet turbine 多喷嘴冲击式水轮机
multi-junction cell 多结叠层太阳电池
multijunction solar cell 多结太阳电池
multijunction space solar cell 空间用多节太阳能电池
multi-kilowatt solar array 多千瓦太阳能阵列
multilayer insulant 多层绝热材料,多层保温材料
multi-layer insulation system 多重隔热系统
multi-layer solar battery 多层太阳能电池
multi-layer solar cell 多层太阳能电池
multi-layer thermal insulation system 多层保温系统,多层热绝缘系统
multilevel converter 多电平变换器
multimeter 多用表
multiphase flow 多相流
multiple basin project 多库方案
multiple cyclone 多管旋风分离器,多管式旋风分离器
multiple effect solar still 多效太阳能蒸馏器
multiple hearth incinerator 多炉膛焚烧炉

multiple heliostat furnace 多级定日镜式太阳能炉
multiple junction cell 多结电池
multiple rotor 多转子
multiple tower wind farm 多塔风电场
multiple tower 多层塔
multiple transparent cover absorber system 多层透明盖板吸热器（板）系统
multiple well system 多井系统
multiple wellhead 多油管井口装置
multiple wind turbine 多风力机机组
multiple-effect rotating-disk wiped-film distiller （太阳能辅助的）多效转盘涂膜蒸馏器
multiple-effect solar still 多效太阳能蒸馏器
multiple-stage planetary gear train 多级行星齿轮系
multiple-tank storage 多箱蓄热，多箱储能
multiplexor 多路器，多路复用器
multipole wound SG 多极绕线屏栅极
multi-port injection 多点喷射
multipurpose development 多目标开发（综合利用）工程
multipurpose transmission line 多用途输电线
multirotor 多元转子，多重转子
multistage adjustable vane propeller pump 多级可调叶轮推进泵
multistage cementing 多级注水泥
multistage digester 多级消化池
multistage fermentation system 多级发酵系统
multistage pump 多级泵
multistage receiver 多级式吸热器
multi-transition solar cell 多转换太阳能电池
municipal refuse 城市垃圾
municipal refuse recycling 城市垃圾循环（利用）
municipal solid waste 城市固体废物
municipal solid waste power generation 城市固体垃圾发电
municipal sorted waste 城市分类垃圾
municipal waste 城市垃圾，城市废物，城市废料
municipal waste disposal 城市废物处理
municipal waste disposal plant 城市垃圾处理厂
municipal water supply 城市供水，城市给水
mutual diffusion 相互扩散
mutual magnetic flux 互感磁通
Myklestad method 莫克来斯塔德法

N

nacelle 机舱
nacelle cover 机舱罩
nacelle machine bed 引擎机床
nacelle yaw 机舱偏航
nadir 天底
nanofiltration 纳滤，纳米过滤
nano-particle 纳米粒子
nanotechnology 纳米技术
naphtha 石脑油，挥发油，粗汽油
narrow gap semiconductor 窄带隙半导体
Natick furnace 纳缔克式太阳能炉
National Electrical Code 美国国家电气规程
National Electrical Safety Code (NESC) 美国国家电气安全代码

National Energy Modeling System 美国国家能源建模系统
National Solar Radiation Data Base 美国国家太阳辐射数据库
national wind resource 国家风力资源
Na-to-steam heat exchanger 钠一汽热交换器
natural ash 固有灰分
natural boundary layer wind 自然界边界层风
natural circulation 自然循环
natural circulation system 自然循环系统
natural clamped converter 自然箝位变换器
natural cooling 自然冷却
natural distribution of air flow 风量自然分配
natural draft cooling tower 自然通风冷却塔
natural draught 自然通风
natural drying 自然干燥
natural environment 自然环境
natural fracture 天然裂缝
natural frequency 固有频率,自然频率
natural gas cogeneration 天然气热电联供系统
natural gas combined cycle 天然气联合循环（发电）
natural gas hydrate 天然气水合物
natural gas vehicle 天然气车辆
natural gas-fired reciprocating engine 天然气往复式发动机
natural geothermal condition 天然地热条件
natural heat flow 天然热流量法（估计地热田能量潜力的一种方法）
natural heat output 天然热输出
natural heat release 天然放热量
natural heat resource 天然热资源
natural heat source 天然热源
natural heated groundwater 自然变热的地下水
natural hot spring 天然热泉
natural hydrothermal reservoir 天然热液储层
natural outflow of heat 天然热流出
natural steam 天然蒸汽
natural steam field 天然蒸汽田
natural water 天然水
natural water level 天然水位
natural wind 自然风
natural wind boundary layer 自然风边界层
natural wind environment 自然风环境
natural wind field 自然风场
natural wind gust 自然阵风
nature fissure 天然裂缝
nautical mile 海里
naval 海洋的
neap range 小潮潮差
neap rise 小潮升
neap tidal current 小潮潮流
neap tide 小潮
near field 近场,近区
near gale 七级风,疾风
near surface groundwater 近地表地下水
near surface wind 近地风
near wake 近尾流
near-ambient pressure operation 近环境压力操作
nearshore 近岸的
nearshore circulation 近岸环流

nearshore current 近岸流
nearshore device 近岸水域装置
nearshore ocean thermal power plant 近岸海洋热能电站
nearshore waters 近岸水域
nearshore zone 近滨带
near-surface rock 近地表岩石
near-surface wind 近地表风
needle probe 针形探头
negative charge 负电荷
negative current 负电流
negative electrode 负极, 阴极
negative electrode plate 负电极板
negative emission 负排放
negative energy 负能量
negative polarity terminal 负极性终端
negative pressure 负压
negative torque 负扭矩
negative voltage 负电压
negligible emission 可忽略排放
N-endowment N掺杂
net calorific value 净热值
net charge 净电荷
net collection efficiency 净集热效率
net coupling 联网（太阳能发电装置）
net effective area of collector 集热器净有效面积
net electric power output 净电功率输出
net energy 纯能量
net energy gain 净能量增益
net energy metering 净能计量
net energy output 净能量产出, 净输出电量
net energy ratio 净能比
net gain of heat 净热增益
net generating station capability 发电站净出力

net generation 净发电量
net head 净水头
net heat 净热
net heat exchanger 净热交换器
net heat gain 净热增益
net heat rate 净热耗率
net longwave radiation 净长波辐射
net metering 电力回馈, 净电量计量法
net power 净功率
net present value 净现值
net pyrradiometer 净（全）辐射表
net radiation flux 净辐射通量
net reference thermal load 净参考热负荷
net shortwave radiation 净短波辐射
net solar radiation 净太阳辐射
net space charge 净空间电荷
net thermal efficiency 净热效率
net thermal power of receiver 吸热器净热功率
net torque 净力矩
net total radiation 净（全）辐射
net weight 净重
network connection point 电网连接点
network impedance phase angle 电网阻抗相角
network splitting 电网解列
neutral air cell 中性空气电池
neutral atom 中性原子
neutral axis 中性轴
neutral conductor 中性（导）线, 中线
neutral current 中线电流
neutral point 中性点
neutral turbulent flow 中性湍流流动
neutralizing agent 中和剂
neutralizing voltage 中和电压
neutron 中子

neutron absorber 中子吸收剂
neutron absorbing material 中子吸收材料
neutron absorption 中子吸收
neutron binding energy 中子结合能
neutron capture 中子俘获
neutron current 中子流
neutron diffraction 中子衍射
new energy conservation technology 节能新技术
new energy development 新能源开发
new energy resource 新能源资源
new energy source 新能源
new wave-energy converter 新型波能转换装置
Newton-meter 焦耳
Newton's Second Law 牛顿第二定律
NiCd battery (Nickel-cadmium battery) 镍镉电池
nickel alloy 镍合金
nickel hydroxide 氢氧化镍
nickel-metal hydride battery 镍氢电池
nighttime wind speed 夜间风速
nitrogen oxide 氧化氮，氮氧化物
nitrogen source 氮源
nitrogen support system 氮保障系统，氮维持系统
nitrogen to phosphorus to potassium ratio 氮磷钾比例
noble gas 惰性气体
nodding duck 点头鸭式（一种波能发电装置）
nodding duck wave energy extractor 点头鸭式波能提取装置
nodding duck wave power generator 点头鸭式波能发电机
nodding-vane pump system 点头叶轮泵系统
node 波节
noise control 噪声控制
noise current 噪声电流
noise energy 噪声能量
noise level 噪声级，噪声电平
noise muffling equipment 消声设备
noise of wind power plant 风电场噪声
noise reduction 噪声抑制
noise shield 噪声屏蔽
noise voltage 噪声电压
no-load 空载，空转
no-load consumption 空载消耗，空转消耗
no-load loss 空载损失
no-load operation 空载运行
no-load power consumption 空载功耗
no-load run 空载运行
no-load speed 空载转速
nominal cross-section 公称横截面
nominal load 额定负载，额定荷载，标称负载
nominal load-bearing capacity 额定承载力
nominal module voltage 组件额定电压
nominal operating cell temperature 电池额定工作温度
nominal power 标称功率
nominal rating 标称定律，额定值
nominal top size 标称最大粒度
nominal voltage 额定电压
nominal wind speed 额定风速
nominal working pressure 额定工作压力
non polluting sea wave energy 无污染海洋波能
non-agitated fuel pile 非搅拌燃料堆

non-biogenic waste 非有机垃圾
noncoherent boundary 非共格界面
noncombustible 非可燃物
noncombustible material 不燃材料
non-combustible waste 非燃性废料，非燃性垃圾
non-commercial energy 非商品能源
non-concentrating collector 非聚光型集热器
non-condensable gas 不可凝气体，不凝结气体
noncondensible gas content 不凝气体含量
non-conducting voltage 截止电压
non-continuous source 非连续能源
non-convective pond 无对流太阳能池
non-crystal magnetic material 非晶态磁性材料
non-crystal solar cell 非晶态太阳电池
non-crystalline semiconductor material 非晶态半导体材料
non-crystalline silicon 非晶硅
non-crystalline silicon solar cell 非晶硅太阳电池
nondeviated absorption 非折射吸收
non-dimensional parameter 无量纲参数，无因次参数
non-dimensional power coefficient 无量纲功率系数，无因次功率系数
non-dimensional wind shear 无量纲风切变
nonelectronic solar tracker 非电子太阳能跟踪机
nonflammable 不可燃的
nonfocusing collector 不聚焦集热器
non-food biomass 非食物生物质

nonfossil fuel 非矿物燃料
non-hydro renewable energy 非水能可再生能源
non-imaging collector 非成像集热器
non-imaging concentrator 非成像聚光器
non-linear aerodynamics 非线性空气动力学
non-linear propagation 非线性传播
non-linear wave interaction 非线性波相互作用
non-load current 空载电流
non-luminous flame 无光焰
non-magnetic material 非磁性材料
non-mechanical separation process 非机械分离工艺
non-methane volatile organic compound 非甲烷挥发性有机化合物
non-Newtonian fluid 非牛顿流体
nonpolluting source of energy 无污染能源
non-pressurized solar water heater 不加压太阳能热水器
non-renewable energy source 不可再生能源
nonselective surface 非选择性表面
non-sinusoidal alternating voltage 非正弦交流电压
nonsolar auxiliary 非太阳能辅助设备
nonsolar auxiliary system 非太阳能辅助设备系统
nonsolar energy 非太阳能
nonsolar energy cost 非太阳能成本
nonsolar energy source 非太阳能源
nonsolar energy subsystem 非太阳能子系统
nonsolar heating subsystem 非太阳能加热子系统

nonsolar system 非太阳能系统
non-specular conductor 非镜面导体
nonstoichiometric compound 非化学计量化合物，非整比化合物
non-thermal energy 非热能
non-tracking concentrator 非跟踪聚光器
nontracking CPC 非跟踪式复合抛物面集热器
non-tracking photovoltaic concentrator 非（无）跟踪太阳能聚光器
non-transferred arc 非转移弧
non-uniform distribution 不均匀分布
non-uniform ground fluid 非均匀地下水
non-uniform rotor blade 变速叶轮叶片
non-working flank 非工作齿面
nonzero lift coefficient 非零升力系数
normal (Gaussian) distribution 正态分布
normal braking system 正常制动系
normal condition 正常状态
normal current 正常电流
normal direct solar irradiance 法向直（接日）射辐照度
normal duty 正常负荷
normal fault zone 正断层带
normal force 正向力，法向力
normal geothermal gradient 正常地热梯度
normal geothermal heatflow 正常地热热流量
normal ground water 正常地面水
normal hydrogen eletrode 标准氢电极
normal incidence calibrating method 准直标定法
normal incidence pyrheliometer 直接辐射计

normal incidence pyroheliometer 垂直入射太阳能热量计
normal load 额定负荷
normal maintenance operating plan 正常检修运行方式
normal operation 正常运行
normal pitch 法向齿距
normal probability density function 正态概率密度函数
normal radiation 法向辐射
normal shutdown 正常关机
normal temperature and pressure 常温常压
normal terrestrial heat flow 正常大地热流量
normal value for terrestrial heat flow 大地热流量正常值
normal wind 正常风
normal-plate anemometer 垂直叶板风速计
north quaternity pattern 北方"四位一体"模式
Norton's equivalent circuit 诺顿电路
nose 导流罩
nose cone 整流罩
not in my backyard 邻避效应，邻避主义
nozzle 喷嘴
nozzle of mud drilling bit 泥浆钻头喷嘴
nozzle velocity 喷嘴速度
NS tracking collector 南北向跟踪集热器
N-type base N型基
N-type semiconductor N型半导体，电子型半导体
nuclear electricity 核电

nuclear electron 核电子
nuclear energy 核能
nuclear energy gap 核能隙
nuclear energy level 核能极
nuclear energy plant 核能装置，核电站
nuclear energy surface 核能面
nuclear engine 核发动机
nuclear engineering 核工程
nuclear excitation 核激发
nuclear excitation energy 核激发能
nuclear fission 核裂变，核子分裂
nuclear fission chain reaction 核裂变链式反应
nuclear force 核力
nuclear fuel 核燃料
nuclear fuel cycle 核燃料循环
nuclear fuel resource 核燃料资源
nuclear fuelled boiler 核锅炉
nuclear furnace 核反应堆
nuclear fusion 核聚变，核子融合
nuclear grade 核级
nuclear heat 核热量
nuclear installation 核设施
nuclear island 核岛，核厂房及设备
nuclear level 核能级
nuclear level density 核能级密度
nuclear level spacing 核能级间距
nuclear magnetic resonance thermometer 核磁共振温度计
nuclear material 核材料
nuclear material control 核材料控制
nuclear material control system 核材料控制系统
nuclear matter 核物质
nuclear particle 核粒子
nuclear photoeffect 核光电效应
nuclear photoelectric effect 核光电效应
nuclear potential energy 核势能，核位能
nuclear power 核动力
nuclear power complex 核动力装置
nuclear power industry 核能工业，原子能工业
nuclear power plant 核动力站，核站
nuclear power production 核能生产
nuclear power station 核电站，核动力站
nuclear power unit 核动力装置
nuclear reaction energy 核反应能
nuclear reactor 核反应堆
nuclear reactor safety 核反应堆安全性
nuclear rotation band 核转动能带
nuclear rotational energy 核转动能
nuclear rotational level 核转动能级
nuclear safety criterion 核安全准则，核安全标准
nuclear safety limit 核安全限额，临界安全限额
nuclear siting policy 核电站选址方针
nuclear steam electric plant 核电站
nuclear thermohydrodynamic instability 核—热流体动力不稳定性
nucleation energy 形核功
nucleation rate 形核率
nucleonic 核子的
nucleonic component 核子组成，核子成分
nucleonic configuration 核子位形
nucleonic excitation 核子激发
nucleonic motion 核子运动
nucleonic separation energy 核子分离能

nucleon-nucleon collision 核子—核子碰撞
nucleon-nucleon force 核子—核子力
nucleon-nucleon scattering 核子—核子散射
nucleus 晶核
nucleus growth 晶核生长
number of blade 叶片数
number of sunshine hours 日照时数
number of teeth 齿数
numerical modelling 数字模型,数字建模
numerical simulation 数值模拟
nutrient content 养分含量,营养成分

O

objective function value 目标函数值
oblique flow 斜流
oblique ray 倾斜光线
oblique wind 斜向风
obliquity 倾斜,斜度
obliquity of the ecliptic 黄赤交角
observation boat 观测船,测量船
observation buoy system 观测用浮标系统
observation desk 观测台,试验台
observation door 观测孔
observed current 实测流
observed maximum wind velocity 实测最大风速
observed wind speed 观测风速
obstacle 障碍物
obstruction height 壅高(海侧流通口处的水位与实际潮位的差值,是正向发电时的水头损失)
obtuse angle 钝角
ocean current 海流,洋流

ocean alternative energy 海洋新能源,海洋替代能源
ocean basin 海洋盆地
ocean bed 洋底,洋床
ocean circulation 海洋环流,大洋环流
ocean climate 海洋性气候
ocean cold-water-mass 海洋冷水团
ocean current 海流
ocean current energy 海流能
ocean current generation 海流发电
ocean current power generation 海流发电
ocean current system 大洋流系
ocean data station 海洋资料站
ocean data station buoy 海洋资料站浮标
ocean data transmitter 海洋资料发送器
ocean depot 海洋油库
ocean development industry 海洋开发产业
ocean disposal 海洋处置
ocean dumping 海洋倾(倒)废(物)
ocean dynamics 海洋动力学
ocean energy 海洋能
ocean energy conversion factor 海洋能转换因子
ocean energy conversion technique 海洋能转换技术
ocean energy converter 海洋能转换器
ocean energy development 海洋能开发
ocean energy exploitation 海洋能源开发
ocean energy farm 海洋能地段
ocean energy from salinity gradient 海洋盐差能

ocean energy source 海洋能源
ocean energy system 海洋能系统
ocean engineering 海洋工程
ocean engineering system development 海洋工程系统开发，海洋工程系统研制
ocean environment 海洋环境
ocean environmental energy source 海洋环境能源
ocean exploitation 海洋开发
ocean farm 海洋农场
ocean floor 海底，洋底
ocean floor drilling 海底钻井
ocean floor-ocean interface 海底—海洋界面
ocean gyre 海洋环流圈
ocean heat flow 海洋热流
ocean level 海洋水位
ocean loading strain tide 海洋负荷应变潮汐
ocean outfall 入海河口
ocean prediction center 海洋预测中心
ocean resource 海洋资源
ocean resource development 海洋资源开发
ocean scarp 海洋悬崖
ocean sediment 海洋沉积物
ocean sited power plant 海上动力站，海上发电站
ocean surface 海洋表面
ocean surface current 洋面海流
ocean surface temperature 洋面温度
ocean surface wave 海洋表面波
ocean swell 大洋涌浪，海洋涌浪
ocean temperature 海洋温度
ocean temperature difference 海洋温差
ocean temperature energy circulation 海洋温差发电
ocean temperature gradient system 海洋温差系统
ocean (sea) temperature gradient power station 海水温差电站
ocean thermal conversion 海洋热转换
ocean thermal difference 海洋温差
ocean thermal difference power plant 海洋温差电站
ocean thermal electric conversion 海洋热转换
ocean thermal energy 海洋热能，海洋温差能
ocean thermal energy conversion 海洋热能转换
ocean thermal energy conversion 海洋热能转换
ocean thermal energy conversion power plant 海洋热能转换发电站
ocean thermal energy conversion power system 海水温差发电系统
ocean thermal energy conversion unit 海洋热能转换装置
ocean thermal gradient 海洋热梯度，海洋温差
ocean thermal gradient electricity generating system 海洋温差发电系统
ocean thermal gradient energy 海洋温差能
ocean thermal gradient power generation 海洋温差发电
ocean thermal gradient system 海洋温差系统，海洋温差装置
ocean thermal gradient 海洋热梯度
ocean thermal power 海洋热能
ocean thermal power generation 海洋温差发电

ocean thermal power plant 海洋热能发电厂
ocean thermal resource 海洋热能资源
ocean thermocline 海洋斜温层,海洋跃层
ocean tidal movement 海洋潮汐动量
ocean tidal power 海洋潮汐能
ocean tide 海洋潮汐
ocean turbine system 海洋涡轮系统
ocean underwater environment 海洋水下环境
ocean volume 海洋水域
ocean water 海水
ocean wave 海浪,海洋波浪
ocean wave electric generator 海洋波浪发电机
ocean wave energy 海洋波浪能
ocean wave energy absorbing structure 海洋波能吸收结构物
ocean wave energy conversion 海洋波能转换
ocean wave energy resource 海洋波浪能资源
ocean wave energy utilization 海洋波浪能利用
ocean wave field 洋浪场
ocean wave lens 海洋波透镜
ocean wave mixing 海水混合
ocean wave power 海洋波浪能
ocean wave power utilization 海洋波能利用
ocean weather station 海洋气象站
ocean wind 洋风
ocean wind power system 海上风能系统
ocean-based solar plant 海底太阳能电站
ocean-current 海流
ocean-current energy 海流能
oceaneering 海洋工程
ocean-going heaving buoy 远洋垂直浮标
oceanic climate 海洋性气候
oceanic crust 海洋地壳
oceanic current 洋流
oceanic cycle 海洋循环
oceanic delta 海洋三角洲
oceanic general circulation model 海洋环流模式
oceanic gyre 海洋环流
oceanic hydrothermal system 海洋热液系统
oceanic lithosphere 海洋地壳
oceanic stratosphere 海洋同温层
oceanic thermal gradient 海洋热度梯
oceanic tide 海洋潮汐
oceanic troposphere 海洋对流层
oceanic turbulence 大洋湍流,海洋湍流
oceanic vessel 海洋船
oceanic wind power 海洋风力
oceanic wind power resource 海洋风力资源
oceanity 海洋度
oceanodromous migration 海洋洄游
oceanographer 海洋学家
oceanographic (al) current meter 海流计,海洋流速仪
oceanographic international total organic carbon analyzer 海洋学国际总有机碳分析计
oceanographic research buoy 海洋研究浮标
oceanographic research ship 海洋调查船

oceanography 海洋学
oceanology 海洋学,海洋开发技术
oceanophysics 海洋物理学
oceanward 向洋,向洋侧,向洋方
ocean-wave field 洋浪场
ocean-wind system 海洋能风系统
octane rating 辛烷值
odd function 奇函数
Odeillo furnace 奥德罗太阳能炉
off leakage current 关态漏电流
off peak energy 基荷电能
off peak load 非高峰荷载
off standard voltage 非标准电压
off-axis wind 偏轴风
off-design operation 非设计工况操作
off-design performance 非设计性能
off-gas 尾气,废气
off-gas pump 抽气泵
off-grid electricity 离网电力
off-grid electricity generation 离网发电
off-grid home power system 独立家庭电源系统
off-grid PV system 独立光伏发电系统
off-grid wind power generator 离网型风力发电机
off-grid wind-electric system 离网风电系统
off-line method 脱机法
off-load period 卸荷期,非载荷期
off-loading 卸荷,卸料
off-peak 非尖峰的
off-peak electricity 避峰用电,非高峰用电
off-sea wind 离海风
offset current 失调电流
offset voltage 补偿电压,失调电压
offshore application 海上应用

offshore aquaculture 海洋养殖
offshore area 近海区
offshore breeze 离岸风,海风
offshore device 离岸装置,海上设备
offshore discharge 近海排放
offshore drilling 近海钻井
off-shore oil rig 近海石油平台,海上平台
offshore platform 海上平台
offshore structure 近海结构,海域结构物,离岸工程结构物
offshore substation platform 海上变电站(平台)
offshore wind 离岸风
offshore wind energy 海上风能
offshore wind farm 近海风电场
offshore wind power 海上风力发电,海上风电
offshore wind power plant 近海风力发电装置
offshore wind power project 海上风力发电工程
offshore wind power system 海上风力发电系统,近海风能系统
offshore wind shear 离岸风切变
offshore wind station 近海风力发电站
offshore wind-energy conversion system 离岸风能转换系统
ohm loss 欧姆损耗
ohmic contact 欧姆接触
ohmic heating 欧姆加热
ohmic loss 欧姆损耗
ohms 欧姆
oil and gas co-production 油气联产
oil burner 油燃烧器
oil consumption 油耗量
oil controlled valve 液动阀,油动阀
oil cooler 油冷却器,机油冷却器

oil crop 油料作物
oil emulsion 油乳胶
oil extraction 油料提取
oil fuel trip 油燃料跳闸
oil furnace 油炉
oil head 受油器
oil lance 油喷枪
oil palm 油棕榈
oil seal 油封
oil streak 油纹
oilfield 油田
oilfield brines 油田卤水
oil-fired furnace 燃油锅炉，燃油炉
oil-immersed type transformer 油浸式变压器
oil-in-place 地质储量
oilseed crop 油料作物
oldham coupling 滑块连接
olefin 烯烃
oleo-duct 输油管道，输油管线
olive residue 橄榄残渣
olivine 橄榄石
omni directional wind 全向风
on sun 一个太阳值（自然太阳辐照的最大值）
on-board power electronics 车载电力电子设备，车载电力电子技术
once-through boiler 直流锅炉
once-through circulation 直流流动
once-through cooling 直流冷却
once-through system 直流系统
oncoming flow 迎面流
oncoming turbulence 来流湍流
oncoming wind 迎面风
one axis tracking solar concentrator 单轴跟踪太阳能聚光器

one effect evaporator 单级蒸发器，单效蒸发器
one electron theory 单电子理论
one-axis tracker 太阳能单轴跟踪器，单轴太阳跟踪器
one-dimensional momentum theory 一维动量理论
one-one exchanger 单壳单程热交换器
one-pipe steam system 单管蒸气系统
one-point spectrum 一点谱
one-tank system 单箱系统
one-to-one control 单个控制
one-to-one control mode 一对一控制方式
one-two exchanger 单壳双程热交换器
one-way operation 单向发电（该潮汐电站只能在涨潮或落潮时发电）
one-way system 单向系统
on-grid wind power generator 并网型风力发电机
on-load indicator 有载指示灯
on-load operation 有载运行
on-off valve 双位阀，截止阀
on-orbit power generation 轨道太阳能发电
on-peak energy 峰荷电能
Onsager reciprocal relation 昂萨格倒易关系
onshore 陆上的
onshore device 陆上装置
onshore wind 向岸风
on-site energy conversion 现场能量转换
on-site energy storage 就地储能
on-site generation 现场发电
on-site power system 场内电源系统

on-state current 开态电流
oozing 渗漏
opal 蛋白石
opaque cloud cover 不透明云遮系统
opaque cover 不透明盖板
opaque material 不透明材料
open bay 敞湾
open burning 开放焚烧，露天焚烧
open circuit 开路，断开电路
open circuit OTEC plant 开式海洋温差电站
open circuit photovoltage 开路光生电压
open circuit voltage 开路电压
open current 开路电流
open cycle 开式循环
open cycle ocean thermal energy conversion 开式循环海洋热能转换
open cycle system 开式循环系统
open cycle turbine 开式循环涡轮机
open drip proof 开式防滴型
open ended wind tunnel 开路式风洞
open floor and roof wind tunnel 无上下壁风洞
open flow flat plate collector 开口式液体单板（太阳能）集热器
open hearth steel 平炉钢
open loop air receiver 开口式空气吸热器
open loop application 开环应用程序
open loop cycle 开放式循环周期
open ocean 公海，外海
open power plant energy from salinity gradient 开式盐差能动力装置
open site 开阔地带
open system 敞开系统
open trough solar water heater 开槽式太阳能热水器
open tube deposition 开管淀积
open water 无冰水面，开阔水面
open-air climate 露天气候
open-circuit characteristic 开路特性
open-circuit operation 开路运行
open-circuit reactance 开路电抗
open-circuit voltage 开路电压
open-cycle 开放式（海洋发电厂）
open-cycle engine 开放式循环动力机，开放式循环发动机
open-cycle ocean thermal difference power plant 开式循环海洋温差电站
open-cycle ocean thermal energy conversion unit 开式循环海洋热能转换装置
open-cycle OTEC 开式循环海洋热能转换
open-cycle sea thermal plant 开式循环海洋热能（转换）装置
opened-cycle system 开放式循环系统
open-ended pipe 开口管
opening angle 开敞角
open-loop active solar system 开环主动式太阳能系统
open-loop air-preheating system 开放式空气预热系统
open-loop collector 开放式集热器
open-loop control system 开环控制系统
open-loop domestic water heater 民用开环太阳能热水器
open-loop system 开式循环系统，直接加热系统
open-phase protection 断相保护
operating bias 工作误差
operating condition 工况

operating cost　经营成本，生产费用，营运费，营运成本
operating current　工作电流
operating distance of solar cooker　太阳灶操作距离
operating height of solar cooker　太阳灶操作高度
operating plan after accident　事故后运行方式
operating point　工作点
operating regime　操作规程
operating state　操作状态，工作状态
operating temperature　工作温度，操作温度
operating temperature of solar cell　太阳电池的工作温度
operating voltage　工作电压
operation and maintenance　运行与维护
operation management　运行管理
operational characteristic　运行特性
operational modulation　运行调节
operational wind speed　运行风速
opposing wind　逆风，迎面风
opposite flow　逆流
optic axis　光轴
optical　光学的，视觉的
optical (quantity) sensor　光学（量）传感器
optical (quantity) transducer　光学（量）传感器
optical absorption　光学吸收
optical absorption constant　光学吸收常数
optical absorption edge　光学吸收限
optical absorption spectrum　光学吸收谱

optical air mass　大气光学质量
optical booster　（太阳能）光学增温器
optical capacity　光学容量
optical coating　光学涂料
optical constant　光学常数，光学恒量
optical contact　光接触
optical coupling　光学耦合
optical densitometer　光密度计
optical density　光密度
optical distance　光程
optical efficiency　光学效率，光效率
optical electron　光学电子
optical energy　光能
optical energy gap loss　光能隙损失
optical filter　滤光器
optical grid　光栅
optical measurement　光学测量，光学法测量
optical orientation　（集热器）最佳朝向
optical power meter　光功率计
optical pyrometer　光学高温计
optical scanning technique　光学扫描技术
optical sensor　光学传感器
optical system of furnace　太阳炉光学系统
optical thickness　光学厚度
optical wind speed sensor　光学风速传感器
optically coated glass　光学镀层玻璃
optics　光学，光学系统
optimal blade design　叶片优化设计
optimal control　最优控制
optimal orientation　最佳朝向
optimal power　最优功率
optimal steam temperature　最佳蒸汽温度

optimal tip speed ratio 最佳叶尖速比
optimization criterion 最优化准则
optimum angle of inclination 最佳倾角
optimum collector area 最佳集热器面积
optimum collector orientation 集热器最佳方向
optimum collector temperature 集热器最佳温度
optimum concentration ratio 最佳聚光比
optimum efficiency 最佳效率
optimum facing (集热器)最佳朝向
optimum flow rate 最佳流速,最佳流率
optimum load 最佳负载,最佳负荷
optimum operating current 最佳工作电流
optimum operating voltage 最佳工作电压
optimum output 最佳出力,最佳输出
optimum rated wind speed 最优额定风速
optimum refractive index 最佳折射系数
optimum speed to windward 最佳迎风速度
optimum tilt angle 最佳倾角(一定时间内固定式太阳能装置获得太阳辐射能最大时与地面的倾角)
optimum tip 最优叶尖
optimum tip speed ratio 最优叶尖速度比
optimum useful life 最佳使用年限
optional test condition 任选测试条件
optislip 转子电流控制结构,斩波调阻结构
optoelectronic coupler 光电耦合器

optoelectronic property 光电性质
optoelectronics 光电子学
orbiting solar power station 轨道太阳能站
orbiting solar reflector system 太阳能轨道反射器系统
organic biodegradable matter 有机生物降解物质
organic content 有机物含量,有机质含量
organic dry matter 有机干物质
organic fluid 有机流体
organic food farm 有机食品农场
organic fraction of municipal solid waste 城镇有机垃圾
organic fraction of the MSW 城市生活有机垃圾
organic household waste 有机生活垃圾
organic loading rate 有机负荷率
organic matter 有机质,有机物
organic photovoltaic cell 有机太阳能电池
organic Rankine cycle 有机朗肯循环
organic Rankine cycle (ORC) technology 有机朗肯循环技术
organic semiconductor solar cell 有机半导体太阳电池
organic waste 有机废物,生化废料
organism 生物体,有机物,有机体
orientation 方向,方位
orientation device 定向设备
orientation drive 迎风驱动装置
orientation mechanism 迎风机构
orientation of collector 集热器的朝向
orientation of the rotor 风轮迎风
oriented array 定向方阵

orifice 孔口，注孔
orifice meter 孔板流量计
origin of geothermal heat 地热成因
origin of thermal spring 地热起源
original equipment manufacturer 原始设备制造商
original geothermal gradient 初始地热梯度
original rock temperature 原始岩石温度
original thermal condition 原始热状态
orogenic zone 造山带
orographic upward wind 地形性上升风
Orsat gas analysis 奥尔氏气体分析
orthographic projection 正射投影，直角投影
oscillating airfoil 振动翼
oscillating body 振荡体
oscillating cable WECS 电缆振荡式风能转换系统
oscillating current 振荡电流
oscillating force 振荡力，振动力
oscillating hydrofoil 振动水翼
oscillating kiln 振动窑
oscillating shaft 摆动轴
oscillating vane 摆动叶片
oscillating water column 振荡水柱
oscillating water column system 振荡水柱系统
oscillating water column wave energy converter 振荡水柱式波能转换装置（由波浪运动驱动固定在岸边或半潜在海面的腔体内的水柱上下振荡，压迫空腔内的空气，产生往复气流，推动空气涡轮机发电的装置）
oscillating wave 振荡波

oscillating wave surge converter 振荡浪涌转换器
oscillating WECS 振荡式风能转换系统
oscillating wing WECS 机翼振荡式风能转换系统
oscillation 振动，振幅
oscillator 振荡器
oscillatory 振荡的，振动的
oscillatory current 振荡电流
oscillatory flow 脉动流，振荡流，摆动流，振荡流动
oscillatory line turbine 摆线式透平
oscillatory motion 振荡运动，振动
oscillatory voltage 振荡电压
oscillatory wave 振荡波
oscillograph 示波器
oscillograph trace 示波图
oscillogram 波形图
oscillographic testing 示波测试
osmosis 渗透作用
osmosis principle 渗透原理
osmotic energy conversion 渗透能量转换
osmotic flow 渗透流
osmotic heat pipe 渗透热管
osmotic potential 渗透势
osmotic pressure 渗透压
osmotic pressure difference 渗透压差
osmotic pressure energy 渗透压能
osmotic pressure gradient 渗透压梯度
osmotic pressure salinity gradient power conversion 渗透压式盐浓度差能发电
osmotic pump 渗透泵
Otake geothermal field （日本）大岳地热田
OTEC apparatus 海洋热能转换装置

OTEC application 海洋热能转换应用
OTEC closed-cycle system 海洋热能转换闭式循环系统
OTEC compact heat exchanger 海洋热能转换致密换热器
OTEC control function 海洋热能转换控制功能
OTEC design process 海洋热能转换设计过程，海洋热能转换设计方法
OTEC development 海洋热能转换开发
OTEC economics 海洋热能转换经济学
OTEC evaporator 海洋热能转换汽化器
OTEC gazing plant 海洋热能转换放牧场
OTEC heat exchanger 海洋热能转换热交换器
OTEC heat transfer 海洋热能转换传热
OTEC lift cycle 海洋热能转换提升循环
OTEC methanol plantship 海洋热能转换甲醇电站船
OTEC module 海洋热能转换组件
OTEC open-cycle system 海洋热能转换开式循环系统
OTEC pilot plant pipe system 海洋热能转换中试管道系统
OTEC plant 海洋热能转换电站
OTEC plate-type heat exchanger 海洋热能转换板式换热器
OTEC platform 海洋热能转换平台
OTEC power cycle 海洋热能转换动力循环
OTEC power generation 海洋热能转换发电
OTEC power integration 海洋热能转换电功率集成化
OTEC power plant 海洋热能转换电站
OTEC power system 海洋热能转换发电系统
OTEC program 海洋热能转换计划
OTEC pump 海洋热能转换泵
OTEC shell 海洋热能转换（发电）壳
OTEC shell-and-tube evaporator 海洋热能转换壳一管汽化器
OTEC station 海洋热能转换电站
OTEC steam-lift cycle 海洋热能转换蒸汽一提升循环
OTEC technology 海洋热能转换技术
OTEC test platform 海洋热能转换试验平台
OTEC energy 海洋热能转换热能
OTEC resource 海洋热能转换资源
OTEC tower 海洋热能转换塔
OTEC tube heat-exchanger 海洋热能转换用管式换热器
OTEC workshop 海洋热能转换工场
OTEC-CWP computer-aided design methodology 海洋热能转换冷水管计算机辅助设计法
OTEC-CWP design 海洋热能转换冷水管设计
OTEC-CWP field experiment 海洋热能转换冷水管现场试验
Otto cycle engine 四冲程循环发动机，奥托循环发动机
Otto engine 奥托发动机
out of phase current 异相电流，不同相电流
out of phase voltage 异相电压
outage rate 停电率
outdoor air temperature 露天气温
outdoor boiler 露天锅炉
outdoor design condition 室外设计条件
outdoor design temperature 室外设计温度
outdoor equipment 露天装置

outer chute of water 外层水流
outer continental shelf 陆棚外域，大陆架外缘，外陆架
outer cover （集热器）外层盖板
outer packing 外包装
outer photoeffect 外层光效应
outer ring 外环
outer rotor 外转子，外齿轮
outerspace solar spectrum 外层空间太阳光谱
outgoing current 输出电流
outlet 出口
outlet steam 出口蒸汽
outlet temperature of collector 集热器出口温度
out-of-phase 反相位
output 产能
output characteristic of WTGS 风力发电机组输出特性
output coupling 输出连接
output current 输出电流
output impedance 输出阻抗
output of collector 集热器输出（能量）
output of steam 蒸汽产量
output power 输出功率
output shaft 输出轴
output temperature 输出温度
output torque 输出转矩，输出力矩，输出扭矩
output voltage 输出电压
output-terminal current 输出端电流
oven dry wood 烘干的木材
over circuit 过电流
over current protection 过电流保护
over power 过载功率
over size particle 筛上物颗粒
over voltage protection 过电压保护

overall efficiency of solar energy system 太阳能系统总效率
overall energy 总能量
overall heat transfer 总传热
overall heat transfer coefficient 总传热系数
overall photosynthetic efficiency 总光合效率
overall plant efficiency 水电站总效率
overall solar heat gain 总太阳热能增益
overall storage heat loss factor 热储存总损失系数
overall thermal resistance 总热阻
overexcitation 过励磁，过度激发
overfire air 燃尽风
over-fire boiler 火焰炉
overhead crane 高架起重机
overhead power line 高架输电线
overhead transmission line 架空输电线，高架输电线路
overhead travelling crane 桥式起重机
overhead wire 高架电线
overlapped glass plate 重叠玻璃板
overlapped glass air heater 重叠玻璃空气加热器
overlapped glass plate air heater 重叠玻璃平板型空气加热器
overload forward current 过负载正向电流
overnormal wind 超常风
overshot 水流冲击（潮轮）上部而转动的
overspeed control 超速控制
overthrust 上冲断层，逆掩断层，逆冲断层
overtopping 漫溢
overturning moment 倾覆力矩

overturning wind speed 倾覆风速
over-voltage 过电压
Ovonic 奥夫尼克（将热能或太阳光直接转换为电能的装置，由奥夫辛斯基发明。装置内有特殊的从非导电状态变成半导体状态的玻璃成分）
ovonic device 热（光）电转换器件，光电转换装置
oxalate 草酸盐
oxidation catalyst 氧化催化剂
oxidation reaction 氧化反应，氧化作用
oxidation refining 氧化精炼
oxidation zone 氧化区，氧化带
oxidation-reduction potential 氧化还原电位
oxidation-reduction system 氧化还原体系
oxidative phosphorylation 氧化磷酸化作用
oxide fuel cell power generator 氧化物燃料电池电力发电机
oxide layer 氧化膜，氧化层
oxide vapor pressure 氧化物蒸汽压
oxidizing agent 氧化剂
oxyacetylene torch 氧乙炔炬
oxygen content 氧含量
oxygen depleted environment 无氧环境
oxygen enrichment 富氧
oxygen gasification 氧化的气化（作用）
oxygen gasifier 氧气气化炉
oxygen remover 除氧器
oxygen-steam system 氧—蒸汽（反应）系统
oxyhydrogen voltameter 氢氧电量计，气电量计
ozone content 臭氧含量

ozone depletion potential 臭氧消耗潜势，臭氧耗损潜能值

P

packed bed 填充床，填充层
packed bed storage 填充床储热
packed distillation column 填充蒸馏塔
packed-bed scrubber 填充床洗涤器
packer 封隔器
padding 填料，跟踪
paddle 桨
paddle blade 平叶片
paddle wheel fan 叶轮风机
paddle-wheel 明轮，蹼轮
pad-mounted transformer 垫板上变压器
paint for biogas digester 沼气池涂料
paired-basin scheme 双库方案
paleomagnetics 古地磁
palladium tube 钯管
palm fibre 棕丝
panel bed filter 板层过滤器
panel control 配电盘控制
panel method 小样本连续研究法，盘区开采法，板块法
panel protection diode 面板保护二极管
paper processing plant 造纸厂
parabola 抛物线
parabolic aluminized reflector 抛物型渗铝反射器
parabolic aluminum mold 抛物型铝模
parabolic collector 抛物型集热器
parabolic concentrator 抛物面聚光器
parabolic cylinder 抛物柱面
parabolic cylinder collector 抛物柱面集热器

parabolic cylinder mirror collector 抛物柱面反射镜集热器
parabolic cylinder solar reflector 抛物柱面太阳能反射器
parabolic dish 抛物面反射镜，抛物面反射器，抛物柱面反射器
parabolic dish collector 抛物面碟形集热器
parabolic dish concentrator 抛物面碟式聚光器
parabolic dish solar concentrator system 抛物碟型太阳能聚光器系统
parabolic dish solar thermal power generation system 抛物碟式太阳热发电系统
parabolic dish technology 抛物面碟技术
parabolic equation 抛物线方程
parabolic focusing collector 抛物线型聚焦集热器
parabolic Fresnel mirror 抛物面菲涅耳反射镜
parabolic heliostat 抛物面定日镜
parabolic mirror 抛物镜，抛物柱面镜
parabolic mirror collector 抛物柱面镜集热器
parabolic point focus collector 抛物线状点聚集集热器
parabolic reflecting surface 抛物型反射面
parabolic reflector 抛物面反射镜，抛物线形反射镜
parabolic shaped collector 抛物面形集热器
parabolic support 抛物线形支架
parabolic trough 抛物槽
parabolic trough concentrator 抛物面槽式聚光器
parabolic trough field 抛物槽场
parabolic trough power plant 抛物槽式太阳能电厂
parabolic trough receiver tube 抛物面槽式吸热管
parabolic trough reflector 抛物线形槽反射器
parabolic trough solar power 抛物面槽式太阳能热发电
parabolic trough solar thermal power generation system 抛物面槽式太阳热发电系统
parabolic trough surface 抛物槽式表面
parabolic trough technology 抛物面槽技术
parabolic-dish collector 旋转抛物面集热器，抛物盘集热器
parabolic-mirror focusing solar collector 抛物柱面镜聚焦太阳能集热器
parabolic-trough collector 槽形抛物面集热器
paraboloid collector 抛物面集热器
paraboloidal concentrating solar collector 抛物面太阳能聚光集热器
paraboloidal concentrator 抛物面聚光器
paraboloidal dish collector 抛物面碟形集热器
paraboloidal dish solar collector 抛物面盘状太阳能集热器
paraboloidal photoelectrochemical system 抛物面光电化学系统
paraboloidal solar concentrator 抛物面太阳能聚光器
paraboloidal trough 抛物面（聚光）槽
paraffin wax heat storage 石蜡储热
parallel auxiliary 并联辅助设备

parallel co-firing	平行混燃，平行共燃
parallel combustion	平行燃烧
parallel configuration	并联布置
parallel flow	平行流，顺流
parallel flow heat exchanger	平行流动热交换器，顺流热交换器
parallel impedance	并联阻抗
parallel resonance	并联谐振
parallel shaft	平行轴
parallel stream	平行流
parameter requirement	参数要求
parameter sensitivity	参数灵敏度
parasitic back loss	（集热器）背面附加热损（失）
parasitic current	寄生电流
parasitic energy	附加热量
parasitic heat leak	寄生热泄漏，附加热泄漏
parasitic heat loss	寄生热损失，附加热损失
parasitic load	附加载荷
parasitic power	寄生功率，寄生能源
parasitic power loss	寄生功率损耗
parasitic rear loss	（集热器）背面附加热损（失）
parasitic resistance	寄生电阻，寄生阻力
parasitic voltage	寄生电压
parastic pressure drop	附加压降
parked wind turbine	风力机停机
parking	停机
parking brake	停机制动
part load	部分负荷
part load behavior	低负荷性能
part load efficiency	低负荷效率
part of collecting system	集热系统构件
partial oxidation	部分氧化
partial safety factor	分项安全系数
partial throwing	部分倒伏状（V级植物风力指示）
partially closed cycle	半闭式循环
particle bed storage	颗粒床储热
particle density	颗粒密度
particle dimension	颗粒维数
particle flux	粒子通量
particle motion convertor	波粒子运动能量转换器
particle receiver	粒子吸热器
particle separation	颗粒分离
particle separator	粒子分离器
particle size	颗粒粒径
particle size distribution	粒度分布
particle size reduction	粒度破碎
particle storage bed	粒子蓄热床
particleboard residue	微块板残渣
particulate	微粒，微粒物质
particulate emission	颗粒排放，颗粒物排放
particulate matter	微粒物质
particulate radiation	粒子辐射，微粒辐射
part-load operation	部分负荷运转
Pascal	帕斯卡（压力单位）
Pascal's law	帕斯卡定律
pass band	通带
passivated emitter solar cell	钝化发射极太阳电池
passivating	钝化
passivation film	钝化膜
passivator	钝化剂
passive collector	被动式集热器
passive component	被动元件，无源元件
passive cooling	被动式冷却

passive cooling in greenhouse 温室被动式冷却
passive heating system 被动加热系统
passive solar array 被动式太阳能阵列，无源太阳能阵列
passive solar building 被动式太阳能建筑
passive solar design 被动式太阳能设计
passive solar energy 被动式太阳能，无源太阳能
passive solar heat 被动式太阳热
passive solar heatlng 被动式太阳能供热，无源太阳能供暖
passive solar heating system 被动式太阳能供热系统，无源太阳能供暖系统
passive solar house 被动式太阳房
passive solar system 被动式太阳能系统
passive stall 被动失速
passive storage wall 被动式蓄热墙
passive system control 非能动控制调节系统
passive thermosyphon system 被动热虹吸系统
passive type of solar system 被动式太阳能系统，无源太阳能装置
passive water heating system 被动式水加热墙系统
passive yaw 被动偏航
passive yawing 被动偏航
pasture crop 牧草作物
path length 轨迹长度，路程长度，行程长度，迹线长度
pathway 输导通道
Pauli principle 泡利不相容原理
Pauling's rule 鲍林规则
pause mode 暂停模式

payload 净负荷，有效负荷
payload capacity 额定载重量
PE converter 电力电子变流器
peak allowable transient voltage 允许的峰值瞬变电压
peak capacity 高峰容量，峰容量
peak current 峰值电流
peak efficiency 最高效率
peak force 峰值力
peak forward anode voltage 峰值正流向阳极电压
peak forward voltage 最大前向电压，最大正向电压
peak frequency 峰值频率
peak gust 最大阵风
peak heat flux 热通量峰值
peak insolation 日照峰值
peak inverse current 最大逆电流，反峰值电流
peak load 最大负荷，峰值负荷
peak load effective duration factor 峰值功率持续时间因素
peak load hydropower plant 峰荷水电站
peak load operation of tidal power station 潮汐电站最大负载运行
peak load regulation solar thermal power plant 太阳能热发电调峰电厂
peak load plant 峰荷电厂
peak load power 峰值负荷功率，最大负荷功率
peak load power plant 尖峰负荷发电厂
peak load station 尖峰负荷电站
peak period 波峰周期，尖峰周期，峰期
peak plate current 最大屏极电流，最大阳极电流

peak point 峰值点
peak point current 峰点电流
peak point voltage 峰点电压
peak power consumption 高峰能耗
peak power point 峰值功率点
peak power tracker 峰值功率跟踪器
peak power tracking photovoltaic power system 光伏发电最大功率峰值跟踪系统
peak power tracking scheme 峰期功率跟踪方案
peak pulse voltage 脉冲峰压
peak reverse voltage 峰值反向电压
peak solar radiation 太阳能峰值辐射（量）
peak sun hours 峰值日照时数
peak temperature 峰值温度
peak to peak current 峰间电流
peak to peak voltage 峰间电压
peak to valley ratio 峰谷比
peak unit 调峰机组
peak voltage 峰值电压
peak wind speed 最大（瞬时）风速
peaking 调峰
peaking capability 尖峰能力
peaking power plant 尖峰负荷电厂
peaking service 尖峰负荷运行
peaking unit 峰荷发电机组
peak-load generating unit 峰值负荷发电机组
peak-shaving device 调峰设备
peat 泥煤，泥炭
peat briquette 泥炭煤球
pebble 小卵石，小圆石
pebble bed 卵石床
pebble bed storage 蓄热卵石床
pebble heater 粒状载热体加热器，卵石加热器

pebble-bed receiver 球床吸热器
peephole 观测孔，窥视孔
pelagic 浮游的，远洋的，深海的
pellet boiler 团矿锅炉
pellet burner 颗粒燃烧机
pellet press 压粒机
pellet silo 颗粒筒仓
pellet stove 颗粒壁炉
pelletisation 粒化
pen recorder 笔式记录器
pendulum anemometer 摆式风速表
pendulum mass 摆锤
pendulum wave energy device 摆式波浪能装置
pendulum wave power device 摆式波浪发电装置
penetration current 渗透电流
penetrometer 针入度仪，贯入度仪
penstock 水渠
per rev load 每转负荷
per unit span 单位跨度
percent possible sunshine 阳光百分率
percentage of air space 通风截面比
percentage of sunshine 日照百分率
percolation 过滤，浸透
perennial plant 多年生植物
perfect collector 理想集热器
perfect combustion 理想燃烧
perfect mirror surface 理想反射镜面
perfect optical system 理想光学系统
perfect radiation 完全辐射
perfect radiator 完全辐射体
perfect reflection 完全反射，理想反射
perfect tracking mirror 理想跟踪反射镜
perfectly black body 理想黑体，完全黑体

perfectly stirred reactor 搅拌反应器，均匀搅拌反应器
perforated distributor 多孔分配板，多孔布风板
perforated pipe 多孔管，穿孔管
perforation 穿孔
performance 性能
performance analysis 性能分析
performance curve of pump 泵特性曲线
performance curve of wind turbine 风力机性能曲线
performance index 性能指数
performance indicator 绩效指标，业绩指标
performance parameter 性能参数
performance rating 额定性能
performance test of unit 机组性能试验
perigean spring tide 近地点大潮
perigean tide 近地点潮（指月球运行到近地点附近时潮差随之增大的潮汐）
perigee 近地点
perihelion 近日点
period of duty 运行时间，运行期
periodic duty 周期性负荷，周期性工作
periodic fluctuating windspeed 周期脉动风速
periodic heat transfer 周期性传热
periodic source testing 周期性源测试
periodic vibration 周期振动
periodicity 周期性，定期性，频率
permanent load 永久负荷
permanent magnet 永久磁铁
permanent magnet alternator 永磁发电机
permanent magnet DC machine 永磁直流电机

permanent magnet generator 永磁发电机
permanent magnet pole 永磁极
permanent magnet square-wave motor 永磁方波电机
permanent magnet synchronous generator 永磁同步发电机
permanent mooring 永久系泊（系统）
permanent thermocline 海洋主温跃层
permanent wind 恒定风
permeability 渗透性，透磁率，导磁系数
permeable fault 渗透断层
permeable layer 透水层
permeable linking 渗透联系
permeable membrane 渗透膜
permeable rock 可渗透岩层，渗水岩，透水岩
permissible load 许用负荷，容许负荷
permittance current 电容性电流
perpendicular 垂直的，垂线，垂直的位置
perpendicular force 正交力
persistent current 持久电流
persistent organic pollutant 持久性有机污染物
persistent wind 持续风
perspective view of drilling assembly 钻井装置透视图
perturbation 微扰
perunit reactance 单位电抗
petrographic analysis 岩石分析
petroleum coke 石油焦
petroleum plant 石化厂
petroliferous basin 含油气盆地
phase 波动幅度，振幅，相位
phase advance 相位超前

phase angle 相角，相位角，相移角
phase boundary 相界
phase boundary crosslinking 相界交联
phase boundary line 相界线
phase boundary reaction 相界反应
phase change 相变
phase change energy storage 相变储能
phase change material 相变物质
phase change material storage unit 相变蓄热单元
phase change medium 相变介质
phase change section 相变区
phase change storage 相变蓄能
phase change storage container 相变蓄热容器
phase change storage substance 相变蓄热物质
phase change storage system 相变储存系统
phase change thermal storage 相变蓄热
phase coil 相位线圈
phase control 相位控制
phase control device 相位控制设备
phase control factor 相位控制因素
phase controlled rectifier 相控整流器
phase controller 相位补偿器
phase conversion 相位变换
phase coupling effect 相位耦合
phase equilibrium 相平衡
phase fault protection 相间短路保护
phase lag 相位滞后
phase liquid 液相
phase load 相负荷
phase modifier 调相器
phase transition 相变

phase velocity 相速度
phase voltage 相电压
phase-change heat storage element 相变储热元件
phase-change material 相变（储热）材料
phase-change material storage 相变材料储存
phase-change medium 相变介质
phase-change solar system 相变太阳能系统
phase-change solar water heater 相变太阳能热水器
phase-change-transition 相变转换，相变转化
phase-changing salt 相变盐
phasor diagram 相量图，矢量图
phasor voltage 相量电压
phenolics 酚醛塑料
phosphate 磷酸盐
phosphate treatment 磷化处理，磷酸盐处理
phosphoric acid fuel cell 磷酸燃料电池
phosphors and fluorescent material 磷光体和荧光材料
phosphorus 磷
photic zone （海洋）透光层，透光区
photistor 光敏晶体管
photo actinic action 光化作用
photo cell 光电池，光电管
photo chemical vapor deposition 光化学气相沉积
photo effect 光子效应
photo electronics chemical solar cell 光电子化学光电池板
photo emissive cell 光电发射管，外光电效应光电管

photo thermal energy conversion 光热能量转换
photoabsorption 光吸收
photoabsorption coefficient 光吸收系数
photoactor 光电变换器件
photocell 光电管,光电池,光电元件
photochemical absorption 光化吸收
photochemical cell 光化学电池
photochemical cell energy conversion unit 光化学电池能量转换装置
photochemical conversion 光化学转换
photochemical conversion efficiency 光化学转换效率
photochemical diode 光化学二极管
photochemical electric power 光化学发电
photochemical energy conversion 光化学能量转换
photochemical inversion 光化转化
photochemical oxidation potential 光化学氧化潜势
photochemical reaction 光化(学)反应
photochemistry 光化学
photocomposing machine 光电组合机
photoconduction 光电导
photoconductive cell 变阻光电池
photoconductive current 光电导电流
photoconductive device 光电导器件
photoconductive effect 光电导效应
photoconductive resistance 光敏电阻
photoconductive response 光电导响应
photoconductive sensor 光导式传感器
photoconductive transducer 光导式传感器
photoconductivity 光电导率,光电导性
photoconductor 光导体
photoconversion efficiency 光转化效率,光电转换效率
photocurrent 光生电流,光电流
photodiffusion voltage 光致扩散电压
photodiode 光电二极管
photodiode dark current 光电二极管暗电流
photoeffect 光电效应
photoelastic stress visualization 光弹性应力可视化
photoelectric 光电的,光电效应的
photoelectric absorption 光电吸收
photoelectric absorption coefficient 光电吸收系数
photoelectric barrier layer cell 有阻挡层的光电池
photoelectric cathode material 光电阴极材料
photoelectric constant 光电常数,光电效应常数
photoelectric conversion efficiency 光电转换效率
photoelectric current 光电流
photoelectric device 光电器件
photoelectric effect 光电效应
photoelectric emission 光电发射
photoelectric lighting controlling 光电照明控制
photoelectric material 光电材料
photoelectric phenomenon 光电现象
photoelectric photometer 光电光度计
photoelectric power converter 光电换能器
photoelectric sensitivity 光电灵敏度
photoelectric spectrophotometer 光电分光光度计

photoelectric switch 光电开关
photoelectric thermometer 光电温度计
photoelectric threshold energy 光电阀值能量
photoelectric transducer 光电换能器
photoelectric yield 光电子产额
photoelectricity 光电，光电现象，光电学
photoelectricity conversion 光电转换
photoelectrochemical cell 光电化学电池
photoelectrochemical effect 光电化学效应
photoelectrochemical energy conversion 光电化学能转换
photoelectrochemistry 光电化学
photoelectrode 光电极
photoelectrolysis 光电解（作用）
photoelectromotive force 光电电动势
photoelectron 光电子
photoelectrosynthesis 光电合成（作用）
photoelectrosynthetic reaction 光电合成反应
photoenergy 光能
photogalvanic process 光电过程
photogenerated current 光生电流，光电流
photoionization 光电离
photojunction battery 光电结电池
photomagnetic coupling 光磁耦合
photomagnetoelectric effect 光电磁效应
photomontage 集锦照片，集锦照相
photomultiplier 光电倍增管（器）
photomultiplier cell 光电倍增器电池
photon 光子
photon enhanced thermionic emission 光子增强型热离子辐射

photoreactive chlorophyll 光反应性叶绿素
photoresistance 光敏电阻
photoresistance and photoconductive detector material 光敏电阻和光电导探测器材料
photoresistor 光敏电阻器
photosensitive material 光敏材料
photosphere 光球，光球层
photoswitch 光控开关
photosynthesis 光合作用
photothermal conversion 光热（直接）转换
photothermal converter 光热（直接）转换器
photothermal device 光热（转换）装置
photothermoelectric effect 光热电效应
phototransistor 光电三极管
phototube 光电管
photovaristor 光敏电阻
photovoltage 光生电压
photovoltaic 光电的
photovoltaic array 光伏阵列，光伏方阵
photovoltaic cell 光电电池
photovoltaic demonstration plant 光伏示范电厂
photovoltaic panel （太阳能）光伏电池板
photovoltaic action 光伏作用，光电作用
photovoltaic array 光电池阵
photovoltaic building material 光伏建筑材料
photovoltaic cathodic protection 光伏阴极保护

photovoltaic cell 光电池
photovoltaic charger 光伏充电器
photovoltaic collector 光电池组件，光电池阵
photovoltaic concentrating array 太阳能聚光电池阵
photovoltaic concentrator array 聚光太阳电池方阵
photovoltaic concentrator array field 聚光太阳电池方阵场
photovoltaic concentrator module 聚光太阳电池组件
photovoltaic concentrator system 光电聚光器系统
photovoltaic conversion 光伏转换，光电转换
photovoltaic conversion of solar energy 太阳能光电转换
photovoltaic conversion system 光电转换系统
photovoltaic converter 光电转换器
photovoltaic detector 光伏探测器
photovoltaic device 光伏器件，光电装置
photovoltaic device technology 光电装置技术，光伏装置技术
photovoltaic effect 光电效应，光伏效应
photovoltaic electric power generation 光电发电
photovoltaic energy 光电能
photovoltaic energy system 光电能系统
photovoltaic flat plate 平光电池板
photovoltaic flat plate module 平板光电池组件
photovoltaic generator 光电发生器

photovoltaic micropower system 光伏微功率系统
photovoltaic module 光伏组件，太阳能电池组件
photovoltaic panel 光伏板，光电板
photovoltaic panel wall 光伏幕墙
photovoltaic power 光伏发电，光电
photovoltaic power conversion 光电功率转换
photovoltaic power generating unit 光电发电装置
photovoltaic power generation 光电发电
photovoltaic power plant 光电电站
photovoltaic powered pump 光电动力泵
photovoltaic process 光电过程
photovoltaic radiation detector 光电辐射探测器
photovoltaic sensor 光伏式传感器
photovoltaic silicon pyrheliometer 硅光电池直接日射计
photovoltaic solar cell 太阳能光电电池
photovoltaic solar energy conversion 光伏太阳能转换，太阳能光电转换
photovoltaic solar energy conversion device 光电太阳能转换装置，光伏太阳能转换装置
photovoltaic system 光伏发电系统
photovoltaic system technology 光电系统技术
photovoltaic thermal system 光电热能（综合利用）系统
photovoltaic transducer 光伏式传感器
photovoltaic water pump 光伏水泵
photovoltaic system 光伏系统

photovoltaics 太阳光电，太阳能光电板
photovoltaic-thermal system 光伏光热系统
photronic cell 光电池
phreatic eruption 潜水喷发
physical dimension 实际尺寸，外形尺寸
physical impediment 物理障碍
physical model test 物理模型试验
physical oceanography 物理海洋学，海洋物理学
physical process in bio-conversion technology 生物转化之物理程序
physical property type sensor 物性型传感器
physical property type transducer 物性型传感器
physical prospecting 物理勘探
physical sensor 物理量传感器
physical transducer 物理量传感器
physical transmission right 物理输电权
physical vapor deposition 物理气相沉积
physico-chemical technology 物化技术
physiognomy 地貌
physiographic feature 地文特征
pick off 敏感元件，传感器
pico hydro 微水电
picoplankton 超微型浮游生物，微型浮游生物
piecewise polynomial 分段多项式
pier 码头，桩
Pierson-Moskowitz spectrum 皮尔逊莫斯科维茨谱，皮尔逊光谱
piezocrystal 压电晶体
piezoelectric effect 压电效应
piezoelectric pickup 压电传感器
piezoelectric polymer 压电高分子
piezoelectricity 压电（现象）
piezometer 压强计，压力计，测压计
piezoresistivity 压电电阻率
piezoresistor 压敏电阻
pile 热电堆，反应堆
pilot balloon 测风气球
pilot exciter 引示励磁机，辅助励磁机，副励磁机，导频励磁器
pilot flame 引燃火焰，起火焰
pilot injection 引燃喷射，前导喷射
pilot voltage 标准电压
pilot wind power plant 风力发电试验场
pin 柱销
pinch point temperature difference 节点温差
pinion 小齿轮
pinion gear 小齿轮，行星齿轮
pipe camera inspection system 管道摄像检查系统
pipe fitter 管道安装工，管装工
pipe receiver 管式吸热器
pipe size 管径
pipeline network 管网
pipeline pump 管道泵
pipework 管道工程
pipework system 管道工程体系，管道工程系统
piping layout 管道布置
piston 活塞
piston pump 活塞泵
piston ring 活塞环
piston steam engine 活塞式蒸汽机，往复式蒸汽机
piston valve steam engine 活塞阀蒸汽机

piston wave power device 活塞式波浪力发电装置
piston wavemaker 活塞造波机
pitch 伏角，倾斜
pitch angle 节面角（锥齿轮的），螺旋角，桨距角
pitch bearing 变桨轴承
pitch circle 节距圈
pitch control 桨距控制，螺距调节
pitch controlling 节距控制
pitch motor 变桨距电机
pitch point 节点
pitch power 螺距力
pitch regulated 桨距调节
pitch regulation 变桨调节
pitch response 俯仰反应
pitching 倾斜的，护坡，海漫，吊桩定位，俯仰，前后
pitching moment 俯仰力矩
pitching momentum coefficient 变桨动量系数
pitot tube 空速管，皮托管
pitot tube anemometer 皮托管式风速计
pitting corrosion 点腐蚀，坑蚀，孔蚀
planar 平板式，平面性
planar booster reflector 平面加强反射镜
planar mirror 平面镜
planar rotation 平面旋转
Planck constant 普朗克常量，普朗克常数
Planck's law 普朗克定律
plane 翼面，平面
plane black surface 黑体平板表面
plane of rotation 转动面
plane receiver with plane reflector 平面反射式平板集热器

plane wave 平面波
planet carrier 行星架
planet gear 行星齿轮
planetary boundary layer 行星边界层
planetary gear drive mechanism 行星齿轮传动机构
planetary gear train 行星齿轮系
planetary gearbox 行星齿轮箱，行星减速箱
plankton 浮游生物
plankton net 浮游生物网
plankton pulse 浮游生物消长
plano-concave lens 平凹镜
plano-convex lens 平凸镜
plant availability 设备利用率，电厂可用率
plant capacity 装置功率，装机容量
plant lifetime 电厂寿命
plant load 电厂负荷，电站负荷
plant outage 电厂停运
plant size 工厂规模，电站容量
plant stoppage 电厂故障
plasma 等离子体
plasma chamber 等离子室，电浆腔室
plasma column 等离子体柱
plasma etching 等离子腐蚀
plasma incinerator 等离（子体）焚烧炉
plasma jet 等离子流，等离子体射流
plasma magnetohydrodynamics power generation 等离子体磁流体发电
plasma reactor 等离子体反应器
plasma torch 电浆炬，等离子体焰炬
plasma-arc gasification 等离子电弧气化
plastic bag solar water heater 塑料袋式热水器

plastic collector 塑料集热器
plastic crystal 塑料晶体，塑晶
plastic digester 塑料沼气池
plastic film 塑料薄膜
plastic film absorber 塑料薄膜吸收体
plastic film cover 塑料薄膜顶盖
plastic Fresnel lens 塑料菲涅尔透镜
plastic heat exchanger （海洋热能转换）塑料换热器
plastic parabolic reflector 塑料抛物面反射镜
plastic pillow water heater 塑料枕形热水器
plastic piping 塑料管路
plastic solar cooker 塑料太阳灶
plastic solar energy heat collector 塑料太阳能集热器
plastic solar still 塑料太阳能蒸馏器
plastic solar still with sandbag 砂袋式塑料太阳能蒸馏器
plastic solar water heater 塑料太阳能热水器
plastic-glass laminate 塑料-玻璃叠片
plate 散热片
plate anemometer 压板风速表
plate clutch 盘式离合器
plate heat exchanger 板式换热器
plate tectonics 板块构造
plate-fin type heat exchanger 板翅式换热器
plate-type heat exchanger 板式换热器
platform region 地台区域
plug flow 活塞流，平推流
plug flow digester 活塞流连续消解器
plug flow reactor 塞流反应器
plug-flow anaerobic fermenter 塞流厌氧发酵罐
plug-flow digester 塞流式沼气池
plug-in hybrid electric vehicle 插电式混合动力车
plunging breaker 卷碎波，卷浪
pluton 深成岩体
plywood residue 胶合板残渣
pneumatic chamber 震荡水柱
pneumatic conveying 气动输送，风力输送
pneumatic conveying drying 风力输送干燥，风气动输送干燥
pneumatic conveying system 气动输送系统，风力输送系统
pneumatic conveyor 气动运输机
pneumatic ocean wave energy conversion 风动海洋波能转换
pneumatic pick 风镐
pneumatic power 气动功率，气动动力
pneumatic reservoir 气压缸
pneumatic system 气动系统，风动系统
pneumatic wave-energy converter 气动式波能转换器
pneumatic-reciprocating compressor 气动往复式压气机
point absorber 点吸收器（波能装置）
point concentrating system 点聚光系统
point contact type solar cell 点接触型太阳能电池
point focus collector 点聚焦集热器
point focus concentrating mirror system 点聚焦反射镜系统，点聚焦反射镜装置
point focus concentration 点聚焦聚光
point focus concentrator 点聚焦聚光器
point focus configuration 点聚焦结构

point focus system 点聚焦系统，点聚焦装置
point focusing reflector 点聚焦反射器，点聚焦反射板
point mass 点质量
point of attachment 附着点
point of common coupling 公共连接点，公共耦合点
point of interconnection of wind farm 风电场并网点
point source 点源，点污染源，点放射源
point-focus collector 点聚焦集热器
point-focus concentrator 点聚焦聚光器
point-of-power utilization 点式用电
polar axis fully tracking mounting 极轴全跟踪装置
polar form 点斜式
polar front jet 极锋气流
polar molecule 极性分子
polar moment 极距
polar mounted collector 装有极轴的集热器
polar mounting （全跟踪集热器的）极轴装置
polar solar tracking 极轴太阳跟踪
polarization current 极化电流
polarizing voltage 极化电压
pole 电极
pole tower 杆塔
pole-changing motor 变极式感应电动机
pole-mounted panel 立柱式太阳能支架
poleward 向南极的，向极的
polishing scrubber 抛光洗涤器
polishing separator 抛光分离器
pollution control department 污染控制部
pollution control system 污染控制系统
pollution control test 污染控制测试
pollution prevention 污染防治，污染预防
pollution-free electrical power 无污染电源
pollution-free engine 无污染发动机
polychlorinated biphenyl 多氯联苯
polychromatic 多波的
polychromatic wave 多色波
polycrystalline 多晶的
polycrystalline block 多晶硅锭
polycrystalline coll 多晶电池片
polycrystalline diamond compact 聚晶金刚石复合片
polycrystalline ingot 多晶硅锭
polycrystalline silicon 多晶硅
polycrystalline silicon film solar cell 薄膜多晶硅太阳能电池
polycrystalline silicon ingot 多晶硅锭
polycrystalline silicon module 多晶硅组件
polycrystalline silicon solar cell 多晶硅太阳能电池
polycrystalline solar cell 多晶太阳电池
polycrystalline structure 多晶结构
polycrystalline wafer 多晶硅片
poly-crystalline-silicon solar cell 多晶硅太阳能电池
polycrystallining 多晶性
polycrystallinity 多晶结晶度
polycyclic aromatic hydrocarbon 多环芳烃，稠环芳烃，多环芳香烃
polyester 聚酯
polyester resin 聚酯树脂
polyethylene pipe 聚乙烯管
polyimide 聚酰亚胺

polylactic acid plastic 聚乳酸塑料
polymaleic acid 聚马来酸
polymer 聚合物
polymer film solar collector 聚合物薄膜太阳能集热器
polymer fuel cell 聚合燃料电池
polymer insulating material 高分子绝缘材料
polymer phenyl propane 聚合物苯基丙烷
polymer semiconductor cell 聚合物半导体电池
polymer semiconductor solar cell 聚合物半导体太阳电池
polymorphic 多晶的，多形的，多态的
polymorphic transformation 同质多晶转变
polymorphism 多晶形
polynuclear aromatic compound 多环芳香族化合物
polyphase current 多相电流
polysilicon 多晶硅
polysulfide bromide battery 聚硫化溴电池
polyvinyl chloride 聚氯乙烯
pond system 池塘系统
pontoon 浮舟
poplar 白杨，白杨木
populus 杨树
pore 孔隙，气孔
pore pressure 孔压
pore space 孔隙
porosity 孔隙率，孔隙度
porosity percent 百分孔隙率
porous absorber-type solar air heater 多孔吸热器型太阳能空气加热器
porous absorbing head 多孔吸热顶盖
porous disc 多孔过滤板
porous layer 多孔岩层，多孔质层
porous matrix 多孔基体，多孔基质
porous matrix-type air heater 多孔阵型空气加热器
porous media 多孔介质，孔隙介质，漏失通道
porous rock 多孔岩石
portable absorption-desorption cooling unit 便携式吸热—解吸制冷装置
portable power generator 便携式发电机
portable solar power generation system 便携式太阳能发电系统
position feedback 位置反馈，定位反馈
position indicator 位置指示灯
position of tide 潮位
position sensor 位置传感器
position vector 位置矢量
positive back current 正离子反向电流
positive current 正电流
positive displacement pump 正排量泵
positive electrode plate 正电极板
positive electrode, anode 正极，阳极
positive output terminal 正极输出端
positive output terminal nut 正极输出线接线柱螺帽
positive output terminal torque 正极输出线接线柱扭矩
positive polarity dc voltage 正极性直流电压
positive polarity voltage 正极电压
positive primary photoelectric current 正原始光电流
positive ray current 阳极射线电流
positive voltage 正电压
positron 正电子

possible installed capacity estimate 可用装机容量评估
possible resource estimate 可用资源评估
post-combustion 后燃
post-combustion catalytic control system 后燃催化控制系统
post-digestion processing unit 消化后处理器，消化后处理机组
post-stall airfoil 失速后翼型
potable water 饮用水
potassium 钾
potassium acetate 醋酸钾，乙酸钾
potassium hydroxide electrolyte 氢氧化钾电解质
potential 电位，电势
potential barrier 势垒，位垒
potential difference 电位差，电势差
potential energy 势能，位能
potential energy of column of water 水柱位能
potential energy storage system 势能储能系统
potential environmental impact 环境影响潜值
potential flow 位流
potential gradient 位能梯度，电位梯度
potential heat 潜热
potential jump 电位跃变
potential of wind energy 风能潜力
potential pollutant 潜在污染物
potential pollution from use of energy resource 能源利用的潜在污染
potential site for wind machine 风力机潜在场址
potential tidal power resource 潮汐能资源蕴藏量
potential wind power site 潜在风能场址
potentially available wind power 潜在可用风能
potentiometer 电位计
power 功率
power amplifier 功率放大器
power and energy compilation table 电力和能源编制表
power and heating plant 热电站，热电厂
power angle 功率角，功角
power balance 功率平衡
power boiler 动力锅炉，电站锅炉
power cable 电力电缆
power chain 动力传动系，动力传动机构
power circle diagram 功率圆图
power coefficient 动力系数，功率系数
power conditioner 功率调节器
power conditioning 电源调节，功率调节
power conditioning unit 功率调节器
power connection and control unit 电源连接和控制单元
power consumption 功率消耗
power contactor 电源接触器
power content 电储量
power control 功率控制
power control algorithm 功率控制算法
power conversion efficiency 功率转换效率，能量转换效率
power conversion unit 能量转换装置
power converter 电源转换器，功率变流器
power current 功率电流
power curve 功率曲线

power cycle 动力循环，电力循环
power delivery 功率输出，电力传输
power delivery cable 功率输出电缆
power density 功率密度，动力密度，能量密度
power density diagram 功率密度图
power density distribution 功率密度分布
power density spectrum 功率密度频谱
power diode 电力二极管
power electronic 电力电子学
power electronic converter 电力电子变换器
power electronic device 电力电子器件
power electronics 电力电子
power equation 功率方程
power extraction 获能
power extraction coefficient 获能系数
power factor 功率因数
power factor control capacitor 功率因数控制电容器
power factor correction capacitor 功率因数矫正电容器
power fatigue 功率疲劳
power fence 发电栅（风力发电的一种新设想）
power flow 电力潮流，功率流
power flow meter 功率流计
power fluctuation 功率波动
power flux 功率通量
power frequency electrical field 工频电场
power frequency magnetic field 工频磁场
power generation 发电
power generation by a solar chimney 太阳烟囱发电
power generation by combustion 燃烧发电
power generation cycle 发电循环
power generation cycle efficiency 发电循环效率
power generation from sea current 海流发电
power generation from tide 潮汐发电
power generation mix 发电结构
power generation using steam turbine 汽轮机发电
power generator 发电机组
power grid 电网
power house 动力室，发电站，发电间
power island 动力岛，电源岛
power law 幂定律
power law exponent 幂律指数
power law for wind shear 风切变幂律
power law profile 幂律廓线
power line 输电线
power mains 输电干线
power matrix 功率矩阵
power meter 功率计
power metering 电能计量
power output 输出功率，功率输出，电源输出
power output prediction 输出功率预测
power park 发电厂区
power performance 功率系数
power plant ash 电厂灰
power plant basics 电厂基础
power plant of energy from salinity gradient 盐差能动力装置
power plant of energy from salinity gradient in the wizened salt lake 干涸盐湖盐差能动力装置
power plant of energy from salinity gradient of mixer in the inside 内混式盐差能动力装置

power plant of energy from salinity gradient of mixer in the outside 外混式盐差能动力装置
power pool 联合电力系统，电力联营（机构）
power purchase agreement 购电协议
power quality 电能质量
power quality and reliability (PQR) 电能质量与供电可靠性
power quality control 电能质量控制
power quality evaluation 电能质量评估
power quality monitor 电能质量监测仪，电能质量监控器
power rating 额定功率
power regulation and storage unit 功率调节和存储单元
power semiconductor 功率半导体
power separator 电力分选机
power shaft 动力轴，传动轴
power spectral density 功率谱密度
power spectral density function 功率谱密度函数
power spectrum 功率谱
power spectrum of rotor thrust 风轮推力功率谱
power spectrum of tip deflection 叶梢上翘功率谱
power spike 功率峰值
power station 电站
power station boiler 电站锅炉
power station with reservoir 堤坝式水电站
power substation 变电站
power subsystem 发电子系统
power supply reliability 供电可靠性
power take-off 动力输出装置，分动箱
power take-off system 动力输出装置

power train 动力传动系统，动力传动机构
power transformer 电力变压器，功率变换器
power transistor 功率晶体管
power transmission 电力传输
power transmission and distribution equipment 输配电设备
power unit using solar pond 太阳池发电装置
power versus wind speed curve 功率—风速曲线
powerbuoy 能量浮标
powerhouse 动力室
power mixer 功率混合
practical efficiency of a solar array 太阳能方阵的实际效率
practical module efficiency 组件实际效率
Practical Ocean Energy Management Systems 美国应用海洋能管理系统
practical resource 实践资源
precambrian bedrock 前寒武纪岩床
precession axis 进动轴
precipitation 沉淀，沉淀作用，沉淀物，降水
precipitation level 降雨量
precipitator 除尘器
pre-coating 预涂层
predefined operation 预定义操作
predicted energy production 预计发电量
pre-digestion processing unit 预消化处理器，预消化处理机组
predominant wind direction 主风向
preheat sequence 预热序列
preheater 预热器，预热炉

preheating 预热
preliminary reconnaissance 预先勘察
premixed flame 预混火焰
preset limit 预定极限
press cake 滤饼
pressing aid/additive 压缩辅料
pressure accumulator 储压器，压力回收锅炉
pressure anemometer 压力风速表
pressure angle 压力角
pressure build-up in well 井内压力恢复，井内压力升高
pressure chamber 压力室，压力舱
pressure control valve 压力控制阀
pressure differential 压差，压力差
pressure drop 压力下降
pressure drop calculation 压降计算
pressure element 压力元件，测压元件
pressure exchanger 压力交换器
pressure fluctuation 压力脉动，压力波动
pressure gauge 压力表
pressure gradient 压力梯度，气压梯度
pressure gradient force 气压梯度力
pressure head 压（力）头，承压水位，测压水头
pressure of wind 风压
pressure oscillation 压力波动，压力振荡
pressure override valve 超压阀
pressure plate 压板
pressure range 压力范围
pressure reducing valve 减压阀，降低压力阀
pressure retarded osmosis 压力延迟渗透
pressure sensor 压力传感器

pressure side 压力面
pressure stabilization 压力稳定作用
pressure swing absorption 变压吸附
pressure tap 测压孔
pressure transducer 压力传感器
pressure transient 压力瞬变
pressure transient analysis 压力瞬时分解
pressure tube anemometer 压力管风速表
pressure wave energy conversion device 压力波能转换装置
pressure-retarded osmosis 正渗透
pressure-temperature safety device 压力温差安全器
pressurization 增压，加压，气密
pressurized acid digestion 压力封闭酸溶
pressurized air 压缩空气，增压空气
pressurized air receiver 承压式空气吸热器
pressurized system 预压系统
prevailing current 盛行流
prevailing wind power direction 风能盛行方向
prevailing wind 主导风，盛行风
prevailing wind direction 主导风向
primary air 一次风，一次空气
primary air control 一次风控制
primary air distribution 一次风分布
primary air heater 第一级空气预热器
primary air outlet 一次风扩口
primary air pressure control system 一次风压力控制系统
primary air ratio 一次风比率
primary air supply 一次空气供应，初级空气供应

primary bioenergy supply 一次生物质能供应量
primary burner 主燃烧器
primary circuit 一次回路，前置回路
primary combustion 一次燃烧，初级燃烧
primary combustion chamber 主要燃烧室
primary combustion zone 主燃区
primary cooling circuit 初级冷却回路
primary current 一次电流
primary energy 一次能源
primary energy mix 一次能源构成
primary energy saving index 初级能源节约指数
primary energy source 一次能源
primary fission neutron energy 初裂变中子能量
primary fluid 一次热媒
primary fuel 原生燃料
primary generator 主发电机
primary material 原材料
primary photoelectric current 原光电流
primary power source 一次能源
primary standard pyrheliometer 一等标准直接日射表
primary standard solar cell 一级标准太阳电池
primary voltage 一次电压
primary wave energy conversion device 波能一次转换装置
primary winding 一次绕组，初级绕组，原边绕组
prime mover 原动机
principal geothermal zone in the world 世界主要地热带
principal radiating surface 基准发射面
principal stress 主应力
principal theorem 基本原理
principle of converting sunlight 阳光转换原理
principle of hourglass 沙漏原理（集热器跟踪太阳的一种原理）
principle of operation 工作原理，操作原理
print electrode 印刷电极
prism solar concentrator 棱镜太阳能聚光器
prismatic window 棱镜窗
probabilistic model 概率模型
probability density function 概率密度函数
probability distribution function 概率分布函数
probability distribution of wind speed 风速的概率分布
probable maximum extratropical storm 可能最大温带风暴
probable maximum hurricane 可能最大飓风
probable maximum storm surge 可能最大风暴潮
probable maximum tropical cyclone 可能最大热带气旋
process control system 过程控制系统
process flexibility 过程柔性
process flowsheet 工艺流程图，生产流程图
process heat 工业用热，工艺用热
process heat system 过程加热系统
process heater 过程加热器
process integration 联合工艺
process residue 加工废料，工艺废料

process steam 生产用蒸汽，工艺蒸汽
process unit 加工单位，工艺设备，工艺装置
producer gas 发生炉煤气
producing zone 生产层
product size 产品尺寸
production casing 生产套管
production decline curve 产量递减曲线
production decline rate 产量递减率
production fluid 产液
production pressure 生产压差
production strategy 生产策略
production tax credit 生产税抵免
production type wind tunnel 生产性风洞
production well 生产井
production-scale deep drilling 生产规模深孔钻井
productivity index decline 生产力指数下降
profitability screening chart 盈利筛选图表
programmable control 可编程序控制
programmable logic controller 可编程逻辑控制器
Programme for the Endorsement of Forest Certification 森林认可认证计划
project cost budget 工程预算
project delay 工程延期
project final acceptance 工程竣工验收
project final account 工程决算
project financial assessment 工程财务评价
project handover for operation 工程移交生产验收
project investment management 工程投资管理
project management information system 工程管理信息系统
project national economic assessment 工程国民经济评价
project planning management 工程计划管理
project quality management 工程质量管理
project technical management 工程技术管理
projected area of blade 叶片投影面积
projectile drilling 弹丸凿岩，抛物体凿岩
projection 发射，投射，投影（图），抛射物
projection angle 投射角
projection drawing 投影图
propagating wave 发散波，行进波
propagation curve 传播曲线
propagation direction 传播方向
propane 丙烷
propane boiler 丙烷锅炉
propane burner 丙烷燃烧器
propane compressor 丙烷压缩机
propane condenser 丙烷冷凝器
propane exhaust pipe 丙烷排气管
propane purge tank 丙烷清洗舱
propane refrigeration unit 丙烷冷冻设备
propane turbine 丙烷涡轮机
propane-vapor pipe 丙烷蒸汽机
propanol 丙醇
propeller 螺旋桨
propeller anemometer 螺旋桨式风速计
propeller blade 螺旋桨叶片
propeller fan 螺旋式风机

propeller turbine 螺旋桨式涡轮机
propeller type anemometer 螺桨式风速计
propeller type fan 轴流式风机
propeller type rotor 螺旋桨型风轮（转子）
propeller type wind machine 螺旋型风力机
propeller type windmill 螺旋桨型风车
proportional-integral (PI) controller 比例积分调节器
proportional-integral-derivative controller 比例积分微分调节器
proportionality 相称性
proposed tidal power scheme 计划中的潮汐发电方案
proppant 支撑剂
propylene glycol 丙二醇
prospect evaluation 勘探评估
prospect for geothermal resource 地热资源勘探
protected against heavy seas 防海浪的
protection 保护
protection circuit 保护电路
protection circuit breaker 短路保护器
protection level 保护等级
protection system 保护系统
protective circuit 保护电路
protective earthing 保护接地
protective relay 保护继电器
protein degradation 蛋白质降解
proton exchange membrane fuel cell 质子交换膜燃料电池
prototype wind speed 原型风速
protype testing 样机试验
proximate analysis 工业分析，近似分析

proximity effect 邻近效应
pseudo-eutectic 伪共晶
P-type semiconductor P 型半导体，空穴型半导体
P-type silicon P 型硅
Public Utilities Regulatory Policies Act of 1978 (PURPA) 1978 年美国公共事业监管政策法案
pulling out of synchronism 牵出同步
pull-out torque 牵出转矩
pulp 纸浆
pulp buffering tank 纸浆缓冲罐
pulp mill 纸浆厂
pulsating current 脉动电流
pulsating direct current 脉动直流电流
pulsating voltage 脉动电压
pulsation 脉动
pulsator 搅拌器，震动器
pulsatory current 脉动电流
pulse bandwidth 脉冲带宽
pulse energy 脉冲能量
pulse firing technology 脉冲燃烧技术
pulse frequency modulation 脉冲频率调制，脉频调制
pulse jet cleaning 脉冲布袋清灰
pulse signal 脉冲信号
pulse solar simulator 脉冲式太阳模拟器
pulse stream 脉冲流
pulse threshold energy 脉冲阈能
pulse voltage 脉冲电压
pulse wavelength modulation 脉波长调制
pulse width modulation 脉冲宽度调制
pulse-amplitude modulation 脉幅调制
pulsed laser deposition 激光溅射沉积，脉冲激光沉积

pulsed wire anemometer 脉冲热线风速计
pulse-jet bag filter 脉冲喷吹袋式除尘器
pulse width modulated converter 脉冲宽度调制转换器
pulse width modulated thyristor 脉冲宽度调制晶体闸流管
pultruded FRP 拉挤玻璃钢
pultrusion 拉挤成型，挤压成型
pulverization 粉化
pulverized biofuel 粉碎的生物质燃料
pulverized coal boiler 煤粉锅炉
pulverized fuel 粉末燃料
pulverized fuel combustion 粉末燃料燃烧
pulverized fuel combustion system 粉末燃料燃烧系统
pulverized fuel combustor 煤粉炉膛
pump capacity 泵容量
pump impeller 泵轮，泵推动器
pump power 泵耗电率，泵浦功率
pump speed 泵速，泵速率，泵冲数
pump turbine 泵水轮机
pumped hydro energy storage 抽水储能
pumped storage 抽水蓄能
pumped storage hydropower station 抽水蓄能电站
pumped storage power resource 抽水蓄能资源
pumped storage set 抽水蓄能机组
pumped storage station 抽水蓄能电站
pumped storage unit 抽水蓄能机组
pumped-hydro storage 抽水蓄能
pumping load 抽水量
pumping loss 泵气损失

pumping station 泵站
pumping technology 泵汲技术
pump-jet turbine 泵喷涡轮机
punched screen 冲孔钢筛板
purchasing power parity 购买力平价
pure pumped storage hydropower station 纯抽水蓄能电站
pure sine wave 纯正弦波
purified silicon 提纯硅
purpose grown energy crop 专用能源作物
pusher furnace 推杆式炉
putting into operation in compliance with standard 达标投产
PV array 光伏电池组件
PV charge controller 太阳能充电控制器
PV module 光伏组件
PV power system for GSM base station 移动通信基站太阳能电源系统
PV ribbon 光伏焊带
PV sheet 光伏板，太阳能光伏板
PV string 光伏组列，光伏组件
PV system 光伏发电系统
PV array support structure 光电阵列支持结构
PV thermal hybrid CPC module CPC聚光型混合光伏光热组件，光伏光热混合CPC组件
pylon 塔架，高压线铁塔
pyramidal optical condenser 棱锥光学聚光镜
pyranograph 总日射计
pyranometer 总日射表，日射强度计
pyranometer network 辐射强度计网络
pyrex 硼硅酸玻璃，耐热玻璃
pyrgeometer 地球辐射表，大气辐射强度计，地面辐射强度计

pyrheliometer 直接日射表，日温计，太阳热量计，日射强度计
pyrheliometry 直接日射学，太阳热量学
pyroelectric crystal 热电晶体
pyroelectric effect 热电效应
pyroelectric material 热释电材料
pyroelectricity 热电性
pyroheliometer 太阳热量计
pyrolytic gasifier 热解气化炉
pyrolysis 裂解，高温分解，热解，热裂
pyrolysis gas 裂解气
pyrolysis gasification 热解气化
pyrolysis oil 裂解油
pyrolysis product 热解产物
pyrolytic cracking 热裂（作用），高温裂化
pyrolytic gasification 热解气化,高温缺氧气化
pyrolytic oil 热解油
pyroxene 辉石
pyrradiometer 全辐射表

Q

q-axis reactance 横轴电抗
quadrature current 正交电流
quadrature-axis 正交轴线
quadrature-axis magnetizing reactance 横轴磁化电抗
quality of consumption 用电质量
quality of geothermal fluid 地热流体质量
quality of supply 供电质量
quantitative differential thermal analysis 定量差热分析
quantity of air flow 风量
quantity of radiant energy 辐射能性质

quantity of solar energy 太阳能量
quantometer 光量计，辐射强度测量计，光谱分析仪
quantum dot solar cell 量子点太阳能电池
quantum dot 量子点
quantum effect 量子效应
quantum energy 量子能量
quantum exploit 量子开采
quantum level 量子能级
quantum number 量子数
quarrystone 粗石，荒料
quartz 石英
quartz clock powered by solar cell 太阳能电池石英钟
quartz crucible 石英坩埚
quartz crystal 石英晶体,石英岩,石英
quartz crystal resonator 石英晶体谐振器
quartz solubility 石英溶解度
quartz thermometer 石英温度计
quartzose sand 石英砂
quasi paraboloidal trough concentrator 准抛物槽聚光器,准抛物槽聚光式集热器
quasi static wind load 准静态风载
quasi stationary current 似稳电流,准稳电流
quasi steady source of energy 准稳态能源
quasi-renewable energy form 准可再生能源形态
quasi-steady state 准稳态,（准）定态
quaternary volcanic center 第四纪火山中心（它与地热田有密切关系）
quench 淬火,淬火冷却

quench process 淬火工艺
quench tower 急冷塔，急冷反应塔
quenching heat exchanger 急冷热交换器
quick startup 快速启动
quicklime 生石灰，氧化钙
quiescent current 静态电流，无信号电流
quiescent value 空载值
quieter operating engine 无声引擎
quill 套管，衬套

R

race rotation 空转
race time 空转时间
racing 空转
rad 拉德（辐射剂量单位）
radex 放射性气体（物质）排除
radgas 放射性气体
radiac 辐射仪
radiac instrument 辐射仪表
radiacmeter 剂量计
radial acceleration 径向加速度
radial axis 径向轴
radial blade fan 径向叶片风机（扇）
radial blanket 径向再生区
radial breeder 径向增殖区
radial buckling 径向曲率
radial chromatography 辐射状色谱法，圆形色谱法
radial coordinate 径向坐标
radial distribution 径向分布
radial extraction 径向引出
radial field 径向场
radial flow 径向流动
radial flow compressor 径流式压气机
radial flow pump 径流泵
radial flow turbine 径流涡轮机
radial flux 径向通量
radial focusing 径向聚焦
radial force 径向力
radial gap induction machine 径向间隙感应机
radial heat pipe 径向热管
radial interference 径向干扰
radial load 径向荷载
radial mixing 径向混合
radial oscillation 径向振荡
radial peaking hot spot factor 径向峰值热点因子
radial pin coupling 径向销连接
radial quantum number 径向量子数
radial shuffling of fuel 燃料径向倒换
radial stress 径向应力
radial vane 径向叶片
radial velocity 径向速度
radial wave function 径向波函数
radially bare 径向无反射层的
radiance exposure 曝辐（射）量
radiant absorbed dose 辐射吸收剂量
radiant absorption 辐射吸收
radiant body 辐射体
radiant capacity 辐射本领
radiant efficiency 辐射效率
radiant electric heater 辐射电热器
radiant emissivity 辐射发射率
radiant emittance 辐射发射率，辐射能流密度
radiant energy 辐射能
radiant energy density 辐射能量密度
radiant energy exchange 辐射能量交换
radiant energy flow 辐射能量流
radiant energy flux 辐（射能）通量

radiant energy from sun 太阳辐射能
radiant energy in sunlight 阳光辐射能
radiant exchange 辐射换热
radiant exitance 辐（射）出（射）度
radiant exposure 辐射曝露量
radiant flux 辐射通量
radiant flux density 辐射通量密度
radiant heat exchange 辐射热能交换
radiant heat flow 辐射热流
radiant heat flux 辐射热通量
radiant heat of anode 阳极辐射热
radiant heat transfer 辐射传热
radiant heat transfer coefficient 辐射放热系数
radiant heating system 散热片式采暖系统
radiant intensity 辐射强度
radiant interchange 辐射热交换
radiant panel 散热片
radiant power 辐（射）功率
radiant ray 辐射射线
radiant reheater 辐射式再热器
radiant section 辐射区段
radiant solar energy 辐射太阳能
radiant state 辐射态
radiant tube 辐射管
radiant-type heater 辐射型加热器
radiate pine 辐射松
radiated angle 辐射角
radiated atom 辐射原子
radiated energy 辐射能
radiating body 辐射体
radiating capacity 辐射本领
radiating element 辐射元件
radiating fin 散热片
radiating flange 散热片凸缘
radiating gill 散热片
radiating heat 辐射热
radiating power 辐射能
radiating rib 散热片
radiating surface 辐射表面
radiation 辐射
radiation absorption 辐射吸收
radiation absorption medium 辐射吸收介质
radiation absorptivity 辐射吸收率
radiation accident 辐射事故
radiation action 辐射作用，辐射影响
radlation alarm system 辐射警报系统
radiation angle 辐射角
radiation area 辐射区
radiation attenuation 辐射衰减
radiation back reaction 辐射逆反应
radiation background 辐射本底
radiation balance flux concentration 辐射平衡通量聚光度
radiation biochemistry 辐射生物化学
radiation botany 辐射植物学
radiation breakdown 辐射损伤
radiation capture 辐射俘获
radiation center 辐射中心
radiation chemical reaction 辐射化学反应
radiation chemical synthesis 辐射化学合成
radiation chemical yield 辐射化学产额
radiation chemistry 辐射化学
radiation coefficient 辐射系数
radiation constant 辐射常数
radiation control 辐射管理，辐射控制
radiation control area 辐射管理区域
radiation cooling 辐射冷却
radiation corrosion 辐射腐蚀
radiation cross section 辐射截面

radiation damage reaction 辐射损伤反应	radiation interference 辐射干扰
radiation damage susceptibility 辐射损伤敏感区	radiation ionization 辐射电离
radiation damping 辐射阻尼	radiation ionization chamber 辐射电离室
radiation decomposition rate 辐射分解速率	radiation leakage 辐射泄漏
radiation decrement factor 辐射衰减因数	radiation level 辐射水平
radiation destruction 辐射破坏	radiation loss 辐射损失
radiation detection 辐射探测	radiation monitoring system 辐射监测系统
radiation dosimeter 辐射剂量计	radiation polymerization 辐射聚合
radiation effect reactor 辐射效应研究（反应）堆	radiation power 辐射功率
radiation emitter 辐射体	radiation preservation 辐射保藏
radiation energy 辐射能量	radiation pressure 辐射压力
radiation exchange factor 辐射交换系数	radiation problem 辐射问题
radiation excitation 辐射激发	radiation processing reactor 辐射处理堆
radiation factor 辐照系数	radiation product 辐射产物
radiation field 辐射场	radiation protection 辐照防护
radiation flux density 辐射通量密度	radiation protection device 辐射防护装置
radiation force 辐射力	radiation quantity 辐射量
radiation front 辐射阵面	radiation reaction 辐射反应
radiation hardened 抗辐射的	radiation resistance 耐辐射性,辐射阻力
radiation heat exchange 辐射热交换	radiation resistant 耐辐射的
radiation heat loss 辐射热损失	radiation safety 辐射安全
radiation heat transfer 辐射换热	radiation scattering 辐射散射
radiation heating surface 辐射受热面	radiation selfdecomposition 辐射自分解
radiation impedance 辐射阻抗	radiation sensitivity 辐射灵敏度
radiation indictor 辐射指示器	radiation sensor 辐射敏感器
radiation induced creep 辐射感生蠕变	radiation shield 辐射屏蔽
radiation induced genetic effect 辐射诱发遗传效应	radiation shielding 辐射屏蔽
radiation injury 辐射损伤	radiation source 辐射源
radiation instability 辐射不稳定性	radiation stability 辐射稳定
radiation intensity 辐射强度	radiation stable polymer 辐射稳定聚合物

radiation surface 辐射表面
radiation threshold energy 辐射阈能
radiation tolerance level 允许辐射量
radiation tolerance status 辐射耐受能力
radiation vector 辐射矢量
radiation wave force 辐射波浪力
radiation width 辐射宽度
radiation window 辐射窗
radiation yield 辐射产额
radiation zone 辐射区
radiationless 无辐射的
radiationmeter 辐射计
radiation-type density gauge 辐射型密度计
radiative absorption 辐射吸收
radiative angle factor 辐射角系数
radiative capture rate 辐射俘获率
radiative collision 辐射碰撞
radiative cooling 辐射冷却
radiative heat loss 辐射热损失
radiative heating 辐射加热
radiative loss 辐射损失
radiator 散热器
radical 自由基，基
radio inference 无线电干扰
radioactive decay 放射性衰变
radioactive waste 放射性废物
radiometer 辐射表
radiometric technique 辐射测量技术
radius of blade section 叶片剖面半径
raft 筏
raft energy generator 浮箱波能发电机
raft unit 浮箱单元
rainfall regime 雨量型，降雨型
rainflow cycle counting 雨流计数法，塔顶法

rainstorm 暴风雨
ramp 坡板
ramp rate 缓变率
Rance tidal power station （法国）朗斯潮汐电站
random failure 随机故障
random manner 随机状态
random synchronizing 不规则同步
random vibration 随机振动
range of effective wind speed 工作风速范围
Rankine cycle 朗肯循环
Rankine cycle engine 朗肯循环发动机
rapid thermal annealing 快速退火
Raschig ring 拉西环
raster 光栅
rate control 速率控制
rate of change of momentum 动量变化率
rate of deposition 沉积速度
rate of discharge 流量
rate of evaporation 蒸发量，蒸发强度
rate of flow 流量
rate of geothermal utilization 地热利用率
rate of heat flow 热流强度
rate of heat generation 热发生速率
rate of heat liberation 放热速率
rate of house power 厂用电率
rate of solar radiation 太阳辐射速率
rated capacity 额定容量，额定功率
rated condition 额定工况
rated current 额定电流，规定电流
rated electrical output 额定出力
rated electric power of turbine-generator unit 汽轮发电机组额定发电功率
rated head 额定龙头，额定水头

rated horsepower 额定功率
rated impedance 额定阻抗
rated load 额定负载
rated load operation 额定负荷运行
rated load torque 额定转矩
rated operational voltage 额定工作电压
rated output capacity 额定输出容量
rated output current 额定输出电流
rated power 额定功率
rated power density 额定功率密度
rated power factor 额定功率因素
rated power output 额定输出功率
rated power supply 额定供电
rated reactive power 额定无功功率
rated specific speed 额定比转速
rated speed 额定转速
rated tip-speed ratio 额定叶尖速比
rated turning speed of rotor 风轮额定转速
rated value 额定值
rated voltage 额定电压
rated waste treatment capacity 额定垃圾处理量
rated wind speed 额定风速
rated wind velocity 额定风速
rates of penetration 钻进速度
rating 额定值
ratio of over load 过载度
ratio of tip-section chord to root-section chord 叶片根梢比
ratio of transformer capacity to load 变电容载比
rational use energy 合理使用能源
Rayleigh scattering 瑞利散射
Rayleigh distribution 瑞利分布
raytracing 光线追踪,光线跟踪

reactance 电抗
reactance voltage 电抗电压
reaction chamber 反应室,反应箱,分馏塔
reaction hydro-turbine 反动式水轮机
reaction steam turbine 反动式汽轮机
reaction turbine 反动式涡轮机,反动式汽轮机
reactive control 被动控制,反应式控制
reactive current 无功电流,电抗性电流
reactive fuel 活性燃料,反应性燃料
reactive power 无功功率
reactive power compensation unit 无功功率补偿单元
reactive power compensator 无功功率补偿器,无功功率补偿装置
reactive power control 无效功率控制
reactive power of wind farm 风电场无功功率
reactive silica 可溶性硅
reactor core 反应堆芯
ready for operation 准备运行
reagent 试剂
real time 实时
real-time outdoor exposure testing 户外实时寿命试验
real-time system 实时系统
reburning 再燃烧
reburning chamber 二次燃烧室
received energy 接受能量
received heat 受热量,传热量
receiver 吸热器,接收器
receiver aperture 吸热器采光口
receiver efficiency 吸热器效率
receiver peak flux density 吸热器峰值辐射通量密度

receiver spillage 吸热器溢出
receiver system 吸热器系统
receiver vent and drain system 吸热器疏水放气系统
receiving end voltage 受电端电压
receiving surface 接收面
recent data 最近数据
recharge current 再充电电流，过充电电流
rechargeable alkali manganese battery 可充碱锰电池
rechargeable battery 蓄电池
reciprocal density 倒易密度
reciprocating engine 活塞式发动机，往复式动力机
reciprocating engine system 往复式动力系统
reciprocating grate 往复炉排
reciprocating mechanical grate 往复式机械炉排
reciprocating motion 往复运动
recirculation current 再循环电流
recirculation region 回流区
recirculation zone 循环区，回流区
reclaimed stock 回收浆料
reclaimed water 回用水
recombination current 复合电流
reconnaissance tool 勘测工具
recording anemometer （自动）记录风速计
recording pyrheliometer 太阳热量记录仪
recording thermometer 记录温度计（带有遥感球）
recording wind vane 自记风向计
recoverable heat 可再生热
recoverable resource 可回收资源，可采资源
recoverable wave energy resource 可回收波能资源
recovered construction wood 回收的建筑木材
recovered energy 回收能量
recovery boiler 回收锅炉
recovery efficiency 回收效率，采收率
rectangular three-dimensional grid 三维矩形栅格
rectangular tube 矩形管
rectangular two-dimensional grid 二维矩形栅格
rectangular weir 矩形堰
rectifier 整流器
rectifier harmonic 整流器谐波
rectifier wave power plant 整流型波能发电站
recuperated system 复原系统
recuperator 复热器，同流换热器
recurrence interval 重现期
recycled fuel 再循环燃料
recycling rate 回收率
red tide 赤潮
red clay plastics 红泥塑料
redox system 氧化还原体系
redox flow battery 氧化还原液流电池
reducing agent 还原剂
redundance 冗余
reference area of the rotor 风轮基准面
reference curve 参考曲线
reference device 基准器件（是一种以标准太阳光谱分布为依据，用来测量辐照度或校准太阳模拟器辐射度的光伏器件）
reference distance 基准距离

reference height 基准高度
reference roughness length 基准粗糙长度
reference station 基准站，参考站
reference wind speed 参考风速
refinement 提纯
refining 晶粒细化，精炼
reflectance 反射，反射比，反射系数，反射率
reflected light 反射光
reflected intensity 反射强度，反射光强度
reflected radiation 反射辐射
reflected solar radiation 反射日射
reflected wave 反射波
reflecting insulating cover 反射绝热盖
reflecting surface 反射面，反射表面，反光表面
reflection 反射
reflection coefficient 反射系数
reflection factor 反射因数
reflection survey 反射波法勘查
reflective aluminum coated 反射涂铝纸
reflective blind 反射帘
reflective film 反射薄膜
reflective loss 反射损失
reflective power 反射能力
reflective scattering 反射散射
reflective scattering surface 反射散射表面
reflective solar collector 反射式太阳能集热器
reflective solar device 反射式太阳能装置
reflective surface 反射面
reflective surface area 反射面积
reflectivity 反射率，反射系数，反射性

reflectivity factor 反射率因子
reflectivity for sunlight 阳光反射率
reflectometer 反射计，反射比表
reflector 反射体，反射器，反射层
reflector area 反射器面积
reflector booster 反射板增温器
reflector collector 反射集热器
reflector control 反射器控制
reflector-shade 反射型遮热板
reflector-type solar cooker 反射器型太阳能灶
reformed fuel 改进燃料
refracted wave 折射波
refracting solar concentrator 折射式太阳能聚光器
refracting surface 折射面
refraction 折射，折射角
refraction coefficient 折射系数
refraction survey 折射波法勘探
refractive index 折射率，折光率，折射指数
refractory 耐火材料，耐火的
refractory lining 耐火内衬，耐火衬砌
refractory oxide 耐高温氧化物
refrigerant 制冷剂
refrigerant piping 制冷剂管
refrigerant vapor 制冷剂蒸汽
refrigeration cycle 制冷循环，冷却循环
refrigeration load 制冷负荷
refuse derived fuel 垃圾衍生燃料
refuse feeder 垃圾喂入机
refuse incinerator 垃圾焚烧炉
refuse-derived fuel (RDF) 垃圾衍生燃料
regenerative cell 再生电池
regenerative cycle 回热循环，再生循环

regenerative fuel cell 再生式燃料电池
regenerative heat exchanger 交流换热器
regenerative heating 交流换热
regenerative open cycle 开式回热循环
regenerative reheat cycle 回热再热循环
regenerator 蓄热器，容器式热水器
regional reconnaissance 区域性普查
regional transmission organization 区域输电组织
reglonal wind 区域风
regional wind resource 地区风力资源
registering weather vane 风向计
regular load 正则载荷
regular polygon 正多边形
regular wave 规则波
regulated voltage 稳定电压
regulating bar 调节棒
regulating damper 调节挡板
regulating mechanism 调速机构
regulating mechanism by adjusting the pitch of blade 变桨距调节机构
regulating mechanism of turning wind rotor out of the wind sideward 风轮偏侧式调速机构
regulating voltage 调节电压
reheat 再（加）热，二次加热
reheat boiler 再热锅炉
reheat cycle 再热循环
reheat factor 重热系数
reheat steam 再热蒸汽
reinforced concrete 钢筋混凝土
rejected heat 散失热量
relative biological effectiveness 相对生物效应
relative capture width 相对捕获宽度
relative direction 相对方向
relative humidity 相对湿度
relative permeability 相对渗透率
relative radiometer 相对辐射表
relative spectral response 相对光谱响应
relative spectral sensitivity 相对光谱灵敏度
relative thickness of airfoil 翼型相对厚度
relative wind 相对风
relative wind direction 相对风向
relative wind velocity 相对风速
relative-spectral-sensitivity characteristic 相对光谱灵敏度特性
relay 继电器
relay logic 继电器逻辑
relay output 继电器输出
relay protection 继电保护
relay protection device 继电保护装置
release valve 排出阀,安全阀,释放阀,放泄阀，排气阀
released heat 放出热
releasing current 放出电流
reliability 可靠性
reliability determination 可靠性测定试验
reliability index of generating and transmission 发输电系统可靠性指标
reliability of generating and transmission 发输电系统可靠性
relief valve 安全阀，减压阀
reluctance 磁阻
reluctance synchronous generator 磁阻同步发电机
reluctance torque 磁阻转矩，反应转矩

reluctivity 磁阻率
remaining reserve 剩余储藏量
remnant 残留物
remote collection 间接集热（作用）
remote harnessing of ocean thermal energy 海洋热能远程利用
remote monitoring 远距离监控
remote power 远方动力
remote power generation 远程发电
remote reading dial thermometer 遥测度数刻度盘式温度计
remote-storage (solar water heating) system 分体式（太阳能热水）系统
remote terminal unit (RUT) 远动终端
removable bottom 活动炉底
removal efficiency 去除效率，清除效率
remove iron from geothermal water 地热水除铁
remove sand from geothermal water 地热水除砂
renewability 再生性
renewable biological source 可再生生物能源
renewable diesel 再生柴油
renewable electricity 可再生电
renewable energy 可再生能源
renewable energy credit 可再生能源信贷
renewable energy directive 可再生能源指令
Renewable Energy Law 可再生能源法
renewable energy portfolio 可再生能源组合
renewable energy resource 可再生能源
renewable energy source electricity 可再生能源电力

renewable energy technology 可再生能源技术
renewable gasoline 再生汽油
renewable generation 可再生能源发电
renewable heating oil 可再生取暖油
renewable jet fuel 再生喷气燃料
renewable natural resource 可再生自然资源
renewable ocean energy source 可再生海洋能源
renewable portfolio standard 可再生能源配额制
renewable primary energy 可再生一次能源
renewable solar energy 可再生太阳能
renewable technology 可再生技术，再生技术
Renewable Grid Integration China Week 中国新能源并网接入国际峰会
repair time 修复时间
replenishment 补给
representative year for wind energy resource assessment 代表年（风能资源评估）
reproduction 复现
reradiation loss 再辐射损失
reserve capacity factor 备用容量系数
reserve capacity of power plant 发电厂的储备容量
reserve generating capacity 储备发电容量
reserve service 备用服务（为维持电力系统正常运行，保持系统发电和负荷需求处于平衡状态，由发电厂提供的增、减有功功率后备服务，包括冷备用和热备用两种。冷备用亦称检修备用，常在发电设备检修时启用，热备用用于事故备用和负荷备用）

reservoir 水库，蓄水池，储热层
reservoir behavior 油层动态
reservoir capacity 库容
reservoir characterization 储层表征
reservoir chemistry 热储化学
reservoir decay 热储衰减
reservoir engineer 油藏工程师，储藏工程师
reservoir engineering 热储工程
reservoir fracture 裂缝性油藏
reservoir geology 储层地质学
reservoir management 热储管理
reservoir model 热储模型
reservoir modeling 热储模型
reservoir of geothermal field 地热田储层
reservoir parameter 储层参数
reservoir permeability 储层渗透性
reservoir pressure decline 热储压力衰减
reservoir recovery 储藏采收，储层恢复
reservoir requirement 储层要求
reservoir rock 热储岩体
reservoir simulation 数值模拟，油藏模拟，热储模拟
reservoir temperature distribution 储层温度分布
reservoir volume 储集层体积
reservoir volumetric change 热储容积变动
residential area PV power system 小区太阳能发电系统
residential scale power system 住宅规模电力系统
residential solar power device 住宅用太阳能发电装置，民用住宅太阳能发电装置，家用太阳能发电装置
residential solar array 民用太阳电池方阵
residua (ash and slag) 残渣
residual flow 残流，余流
residual heat 余热
residual liquid 剩余溶液，残余溶液
residual oil content 剩余油含量
residual-heat utilization 余热利用
residue derived fuel 废弃物衍生燃料
residue disposal 残余物处理
residue recycling 残余物回收
resilience 回弹
resistance 电阻
resistance head 阻力损失
resistance heating 电阻加热
resistance loss for generator 发电机铜损
resistance of an earthed conductor 接地电阻
resistance of flow 流阻
resistance to irradiation damage 抗辐照损伤
resistance voltage 电阻电压
resistive load 阻性负载
resistive material for heater 发热电阻材料
resistivity 电阻率，电阻系数，比电阻
resistor 电阻器
resolution 分辨率
resonance 谐振
resonance curve 谐振曲线
resonance filter 共振滤波器
resonance frequency 谐振频率
resonance scattering surface 共振散射表面
resonance voltage 共振电压

resonant converter 谐振变换器
resonant response 共振响应
resonant system 共振系统
resonant tip response 共振叶尖响应
Resource Conservation and Recovery Act (RCRA) （美国）《资源保护与回收法案》
resource potential 资源潜力
resource uncertainty 资源不确定性
response amplitude operator 反应振幅运算子
response gust factor 响应阵风因子
response magnitude 反应量
response time 响应时间
responsibility system of project legal person 项目法人责任制
Ressel rectifier wave energy conversion device 拉塞尔整流波能转换装置
rest energy 静止能量
restrictor 限制器，节流阀
resultant field 复合磁场
resultant force 合力，总反力
resultant wind 合成风
resultant wind direction 合成风向
resulting pressure 最终压力
retainer 护圈
retardation of wind 风力减弱
retarding force 制动力
retreating blade 后行桨叶
return convection 回返对流
return flow 回流
return header 回流集管
return hose 回流软管
return information 返回信息
return line 回流线，回流管路
return manifold 返回集管
return pipe 回流管

return tank 回流箱
return water 回水
returning polar air 回流极地气团
revaporization 二次蒸发，再汽化
revegetation 植被重建
revenue stream 收益源，收入流
reversal of the monsoon 季风转换
reverse current 反向电流，逆流
reverse current heat exchanger 逆流热交换器
reverse cycle 逆循环
reverse electrodialysis 反电渗析
reverse flow 逆流
reverse generation 反向发电
reverse impeller 反向叶轮
reverse induced current 反向感应电流
reverse osmosis 反渗透，逆向渗透
reverse osmosis desalinating 反渗透脱盐
reverse osmosis desalination 逆渗透淡化，逆渗透脱盐
reverse osmosis membrane 反渗透膜
reverse osmosis process 反渗透过程
reverse osmosis separation 反渗透分离
reverse pumping 反向抽水
reverse return 反向回路
reverse return piping 反向回路管道
reverse saturation current 反向饱和电流
reverse thermosyphon 反热虹吸
reversed current heat exchanger 逆流热交换器
reversed electrodialysis (reverse dialysis) 反向电渗析
reversed tide 逆潮

reversed wind 反向风
reversible adiabatic compression 可逆绝热压缩
reversible cell 可逆电池
reversible chemical reaction heat storage 可逆化学反应储热
reversible cycle 可逆循环
reversible process 可逆过程
reversible propeller 活叶螺旋桨
reversible tidal current 往复潮流
reversing operation 反循环操作
reversing valve 换向阀，可逆阀，回动阀
revolution 旋转
revolving fluidised bed 旋转流化床
Reynolds number 雷诺数
rhyolite 流纹岩
ria 河口，溺湾（指狭长的楔形入海口，其形成原因为原本的河谷因海平面上升或地形沉降而被海水入侵）
rib 翼肋
ribbon cell 带状电池
ribbon silicon solar cell 带状硅太阳电池
rice husk 稻壳
ridge impinge 脊线撞击
rig floor 钻台
right angle 直角
right-of-way fee 开路费
rigid beam 刚性梁
rigid body 刚体
rigid coupling 刚性联轴器
rigid rotor 刚性转子，刚接式旋翼，整体转筒
rigid support 刚性支承
rigidity gear 刚性齿轮
rim turbine 轮缘涡轮机

rime 霜凇
rim-generator turbine 全贯流式水轮机
ring closing 合环
ring die pelletizer 环模制粒机
ring earth electrode 环形接地体
ring fracture 环状裂缝
ring gauge 环规
ring gear 环形齿轮，齿圈
ring generator 环生成元
ring opening 解环
ring operation of a part of a network 电网局部环式运行
ring-cam hydraulic pump 环形凸轮液压泵
rip current 离岸流，裂流
ripple current 波纹电流，脉动电流
ripple voltage 脉动电压
riser 电杆
riser cable 吊索
riser pipe 升流管
rising air 上升气流
rising flow 上升流
rising gust 上升风
rising out of synchronism 超出同步
rising tide 涨潮线
river basin 流域
river delta sediment 三角洲沉积物
river energy 河流能量
river power station 径流式电站
river stage 水位
Robinson's anemometer 鲁滨逊风速表
robot weather buoy 机器人气象浮标
robotic claw 机械爪
rock bed heat storage unit 岩石床储热装置
rock catcher 采石机

rock cuttings 岩屑
rock formation 岩层，岩系
rock fracture technique 岩石破碎技术
rock fragments 岩屑
rock layer 岩层
rock pressure 岩石压力
rock property 岩石性质
rock salt 岩盐，石盐，矿盐
rock strata 岩层
rock-water interaction 水岩相互作用
rock-fill breakwater 堆石防波堤
rocking grate 摇动炉排，摇杆炉排
rocking wave power device 摆动式波能装置
rod 连杆
roller 柱销套
roller bearing 滚柱轴承
roller bit 牙轮钻头，旋转钻头
roller conveyor 滚筒（柱）式输送机
roller grate 辊式炉条
roller mill 滚压机
rolling mill furnace 轧机炉
rolling terrain 丘陵区
roof collector （太阳房）顶部集热器
roof mounted collector array 屋顶集热器阵
roof pond collector 屋顶蓄热池集热器
roof solar absorptance 屋顶太阳能吸收率
roof still 屋顶式蒸馏器
roof top solar collector 屋顶太阳能集热器
roof top solar concentrator 屋顶聚光集热器
roof type solar still 屋顶式太阳能蒸馏器
room temperature 室温，常温

room temperature change 室温变化
root airfoil 翼根翼型
root circle 齿根圆
root loss factor 叶根损耗系数
root mean cube wind speed 均立方风速
root mean square value 均方根值
root of blade 叶根
root region 叶根
root vortex 叶根涡流
rossby wave 罗斯贝波，罗斯比波
rossby-gravity wave 罗斯比重力波
rotameter 转子式测速仪
rotary 旋转的，转动的
rotary anemometer 旋转式风速计
rotary arm disconnector 旋转臂式隔离开关
rotary atomizer 旋转雾化，旋转雾化器
rotary cone bit 旋转锥形钻头
rotary converter 旋转变流机，转动换流机
rotary current 旋流
rotary drilling 旋转钻孔，回转钻进
rotary engine 旋转发动机，转轮式发动机
rotary gas heater 回转式烟气换热器
rotary hearth furnace 转底炉，环形加热炉
rotary hydraulic pump 旋转液压泵
rotary kiln 回转窑，回转炉
rotary kiln incinerator 回转式焚烧炉
rotary packer 回转包装机
rotary table 转盘，回转台
rotary valve 旋转阀
rotate 旋转
rotating beam 回旋射束
rotating blade 转动叶片，旋转叶片

rotating coordinate system 转动坐标系
rotating drum 旋转鼓轮，转筒
rotating field 旋转场
rotating grate 旋转栅
rotating guide vane 转动导流叶片
rotating magnetic field 旋转磁场
rotating mass 转动质量
rotating shaft 转轴，旋转轴，回转轴
rotating speed 转速，回转速度
rotating stall 旋转失速
rotating time 纯钻井时间
rotating union 旋转接头
rotating vane 转叶（旋转叶片）
rotation angle 旋转角
rotation axis 转动轴
rotation energy 转动能
rotation frequency 转动频率
rotation of the earth 地球自转
rotational component 旋转分量
rotational energy 转动能
rotational flow 有旋流
rotational frequency 旋转频率
rotational instability 转动不稳定性
rotational kinetic energy 转动动能
rotational level 转动能级
rotational period 旋转周期
rotational sampling 旋转采样
rotational speed 转速，旋转速度
rotational stiffness 转动刚度
rotationally sampled wind velocity 旋转采样风矢量
rotations per minute 每分转数
rotator 转子，旋转体
rotator axis 转子轴，风轮轴
rotor 风轮，转子，水平旋翼，旋转体
rotor angle （发电机）转子角

rotor area 风轮区域
rotor axial thrust 转子轴向推力
rotor azimuth angle 转子方位角
rotor bar 转子导条
rotor bearing 转子轴承
rotor blade 转子叶片
rotor blade theory 风轮叶片理论
rotor blade twist 风轮叶片扭曲
rotor brake 旋翼制动器，转子制动器
rotor cage 转子笼
rotor conductor 转子导体
rotor conductor resistance 转子导体电阻
rotor core 转子铁芯
rotor critical speed 转子临界转速
rotor current 转子电流
rotor diameter 风轮直径
rotor disc 风轮转动圆盘面
rotor disk 转子轮盘
rotor dynamics 转动动力学
rotor eccentricity monitor 转子偏心度监视器
rotor efficiency 转子效率
rotor field 转子磁场
rotor field coil 转子磁场线圈
rotor frontal area 风轮迎风面积
rotor geometry 转子尺寸
rotor hub 风轮轴
rotor inertia 转子惯量，转能惯量
rotor interface 风轮界面
rotor lamination 转子铁芯片
rotor leakage reactance 转子漏磁电抗
rotor locking device 风轮锁紧装置，风轮闭锁装置
rotor loss 转子损耗
rotor mechanical speed 转子机械速度

rotor of low tip speed ratio 低叶尖速度比风轮
rotor of medium tip speed ratio 中叶尖速度比风轮
rotor performance modelling 转子性能模型
rotor plate 动片
rotor pole 转子磁极
rotor power coefficient 风能利用系数
rotor power efficient 风轮功率系数
rotor pre-cone 风轮预锥角
rotor punching 转子冲片
rotor radius 转子半径
rotor resistance 转子电阻
rotor rotation moment 风轮转矩
rotor shaft 风轮轴,转轴
rotor slip 转子转差率
rotor solidity 风轮实度
rotor speed 风轮转速
rotor spindle 转子轴
rotor stress field 转子应力场
rotor swept area 风轮扫掠面积,风轮迎风面积
rotor system 转子系统
rotor thrust 风轮推力
rotor thrust power spectrum 风轮推力功率谱
rotor tilt 风轮斜置角度
rotor torque 风轮力矩
rotor torque fluctuation 风轮扭矩波动
rotor torque standard deviation 风轮扭矩标准差
rotor wake 风轮尾流
rotor whirl mode 风轮涡旋状态
rotor winding 转子绕组
rough sea 大浪
roughness band 粗糙带

roughness coefficient 粗糙度系数
roughness element 粗糙元
roughness length 粗糙长度
round robin experiment 周游实验
round rotor 整圆转子
round trip conversion efficiency 往返转换效率
round trip energy efficiency 循环能源效率
round-bottomed clay vessel 圆底形白土容器
round-bottomed metal vessel 圆底形金属容器
roundtable on sustainable biofuel 可持续生物燃料圆桌论坛
routine test 常规试验
rubber belt conveyor 橡胶带输送机
rubber bumper 橡胶缓冲器
rubber tube 橡皮管
rubbing action 摩擦作用
rule of twelfths 十二分规则
run away speed 极限转速
run condition 运行工况,工作条件
run empty 空转
run of the wind 风程
run off 径流
run on time 运转时间,连续时间
run out load 失载负荷
runner blade servomotor 转轮叶片接力器
runner cone 泄水锥
running current 正常工作电流
running gear 行动装置
running program 运行程序
running temperature 工作温度
runoff 径流
runoff intensity 径流强度

run-of-river power station　径流式发电厂
run-of-river system　径流式系统
rural energy-ecology pattern by biogas　以沼气为纽带的农村能源生态模式
Russel rectifier　鲁瑟尔整流器
Russel rectifier wave power plant　鲁瑟尔整流器型波能电站

S

saccharification　糖化（作用）
safe distance　安全距离
safe life　安全寿命
safe marking　安全标志
safe value　安全值
safer impedance　安全阻抗
safety color　安全色
safety margin　安全界限，安全裕度，安全系数
safety system　安全系统
safety valve　安全阀
safety voltage　安全电压
safety zone　安全区，安全地带，安全区域
sag　垂度
sail assisting　风帆助航
sail wind generator　帆翼风力发电机
sail windmill　帆翼风车
sail wing　帆翼
sail wing blade　帆翼叶片
sailing ship　帆船
sailwing propeller　帆翼式螺旋桨
sailwing rotor　帆翼风轮
sailwing windmill　帆翼风车
salt detector　盐分探测器
salient　扇形地背斜轴
salient pole　凸极
salient pole rotor　凸极转子

salification　积盐
saline　盐水，含盐的，咸的
saline energy　盐渗透能
saline intrusion　海水入侵，盐水入侵
saline matter　盐分
saline sea　盐海，海
saline solution　盐溶液
saline water　海水，咸水
saline water conversion　盐水转化
saline-free water　淡水
salineness　含盐度
salinification　盐化作用，盐渍化
salinity　盐度，盐分，盐性
salinity bridge　盐度电桥
salinity cell　浓淡电池
salinity cell energy conversion　浓淡电池式浓度差能发电
salinity difference power generation　海洋盐浓度差发电
salinity energy　盐（度）差能
salinity gradient　盐（度）差，盐度梯度
salinity gradient energy　（海水）盐（度）差能
salinity gradient energy conversion　浓度差能转换，盐（度）差（度）能转换
salinity gradient　盐度梯度，含盐量梯度
salinity indicator　盐量计，盐液密度计
salinity membrane　盐度差膜
salinity power　盐动力
salinity power device　盐度差发电装置
salinity power plant　盐度差电站
salinity-gradient energy　盐度差能
salinity-gradient energy converter　盐（度）差（度）能转换装置
salinity-gradient vapor-pressure power conversion　蒸气压差式盐度差能发电
salinometer　盐液比重计

salt accumulator 储盐器，蓄盐器
salt battery 盐电池
salt concentration 含盐浓度
salt concentration gradient 盐浓度梯度，盐浓度差
salt content 盐分
salt crust 盐结皮
salt deposit 盐沉积
salt detector 盐分探测器
salt diapir 含盐底辟
salt diffusion in solar pond 太阳池的盐分扩散
salt dome 盐丘
salt fog 盐雾
salt gradient pond 盐梯度太阳池
salt hydrate 水合盐
salt intrusion 盐侵
salt penetration 渗盐
salt rejecting membrane 脱盐膜
salt spray 盐雾
salt water corrosion 盐水腐蚀，海水腐蚀
salt water inlet 咸水入口
salt wind 盐风
salt windmill 盐场风车
Salter duck 索尔特鸭式装置（一种像鸭子在水中上下摆动的波浪发电装置）
salt-heat transfer fluid exchanger 熔盐—传热流体换热器，熔盐换热器
salting liquid 盐溶液
salting liquor 盐水，盐汁
salting-out evaporator 脱盐蒸发器
saltmarsh 盐沼
saltpan 盐田
saltwater battery 海水电池
saltwater intrusion 盐水浸入作用，海水浸入作用

saltwater-activated battery 海水电池
salvage 废物利用
sample 样品
sample circuit 取样电路
sample division 样品缩分
sample preparation 样品制备
sample reduction 样品破碎
sampled system 采样系统，取样系统
sampling of geothermal fluid 地热流体取样
sampling rate 采样率，抽样率
sampling test 抽样试验
sampling unit 采样单元
sampling voltage 取样电压
sand bearing wind 含沙风
sand driving wind 挟沙风
sand moving wind 起沙风
sand pumping 沙泵，泥沙泵，抽沙泵
sand reel 捞沙滚筒
sandbank 沙丘，沙洲，沙滩
sanitary sewer system 污水排放系统
sarcina mobilis 活动性八叠菌
satellite altimeter 卫星高度计
satellite power system 卫星能源系统
satellite solar power station 卫星太阳能发电站
satin film 半透明膜
saturate steam 饱和蒸汽
saturate steam pressure 饱和蒸汽压
saturated adiabatic process 饱和绝热过程
saturated air 饱和空气
saturated Carnot vapor cycle 饱和蒸汽卡诺循环
saturated steam 饱和水蒸气，饱和蒸汽
saturated vapor 饱和蒸汽

saturated vapor pressure 饱和蒸汽压
saturated water 饱和水
saturation 饱和
saturation adiabat 湿绝热线
saturation area 饱和区域
saturation base current 饱和基极电流
saturation characteristic 饱和特性
saturation condition 饱和状态，饱和条件
saturation current 饱和电流
saturation energy 饱和能量
saturation line 饱和线
saturation mixing ratio 饱和混合比
saturation point 饱和点
saturation pressure 饱和压力
saturation specific humidity 饱和比湿
saturation vapor pressure 饱和蒸汽压
saturation voltage 饱和电压
saturation water vapor pressure 饱和水汽压
saturation-adiabatic lapse rate 湿绝热直减率
saturation-adiabatic process 湿绝热过程
saturation-adiabatics 湿绝热曲线
Savonius 萨伏纽斯式
Savonius rotor 萨伏纽斯转子
Savonius windmill 萨伏纽斯风车
sawdust 锯屑
sawmill 锯木厂
sawmill waste 锯木厂垃圾，锯木厂废弃物
scale coefficient 水垢热阻系数
scale deposition 规模沉积
scale deposit 水垢
scale effect 规模效应
scale of thermometer 温（度）标

scale of wind-force 风级
scale parameter 尺度参数，尺度母数，标尺参数
scaler 除垢器，计数器，换算器，定标器
scaling inhibitor 阻垢剂
scaling loss 结垢损失
scaling potential 潜在结垢
scaling relation 比例关系，尺度关系
scanning electron microscopy 扫描电子显微镜
scanning voltage 扫描电压
scatter diagram 散布图，分布图，散射图，离散图形
scattered amplitude 散射振幅
scattered beam 散射束
scattered electron 散射电子
scattered field 散射场
scattered light 散射光
scattered particle 散射粒子
scattered pattern 散射条纹
scattered radiation 散射辐射，散射线，散辐射
scattered ray 散射线
scattered reflection 漫反射
scattered wave 散射波
scatterer 散射体
scattering 散射
scattering absorption coefficient 散射吸收系数
scattering amplitude 散射幅度
scattering angle 散射角
scattering area 散射面积，扩散面积
scattering area coefficient 散射面积系数
scattering attenuation coefficient 散射衰减系数

scattering coefficient 散射系数，漏损系数
scattering collision 散射碰撞
scattering cross section 散射截面
scattering effect 散射效应
scattering factor 散射因子
scattering force 散射力
scattering function 散射函数
scattering interference 散射干涉
scattering law 散射定律
scattering loss 散射损失
scattering mechanism 散射机理
scattering medium 散射介质
scattering parameter 散射参数
scattering particle 散射粒子
scattering reaction 散射反应
scattering solar irradiance 散射（日射）辐照度
scattering solar radiation 散射日射
scattering state 散射状态
scattering threshold 散射阈
scattering vector 散射矢量
scattering-in 内散射
schematic 原理图
Scherbius machine 谢尔比斯电机
Scherbius variable-speed drive 谢尔比斯变速传动
Schottky barrier cell 肖特基势垒电池
Schottky diode 肖特基二极管
Schottky solar cell 肖特基太阳电池
screen 筛分机
screen display 屏幕显示
screen printing 丝网印刷术
screen printing process 丝网印刷工艺
screw conveyor 螺旋输送机
screw feeder 螺旋加料器
screw rolling mill 螺旋轧机

scrubber 洗涤器，辐射防护
scrubber tank 洗气罐
scum 浮垢
scum layer 浮渣层
sea bed 海床，海底
sea breeze 海风
sea breeze front 海风锋
sea chart 海图
sea cliff 海崖，海蚀崖
sea condition 海况
sea depth transducer 深海换能器
sea floor 海床
sea land wind 海陆风
sea level elevation 海拔高度
sea main thermocline 海洋主温跃层
sea map 海图
sea ridge 海脊
sea solar power 海洋太阳能
sea space constraint 海洋空间约束
sea spectra 海浪谱，海洋光谱
sea state 海况
sea state spectrum 波浪谱
sea surface roughness 海面粗糙度
sea surface temperature 海面温度
sea thermal power generation 海洋热能发电
sea thermal power plant 海洋热能电站
sea trial 海上试航
sea water battery 海水电池
sea water by distillation 海水蒸馏
sea water condenser 海水冷凝器
sea water corrosion 海水腐蚀
sea water desalination 海水淡化
sea water desalting plant 海水淡化装置
sea water osmotic pressure energy 海水渗透压能
sea water primary battery 海水原电池

sea water pump 海水泵
sea water self purification 海水自净
sea water temperature difference power generation 太阳能海水温差发电
sea wave device 海浪装置
sea wave electric power system 海浪发电系统
sea bed mounted system 海床安装系统
seafloor spreading 海底扩展
seaflow 洋流
sea-land breeze 海陆风
sealed system 密封系统
sea level pressure 海平面气压
seamless pipe 无缝管
seamless power 无缝功率
seamount 海底山
search algorithm 搜索算法
seasonal control （太阳能的）季节控制
seasonal efficiency （太阳能利用的）季节效率
seasonal heating load 季节性供热负荷
seasonal operating plan 季运行方式
seasonal performance factor 季节性能系数
seasonal resource 季节性资源（指风力资源）
seasonal storage 季节性蓄能
seasonal storage of energy 能量的季节储存
seasonal total beam radiation （太阳能）直接辐射季总量
seasonal variation 季节性变化
seasonal variation of sunlight 阳光季节性变化
seasonal weather pattern 季节性气候类型

seasonal wind 季风
seasonally averaged daily directnormal solar flux 季平均法线太阳能日通量
seawall 海塘，防波堤
seaward slope 向海坡面
seawater air conditioning 海水空调
seawater corrosion 海水腐蚀
seawater osmotic pressure 海水渗透压能
seawater osmotic pressure energy 海水渗透压能
seawater salinity gradient power generation 海水盐差发电
seawater thermometer 海水温度表
seaweed 海藻，海草
second battery 二次电池
second class pyranometer 二级（工作）总日射表
second class pyrheliometer 二级（工作）直接日射表
second class pyrradiometer 二级（工作）全辐射表
second generation device 改进型发电设施，第二代装置
second law of thermodynamics 热力学第二定律
second water-pumping windmill 第二级提水风车
secondary air 二次空气，二次风
secondary air injection 二次空气喷射
secondary air supply 二次空气供应
secondary battery 二次电池，蓄电池
secondary cell 二次电池，蓄电池
secondary circulation 二次环流，次环流
secondary combustion air 次级燃烧空气
secondary combustion chamber 后燃烧室

secondary concentration 二次聚光
secondary concentrator 二次聚光器
secondary current 二次电流
secondary emission 二次发射，次级发射
secondary emission cell 二级发射光电管
secondary energy 二次能源
secondary energy source 二次能源
secondary exchanger 二次（离子）交换器
secondary fermenter 二级发酵罐
secondary fluid 二次热媒
secondary heat exchanger 二次热交换机
secondary mineral 次生矿物
secondary permeability 次生渗透性
secondary pollutant 次生污染物（质），二次污染物（质）
secondary radiation 二次辐射
secondary raw material 次生原料
secondary reflection 二次反射
secondary separation 二次分离
secondary standard pyranometer 二等标准总日射表
secondary standard pyrheliometer 二等标准直接日射表
secondary standard pyrradiometer 二等标准全辐射表
secondary standard solar cell 二级标准太阳电池
secondary steam 二次蒸汽
secondary treatment 二次处理
secondary voltage 二次电压
secondary wave energy conversion device 波能二次转换装置
secondary wind direction 次级风向
secondary wind direction effect 次生风效应
secondary winding 二次绕组，次级绕组
section drag coefficient 翼型阻力系数
secure bay 安全海湾
security coupling 安全联轴器
sediment 沉淀物，沉积物
sediment transport 泥沙输移，泥沙输送，沉积物迁移，沉积物搬运
sedimentary deposit 积淀
sedimentary layer 沉积层
sedimentary rock 沉积岩
sedimentation 沉降，沉淀
sedimentation tank 沉淀池，沉积槽
sedimentation unit 沉积装置
seepage 渗透（现象）
seesaw rotor 翘板式风轮
segmented mirror 瓦状（排列）反射镜
seiche 湖面波动，湖震，假潮
seismic activity 地震活动
seismic method 地震勘探法
seismic reflection 地震反射，地震波反射
seismic refraction survey 震波折射测勘
seismic refraction technique 地震折射技术
seismic sea wave 地震海浪
seismic velocity 地震速度
seismicity 地震活动性
seismogram 地震波曲线
selective absorption 选择性吸收
selective black paint 选择性黑色涂料
selective catalytic reduction 选择性催化还原

selective coating 选择性涂层
selective diffusion 选择性扩散
selective film 选择性薄膜
selective greenhouse 选择性温室
selective heat surface 热选择性表面
selective mirror 选择性反射镜
selective multilayer surface 多层选择性表面
selective non-catalytic reduction 选择性非催化还原
selective permeability 选择性渗透
selective radiation 选择辐射
selective receiver （太阳能）选择性接收器
selective reflection 选择反射
selective reflection plate 选择性反射板
selective reflector 选择性反射器
selective scattering 选择散射
selective scattering coefficient 选择散射系数
selective surface 选择性表面
selective surface absorber 选择性面吸收器
selective transmission 选择性透射
selective transmission film 选择性透射膜
selective transparent film 选择性透明膜
selectivity of radiation 辐射选择性
selenium cell 硒光电池
selenium oxide photovoltaic cell 氧化硒光电池，氧化硒光伏打电池
selenium photocell 硒光电池
self irradiation 自辐射
self leveling 自动找平，自动调平

self orientation 自迎风
self regulation 自调整
self spring 自拉
self-acting control 自行控制
self-balancing hot wire anemometer 自平衡热线风速计
self-cleaning membrane 自动滤净膜
self-commutated inverter 自换相逆变器
self-contained cell （太阳能）自给电池
self-discharge 自放电，自身放电
self-energy 自能
self-excitation 自励
self-excitation capacitance 自励电容
self-excited generator 自励发电机
self-ignition 自燃
self-ignition point 自燃（着火）点
self-orientating （风轮）自动调向，自动对风
self-potential 自然电位
self-potential method 自然电位法，自然电场法
self-propelled baler 自走式捡拾压捆机，自力推进压捆机
self-propelled pelletizer 自走式制粒机，自力推进制粒机
self-purging 自动净化
self-recording hygrometer 自记湿度表
self-recuperative burner 自动预热烧嘴
self-registering anemometer 自记风力表
self-registering barometer 自记气压表
self-registering thermometer 自记温度表
self-regulating windmill 自动调节风车
self-starting capability 自动起动

self-startup 自起动
self-synchronization 自同步
semi-aerodynamic shape 半气动形状，半流线型
semiautomatic valve 半自动阀
semi-closed cycle gas turbine 半闭式循环燃气轮机
semi-closed-jet wind tunnel 半闭口式风洞
semiconducting material 半导体材料
semiconductive photovoltaic refrigerator 半导体光伏冰箱
semiconductor 半导体
semiconductor alloy 半导体合金
semiconductor cell 半导体电池
semiconductor coating 半导体膜，半导体层
semiconductor electrode 半导体电极
semiconductor grade silicon 半导体级硅
semiconductor junction 半导体结
semiconductor layer 半导体层
semiconductor photoelectrode 半导体光电极
semiconductor power converter 半导体功率转换器
semiconductor rectifier 半导体整流器
semiconductor solar cell 半导体太阳电池
semiconductor thin film junction 半导体薄膜结
semiconductor-electrolyte interface 半导体—电解液界面
semiconductor-liquid junction 半导体—液体结
semiconductor-liquid solar cell 半导体—液体太阳电池

semiconverter 半额电能转换器
semicrystalline 半晶质，半晶质的
semidiurnal 每十二小时一次的
semidiurnal flow 半日分潮流
semidiurnal tide 半日潮
semi-dry flue gas desulphurization 半干法烟气脱硫
semi-dry process 半干法，半干压工艺
semi-dry scrubber 半干洗涤器，半干除尘器
semi-featuring propeller 半活叶螺旋桨
semi-flash conditions 半闪蒸条件
semilog analysis 半测井图分析
semilog straight line 半对数直线
semi-open-jet wind tunnel 半开口式风洞
semi-osmotic membrane 半透膜
semipermeable membranes 半透膜
semipermeable membrane 半透膜
semi-submerged concrete caisson 半潜式混凝土沉箱
semi-submerged cylindrical section 半潜圆柱段
semi-submersible 半潜式
semi-thermal area 半热区
sending end voltage 发送端电压
sensible heat 显热，可感热
sensible heat energy storage 显热储存
sensible heat storage 显热储存（利用物质在温度升高/降低时吸收/放出热量的性质储存/释放热能）
sensitive load 敏感负荷
sensitivity 灵敏度
sensitivity analysis 灵敏度分析
sensitivity of following wind 调向灵敏性
sensor 传感器

sensor cable 传感器电缆
sensor coverage 探测覆盖
sensor input 传感器输入
sensorless drive system 无转轴侦测元件的驱动系统
separate collection 分类收集
separate collector 分离式集热器
separate combustor 独立燃烧器
separate gasholder digester 分离式沼气池
separate refrigeration load 分散式制冷负荷
separated collector 独立式太阳能集热器,分立式太阳能集热器
separation pressure 分离压力
separator 分离器,隔离膜
separator vessel 分离器容器
septic tank 化粪池
sequential batch anaerobic composting 序批式厌氧堆肥
sequential order of the phase 相序
serial correlation 序列相关
serial resistance 串联电阻
series block 串联组件
series circuit 串联电路
series coil 串联线圈
series configuration 串联布置
series connected 串联的
series connected cell 串联电池
series connection 串联
series diode 串联二极管
series of cells 电池组
series of inter-connected rafts 相互连接的筏串
series resistance 串联电阻
series resistance of solar cell 太阳电池的串联电阻

series resonance 串联谐振
series-connected 串联的
series-connected system 直流系统
service conditions 操作条件,使用情况,工作情况
service life 使用寿命,使用期限
service load 工作负载,实用载荷
service main 接户线
service temperature 使用温度,工作温度
service voltage 额定工作电压
serviceability limit state 使用极限状态
serviceable conditions 工作情况,运行情况
servo actuator 助力器,伺服机构
servomotor 接力器
servomotor foundation plate 接力器基础板
servomotor stroke 接力器行程
set point 设定点
set point tracing 照射方向自动跟踪
set point valve 设定值
setter 定型机
setting angle of blade 叶片安装角
setting pond 沉淀池
setting pressure 设定压力
settling chamber 沉降室
settling pit 沉降坑,沉淀槽,沉沙池
settling tank 沉降池,固体沉降槽
setup control 设置控制,功能调节
sewage disposal system 污水处理系统
sewage oxidation pond 污水氧化池,污水氧化塘
sewage sludge 污水污泥
sewage sludge bed 污泥床
sewage treatment 污水处理
sewage treatment plant 污水处理厂

sewage water 污水
sewer system 污水管道系统
shade disk 遮光片
shade ring 遮光环
shading 遮蔽，阴影，明暗
shading factor 阴影因子
shading loss 阴影损失
shadow 阴影，遮蔽，静区，盲区
shadow band 遮光带
shadow effect 荫蔽效应，阴影效应
shaft 轴
shaft furnace 竖炉，竖式炉
shaft mine 竖井开采矿
shaft power 轴功率
shaft seat 轴座
shaft speed control 转速控制
shaft torque 轴扭矩
shaker 振动器，混合器
shallow bore hole 浅孔
shallow crustal structure 浅层地壳构造
shallow diffused cell 浅扩散电池
shallow ditch 浅沟，小沟
shallow earthquake 浅层地震
shallow geothermal energy 浅层地热能
shallow resistivity survey 浅层电阻率测量
shallow sea deposit 浅海沉积
shallow subsurface 地表浅层
shallow subsurface temperature 地表浅层温度
shallow surface thermometry 地表浅层温度测量
shallow temperature surveying 浅层温度测量
shallow water 滩海，浅水，浅海
shallow water wave 浅水波

shallower well 浅井
shape parameter 形状参数，谱轮廓参量
sharp-nosed blade 尖头叶片
shear flow 剪切流
shear loading 剪切荷载
shear of wind 风的切变
shear stress 剪切力，剪切应力，剪应力
shear wave 剪切波，横波
shear wind 切变风
sheave 槽轮
sheet bar furnace 板坯加热炉
sheet resistance 薄层电阻
sheet resistivity 薄层电阻率
sheet slot 钣金槽
shell evaporator 封闭式蒸发器
shell-and-tube 壳管式
shell-and-tube heat exchanger 壳管热交换器
shelter belt 防风林带，防护林带（保护农作物的）
shelter thermometer 百叶箱温度表
sheltered area 避风水域，遮蔽区
sheltered bay 避风（河）湾
shielding 屏蔽，防护
shielding gas 保护气体
ship propeller 船用螺旋桨，船用推进器
ship-shape hull 船型壳体
shoaling 浅滩，浅水区
shock absorber 减震器，缓冲器
shock blow stress 冲击应力
shock load 冲击荷载
shock voltage 电压脉冲
shoreline 海岸线，海岸线地带
short circuit current 短路电流

short circuiting 短路效应
short diode 短二极管
short duration gust 短期阵风
short duration loading 短期荷载
short fiber 短纤维
short rotation 短期轮作
short rotation forest 短轮伐期林
short rotation poplar 短期轮作杨树
short wave sunshine 短波阳光射线
short wavelength radiation 短波辐射
short ton 短吨
short-circuit characteristic 短路特性
short-circuit current 短路电流
short-circuit current density 短路电流密度
short-circuit flow 短路流
short-circuit operation 短路运行
short-circuit ratio 短路比
short-circuit voltage 短路电压
short-focus concentrator 短焦距聚光器
shorting voltage 击穿电压，闪路电压
short-term cut-out wind speed 短时切出风速
short-term exposure limit 短期暴露限值
short-term thermal storage 短期蓄热
short-term wind data 短期风资料
shortwave radiation 短波辐射
shortwave sharp cut-off color glass filter 短波端锐截止型有色玻璃滤光片
shortwave length radiation 短波辐射
shovel 挖掘机
shredded biofuel 破碎的生物质燃料
shroud 护罩
shroud aerogenerator 折流板风能发电机
shroud force 折流力
shrouded impeller 闭式叶轮
shrouded propeller 侧板螺旋桨，折流板螺旋桨
shrouded rotor 有罩风轮
shrouded turbine 闭式涡轮，带冠涡轮，闭式涡轮机
shrouded wind turbine 集风罩式风力机
shrouded windmill 折流板风力机
shrouded-to-rotor-area ratio 折流板—风轮面积比
shunt 分流器
shunt capacitor 并联电容器，分路电容器，旁路电容器
shunt impedance 分路阻抗
shunt power 并联功率
shunt resistance 并联电阻
shunt resistance of solar cell 太阳能电池的并联电阻
shut off valve 关闭阀，节流阀
shutdown 关机
shutdown for wind turbine 风轮机关机
shutdown wind speed 停车风速
shutoff valve 截止阀
SI unit 国际标准单位
SiC sleeve 碳化硅衬套
side gust 侧阵风
side headwind 侧逆风
side leading wind 侧顺风
side length 边长
side load 横向负荷
side stream 侧流，旁流，分流
Siemens process 西门子法（生产多晶硅的一种方法）
sieve-plate 筛板

signal circuit 信号电路
signal conditioner 信号调整器
signal modifier 信号调节器
signal-to-noise ratio (SNR) 信噪比
significant wave height 有效波高
silane method 硅烷法
silencer 消音器
silica aerogel 氧化硅气凝胶
silica brick 硅石，硅砖
silica extraction 二氧化硅萃取
silica-alumina cracking catalyst 硅铝裂化催化剂
silica-free brine 无硅盐水
silicate compound 硅酸盐化合物
siliceous sediment 硅质岩沉积
silicon block 硅块
silicon crystal 硅晶体
silicon ingot 硅锭
silicon nitride 氮化硅
silicon oil 硅油
silicon photocell 硅光电池
silicon photocell equi-energy response 硅光电池等能量响应
silicon photovoltaic cell 硅光伏电池
silicon photovoltaic device 硅光电装置，硅光伏装置
silicon P-N junction photocell （太阳能转换为电能的）P-N 结硅光电池
silicon polycrystal 多晶硅
silicon ribbon 硅带
silicon ribbon solar cell 硅带太阳电池
silicon rubber adhesive 硅胶粘合
silicon solar cell 硅太阳能电池
silicon solar energy converter 硅太阳能转换器
silicon wafer 硅片
silicon-controlled rectifier 晶闸管整流器
silicone 硅有机树脂
silicone weather shed 硅遮雨缘
silicon-solar-cell parameter 硅太阳能电池参数
silt 淤泥，泥沙
siltation 淤积，聚积
siltstone 砂泥岩
silver-disk pyrheliometer 银盘式直接日射计
silvered polymer mirror 镀银聚合物反射镜
silviculture 森林培育学，造林学
simple circuit 简单回路
simple cycle microturbine 简单回路微型燃气轮机
simple solar cooker 简易太阳灶
simple solar room-heater 简易太阳房采暖器
simplex transmission 单向传输
simultaneity factor 同时系数
sine response 正弦响应
sinewave 正弦波
single basin project 单库方案
single-basin double-effect scheme 单库双作用方案
single-basin double-effect scheme with pumping 单库加泵水双向方案
single-basin single-effect scheme 单库单作用方案
single-basin single-effect tidal power plant 单库单向式潮汐电站
single bladed rotor 单叶片风轮
single bladed windmill 单叶片风车
single board computer 单板计算机
single bore hole 单孔
single bore-hold technology 单井筒技术

single bore-hole technology 单井眼技术
single crystal cell 单晶电池
single crystal silicon solar cell 单晶硅太阳能电池
single crystalline silicon 单晶硅
single crystalline silicon solar cell 单晶硅太阳能电池
single day tide 全日潮
single direction decomposition 单向分解
single effect absorption chiller 单效吸收式制冷机
single effect evaporator 单级蒸发器，单效蒸发器
single effect solar still 单级太阳能蒸馏器
single fluid cell 单液电池
single glass cover 单层玻璃盖板
single glass covered flat plate collector 单层玻璃盖板平板集热器
single glazed black painted collector 单层透明盖板黑吸收表面集热器
single glazed collector 单层透明盖板集热器
single glazed flat plate air collector 单层透明盖板空气型集热器
single glazing clear glass 单层透明玻璃
single glazing heat-absorbing glass 单层吸热玻璃
single glazing reflecting glass 单层反射玻璃
single heliostat furnace 单定日镜太阳能炉
single layer glass cover flat plate collector 单层玻璃盖板平面集热器
single loop 单回路
single loop system 单回路系统

single pass heat exchanger 单程热交换器
single phase flow 单相流动
single planetary gear train 单级行星齿轮系
single P-N junction constant energy gap solar cell 单 P-N 结恒能带宽度太阳能电池
single P-N junction form energy gap solar cell 单 P-N 结均匀能带宽度太阳能电池
single point mooring system 单点系泊系统
single pool 单库
single pool one way tide working installation 单库单向潮汐工程设施
single reheat cycle 一次再热循环
single rope 单绳
single shaft 单轴
single solar cell 单体太阳电池
single squirrel cage rotor 单鼠笼式转子
single stage digester 单级消化池
single step of surface relief 单台阶式地表起伏
single tank storage 单箱蓄热
single thermal storage tank 蓄热单罐
single tubular collector 单管集热器
single weather rotor 单朝向风轮，定向风轮
single-basin scheme 单库方案
single-basin unidirectional power production scheme 单库单向潮汐能生产方案
single-basin-single effect scheme 单库单向方案

single-bladed propeller 单叶片螺旋桨
single-bladed wind turbine 单叶片风力机
single-cover collector 单层盖板集热器
single-cycle combustion turbine unit 单循环燃气轮发电机组
single-effect single-basin scheme 单向单库方案
single-effect straflo turbine 单效贯流涡轮机
single-effect tidal plant 单向潮汐电站
single-effect unit 单向装置
single-glazed collector 单层玻璃集热器
single-junction solar cell 单结太阳能电池
single-junction space solar cell 太空（用）单结太阳能电池
single-pane glass 单窗格玻璃
single-phase current 单相电流
single-phase flow 单相流
single-phase generator 单相发电机
single-phase pressure 单相压力
single-phase system 单相系统
single-phase value 单相位值
single-pool ebb-generation system 单库退潮发电系统
single-pool system 单库系统
single-pool tidal power project 单库潮汐能工程
single-pool two way tide working installation 单库双向潮汐功能设施
single-pool two-way system 单库双向系统
single-side 单面的
single-sided with surface magnet 单面表面磁铁

single-speed pump 单速泵
single-stage combustion 单级燃烧
single-stage gearbox 单级变速箱
single-tank solar-gas water heater 单罐太阳能热水器
single-valve filter 单阀滤池
sink temperature 水槽温度
sinter 泉华
sintering process 烧结法
sintering temperature 烧结温度
sinusoidal circuit 正弦电路
sinusoidal current 正弦电流，正弦式电流
sinusoidal input 正弦输入
sinusoidal signal 正弦信号
sinusoidal variation 正弦变化
sinusoidal voltage 正弦电压
sinusoidal wave 正弦波
sinusoidal wave generator 正弦波发生器
sinusoidal waveform 正弦波形
siphon drainback system 虹吸回排系统
SIS solar cell (semiconductor-insulator-semiconductor solar cell) 半导体—绝缘体—半导体太阳能电池
site accessibility 场地通达度
site assessment 现场评估
site control 现场控制
site control point 现场控制点
site electrical facilities 风场电气设备
site exploration 基址探测
site monitoring 工地监测
site planning 场地规划
site potential 厂址潜势
site substation 变电站
site-built collector 内装式集热器
site-specific geology 特定场地地质

sitting 选址
six-phase winding 六相绕组
size analysis sample 粒度分析样品
size reduction factor 尺度折减系数
size-reduction 破碎
skew distribution 偏斜分布
skew wind 斜风
skewed wake 歪斜流
skin effect 表皮效应，集肤效应
skin factor 表皮系数
skin friction drag 表面摩擦阻力
(effective) sky temperature 天空（有效）温度
sky radiation 天空辐射
slack tide 憩流
slack water 憩流，平潮
slag 炉渣，熔渣
slag deposit 结渣
slag formation 造渣
slag particle 渣粒
slag resistance 抗渣性
slag tap 放渣口
slagging 出渣，排渣
slagging Lurgi gasifier 液态排渣鲁奇气化炉
slagging propensity 结渣性
slick line 平直管线
slide valve 滑阀
sliding pressure operation of turbine 汽轮机滑压运行
sliding shoe 滑动制动器
slight air 软风
slimhole 小直径井，小井
sling hygrometer 摆动湿度计
sling psychrometer 摆动湿度计
sling thermometer 旋转温度表
slip 转差率

slip band 滑移带，滑脱带
slip flow 滑流
slip frequency 滑差频率
slip line 滑移线
slip power 转差功率
slip ring 集电环，滑环
slip system 滑移系
slipping rotor 滑动转子
slipstream 滑流，尾流
slipstream contraction 滑流收缩
slipstream velocity 滑流速度
slope convection 弱对流
slope down wind 下坡风
slope stability 边坡稳定性
slope up wind 上坡风
slope wind 坡风
slotted liner 割缝衬管
slow running rotor 低速运行风轮
slow sand filter 慢滤池
slow speed rotor 低速风轮
slow-burning fuel 耐火燃料
sludge 污泥
sludge degradation and stabilization 污泥降解和稳定
sludge dewatering 污泥脱水
sludge digestion pond 污泥消化池
sludge index 污泥指数
sludge treatment 污泥处理
sludge volume index 污泥体积指数
slug flow 段塞流体
sluice 水闸，流槽
sluice flow 闸流，孔流
sluice valve 闸式阀
sluiceway 闸沟，泄水道
slurry exchange 更换砂浆
slurry pit 泥浆池
slurry pump 泥浆泵，浆液泵

small container of phase-changing salt 小型相变盐蓄能器
small hydro 小水电
small on-grid PV power station 小型并网光伏电站
small reactance 小电抗
small rim angle tracking collector 小边界角跟踪式集热器
small wind energy system 小型风能工程
small wind system 小型风力发电系统
small wood 小木头
small-scale WECS 小型风能转换装置
small-size wind machine 小型风力机
smart control 智能控制
smart rotor blade 智能旋翼桨叶
Smith-Putnam machine 史密斯—普特南（试验风力发电）机
Smith-Putnam wind generator 史密斯—普特南风力发电机
Smith-Putnam wind turbine 史密斯—普特南风力涡轮机
smooth flow 平滑流
smooth sea 微浪
smooth wind 平滑风
sneaker wave 大型近岸波
Snell's law 斯涅尔定律
snow load 雪载
snow trap 集雪器
snow-melt photography 融雪摄影
soaking zone 均热区，均热段
soapstone 皂石，滑石
social environment 社会环境
sock 风向袋
sodium 钠
sodium chloride 氯化钠
sodium chloride water 氯化钠水
sodium ion 钠离子
sodium nitrate 硝酸钠
soft coal 烟煤
soft energy source 软能源
soft mode 软模式
soft radiation 软辐射
soft start 缓起动
soft starter 软起动器
softening 软化
softening point 软化点
soft-start circuit 软启动电路
software logic 软件逻辑
software platform 软件平台
softwood 软木材，软木，针叶树
soil amendment 土壤改良剂
soil carbon sequestration 土壤（有机）碳封存
soil conditioner 土壤改良剂，土壤调理剂
soil mineral 土壤矿物
soil temperature measurement 土壤温度测量
soil thermal diffusivity 土壤导温率
soil thermograph 土壤温度计
soil thermometer 土壤温度表
soil warming 土壤加温
soil warming area 土壤加温区
Solapak collector panel 索拉派克集热器板
solar 太阳的，日光的
solar (irradiance) simulator 太阳（辐照度）模拟器
solar minimum 太阳活动极小期
solar absorbing surface 太阳能吸收表面
solar absorbtivity 太阳能吸收率
solar absorptance 太阳（能）吸收率
solar absorption cooler 太阳能吸收冷

却器
solar absorption cooling 太阳能吸收式制冷
solar absorption and dehumidification system 太阳能吸收去湿系统
solar absorption factor 太阳能吸收因数
solar absorption index 太阳能吸收指数
solar absorptivity 太阳（能）吸收率，阳光吸收率
solar access 太阳光通路
solar activity 太阳活动
solar age 太阳能时代
solar air collector 太阳能空气集热器
solar air conditioner 太阳能空调机
solar air conditioning 太阳能空调
solar air conditioning cycle 太阳能空调循环
solar air heater 太阳能空气加热器
solar air heating collector 太阳能空气加热集热器
solar air heating system 太阳能空气加热系统
solar air mass 太阳大气光学质量
solar air system 太阳能空气系统
solar air-heating system 太阳能空气加热系统，太阳能干燥装置
solar altitude 太阳高度
solar altitude angle 太阳高度角
solar angle of incidence 太阳入射角
solar angle simulator 太阳角模拟器
solar annual tide 太阳周年潮
solar antapex 太阳背点
solar apex 太阳向点
solar application 太阳能应用
solar architecture 太阳能建筑
solar array 太阳能电池阵，太阳能装置阵列
solar assisted absorption 太阳能辅助吸收
solar assisted absorption cooling system 太阳能辅助吸收冷却系统
solar assisted cooling 太阳能辅助冷却
solar assisted gas energy system 辅以太阳能的气体能量系统
solar assisted heat pump system 太阳能辅助热泵系统
solar assisted liquid pressurized biogas digester 太阳能水压式沼气池
solar atmospheric tide 太阳大气潮
solar attenuation loss 太阳能传输损失
solar automobile 太阳能汽车
solar azimuth 太阳方位，太阳方位角
solar azimuth angle 太阳方位角
solar balloon 太阳能空气气球
solar battery 太阳（能）电池
solar battery cell 太阳能电池
solar beam spread 太阳光散
solar bioconversion efficiency 太阳能生物能转换效率
solar bioconversion process 太阳能生物能转换过程
solar biological process 太阳能—生物能（转换）过程
solar biotechnology 太阳能生物技术
solar boiler auxiliary boiler 太阳能锅炉辅助锅炉系统
solar boiler-steam turbine system 太阳能锅炉—蒸汽轮机系统
solar breeder 太阳能反应堆
solar building 太阳能建筑
solar burst 太阳爆发，太阳暴
solar calculator 太阳能计算器
solar capacity 太阳能出力，太阳能容量

solar capital 太阳能投资
solar car 太阳能汽车
solar cavity receiver 太阳能吸热器
solar cell 太阳能电池
solar cell absorption efficiency 太阳能电池吸收效率
solar cell area 太阳电池面积
solar cell array 太阳能电池方阵
solar cell basic plate 太阳电池底板
solar cell battery 太阳能电池组
solar cell for satellite 卫星用太阳能电池
solar cell material 太阳电池材料,太阳能电池材料
solar cell module 太阳能电池组件
solar cell module area 太阳电池组件面积
solar cell module for residential house 住宅用太阳能电池
solar cell module surface temperature 太阳电池组件表面温度
solar cell panel 太阳能电池板
solar cell structure 太阳能电池结构
solar cell subarray 太阳电池子方阵
solar cell theory 太阳能电池理论
solar central receiver electric generation system 太阳能中心接收器发电系统
solar charge controller 太阳能充电控制器
solar chimney 太阳能烟囱
solar chimney-stack power generation 抽风式太阳能电站
solar close loop control system 太阳闭环控制系统,太阳能闭环控制系统
solar clothes 太阳能服装
solar coating 太阳能涂层,太阳能吸收涂层
solar collection area 太阳能集热面积

solar collection efficiency 太阳能集热效率
solar collector 太阳能集热器
solar collector assemblies 太阳能集热器组件
solar collector flow rate 太阳能集热器流体流速
solar collector hot water storage system 太阳能集热器储热水系统
solar collector on roof 屋顶太阳能集热器
solar collector orientation 太阳能集热器方向
solar collector panel 太阳能采集板
solar collector slope 太阳能集热器坡度
solar collector thermal power system 太阳能集热器热力发电系统
solar collector tilt 太阳能集热器斜度
solar collector tracking device 太阳能集热器追踪装置
solar collector tube 太阳能集热器吸收管
solar component tide 太阳因子潮
solar compression cooling 太阳能压缩式制冷
solar concentrating collector 聚焦性太阳能集热器
solar concentration 太阳能聚光
solar concentration factor 太阳能聚光系数
solar concentration ratio 太阳能聚光比
solar concentrator 太阳聚光器,太阳能聚光器
solar constant 太阳常数
solar contribution 太阳能供热量
solar control system 太阳能控制系统

solar controller 太阳能控制器
solar conversion efficiency 太阳能转换效率
solar conversion method 太阳能转换法
solar conversion technology 太阳能转换技术
solar converter 太阳能转换器
solar cooker 太阳灶
solar cooking 太阳能烹饪
solar cooling 太阳能制冷
solar cooling application 太阳能冷却应用
solar cooling house 太阳能冷却机房
solar cooling system 太阳能制冷系统
solar cooling technology 太阳能冷却（制冷）系统
solar cooling-heating system 太阳能冷却—加热（取暖）系统
solar corona 日冕
solar corpuscular 太阳微粒
solar cosmic ray 太阳宇宙射线
solar courtyard light 太阳能庭院灯
solar crude 太阳能总量
solar cycle 太阳活动周期
solar day 真太阳日
solar decentralized system 太阳能分散收集系统
solar declination 太阳赤纬，太阳偏角，太阳方位角，太阳倾斜角
solar degradation 太阳能材料老化，太阳能降解
solar dehumidification 太阳能除湿
solar dehydration facility 太阳能脱水设备
solar dehydrator 太阳能脱水器
solar desalination 太阳能海水淡化

solar desalination system 太阳能海水淡化系统
solar desalting 太阳能脱盐
solar detoxification 太阳能去毒
solar device 太阳能装置
solar diode 太阳能二极管
solar dish 太阳能聚光器
solar dish-Stirling engine 碟式—斯特林太阳能热发电
solar disinfection 阳光消毒法，日光消毒法
solar disk 日面
solar dispersion 太阳色散
solar distance 日（地）距
solar distillation 太阳能蒸馏
solar distillation plant 太阳能蒸馏厂
solar distillator 太阳能蒸馏器
solar distilling operation 太阳能蒸馏作业
solar distribution 太阳能分布
solar document bag 太阳能公文包
solar domestic hot water heater 家用太阳能热水器
solar domestic water heating 太阳能家用水加热
solar domestic water-heater 太阳能家用热水器
solar drier 太阳能干燥器
solar driven power generator 太阳能发电机（装置）
solar driven refrigeration unit 太阳能制冷设备（装置）
solar dryer 太阳能干燥器
solar drying experiment 太阳能烘干试验
solar drying process 太阳能烘干过程

solar drying system 太阳能干燥系统，太阳能烘干系统
solar ebb 太阳能活动低潮
solar eclipse 日食
solar economic analysis 太阳能经济分析
solar economics 太阳能经济学
solar economy 太阳能经济
solar ejector cooling 太阳能喷射式制冷
solar electric cell 太阳能电池
solar electric conversion 太阳能一电能转换
solar electric direct conversion 太阳能光发电
solar electric energy system 太阳能电能系统
solar electric equipment 太阳能发电设备
solar electric power generation 太阳能发电
solar electric source 太阳能电源
solar electric system 太阳能电力系统
solar electricity 太阳能发电，太阳能电力
solar electric-thermal flat-plate collector 太阳能电—热式平板集热器
solar elevation 太阳高度角
solar elevation angle 太阳高度角
solar energy 太阳能
solar energy application 太阳能利用
solar energy benefit guarant 太阳能效率保证（太阳能热水系统产权方和太阳能热水工程实施责任方之间的一种合同）
solar energy boiler 太阳能锅炉
solar energy charging light 太阳能充电灯

solar energy chemical heat pump 太阳能化学热泵
solar energy collection 太阳能收集
solar energy collector 太阳能集热器
solar energy computer model 太阳能计算机模型
solar energy concentrator 太阳能集热器
solar energy conversion 太阳能转换
solar energy conversion device 太阳能转换装置
solar energy conversion facility 太阳能转换设备
solar energy conversion system 太阳能转换系统
solar energy conversion technology 太阳能转换技术
solar energy converter 太阳能转换器
solar energy cost 太阳能成本
solar energy courtyard light 太阳能庭院灯
solar energy delivery 太阳能供给量
solar energy device 太阳能装置
solar energy distribution 太阳能分布
solar energy engine 太阳能发动机
solar energy engineering 太阳能工程
solar energy equipment 太阳能设备
solar energy flux 太阳能流密度，太阳能通量
solar energy for heating 供暖用太阳能
solar energy fridge 太阳能冰箱
solar energy generating system 太阳能发电系统
solar energy heat power station 太阳能热电站
solar energy heat pump 太阳能热泵
solar energy heat pump system 太阳能热泵系统

solar energy industry 太阳能工业
solar energy laboratory 太阳能实验室
solar energy lawn light 太阳能草坪灯
solar energy material 太阳能材料
solar energy measuring equipment 太阳能测量设备
solar energy output 太阳能输出（量）
solar energy panel 太阳能收集板
solar energy planning 太阳能规划
solar energy photothermal conversion unit 太阳能的光热转换器
solar energy polyresin crafts 太阳能树脂工艺品
solar energy power system 太阳能动力系统
solar energy resource distribution 太阳能资源分布
solar energy resource 太阳能资源
solar energy society 太阳能学会
solar energy source 太阳能源
solar energy storage 太阳能储存，太阳能储存器
solar energy street lamp 太阳能路灯
solar energy supply 太阳能供应（量）
solar energy system design 太阳能系统设计
solar energy system thermal process 太阳能热能转换过程，太阳能热过程
solar energy technology 太阳能工艺学
solar energy thermal collector 太阳能热量收集器，太阳能集热器
solar energy thermoionic conversion system 太阳能热离子转换系统
solar energy through biological system 通过生物系统的太阳能
solar energy traffic light 太阳能交通灯
solar energy traffic signal 太阳能交通信号灯
solar energy transmission 太阳能传输
solar energy use 太阳能利用
solar energy utilization 太阳能利用
solar energy utilizing device 太阳能利用装置
solar energy water heater 太阳能热水器
solar energy-chemical energy conversion unit 太阳能—化学能转换装置
solar engine 太阳能发动机
solar engineer 太阳能工程师
solar engineering 太阳能工程
solar era 太阳能时代
solar evaporation 太阳能蒸发（作用）
solar eyeball collector 眼式太阳能集热器
solar farm 太阳能农场
solar farm plant 太阳能发电厂
solar feasibility 太阳能的可利用性
solar field size 集热器阵列面积
solar film 太阳能薄膜
solar firing 太阳能点火
solar flare 耀斑，太阳闪射
solar flare measuring system 太阳能闪光测量系统
solar flashlight 太阳能手电筒
solar flat-plate collector 平板型太阳能集热器
solar flat-plate collector of flat-pipe absorber 扁管式太阳能平板集热器
solar flux 太阳辐射通量
solar flux amplification 太阳能通量放大率（倍数）
solar flux collector 太阳能（通量）集热器（收集器）

solar flux density 太阳能量密度
solar flux map 太阳能通量图
solar focusing collector 聚焦性太阳能集热器
solar fraction 太阳能指数
solar fuel 太阳能燃料
solar furnace 太阳(能)炉,太阳(能)灶
solar gain 太阳能获得量,太阳能增益
solar garden light for cities 城市太阳能庭院灯
solar garden light for towns 乡镇太阳能庭院灯
solar gas receiver 太阳能气体接收器
solar generating station 太阳能发电站
solar generator 太阳能发电机
solar grade silicon 太阳能级硅
solar grade silicon cell 太阳能级硅电池
solar greenhouse 太阳能温室
solar halo 日晕
solar heat 太阳热能
solar heat collection system 太阳能集热系统
solar heat collector 太阳能热量收集器
solar heat control window 太阳热能控制窗
solar heat engine 太阳能热机
solar heat exchanger 太阳能热交换器,太阳能换热器
solar heat exchanger drive 太阳热能驱动(装置)
solar heat gain 太阳能获得量,日照得热量,太阳热增益
solar heat power generation 太阳热发电
solar heat pump 太阳能热泵
solar heat pump system 太阳能热泵系统
solar heat rate 太阳能热率
solar heat storage 太阳能储热(器)
solar heat system 太阳能加(供)热系统
solar heated green house 太阳能采暖暖房
solar heated swimming pool 太阳能加热游泳池
solar heated thermionic converter 太阳能加热热离子转化器
solar heater 太阳能加热器
solar heating 太阳能供暖,太阳能供热,太阳能加热
solar heating and cooling 太阳能供热和冷却
Solar Heating and Cooling Demonstration Act 太阳能供热和冷却示范法律
solar heating and cooling system 太阳能供热和冷却系统
solar heating and hot water system 太阳能供热和热水系统
solar heating building 建筑物太阳能供热
solar heating collector 太阳能加热集热器
solar heating economics 太阳能供热经济学
solar heating equipment 太阳能供热设备
solar heating experiment 太阳能供热试验
solar heating installation 太阳能加热装置,太阳能取暖设备
solar heating orientation 太阳能加热取向,太阳能加热方向

solar heating system 太阳能供热系统，太阳能供暖系统，太阳能加热系统
solar heating technology 太阳能供热技术，太阳能加热技术
solar heating-cooling system 太阳能供暖—冷却系统
solar heat-storage pond 太阳能储热池
solar height 太阳高度（角）
solar home system 太阳能家庭发电系统
solar home without a southern exposure 不向阳太阳房
solar hot box 太阳能热箱
solar hot water 太阳能热水
solar hot water bath house 太阳能热水浴室
solar hot water heater 太阳能热水器
solar hot water system 太阳能热水系统
solar hot-air engine 太阳能热气机
solar hot-water heater 太阳能热水器
solar hour angle 太阳时角
solar house 太阳房，太阳能房
solar house heating 太阳能房供暖
solar humidification 太阳能热增湿（淡化）法
solar hydraulic engine 太阳能水力发动机
solar hydrogen energy system 太阳—氢能系统
solar hydrogen production 太阳能制氢
solar image 太阳映像
solar incidence 太阳入射角
solar incidence angle 太阳入射角
solar industrial process heat 太阳能工业加热

solar industry 太阳能工业，太阳能产业
solar infrared radiation 太阳红外辐射
solar input 太阳能输入
solar insolation 太阳照辐射量
solar insolation data 太阳辐射数据，太阳日照数据
solar installation 太阳能装置
solar intensity 太阳强度
solar interference boundary 太阳干涉边界
solar interference contour 太阳干涉等值线
solar investment 太阳能投资
solar irradiance 太阳辐射照度
solar irradiance error estimate 太阳辐照误差估计
solar irradiation 太阳辐照
solar irrigation pump 太阳能灌溉泵
solar irrigation pumping system 太阳能灌溉系统
solar isoflux line 太阳能等通量线
solar isoflux map 太阳能等通量图
solar lamp 太阳能灯具
solar laser 太阳能激光器
solar laser power station for space use 空间用太阳能激光动力站
solar lawn lamp 太阳能草坪灯
solar lawn light 太阳能草坪灯
solar lawn light for suburb 郊区太阳能草坪灯
solar liquid collector 太阳能液体集热器
solar liquid system 太阳能液体系统
solar load fraction 太阳能负载部分
solar load ratio method （性能估计的）太阳能负荷比法
solar longitude 日面经度
solar low grade heat 低品位太阳能热

solar luminosity 太阳光度
solar machine 太阳能机械
solar magnetic cycle 日磁周期
solar magnetic fluid power generation 太阳能磁流体发电
solar magnetism 太阳磁,太阳磁学
solar magnetograph 太阳磁像仪
solar mass 太阳质量
solar mass wall 太阳能(玻璃隔板)集热墙
solar material 太阳能材料
solar maximum 太阳活动高峰期
solar maximum temperature 最高日射温度
solar measuring instrumentation 太阳测量仪表
solar mechanical process 太阳能机械能(转换)过程
solar microwave relay station 太阳能微波中继站
solar module 太阳能电池模块,光伏模块
solar multiple 太阳倍数
solar neutrino 太阳中微子
solar noise 太阳(射电)噪声
solar noon 太阳正午
solar noon incidence angle 正午太阳入射角
solar optical fiber lighting 太阳光纤照明
solar optical material 太阳能光学材料
solar oriented architecture 朝阳建筑
solar paddle 太阳能电池叶片
solar panel 太阳能电池板,太阳能收集板,平板型太阳能集热器,太阳电池帆板,太阳翼,聚光板
solar panel test stand 太阳能收集板实验架

solar parabolic collector 太阳能抛物面集热器
solar parabolic dish power generation 盘式太阳能电站
solar parabolic trough power generation 槽式太阳能电站
solar parallax 太阳视差
solar photochemical conversion 太阳能光化学转换
solar photochemical process 太阳能光化学过程
solar photochemical system 太阳能光化学系统
solar photochemical technology 太阳能光化学技术
solar photoelectric collector 太阳能光电收集器
solar photon 太阳能光子
solar photosynthesis 太阳光合(作用)
solar photovoltaic cell 太阳能光伏电池
solar photovoltaic conversion 太阳能光电转换
solar photovoltaic conversion efficiency 太阳能光电转换效率
solar photovoltaic conversion technology 太阳能光电转换技术
solar photovoltaic energy system 太阳光伏能源系统
solar photovoltaic fence 太阳能电围栏
solar photovoltaic module 太阳能光电池组件
solar physicist 太阳(能)物理学家
solar physics 太阳(能)物理学
solar physiotherapy 太阳能理疗
solar plane 太阳能飞机
solar plastic 太阳能塑料

solar pond collection efficiency 太阳池集热效率
solar pond collector 太阳能水池集热器
solar pond efficiency 太阳池效率
solar pond plant 太阳池发电厂
solar pond power generation 太阳池发电站
solar pond thermal efficiency 太阳池热效率
solar pond thermal energy 太阳池热能
solar pool power generation 太阳池发电
solar position angle 太阳定位角
solar power 太阳能热动力
solar power generation 太阳能发电
solar power charger 太阳能充电器
solar power density 太阳能量密度
solar power direct conversion system 太阳能直接转换系统
solar power flat plate module 太阳能电池板组件
solar power heat engine 太阳能热机
solar power heat engine system 太阳能驱动的热机系统
solar power machine 太阳能动力机
solar power module's power output 太阳能电池组件输出功率
solar power plant 太阳能发电厂
solar power pump 太阳能动力泵
solar power satellite 太阳能电力卫星
solar power satellite station 太阳能卫星站
solar power source 太阳能源
solar power station 太阳能站
solar power system 太阳能供电系统
solar power thermal conversion system 太阳能热转换系统

solar power tower 塔式太阳能电站
solar power tower plant 太阳能收集塔发电厂
solar power tower system 太阳能收集塔系统
solar power unit demonstrator 太阳能源装置示范产品
solar power water pump 太阳能水泵
solar powered aeroplane 太阳能飞机
solar powered cooling 太阳能冷却,太阳能制冷
solar powered engine 太阳能发动机
solar powered facility 太阳能供电设备,太阳能供电装置
solar powered irrigation 太阳能动力灌溉
solar powered light beacon 太阳能灯标
solar liquid metal magnetohy dro-dynamic power generation system 太阳能液态金属磁流体发电系统
solar powered pump 太阳能泵
solar powered satellite station 太阳能卫星站
solar powered vehicle 太阳能车
solar preheating system 太阳能预热系统
solar preheating 太阳能预热
solar prepared hydrogen 太阳能制氧
solar probe 太阳能探测器
solar production of hydrogen 太阳能制氢
solar project for new villages 太阳能建设新农村工程
solar propulsion 太阳能推进
solar pump 太阳能泵

solar pumping device 太阳能抽运装置
solar PV power system for mobile communication signal station 移动通信信号站太阳能发电系统
solar radar 太阳能雷达
solar radiation 太阳辐射
solar radiation absorber 太阳能辐射吸收器
solar radiation concentrator 太阳辐射聚光器
solar radiation data 太阳辐射数据
solar radiation energy 太阳能辐射能
solar radiation error 太阳能辐射误差
solar radiation flux 太阳辐射通量
solar radiation instrument 太阳辐射测量仪器
solar radiation intensity 太阳照射强度
solar radiation map 太阳能辐射资源地图
solar radiation matching 太阳辐射匹配
solar radiation measurement 太阳辐射测量
solar radiation meter 太阳辐射计
solar radiation observation 太阳能辐射观测，日射观测
solar radiation rate 太阳辐射速率
solar radiation record 太阳辐射记录
solar radiation simulation 太阳辐射仿真，太阳辐射模拟
solar radiation spectrum 太阳辐射光谱
solar radiation thermometer 日射温度表
solar radio burst 太阳射能爆发
solar radio electron 太阳射线电子
solar radio noise 太阳射电噪声
solar radio wave 太阳射电波

solar radiometry 太阳辐射测量
solar radius 太阳半径
solar receiver 太阳能接收器，太阳能吸热器
solar receiver support tower 太阳能吸热器支撑塔
solar rechargeable pack 太阳能充电包
solar recorder 太阳能记录器
solar reflectance 太阳光反射系数，太阳光反射比
solar reflectance index 太阳能反射指数
solar reflecting surface 太阳能反射表面
solar reflectivity 太阳（能）反射率
solar reflector material 太阳能反光材料
solar refrigeration 太阳能制冷
solar refrigeration system 太阳能制冷系统
solar refrigerator 太阳能冷冻机，太阳能冰箱
solar regenerative system 太阳能再生系统
solar regulator 太阳能控制器
solar residential system 太阳能住宅系统
solar resource 太阳能资源
solar rights 太阳能使用权
solar rocket 太阳能火箭
solar roof 太阳能屋顶
solar roof collector 房顶太阳能集热器
solar roof pond system 房顶池太阳能装置
solar room-heater 太阳房采暖器
solar sail 太阳帆，光帆

solar sailer 太阳能帆船
solar salt 日晒盐
solar salt power 日晒盐度差能,太阳盐度差能
solar salt-water conversion 太阳能咸水转换,太阳能咸水淡化
solar satellite 太阳能卫星
solar saving 太阳能保存
solar science 太阳能科学
solar scientist 太阳能科学家
solar sea power plant 海洋太阳能电站
solar sea thermal power plant 太阳能海洋热电站
solar sea water still 太阳能海水蒸馏器
solar selective absorber 太阳能选择性吸收器
solar selective absorbing surface 太阳能选择性吸收表面
solar selective coating 太阳能选择(性)涂层
solar selective film 太阳能选择性薄膜
solar selective receiver 太阳能选择性接收器
solar selective surface 太阳能选择性表面
solar sensor 太阳能传感器
solar shading estimator 太阳覆盖层测量器
solar shield 阳光防护窗
solar signal lamp 太阳能信号灯
solar silicon 高纯度硅(光伏发电)
solar simulator 太阳能模拟器,太阳能仿真器
solar sorption cooling 太阳能吸附式制冷
solar source 太阳源
solar source in space 空间用太阳能源

solar space heater 太阳能加热装置,太阳能供热装置
solar space heating 太阳能空间供热,区域太阳能供暖
solar space heating facility 太阳能空间供热设备,太阳能区域供暖设备
solar space power plant 太阳能空间电站
solar spectral intensity 太阳光谱强度
solar spectrograph 太阳摄谱仪
solar spectrum 太阳光谱
solar spot 太阳能聚光灯,太阳黑点,日斑
solar spot light 太阳能聚光灯
solar steam 太阳发射微粒流
solar steam engine 太阳能蒸汽发动机
solar steam generating station 太阳能蒸汽发电站
solar steam plant 太阳能蒸汽发电厂
solar still 太阳能蒸馏器
solar still with external condensation 外冷凝式太阳能蒸馏器
solar Stirling hot-air engine 太阳能斯特林热空气发动机
solar storage tank 太阳能储罐
solar storage temperature 太阳能蓄热温度
solar storm 太阳风暴
solar stove 太阳能灶,太阳能炉
solar stream 太阳微粒流
solar street lamp 太阳能路灯
solar streetlight for rural road 乡镇公路太阳能路灯
solar strip concentrator 太阳能条型聚光器
solar swimming pool heater 太阳能游泳池加温装置

solar system 太阳能系统，太阳系
solar system capital 太阳能系统基本投资
solar system controller 太阳能系统控制器（装置）
solar system cost equation 太阳能系统费用（成本）方程
solar system economic optimization 太阳能系统最优经济性选择
solar system efficiency 太阳能系统效率
solar system future cost 太阳能系统的未来成本
solar system initial cost 太阳能系统的初始成本
solar tank 太阳能储存（水）箱
solar technology 太阳能技术
solar telephone 太阳能电话
solar telescope 太阳能望远镜
solar temperature 太阳温度
solar temperature difference power generation 太阳能温差发电
solar terrestrial physics 日地物理学
solar terrestrial space 日地空间
solar thermal and heat storage system 太阳热能与蓄热系统
solar thermal capacity 太阳能热容量
solar thermal central receiver system 集中式太阳能热发电系统
solar thermal collection system 太阳能集热系统
solar thermal collector 太阳能集热器
solar thermal collector design 太阳能集热器设计
solar thermal conversion 太阳能—热能转换
solar thermal distributed power system 分散式太阳能热发电系统
solar thermal electric power generation 太阳能热发电
solar thermal electricity 太阳能热发电
solar thermal energy 太阳热能
solar thermal energy storage 太阳能热储存
solar thermal panel 太阳能板
solar thermal power 太阳能热发电
solar thermal power facility 太阳能热力发电设备
solar thermal power generation 太阳能热力发电
solar thermal power generation system 太阳能热发电系统
solar thermal power generation system with tower and heliostat plant 塔式定日镜装置的太阳能发电系统
solar thermal power installation 太阳能热力发电装置
solar thermal power plant 太阳能热力发电站
solar thermal power plant capacity 太阳能热发电厂容量
solar thermal power pump 太阳能热力泵
solar thermal power station 太阳能光热发电站，太阳能热发电厂
solar thermal power system 太阳能光热发电系统，太阳能热力发电系统
solar thermal system 太阳能热（收集）系统
solar thermionic generator 太阳能热离子发电机
solar thermoelectric converter 太阳能热电转换器
solar thermal electric conversion 太阳

能热—电转换
solar thermoelectric generator 太阳能热电发电机
solar thermoelectric system 太阳能热电系统
solar thermoionic power generation 太阳能热离子发电
solar tide 太阳潮汐，日潮
solar time 太阳时
solar toothbrush 太阳能牙刷
solar total energy system 太阳能全能量系统
solar total irradiance 太阳总辐射
solar total radiation 太阳总辐射，总日射量
solar total-energy-system test facility 太阳能全能系统试验设施
solar tower 太阳能塔
solar tower power generation 塔式太阳能电站
solar town 太阳城
solar tracker 太阳跟踪器
solar tracker with shade disk kit 自动遮光装置
solar tracking concentrator 太阳能跟踪聚光器
solar tracking device 太阳能跟踪装置，太阳能自动跟踪装置，跟踪式太阳能装置
solar tracking water heater 跟踪式太阳能热水器
solar traffic sign 太阳能交通信号灯
solar transformity 太阳能转换率
solar transmission rate 太阳（能）透射率
solar transmissivity 太阳（能）透射率
solar transmittance 太阳（能）透射率

solar transport system 太阳能输送系统
solar turbo power plant 太阳能涡轮发电机
solar turboelectric drive 太阳能—涡轮发电驱动（装置）
solar type reaction 太阳型反应
solar ultraviolet radiation 太阳紫外线辐射
solar umbrella 太阳伞
solar utilizaiton 太阳能利用
solar utilization rate 太阳能利用率
solar utilization technique 太阳能利用技术
solar vapor generator 太阳能蒸汽发生器
solar vector 太阳矢经
solar village 太阳村
solar wall 太阳能集热墙
solar wall with two-sides 双面太阳墙
solar water boiler 太阳能热水锅炉
solar water collector 太阳能热水器
solar water distiller 太阳能净水器
solar water heater 太阳能热水器
solar water heating system 太阳能热水系统
solar water preheat system 太阳能热水器预热系统
solar water preheated 太阳能水预热器
solar water pump 太阳能水泵
solar water pump and fountain 太阳能水泵及喷泉
solar wind 太阳风，日射微粒流
solar window 太阳能窗口
solar yard lamp 太阳能庭院灯
solar zenith angle 太阳天顶角

solar-assisted heat pump 太阳能热泵，太阳能辅助热泵
Solarator collector panel 索拉雷托集热器板
Solarblak coating 索拉布莱克涂料
solar-cell efficiency 太阳能电池效率
solarcell power generation system 太阳能电池发电系统
solar-distilled water 太阳能蒸馏水
solar-energized cooling 太阳赋能冷却
solar-energized mode 太阳赋能模式
solar-energy device 太阳能转换装置
solar-excited laser 太阳能激光器
solar-generated power 太阳能发电
solar-generated thermal energy 太阳热能
solar-heat collector 太阳能集热器
solar-hydrogen economy 太阳能制氢经济
solar-hydrogen energy system 太阳能—氢能系统
solarigram 日射总量曲线
solarigraph 日射总量计
solarimeter 总日射表，太阳辐射强度计，日射总量表
solar-induced-current 太阳能感生电流
solarium 日光浴室
solarization 日晒作用，曝晒作用，阳光辐射作用，暴晒作用，负感现象
solar-only system 太阳能单独系统
solar-operated device 太阳能装置
solar-operated hot-air engine 太阳能热气机
solar-operated photovoltaic unit 太阳能光伏打装置，太阳能光电装置
solar-operated steam engine 太阳能蒸汽机
solar-operated turbine 太阳能涡轮机
solar-plus-supplementary system 太阳能带辅助热源系统
solar-power-driven fan 太阳能驱动的风扇，太阳能驱动的风机
solar-powered water pump 太阳能水泵
solar-powered electric-generating station 太阳能发电站
solar-powered heat engine 太阳能热机
solar-powered orbiting generating station 空间太阳能发电站，轨道太阳能发电厂，卫星太阳能发电站
solar-powered steam engine 太阳能蒸汽机
solar-produced electricity 太阳能发电
solar-thermal central receiver system 太阳能—热能集中收集系统
solar-thermal collector 太阳能热能集热器（收集器）
solar-thermal conversion facility 太阳能—热能转换设备
solar-thermal conversion system 太阳能—热能转换系统
solar-thermal distributed power station 太阳能—热能分布式发电厂
solar-thermal distributed receiver system 太阳能—热能分散收集系统
solar-thermal electric conversion 太阳能—热能的电能转换
solar-thermal electric energy conversion system 太阳能—热能电能转换系统
solar-thermionic system 太阳能—热离子系统
solar-to-biomass conversion efficiency 太阳能—生物质转换效率

solar-wind energy system 太阳能—风能系统
solar-wind energy hybrid system 太阳能风能混合系统
solar-wind power generator 太阳能—风能发电装置
solenoid 通电螺线管
solenoid valve 电磁阀
solenoidal magnetic field 螺线管磁场
solid biofuel 固体生物质燃料，固体生物燃料
solid biomass fuel 固体生物质燃料
solid coupling 刚性联轴节
solid diffusion 固态扩散
solid forming 固化成形
solid fossil fuel 固体化石燃料
solid fuel 固体燃料
solid fuel cell 固态燃料电池
solid fuel kinetics 固体燃料动力学
solid impurity 固体杂质
solid material storage unit 固体材料蓄热单元
solid oxide fuel cell 固体氧化物燃料电池
solid phase diffusion 固相扩散
solid phase heat storage 固相蓄热，固相储能
solid phase storage medium 固相蓄热介质，固相储能介质
solid phase thermal storage 固体相蓄热，固相储能
solid polymer electrolyte 固体聚合物电解质
solid recovered fuel 固体回收燃料，固体再生燃料
solid rock 固态岩石
solid solubility 固溶性，固溶度

solid state circuit 固体电路
solid state reaction 固态反应
solid state rectifier 固态整流器
solid storage 固相蓄热，固相储能
solid storage material 固相蓄热物质，固相储能物质
solid volume 固体体积
solid waste 固体垃圾，固体废物
solid-gas phase change heat storage 固—气相变储热
solidification 固化，凝固
solidification refining 凝固精炼
solidity 实度
solidity loss 实度损失
solidity ratio 稠度，固性比
solid-liquid equilibrium 固液平衡
solid-liquid phase change storage 固液相变蓄热
solid-liquid storage system 固液储能系统
solid-oxide fuel cell 固态氧化物燃料电池
solids retention time 固体停留时间
solid-solid diffusion 固—固扩散
solid-state inverter 固态逆变器
solid-state particle 固态粒子
sollar pest killer 太阳能灭虫器
solsticial tide 至点潮
soluble mineral 可溶矿物质
soluble monomer 可溶性单体
soluble salt 可溶性盐，可溶盐，易溶盐
solute 溶质
solution viscosity 溶液黏度
solvent 溶剂
sonic anemometer 声波风速计
sonic detection and ranging 声波探测和测距

sonic soot-blower 声波吹灰器
soot 烟灰，碳烟
soot blower 吹灰器
soot blowing 吹灰
sophisticated heat exchanger 高级热交换器
sorbent 吸附剂，吸着剂
sorghum 高粱，蜀黍
sorption with activated carbon 活性炭吸附
sorting cabin 分类仓
sound pressure level 声压级
sounding station 探测站
source impedance 电源阻抗
source of geothermal energy 地热能源
source of low temperature heat 低温热源
source of pollution 污染源
source of power supply 电源
source separated municipal solid waste 源分类城市生活垃圾
source temperature 热源温度
source voltage 源电压
source well 水源井
source-rock-fluid interface 源岩流体界面
source-separated waste 源分类垃圾
south equatorial current 南赤道洋流
south swine-biogas-fruit pattern 南方"猪沼果"模式
southerly-facing flat collector 朝南平面集热器
south-facing glazed window 向阳玻璃窗
south-facing vertical solar collector 向阳立式太阳能集热器

space charge 空间电荷
space charge density 空间电荷密度
space charge effect 空间电荷效应
space charge layer 空间电荷层
space charge region 空间电荷区
space cooling 空间制冷，空间冷却
space dimension for maintenance 维修空间尺寸
space energy system 空间能源系统，宇宙能源系统
space environment effect 空间环境效应
space harmonics 空间谐波
space heater 空间加热器，局部供热装置
space heating with solar energy 太阳能空间加热，太阳能区域供暖
space power system 空间能源系统，宇宙能源系统
space solar cell 空间太阳能电池
space solar energy 空间太阳能，宇宙太阳能
space solar energy station 空间太阳能发电站
space solar power station 空间太阳能发电站
space solar power system 空间太阳能系统
space vector 空间矢量
spaceborne measurement 星载测量
spallation drilling 分裂凿岩，断裂凿岩，破碎凿岩
spalling 散裂，剥落
spalling resistance 抗散裂强度
spanwise airfoil distribution 沿翼展分布
spanwise flow 展向流动，跨向流动
spar 翼梁

spar buoy 栓行浮标，杆状浮标
spare pump 备用泵，库存泵
spark gap 火花隙，电花隙
spark ignition 火花点火
spark ignition engine 点火式发动机，点燃式发动机
spark plug 火花塞
sparking voltage 击穿电压，放电电压
spark-over voltage 放电电压
spatial distribution 空间分布
spatial variation 空间变化
special polymer 特种聚合体
specialized control system 专用控制系统
specific capacity 单位产量
specific denitrification rate 反硝化速度
specific discharge 比流量，单位流量
specific geofluid consumption 地热流体消耗
specific gravity cell 比重电池
specific heat 比热
specific heat at constant pressure 定压比热
specific heat at constant volume 定容比热
specific heat capacity 比热容
specific heat of hot fluid 热流体比热
specific power output 比功率输出
specific production index 单位生产率，单位产出率
specific rated capacity 额定功率
specific resistance 电阻率
specific speed 比转速
specific surface area 比表面积
specific value 比值
specific volume 比容

specification 规格
specified load 额定负载
specified performance 设计性能
spectral analysis 频谱分析，光谱分析
spectral bandwidth 光谱带宽
spectral concentration of irradiance 辐（射）照度的光谱密集度
spectral distribution 谱分布
spectral distribution of sun's energy 太阳能光谱分布
spectral emittance 光谱发射，光谱发射率
spectral energy distribution 谱的能量分布
spectral energy distribution curve 光谱能量分布曲线
spectral gap 谱隙，谱空缺
spectral hardening 谱硬化
spectral irradiance 光谱辐照度$[E_\lambda]$
spectral irradiance distribution 光谱辐照度分布
spectral line 光谱线
spectral mismatch 光谱失配
spectral moment 谱矩
spectral photo irradiance 光谱光子辐照度
spectral photon irradiance 光谱光子辐照度$[N_\lambda]$
spectral pyranometer 分光总日射表
spectral resolution 光谱分辨率
spectral response （太阳能电池）光谱响应
spectral response of photoconductivity 光电导的光谱响应
spectral responsivity 光谱响应率
spectral sensitivity 光谱灵敏度，频谱灵敏度，光谱感光度

spectral shape 频谱形状，谱形
spectral signal 谱信号
spectral solar irradiance 太阳光谱辐照度
spectral-sensitivity characteristic 光谱灵敏度特性
spectrophotometric method 分光光度法
spectroscopic carbon rod 光谱碳棒
spectroscopy 光谱学，波谱学，分光镜检查
spectrum analysis 光谱分析
spectrum of a radiation 辐射光谱
spectrum splitting 频谱分裂
specular conductor 镜面导管
specular reflectance 镜面反射比
speed control 速度控制，转速控制
speed control with frequency signal 转速频率控制
speed governor 调速器
speed increaser 增速器
speed increasing gear pair 增速齿轮副
speed increasing gear train 增速齿轮系
speed increasing ratio 增速比
speed mode 速度模式
speed reduction gear 齿轮减速器
speed sensor 速度传感器，测速装置
speed stabilizing control 稳速控制
speed adjusting controller 调速控制器
spent fuel 乏燃料
spheral panel 球形面板
spherical collector 球形集热器
spherical concentrator 球面聚光器
spherical silicon solar cell 球型硅太阳电池

spill containment system 泄漏围堵系统
spill valve 溢流阀，旁流阀
spilling breaker 崩碎波，崩顶碎波
spillway 溢洪道，泄洪道
spillway draining system 溢洪道排水系统
spindle interface 主轴接口
spiral concentrator 螺旋聚光器
spiral concentrator cooker 螺旋聚光（太阳能）灶
spiral dynamic pump 螺旋动力泵
spiral turbine 蜗壳式叶轮机
spiral-type point focusing concentrator 螺旋型点聚焦聚光器
splat cooling 急冷技术
spline hydraulic pump 花键液压泵
splined coupling 花键连接
split Savonius wind turbine 开裂式萨沃纽斯型风轮机
split spectrum cell 分光谱电池
split spoon 对开式取土勺，劈管
splitting 解列
splitting point 解列点
spoiler 扰流器
spoiling flap 阻尼板
spoke 轮辐，（车轮的）辐条
spot downhole temperature 井下点温
spot wind 定点风
spotted gum 斑桉
spout 喷泉
spray chamber 喷雾室
spray dryer 喷雾干燥器，喷雾干燥机
spray dryer absorber 喷雾干燥吸收器，喷雾干燥减震器
spray flow 喷涂流体
spray gun 喷枪，喷涂枪，喷漆枪

spray humidifier 喷雾式加湿器，喷雾增湿器
spray nozzle 喷雾嘴
spray tower 喷淋塔
spray type heat exchanger 喷射式热交换器
spread voltage 分布电压
spreader stoker 抛煤机炉排
spreader stoker boiler 抛煤机锅炉
spring 弹簧
spring constant 弹簧常数
spring overturn （湖水）春季对流
spring pressure 弹簧压力
spring return 弹簧回位
spring tide 潮差
spruce 云杉
SPS (satellite power system) ground receiving station 卫星能源地面接收站
SPS (satellite power system) space transportation system 卫星能源空间传输系统
spur gear 直齿圆柱齿轮
spurious response rejection ratio 杂散响应抑制比
sputtering 喷溅涂覆法，真空镀膜
square law voltmeter 平方律特性电压
square value 平方值
square-cube law 平方—立方定律
squirrel cage induction generator 鼠笼感应（异步）电机
squirrel cage repulsion motor 鼠笼式推斥电动机
squirrel cage rotor (cage rotor) 鼠笼式转子
squirrel-cage 鼠笼式
SRTA (satationary reflector tracking absorber) collector 固定反射器跟踪吸收器式太阳能集热器
S-shaped rotor S型转子
stability limit 稳定极限
stability of following wind 调向稳定性
stability of motion 运动稳定性
stabilizer 稳定剂，稳定器
stable equilibrium point 稳定平衡点
stable manifold 稳定流形
stable stratification 稳定层结
stack loss 废气损失，烟道损失
stack sampling 烟道取样，烟道采样
stack solar cell 堆积光伏发电板
stack temperature 锅炉炉身温度
stack testing 栈测试
stacked solar cell 叠层太阳电池（级联太阳电池）
stacked volume 堆积体积
Staebler-Wronski effect 光致效应，本证衰减
stage test 分层试验，分段试验
staged combustion 分级燃烧
staged-air combustion 空气分级燃烧
staggered finned solar air heater 错列肋片太阳能空气加热器，错列翅片太阳能空气加热器
staging area 备料区，暂存区
stagnant layer 滞止层
stagnant wake 滞止尾流
stagnant water 停滞水，滞水
stagnation 闷晒
stagnation condition 滞止状态
stagnation temperature 闷晒温度
stagnation test 空晒（集热器内只有空气而不充入其他载热工质时接受太阳辐射的工作状态）
stainless steel body 不锈钢体
stall 矿坑
stall control 失速控制

stall control rotor 失速控制风轮
stall development region 完全失速区
stall point 失速点
stall proof 防失速的
stall region 失速区
stall regulated 失速调节
stall regulated rotor 失速式限速风轮
stall regulation 失速控制
stall torque 失速转矩
stalled area 失速区
stalled blade 失速叶片
stalled condition 失速状态
stalled flow 失速气流
stalling 失速
stalling load 停转负载
stand alone system 单机系统
stand-alone electric generation unit 独立发电单元
stand-alone photovoltaic power generation 独立光伏电站
stand-alone photovoltaic system 独立光电系统，独立光伏打系统
stand-alone power system 离网型发电系统
stand-alone PV power system 独立式光伏发电系统
stand-alone remote power generation 独立远程发电
stand-alone solar thermal system 独立式太阳能光热发电系统
stand-alone system 单独系统
stand-alone wind farm 离网型风电厂
stand-alone wind system 独立风力系统
stand-alone PV system 独立光伏发电系统
standard air pressure 标准大气压

standard atmospheric state 标准大气状态
standard cell 标准电池
standard cubic feet 标准立方英尺
standard deviation 标准偏差
standard deviation of wind velocity 风速标准差
standard dimension ratio 标准尺寸比
standard irradiance curve 标准辐照度曲线
standard light source 标准光源
standard liquid heating system 标准（太阳能）液体供热系统，标准（太阳能）液体加热系统
standard liquid system 标准液体系统
standard radiometer 标准辐射表
standard rating cycle 标准循环
standard sea level 标准海平面
standard solar cell 标准太阳电池
standard solution 标准溶液
standard source 标准光源，标准信号源
standard surface 标准表面
standard temperature and pressure 标准温度和压力
standard test conditions 标准测试条件（太阳电池的标准测试条件）
standard thermometer 标准温度表
standard uncertainty 标准误差
standard voltmeter 标准电压表
standard water 基准水，标准水
standardization of solar radiometer 太阳辐射计标准化
standardized grids 标准化电网
standardized wind speed 标准风速
stand-by battery 备用电池组
stand-by capacity 备用容量

stand-by condition 备用状态
stand-by diesel generator 备用柴油发电机
stand-by heat 备用热
stand-by heat loss 储存热损失
stand-by power 维持功率，备用电源
stand-by pump 备用泵
stand-by source of power 备用动力源，备用电源
stand-by station 备用电站，尖峰负荷电站
standing column 独立柱，站柱
standing column well 站列井
standing current 稳定电流，驻流
standing wave 驻波
standpipe 支撑管
standstill 停机
star connection 星形连接
star valve 星形阀
star voltage 相电压（星形接线）
star winding 星形绕组
starch 淀粉
starch crop 淀粉作物
starch-based ethanol 淀粉基乙醇
star-delta starter 三角形起动器
start relay 起动继电器
start up 起动
start up operation 起动运行
start up of turbine-generator unit 汽轮发电机组起动
starter gear 起动齿轮，起动装置
starting voltage 起动电压
starting wind speed 起动风速
starting current 起动电流
starting inrush current 启动冲击电流
starting loss 起动损失
starting motor 起动电动机

starting resistance 起动阻力
starting signal 起动信号
starting torque 起动力矩，起动扭矩
starting torque coefficient 起动力矩系数
starting transformer 起动变压器
starting transient 起动瞬变量
starting under windmilling conditions 风状态起动条件
starting wind speed 起动风速
starting wind velocity 起动风速
start-up burner 点火燃烧器，启动燃烧器
start-up cycle 起动系统，起动周期
start-up device 起动装置
start-up wind speed 起动风速
state information 状态信息
state of sea 海面状况
state space 状态空间
state-observer 状态观测器
state-of-charge 荷电状态
state-of-the-art power plant 现代电厂，最先进的发电厂
static bus impedance 静态总线阻抗
static concentrator 静态集光器
static electricity 静电
static free propeller 不受气流扰动的螺旋桨
static load testing 静载荷试验
static polymeric drilling fluid 静态聚合物钻液井
static pressure 静压力
static pressure hole 静压孔
static project cost estimate 静态工程投资概算
static strength 静强度
static synchronous compensator 静态型同步补偿器

static var compensator 静态无功补偿器
static var controller 静态无功控制器
static water 静水，静态水体
station absorber 固定吸尘器
station capacity 电站功率
stationary armature 固定电枢
stationary battery 固定电池组
stationary collector 固定集热器
stationary compound parabolic concentrator 固定式半球面聚光（太阳能）集热器
stationary concentrator 固定式聚光器
stationary flow 稳流，稳定流，稳态流
stationary honeycomb solar pond 蜂窝状固定太阳池
stationary pollution source 固定污染源
stationary random load 平稳随机负荷
stationary receiver tube 固定接受管，固定集热管
stationary reflector tracking-absorber spherical mirror collector 跟踪吸收器式固定球面反射镜集热器
stationary sensor 固定传感器
stationary spherical reflector tracking absorber solar collector 跟踪吸收器式固定球面反射太阳能集热器
stationary surface water 固定表面水
stationary vertical collector 固定立式集热器
statistical discrete gust method 孤立阵风统计方法
stator 定子
stator blade 静叶片，定子叶片
stator conductor 定子导线，定子导体
stator conductor resistance 定子导体电阻

stator core 定子铁芯
stator field 定子磁场
stator flux 定子磁链
stator flux wave 定子磁链定向波
stator flux wave speed 定子通量波速度
stator frequency 定子频率
stator housing 定子外壳
stator leakage reactance 定子漏抗
stator magnetic field 定子磁场
stator pole 定子杆
stator resistance 定子电阻
stator slot 定子槽
stator winding 定子绕组
status of bioelectricity in China 中国生物质发电的产业现状
status of hydropower development 水电发展现状
status of wind power in the world 国际风电产业化现状
stay ring 座环
steady current 稳定电流
steady load 恒定负载
steady solar simulator 稳态太阳模拟器
steady state 稳态
steady state aerodynamics 稳态空气动力学
steady state availability 稳态可用度
steady state operation 稳态运行
steady state stability limit 静态稳定限度，静态稳定极限
steady state value 稳态值，恒定值
steady uniform flow 稳定等速流动
steady wind 常风，稳风
steady wind load 定常风载
steady-state condition 稳态条件，稳态工况，稳态状况

steady-state current 稳定电流，稳态电流
steady-state heat loss 稳态热损失
steady-state heat transfer 稳定传热
steam accumulator 蒸汽蓄热器
steam bleed 抽汽
steam boiler 蒸汽锅炉
steam capacity 蒸发量
steam conductivity 蒸汽导电度
steam consumption 蒸汽量，汽耗
steam content 含汽量
steam cycle 蒸汽循环
steam deposit 蒸汽储量
steam drive 蒸汽驱动
steam driven 蒸汽驱动的
steam drum 上汽包，汽鼓，蒸汽鼓筒
steam drying 蒸汽干燥
steam dumping 排汽
steam field 蒸汽田
steam flashing system 蒸汽闪蒸系统
steam flow 蒸汽流量
steam flow meter 蒸汽流量表
steam flow rate 蒸汽流量率
steam gathering system 蒸汽采集系统
steam gauge 蒸汽压力表
steam generation system 蒸汽发生系统
steam generation unit 蒸汽发生装置
steam generator 蒸汽发生器
steam generator block 锅炉机组
steam heating system 蒸汽加热系统
steam jet 蒸汽喷射，蒸汽射流，蒸汽喷流，蒸汽喷嘴
steam jet ejector 蒸汽喷射器
steam jet pump 射汽泵
steam kettle 汽锅
steam lift 蒸汽提升器

steam lift pump 蒸汽提升泵
steam line 蒸汽管道
steam load 蒸汽负荷
steam main 蒸汽总管
steam out-put 蒸发量
steam parameter 蒸汽参数
steam phase 汽相
steam piston engine 活塞式蒸汽机，往复式蒸汽机
steam pocket 蒸汽包，汽袋
steam power plant 蒸汽动力装置，蒸汽发电厂，热电站
steam power station 蒸汽电站
steam pressure 蒸汽压力
steam processing 蒸汽处理
steam production 蒸汽产量
steam production rate 蒸汽生产率
steam production well 蒸汽生产井
steam purification 蒸汽净化
steam purity 蒸汽纯度
steam quality 蒸汽质量，蒸汽品质，蒸汽干度
steam radiator 蒸汽散热器
steam receiver 蒸汽吸热器
steam reformer 蒸汽转化炉，蒸汽转化器
steam reforming 蒸汽改质，蒸汽转化
steam reforming process 蒸汽改质过程，蒸汽转化过程
steam reheater 蒸汽回热器
steam relieving capacity 蒸发量
steam reservoir 蒸汽储层
steam separator 汽水分离器，蒸汽分离器
steam table 水蒸气表，蒸汽表
steam temperature 蒸汽温度
steam throughput 蒸汽流量

steam turbine 蒸汽轮机
steam turbine generator 汽轮发电机
steam turbine plant 汽轮机装置
steam type airheater 蒸汽加热空气预热器
steam vapor 水蒸气
steam vent 汽孔，蒸汽裂口
steam water cycle 汽水循环
steam water separation 汽水分离
steam well 蒸汽井
steam zone 蒸汽带
steam-electric generating station 蒸汽电站
steam-electric plant 火力发电厂
steaming ground 冒汽地面
steaming rate 蒸发率
steaming zone 蒸发区
steam-only reserve 蒸汽储层
steam-to-steam heat exchanger 汽—汽热交换器
steam-turbine-driven electrical generator 蒸汽涡轮机驱动的发电机，蒸汽涡轮发电机
steam-water mixture 汽水混合物
steamy 蒸汽的，蒸汽似的
steel alkaline battery 碱性蓄电池，铁镍蓄电池组
steel bushing 钢套
steel pipe at wellhead 井口钢管
steel rod 钢条，钢棍
steel tube 钢管
steepness 陡度
steering current 引导气流
steering gear 转向装置，转向齿轮
steering mechanism 转向机械
steering shaft 转向轴
steering system 转向系统
steering tool 导向工具
steering worm 转向蜗杆
Stefan-Boltzmann constant 斯蒂芬—波尔兹曼常数
Stefan-Boltzmann's law 斯蒂芬—波尔兹曼定律
stemwood 杆木
stemwood chip 杆木片
stemwood log 原木干材
step current 阶跃电流
step response 阶跃响应
step voltage 跨步电压
step-by-step control 逐级控制
step-down side 低压端，低压侧
stepless control 无阶调节，连续调节
stepper motor 步进电机
step-up gear 增速齿轮，增速传动装置
step-up transformer 升压变压器
stepwise controllable apparatus 逐步控制装置
STES (solar total energy system) 太阳能全能量系统
sticking valve 黏滞阀
stiff rotor 刚性转子
stiffness 刚度
still air 无风
still gas 蒸馏气体
still-water level 静水位，静水面
stimulation of geothermal energy resource 地热能源激励法
stimulation of geothermal system 地热系统激励
Stirling cycle 斯特林循环
Stirling engine 斯特林机
Stirling receiver 斯特林吸热器
Stirling regenerator 斯特林机回热器
Stirling solar hot air engine 斯特林太

阳能热空气发动机
Stirling thermal motor 斯特林电动机
Stirling-cycle engine 斯特林循环引擎
stirrer 搅拌器
stochastic load 随机荷载
stochastic method 随机性法
stochastic optimal control 随机最优控制
stochastic response 随机反应,随机响应
stochastic rotor thrust fluctuation 随机风轮推力波动
stochastic tower bending moment 随机塔架弯矩
stochastic wind field simulation 随机风场模拟
stochastic wind load 随机风载
stoker combustion 炉排燃烧
stoker combustor 炉排燃烧室,炉排燃烧器
stoker grate 加煤机炉排
Stokes drift 斯托克斯漂移
Stokes wave 斯托克斯波
stop valve 截止阀,断流阀
stop wind speed 停止风速
stop-gap emergency 临时紧急措施
storable 可储存的
storage basin 蓄水池
storage battery plate 蓄电池板
storage bed 蓄热床,蓄能床
storage bin 储料仓,储料箱,储存箱
storage capacitance 储藏容量
storage capacity 蓄热容
storage cell 蓄电池
storage cost (太阳能)储存成本(费用)
storage density 蓄热密度
storage effect 储热作用,储能作用
storage electrode 储能电极

storage energy density 储能密度
storage energy system 储能系统
storage facility 储藏设施,存储设备
storage floor 蓄热地板
storage fluid flow rate 蓄热流体流速
storage fluid heat capacity 蓄热流体热容量
storage hopper 储料斗
storage limit 蓄热极限,储能极限
storage loss 蓄能损失,储能损失
storage mass 蓄热质量
storage material 储热材料
storage of energy from solar energy 太阳能的能量储存
storage particle bed 蓄热粒子床
storage pond 蓄水池
storage pool 储存池
storage porosity 封闭孔隙度
storage power station 蓄能电站
storage reservoir 蓄水库,储水池
storage rock bed 蓄热岩石床
storage roof 蓄热屋顶
storage station 储热站,蓄能站
storage subsystem 储热子系统
storage system 储热系统,蓄能系统
storage tank 储存池,储水箱
storage tank regulation 储罐法规,特殊废弃物条例
storage temperature 储热温度
storage temperature peak 储存温度峰值
storage temperature swing 储存温度波动
storage utilization factor 储能利用因子
storage vacuum tube 储热式真空(集热)管

storage water heater 蓄热器，容器式热水器
storage-coupled 储能投运
stored energy 储能
stored energy system of wind power 风力发电储能系统
stored heat 储热（法）（估计地热田能量潜力的一种方法）
storm 风暴，暴雨，狂风
storm drum 风暴风量筒
storm duration 风暴期，暴雨历时
storm surge forecast 风暴潮预报
storm tide 风暴潮
storm warning 风暴警报
storm wave 风暴浪
storm-swept 受风暴破坏的
straight finned solar air heater 直列肋片太阳能空气加热器，直列翅片太阳能空气加热器
straight flow turbine 全贯流式水轮机
straight-bladed vertical axis wind turbine 直线翼垂直轴风力机
straight-flow turbine 直流式水轮机
straight-line plot 直线图
strain 应变
strain gauge 应变片，应变计
Strangford Lough SeaGen 斯特兰福特湾海洋发电站
strategic environmental assessment 战略环境评价
strategic environmental performance indicator 战略环境绩效指标
stratification 层理，成层
stratified downdraft gasifier 分层下吸式气化炉
stratified flow 分层次流（体）
stratified fluid 分层流体
stratified reservoir 层状热储
stratified-bed gasifier 分层床气化炉
stratigraphic section 地层剖面
stratigraphic thickness 地层厚度
stratigraphic-bound reservoir 沉积盆地型层状热储地热
stratigraphy 地层学，地层中的岩石组成
stratocumulus cloud 层积云
stratum structure 地层结构
straw 秸秆
straw combustion 秸秆燃烧
straw shredder 切碎机
stray inductance 杂散电感
stream erosion 河流侵蚀
stream tube 流管
streamline 流线，流线型
streamtube 流管
stress corrosion 应力腐蚀
stress cycle 应力周期
stress field 应力场
stress measurement 应力测量
stress profile 应力剖面
strip mine 露天矿
strip theory 切片理论
strip-cleaning of silicon wafer 硅片剥离清洗
stroboflash 频闪
strong breeze 六级风，强风
strong current control 强电控制
strong gale 九级风，烈风
strong mechanical resistance 抗机械强度高
structural dynamics 结构动力学
structural load 结构载荷
structural stiffness 结构刚度
structural support 结构支撑
structure of melt 熔体结构

stub pole 主梁
stud-bolt 双头螺栓
sub-bituminous coal 次烟煤
subcooled liquid 过冷液体,再冷却液
subcritical refrigerating cycle 亚临界制冷循环
submarine fresh water spring 水下冷水泉
submarine power cable 海底电缆
submarine river 海底流
submerged ash conveyor 淹没灰输送机
submerged pressure differential 海底压力差动式(多安装在近岸,固定于海底,由波浪造成的海水起伏带动差动器进行发电)
submersible 能潜水的,可沉入水中的
submersible motor 潜入式电动机
submersible pump 潜水泵
submersible turbine pump 潜油涡轮泵
submersible type sump pump 可潜式排水泵
subsample 分样
subsea cable 海底电缆
subsidence 沉降
subsolar point 日下点
substation 变电站
substitutional solid solution 置换型固溶体
substrate 基质,基片,基板,底层
substrate degradation 基质降解
subsurface aquifer 浅表面含水层
subsurface current 次表层流,面下流
subsurface fracturing 浅表层断裂
subsurface information 井下资料
subsurface structure 地下构造,浅埋结构
subsurface temperature 井下温度,地下温度,次表层温度
subterranean molten rock 地下熔岩
subterranean steam field 地下蒸汽田
subtransient reactance 起始瞬态电抗,次瞬态电抗
successive overrelaxation 连续超松弛
suction anemometer 吸管式风速表
suction jet 抽吸射流
suction line 吸引管线
suction pit 吸水坑,泥浆池
suction side 吸力面
sugar beet 甜菜
sugar cane production plant 甘蔗生产工厂
sugar crop 糖类作物
sugar-based diesel 糖基柴油
sugar-based ethanol 糖基乙醇
sulfide 硫化物
sulfide stress cracking 硫化物应力破裂
sulfidic ore 硫化物矿石
sulphide ore 硫化物矿石
sulphur content 含硫量,硫含量
summer cooling system 夏季制冷系统
summer heat gain 夏季热量增益
summer shade 夏季遮荫
sun 太阳
sun angle 太阳高度角,日照角
sun collector 太阳能收集器(集热器)
sun compass 太阳罗经
sun drying peat 太阳能干燥泥炭
sun effect 日光效应
sun equivalent hour 太阳能的等价小时
sun field angle 太阳张角

sun gear 太阳轮
sun hour 太阳小时
sun path from east to west 太阳自东向西的轨迹
sun shield 遮阳板
sun simulator 太阳模拟器
sun spectrum 光谱
sun tracker 太阳跟踪仪（器）
sun-earth distance 太阳—地球距离，日地距离
sun-following servosystem 太阳跟踪伺服系统
sun-following surface 太阳跟踪表面
sunfuel 日光燃料
sun-generated electric power 太阳能电源
sun-heat absorber 太阳能接收器
sun-illuminated area 日照面积，日照区域
sunlight focusing instrument for physiotherapy 阳光聚焦理疗仪
sunlight panel 阳光板（双层透明聚碳酸酯板。它是一种具有透光、高强、轻质、隔热、阻燃、耐候等优良性能的新型建筑材料，可以用作太阳能装置的透明盖板）
sunlight pattern （随季节与时间变化的）阳光（照射）型式
Sun-Light Plan 阳光计划（日本的一项全国性计划，其目标是开发除原子能以外的所有新能源）
sunlight power generation 太阳光发电
sunlight wind machine 日照型风力机
sunlight-transparent sheet 透光板
sun-path diagram 太阳行程图
sun-resistant plastic 耐晒塑料
sunrise 日出

sunset 日没
sunshine 日照
sunshine duration 日照时数
sunshine recorder 日照记录仪
Sunstor solar water heater 森斯托太阳能热水器
Sunstrap flat plate collector 森特拉普平板型集热器
sun-synchronous orbit 太阳同步轨道
sun-tracking controller 太阳跟踪控制器
sun-tracking mirror 太阳能定日镜
sun-tracking photovoltaic power generation 跟踪太阳光发电
sun-tracking structure 太阳跟踪系统结构
supercapacitor 超级电容器
supercharged boiler 增压锅炉
supercharger 增压器
superconducting coil 超导线圈
superconducting magnetic energy storage 超导磁储能
superconducting magnetic energy storage system 超导磁储能系统
superconductive material 超导材料
supercooling melt 过冷液体
supercritical boiler 超临界锅炉
supercritical cycle 超临界循环
supercritical fluid 超临界流体
supercritical pressure 超临界压力
supercritical steam condition 超临界蒸汽状态
supercritical steam pressure 超临界蒸汽压力
super-duty refractory 特级耐火材料
superheated 过热的，过热蒸汽，过热状态

superheated fluid 过热流体
superheated gas 过热气体
superheated steam 过热蒸汽
superheated steam dryer 过热蒸汽干燥机
superheated steam locomotive 过热蒸汽机车
superheated vapor 过热蒸汽
superheated water 过热水
superheater 过热器,过热炉,过热设备
superheater coil 过热蛇形管
superposition 叠加,重合
supersaturated steam 过饱和蒸汽
supersonic-flow anemometer 超音速(流)风速表
supervisory control and data acquisition system 数据采集与监控系统
supervisory control system 监控系统
supervisory controller 监控器
supplemental heat 补充热量
supplementary cooling equipment 辅助冷却设备
supplementary heat 额外热量
supply chain 供应链,供给链,供需链
supply header 供应集管
supply interruption cost 停电费用
supply line 补给线,运输线,供应线
supply voltage 电源电压
supply well 供水井
support post for glass plate 玻璃板支柱
support structure for wind turbine 风机支撑结构
supporting hub 支撑轮毂
surf zone 碎波带,碎波区
surface albedo 地表反照率
surface aluminated mirror 镀铝镜
surface casing 表层套管,地面套管

surface coating 表面涂层
surface condenser 表面式凝汽器,表面式冷凝器,表面冷却器,间壁冷凝器
surface corrosion 表面腐蚀
surface debris 叶面残片
surface diffusion 表面扩散
surface elevation 表面高度,水面标高
surface emissivity 表面发射率,表面辐射率
surface energy 表面能
surface evaporation 表面蒸发
surface evaporative condenser 表面蒸发式冷凝器
surface geomorphology 表面地貌
surface grinding 表面磨削
surface heat exchanger 表面式换热器
surface manifestation 地表显示
surface mine 露天矿
surface modification 表面改质,表面改性
surface observation 地表观察
surface piercing foil 穿面式水翼
surface relief 地表起伏
surface roughness 地面粗糙度
surface roughness effect 表面粗糙效应
surface site 地面现场
surface specific heat 地表比热
surface technology 地面技术
surface temperature 表面温度
surface temperature anomaly 地表温度异常
surface temperature value 表面温度值
surface tension 表面张力
surface tension effect 表面张力效应
surface thermal activity 地表热活动
surface thermal phenomena 表面热现象

surface water discharge area 地表水排泄区
surface water heat pump 地表水热泵
surface water intake area 地表水补给区
surface water loop 表面水循环
surface wave dispersion 面波频散
surface wind 地面风，地表风
surface-water heat pumps 地表水热泵系统
surfactant 表面活性剂
surficial sediment 表层沉积物
surge 大浪，巨浪
surge current 浪涌电流
surge protective device 浪涌保护器
surge protective device assembly 组合式浪涌保护器
surge suppressor 电涌保护区
surge tank 缓冲槽（罐）
surging breaker 激碎波，上涌破浪
surging-wave energy convertor 涌浪能量转换器
surplus activated sludge 剩余活性污泥，废活性污泥
surrounding air 周界风
surrounding air speed 环境风速
surrounding flow 环境环流
survivability 耐受性
survival wind speed 安全风速
suspended particle gasifier 悬浮颗粒气化炉
suspended solid 悬浮固体
suspension combustion 悬浮燃烧，火室燃烧
suspension firing 悬空燃烧
sustainable capacity 可持续容量
swash-plate 转盘
swather 割谷机
sweep current 扫描电流
sweet potato tree 番薯树
sweet sorghum 甜高粱
sweet water 淡水
swell 海涌，浪涌
swept area 后掠面积
swept area of rotor 风轮扫掠面积
swept volume 波及体积
sweptback blade 后掠桨叶
sweptback vane 后掠叶片
sweptback wing 后掠翼
swimming pool solar system 游泳池用太阳能系统
swing frame 回转框架
switch 开关
switch board 配电盘
switched reluctance 开关磁阻
switched reluctance generator 开关磁阻发电机
switched reluctance machine 开关磁阻电机
switched-off time 关闭时间
switchgear 开关设备，接电装置
switchgrass 柳枝稷
switching circuit 开关电路，转换电路
switching operation 切换运行
switching quadrant 开关象限
switchover standby power supply 备用电源自投（变电站备用电源自投入。当主供电源发生故障瞬间，自动控制开环充电线路或母线联络断路器即合闸，变电站恢复供电）
swivel 旋转
swivelling vane 回转叶片，风标
symmetric aerofoil turbine 对称翼型透

平，威尔斯透平
symmetrical airfoil 对称翼型
symmetrical tensor 对称张量
synchronization of two systems 两系统同步
synchronized sampling technology 同期采样技术
synchronizing breaker 同步断路器
synchronizing circuit 同步电路
synchronizing coefficient 同步系数
synchronizing lamp 同步指示灯
synchronizing power 同步功率
synchronous condenser 同步调相机，同步补偿机
synchronous generator 同步发电机
synchronous impedance 同步阻抗
synchronous machine 同步电机
synchronous motor 同步电机
synchronous reactance 同步电抗
synchronous speed 同步转速
synchronous starting 同步起动
synchronous voltage 同步电压
synchroscope 同步示波器
synergy 协同作用，增效
syngas 合成气
synodic month 朔望月
synthesis gas 合成气
synthesis route 合成路线
synthetic aperture radar 合成孔径雷达
synthetic diesel 合成柴油
synthetic fiber 人造纤维
synthetic gasoline 合成汽油
synthetic jet 合成射流
synthetic weather factor 综合气象因数
syntrophy 互养，共栖
system dynamics 系统动力学
system check 系统检验，系统检查

system clock 系统时钟
system dynamics 系统动力学
system fault 系统故障
system flushing 系统冲洗
system identification 系统识别
system meter 系统仪表
system parameter 系统参数
system power consumption 系统功耗
system sizing 系统规模
system software 系统软件
system stability 系统稳定性
system stiffness 系统刚度
system with effectively earthed neutral 中性点有效接地系统
syzygy tide 朔望潮

T

table-shaped turbine 桌型水轮机，T型水轮机
tail boom 尾撑
tail pulley 尾滑轮
tail rotor 尾桨
tail vane 尾舵
tailing edge 后缘
tailrace 尾水渠
tall oil 妥尔油，液体松香
tall tower 高塔
tallow 牛脂，兽脂，动物脂油
tan sandstone 棕黄色砂岩，棕褐色砂岩
tandem 串联
tandem converter 串联变换器
tandem solar cell 叠层太阳电池（级联太阳电池）
tandem solar cell 串接太阳电池
tangential flow induction factor 切向风流诱导因子

tangential force 切向力，切线力
tangential hydraulic compressor 正切液压压缩机
tangential stress 切线应力
tangential velocity 切向速度
tangentially fired boiler 切向燃烧锅炉
tank capacity 储水量
tank immersion heater 浸入罐体式加热器
tank testing 储罐检测，液舱试验
tank voltage 槽电压
tanker 油轮，运油飞机
tap position information 分接头位置信息
taper 锥度
tapered channel wave power station 收缩水道波流式电站
tapered roller bearing 圆锥滚子轴承
tapered waveguide 递变波导器，锥形波导，递变截面波导
tar-cracking gasifier 焦油裂解气化炉
target-aligned tracking 目标一定位跟踪
tax abatement 减税
tear resistance 抗扯力
tearing force 撕力
technical brake 机械制动
technical characteristic 技术特性
technical feasibility 技术可行性
technical parameter 技术参数
technical potential 技术潜势
technical resource 技术资源
technical specification 技术规范
tectofacies map 构造相图
tectonic plate 构造板块
tectonic setting 地质构造，构造背景
tectonic stress 构造应力

teetered hub 翘板型轮毂
teeth mesh 轮齿啮合
teflon film 聚四氟乙烯薄膜
telecommunication cable 通信电缆
telemonitoring 远程监视
teleseismic delay 远震滞后
teleseismic P-wave delay 远震 P 波滞后
teleseism 远震
tellerette 泰勒填料
temperate zone 温带
temperature effect 温度效应
temperature at earth's surface 地表温度
temperature average 平均温度
temperature barrier 挡热层，温度屏障
temperature booster 加热器，增热器
temperature coefficient 温度系数
temperature compensation 温度补偿
temperature condition 温度条件
temperature conductivity 导热性,热导率
temperature control loop 温度控制回路
temperature controller 温度控制器,温度调制器
temperature cycling testing 温度循环试验
temperature difference 温差
temperature difference energy 温差能
temperature difference heat engine 温差热机
temperature differential 温差
temperature diffusivity 温度扩散率
temperature distortion 受热变形
temperature disturbance 温度扰动
temperature drift 温度漂移

temperature equilibrium 温度平衡
temperature excess 温差
temperature factor 温度因数
temperature feedback 温度反馈
temperature field 温度场
temperature gradient 温度梯度
temperature gradient chromatography 温度梯度色谱法
temperature gradient hole 温度梯度孔
temperature gradient mass transfer 温度梯度质量迁移
temperature gradient measurement 温度梯度测量
temperature gradient method 温度梯度法
temperature gradient solution growth 温度梯度溶液生长
temperature gradient zone melting 温度梯度逐区熔化，温度梯度区域精炼
temperature gradient zone refining 温度梯度区域精炼
temperature head 温压
temperature history 温度历史
temperature imbalance 温度失衡
temperature increase in soil 土壤增温
temperature indicator 温度指示器
temperature instability 温度不稳定性
temperature inversion 逆温，逆温现象
temperature lapse rate 温度递减率
temperature log 温度测井图
temperature logger 温度记录器
temperature loss 热损失
temperature maintenance 温度维持
temperature measuring station 温度测位
temperature of geothermal fluid 地热流体温度
temperature of hot source 热源温度
temperature of hot spring 热泉温度
temperature pickup 温度传感器
temperature probe 温度传感器，测温探头
temperature profile 温度曲线
temperature regulator 温度调节器
temperature resistance 耐热性
temperature response 温度响应
temperature rise 温升，温度上升
temperature rise in soil 土壤温升
temperature scale 温标
temperature sensor 温度传感器
temperature shock 热冲击，温度骤变
temperature slope 温度梯度
temperature stability 温度稳定性
temperature stratification 温度层结
temperature survey 地热测量
temperature swing 温度波动
temperature transmitter 温度传热器，温度发送器
temperature-humidity index 温度—湿度指数
temperature-sensing equipment 温感设备
temperature-sensitive material 温敏材料
templifier 温度放大器
temporary facility for construction 施工临时设施
tensile energy 断裂能
tensile failure 拉伸破坏，拉伸断裂
tensile force 张力，拉力
tensile strength 抗张强度
tensile stress 拉伸应变
tensile structure 张力结构
tensile test 抗拉试验，拉力试验
tensility 延性

tension axis	张力轴
tension load	张力荷载
tensioned membrane	张拉膜结构
tensioned structure	张拉结构
tensor description	张量描述
tentative design	初步设计
terawatt (TW)	太瓦（兆兆瓦）
terawatt hour (TWhr)	太瓦小时
terminal	接线端
terminal moraine	终碛
terminal voltage	终端电压，端电压
terminator	终止子
terminator device	漫溢设备
terrain category	地形分类
terrain downwash	地形诱导下洗
terrain height	地形高度
terrain induced mesoscale system	地形性中尺度系统
terrain roughness	地形粗糙度
terrestrial array	地面方阵
terrestrial beam radiation	地面直接辐射量
terrestrial concentrator array	地面聚光器方阵
terrestrial current	地电流
terrestrial deposit	陆地沉积
terrestrial environment	地面环境
terrestrial flat plate array	地面平板方阵
terrestrial geothermal fluid	陆相地热流体
terrestrial heat	地热
terrestrial heat balance	地热平衡
terrestrial heat capacity	地热容量
terrestrial heat content	地热含量
terrestrial heat emission	地热放射
terrestrial heat energy	地热能
terrestrial heat flow	大地热流
terrestrial heat flow control	地热流量控制
terrestrial heat flow data	地热流量数据
terrestrial heat flow gauge	地热流量计
terrestrial heat flow meter	地热流量计
terrestrial heat flux	地热流量
terrestrial heat power station	地热电站，地热发电厂
terrestrial heat source	地热源
terrestrial heat storage	地热储存
terrestrial insolation	地面日照，地面太阳能辐射
terrestrial irradiance	地球辐照度，地面辐照度
terrestrial irradiance spectra	地面辐射光谱
terrestrial photovoltaic system	地面太阳能电池系统
terrestrial radiation	地球辐射
terrestrial refraction	地面折射
terrestrial solar cell array	地面太阳电池方阵
terrestrial solar-powered device	地面太阳能供电设备
tertiary circulation	三级环流，再次环流
tertiary pump	第三级热泵
tertiary tuyere	三级风口
test data	试验数据
test method	测试方法
test on bed	台架试验
test portion	试样量
test rig	测验台，试验装置
test site	试验场地
test well	探勘井
testbed	试验台
tethered balloon	系留气球

textured cell 绒面电池
textured silicon solar cell 绒面硅太阳电池
textured solar cell 绒面太阳电池
thalassothermal energy 海洋热能
thallium-sulfide photovoltaic device 硫化铊光电装置，硫化铊光伏打装置
thalweg 海谷底线，海谷深泓线，最深谷底线
thatched Dutch windmill 荷兰草屋风车
thatched windmill 草屋风车
the family of airfoil 翼型族
theorem 定理，原理
theoretical concentration 理论聚光
theoretical concentration ratio 理论聚光比
theoretical efficiency 理论效率
theoretical maximum value 理论最大值
theoretical resource 理论资源
thermal 热的，热量的，上升暖气流
thermal absorber 吸热器
thermal absorption 热量吸收，热中子吸收
thermal acceptor 热接收器
thermal activation crosssection 热（中子）激活截面
thermal analysis 热分析
thermal annealing 热退火
thermal baffle 绝热隔板
thermal barrier 绝热层，保温层，热障
thermal behavior 热性能
thermal boundary layer 热边界层
thermal capacitivity 热容率
thermal capacity 热容量
thermal capacity of storage 储热能力，储热容量
thermal centre 热中心
thermal chimney 热风筒，热烟气管
thermal circuit 热回路
thermal cleaning bath 地热清洗剂
thermal coefficient 热系数
thermal collector 集热器
thermal column 热柱
thermal compensation 温度补偿，热补偿
thermal condition 热工况
thermal conduction 热传导，导热性
thermal conductivity 热导率，热导系数
thermal conductivity detector 热导检测器，热导率探测器
thermal conductivity of rock 岩石热导体
thermal conductor 导热体
thermal contact 热触点
thermal convection 热对流
thermal convection current 热对流
thermal conversion 热转换
thermal conversion device 热转换装置
thermal converter 热中子转换反应堆
thermal cracking 热裂解，热裂化
thermal current 热电流，热流
thermal cycle 热循环
thermal decline 热衰减
thermal decomposition 热分解，热解，热分解法
thermal defect 热缺陷
thermal defector 热探测器
thermal degradation 热降解
thermal delay 热延时
thermal depolymerization 热解聚
thermal detector 热探测器
thermal diffusion 热扩散

thermal diffusion coefficient 热扩散系数
thermal diffusivity 热扩散率,热扩散系数
thermal discharge 放热
thermal dose 热中子剂量
thermal drop 热降
thermal drying 热干化
thermal drying of biomass 生物质热干化
thermal effect 热效应
thermal effectiveness 热力经济性
thermal efficiency 热效率
thermal efficiency of waste incineration boiler 余热锅炉热效率
thermal efficiency ratio 热效比
thermal emissivity 热辐射率,热发射率
thermal emittance 热发射(率)
thermal energy flux 热能通量
thermal energy from sea 海洋热能
thermal energy neutron 热(能)中子
thermal energy of ocean 海洋热能
thermal energy storage 热能储存
thermal energy storage capacity 储热容量
thermal energy storage system 热能储存系统
thermal engine 热机
thermal envelope 热封套
thermal equilibrium 热平衡
thermal equipment protection 热工保护
thermal equivalent 热当量
thermal etching 热侵蚀
thermal evaporation 热蒸发
thermal exchange 热交换
thermal excitation energy 热激发能量

thermal expansion 热膨胀
thermal expansion coefficient 热膨胀系数
thermal factor 保热系数
thermal flow measurement technique 热流测量技术
thermal flow-reversal reactor 热力双向流反应器
thermal gasification 热气化
thermal generator 火力发电机
thermal gradient 热能梯度,热梯度,地温梯度
thermal hydraulic modeling approach 热工水力建模方法
thermal hysteresis 热滞后
thermal imaging system 热像仪
thermal inertia 热惯性
thermal infrared technique 热红外技术
thermal insulation 热绝缘,隔热
thermal insulation shutter 绝热百叶窗
thermal intensity 热强度
thermal inversion 逆温
thermal inversion layer 逆温层
thermal island 热岛
thermal lag 热惯性,热传导迟缓
thermal leakage coefficient 热漏泄系数
thermal load 热负荷
thermal loss 热损失
thermal loss from collector 集热器热损失
thermal machine 热力发电机
thermal map of sea surface 海面热图
thermal mapping 热测绘图
thermal metamorphism 热变质
thermal mismatch 热失配

thermal mixing 热混合
thermal model 热模型
thermal motion 热运动
thermal network 热网络
thermal noise 热噪声
thermal oil 热采原油
thermal output 热功率
thermal oxidation 热氧化，热氧化作用
thermal parameter 热力指数，热力参数
thermal performance 热力特性
thermal plasma 热等离子体
thermal poison cross section 热害截面
thermal pollution 热污染
thermal potential 热势
thermal power 火力发电，热功率
thermal power generation plant 热发电厂
thermal power station 热电站
thermal pressure sealing 热压封
thermal probe 热探针，测温探针
thermal process protection 热工保护
thermal profile 热辊型，热剖面
thermal property 热性能，热特性
thermal pump 热泵
thermal radiation 热辐射
thermal radiator 热辐射器
thermal rating 热功率
thermal receiver 热接收器
thermal refraction index coefficient 折射率温度系数
thermal relaxation 热松弛
thermal relay 热继电器
thermal reservoir 地热储层
thermal resistance 热阻，热敏电阻，热变电阻

thermal resistivity 热阻率
thermal resistor 热敏电阻
thermal resource 热资源，热量资源
thermal response 热响应，热反应
thermal rising 热抗计，热泡抗计
thermal scattering 热散射
thermal sensitivity 热灵敏性
thermal shield 热屏蔽
thermal shielding 热屏，热防护层
thermal shock 热冲击，温度急增
thermal shock parameter 热震参数，热冲击系数
thermal shock resistance 抗热震阻力，抗热冲击阻力
thermal simulation 热模拟
thermal skin effect 热肤效应
thermal source 热辐射源，热源
thermal spalling 热剥落
thermal spraying 热喷涂
thermal spring 热泉，温泉
thermal stability 热稳定性
thermal steering 热控制
thermal storage 热储存，蓄热
thermal storage battery 蓄热器组，储汽器组
thermal storage capacity 蓄热能力，蓄热量
thermal storage device 蓄热器
thermal storage heat supply system 蓄热供热系统
thermal storage heater 蓄热式加热器
thermal storage material 储热材料
thermal storage of energy 蓄热储存
thermal storage rock bed 储热岩石床
thermal storage roof pond 储热层屋面池
thermal storage system 储热系统

thermal storage unit 储热装置，蓄热装置
thermal storage wall 储热壁
thermal strain 热变形，热应变
thermal stratification 热成层作用
thermal stress 热应力
thermal structure 温度结构
thermal subsea power source 海底热能源
thermal tar cracking 焦油热裂解
thermal time constant 热时间常数
thermal time delay relay 热延时继电器
thermal tower 热塔
thermal tracer 热示踪物
thermal transition 热跃迁
thermal transmission 热传递
thermal transmittance 热传递系数
thermal treatment 热处理
thermal unit 热量单位
thermal utilization 热利用
thermal vacuum 热真空
thermal vacuum cycling 热真空循环
thermal valve 热力阀，热动式调节阀
thermal velocity 热速度
thermal vibration 热振动
thermal voltage 热电压
thermal waste 热能废物
thermalchemical property of biogasification 生物质汽化的热化学特性
thermalization 热（能）化
thermal-lock 热锁
thermally active area 热显示区
thermally active zone 热显示区
thermally-conductive coating 热导电涂层
thermal-photovoltaic solar collector 热光电太阳能集热器

thermalstorage wall 储热墙，蓄热墙
thermistor 热调节器，热敏电阻
thermoacoustic-Stirling heat engine 热声斯特林热机
thermochemical biomass conversion technology 热化学生物质转换技术
thermochemical conversion 热化学转化
thermochemical conversion technology 热化学转换技术
thermochemistry reactor receiver 热化学吸热反应器
thermocline 温跃层，斜温层，变温层
thermocline thermal storage system 斜温层蓄热系统
thermocouple 热电偶
thermodiode solar plate 热二极管太阳板
thermodynamic calculation 热力学计算
thermodynamic coefficient 热扩散系数
thermodynamic conversion efficiency 热能转换效率
thermodynamic cycle 热力循环
thermodynamic cycle efficiency 热力学循环效率，热力循环效率，循环热效率
thermodynamic effectiveness 热力经济性
thermodynamic efficiency 热力学效率
thermodynamic equilibrium 热力平衡
thermodynamic force 热动力
thermodynamic irreversibility 热力不可逆性
thermodynamic loss 热力损失
thermodynamic parameter 热力学参数
thermodynamic potential 热力势
thermodynamic process 热力过程

thermodynamic property 热力学性质
thermodynamic scale 热力学温标
thermodynamic scale of temperature 热力学温标
thermodynamic surface 热力面
thermodynamics 热力学
thermodynamics of solid 固体热力学
thermoelectric 温差电，热电的
thermoelectric actinometer 热电日射表
thermoelectric cell 温差电偶，热电式感温元件
thermoelectric conversion 热电转换（作用）
thermoelectric conversion of solar energy 太阳能热电转换
thermoelectric conversion of solar radiation 太阳辐射热电转换
thermoelectric conversion system 热电转换系统
thermoelectric cooling 温差电制冷
thermoelectric effect 温差电效应
thermoelectric generator 热电发电机
thermoelectric inversion 温差电反转
thermoelectric junction 温差电偶接头
thermoelectric material 热电材料
thermoelectric metal 热电金属
thermoelectric neutral point 温差电中立点
thermoelectric pyrheliometer 热电直接日射计
thermoelectric series 温差电势率
thermoelement 温差电偶
thermograph 温度记录器，热录像仪
thermographic stress visualization 温度应力可视化
thermogravimetric analysis 热重分析
thermohaline 热盐的，温盐的

thermohaline circulation 温盐环流，热盐环流
thermoluminescence 热致发光的
thermomagnetic 热磁，热磁的
thermomechanical conversion efficiency 热机转换效率
thermomechanical effect 热机械效应
thermometric liquid 测温液（体）
thermometric scale 温标
thermometrograph 温度记录器
thermophilic anaerobic bacteria 嗜热厌氧细菌
thermophilic digester 高温发酵池，嗜热发酵池
thermophilic temperature optimum 最佳中温
thermophoresis 热泳，热迁移
thermophotonic converter 热光子转换器
thermophotovoltaic converter 热光电转换器
thermopile 热电堆
thermoselect technology 热选技术
thermosiphon 热虹吸管
thermosiphon cooling 热虹吸管冷却
thermosiphon solar water heating system 虹吸吸管太阳能热水系统
thermosiphon system 热虹吸系统
thermosiphoning action pump 热虹吸作用水泵
thermosiphoning air collector 热虹吸空气集热器
thermosiphoning solar collector 虹吸吸管太阳能集热器，差温环流式太阳能集热器
thermostat 恒温器，自动调温器
thermosiphon solar hot-water device 虹吸吸管太阳能热水装置

thermosiphon solar hot-water system 虹吸吸管太阳能热水系统
thermosiphon solar water heater 自然循环太阳热水器
thermosiphon water collector 热虹吸水收集器
Thevenin equivalent circuit 戴维南等效回路
Thevenin equivalent parameter 戴维南等值参数
Thevenin's equivalent circuit 戴维南等效电路
thickener 增稠剂
thickness 厚度
thickness function 厚度函数
thickness measurement 厚度测量
thickness of airfoil 翼型厚度
thickness variation 厚度变化，厚度偏差，厚度误差
thin film 薄膜
thin film solar cell 薄膜太阳能电池
thin lens equation 薄透镜成像公式
thin shell structure 薄壳结构
thin-film polycrystalline silicon solar cell 薄膜多晶硅电池
thin-film silicon solar cell 硅基薄膜太阳电池
thin-film solar cell material 薄膜太阳能电池材料
thinning residue 削磨残渣
third generation device 三代发电装置，第三代装置
thixotropic property 触变性能
Thomason solar house 汤姆逊太阳能房
Thomason-type trickle collector 汤姆逊型滴流式集热器

three circles calculative design method 三圆计算作图法
three dimension stalling effect 三维失速效应
three phase AC power 三相交流电
three phase commutator machine 三相换向电动机
three phase load 三相负载
three phase machine 三相电机
three phase non-salient synchronous machine 三相隐极同步电机
three phase salient synchronous machine 三相凸极同步电机
three-blade wind turbine 三叶风力涡轮机
three-blade wind wheel 三叶风轮
three-blade windmill 三叶片风车
three-bladed upwind turbine 三叶逆风涡轮机
three-bladed wind generator 三叶风力发电机
three-dimensional boundary layer effect 三维边界层效应
three-in-one digester 三结合沼气池
three-layer solar energy converter 三分子层太阳能转换器
three-phase AC system 三相交流系统
three-phase current 三相电流
three-phase grid system 三相并网
three-phase induction motor 三相感应电动机
three-phase machine 三相电机
three-phase power conversion 三相功率转换
three-phase winding 三相绕组

three-pool system 三库系统
three-way catalyst system 三元催化剂系统
threshold energy 临界能量
threshold limit value 阈限值
threshold wind speed 启动风速
throttle （机器的）风门，节流阀
throttle valve 节流阀
throttling loss 节流损失
throughput rate 生产率，通过速度
thrust 推力，冲断层
thrust bearing 止推轴承
thrust belt 冲断带
thrust coefficient 推力系数
thrust fault 逆断层，冲断层，逆冲断层
thrust loading 推力负载
thrust plane 冲断面
thrust plate 冲掩体，推力板
thrust ripple 推力波动
thrust value 推力值
thunderstorm 雷暴
thyratron 晶闸管
thyristor 晶闸管，闸流体
thyristor controlled reactor 晶闸管控制电抗器
thyristor controlled series compensation 晶闸管可控串联补偿
thyristor element 硅整流元件
thyristor firing circuit 晶闸管触发电路
thyristor level 晶闸管级
thyristor module 晶闸管组件
thyristor protective circuit 晶闸管保护电路
thyristor rectifier 晶闸管整流器，晶体闸流管逆变器
thyristor soft-start unit 晶闸管软启动装置
thyristor trigger circuit 晶闸管触发电路
thyristor valve 晶闸管阀，晶闸管换流阀
thyristor controlled inductance 晶闸管控制电感
tidal 潮汐的
tidal acceleration 潮汐加速度
tidal accommodation 潮汐码头
tidal action 潮汐作用
tidal age 潮龄
tidal amplitude 潮振幅，潮差，潮（沙波）幅
tidal atlas 潮汐图
tidal average power 潮汐能平均功率
tidal backwater 潮汐回水
tidal ball 报潮球，验潮仪
tidal bank 潮滩
tidal barrage 防潮堰堤，拦潮坝，潮汐堰坝
tidal barrier 拦潮堤
tidal basin 潮汐湖，潮汐盆地，蓄水池
tidal bay 潮汐湾
tidal benchmark 验潮水准点
tidal bore 涌潮，潮浪，潮津波
tidal bulge 潮汐隆起
tidal channel 感潮水道
tidal climate 潮汐气候
tidal compartment 潮区，有潮区
tidal compensation 潮汐补偿
tidal component 分潮
tidal constant 潮汐常数
tidal constituent 分潮，潮汐组分
tidal correction 潮汐订正，潮汐校正
tidal creek 进潮口，潮沟
tidal current 潮流

tidal current curve 潮流曲线
tidal current difference 潮流速度差
tidal current energy 潮流能
tidal current table 潮流表
tidal curve 潮位曲线
tidal cycle 潮汐周期
tidal datum 潮位基准面
tidal datum plane 潮汐基准面
tidal deformation 潮汐变形
tidal delta 潮汐三角洲
tidal difference 潮差
tidal dissipation 潮能耗散
tidal divide 分潮岭
tidal double ebb 双低潮
tidal double flood 双高潮
tidal ebb 落潮，退潮
tidal electric power generation 潮汐发电
tidal electric station 潮汐电站
tidal electrical energy 潮汐电能
tidal embankment 挡潮堤
tidal energy 潮汐能
tidal energy application 潮汐能应用
tidal energy conversion 潮汐能转换
tidal energy conversion device 潮汐能转换装置
tidal energy conversion system 潮汐能转换系统
tidal energy converter 潮汐能量变流器
tidal energy development 潮汐能开发
tidal energy device 潮汐能发电装置
tidal energy extraction 潮汐能提取
tidal energy integration 潮汐能综合利用
tidal energy output 潮汐能输出
tidal energy production 潮汐能生产

tidal energy scheme 潮汐能方案
tidal energy tapping scheme 潮汐能开发方案
tidal energy utilization 潮汐能利用
tidal epoch 潮相迟角
tidal estuary 潮汐河口，有潮河口，感潮河口
tidal excursion 潮程
tidal fall 落潮，退潮
tidal fence 潮汐篱笆，潮汐栅栏式结构
tidal flat 潮滩，潮汐平地
tidal flat deposit 潮滩沉积
tidal flood strength 最大涨潮流速
tidal flood 涨潮
tidal flow 潮流
tidal flush 潮流冲刷作用，潮汐冲刷作用
tidal flushing 潮流冲刷
tidal force 潮汐力
tidal friction 潮汐摩擦
tidal friction energy 潮汐摩擦能
tidal gage 测潮仪
tidal gaging station 潮汐站，验潮站
tidal gate 拦潮闸门，防潮闸门
tidal gauge 验潮仪，潮位水尺
tidal generating plant 潮汐能发电厂
tidal generation force 引潮力
tidal generator 潮汐发电机
tidal harbor 有潮港
tidal height 潮位
tidal hour 潮时
tidal hydraulics 潮汐水力学
tidal indicator 示潮器
tidal influx 进潮量，潮水量
tidal inlet 进潮口
tidal instability 潮汐不稳定性
tidal interval 潮汐间隙

tidal kinetic energy 潮汐动能	计
tidal lag 潮滞，潮时滞后	tidal power project 潮汐能发电工程，潮汐发电计划
tidal land 沿岸带，潮间地	
tidal level 潮位	tidal power review board 潮汐能审查委员会
tidal lift 涨潮	
tidal limit 潮区界限，涨潮界	tidal power scheme 潮汐发电站开发方案，潮汐发电方案
tidal load 潮压	
tidal lock 挡潮闸	tidal power station 潮汐发电站
tidal mark 潮标，潮痕	tidal power station availability 潮汐发电站可利用率
tidal mashland 潮漫滩	
tidal mechanism 潮汐（发生）仪	tidal power technology 潮汐能技术
tidal meter 潮汐仪	tidal power unit 潮汐发电机组
tidal mill 潮汐磨坊	tidal power utilization 潮汐能利用
tidal movement 潮流运动	tidal prism 纳潮量，进潮量
tidal oscillation 潮汐振荡	tidal progressive wave 潮汐推进波
tidal outlet 出潮口	tidal race 强潮流，潮汐流
tidal period 潮汐周期	tidal raising force 引汐力，涨潮力
tidal potential 潮汐势，潮汐位	tidal range 潮差，潮汐变化范围，潮位变幅
tidal potential energy 潮汐位能	
tidal power 潮汐发电，潮汐力，潮汐功率，潮汐能	tidal range resource 潮差资源
	tidal rapid 强潮流，潮汐激流
tidal power barrage 潮汐发电坝，潮汐坝	tidal reach 感潮河段
	tidal regime 潮流特性，潮汐状况
tidal power development 潮汐能开发	tidal regulation 潮汐调节，潮汐控制
tidal power engineering 潮汐能工程	tidal reservoir 潮汐水库
tidal power generating plant 潮汐发电站	tidal resonance 潮汐共振
	tidal rip 激潮，潮头卷浪
tidal power generation 潮汐发电	tidal rise 潮升
tidal power generator 潮汐发电机	tidal river 有潮河，潮水河，潮水河流
tidal power generator set 潮汐发电机组	tidal sand ridge 潮汐沙脊
	tidal scheme 潮汐发电工程，潮汐发电方案
tidal power house 潮汐动力房	
tidal power peak 潮力峰值	tidal scour 潮流冲刷，潮水冲刷
tidal power planning 潮汐发电规划	tidal sluice 挡潮闸
tidal power plant 潮汐发电厂，潮力发电厂	tidal stand 平潮
	tidal station 验潮站，潮位站
tidal power plant design 潮汐发电站设	tidal stream device 潮流装置

tidal stream generator	潮汐流发电机
tidal stream technology	潮汐流技术
tidal stream turbine	潮汐流涡轮机
tidal stretch	潮汐区
tidal surge	潮汐起伏，涌潮，大潮汛
tidal table	潮汐表
tidal turbine	潮汐发电水轮机
tidal turbine pump station	潮汐水轮泵站
tidal undulation	潮振动，潮位升降
tidal variation	潮汐变化，潮高变化
tidal water	有潮水域，感潮水域
tidal water furrow	潮水沟
tidal water level	潮位
tidal water volume	潮汐水量
tidal waterway	潮汐河水道，感潮水道，潮汐航道
tidal wave	潮波，潮汐波
tidal wedge	潮汐楔，楔形潮沟口
tidal weir	潮堰，潮堤
tidal wind	潮风
tidal zone	潮汐带
tidal-generating force	成潮力
tidalmeter	测潮表
tidal-power generator	潮汐发电机
tidal-powered electric generating plant	潮汐发电站
tide	潮，潮汐
tide ball	报潮球
tide blow	激潮
tide box	潮水箱（水工模型用的）
tide bulge	潮涨
tide channel	潮沟
tide control apparatus	潮汐控制仪
tide current	潮流
tide curve	潮汐曲线，潮位曲线
tide cycle	潮汐周期
tide embankment	防潮堤
tide force	引潮力
tide gate	防潮闸，挡潮闸
tide gauge	验潮仪
tide gauge telemetry	遥测验潮仪
tide gauge well	验潮井
tide head	有潮界限，感潮界限
tide interval	潮流间隙
tide lag	潮滞
tide level	潮位
tide level of water	潮汐水位
tide mark	潮标，潮痕
tide mill	潮汐磨坊
tide mill schematic	潮轮原理图
tide of long period	长周期潮汐
tide phase	潮汐相位
tide pool	蓄潮池
tide power generating station	潮汐发电站
tide power plant	潮汐发电站
tide predicting machine	潮汐自动推算机
tide producing force	引潮力
tide race	急潮流
tide raising force	涨潮力
tide range	潮差
tide recorder	自记验潮仪，自记潮位仪
tide reproducing force	引潮力，起潮力
tide spectrum	潮汐谱
tide staff	水尺，验潮标
tide stage	潮位
tide station	验潮站
tide surge	涌潮，潮汐起伏
tide table	潮汐表
tide water level	潮水位
tide wave	潮波
tide well	验潮井

tide-free 无潮汐的
tide-gate 挡潮闸
tide-gauge 测潮计，潮位计
tide-generating force 引潮力，生潮力
tide-generating mechanism 潮浪发生器
tide-generation force 引潮力
tideland 潮间地
tideless 无潮的
tideless coast 无潮海岸
tideline 涨潮线，逐浪
tide-meter 潮汐计
tide-motor 潮汐发动机
tide-plant 潮汐发电站
tide-pole 验潮杆
tide-predicting machine 潮汐预测机
tideway 潮流，潮水河
tide-worn 潮汐冲毁的
tidology 潮汐学
tier conveyor drier 多层输送带式干燥机
tie-rod ring 连杆环，系杆环
Tigris 底格里斯河
tiled stove 花砖火炉
tilt angle 仰角，倾角
tilt angle of collector 集热器倾角
tilt angle of rotor shaft 风轮轴仰角
tilted collector 倾斜式集热器
tilted device 倾斜装置
tilted flat-plate collector 倾斜平板型集热器
tilted reflector booster 斜面反射板增温器
tilted solar system 倾斜式太阳能系统
tilted still 倾斜式蒸馏器
tilted thermosiphoning solar collector 倾斜式热虹吸太阳能集热器

tilted tray solar still 阶梯置放碟式太阳能蒸馏器
tilted tubular collector 倾斜状管式集热器
tilted water-heating tray 倾斜式热水槽
tilted wick 倾斜衬里
tilted wick type solar still 斜置绒布式太阳能蒸馏器
tilting flume 活动水槽
time constant 时间常数
time domain 时域，时间域
time domain analysis 时域分析
time domain and frequency domain 时域与频域
time domain electromagnetics 时域电磁学
time domain modelling 时间域模型
time history of wind speed 风速时间历程
time of use 使用时段
time-of-use power price （谷峰）分时电价
time response （光电电池）时间响应
time synchronization 时间同步
time-invariant flow 非时变流
time-varying system 时变系统
timing circuit 定时电路
tin doped indium oxide 氧化锡铟
TiO_2 nanometre solar cell 二氧化钛纳米太阳电池
tip airfoil 翼尖翼型
tip brake 尖端制动，翼尖刹车
tip circle 齿顶圆
tip deflection （螺旋桨）叶梢上翘
tip feathering 叶尖顺桨
tip leakage 叶（片）顶（部）漏泄
tip leakage loss 叶顶漏泄损失

tip loss 叶尖损失
tip loss factor 叶尖损失因子
tip of blade 叶尖（叶片）
tip radius （风力涡轮机）叶尖半径
tip section pitch 叶尖段桨距
tip speed 叶尖速度，（桨叶）梢速，（叶片）端速
tip speed ratio 叶尖速比
tip stall 叶尖脱流，叶尖失速
tip vane augmented wind turbine 叶尖舵片增力型风轮机
tip velocity ratio （风力涡轮机）叶尖速度比
tip vortex 翼梢旋涡
tip vortice 叶尖涡旋
tipper truck 翻斗卡车，自动倾卸车
tipping floor 卸料间
tire derived fuel 轮胎衍生燃料
tissue 生物组织，织物
toggle pump 肘杆泵
ton per year 吨/年
tonality 音值
ton per day 日产出量（吨）
toolset 工具集
tooth depth 齿高
tooth flank 齿面
tooth space 齿槽
tooth thickness 齿厚
topographic condition 地形条件
topographic effect 地形效应
topographic feature 地形特征
topographical factor 地形因素
topography 地形学，地势
topsoil 表层土，上层土
tornado tower 飓风塔
tornado-type wind turbine 旋风式风力涡轮机

toroidal concentrator 超环面聚光器，轮胎面聚光器
toroidal surface 超环面，轮胎面
torque 扭矩
torque control 转矩控制，转速控制
torque converter 转矩变换器
torque damper 扭矩减震器
torque fluctuation 转矩波动
torque force 扭力
torque level 扭矩水平
torque meter 扭矩计
torque pickup 扭力计
torque pulsation 转矩脉动
torque transducer 扭矩传感器
torque-speed characteristic 转矩一速度特性
torrefaction 烘焙，干燥
torsion constant 扭转常数
torsion deflection 扭转变形
torsion elasticity 扭转弹性
torsion indicator 扭力计
torsional strength 扭转强度，抗扭强度
torsional load 扭转荷载
torsional rigidity 扭转刚度
torsional stall flutter 扭转失速颤振
torsional strain 扭转应变
torsional stress 扭应力，扭转应力
torsional system 扭转系统
torsional vibration 扭转振动
torus generator 圆环发电机
total solar flux 总太阳能通量
total amount of solar energy 太阳能总量
total annual solar radiation 全年太阳辐射总量
total daily solar radiation 日太阳辐射总量

total dissolved solid 总溶解固体量
total dissolved solid content of steam 蒸汽的溶解固体总含量
total efficiency of hydroelectric power station 潮汐电站总效率
total energy of ocean thermal 海洋热能总能量
total flow system for electric power production from geothermal energy 全流式地热发电系统
total generating capacity 总发电容量
total harmonic distortion 总谐波失真
total incident radiation 入射辐射总量
total insolation 日射总量
total installed capacity 总装机容量
total irradiance 总辐照度
total irradiation 总辐射
total moisture 全水分
total organic carbon 总有机碳，有机碳总量
total power 总功率
total power available in the wind 风可利用的总功率
total power consumption 总功率消耗
total power density 总动力密度
total power dissipation 全功耗
total power generating capacity 总发电容量
total power loss 总功率损耗
total power of station 电站总功率
total project cost 工程总投资
total radiation 总辐射
total solar energy incidence 太阳能总入射量
total solar photovoltaic energy system 全太阳能光电能系统
total solar radiation 太阳能总辐射量
total solar reflectance 太阳能总反射比
total solid content 总固体含量
total solids 总固体量
total sulfur 全硫
total suspended solid 总悬浮固体，总悬浮固形物
totally enclosed fan cooled 全封闭风扇冷却
touch voltage 接触电压
toughened glass 钢化玻璃
towed machine 牵引机
tower 塔架
tower array （风力涡轮机）塔架阵列
tower cabinet 塔基机柜
tower base fore-aft bending moment 塔架底部前后弯矩
tower bending 塔弯曲
tower evaporator 塔式蒸发器
tower focus power plant 塔聚焦式发电厂
tower for wind turbine generator system 风电机组塔架
tower height 塔架高度
tower loading 塔架荷载
tower logic 塔逻辑
tower motion （风力涡轮机）塔架运动
tower natural frequency 塔架自振频率
tower section 筒体
tower shadow 塔影
tower shadow loading 塔影荷载
tower stiffness 塔架刚度
tower top weight 塔顶重量
tower vibration 塔振动
towing tank 拖曳水池，拖曳水槽，船模试验池，船模试验槽
town gas 民用煤气，民用燃气
toxic equivalent quantity 毒性当量

toxicity equivalent 毒性当量
toxicological impact category 毒性影响类别，毒性影响分类
trace compound 微量化合物
trace element 微（痕）量元素
trace gas 微量气体，痕量气体
trace metal 微量金属，痕量金属
tracer test 示踪试验
tracking 跟踪
tracking accuracy (or error) 跟踪准确度（或跟踪误差）
tracking accuracy controller 跟踪准确度控制器
tracking array 追踪阵列
tracking collector 跟踪集热器
tracking concentrator 跟踪式聚光器
tracking error 跟踪误差
tracking heliostat 跟踪式定日镜
tracking mirror 跟踪反射镜
tracking precision 跟踪精确度
tracking receiver 跟踪接收器
tracking reflector 跟踪反射器
tracking solar collector 跟踪式太阳能集热器
tracking system 跟踪机构（实现对太阳跟踪的装置）
trade wind 信风
traditional oil 传统原油
trailing cable 拖拽缆
trailing edge 翼片后缘，叶片后缘
train of gears 齿轮系
transducer 换能器
transesterification 酯基转移，酯交换反应
transfer by conduction 热导换热
transfer fluid 传热流体
transfer function 传递函数

transfer ration 传递比
transfer switch 转接开关
transferred arc 转移弧
transformer 变压器
transformer platform 变压器平台
transgenic yeast strain 转基因酵母菌株
transient coefficient 非稳态系数
transient current 瞬态电流
transient electrical performance test 瞬态性能测试
transient electromechanical torque 瞬态机电扭矩
transient force 瞬时力
transient load 瞬时荷载
transient performance 瞬态性能
transient reactance 瞬态电抗
transient stability limit 暂态稳定边界，瞬态稳定边界，动态稳定极限
transient technique 不稳态渗流技术
transient torque 瞬时转矩
transient voltage 瞬时电压
transient 瞬变现象，瞬变电流（或电压）
transistor 晶体管
transitional rock site 过渡性岩石区
translator 转换器
transmission 透射
transmission accuracy 传动精度
transmission and distribution system 输电和配电系统
transmission cost 传输成本
transmission distance 传输距离，传真距离
transmission error 传动误差
transmission factor 透射因数
transmission grid 输电网
transmission line 传输线

transmission line pole 电线杆
transmission link 传输线路
transmission loss 传输损耗
transmission pipeline 输送管道
transmission power cable 传输电力电缆
transmission ratio 传动比
transmission system 传输系统
transmissivity 透射率
transmittance 透射比，透射率，透射系数
transmittance-absorptance product 透射吸收产品
transmitter 传送器
transparent conduction oxide 透明传导氧化物
transparent conductive oxide 透明导电薄膜，导电氧化物薄膜，透明导电氧化物
transparent cover 透明盖层
transparent cover plate 透明盖板
transparent insulation material 透明绝热材料
transparent mylar 透明聚酯薄膜
transparent plastic roof 透明塑料顶盖
transparent plastics 透明塑料
transparent quartz window 透明石英窗
transparent sun-resistant plastics 耐晒透明塑料
transpired honeycomb 发泡蜂窝（结构）
transportation pipe 输送管
transversal flux 横向磁通
transverse flux machine 横向磁场电机
transverse flux permanent magnet 横向磁场永磁
transverse flux permanent magnet machine 横向磁场永磁电机
transverse lamination 横向层压（薄板）
transverse oscillator 贯轴振荡器
trap reservoir 地压型热储
trash storage bunker 垃圾储料仓
trashrack 拦污栅
travel time 传播时间
traveling block 移动滑车，游动滑车
traveling grate stoker 活动炉排加煤机
traveling-wave thermoacoustic heat engine 行波热声热机
travelling grate 移动炉排
travelling grate furnace 移动炉排炉
travertine 钙华
treatment of combustible waste 可燃性废料处理
tree section 树段
tremie pipe 混凝土导管
trend analysis 趋势分析
trial phase 试运行阶段
tributary 支流
trichloromonofluoromethane 三氯氟代甲烷（太阳热间接转换为电能的一种工质）
trickle charge 浮接充电
trickle collector 涓流集热器
trickling filter 生物滤池
trigeneration 三联产，热电冷三联产
trip relay 跳闸继电器
trip-free 自由脱扣
triple glass 三层玻璃
triple junction compound solar cell 三结复合太阳电池
triple junction solar cell 三结太阳电池

triple valve 三通阀
tripod piled structure 三脚架结构
triptyl 三蝶烯基
triticale 黑小麦
tritium 氚，超重氢
tritium content 含氚量
trituration 研碎，磨碎，粉状
Trombe wall 特朗伯被动式太阳能墙
Trombe wall solar house 特朗伯墙式太阳房
trommel 矿石筛
trommel screen 滚筒筛，转筒筛，圆筒筛
trommel screen sieve 滚筒分离筛，圆筒分离筛
tropic tide 回归潮
tropical water 热带水
tropical ocean region 热带海洋区域
troposphere 对流层
trough 水槽，低谷期，饲料槽，低气压
trough angle 槽角
trough collector 槽形集热器
trough idler 槽形托辊
trough of the wave 波槽
trough solar energy power generation 槽式太阳能发电
trough type solar water heater 槽式太阳能热水器
troughlike concentrator 类槽形聚光式太阳能集热器
trough-type concentrator 槽形聚光器
true solar time 真太阳时
truncated Fourier series decomposition method 截位傅里叶序列分解法
trunnion 耳轴
trunnion pin 炮耳轴
tsunami 海啸

tube anemometer 管状风速表
tube ball mill 管式球磨机
tube bank 排管
tube bundle dryer 管束烘干机，管束干燥机
tube cleaner 洗管器，管内清垢
tube fin pitch 管子鳍片节距
tube furnace 管式炉
tube plate 管板
tube plate packing 管子管板之间的密封
tube section 塔体
tube sheet 管板
tube spacing 管距，管排配置
tube tower 管塔
tube turbine 管道式水轮机
tube wall thickness 管壁厚度
tube-in-sheet water heater 管壳式热水器
tube-in-plate absorber 套管板式吸收器
tube-in-plate configuration 套管板配置
tube-in-sheet evaporator 管壳式蒸发器
tube-type collector 管式集热器
tube-type heat exchange 管式热交换
tube-type turbine 管道式水轮机
tubular absorber 管状吸收器
tubular collector 管状集热器
tubular evaporator 管式蒸发器
tubular linear induction motor 管形直线感应电动机
tubular permanent magnetic linear synchronous motor 圆筒形直线永磁同步电机
tubular receiver 管式接收器
tubular solar energy collector 管式太阳能集热器

tubular steel 铁架风塔
tubular steel tower 钢制塔筒，钢结构塔筒
tubular synchronous machine 圆筒形同步发电机
tubular tower 圆筒式塔架
tufa 凝灰岩
tungsten carbide 硬质合金，碳化钨
tungsten halogen lamp 卤钨灯
tungsten-inert gas welding 钨极惰性气体保护电弧焊
tunnel kiln 隧道窑
turbine aerator 涡轮充气器
turbine blade 涡轮叶片，透平叶片
turbine blade characteristic 涡轮机叶片特性
turbine blade technology 涡轮叶片技术
turbine bypass 涡轮机旁通管
turbine discharge 涡轮机流量
turbine dynamometer 水轮机测功机
turbine efficiency 涡轮效率，透平效率
turbine exhaust 汽轮机排汽
turbine exit 涡轮出口
turbine for tidal power 潮汐发电用涡轮机
turbine gate 水轮机闸门
turbine generator 涡轮发电机，汽轮发电机，透平发电机
turbine head cover 水轮机顶盖
turbine inlet bend 水轮机进水弯管
turbine inlet temperature 涡轮机进口温度
turbine outage 涡轮机停机
turbine performance 涡轮性能
turbine pump 涡轮泵
turbine rating 涡轮定级，透平额定功率

turbine room 汽轮发电区
turbine rotor shaft 涡轮转轴
turbine shaft 涡轮轴，水轮机轴
turbine shroud 涡轮壳体
turbine technology 汽轮机技术，涡轮技术
turbine tower 风力机塔架
turbine wake 风力机尾流
turbine windmill 涡轮式风车,涡轮式风力发动机
turbine with rotary wing 旋翼涡轮机
turbine with solar energy 太阳能涡轮机
turbine generator 汽轮发电机
turbine generator system 汽轮发电机系统
turbine-type machine 汽轮型电机
turbo generator unit 涡轮发电机装置
turbo rotor 涡轮转子
turbocharger 涡轮增压器
turbo-expander 涡轮膨胀机,透平膨胀机
turbofan 涡轮风机
turbogenerating 涡轮发电
turbo-generator 涡轮发电机,汽轮发电机
turbo-generator set 涡轮发电机组,汽轮发电机组
turbomachinery 涡轮机械
turboset 汽轮机组
turbosupercharger 涡轮增压机
turbulance intensity 湍流强度
turbulence 湍流，狂暴，紊流
turbulence buffeting 湍流抖振
turbulence intensity 湍流强度
turbulence intensity model 紊流强度模型

turbulence inversion 湍流逆温
turbulence level 湍流度
turbulence resistance 湍流阻力
turbulence scale parameter 湍流尺度参数
turbulence wind 脉动风
turbulent bursting 湍流猝发
turbulent convection 湍流对流
turbulent flow 湍流,紊流
turbulent fluctuation 湍流脉动
turbulent flux 湍流通量
turbulent friction 湍流摩擦
turbulent gust 湍流阵风
turbulent intensity 湍流强度
turbulent reattachment 湍流再附
turbulent shear 湍流剪切
turbulent viscosity 湍流黏性
turbulent wind 湍流风
turbulivity 湍流度,湍流系数
turgo turbine 斜击式水轮机
turndown ratio 调节比
turning direction of rotor 风轮转动方向
turning vane 导向叶片,导流板
turn-key photovoltaic system "交钥匙"光伏系统
turns ratio 匝数比
turret lathe 六角车床
tuyere 鼓风口,风口
TV tower 电视塔
twin rotor 双旋翼
twin rotor capacitor 双动片电容器
twin rotor system 双旋翼系统
twin semi-submersible steel vessel 双体半潜式钢船(温差电站的一种装置)
twist 捻度
twist angle of blade 叶片扭角

twist blade 扭转叶片
twist distribution 捻度分布,扭转分布
twist grain boundary 扭转晶界
twist of blade 叶片扭角
twisting stress 扭应力
two blade rotor 二叶片风轮
two dimensional rotation 二维旋转
two layered ground system 双层地下系统
two phase operation 两相运行
two pool continuous 双库连续
two pool intermittent 双库间断
two pool system 双库系统
two-axis oriented array 双轴定向方阵
two-axis sun sensor 双轴阳光传感器
two-axis tracking dish 双轴跟踪碟形抛物面反射器
two-axis tracking solar concentrator 双轴跟踪太阳能聚光器
two-axis tracker 太阳能双轴跟踪器
two-blade horizontal rotor 双叶水平转子,双叶水平风轮
two-blade rotor 双叶转子
two-bladed propeller 双叶螺旋桨
two-bladed wind station 双叶风力发电站
two-bladed wind turbine 双叶风力涡轮机
two-cycle engine 二冲程发动机
two-dimensional absorber 二维吸收器
two-dimensional concentrator 二维聚光器,二维聚光式集热器
two-dimensional CPC 二维复合抛物面集热器
two-dimensional crystal lattice 二维晶体点阵
two-dimensional drag coefficient 二维

阻力系数
two-dimensional receiver 二维接收器
two-dimensional reflector 二维反射器
two-loop profiling method 双回路剖面法
two-loop sounding method 双回路测深法
two-pass solar air heater 双路太阳能空气加热器
two-phase aquifer 双相含水层，二相含水地带
two-phase cooling 双相冷却
two-phase flow 双相流
two-phase fluid 双相流体
two-phase mixture 双相混合物
two-phase thermal storage 双相蓄热，双相热储存
two-phase value 双相位值
two-pool 双库
two-pool plan 双池式布置（潮汐发电）
two-source evaporation 双源蒸发
two-stage gas turbine 双级式燃气轮机
two-stage turbine 双级透平
two-stage vacuum pump 两级真空泵
two-tank storage system 双箱储能系统
two-value capacitor motor 双值电容式电动机
two-way isolating valve 双向隔离阀
type of collector fluid 集热器中流体种类
type of tide 潮型
typhoon 台风

U

UAUB anaerobic reactor UAUB 厌氧反应器
U-bend U 形弯头
U-bend test U 形弯曲测试
U-loop U 形管圈
ultimate analysis 元素分析
ultimate consumer 终端用户
ultimate limit state 极限限制状态，最大极限状态
ultimate load 极限荷载
ultimate strength 极限强度
ultrafiltration 超滤
ultra-low head turbine 超低压头涡轮机
ultrasonic thermometer 超声波温度计
ultra-supercritical power plant 超临界电厂
ultra-thin solar cell 超薄太阳电池
ultraviolet 紫外线
ultraviolet absorption 紫外线吸收
ultraviolet absorption spectrum 紫外线吸收光谱
ultraviolet band 紫外线谱带
ultraviolet energy 紫外线能量
ultraviolet lamp 紫外线灯
ultraviolet laser 紫外激光器
ultraviolet light 紫外光
ultraviolet light sensor 紫外光传感器
ultraviolet light transducer 紫外光传感器
ultraviolet pyranometer 紫外总日射表
ultraviolet radiation 紫外线辐射，紫外线照射
ultraviolet ray 紫外线
ultraviolet spectroscopy 紫外线光谱法
uncertainty in measurement 测量误差
unconsolidated aquifer material 松散含水层材料

unconsolidated aquifer 未固结含水层
unconsolidated deposit 松散沉积物
unconsolidated sediment 松散沉积物，疏松沉积物
unconsolidated sediment 疏松沉积
undamped vibration 无阻尼振动
underdamped oscillation 欠阻尼振荡
underexcitation 励磁不足，欠励磁
underfeed stoker 下部加料锅炉
underfire air 一次空气
underground copper grid 接地铜网格
underground heat exchanger 埋地换热器
underground power station 地下发电站
underground pumped hydrostorage 地下抽水储能
underground reservoir 地下热储
underground steam 地下蒸汽
underground storage 地下储能
underground thermal energy storage system 地下热能储存系统
underreamer 管下扩眼器，管眼扩大器，扩孔器
underrun 潜流
undershot 水流冲击（潮轮）下部而转动的
undertow 退波，（海面下的）下层逆流
under-voltage 欠电压
underwater construction 水下施工，水下建筑
underwater manipulator 水下机械手，水下操作器
underwater mill 水中水车
underwater thermal powerplant 水下热电站
underwater turbine 水下风车

undisturbed flow 稳态气流
uneven heating 不均匀加热
unfavorable mobility 不利流动比
unglazed collector 无透明盖板集热器，无上釉集热器
uniaxial loading 单轴负载
unidirectional 单向的
uni-directional flapper 单向瓣阀，单向拍动板
uni-directional ocean-current energy source 单向海流能源
uni-directional turbine 单向涡轮机，单向透平
unified power flow controller 统一潮流控制器
unified power quality conditioner 统一电能质量控制器
unified voltage controller 统一电压控制器
unijunction solar cell 单结（太阳能）电池
uninterruptible power supply 不间断电源
unit 单位
unit capacity 单机容量，单机出力
unit cell 晶胞
unit centralized control 单元集中控制
unit control 单元控制
unit control room 单元控制室
unit efficiency 机组效率
unit heater 成套加热器
unit start-up and commissioning 整套启动试运
unit vector 单位矢量
unit wattage 单体（位）功率
universal coupling 万向联轴器

universal time 世界时
unleaded blend 未加铅的掺合燃料
unleaded fuel 无铅燃料
unloading valve 释荷阀
unloading well 自喷井
unreacted biomass 未反应的生物质
unreacted cellulose 未反应的纤维素
unreacted charcoal 未反应木炭
unrecuperated microturbine 不可复原式微型燃气轮机
unsaturated polyester 不饱和聚酯
unsaturation 不饱和度
unshrouded rotor 无罩风轮
unstable thermal interface 非稳定热界面
unsteady aerodynamics 非定常空气动力学
unsteady current 不稳定电流
unsteady flow 不稳定流动
unsteady load 不稳定负荷
untubed well 未下油管的井
untwist 解缆
unusual storm conditions 异常暴(风)雨状况
up wind 上风向
updraft 上升气流
up-draught stove 鼓风炉
upflow anaerobic sludge blanket 上流式厌氧污泥床
up-flow anaerobic sludge blanket reactor 升流式厌氧污泥床反应器
upflow tube 溢流管
upflow zone 上流区
upgraded metallurgical silicon 升级冶金级硅
upper air wind 高空风
upper limit LHV of waste 焚烧炉上限垃圾低位热值
upper limit waste treatment capacity for incinerator 焚烧炉上限垃圾处理量
UPS system 不间断电源系统
upstream 逆流,向(或在)上游
upstream end 进汽端
upstream wind 逆风
upwelling zone 上升流带
upwind 上风向
upwind configuration 上风向布置
upwind direction 上风向
upwind propeller 逆风螺旋桨
upwind rotor 上风式风轮
upwind turbine 迎风风力机
up-wind wind turbine 上风向风力机
upwind wind turbine generator system 上风向式风电机组
uranium extraction from seawater 海水提铀
uranium from sea water 海水提铀
urban waste 城市垃圾
urban wood waste 城市木材垃圾,城市木材废料
usable heat collection 可用集热,有效集热
use of geothermal energy 地热能利用
used wood 用过的木材
useful energy 有效能,有用能
useful life 使用寿命
useful power 有效功率,可用功率
useful solar heat 有效太阳热能,有用太阳热能
utility boiler 电站锅炉
utility cost 公用工程费用,效用成本,效益成本
utility electricity 公用电力,工业用电力

utility grid 公用电网
utility line 公用电缆
utility natural gas 公用天然气
utility power plant 公用事业电站
utilization for geothermal energy 地热能利用
utilization hour of generation 发电设备利用小时
utilization of solar energy 太阳能利用
utilization of tidal power 潮汐力利用
utilization rate of wind energy 风能利用率
U-tube U形管
U-tube heat exchanger U形管热交换器
U-type vacuum tube U形管式真空集热管
UV radiation 紫外线，紫外辐射

V

V dip 电压跌落
V fault 电压故障
V pre-fault 故障前电压
vacancy concentration 空位浓度
vacancy diffusion 空位扩散
vacancy mechanism 空位机理
vacant defect 空位缺陷
vacant lattice site 晶格空位，阵点空位
vacant shell 空壳层
vacuum chamber 真空室，压力室
vacuum circuit breaker 真空断路器
vacuum degasifier 真空脱气器
vacuum evaporated film 真空蒸镀薄膜
vacuum filtration 真空过滤
vacuum limit 极限真空

vacuum melting method 真空熔炼法
vacuum mixer 真空搅拌机，真空混合器
vacuum pump 真空泵
vacuum refining 真空精制
vacuum tube collector 真空管集热器
vacuum tube solar collector 真空管式太阳能集热器
vacuum tubular glass solar collector 玻璃真空管太阳能热水器
valence band 价带
valence electron 价电子
validation 实用性
valley floor 谷底
valve regulated lead acid battery 阀控铅酸电池
Van Allen radiation belt 范艾伦辐射带
vanadium oxide 氧化钒
vanadium redox battery 钒氧化还原液流电池
vane 风向标，风信旗，叶片
vane anemometer 叶轮风速计
vane vortex generator 叶片涡流发生器
vapor 蒸汽
vapor compression 蒸汽压缩
vapor compression equipment 蒸汽压缩装置
vapor deposition 气相沉积
vapor growth 气相生长
vapor heated evaporator 蒸汽加热蒸发器
vapor phase 气相
vapor phase deposition 气相沉积
vapor phase diffusion 气相扩散
vapor phase epitaxy 气相外延
vapor phase growth 气相生长
vapor plume 蒸汽羽烟

vapor pressure 蒸汽压,蒸汽压力
vapor quality 蒸汽质量
vapor-dominated geothermal field 蒸汽地热田
vapor-dominated geothermal resource 蒸汽型地热资源
vaporization efficiency 汽化效率
vaporization energy 汽化能
vaporization heat 汽化(潜)热
vaporizer 汽化器,蒸发器
vaporizing point 沸点,蒸发温度
variable air volume 变风量
variable chord blade 变截面叶片
variable flow 变速流,异形流动
variable flow system 变流量系统,变水量系统
variable frequency 变频
variable frequency drive starter 可变频率驱动式起动机
variable frequency drives 变频调速,变频器
variable frequency power 变频电源
variable geometry type wind turbine 可变几何翼型风力机
variable pitch blade 变距桨叶,变螺距叶片
variable pitch propeller 可调螺距螺旋桨,变距螺旋桨
variable pitch turbine 可调螺距水轮机
variable propeller 调距螺旋桨
variable reflective coating 可变反射层,可变反射膜
variable reflective layer 可变反射层
variable rotor speed 叶轮变速
variable specific heat 可变比热
variable speed constant frequency wind turbine generator system 变速恒频风电机组
variable speed damper 变速消振器
variable speed gear 变速齿轮,变速装置
variable speed rotor 变转速风轮
variable speed-constant frequency system 变速恒频发电系统
variable speed-constant frequency WTGS 变速恒频风电机组
variable voltage 变压
variable wind 不定风
variable-frequency alternator 变频交流发电机
variable-pitch rotor 变螺距转子
variable-rpm operation 变速运行
variable-speed control 变速调节
variable-speed direct drive 变速直驱
variable-speed generator 变速发电机
variable-speed mechanism 变速机械装置
variable-speed operation 变速运转
variable-speed tidal mill 变速潮力磨坊
variable-speed wind turbine 变速风力发电机
variable-speed-constant-frequency 变速恒频
varistor 变阻器
varying field 变化场
varying load 变动负荷,变动负载
varying underflow 变底流
VAV air system 变风量空气系统
vector 矢量
vector control speed system 矢量控制调速系统
vector equation 向量方程
vector sum 矢量和,向量和
vector wind field 向量风场

vee-corrugated absorber　V形波纹吸收器
vee-corrugated absorbing surface　V形波纹吸收表面
vee-corrugated specular surface　V形波纹镜表面
veering wind　顺转风
vee-trough reflector　V形槽反射器（镜）
vee-trough vacuum tube receiver　V形槽真空管接收器
vegetable waste　蔬菜废弃物
vegetal matter　植物物质
vegetation indicator　植被指数
velocity　速度，速率，流速
velocity amplitude　速度幅值
velocity deficit　速度不足
velocity distribution　流速分布
velocity head　速度落差，速位差
velocity in the rotor plane　风轮平面内速度
velocity of light　光速
velocity pressure　风速压力，速度压力，动压力
velocity scale　速度比例
velocity signal　速率信号
venetian blind collector　软百叶帘集热器
vent condenser　空气凝汽器
vent for summer use　夏季用通风口
vent stack　（垂直）排气筒，排气通道
vent valve　排气阀，通风阀，排水阀，排泄阀
vented system　开口系统
ventilating duct　导风筒
ventilating fan　通风机
ventilation air　通风气流
ventilation capacity　通风量
ventilation coefficient　通风系数
ventilation duct　通风管道
ventilation load　换气负荷
ventilation rates　通风率
ventilation resistance　通风阻力
ventilation system　通风系统，通排风系统
ventilator　通风设备，换气扇
Venturi flow measuring element　文丘里测风装置
Venturi meter　文丘里流量计
Venturi scrubber　文丘里洗涤器，文丘里除尘器
Venturi tube　文丘里流量计
Venturi type wind turbine　文丘里管式风电机
verification　检验
vernier gauge　游标尺
vernier hybrid permanent magnet machine　游标混合永磁电机
vertex　顶点，角顶，台风转向点
vertical　纵的
vertical lathe　立式车床
vertical anemometer　铅直风速表
vertical axial wind turbine　垂直轴式风车
vertical axis　垂直轴
vertical axis configuration　竖轴结构
vertical axis rotor　竖轴风轮
vertical axis tidal power turbine　垂直轴潮汐能水轮机
vertical axis tidal stream turbine　垂直轴潮流水轮机
vertical axis turbine　垂式涡轮机，（海流发电用）竖轴式水轮机
vertical axis wind machine　垂直轴风力机

vertical axis wind rotor 竖轴式风机转子，竖轴式风机叶轮
vertical axis wind turbine 竖轴风机（风车）
vertical axis wind turbine aerodynamic performance 竖轴式风力涡轮机空气动力性能
vertical axis wind turbine rotor 竖轴式风力涡轮机转子，竖轴式风力涡轮机叶轮
vertical blind 垂直百叶窗
vertical borehole ground-coupled heat pump 垂直埋管地源热泵系统
vertical centrifugal chemical pump 立式离心化工泵
vertical circle 地平经圈
vertical closed loop 垂直闭合回路
vertical collector 立式集热器
vertical combustion chamber 垂直燃烧室
vertical composting unit 垂直堆肥装置
vertical dimension 垂直距离
vertical eddy 纵向涡流
vertical flap 垂直活板
vertical flute-tube heat exchanger 立式槽—管型换热器
vertical ground heat exchanger 竖直地埋管换热器
vertical junction solar cell 垂直结太阳电池
vertical layout of plant 厂区竖向布置
vertical loop 垂直回路
vertical loop biological reactor 立环式生物反应器
vertical multijunction solar cell 垂直多结太阳电池
vertical plane 垂直平面，垂直面
vertical profile 垂直廓线，垂直剖面
vertical sail-type mill 垂直帆式风车
vertical seismic profiling 垂直地震剖面法，垂直测震剖面法
vertical stress 垂直应力，竖向应力
vertical turbine pump 竖式叶轮泵
vertical type wind turbine 立轴风力机
vertical wind profile 垂直风廓线
vertical wind shear 垂直风切
vertical-axis hydraulic machine 垂直轴液压机器
vertical-axis machine 竖轴式风机
vertical-axis rotor-type wind turbine 竖轴式叶轮型风力涡轮机
vertical-axis turbine 竖轴风力涡轮机
vertical-axis wind turbine 垂直轴风力发电机
vertical-axis windmill 竖轴风车
vertically oriented fracture 垂直裂缝
vertical-shaft windmill 竖轴式风车
vertical-spout evaporator 垂直喷口蒸发器
vertical-tube evaporator 立管蒸发器
Vestas 维斯塔斯
Vestas wind system 维斯塔斯风力系统公司
V-groove collector V形槽集热器
vibrating grate 振动炉排
vibration absorber 减震器
vibration damping 振动阻尼
vibration energy 震荡能
vibration isolator 减震器
vibration level 震动水平，振动级别
vibration source 振动源
vibration suppression 减震
vibratory feeder 振动给料机

vibratory motion 振动	vitreous state 玻璃态
vicinity 邻近地区	vitrified slag 玻璃状炉渣
viewing angle 角度	void boundary 空穴边界
village electrification 农村电气化	void coefficient 空穴系数
vinyl ester 乙烯基酯	void effect 空穴效应
violent storm 暴风	void ratio 孔隙比,孔隙率
violet cell 紫电池	volatile 挥发物
violet solar cell 紫光太阳电池	volatile combustion 挥发性燃烧
virgin pulp 原生纸浆	volatile component 挥发性组分
viscid lava 黏质熔岩,半流体熔岩	volatile compound 挥发性化合物
viscose residue 黏液残渣	volatile content 挥发物含量
viscosimeter 黏度计	volatile evaporation 挥发性蒸发
viscosity 黏性,黏度	volatile fatty acid 挥发性脂肪酸
viscosity coefficient 黏性系数	volatile fraction 挥发性馏分
viscosity effect 黏度效应,黏滞效应	volatile fuel 挥发性燃料
viscosity index 黏度指数	volatile matter 挥发物
viscosity resistance 黏性阻力	volatile organic compound 挥发性有机化合物,挥发性有机物
viscous dissipation 黏性耗散,黏滞耗散,黏滞扩散	volatile solid 挥发性固体
viscous drag 黏性阻力	volatility 挥发性,挥发度
viscous effect 黏滞效应,流体黏滞性效应	volatilization 挥发
	volcanic rent 火山裂缝
viscous flow 黏性流	volcanic rock 火山岩
viscous flow sintering mechanism 黏性流动烧结机制	volcanism 火山活动
	Volochine pyranometer 沃罗钦总日射计
viscous fluid 黏性流体	
viscous force 黏性力	voltage 电压
viscous friction 黏滞摩擦	voltage anomaly 电压异常
viscous layer 黏性层	voltage change factor 电压变化系数
viscous liquid 黏性液体	voltage divider 分压器
viscous wake 黏性尾流	voltage drop 电压降
visible light 可见光	voltage feedback system 电压反馈系统
visible light transducer 可见光传感器	
visible radiation 可见辐射	voltage flicker 电压闪变
visible spectrum 可见光谱	voltage fluctuation 电压波动
visual appearance 外观	voltage level 电压电平
visual impact 视觉冲击	voltage magnitude 电压值

voltage monitoring 电压监视
voltage regulation 稳压，电压调整，电压变动率
voltage regulator 电压调节器
voltage restoration 电压恢复
voltage ride-through capability 电压穿越能力
voltage sag 电压下降
voltage source converter 电压源换流器
voltage source converter based high voltage direct current 电压源换流器高压直流输电
voltage source inverter 电压源型逆变器
voltage temperature coefficient 电压温度系数
voltage temperature coefficient of a solar cell 太阳电池的电压温度系数
voltage to earth 对地电压
voltage transformer 电压互感器，变压器
voltage transient 电压瞬变
voltage vector 电压矢量
voltage wave 电压波
voltage waveform 电压波形
volt-ampere 伏安
volts/hertz ratio 伏特/赫兹比
volume defect 体缺陷
volume of ebb 落潮潮量
volume of flood 涨潮潮量
volume shrinkage 体积收缩
volume stability 体积稳定性
volumetric compressibility factor 容积压缩系数
volumetric flow rate 体积流量，容积流量

volumetric receiver 容积式吸热器
volumetric Stirling receiver 容积式斯特林吸热器
vortex 漩涡
vortex augmentor 旋涡增力型装置
vortex concentrator device 旋涡聚集式风能装置
vortex cone 涡锥
vortex core 涡核
vortex decay 旋涡衰减
vortex flow 涡流
vortex generator 涡流发生器
vortex induced oscillation 涡致振荡
vortex induced response 涡致响应
vortex shedding 涡漩滑泻，漩涡泄离，漩涡脱离
vortex shedding converter 漩涡脱离转换器
vortex sheet 涡流层
vortex street 涡列
vortex tail 涡迹
vortical flow 涡流
vortices 涡流
V-trough collector V形槽集热器
V-trough solar concentrator V形槽太阳能聚光器

W

wading bird 涉水鸟，涉禽
wafer 晶片，薄片
wafer cleaning 晶圆清洗，晶片清洗
wafer cutting 切片
wafering 压块，（硅等的）切片
waiting situation 等候工况
wake axial induced flow 轴向尾流诱导（扰动）气流
wake blockage 尾流阻塞

wake buffeting 尾流抖振	wake-induced vibration 尾流致振
wake capture 尾流捕获	walking beam 步进梁，活动梁
wake cavity region 尾流空穴区	walking beam furnace 步进式加热炉
wake circulation 尾流环量	wall barrier 挡板
wake closure 尾流封闭区	wall collection 墙式集热
wake decay 尾流衰减	wall collector 墙式集热器，集热墙
wake defect 尾流损失	wall fired boiler 墙式锅炉
wake drag 尾流阻力	wall loss 壁热损失
wake effect 尾流效应	wall orientation 屋墙朝向
wake effect loss 尾流效应损失	wall pressure 壁压
wake energy 尾流能量	wall rock 围岩
wake entrainment 尾流卷挟	warm air 暖空气
wake excitation 尾流激励	warm spring 温泉
wake excited crosswind response 尾流横风响应	warm water flow rate 温水流量
wake expansion 尾流膨胀，尾流扩展	warm-water discharge 温水排放，暖水排放
wake flutter 尾流颤振	warm-water inlet 温水进口，暖水进口
wake galloping 尾流驰振	warm-water intake 温水入口，暖水入口
wake geometry model 尾流几何模型	warm-water intake pipe 温水进水管，暖水进水管
wake interaction 尾流交互作用	warm-water pipe 温水管，暖水管
wake interference 尾流干扰	washer 洗涤器，垫圈，垫片
wake loss 尾流损失	washout valve 冲刷阀
wake momentum thickness 尾流动量厚度	waste activated sludge 剩余污泥，剩余活性污泥，废活性污泥
wake oscillator model 尾流振子模型	waste biomass 生物质废料
wake rotation 尾流旋转	waste burning 垃圾燃烧，垃圾焚烧
wake Strouhal number 尾流斯特鲁哈数	waste burning power generation 垃圾焚烧发电
wake suction 尾流吸力	waste classification 垃圾分类
wake traversing 尾流游测	waste classification and management 垃圾分类和管理
wake turbulence 尾流流度	waste crane 垃圾起重机
wake vortex 尾流旋涡	waste discharge standard 废物排放标准
wake vortex resonance 尾流旋涡共振	waste disposal 垃圾处理，废物处理
wake vortex system 尾流涡系	
wake vorticity 尾流涡量	
wake-induced velocity 尾流诱导速度，尾流扰动速度	

waste disposal infrastructure 垃圾处理设施
waste energy 垃圾能
waste feed chute 垃圾喂入槽
waste feedstock 废物原料
Waste Framework Directive （欧盟）《废物框架指令》
waste fuel 燃料渣，废燃料
waste gasification 垃圾气化
waste heat 余热，废热
waste heat boiler 废热锅炉，余热锅炉
waste heat recovery 余热回收，废热利用
waste hierarchy 废物分级
waste incineration 垃圾焚烧，垃圾焚化，废物焚化
waste incineration boiler 垃圾焚烧余热锅炉
Waste Incineration Directive （欧盟）《垃圾焚烧指令》
waste management 垃圾管理，废物管理，废弃物管理
waste management hierarchy 层次式废物管理方法
waste material energy 废料能
waste pretreatment 垃圾预处理
waste reduction algorithm 减废算法，废物减量化算法
waste residue treatment 废渣处理
waste separation 垃圾分选
waste stream 废物源流
waste to energy 垃圾发电
waste wood 废木材，废木
waste-gas loss 废气损失，烟道损失
waste-heat recovery 废热回收
waste-to-energy 废物再生能源，垃圾焚烧发电

waste-to-energy plant 垃圾焚烧发电厂
waste-to-energy power plant 垃圾焚烧发电厂
wastewater treatment plant 污水处理厂
water ocean current plant 水下海流发电装置
water availability 水分有效性
water bag collector 水袋式集器
water bag house 水袋式太阳房
water balancing 水平衡
water bearing layer 含水层
water buffer 自来水缓冲器
water chamber 水室，水夹套
water chemistry 水化学
water column 水柱，水塔，水层
water column structure 水体结构
water content 含水量
water cooled grate 水冷炉排
water cooling 水冷却
water cycle 水循环
water deflection part 导水机构
water delivery rate 供水速率
water delivery temperature 供水温度
water density 水密度，水质稠密度
water discharge 流量
water disposal system 水处理系统
water draw-off rate 取水流量
water emulsion 水乳胶
water erosion 水蚀
water evaporator 水蒸发器
water filter 滤水池，滤水器
water flux 水通量，水溶性焊剂
water gate 水闸
water guide mechanism 导水机构
water hammer 水锤，水击

water head of driving a tidal plant 潮汐电站发电水头
water impinge 水冲击
water inlet 进水口
water intake 进水口
water intake intensity 吸水强度
water intrusion 进水
water jetting 水冲法，射水法
water level behind barrage 坝后水位
water line 水位线，水线
water misting system 水雾化系统
water oozing 渗水，漏水
water outlet 出水口
water pond with black bottom 黑底水池
water potential 静水压，水位，水势
water pressure 水压，水压力
water proof junction box 防水接线盒
water proof machine 防水式电机
water regime 水文状况
water saturated debris 饱和岩屑
water source heat pump 水源热泵
water storage system 储水能系统
water supply 水源
water swivel 旋环，水旋转接头
water table 地下水位，潜水面，泄水台
water tank 水槽
water tower storage 水塔贮能
water tube boiler 水管式锅炉
water tunnel 水洞，水槽，输水隧道
water turbine 水轮机
water turbine blade 水轮机叶片
water turbine driven generator 水轮发电机
water turbine generator set 水轮发电机组
water turbine pump 水轮机泵
water vapor 水蒸气
water vapor absorption 水汽吸收
water vapor content 水汽含量
water vapor density 水汽密度
water vapor pressure 水汽压
water velocity 水流速度
water-boiling thermal efficiency of solar cooker 太阳灶煮水热效率
water-circulating system 水循环系统
water-cooled combustion chamber 水冷式燃烧室
water-cooled photovoltaic-thermal cogeneration system 水冷式太阳能热电联产系统
water-cooled system 水冷系统
water-cooling 水冷，水冷却法
water-heating system 水加热系统
waterline 吃水线（海陆边界），水印横线
waterplane area 水线（水平面）
waterplane moment 水线面力矩
water-power engineering 风力工程
waterproof 防水
waterproof layer 防水层
water-proofing of generator 发电机防水
water-pumping assembly 抽水装置
water-pumping windmill 泵水风力机，抽水风力机
water-rock storage bin 水—岩石储能仓
water-source heat pump 水源热泵
water-to-water heat pump 水水热泵
water-turbine generator set 水轮发电机组
water-type heat storage 水介质蓄热器
waterwall 水墙，水冷壁

waterway 航道,水路,排水沟
waterwheel 水车,水轮
watt peak 峰瓦(太阳电池的峰值功率)
watt per square meter 瓦特/平方米
wave 波浪
wave abrasion 波蚀
wave action 波浪作用,波作用,激活作用
wave activated generator 波浪发电机
wave activated generator group 波浪发电机组
wave activated power generation 波浪力发电
wave activated power generator 波浪发电机
wave advance 波浪传播,波浪推进
wave age 波龄
wave amplitude 波振幅
wave as energy source 波浪能源
wave attack 波浪冲击,波浪袭击
wave base 浪基面,波底(静水中水面波动扬不起沉淀物的深度)
wave basin 波浪水池,造波水池
wave breaking 波破碎
wave breaking revetment 防波堤,消波堤
wave built terrace 波成阶地
wave climate 波候,波象
wave clutter 波浪回波干扰,海面杂乱回波
wave condition 海浪
wave contouring raft 波面筏(一种波浪发电设备)
wave crest 波峰
wave current 波浪流
wave delta 浪成三角洲
wave device 波能装置

wave direction 波向,浪向
wave dispersion 波频散
wave dragon 龙波
wave drift 波浪漂移
wave drift current 波成流
wave drift damping 波浪漂移阻尼
wave drift force 波浪漂移力
wave dynamics 波浪动力学
wave electric power system 波浪发电系统
wave energy 波能
wave energy absorbing device 波能吸收装置
wave energy absorbing floater 波能吸收浮子
wave energy absorption 波能吸收
wave energy air turbine 波能空气透平
wave energy attenuator 波能衰减器
wave energy balance 波能平衡
wave energy barge 波能(转换)驳船
wave energy coefficient 波能系数
wave energy conversion 波能转换
wave energy conversion buoy 波能转换浮筒
wave energy conversion concept 波能转换概念
wave energy conversion device 波能转换装置
wave energy conversion efficiency 波能转换效率
wave energy conversion float 波能转换浮体
wave energy conversion scheme 波能转换方案
wave energy conversion terminal 波能转换终端

wave energy converter 波浪发电机, 波浪能吸收转换机, 波浪能转换器
wave energy converter array 波能转换器阵列
wave energy device 波能装置
wave energy distribution 波能分布
wave energy extraction 波浪能萃取
wave energy extractor 波能萃取器
wave energy flux 波能通量
wave energy growth 波能成长
wave energy in random sea 随机海浪中的波能
wave energy origin 波能原点
wave energy park 波浪发电场
wave energy platform 波能(转换)平台
wave energy recovery system 波能回收系统
wave energy rectifier 波能检波器
wave energy rose diagram 波能玫瑰图
wave energy sink 波能槽
wave energy spectrum 波(浪)能(量)谱
wave energy technology 波(浪)能技术
wave energy terminator 波能终止器
wave energy transmission 波能传递
wave energy utilization 波能利用
wave energy-density spectrum 波能密度谱
wave equation 波动方程
wave farm 波浪发电场
wave flow 波能流动,波动流体
wave flume 波浪水槽
wave focusing 聚波,波能聚焦
wave force 波浪力
wave forecast 海浪预报

wave frequency 电波频率,波动频率
wave front angle 波前冲角
wave front method 波前法
wave front reconstruction 波前再现
wave generator 波浪发电机
wave group 波组,波群
wave height 波高
wave height gauge 浪高仪
wave hollow 波谷
wave hub 波中心
wave load 波浪荷载
wave loading 波浪载荷
wave magnitude 波级
wave motor 波力发动机
wave of explosion 爆发波
wave of oscillation 旋转波,摆动波
wave of translation 推进波,移动波
wave packet 波包,波群
wave parameter 海浪要素,波浪要素,波参数
wave period 波周期,波浪周期
wave physics 波动物理学
wave planation 波浪均夷作用
wave power 海浪能,波浪能,波浪力
wave power absorber 波能吸收器
wave power absorption 波能吸收
wave power air-turbine generator 波能汽轮发电机
wave power boat 波力驱动船
wave power breakwater 拦浪发电站
wave power buoy project 波能浮标工程
wave power conversion 波能转换
wave power device 波能装置
wave power farm 波浪发电场
wave power generation 波浪发电,波力发电

wave power generator 波浪发电机,波力发电机
wave power machine 波能发电机
wave power recovery device 波能回收装置,波能利用装置
wave power recovery system 波能回收系统,波能利用系统
wave power station 波能发电站,波浪能电站
wave power unit 波能装置,波浪发电装置
wave powered buoy 波能供电浮筒
wave powered device 波能装置
wave powered generator 波能发电机
wave pressure 波浪压力,波压
wave profile 波形
wave propagation 波传播
wave propelled boat 波浪推进小船
wave protection 波浪防护
wave pump 波力泵
wave radar 连续波雷达
wave ray 波线,波向线,波射线
wave recorder 海波计,波浪计,自记测波仪
wave reflection 波浪反射
wave refraction 波浪折射
wave resistance 波浪阻力
wave resource 波浪资源
wave run-up 波浪爬高
wave scale 波级
wave scatter diagram 波浪分布图
wave set up 波浪组合
wave setup 波增水
wave shoaling 波浪浅化
wave source 波源
wave spectrum 波谱
wave spectrum 海浪谱

wave steepness 波浪陡坡,波陡
wave steepness ratio 波陡比
wave surface 波面
wave tank 造浪水池,波浪槽
wave train 波列
wave trough 波槽
wave turbine 浪轮机,波力透平
wave turbulence 波湍流
wave type 波浪类型
wave velocity 波速,波速度
wave winding 波形绕组
wave-activated air turbine generator 空气涡轮波力发电机
wave-by-wave tuning 逐波调优
wave-conversion device 波浪转换设备
wave-current interaction 波流相互作用
wave-cut platform 浪蚀台
wave-cut terrace 浪蚀阶地
wave-energy conversion cost-ineffective 波能转换无效费用
wave-farm 波浪场(是将波浪力转换为压缩空气来驱动空气透平发电机发电的场所)
wavefront 波前,波阵面
wavefront length 波阵面长度,波前长度
wave-generated power 波浪产生的电力
wavelength 波长
wavelength dependent reflected intensity 波长依赖性反射强度
wavelength dependent transmitted intensity 波长依赖性透射强度
wavelength interval 波长间隔
wavelength selectivity of collector surface 集热器表面的波长选择性
wavemaker 造波机

wave-mud interaction 波浪泥沙交互作用
wavenumber 波数
wave-power park 海能发电厂
wave-powered desalination 水波动力脱盐
wave-powered generator 波能汽轮发电机
wave roller 波形辊
wavescan 波形扫描
wave-sediment interaction 波浪沉积交互作用
weak current control 弱电控制
weak gas drive 弱气驱
weak grid 弱电网
wear rate 磨损率
wear-out failure 磨损故障
weather 气象,天气,风化,气候
weather balloon 气象气球
weather bureau 气象局
weather code 天气电码
weather condition 气象条件
weather data 天气资料
weather element 气象要素
weather parameter 气象参数
weather shed 遮雨缘,雨棚
weather vane 风向标
weather-chart 天气图
weathercock 风向标
weathered coal 风化煤
weathered rock 风化岩
weathering 风化作用
weathering crack 风化裂缝
weathering crust 风化壳
weathering fissure 风化裂隙
weathering of rock mass 岩体风化
weatherproof 防风雨的,全天候的

Weibull distribution 威布尔分布
Weibull parameter 威布尔参数
Weibull probability distribution 威布尔概率分布
weight of blade 叶片重量
weight percent 重量百分数
weight ratio 重量比
weight to power ratio of a solar array 太阳电池方阵的重量—功率比
weighted sound pressure level 声级
weighting material 加重料,填充物
weir head 堰上水头
weir notch 堰缺口
welder 焊工,焊机
well 钻井
well blowout 井喷
well casing 钻孔套管,井管
well completion 成井
well deviation 钻孔弯曲,井斜
well diameter 钻孔直径
well fluid 井内流体
well log 测井曲线,钻井日志,钻井记录
well logging 测井,钻井测试
well pad 井场
well performance 地热井性能
well production 井产量
well response 地热井响应
well screen 井管滤网
well spacing 布井,井距
well stimulation 油气井增产措施
well testing 试井
well water heat pump 井水源热泵
wellbore 钻井孔
wellbore cleaning process 井眼清洗过程
wellbore collapse 井壁坍塌
wellbore diameter 井眼直径

wellhead equipment 井口设备
wellhead pressure 井口压力
wells turbine 威尔斯水轮机
westerlies 西风带
westerlies circle 西风带环流
westerlies rain belt 西风多雨带
western windmill 西部风车
wet bark 湿树皮
wet basis 湿基，湿量基准
wet deposition 湿沉降
wet digestion 湿法分解
wet etching 湿法腐蚀，湿法刻蚀
wet fermentation 湿发酵
wet flue gas treatment 湿式烟气处理
wet fuel 液体燃料
wet mechanical pretreatment 湿式机械预处理
wet process waste 湿处理废物
wet scrubber 湿式除尘器
wet steam 湿蒸汽
wet steam field 湿蒸汽田
wet steam reservoir 湿蒸气储藏
wet weight 湿重，含水重量
wet-bulb depression 湿球温降
wet-bulb temperature 湿球温度
wet-pad evaporative cooler 湿垫蒸发冷却器
wettable biomass 可湿性生物质
wheel gear 大齿轮
wheel loader 轮式装载机
white body 白体
white squall 无形飑，晴天突起的（热带）暴风
whitecapping （波峰有白色碎浪的）白头浪，白帽浪
whole algae hydrothermal liquefaction 全藻类水热液化

whole-tree chip 整树片
wide band gap 宽能带隙
Wien's displacement law 维恩位移定律
winch 绞车
wind abrasion 风蚀，风力侵蚀
wind action 风力作用
wind aloft 高空风
wind and biomass co-firing 风与生物质共燃
wind and drift chart 风力偏流修正表
wind and PV hybrid power system for residential area 居住区风—光互补系统
wind and PV hybrid streetlight 风—光互补路灯
wind angle 风迎角，风角
wind area 迎风面积
wind arrow 风矢
wind assessment 风场评价
wind atlas 风图谱，风能分布图
wind axis system 风轴系，气流坐标系
wind beam 抗风梁
wind belt 风带
wind borne dust 风载尘
wind borne material 风载物质
wind borne sediment 风成沉积物
wind borne snow 风载雪
wind box 风箱
wind box collector 窗口箱式集热器
wind break 防风林，防风墙，风障
wind break barrier 防风墙
wind break fence 防风篱
wind break system 防风林系统
wind change characteristic 风速变化特性

wind characteristic 风特性
wind charger 风力充电机，风力充电器
wind chill factor 风冷指数
wind circulation 风环流
wind classification 风力分级
wind cock 风向标
wind compensation 风力补偿
wind concentrator 聚风器
wind cone 风标
wind convection coefficient 风对流系数
wind conversion device 风能转换装置
wind converter 风电变流器
wind conveyer 风力输送机
wind correction 风力修正
wind corrector 风力修正器
wind corrosion 风蚀
wind data logger 风数据记录器
wind deflection 风力偏移
wind diagram 风向频率图，风向玫瑰图
wind direction 风向
wind direction fluctuation 风向波动
wind direction frequency 风向频率
wind direction indicator 风向指示器
wind direction recorder 风向自记器
wind direction sensor 风向传感器
wind direction shaft 风矢杆
wind direction vane 风向标
wind disturbance 风扰动
wind drag 风阻
wind drag load 风阻荷载
wind driven heat pump 风力热泵
wind driven generator 风力发电机
wind driven power-conversion device 风能转换装置
wind duration 风时
wind dynamic load 动态风载

wind eddy 风涡
wind electric power generation 风力发电
wind electricity 风能电力
wind energy 风能
wind energy and energy security 风能与能源安全
wind energy application 风能利用
wind energy collector 风能收集器
wind energy conversion 风能转换
wind energy conversion system 风能转换系统，风力发电机组
wind energy converter 风能转换器
wind energy density 风能密度
wind energy development 风能开发
wind energy distribution curve 风能频谱图
wind energy flux 风能密度
wind energy monitor 风能监测器
wind energy resource 风能资源
wind energy rose 风能玫瑰图
wind energy system time domain 风能系统时域
wind energy time-domain analyzer 风能时域分析仪
wind energy utilization 风能利用
wind energy utilization coefficient 风能利用系数
wind erosion 风蚀
wind farm 风电场
wind farm layout 风电场布局
wind farm screening chart 风电场筛选图表
wind farm sizing 风电场规模
wind field 风场，风力场
wind flow 风向
wind fluctuation 风性波动

wind force 风力
wind force diagram 风力图
wind force scale 风力等级
wind frame 抗风构架
wind frequency distribution 风频率分布
wind friction 风摩擦
wind gage 风速计
wind gap 风口
wind gauge 风速表，风速器
wind generated noise 风致噪声
wind generated wave 风生（波）浪
wind generating capacity 风电产能
wind generating system 风力发电系统
wind generator 风力发电机
wind generator array 风力发电阵列
wind generator facility 风力发电设施
wind generator group 风力发电机组
wind generator tower 风力发电机塔
wind guarder 防风加强环
wind heating 风力加热
wind indicator 风信标，风力指示器
wind induced heat loss 风致热损失
wind induced heat system 风致热装置
wind induced heater system 风致加热器系统
wind induced load 风致荷载
wind induced oscillation 风致振荡
wind induced pressure 风致压力
wind induced vibration 风致振动
wind load 风荷载
wind load capacity 风载能力
wind load rating 额定风力负荷
wind loading 风荷载
wind machine 风轮机
wind machine design 风力机械设计
wind map 风力地图，风图

wind measurement 风的测定，风速测量
wind measurement mast 测风塔
wind measuring system 风速测量系统
wind monitoring station 风力监察站
wind motor 风力发动机
wind of Beaufort force 蒲氏风力等级
wind of Beaufort scale 2 二级（蒲福）风
wind onset 风增水
wind path 风路，风迹
wind photovoltaic power system 风—光互补电站
wind plant 风电厂
wind porch 风廊
wind power 风力发电，风能，风力
wind power class 风力等级
wind power density 风功率密度
wind power draw water machine 风力提水机组
wind power duration curve 风能频谱图，风能历时曲线
wind power economics 风能经济学
wind power engineering 风力工程
wind power generation 风力发电
wind power grid integration 风电并网
wind power heating 风力制热
wind power industry 风能工业
wind power installation 风力设备
wind power plant 风力发电厂
wind power plant analog (ue) information 风电场模拟信息
wind power plant component 风电场部件
wind power plant management system 风电场管理系统
wind power potential 风能潜势

wind power resource	风力资源
wind power station	风电场
wind power system	风力装置，风电系统
wind power utilization coefficient	风能利用系数
wind power water pumping	风力提水
wind powered aeration system	风力曝气系统
wind powered aerator	风力曝气机
wind powered generator	风力发电机
wind powered heat pump	风力热泵
wind pressure	风压，风压力
wind pressure tap	风压测量孔
wind profile	风廓线，风速分布图
wind profile wind shear law	风廓线风切变律
wind proof design	耐风设计
wind proof performance	耐风性能
wind pump	风力泵，风泵
wind pump water pumping	风力提水
wind recorder	自记测风器
wind reduction efficiency	减风效率
wind regime	风的状况，风况
wind resistance	风阻力
wind resisting truss	抗风桁架
wind resource	风力资源
wind resource assessment	风能资源评估
wind resource assessment program	风力资源评估项目
wind resource map	风力资源地图
wind resources	风能资源
wind response characteristic	风响应特性
wind rib	抗风肋
wind ripple mark	风成波浪
wind rose	风向图，风向玫瑰图
wind rotor	风轮
wind run	风程
wind scale	风级
wind scoop	风穴
wind sea	海浪，波浪，风浪
wind set up	风增水，海面倾斜
wind shadow thermal	风影热，背风区热
wind shaft	风矢杆
wind shear	风切变，风剪力
wind shear exponent	风切变指数
wind shear law	风速廓线，风切变律
wind shield	挡风板
wind sifting	风筛机
wind simulation	风模拟
wind site	风场
wind sleeve	风向袋
wind sock	风向袋
wind span	风载档距，风压径间
wind speed	风速
wind speed deficit	风速不足
wind speed distribution	风速分布
wind speed distribution curve	风速分布曲线
wind speed duration curve	风速历时曲线
wind speed frequency	风速频率
wind speed frequency curve	风频曲线
wind speed histogram	风速直方图
wind speed measurement	风速测量，风速测定
wind speed measurement station	风速测量站
wind speed of test point	测试点风速
wind speed prediction	风速预测
wind speed profile	风速廓线
wind speed profile coefficient	风廓线

指数
wind speed ratio 风速比
wind speed value 风速值
wind speed-duration curve 风速历时曲线
wind spout 龙卷风
wind spun vortex 风动涡旋
wind stand-alone power system 离网型风力发电系统
wind station 风力站(风力发电站)
wind streak 风带
wind stream 气流,迎风气流
wind strength 风力
wind stress 风压应力,风应力
wind swell 海浪
wind system 风系统,风电系统,风能系统
wind technology 风能技术
wind test station 风力测试站
wind tide 风增水,风暴潮
wind tower 风塔
wind tunnel 风洞
wind tunnel turbulence 风道湍流
wind turbine 风力机,风力涡轮机,风力透平
wind turbine acoustic standard 风力涡轮机声标准
wind turbine blade 风力机叶片
wind turbine data acquisition system 风力涡轮机数据采集系统
wind turbine drum 风力涡轮机转鼓
wind turbine dynamics 风力涡轮机动力学
wind turbine efficiency 风力机效率,风力涡轮机效率
wind turbine foundation parameter 风力涡轮机基本参数
wind turbine gearbox 风机齿轮箱
wind turbine generator 风力涡轮发电机
wind turbine generator development 风力涡轮发电机开发
wind turbine generator model 风力涡轮发电机模型
wind turbine generator set 风力涡轮机发电机组
wind turbine generator system 风力发电机组
wind turbine housing 风轮机外壳
wind turbine nacelle 风力涡轮机舱
wind turbine noise 风力涡轮机噪声
wind turbine pad 风机升降台
wind turbine performance 风力涡轮机性能
wind turbine rotor 风轮
wind turbine simulation 风力机模拟
wind turbine spin 风力涡轮机
wind turbine terminal 风力发电机组输出端,风电机端口
wind turbine torque 风力涡轮转矩
wind turbine tower 风机塔架,风力涡轮机塔架
wind turbine wake 风力机尾流
wind turbulence 风湍流
wind vane 风向标
wind vane and anemometer 风向风速计
wind variation 风向变化,风变差
wind vector 风矢量
wind velocity 风速,风矢量
wind velocity fluctuation 风速脉动
wind velocity for windmill 风车风速
wind velocity on elevation basis 风速的高度换算
wind ward rudder 迎风装置

wind water-lifting set 风力提水机组
wind wheel 风轮
wind wheel anemometer 风轮风速计
wind-speed characteristic 风速特性
windage 鼓风,风阻,游隙,风吹损失
windage effect 风阻影响
windage loss 风阻损失,风吹损失,风力损失
windage resistance 风阻
wind-approach angle 来流风迎角,风接近角
wind-axis wind machine 风轴风力机
windblown coal 风化煤
wind-break earth ridge 防风土埂
wind-break forest 防风林
wind-break network 防风林网
windchill 风力降温
wind-diesel photovoltaic system 风力—柴油发电系统
wind-diesel system 风柴系统,柴油风力发电系统
wind-diesel-battery hybrid system 风柴电池混合系统
wind-direction measurement 风向测量,风向测定
wind-direction shaft 风矢杆
winddrift sand 风沙,流沙
wind driven current 风驱流
wind driven device 风力驱动装置
wind driven dynamo 风力发电机
wind driven electrical generator 风力发电机
wind driven generating system 风力发电机组
wind driven generating unit 风力发电机组
wind driven generator group 风力发电机组
wind driven generator set 风力发电机组
wind driven water pump 风力提水机
wind driven wave 风驱波
wind-electric system 风力电力转换系统
wind-electric turbine 风力发电机
wind-electric water pump 风力—电力水泵
wind-electric water pumping system 风力发电抽水装置
winder 卷绕机,绕线器,卷扬机,绞车
wind-farm noise 风电场噪声
wind-force diagram 风力图
wind-generated electricity 风力发电,风生电
wind-generated gravity wave 风生重力波
wind-generated sea wave 风成海浪
wind-generated wave spectrum 波浪谱
wind-generator set 风力发电机组
wind-induced current 风生海流
wind-induced heat system 风致热系统,风致热装置
wind-induced motion 风生运动
wind-induced turbulence 风生湍流
wind-induced wave 风成浪,风浪
winding bend 河湾
winding configuration 线圈形状
winding cylinder 绕线柱
winding drum 卷筒
winding factor 绕组系数
winding overhang 绕组端部
winding pitch 绕组节距
winding pressure arrangement 线圈压紧结构

winding resistance 线组电阻，绕组电阻

winding resistance measurement 绕组电阻测量

winding section 绕组单元

winding temperature 绕组温度，线圈温度

winding wire 绕组线

winding with non-uniform insulation 分级绝缘绕组

winding with uniform insulation 全绝缘绕组

wind-laid deposit 风成沉积

windless 无风的

windless region 无风区

windlop 三角浪，短涌浪

windmeter 风速表

windmill 风车，风力发电机

windmill anemometer 风车式风速表

windmill brake 风车式制动装置

windmill for water lifting 提水风车，抽水风车

windmill generator 风力发电机，风车式发电机

windmill governor 风力调速器

windmill heat generator 风车型生热器

windmill of Okinawa style （日本）冲绳型风车

windmill pump 风车泵，风力泵

windmill sail 风力发动机工作轮叶

windmill size 风车尺寸

windmill turbine （海流发电用）风车式水轮机

windmill type 风车型式

windmill-braking condition 风车制动状态

windmilling action of rotor 叶轮的鼓风作用

windmill-produced noise 风车噪声，风力发电机噪声

window box solar collector 窗框式太阳能集热器

window showcase 窗式橱窗

wind-photovoltaic complementary 风光互补

wind-power capacity 风电装机容量

wind-power electric generating unit 风力发电装置

wind-power extraction system 风力萃取系统

wind-power generation 风力发电

wind-power generation equipment 风力发电设备

wind-power generator 风力发电机

wind-power heat exchanger 风能热交换器，风能换热器

wind-power industry 风能工业

wind-power machine 风力机械，风能机械

wind-power network 风力发电网络

wind-power out 风力出口

wind-power plant 风电厂，风能装置，风力电站

wind-power plant layout 风力电站布置图

wind-power resource 风力资源

wind-power station 风力发电站

wind-power station utilizing solar energy 太阳能风力发电站

wind-power system 风力装置

wind-powered 风力的

wind-powered generator 风力发电机

wind-powered machine 风力机械

wind-powered water pumping system 风力水泵系统
wind-PV hybrid power system 风一光互补电站
wind-PV hybrid system 风压混合系统
wind-PV-battery hybrid system 风一光伏电池混合系统
wind-PV-diesel 风柴电压
wind-PV-diesel hybrid power 风一光伏一柴油混合系统
windrose 风玫瑰
wind-run anemometer 风程表,风程风速计
windsail 风翼板
windscreen 挡风玻璃,风挡
windscreen wiper 风挡刮水器,风挡雨雪刷
windshielding 挡风墙
wind-solar photovoltaic system 风力一光伏发电系统
wind-speed distribution 风速分布(曲线)
wind-speed profile 风速分布曲线
windstorm 风暴
wind-turbine generator set 风力发电机组
wind-turbine noise 风力机噪声
wind-up 终结,结束
windward 上风面(的),向风面(的),迎风面(的),迎风侧
windward drift 向风漂流
windward flood 迎风潮
windward rudder 迎风舵
windward side 向风面
windward slope 迎风坡
windward wall 迎风壁
wind-wave hydraulics 风浪水力学
wind-wave load 风浪载荷,风浪负荷
wind-wave model 风浪模型
windwheel 风轮
windworks 小型高效风力发电站
windy area 多风地区
windy district 多风区
wing 翼缘,叶片
wing chord 翼弦
wing pump 活翼式泵,轮叶泵
wing resistance 翼阻力
wing root chord 翼根弦
wing section 翼截面,翼剖面
wing setting angle 叶片安装角
wing-drift current 吹流
winglet 小翼
winning 回采
Winston collector 温斯顿式集热器
winter collection 冬季集热
winter peak 冬季用电高峰
wire break 断线
wire mesh separator 丝网分离器
wire rope 钢索
wire saw 线锯床,钢丝锯,线切割
wireless two-way remote 无线双向遥控器
wood ash 木灰
wood chip 木片
wood fuel 木质燃料
wood moisture content 木材含水率
wood oil 桐油
wood pellet 木屑颗粒,木质颗粒
wood pellet fuel 木屑颗粒燃料
wood pellet stove 木屑炉
wood processing industry by-product and residue 木材加工业副产品和残渣
wood processing plant 木材加工厂
wood processing residues 木料加工残留物

wood shaving 木刨花
wood stove 柴火炉
wood-fired heating plant 燃木供热设备
woody biomass 木质生物质
woody biomass fuel 木质生物质燃料
woody energy crop 木本能源作物
working earthing 工作接地
working flank 工作齿面
working fluid 工作流体，流体工作介质，加工液
working frequency 工作频率
working gas 工作气体
working principle 工作原理
working standard solar cell 工作标准太阳电池
working-fluid loop 工质回路
working-fluid vapor 工质蒸汽
world radiation reference 世界辐射测量基准
world wave 世界波
worm 蜗杆
worm wheel 蜗轮
wound armature 绕线电枢
wound field 风场，风区
wound rotor 绕线转子
wound rotor asynchronous motor 绕线式异步电动机
wound rotor generator 绕线式发电机
wound rotor induction generator 绕线式感应（异步）发电机
wound rotor synchronous generator 绕线转子同步发电机
wrap-around type solar cell 卷包式太阳电池

X

xylose 木糖

Y

Yangbajing geothermal power plant (China) 羊八井地热电站（中国）
yard waste 庭院垃圾，庭院废物，庭院废弃物
yaw 偏航，偏转
yaw active rotor 主动调向风轮
yaw angle 偏航角
yaw angle of rotor 风轮偏角
yaw axis 偏航轴
yaw axis concentration solar cooker 偏轴聚光太阳灶
yaw base 偏航基座，偏航盘
yaw bearing 偏航轴承
yaw brake 偏航制动器
yaw control 偏航控制
yaw control mechanism 对风控制机构
yaw control system 偏航控制系统
yaw controlling 偏航控制
yaw damper 偏航阻尼器
yaw drive 偏航驱动
yaw error 偏航误差
yaw gear 偏航齿轮
yaw inertia 对风惯性
yaw mechanism 偏航装置
yaw misalignment 偏航角误差，偏航失调
yaw moment 偏航力矩
yaw motor 偏航电机
yaw offset 偏航偏移
yaw passive rotor 被动调向风轮，被动对风风轮
yaw rate 偏航角速度
yaw ring 偏航环
yaw stability 航向稳定性
yaw system 偏航系统

yaw vane 调向尾舵,对风尾舵
yawed flow 偏航气流
yawed wind 斜向风
yawing 偏航,左右摇摆
yawing angle of rotor shaft 风轮偏航角
yawing device 调向装置
yawing driven 偏航驱动
yawing mechanism 偏航机构
yawing moment coefficient 偏航力矩系数
yawing system 偏航系统
yearly operating plan 年运行方式
yeast 酵母
yield strength 屈服强度
yielding information 屈服信息

Z

zenith 天顶
zenith angle 天顶角
zenith distance 天顶距
zero azimuth 零方位角
zero clearance fireplace 零间隙壁炉
zero conductor 零线
zero crossing period 过零周期
zero crossing wave period 零交叉波周期
zero energy house 零能房屋
zero phase 零相位
zero resistance 零电阻
zero-loss collector efficiency 零损失热器效率
zero-speed test 零速度测试
zinc bromine battery 锌溴电池
zinc carbon powder 锌碳粉
zinc-air battery 锌空电池
zirconia 氧化锆
zirconium oxide analyzer 氧化锆分析仪
ZJ solar collector ZJ 型太阳能集热器
Zomeworks solar house 松姆沃克太阳房
zonation 区划,分带性
zone-by-zone load analysis 逐层负载分析
zoned reservoir 带状热储
zone-selected controller 区域选择控制器
zoom factor 缩放系数
zoon time 区时
zymomonas 发酵单胞菌属
zymomonas mobilis 运动发酵单细胞菌

汉英部分

汉英词分

10 的 18 次方焦耳　exajoule (EJ) (unit of measure)
10 分钟最大平均风速　maximum average wind speed for 10 minutes
1978 年美国能源税法案　Energy Tax Act of 1978
1978 年美国公共事业监管政策法案,（美国）《1978 年公共事业监管政策法案》Public Utilities Regulatory Policies Act of 1978 (PURPA)
2004 年德国可再生能源法　Germany's 2004 Renewable Energy Sources Act
3D 地震层析成像　3D seismic tomography
Ⅱ-Ⅵ族太阳电池　Ⅱ-Ⅵ group solar cell
Ⅲ-Ⅴ族太阳电池　Ⅲ-Ⅴ group solar cell
AM0 条件［标定和测试空间用（大气质量等于 0）太阳电池所规定的辐照度和光谱分布］　AM0 condition
AM1.5 太阳光谱　AM1.5 solar spectrum
AM1.5 条件［标定和测试空间用（大气质量等于 1.5）太阳电池所规定的辐照度和光谱分布］　AM1.5 condition
FT 合成　FT synthesis
FT 合成催化剂　FT synthesis catalyst
HRS 整流式　HRS rectifier
Kislaya Guba 潮汐电站　Kislaya Guba Tidal Power Station
LED 数量　LED quantity
LED 手电筒　LED flash light
LED 台灯　LED reading light
MIS 太阳电池　MIS solar cell
n 水槽　n-trough
N 掺杂　N-endowment
n 型半导体　n-type semiconductor
n 型基　n-type base
p-n 结　p-n junction, pnjunction
p-n 结二极管　p-n diode
p-层［空穴较多，掺硼］　p-layer
P 型半导体　p-type semiconductor
P 型硅　p-type silicon
RAM 可充碱锰电池　Rechargeable alkali mangan (RAM) battery
S 型转子　S-shaped rotor
T 型等值电路　equivalent T circuit
UΛUB 厌氧反应器　UAUB anaerobic reactor
Uldolmok 潮汐电站　Uldolmok Tidal Power Station
U 形管　U-tube
U 形管圈　U-loop
U 形管热交换器　U-tube heat exchanger
U 形管式真空集热管　U-type vacuum tube
U 形夹　clevis
U 形弯头　U-bend
U 形弯曲测试　U-bend test
V 形波纹吸收表面　vee-corrugated absorbing surface
V 形波纹镜表面　vee-corrugated specular surface
V 形波纹吸收器　vee-corrugated absorber
V 形槽反射器（镜）　vee-trough reflector
V 形槽真空管接受器　vee-trough vacuum tube receiver
V 形槽集热器　V-groove collector, V-trough collector
V 形槽太阳能聚光器　V-trough solar concentrator
X 平方分布值　square value
ZJ 型太阳能集热器　ZJ solar collector

α(射线的)蜕变能　alpha disintegration energy
β吸收体　beta absorber
γ射线吸收体　gamma absorber

a

阿尔塔蒙特山口风电场(世界最大风电场)　Altamont Pass Wind Farm
阿伏伽特罗定律　Avogadro's Law
阿古拉斯海流(沿着非洲南部东岸向西南流的印度洋洋流)　Agulhas Current
阿基米德波摇摆　Archimedes wave swing (AWS)
阿基米德波摇摆发电机　Archimedes wave swing generator

ai

埃克曼层　Ekman layer
埃克曼螺线　Ekman spiral
埃克曼螺旋　Ekman spiral
埃克曼输送　Ekman transport
埃利斯—里奇法(进行碱度测定的一种方法)　method of Ellis and Ritchie
埃斯特朗直接日射表　Angstrom pyrheliometer
艾伯特水流式日射强度计　Abbot water pyrheliometer
艾伯特银盘式日射强度计　Abbot silver disk pyrheliometer
艾焦耳　exajoule
艾伦内六角扳手　Allen key
艾伦扳手　Allen wrench
爱德华天平(一种测量气体密度的仪器)　Edwards balance
爱迪生电池　Edison battery
爱迪生电气协会　Edison Electric Institute
爱因斯坦方程　Einstein's equation
爱因斯坦能　Einstein energy
爱因斯坦质能关系式　Einstein's mass energy relation

an

安(培小)时　Ampere-hours (Ah)
安德森周期　Anderson cycle
安娜波利斯潮汐电站　Annapolis tidal power station
安排　disposal
安培定律　Ampere's law
安培环路定律　Ampere's circuital law
安培小时　Ah
安全标志　safe marking
安全地带　safety zone
安全电压　safety voltage
安全阀　safety valve, relief valve
安全风速　survival wind speed
安全海湾　secure bay
安全界限　safety margin
安全距离　safe distance
安全联轴器　security coupling
安全门　emergency door
安全区　safety zone
安全区域　safety zone
安全色　safety color
安全寿命　safe life
安全系数　safety margin, factor of safety
安全裕度　safety margin
安全栅　guard grating
安全值　safe value
安全阻抗　safer impedance
安山石　andesite
安息角　angle of repose
安匝　ampere-turns
安装　installation, mount
安装费(从用包装箱运输整套太阳能装

置，到使其开始运转的全部费用，是太阳能成本和一次费用的组成部分） installation charge
安装高度　installation height
安装集热器程序　collector mounting procedure
安装图　installation drawing
桉属植物　eucalyptus
桉树　eucalyptus
氨　ammonia
氨保障系统　ammonia support system
氨计量泵　ammonia measuring pump
氨计量箱　ammonia measuring tank
氨气制冷机　ammonia machine
按钮　knob
按批投料　batch-fed
按容积的流量　volumetric flow rate
按体积计算百万分之一（气体浓度单位） parts per million by volume (PPMV)
按消耗量收费（根据实际消耗量收取公用事业费用，与按需用量收费不同） consumption charge
按需用量收费（公用事业部门根据可能的需用量，而不是按实际消耗量收取费用） demand charge
暗电极　dark electrode
暗电流　dark current
暗钉眼　countersink
暗度　degree of darkening
暗特性曲线　dark characteristic curve

ao

凹槽　chamfer, groove
凹面　concave
凹面镜　concave mirror
凹面透镜　concave lens
凹凸透镜　concave-convex lens
凹形转子　female rotor
奥德罗太阳能炉　Odeillo furnace
奥尔氏气体分析　Orsat gas analysis
奥夫尼克（将热能或太阳光直接转换为电能的装置，由奥夫辛斯基发明。装置内有特殊的从非导电状态变成半导体状态的玻璃成分） Ovonic
奥马特技术公司　Ormat Technologies
奥氏甲烷芽孢杆菌　methanobacillus omelianskii
奥氏体（碳丙铁）　austenite
奥托发动机　Otto engine
奥托循环发动机　Otto cycle engine

ba

八级风　fresh gale, gale
拔钉锤　claw hammer
拔丝机　drawing machine
钯管　palladium tube
坝后水位　water level behind barrage
坝失事　dam failure

bai

白炽灯　incandescent lamp
（波峰有白色碎浪的）白帽（头）浪 whitecapping
白体　white body
白杨　poplar
白云石　dolomite
白云土　dolomite
白云质灰岩　dolomite lime
百分孔隙率　porosity percent
百万分之　parts per million (ppm) (unit of measure)
百万公吨　million metric ton
百万公吨油当量　million tons of oil

equivalent (MTOE)
百万公顷　million hectares (Mha) (unit of measure)
百万千瓦小时　gigawatt hour (GWhr)
百万英热单位　million BTU (mmBTU)
百叶窗　blind
百叶箱温度表　shelter thermometer
柏氏回路　Burgers circuit
柏氏矢量　Burgers vector
摆锤　pendulum mass
摆动波　wave of oscillation
摆动流　oscillatory flow
摆动湿度计　sling psychrometer, sling hygrometer
摆动式波能装置　rocking wave power device
摆动叶片　oscillating vane
摆动轴　oscillating shaft
摆幅　amplitude of oscillation
摆式波浪发电装置　pendulum wave power device
摆式波力装置　pendulum wave energy device
摆式风速表　pendulum anemometer
摆线式透平　oscillatory line turbine
摆振弯曲　flapwise bending

ban

班脱岩　bentonite
斑桉　spotted gum
板层过滤器　panel bed filter
板翅式换热器　plate-fin type heat exchanger
板块构造　plate tectonics
板块法　panel method
板块内热点　intraplate hot spot
板坯加热炉　sheet bar furnace

板式换热器　plate heat exchanger, plate-type heat exchanger
板状吸收体　absorber plate
钣金槽　sheet slot
半闭口式风洞　semi-closed-jet wind tunnel
半闭式循环　partially closed cycle
半闭式循环燃气轮机　semi-closed cycle gas turbine
半闭式循环燃气轮机装置　semi-closed gas turbine plant
半波区　Fresnel zone
半测井图分析　semilog analysis
半潮　half tide
半潮位　half tide level
半导体　semiconductor
半导体薄膜结　semiconductor thin film junction
半导体材料　semiconducting material
半导体层　semiconductor layer, semi-conductor coating
半导体电池　semiconductor cell
半导体电极　semiconductor electrode
半导体—电解液界面　semiconductor-electrolyte interface
半导体功率转换器　semiconductor power converter
半导体光电极　semiconductor photoelectrode
半导体光伏冰箱　semiconductive photovoltaic refrigerator
半导体合金　semiconductor alloy
半导体级硅　semiconductor grade silicon
半导体结　semiconductor junction
半导体—绝缘体—半导体太阳能电池　SIS solar cell (semiconductor-insulator-

半导体膜　semi-conductor coating
半导体太阳电池　semiconductor solar cell
半导体—液体结　semiconductor-liquid junction
半导体—液体太阳电池　semiconductor- liquid solar cell
半导体阴极金属心　mandrel
半导体闸流管　thyristor
半导体整流器　semiconductor rectifier
半对数直线　semilog straight line
半额电能转换器　semiconverter
半风潮差　amplitude of wind tide
半干除尘器　semi-dry scrubber
半干法　semi-dry process
半干法烟气脱硫　semi-dry flue gas desulphurization
半干洗涤器　semi-dry scrubber
半干压工艺　semi-dry process
半活叶桨旋桨　semi-featuring propeller
半晶质　semicrystalline
半晶质的　semicrystalline
半镜（一种防热玻璃，有透光、反射双重功能）　half mirror
半开口式风洞　semi-open-jet wind tunnel
半流体熔岩　viscid lava
半路地　halfway
半气动形状　semi-aerodynamic shape
半潜式　semi-submersible
半潜式混凝土沉箱　semi-submerged concrete caisson
半潜圆柱段　semi-submerged cylindrical section
半球光谱发射率　hemispherical spectral emissivity
半球光谱辐射功率　hemispherical spectral emissive power
半球光谱辐射能力　hemispherical spectral emissive power
半球向反射比　hemispherical reflectance
半球向辐射　hemispherical radiation, hemispherical solar radiation
半球向日射辐照度　hemispherical solar irradiance
半球形碗形反射镜　hemispherical bowl mirror
半球形中空圆筒　hemispherical hollow cylinder
半球总辐射率　hemispherical total emissivity
半热区　semi-thermal area
半日潮　semi-diurnal tide, semidiurnal tide
半闪蒸条件　semi-flash conditions
半渗透膜　semi-permeable membrane
半寿期　half value period
半衰期　half life
半双工传输　half-duplex transmission
半透明盖板　glazing cover
半透明膜　satin film
半透膜　semi-osmotic membrane, semipermeable membrane, semipermeable membrane
半纤维素　hemicellulose
半纤维素酶　hemicellulase
半盐水　brackish water
半圆锉　half-round file
半圆凿　gouge
半圆柱形集热器　hemicylindrical collector
半正弦波　half sine wave
半直驱式风电机组　multibrid technology wind turbine generator system, multibrid technology WTGS
半周期区　Fresnel zone

半轴　half shaft
半自动阀　semiautomatic valve
伴热　heat tracing

bao

薄层电阻　sheet resistance
薄层电阻率　sheet resistivity
薄壳结构　thin shell structure
薄泥浆　grout
薄片　slice
薄透镜成像公式　thin lens equation
饱和　saturation
饱和比湿　saturation specific humidity
饱和点　saturation point
饱和电流　saturation current
饱和电压　saturation voltage
饱和度　degree of saturation
饱和混合比　saturation mixing ratio
饱和基极电流　saturation base current
饱和绝热过程　saturated adiabatic process
饱和空气　saturated air
饱和能量　saturation energy
饱和区域　saturation area
饱和水　saturated water
饱和水汽　saturated vapour
饱和水汽压　saturation water vapour pressure
饱和水蒸气　saturated steam
饱和特性　saturation characteristic
饱和条件　saturation condition
饱和线　saturation line
饱和效应　effects of saturation
饱和压力　saturation pressure
饱和岩屑　water saturated debris
饱和蒸汽　saturated vapor, saturated steam
饱和蒸汽卡诺循环　saturated Carnot vapor cycle
饱和蒸汽压　saturated vapour pressure, saturation vapour pressure, saturate steam pressure
饱和状态　saturation condition
保持器　cage
保护　protection
保护等级　protection level
保护电路　protection circuit
保护电容器　capacitor for voltage protection
保护继电器　protective relay
保护角　angle of shade
保护接地　protective earthing
保护气体　shielding gas
保护系统　protection system
保护栅　guard grating
保热系数　thermal factor
保温材料　heat shield material
保温层　thermal barrier
保温的　heat insulated
保温浇注料　insulating castable
保温力　heat retention
保温炉　heat preserving furnaces
保温套　insulating sleeving
保温性能　heat retention, heat insulating property
保真度　fidelity
报潮球　tidal ball
报警抑制　alarm cut out
鲍林规则　Pauling's rule
暴风　violent storm
暴风浪　storm wave
暴风雨　rainstorm
暴晒作用　solarization
暴雨　storm

暴雨历时　storm duration
爆发波　wave of explosion
爆破孔钻井　blasthole drilling
爆炸隔膜　bursting disc
曝辐（射）量　radiance exposure
曝光计　actinometer
曝气池　aeration tank
曝气浮选　dispersed air floatation
曝气泥　aerate mud
曝晒　insolation
曝晒蒸发　solar evaporation
曝晒作用　solarization

bei

（带有太阳能热量计的）杯式风速计　cuptype anemometer
杯形气流计　cup anemometer
北方"四位一体"模式　north quaternity pattern
贝茨定律（贝茨指出：理论上机械设备能够提取的风能最大值为总量的 59.3%。实际上这一效率的极限值约为 40%）　Betz' Law
贝茨功率系数　Betz power coefficient
贝茨极限　Betz Limit
贝茨理论　Betz theory
贝茨效率　Betz's efficiency
贝尔鞍形填料　Berl saddle
贝尔实验室　Bell Laboratories
贝塞麦转炉　Bessemer converter
贝叶斯方法　Bayesian directional method
备料　feed preparation
备料区　staging area
备用泵　spare pump, standby pump
备用柴油发电机　standby diesel generator
备用电池组　stand-by battery
备用电源　stand-by power, stand-by source of power, backup power
备用电源自投（变电站备用电源自投入，当主供电源发生故障瞬间，自动控制开环充电线路或母线联络开关即合闸，变电站恢复供电）　switchover standby power supply
备用电站　standby station
备用动力源　stand-by source of power
备用发电机　back-up generator
备用发电机组　Backup Generator
备用服务（为维持电力系统正常运行，保持系统发电和负荷需求处于平衡状态，由发电厂提供的增、减有功功率后备服务。包括冷备用和热备用两种服务，冷备用亦称检修备用，常在发电设备检修时启用，热备用用于事故备用和负荷备用）　reserve service
备用负载　dump load
备用供电设备　emergency electric supply unit
备用加热器　back up heater
备用能源系统（常指太阳能供热系统的备用系统。能在太阳能系统停止工作时，提供全部需求的能量）　backup energy system
备用热　standby heat
备用容量　standby capacity
备用容量系数　reserve capacity factor
备用装置　back-up unit
备用状态　stand-by condition
背（电）场太阳电池　back surface field effect solar cell
背场背反射电池　back surface reflection field cell
背场背反射太阳电池　back surface reflection and back surface field solar cell

背场太阳电池　back surface field (BSF) solar cell
背反射电池　back surface reflection cell
背反射太阳能电池　back reflection solar cell
背风处　leeward
背风面　lee side
背风区热　wind shadow thermal
背封　back-seal
背景辐射　background irradiance
背景响应　background response
背景颜色　background colour
背面场　backsurface field
背面场电池　backsurface field cell
背面场效应太阳能电池　backsurface field effect solar cell
背面反射器　back surface reflector (BSR)
背面附加热损(失)　parasitic back loss, parasitic rear loss
背面绝热传导率　back insulation thermal conductivity
背散射　back scatter
背斜　anticline
背斜构造　anticlinal structure, anticline structure
背斜模型　anticline model
背压　back pressure
背压式汽轮机　back pressure turbine
背压装置　back-pressure plant
钡燃料电池（使用钡和氧或钡和氯，把化学能转变为电能的燃料电池）　barium fuel cell
倍压器　doubler
被动对风风轮　yaw passive rotor
被动加热系统　passive heating system
被动控制　reactive control

被动偏航　passive yaw
被动热虹吸系统　passive thermosyphon system
被动失速　passive stall
被动式集热器　passive collector
被动式冷却　passive cooling
被动式水加热墙系统　passive water heating system
被动式太阳房　passive solar house
被动式太阳能　passive solar, passive solar energy
被动式太阳能供热系统　passive solar heating system
被动式太阳能建筑　passive solar building
被动式太阳能设计　passive solar design
被动式太阳能系统　passive solar system, passive type of solar system, passive system
被动式太阳热　passive solar heat
被动式蓄热墙　passive storage wall
被动太阳能　passive solar
被动调向风轮　yaw passive rotor
被动系统　passive system
被动元件　passive component
被吸收功率　absorbed power

ben

本生灯　Bunsen burner
本体压载泵　hull-ballast pump
本征硅　intrinsic silicon
本征扩散系数　intrinsic diffusion coefficient
本征频率　eigen frequency
本征热缺陷　intrinsic thermal defect
本征矢量　eigenvector
本征填充因数（理论填充因数）　intrinsic fill factor (theoretical fill factor)
本征转换效率（本征效率）intrinsic conv-

ersion efficiency (intrinsic efficiency)
本证衰减　Staebler-Wronski effect

beng

崩边　edge chip
崩顶碎波　spilling breaker
崩碎波　spilling breaker
泵冲数　pump speed
泵耗电率　pump power
泵汲技术（一种提高单库双向电站发电效能的技术。其工作过程是：在退潮发电结束后，用泵抽水，降低库面的水位，以增加涨潮发电的水头。由于泵汲是在低水头下进行的，而后的发电则是在高水头下发生的，所以提高水头增加的发电量远大于抽水的耗电量，因而可以得到很大的净能量收益）pumping technology
泵轮　pump impeller
泵喷涡轮机　pump-jet turbine
泵浦功率　pump power
泵气损失　pumping loss
泵容量　pump capacity
泵水风力机　water-pumping windmill
泵水轮机　pump turbine
泵速　pump speed
泵速率　pump speed
泵特性曲线　performance curve of pump
泵推动器　pump impeller
泵站　pumping station

bi

比表面积　specific surface area
比恩法　method of bins
比功率输出　specific power output
比较矩阵　comparative matrix
比较性漏电指数　comparative tracking index
比例关系　scaling relation
比例积分调节器　proportional-integral controller
比例积分微分调节器　proportional-integral-derivative controller
比流量　specific discharge
比热　specific heat
比热容　specific heat capacity
比容　specific volume
比容积　specific volume
比色温度计　colorimetric thermometer
比特　bit
比值　specific value
比重电池　specific gravity cell
比重液电池　gravity battery
比转速　specific speed
笔式记录器　pen recorder
毕奥—萨伐尔定律　Biot-Savart law
闭阀　closed valve
闭合电路　closed circuit
闭合环路　closed loop
闭合回路　complete circuit
闭合回线　closed loop
闭合式循环周期　closed loop cycle
闭环，闭合回路　closed loop
闭环大地耦合　closed loop earth coupling
闭环地热换热器　closed loop geothermal heat exchanger
闭环控制系统　closed-loop control system
闭环系统　closed loop system
闭环系统结构　closed-loop system configuration
闭环蒸汽系统　closed loop steam system

闭井压力　closed-in pressure
闭路冷却器　closed circuit cooler
闭路式循环动力机　closed cycle engine
闭路太阳能供暖系统　closed-loop solar heating system
闭路循环海洋热能　closed cycle ocean thermal
闭路循环海洋热能电站　closed-cycle sea thermal power plant
闭路循环海洋温差电站　closed-cycle sea thermal plant
闭路循环塔　closed-cycle-tower
闭路循环透平系统　closed-cycle turbine system
闭式汽轮机　closed gas turbine
闭式燃气轮机循环　Ackert-Keller cycle
闭式涡轮　shrouded turbine
闭式涡轮机　shrouded turbine
闭式循环　closed cycle
闭式循环海水温差发电系统　closed-cycle ocean thermal energy conversion system
闭式循环海洋热能转换　closed cycle ocean thermal energy conversion
闭式循环燃气轮机电厂　closed-gas-turbine power station
闭式循环燃气轮机装置　closed cycle gas turbine installation
闭式循环系统　closed cycle system
闭式盐差能动力装置　closed power plant of energy from salinity gradient
闭式叶轮　shrouded impeller
闭式蒸发式冷却器　closed circuit evaporative cooler
壁内空气空间　air space in wall
壁热损失　wall loss/ furnace skin loss/ furnace surface loss
壁压　wall pressure
避风（河）湾　sheltered bay
避风水域　sheltered area
避峰用电　off-peak electricity
避雷（装置）　lightning protection
避雷针　lightning rod

bian

边波　edge wave
边角料　edgings
边界层　boundary layer
边界层控制　boundary layer control
边界流　boundary current
边界能量　end point energy
边界条件　end condition
边界要素法　boundary element method
边坡稳定性　slope stability
边缘　flange
边缘波　edge wave
边缘固定系统（使太阳能集热器各层嵌板的边缘固定就位的金属槽道）　edge retaining system
边缘热损失　edge loss
边缘位移　edge dislocation
边长　side length
编码　encode
扁斧　adze
扁管式太阳能平板集热器　solar flat-plate collector of flat-pipe absorber
变底流　varying underflow
变电容载比　ratio of transformer capacity to load
变电所　substation
变电站　site substation
变动负荷　varying load
变风量　variable air volume

变高风 allohypsic wind
变更 alteration
变化场 varying field
变换器拓扑结构 converter topologies
变换设备 conversion device
变换效率 inverter efficiency
变换装置 conversion device
变极式感应电动机 pole-changing motor
变桨动量系数 pitching momentum coefficient
变桨距电机 pitch motor
变桨距调节机构 regulating mechanism by adjusting the pitch of blade
变桨距转子 variable-pitch rotor
变桨调节 pitch regulation
变桨轴承 pitch bearing
变节距水轮机 variable-pitch turbine
变截面叶片 variable chord blade
变距桨叶 variable pitch blade
变距螺旋桨 variable pitch propeller
变距水轮机 variable-pitch turbine
变流量系统 variable flow system
变流器 inverter
变螺距叶片 variable pitch blade
变频 variable frequency
变频电源 variable frequency power
变频交流发电机 variable-frequency alternator
变频器 frequency converter, frequency inverter
变频器柜 cabinet converter
变频调速 variable frequency drives
变水量系统 variable flow system
变速 gear shift
变速潮力磨坊 variable-speed tidal mill
变速齿轮 variable speed gear, changer speed gear

变速发电机 variable-speed generator
变速风力发电机 variable-speed wind turbine
变速杆 gear lever
变速恒频 variable-speed-constant-frequency
变速恒频风电机组 variable speed constant frequency wind turbine generator system, variable speed-constant frequency system
变速机构 gear shifter
变速机械装置 variable-speed mechanism
变速流 variable flow
变速调节 variable-speed control
变速箱 changer speed gear
变速消振器 variable speed damper
变速叶轮叶片 non-uniform rotor blade
变速运行 variable-rpm operation
变速运转 variable-speed operation
变速直驱 variable-speed direct drive
变速装置 variable speed gear
变位齿轮 gears with addendum modification
变温层 thermocline
变形 deformation
变形比 deformation ratio
变形浮箱（一种利用波浪能的设备） deformable raft
变性燃料乙醇 denature fuel ethanol
变压 variable voltage
变压风 allobaric wind
变压器 voltage transformer, transformer
变压器平台 transformer platform
变压吸附 pressure swing absorption
变质水 metamorphic water
变质岩 metamorphic rock
变阻测辐射热表 bolometer

变阻光电池　photoconductive cell
变阻器　varistor
便携式发电机　portable power generator
便携式太阳能发电系统　portable solar power generation system
便携式吸热—解吸制冷装置　portable absorption-desorption cooling unit

biao

标称定律　nominal rating
标称负载　nominal load
标称功率　nominal power
标称最大粒度　nominal top size
标尺参数　scale parameter
标定（获得标准太阳电池的方法或手段称为标定）　calibrating
标定值　calibration value
标度　calibration
标准　criterion
标准液体供热系统　standard liquid heating system
标准表面　standard surface
标准测试条件（太阳电池的标准测试条件）　standard test condition
标准尺　gauge
标准尺寸比　standard dimension ratio
标准大气压　standard air pressure
标准大气状态　standard atmospheric state
标准电池　standard cell
标准电压　pilot voltage
标准电压表　standard voltmeter
标准风速　standardized wind speed
标准辐射表　standard radiometer
标准辐照度曲线　standard irradiance curve
标准光源　standard light source
标准海平面　standard sea level
标准化电网　standardized grid
标准空气污染物　criteria air contaminant
标准立方英尺　standard cubic feet
标准偏差　standard deviation
标准氢电极　normal hydrogen electrode
标准溶液　standard solution
标准水　standard water
标准太阳电池　standard solar cell
标准温度表　standard thermometer
标准温度和压力　standard temperature and pressure
标准误差　standard uncertainty
标准循环　standard rating cycle
标准液体系统　standard liquid system
表层沉积物　surficial sediment
表层套管　surface casing
表层土　topsoil
表观风速　apparent wind speed
表面波雷达　wave radar
表面粗糙效应　surface roughness effect
表面地貌　surface geomorphology
表面发射率　surface emissivity
表面辐射率　surface emissivity
表面腐蚀　surface corrosion
表面改性　surface modification
表面高度　surface elevation
表面活性剂　surfactant
表面扩散　surface diffusion
表面冷却器　surface condenser
表面摩擦阻力　skin friction drag
表面磨削　surface grinding
表面能　surface energy
表面热现象　surface thermal phenomena
表面式换热器　surface heat exchanger
表面式冷凝器　surface condenser

表面式凝汽器　surface condenser
表面水循环　surface water loop
表面涂层　surface coating
表面温度　surface temperature
表面温度值　surface temperature value
表面展扩热交换器　extended surface exchanger
表面张力　surface tension
表面张力波　capillary wave
表面张力效应　surface tension effect
表面蒸发　surface evaporation
表面蒸发式冷凝器　surface evaporative condenser
表皮系数　skin factor
表皮效应　skin effect
表现黏度　apparent viscosity
表征测试　characterization test

bing

冰雹　hail
冰川沉积　glacial deposit
冰川沉积矿床　glacial outwash deposit
冰川地基岩冲刷　glacially scoured bedrock
丙醇　propanol
丙酮—丁醇—乙醇工艺　acetone-butanol-ethanol
丙二醇　propylene glycol
丙烷　propane
丙烷锅炉　propane boiler
丙烷冷冻设备　propane refrigeration unit
丙烷冷凝器　propane condenser
丙烷排气管　propane exhaust pipe
丙烷清洗舱　propane purge tank
丙烷燃烧器　propane burner
丙烷涡轮机　propane turbine
丙烷压缩机　propane compressor
丙烷蒸汽管　propane-vapor pipe
并联布置　parallel configuration
并联电容器　shunt capacitor
并联电阻　shunt resistance
并联辅助设备　parallel auxiliary
并联功率　shunt power
并联谐振　parallel resonance
并联运行　interconnected operation
并联阻抗　parallel impedance
并列　interconnection
并入线路　cutting in
并生藻类　adnate algae
并网变速风力发电系统　grid-connected variable speed wind power system
并网的　grid-connected
并网发电　grid-connected power generation
并网发电机设备　grid-connected power generation
并网光伏电站　grid connected photovoltaic power generation
并网太阳能光伏发电系统　grid-connected solar photovoltaic system
并网型风力发电机　on-grid wind power generator
并向流　co-current flow

bo

波（浪）能技术　wave energy technology
波（浪）能（量）谱　wave energy spectrum
波包　wave packet
波参数　wave parameter
波槽　wave trough, trough of the wave
波成流　wave drift current
波传播　wave propagation
波登管　Bourdon tube
波登管式压力计　Bourdon gauge,

Bourdon tube pressure gauge
波底 （静水中水面波动扬不起沉淀物的深度） wave base
波动 fluctuation
波动筏式波浪能发电装置 contouring raft type wave energy conversion system
波动方程 wave equation
波动幅度 phase
波动流体 wave flow
波动频率 wave frequency
波动物理学 wave physics
波陡 wave steepness, wave steepness ratio
波尔原子模型 Bohr atomic model
波峰 wave crest, crest
波峰长度 crest length
波峰高 crest elevation
波峰高度 crest height
波峰周期 peak period
波幅 amplitude of waves
波腹 loop, antinode
波高 wave height
波谷 wave hollow
波候 wave climate
波候，波象 wave climate
波及体积 swept volume
波级 wave scale, wave magnitude
波节 node
波浪 wind sea, wave condition, wind swell, ocean surface wave
波浪槽 wave tank
波浪产生的电力 wave-generated power
波浪场（是将波浪力转换为压缩空气来驱动空气透平发电机发电的场所） wave-farm
波浪沉积交互作用 wave-sediment interaction
波浪冲击 wave attack
波浪推进 wave advance
波浪动力学 wave dynamics
波浪陡坡 wave steepness
波浪发电 wave power generation
波浪发电场 wave power farm, wave energy park, wave farm
波浪发电机 wave activated generator, wave engergy converter, wave generator, wave power generator, wave activated power generator
波浪发电机组 wave activated generator group
波浪发电设备 wave energy converter
波浪发电系统 wave electric power system
波浪发电装置 wave power unit
波浪发生区 area of wave generation
波浪反射 wave reflection
波浪防护 wave protection
波浪分布图 wave scatter diagram
波浪荷载 wave load
波浪回波干扰 wave clutter
波浪计 wave recorder
波浪均夷作用 wave planation
波浪类型 wave type
波浪力 wave power, wave force
波浪力发电 wave activated power generation
波浪流 wave current
波浪能 energy from the wave, wave power
波浪能电站 wave power station
波浪能吸收转换机 wave energy converter
波浪能源 wave as energy source

波浪能转换 wave energy extraction
波浪能转换器 wave energy converter
波浪泥沙交互作用 wave-mud interaction
波浪爬高 wave run-up
波浪漂移 wave drift
波浪漂移力 wave drift force
波浪漂移阻尼 wave drift damping
波浪谱 wind-generated waves spectrum, sea state spectrum
波浪浅化 wave shoaling
波浪水槽 wave flume
波浪水池 wave basin
波浪推进 wave advance
波浪袭击 wave attack
波浪压力 wave pressure
波浪要素 wave parameter
波浪运动能 energy by wave motion
波浪载荷 wave loading
波浪折射 wave refraction
波浪周期 wave period
波浪转换设备 wave-conversion device
波浪资源 wave resource
波浪阻力 wave resistance
波浪组合 wave set up
波浪作用 wave action
波力泵 wave pump
波力发电 wave power generation
波力发电机 wave power generator
波力发动机 wave motor
波力驱动船 wave power boat
波力透平 wave turbine
波粒子运动能量转换器 particle motion convertor
波连 wave velocity
波列 wave train
波龄 wave age
波流相互作用 wave-current interaction

波面 wave surface
波面筏（一种波浪发电设备） wave contouring raft
波能 wave energy
波能（转换）驳船 wave energy barge
波能（转换）平台 wave energy platform
波能槽 wave energy sink
波能成长 wave energy growth
波能传递 wave energy transmission
波能萃取器 wave energy extractor
波能二次转换装置 secondary wave energy conversion device
波能发电机 wave power machine, wave powered generator
波能发电站 wave power station
波能分布 wave energy distribution
波能浮标工程 wave power buoy project
波能供电浮筒 wave powered buoy
波能回收系统 wave power recovery system, wave power recovery system
波能回收装置 wave power recovery device
波能检波器 wave energy rectifier
波能聚焦 wave focusing
波能空气透平 wave energy air turbine
波能利用 wave energy utilization
波能利用系统 wave power recovery system, wave energy recovery system
波能利用装置 wave power recovery device
波能流动 wave flow
波能玫瑰图 wave energy rose diagram
波能密度谱 wave energy-density spectrum
波能平衡 wave energy balance

波能汽轮发电机　wave-powered generator, wave power air-turbine generator
波能衰减装置　wave energy attenuator
波能通量　wave energy flux
波能吸收　wave energy absorption, wave power absorption
波能吸收浮子　wave energy absorbing floator
波能吸收器　wave power absorber
波能吸收装置　wave energy absorbing device
波能系数　wave energy coefficient
波能一次转换装置　primary wave energy conversion device
波能原点　wave energy origin
波能终结器　wave energy terminator
波能终止器　wave energy terminator
波能转换　wave energy conversion
波能转换方案　wave energy conversion scheme
波能转换浮体　wave energy conversion float
波能转换浮筒　wave energy conversion buoy
波能转换概念　wave energy conversion concept
波能转换器　wave energy converter
波能转换器阵列　wave energy converter array
波能转换无效费用　wave-energy conversion cost-ineffective
波能转换效率　wave energy conversion efficiency
波能转换终端　wave energy conversion terminal
波能转换装置　wave energy conversion device

波能装置　wave device, wave energy device, wave power device
波频散　wave dispersion
波破碎　wave breaking
波谱　wave spectrum
波谱学　spectroscopy
波前　wavefront, wave front
波前冲角　wave front angle
波前法　wave front method
波前再现　wave front reconstruction
波前长度　wavefront length
波群　wave group, wave packet
波射线　wave ray
波蚀　wave abrasion
波数　wavenumber
波速度　wave velocity
波特　baud
波特率　baud rate
波湍流　wave turbulence
波纹电流　ripple current
波纹管式　bellow type
波纹型吸热板　corrugated absorber plate
波线　wave ray
波向　wave direction
波向线　wave ray
波形　wave profile
波形辊　waveroller
波形绕组　wave winding
波形扫描　wavescan
波形铁皮　corrugated iron
波形图　oscillogram
波压　wave pressure
波源　wave source
波增水　wave setup
波长　wavelength
波长间隔　wavelength interval
波长依赖性反射强度　wavelength

dependent reflected intensity, wavelength dependent transmitted intensity
波能转换　wave power conversion
波阵面　wavefront, wave front
波阵面长度　wavefront length
波振幅　wave amplitude
波中心　wave hub
波周期　wave period
波组　wave group
波作用　wave action
玻耳兹曼恒量　Boltzmann constant
玻璃　glazing
玻璃板支柱　support post for glass plate
玻璃半导体　glass semiconductor
玻璃厂　glass foundry
玻璃窗　glazed window
玻璃刀　glass cutter
玻璃底板　glass base
玻璃顶盖　glass roof
玻璃顶盖太阳能蒸馏器　glass-roofed solar still
玻璃顶盖吸热　absorption by glass cover
玻璃封接　glass seal
玻璃封装管　glass-encapsulated tube
玻璃盖板　glass cover, glazing cover, glass cover plate
玻璃盖板温度　glass cover temperature
玻璃盖板下面的水袋　glass covered water bag
玻璃盖片　glass cover sheet
玻璃钢　fiberglass reinforced plastic (FRP)
玻璃隔热　glass spacer
玻璃管　glass tube
玻璃扩散源　glass diffusion source

玻璃平板太阳能集热器　flat-glass solar heat collector
玻璃热阻　glass resistance
玻璃态　vitreous state
玻璃透光率　glass transmission
玻璃涂层　glass coating
玻璃纤维　glass fiber
玻璃纤维布　fiberglass cloth
玻璃纤维盖板　fiberglass cover
玻璃纤维结构　fiberglass structure
玻璃纤维绝缘（材料）　fiberglass insulation
玻璃纤维束　glass fiber bundle
玻璃纤维增强材料　fiber-glass reinforcement
玻璃纤维增强混凝土　glass fiber reinforced concrete
玻璃纤维增强热塑性塑料　fiber glass reinforced thermoplastics
玻璃纤维增强塑料　fiberglass-reinforced plastic, fibrous glass reinforced plastics
玻璃真空管集热器　glass evacuated tube collector
玻璃真空管太阳能热水器　vacuum tubular glass solar collector
玻璃状炉渣　vitrified slag
玻意耳定律　Boyle's law
玻意耳—马略特定律　Boyle-Mariotte law
剥落　spalling, exfoliation
剥蚀　abrasion
伯努利方程　Bernoulli's equation
驳船　barge
薄膜　membrane, thin film
薄膜电路　film circuit
薄膜电阻材料　materials for film resistance
薄膜多晶硅电池　thin-film polycrystalline silicon cell
薄膜多晶硅太阳能电池　polycrystalline silicon film solar cell

薄膜砷化镓太阳能电池　film GaAs solar cell
薄膜太阳能电池　thin film solar cell
薄膜太阳能电池材料　thin-film solar cell material
薄膜形貌　film morphology

bu

补偿电路　compensation circuit
补偿电压　offset voltage
补偿风　compensating wind
补偿流　compensating stream
补偿器　equalizer
补偿式绝对辐射表　compensated pyrheliometer
补偿直接日射（强度）计　compensation pyrheliometer
补充能量　complementary energy
补充热量　supplemental heat
补给　replenishment
补给井　diffussion well
补给水　feedwater makeup
补给线　supply line
补燃室　afterburning chamber
捕获截面　capture cross section
不饱和度　unsaturation
不饱和聚酯　unsaturated polyester
不彻底的　halfway
不充电电池　frozen battery
不定风　variable wind
不定形耐火材料　monolithic refractory
不对称电池　asymmetrical cell
不固定的　floating
不规则　irregularity
不规则波　irregular wave
不规则潮汐　anomalistic tide
不规则同步　random synchronizing
不规则星系　irregular galaxy
不规则性　irregularity
不含矿物盐蒸汽　clean steam
不加压太阳能热水器　non-pressurized solar water heater
不间断电源　uninterruptible power supply
不间断电源系统　UPS system
不聚焦集热器　nonfocusing collector
不均匀分布　nonuniform distribution
不均匀加热　uneven heating
不均匀送风　inhomogeneous air supply
不均匀性　inhomogeneity
不可复原式微型燃气轮机　unrecuperated microturbine
不可更新能源　nonrenewable energy source
不可凝气体　non-condensable gas
不可燃的　nonflammable
不可再生能源　non-renewable energy source
不利流动比　unfavorable mobility
不敏感性　immunity
不凝结气体　non-condensable gas
不凝气体含量　noncondensible gas content
不起作用的　inoperative
不燃材料　noncombustible material
不燃灰　incombustible ash
不溶的　insoluble
不受气流扰动的螺旋桨　static free propeller
不太理想的集热器　less perfect collector
不同相电流　out of phase current
不透辐射热性　athermancy
不透明材料　opaque material
不透明盖板　opaque cover
不透明云遮系统　opaque cloud cover

不透水层　confining layer
不完全燃烧　incomplete combustion
不完全燃烧损耗　incomplete combustion loss
不稳定电流　unsteady current
不稳定负荷　unsteady load
不稳定流动　unsteady flow
不稳态渗流技术　transient technique
不向阳的太阳能住宅　solar home without southern exposure
不向阳太阳房　solar home without a southern exposure
不锈钢体　stainless steel body
不正常温升　abnormal temperature rise
布袋除尘器　baghouse filter
布井　well spacing
布拉格定律　Bragg's law
布拉维指数　Bravais indices
布朗运动　Brownian movement
布雷顿发动机　Brayton engine
布雷顿循环　Brayton cycle
布置　disposal, furnish
布置图　layout drawing
步进电机　stepper motor
步进梁　walking beam
步进式加热炉　walking beam furnace
部分倒伏状（V级植物风力指示）partial throwing
部分的　halfway
部分负荷　part load
部分负荷运转　part-load operation
部分氧化　partial oxidation

cai

材料　material
材料产出　material output
材料的热导　conductance of material
材料光学　materials optics
材料合格证　material certificate
材料回收设备　material recovery facility
材料疲劳　material fatigue
材料特性　material property
材料置换　material substitution
采风口　air intake
采光口　aperture
采光面积　aperture area
采光平面　aperture plane
采矿　mineral extraction
采暖　heating
采暖分配系统　heating distribution system
采暖锅炉　heating boiler
采暖期度日数　degree-day during heating period
采暖水　heating water
（采）暖通（风）工程　heating and ventilation engineering
采石机　rock catcher
采收率　recovery efficiency
采样单元　sampling unit
采样率　sampling rate
采样系统　sampled system
彩色检测系统　color-detection system
彩色识别　color identification
菜籽　rapeseed

can

参考风速　reference wind speed
参考曲线　reference curve
参考站　reference station
参数分布模型　distributed-parameter model
参数灵敏度　parameter sensitivity

参数要求　parameter requirement
残留物　remnant
残流　residual flow
残余溶液　residual liquid
残余物处理　residue disposal
残余物回收　residue recycling
残渣　residua (ash and slag)

cao

操控手册　operating and maintenance manual
操纵性　handiness
操作规程　operating regime
操作条件　service condition
操作维护　operation & maintenance
操作温度　operating temperature
操作原理　operating principle
操作状态　operating state
槽壁修正　blockage correction
槽电压　tank voltage
槽电压，电池电压　cell voltage
槽角　trough angle
槽轮　sheave
槽式太阳能电站　solar parabolic trough power generation
槽式太阳能发电　trough solar energy power generation
槽式太阳能热水器　trough type solar water heater
槽式太阳热发电系统　parabolic trough solar thermal power generation system
槽形集热器　trough collector
槽形聚光器　trough-type concentrator
槽形抛物面集热器　parabolic-trough collector
槽形托辊　trough idler
槽型太阳能聚光器　grooved concentrator

草本能源作物　herbaceous energy crop
草本燃料　herbaceous fuel
草本生物质燃料　herbaceous biomass fuel
草酸钙　calcium oxalate
草酸盐　oxalate
草屑　grass clipping

ce

侧板螺旋桨　shrouded propeller
侧风　cross wind
侧腹　flank
测流　side stream
侧面　flank
侧面弯曲　edgewise bending
侧逆风　side headwind
侧顺风　side leading wind
侧向力　lateral force
侧向平移　lateral translation
侧向移位　lateral displacement
侧阵风　side gust
测潮标尺　tide-staff
测潮表　tidalmeter
测潮计　tide-gauge
测潮仪　tidal gage
测风气球　pilot balloon
测风塔　wind measurement mast
测风学　anemography
测风仪　anemometer
测风站　air measuring station
测井　well logging
测井曲线　well log
测力计　dynamometer
测量参数　measurement parameter
测量船　observation boat
测量电路　measuring circuit
测量功率曲线　measurement power

curve
测量扇区　measurement sector
测量速度　measured speed
测量位置　measurement seat
测量误差　uncertainty in measurement
测量周期　measurement period
测试点风速　wind speed of test point
测试方法　test method
测速装置　speed sensor
测温探头　temperature probe
测温探针　thermal probe
测温液　thermometric liquid
测压计　piezometer
测压孔　pressure tap
测压元件　pressure element
测验台　test rig
策划　hatch

ceng

层　layer, level, class
层次式废物管理方法　waste management hierarchy
层积云　stratocumulus cloud
层理　stratification
层理面　bedding plane
层流　laminar flow
层面　bedding plane
层燃　grate firing
层燃锅炉　grate firing boiler
层燃炉　grate furnace
层燃炉技术　grate furnace technology
层压　lamination, lamination process
层压（制品）　laminate
层压板式集热器　laminated panel collector
层压铝钢复合反射层材料　laminated polymer-aluminium-steel reflector material
层状热储　stratified reservoir

cha

叉架式运货车　fork-lift truck
插电式混合动力车　plug-in hybrid electric vehicle
插入　intervene
插入式套筒扳手　box spanner inset
差动保护　differential protection
差分法　difference method
差分方程　difference equation
差热分析　differential thermal analysis
差速齿轮　differential gear
差温环流式太阳能集热器　thermosiphoning solar collector

chai

拆除　dismantle
拆除的木材　demolition wood
拆袋器　bag breaker
拆建废料　construction and demolition waste, construction and demolition debris
拆卸　dismount
柴火炉　wood stove
柴油　diesel oil
柴油电网　diesel-powered grid
柴油发电机　diesel generator
柴油发电机组　diesel genset
柴油发动机　diesel engine
柴油风力发电系统　wind-diesel system
柴油机燃料　diesel fuel
柴油值　diesel number
柴油指数　diesel index

chan

掺锂太阳电池　lithium-doped solar cell
掺杂硅　doped silicon
掺杂剂　dopant
掺杂物　dopant
产甲烷菌　methanogenic bacteria
产甲烷生物　methanogen
产量递减率　production decline rate
产量递减曲线　production decline curve
产能　output
产能和计算　force production and calculation
产能因子　energy capability factor
产品尺寸　product size
产品单位产量可比综合耗能　comparable comprehensive energy consumption for unit output of product
产品单位产量综合耗能（单位产品综合耗能）　comprehensive energy consumption for unit output value
产品分配系统　distribution system
产气量　gas yield
产气率　biogas production rate
产热　heat production
产热能力　heat producing capability
产生热量的　heat producing
产液　production fluid
铲车　fork-lift truck
颤振　flutter

chang

长臂开采　longwall mining
长臂圆规　beam compass
长波辐射　long wavelength radiation, longwave radiation
长波辐射源　long-wave emittance
长波红外辐射　long-wavelength infrared radiation
长波红外线扫描仪　long wave infrared scanner
长堤　causeway
长定子直线同步发动机　long stator linear synchronous motor
长度规　length scale
长距离供热　long distance heating
长距离管道　long distance line
长聚焦集热器　long-focus collector
长期储热（系统）　long-term storage
长期能源替代计划　long-term energy substitution programme
长石　feldspar
长寿命能源　long-life power source
长寿期太阳能阵列　long-life solar array
长重力波　infragravity wave
长周期潮汐　tide of long period
常风　steady wind
常规固体燃料　conventional solid fuel
常规热液　conventional hydrothermal
常规热液扩展　conventional hydrothermal expansion
常规生物柴油　conventional biodiesel
常规试验　routine test
常规太阳电池　conventional solar cell
常温　room temperature
常温常压　normal temperature and pressure
常温耐压强度　cold crushing strength
常现风速　frequent wind speed
厂房　factory building
厂区绿化系数　greening factor of plant area
厂区竖向布置　vertical layout of plant
厂用电率　rate of house power

厂址潜势　site potential
场边界条件　field boundary condiction
场地规划　site planning
场地通达度　site accessibility
场合　instance
场镜　field mirror
场内电源系统　on-site power system
场能　field energy
场线圈　field coil
场效应　field-effect
场效应管　field effect transistor
场增强光电导率　field enhanced photoconductivity
敞开系统　open system
敞湾　open bay

chao

超薄太阳能电池　ultra-thin solar cell
超常风　overnormal wind
超出额定值的电压　overrating voltage
超出同步　rising out of synchronism
超导材料　superconductive material
超导磁储能系统　superconducting magnetic energy storage system
超导线圈　superconducting coil
超低压头潮汐能发电　extra-low head tidal power
超低压头涡轮机　ultra-low head turbine
超高压层　abnormal pressured formation
超环面（轮胎面）　toroidal surface
超环面（轮胎面）聚光器　toroidal concentrator
超级电容器　supercapacitor
超临界电厂　supercritical power plant
超临界锅炉　supercritical boiler
超临界流体　supercritical fluid
超临界循环　supercritical cycle
超临界压力　supercritical pressure
超临界蒸汽参数　supercritical steam condition
超临界蒸汽压力　supercritical steam pressure
超临界状态　above-critical state
超滤　ultrafiltration
超前功率因数　leading power factor
超前无功功率　leading reactive power
超前相　leading phase
超热能　epithermal energy
超声波温度计　ultrasonic thermometer
超声生物效应　biologic effect of ultrasound
超速控制　overspeed control
超微型浮游生物　picoplankton
超压地热系统　geopressured system
超压阀　pressure override valve
超压设施　geopressured application
超音速（流）风速表　supersonic-flow anemometer
超重氢　tritium
朝南平面集热器　southerly-facing flat collector
朝阳建筑　solar oriented architecture
潮　tide
潮（流冲）刷　tidal scour
潮（汐波）幅　tidal amplitude
潮标　tidal mark, tide mark
潮波　tide wave
潮波　tidal wave, tide bulge
潮差　tidal amplitude, tidal difference, tidal range
潮差资源　tidal range resource
潮汐平地　tidal flat
潮程　tidal excursion
潮船坞　tidal basin

潮堤	tidal weir
潮风	tidal wind
潮幅	amplitude of the tide
潮高变化	tidal variation
潮沟	tidal creek, tide channel
潮痕	tidal mark, tide mark
潮间带	intertidal area
潮间地	tidal land, intertidal zone
潮间区域	intertidal zone
潮津波	tidal bore
潮浪	tidal bore
潮浪发生器	tide-generating mechanism
潮篱	tidal fence
潮力发电厂	tidal power plant
潮力峰值	tidal power peak
潮龄	tidal age
潮流	tidal current, tidal current curve, tidal flow, tideway
潮流表	tidal current table
潮流冲刷	tidal flushing
潮流冲刷作用	tidal flush
潮流间隙	tide interval
潮流能	tidal current energy
潮流曲线	tidal current curve
潮流速度差	tidal current difference
潮流特性	tidal regime
潮流运动	tidal movement
潮流装置	tidal stream device
潮路	tide way
潮轮	tide mill
潮轮原理图	tide mill schematic
潮漫滩	tidal mashland
潮能耗散	tidal dissipation
潮区	tidal compartment
潮区界限	tidal limit
潮升	tidal rise
潮湿	humidity

潮时	tidal hour
潮时滞后	tidal lag
潮水冲刷	tidal scour
潮水发电厂	tidal power plant
潮水沟	tidal water furrow
潮水河	tideway, tidal river
潮水量	tidal influx, tidal power prism
潮水位	tide water level
潮水箱	tide box
潮滩	tidal flat, tidal bank
潮滩沉积	tidal flat deposit
潮头	head of tide
潮位	tidal height, position of tide, tidal water level, tide stage
潮位变幅	tital range
潮位基准面	tidal datum
潮位计	tide-gauge
潮位曲线	tidal curve
潮位升降	tidal undulation
潮位水尺	tidal gauge, tide-staff
潮位站	tidal station
潮汐	tide
潮汐（发生）仪	tidal mechanism
潮汐（变化）范围	tidal range
潮汐坝	tidal power barrage
潮汐变化	tidal variation
潮汐变化范围	tital range
潮汐变形	tidal deformation
潮汐表	tidal table
潮汐波	tidal wave, tide bulge
潮汐补偿	tidal compensation
潮汐不稳定性	tidal instability
潮汐常数	tidal constant
潮汐冲毁的	tide-worn
潮汐冲刷作用	tidal flush
潮汐带	tidal zone
潮汐的	tidal

潮汐电能　tidal electrical energy
潮汐电站发电水头　water head of driving a tidal plant
潮汐电站总效率　total efficiency of hydroelectric power station
潮汐电站最大负载运行　peak load operation of tidal power station
潮汐订正　tidal correction
潮汐动力房　tidal power house
潮汐动能　tidal kinetic energy
潮汐发电　tidal electric power generation, tidal power, tidal power generation, power generation from tide
潮汐发电坝　tidal power barrage
潮汐发电方案　tidal power scheme
潮汐发电工程　tidal scheme
潮汐发电功率　tidal power
潮汐发电规划　tidal power planning
潮汐发电机　tidal power generator, tidal-power generator, tidal generator
潮汐发电机组　tidal power unit, tidal power generator set
潮汐发电计划　tidal power project
潮汐发电水轮机　tidal turbine
潮汐发电用涡轮机　turbine for tidal power
潮汐发电站（厂）　tidal power station, tidal power generating plant
潮汐发电站开发方案　tidal power scheme, tidal scheme
潮汐发电站可利用率　tidal power station availability
潮汐发电站设计　tidal power plant design
潮汐发动机　tide-motor
潮汐港池　tidal basin
潮汐功率　tidal power
潮汐共振　tidal resonance
潮汐航道　tidal waterway
潮汐河口　tidal estuary
潮汐河水道　tidal waterway
潮汐湖　tidal basin
潮汐回水　tidal backwater
潮汐基准面　tidal datum plane
潮汐激流　tidal rapid
潮汐脊　tidal sand ridge
潮汐计　tide-meter
潮汐加速度　tidal acceleration
潮汐间隙　tidal interval
潮汐控制　tidal regulation
潮汐控制仪　tide control apparatus
潮汐篱笆　tidal fence
潮汐力　tidal power, tidal force
潮汐力利用　utilization of tidal power
潮汐流　tidal race
潮汐流发电机　tidal stream generator
潮汐流技术　tidal stream technology
潮汐隆起　tidal bulge
潮汐码头　tidal accommodation
潮汐摩擦　tidal friction
潮汐摩擦能　tidal friction energy
潮汐磨坊　tidal mill
潮汐能　tidal energy
潮汐能发电厂　tidal generating plant
潮汐能发电工程　tidal power project
潮汐能发电装置　tidal energy device
潮汐能方案　tidal energy scheme
潮汐能工程　tidal power engineering
潮汐能技术　tidal power technology
潮汐能经济学　economics of tidal power
潮汐能开发　tidal energy development, tidal power development
潮汐能开发方案　tidal energy tapping scheme

潮汐能利用　tidal power utilization, tidal energy utilization
潮汐能量变流器　tidal energy converter
潮汐能平均功率　tidal average power
潮汐能审查委员会　tidal power review board
潮汐能生产　tidal energy production
潮汐能输出　tidal energy output
潮汐能提取　tidal energy extraction
潮汐能应用　tidal energy application
潮汐能转换　tidal energy conversion
潮汐能转换系统　tidal energy conversion system
潮汐能转换装置　tidal energy conversion device
潮汐能资源蕴藏量　potential tidal power resource
潮汐能综合利用　tidal energy integration
潮汐盆地　tidal basin
潮汐谱　tide spectrum
潮汐气候　tidal climate
潮汐区　tidal stretch
潮汐曲线　tide curve
潮汐三角洲　tidal delta
潮汐势　tidal potential
潮汐试验电站　experimental tidal power plant
潮汐水车　tidal mill
潮汐水道　tide way
潮汐水库　tidal reservoir
潮汐水力发电厂　hydrokinetic wave farm
潮汐水力学　tidal hydraulics
潮汐水量　tidal water volume
潮汐水轮泵站　tidal turbine pump station
潮汐水位　tide level of water
潮汐水域　tidal waters

潮汐调和分析　harmonic analysis of tides
潮汐调节　tidal regulation
潮汐图　tidal atlas
潮汐推进波　tidal progressive wave
潮汐湾　tidal bay
潮汐位　tidal potential
潮汐位能　tidal potential energy
潮汐位相　tide phase
潮汐相位　tide phase
潮汐校正　tidal correction
潮汐楔　tidal wedge
潮汐泻湖　tidal lagoon
潮汐泻水　tidal lagoon
潮汐学　tidology
潮汐堰坝　tidal barrage
潮汐预测机　tide-predicting machine
潮汐栅栏　tidal fence
潮汐栅栏式结构　tidal fence
潮汐站　tidal gaging station
潮汐振荡　tidal oscillation
潮汐周期　tidal cycle, tidal period
潮汐状况　tidal regime
潮汐自动推算机　tide predicting machine
潮汐组分　tidal constituent
潮汐作用　tidal action
潮相迟角　tidal epoch
潮型　types of tides
潮压　tidal load
潮堰　tidal weir
潮振动　tidal undulation
潮振幅　tidal amplitude
潮滞　tidal lag, tide lag

che

车床工具　lathe tool
车床工作台　lathe bench

车用乙醇汽油　ethanol gasoline for motor vehicle
车载电力电子技术　on-board power electronics
车载电力电子设备　on-board power electronics

chen

尘埃注入炉　dust injection furnace
尘饼　dust cake
尘土过滤器　dust filter
沉淀　sedimentation, precipitation
沉淀槽　settling pit
沉淀池　setting pond, sedimentation tank
沉淀物　sediment, deposit, precipitation
沉淀作用　precipitation
沉积　deposition, sediment
沉积槽　sedimentation tank
沉积盆地型层状热储地热　stratigraphic-bound reservoir
沉积速度　rate of deposition
沉积速率　deposition rate
沉积物　sediment
沉积物搬运　sediment transport
沉积物迁移　sediment transport
沉积岩　sedimentary rock
沉积装置　sedimentation unit
沉降　sedimentation, subsidence
沉降池　settling tank
沉降坑　settling pit
沉降室　settling chamber
沉井　caisson
沉砂池　settling pit
衬度　contrast
衬套　quill, bush

cheng

成本比率　cost ratio
成本收回时间　marginal cost payment time
成层　stratification
成潮力　tidal-generating force
成分　composition
成分分析，组成分析　compositional analysis
成分过冷　constitutional supercooling
成井　well completion
成品质量检验　final quality control
成酸　acid forming
成酸剂　acid former
成酸物质　acid former
成套电池　battery stack
成套工具　kit
成套项目太阳能光伏系统　turn-key photovoltaic system
承压式空气吸热器　pressurized air receiver
承压水位　pressure plate
承压水压力　artesian pressure
城堡螺母　castle nut
城市废料　municipal waste
城市废物　municipal waste
城市废物处理　municipal waste disposal
城市分类垃圾　municipal sorted waste
城市给水　municipal water supply
城市供水　municipal water supply
城市固体废物　municipal solid waste
城市垃圾　urban wastes, city rubbish, municipal refuse, municipal waste
城市垃圾处理厂　municipal waste disposal plant

城市垃圾循环（利用） municipal refuse recycling
城市木材废料 urban wood waste
城市生活垃圾焚烧法 incineration of municipal solid wastes
城市太阳能庭院灯 solar garden light for cities
城市有机生活垃圾 organic fraction of the MSW
城镇有机垃圾 organic fraction of municipal solid waste

chi

吃水线 （海陆边界） waterline
池式太阳能蒸馏器 basin-type solar still
池塘系统 pond system
持久电流 persistent current
持久极限 endurance limit
持久性有机污染物 persistent organic pollutant
持续风 persistent wind
持续运行 continuous operation
持续运行的闪变系数 flicker coefficient for continuous operation
尺寸 dimension
尺寸检验 dimensional inspection
尺度（定距）准绳（索） guide wire
尺度参数 scale parameter
尺度关系 scaling relation
尺度母数 scale parameter
尺度折减系数 size reduction factor
齿槽 tooth space
齿顶圆 tip circle
齿高 tooth depth
齿根圆 root circle
齿厚 tooth thickness

齿宽 face width
齿轮 gear wheel
齿轮泵 gear pump
齿轮传动 gear transmission
齿轮传动泵 gear driven pump
齿轮的变位 addendum modification on gear
齿轮风机 geared wind turbine
齿轮负荷 gear loading
齿轮副 gear pair
齿轮毂 gear hub
齿轮减速器 speed reduction gear
齿轮电动机 gear motor
齿轮啮合 gear meshing
齿轮切削机 gear cutting machine
齿轮式泵 gear type pump
齿轮水泵 gear water pump
齿轮速比 gear ratio
齿轮系 gear train, train of gears
齿轮箱 gear box, gear case, planetary gearbox
齿轮箱变比 gear box ratio
齿轮箱齿轮比 gearbox gear ratio
齿轮箱冷却风扇 gear fan
齿轮箱轴承 gearbox bearing
齿轮装置 gearing
齿面 tooth flank
齿啮式连接 dynamic coupling
齿圈 girt gear
齿式离合器 dog clutch
齿数 number of teeth
齿隙 backlash
赤潮 red tide
赤道波 equatorial wave
赤道潮 equatorial tide, equatorial tracker
赤道式跟踪器 equatorial tracker
赤道无风带 doldrum

赤铁矿　hematite
赤纬　declination
翅片板吸收器　finned-plate absorber
翅片管　finned pipe
翅片式吸热板　finned absorber plate

chong

冲淡　dilute
冲动　impulse
冲断层　thrust, thrust fault
冲断带　thrust belt
冲断面　thrust plane
冲击　impinge
冲击电流　impulse current
冲击动荷试验　impulse load test
冲击负载　impact load
冲击荷载　shock load
冲击扭矩　impact torque
冲击强度　impact strength
冲击韧性　impact toughness
冲击式分离机　ballistic separator
冲击吸收能量　absorbed striking energy
冲击压力　impact pressure
冲击应力　shock blow stress
冲击载荷　impulsive load
冲击致振　impact induced vibration
冲击阻尼器　impact damper
冲积床　alluvial bed
冲积扇　fan
冲击式汽轮机　impulse turbine
冲角　incidence angle
冲孔钢筛板　punched screen
冲力式涡轮机　impulse turbine
（日本）冲绳型风车　windmill of Okinawa style
冲刷阀　flush valve, washout valve
冲洗　flushing

冲洗阀　flush valve
冲掩体　thrust plate
充斥电解液蓄电池　flooded electrolyte battery
充电电流　charge current
充电电压　charge voltage
充电功率　charge power
充电控制器　charge controller
充电效率　charging efficiency
充满式系统　filled system
充气电阻挡层光电池　gas filled barrier layer cell
充气剂　aerating agent
充气式塑料蒸馏器　inflated plastic still
充气式蒸馏器　inflated still
充气系统　gassy system
充气效率　charging efficiency
充气钻井液　aerated drilling fluid
充气嘴　charging air inlet
充液玻璃温度计　liquid-in-glass thermometer
充溢系统　flooded system
充足的空间　ample space
重叠　lap
重叠玻璃板　overlapped-glass-plate
重叠玻璃空气加热器　overlapped glass air heater
重叠玻璃平板型空气加热器　overlapped glass plate air heater
重复接地　iterative earth
重合　superposition
重现期　recurrence interval

chou

抽风机　induced draft fan
抽风式风扇　induced draft fan
抽风式太阳能电站　solar chimney-

stack power generation
抽风系统　exhaust system, exhausting system
抽气　air bleed
抽气泵　off-gas pump
抽气机　air pump, air ejector, air extractor
抽气机房　air pump room
抽汽　steam bleed
抽汽式汽轮机　extraction turbine
抽砂泵　sand pumping
抽水风车　windmill for water lifting
抽水风力机　water-pumping windmill
抽水量　pumping load
抽水蓄能　pumped-hydro storage
抽水蓄能电站　pumped storage station, pumped storage hydropower station
抽水蓄能电站抽水耗能量　energy absorbed by storage pumping
抽水蓄能电站水库注满周期　fill period of a pumped storage reservoir
抽水蓄能电站综合效率　comprehensive efficiency of pumped storage station
抽水蓄能机组　pumped storage unit, pumped storage sets, pumped-storage unit
抽水蓄能水库注满周期　filling period of a pumped storage reservoir
抽水蓄能循环转换效率　conversion efficiency of a pumped storage cycle
抽水蓄能资源　pumped storage power resources
抽水装置　water-pumping assembly
抽提装置　extraction plant
抽吸射流　suction jet
抽蓄发电厂　pumped storage power station
抽蓄水力储能　pumped hydro energy storage
抽样率　sampling rate
抽样试验　sampling test
抽真空系统　air extraction system
稠度　solidity ratio
稠环芳烃　polycyclic aromatic hydrocarbon
臭氧含量　ozone content
臭氧耗减潜能　ozone depletion potential
臭氧耗损潜能值　ozone depletion potential

chu

出潮口　tidal outlet
出风管　discharge pipe
出风口　air outlet
出灰机　ash extractor
出口　discharge hole
出口　outlet
出口截面积　discharge area
出口热盐罐　molten salt receiver outlet vessel
出口损失　exit loss, discharge loss
出口蒸汽　outlet steam
出口阻力　exhaust resistance
出射剂量　exit dose
出水口　water outlet
出液管　discharge pipe
出渣　slagging
初步设计　tentative design
初级空气供应　primary air supply
初级冷却回路　primary cooling circuit
初级能源节约指数　primary energy savings index
初级燃烧　primary combustion
初级绕组　prime mover
初建费用　initial cost
初裂变中子能量　primary fission neutron energy

初凝时间 initial setting time
初期渗滤液 initial leachate
初清洗 initial cleaning
初始的 initial
初始地热梯度 original geothermal gradient
初始化 initialize
初始压力 initial pressure
初始压力分布曲线 initial pressure profile
初相 first phase
初相角 epoch angle
除尘 dust precipitation
除尘技术 dust precipitation technology
除尘器 flue dust collector, precipitator
除尘系统 dedusting system
除垢剂 deincrustant
除垢器 scaler
除灰 ash disposal, ash removal
除灰器 ash discharger
除灰系统 ash disposal system
除离子水 deionized water
除离子装置 deionization plant
除泥器 desilter
除气 deaeration
除气设备 deaerating unit
除砂器 desander
除湿 dehumidification
除湿量 moisture removal
除雾器 demister, mist eliminator
除芯机 core breaker
除盐装置 demineralizer
除氧 deaeration
除氧器 degasifier, oxygen remover
除氧器减温减压阀 deaerator reducer
除氧水取样器 deaerated water sampling
储备发电容量 reserve generating capacity
储藏采收 reservoir recovery
储藏工程师 reservoir engineer
储藏容量 storage capacitance
储藏设施 storage facility
储层表征 reservoir characterization
储层参数 reservoir parameter
储层地质学 reservoir geology
储层渗透性 reservoir permeability
储层温度分布 reservoir temperature distribution
储层要求 reservoir requirement
储存成本（费用） storage cost
储存池 storage tank, storage pool
储存罐 hold tank
储存热损失 standby heat loss
储存温度波动 storage temperature swing
储存温度峰值 storage temperatured peak
储存系统循环使用寿命（年限） cyclic durability of storage
储存箱 storage bin
储罐法规 storage tank regulation
储罐检测 tank testing
储灰仓 ash storage silo
储集层体积 reservoir volume
储集区 gathering area
储料仓 storage bin
储料存箱 storage bin
储料斗 storage hopper
储能 stored energy
储能电极 storage electrode
储能极限 storage limit
储能利用因子 storage utilization factor
储能密度 storage energy density
储能器 energy storage

储能损失	storage loss
储能投运	storage-coupled
储能系统	storage energy system
储能作用	storage effect
储气罐	gas storage tank
储气器	gas holder
储气系统	air storage system
储汽器组	thermal storage battery
储热（法）（估计地热田能量潜力的一种方法）	stored heat
储热壁	thermal storage wall
储热材料	thermal storage material
储热层	geothermal reservoir
储热层屋面池	thermal storage roof pond
储热介质	heat storage medium
储热介质选择	choice of storage medium
储热经济学	economics of thermal storage
储热能力	ability to store heat, thermal capacity of storage
储热器	heat reservoir
储热墙	thermalstorage wall, heat storage wall
储热容量	thermal energy storage capacity
储热水箱	heat storage tank
储热式真空（集热）管	storage vacuum tube
储热温度	storage temperature
储热系统	storage system, thermal energy storage system, thermal storage system
储热岩石床	thermal storage rock bed
储热站	storage station
储热质量	storage mass
储热装置	thermal storage unit
储热作用	storage effect
储热子系统	storage subsystem
储水池	storage reservoir
储水量	tank capacity
储水能系统	water storage system
储水箱	storage tank
储盐器	salt accumulator
储压器	pressure accumulator
处罚通知书	disposition notice
处理（置）	disposal
触变性能	thixotropic property
触电	electric shock
触规	feeler gauge
触热交换器	contacting heat exchanger
触头	contact

chuan

穿场（法）	cross-field
穿孔	perforation
穿孔管	perforated pipe
穿面式水翼	surface piercing foil
穿透能力	ability to penetrate
穿透系数和吸收系数的乘积	transmittance-absorptance product
氚	tritium
传播方向	propagation direction
传播曲线	propagation curve
传播时间	travel time
传导热损失	conduction heat loss
传递比	transfer ration
传递函数	transfer function
传动比	transmission ratio
传动精度	transmission accuracy
传动链	drive chain
传动误差	transmission error
传动系惯性	drive train inertia
传动系统	drive train
传动轴	power shaft
传动轴扭转柔性	drive shaft torsional

flexibility
传动装置　drive assembly, gearing
传感器　pick off
传感器　sensor
传感器电缆　sensor cable
传感器输入　sensor input
传感元件　detecting element
传热　heat transmission
传热表面　heat transfer surface
传热分析　heat transfer analysis
传热过程　heat transfer cycle
传热回路　heat transfer loop
传热机理　heat transfer mechanism
传热介质　heat transfer medium, heat transfer media
传热矩阵　heat-transfer matrix
传热量　received heat
传热流体　transfer fluid, heat transfer fluid, heat transport fluid
传热率　heat transfer efficiency
传热模型　heat transfer model
传热系数　heat transfer coefficient, heat transfer rate
传热系统　heat transfer system
传热性能　heat-transfer property
传热盐　heat transfer salt
传输成本　transmission cost
传输电力电缆　transmission power cable
传输距离　transmission distance
传输损耗　transmission loss
传输系统　transmission system
传输线　transmission line
传输线路　transmission link
传送　convection
传送带　conveyor belt
传送器　transmitter

传送系统　delivery system
传统形式锅炉　conventional boiler
传统原油　traditional oil
传真距离　transmission distance
船舶设计　marine design
船舶推进(反应)堆　marine propulsion reactor
船舶推进装置　marine propulsion
船舶系缆力　mooring force
船底附生物　marine growth
船模试验槽（池）　towing tank
船形波浪能发电装置　barge type wave energy conversion system
船型壳体　ship-shape hull
船用电缆　marine cable, submarine power cable, subsea cable
船用锅炉　marine boiler
船用螺旋桨　marine propeller
船用螺旋桨　ship propeller
船用气冷(反应)堆　marine gas cooling reactor
船用推进器　ship propeller
船闸　canal lock
串接太阳电池　tandem solar cell
串联电阻　serial resistance
串联　tandem, series connection
串联变换器　tandem converter
串联布置　series configuration
串联的　series connected
串联电池　series connected cell
串联电路　series circuit
串联电阻　series resistance
串联二极管　series diode
串联加热器　cascade heater
串联线圈　series coil
串联谐振　series resonance
串联组件　series block

chuang

窗框式太阳能集热器　window box solar collector
窗式橱窗　window showcase
床料　bed material
创立　foundation

chui

吹风管　blowpipe
吹风机　blower
吹风式冷却器　air-blown cooler
吹管　blowtorch
吹灰　soot-blowing
吹灰器　soot blower
吹流　wing-drift current
吹气射流　blowing jet
垂荡和纵摇波能转换装置　heaving and pitching body wave energy conversion device
垂度　sag
垂式涡轮机　vertical axis turbine
垂线　perpendicular
垂直百叶窗　vertical blind
垂直闭合回路　vertical closed loop
垂直测震剖面法　vertical seismic profiling
垂直的　perpendicular
垂直的位置　perpendicular
垂直地震剖面法　vertical seismic profiling
垂直定向板　guiding fin
垂直堆肥装置　vertical composting unit
垂直多结太阳电池　vertical multijunctions solar cell
垂直帆式风车　vertical sail-type mill
垂直风廓线　vertical wind profile
垂直风切　vertical wind shear
垂直回路　vertical loop

垂直活板　vertical flap
垂直结太阳电池　vertical junction solar cell
垂直距离　vertical dimension
垂直廓线　vertical profile
垂直力　force perpendicular
垂直裂缝　vertically oriented fracture
垂直埋管地源热泵系统　vertical borehole ground-coupled heat pump
垂直模式　heaving
垂直排气筒　vent stack
垂直喷蒸发器　vertical-spout evaporator
垂直平面　vertical plane
垂直剖面　vertical profile
垂直燃烧室　vertical combustion chamber
垂直入射太阳热量计　normal incidence pyroheliometer
垂直入射直接日射计　normal incidence pyrhelicmeter
垂直推力　thrust perpendicular
垂直叶板风速计　normal-plate anemometer
垂直应力　vertical stress
垂直轴　vertical axis
垂直轴潮流水轮机　vertical axis tidal stream turbine
垂直轴潮汐能水轮机　vertical axis tidal power turbine
垂直轴风力发电机　vertical-axis wind turbine
垂直轴风轮　vertical axis rotor
垂直轴式风车　vertical axial wind turbine
垂直轴液压机器　vertical-axis hydraulic machine
槌　gavel
锤式粉碎机　hammer mill

chun

春季对流　spring overturn
纯抽水蓄能电站　pure pumped storage hydropower station
纯净蒸汽　clean steam
纯能量　net energy
纯正弦波　pure sine wave
纯钻井时间　rotating time

ci

词首的　initial
磁场　magnetic field
磁场畸变　magnetic field distortion
磁场绕组　field winding
磁电风速计　magneto anemometer
磁动势　magnetomotive force
磁法（用于地热勘探的一种地球物理方法）　magnetic method
磁放大器　magnetic amplifier
磁粉检验　magnetic particle examination
磁感风杯风速表　cup-generator anemometer
磁化　magnetization
磁化电抗　magnetizing reactance
磁化电流　magnetizing current
磁化率　magnetic susceptibility
磁极　magnetic pole
磁极对　magnetic pole pair
磁控溅射镀膜　magnetron sputtering coating
磁力分离器　magnetic separator
磁力分选机　magnetic separator
磁链　flux linkage
磁链计算　flux linkage calculation
磁流体发电　magnetohydrodynamic power generation
磁路　magnetic circuit
磁路饱和　magnetic saturation
磁敏电阻式编码器　magneto resistor encoder
磁石　magnet
磁体　magnet
磁条　bar magnet
磁铁　magnetic iron
磁通　flux
磁通量　magnetic flux
磁通密度　magnetic flux density
磁通匝连数　flux linkage
磁透镜　magnetic lens
磁芯　magnetic core
磁芯损耗　magnetic core loss
磁性工具　magnetic tools
磁性联轴　magnetic coupling
磁性分离　magnetic separation
磁滞　hysteresis
磁滞损耗　hysteresis loss
磁阻　reluctance
磁阻率　reluctivity
磁阻同步发电机　reluctance synchronous generator
磁阻转矩　reluctance torque
功率表面　power surface
次表层流　subsurface currents
次表层温度　subsurface temperature
次环流　secondary circulation
次级发射　secondary emission
次级风向　secondary wind direction
次级燃烧空气　secondary combustion air
次级绕组　secondary winding
次生风效应　secondary wind direction effect
次生矿物　secondary mineral

次生渗透性　secondary permeability
次生污染物（质）　secondary pollutant
次生原料　secondary raw material
次瞬态电抗　subtransient reactance
次烟煤　sub-bituminous coal
刺激　impulse

cu

粗糙　coarse
粗糙带　roughness band
粗糙度系数　roughness coefficient
粗糙元　roughness element
粗糙长度　roughness length
粗粉煤灰颗粒　coarse fly-ash particle
粗粒物质　coarse material
粗汽油　naphtha
粗砂　coarse sand
粗砂砾　coarse gravel
粗石　quarrystone
粗碎石　coarse gravel
粗调节　coarse control
粗同步　coarse synchronizing
粗絮状物　coarse floc
促动盘　actuator disk
促进　facilitate
醋酸　acetic acid
醋酸钾　potassium acetate
醋酸盐异化（作用）　dissimilation of acretate

cui

催化反应堆　catalytic reactor
催化反应过程　catalytic reaction process
催化改质　catalytic upgrading
催化剂　catalyst
催化剂床　catalyst bed
催化剂中毒　catalyst poisoning
催化焦油裂解　catalytic tar cracking
催化气化反应器　catalytic gasification reactor
催化燃烧焚烧炉　catalytic combustion incinerator
催化燃烧室　catalytic combustor
催化热解　catalytic pyrolysis
催化转化器　catalytic converter
脆性　brittleness
萃取率　extraction rate
萃取塔　extraction column
萃取效率　extraction efficiency
淬火　quench
淬火工艺　quench process
淬火冷却　quench process

cuo

错列翅片太阳能空气加热器　staggered finned solar air heater
错流　crossflow
错流喷雾室　cross-flow spray chamber
错瞄太阳能辐射　misdirected solar radiation

da

搭接　lap, lap joint
达标投产　putting into operation in compliance with standard
达肯方程　Dark equation
达里厄风轮　Darrieus rotor
达里厄汽轮机　Darrieus turbine
达里厄竖轴风能转换系统　Darrieus verticle axis wind-energy conversion system
达里厄型　Darrieus
达里厄型风车　Darrieus windmill
达里厄型风力机　Darrieus machine,

Darrieus type wind turbine
达因风速表　Dines anemometer
达因风压机　Dines pressure anemograph
打包生物质燃料　baled biofuel/bale
打蛋器型风轮机　eggbeater wind turbine
打卵器式叶轮　egg-beater rotor
大坝　dam
大潮　spring-tide
大潮平均高潮位　mean high water spring
大潮汛　tidal surge
大齿轮　wheel gear, bull gear
大地电磁学　magnetotellurics (MT)
大地电流　earth current
大地接触冷却　earth-contact cooling
大地耦合　earth coupling
大地热流　terrestrial heat flow
大地热流量分量　component of terrestrial heat flow
大地热流量正常值　normal value for terrestrial heat flow
大范围天气形势　large-scale weather situation
大风　fresh gale, high wind, gale
大功率　heavy duty
大功率 LED 格栅灯　high power LED grid lights
大功率 LED 路灯　high power LED street lamp
大功率 LED 球泡灯　high power LED bulb lamp
大功率 LED 射灯　high power LED spot light
大功率 LED 隧道灯　high power LED flood lamp
大功率 LED 天花射灯　high power LED ceiling spot light
大功率 LED 洗墙灯　high power LED wall washer lamp
大功率密度　high power density
大规模储热　large-scale storage of heat
大规模能源利用　large scale energy use
大角度晶界　high angle grain boundary
大块石　boulder
大浪　surge, rough sea
大梁式结构　girder structure
大量　array
大陆架　continental shelf
大陆架外缘　outer continental shelf
大陆块　land mass
大陆性季风气候　continental season wind climate
大陆性气候　continental climate
大陆岩石圈　continental lithosphere
大卵石　boulder
大帽钉　clout nail
大面积薄膜光电电池　large-area thin-film photovoltaic cell
大面积薄膜光伏电池　large-area thin-film photovoltaic cell
大面积多晶薄膜　large-area polycrystalline film
大面积流计数器　large-area flow counter
大面积向阳玻璃面　large south-facing glazed area
大气暴露叶片运动　weathering vaning motion
大气边界层　atmospheric boundary layer
大气波预测系统　atmosphere-wave

forecast system
大气采样 air sampling
大气采样测量 air sampling measurement
大气采样器 air sampler
大气采样装置 air sampling rig
大气层 airspace
大气成分 atmospheric components
大气传递 atmospheric transmissivity
大气顶层太阳辐射 extra-terrestrial radiation
大气顶辐射 extraterrestrial radiation
大气风 atmospheric wind
大气辐射 atmospheric radiation
大气辐射强度计 pyrheliometer
大气光化学 atmospheric photochemistry
大气光学厚度 atmospheric optical depth, atmospheric optical thickness
大气光学质量 optical air mass
大气过程 atmospheric processes
大气—海洋温差 air-sea temperature difference
大气化学 atmospheric chemistry
大气环流 atmospheric circulation
大气环流模式 atmospheric circulation pattern
大气浑浊度 atmospheric turbidity
大气监测 air monitoring
大气监测网 air monitoring network
大气监测仪器 air monitoring instrument
大气监测站 air monitoring station
大气监控装置 air monitoring equipment
大气降水 atmospheric precipitation, meteoric water
大气颗粒物 atmospheric particulates
大气空间 airspace
大气排放 airborne emission
大气气体 atmospheric gas

大气热辐射 atmospheric heat radiation
大气热量输送 atmospheric heat transport
大气热平衡 atmospheric heat balance
大气上界日辐射量 daily extraterrestrial radiation
大气上界水平辐射日总量 daily total extraterrestrial horizontal
大气上界太阳能通量 extraterrestrial solar flux
大气衰减 atmospheric attenuation
大气水库系统 atmospheric reservoir system
大气透射 atmospheric transmissivity
大气微粒 aerosol
大气稳定度 atmospheric stability
大气稳定度类别 atmospheric stability class
大气污染 atmospheric pollution
大气污染物 air contaminant
大气吸收 atmospheric absorption
大气向下辐射 atmospheric counter radiation
大气压力 atmosphere pressure, atmospheric pressure
大气压强 atmospheric pressure
大气影响类别 atmospheric impact category
大气质量 air mass
大气质量标准 air quality criteria, air quality standard
大气质量检测 air quality surveillance
大气质量检测网 air quality surveillance network
大气质量监测系统 air quality monitoring system
大气质量指数 air quality index
大气组成 atmospheric composition
大容量 high duty, heavy duty, high capacity

大容量锅炉　high duty boiler
大型储热装置　large heat-storage unit
大型电站能源消费法令　large combustion plant directive
大型风力发电机阵列　large wind generator array
大型浮式电站　large floating power plant
大型建筑　edifice
大型蒸发器　evaporation tank
大型钻　large power drill
大漩涡　maelstrom
大洋环流　ocean circulation
大洋流系　ocean current system
大洋湍流　oceanic turbulence
大洋涌浪　ocean swell
大圆石　boulder
（日本）大岳地热田　Otake geothermal field
大直径管　broad diameter tube

dai

（风能资源评估）代表年　representative year for wind energy resources assessment
代数运算　algebraic manipulation
带边能量　band edge energy
带电粒子　charge carrier, charged particle
带电粒子辐射　charged particle radiation
带电流体风力发电机　electrofluid dynamic wind driven generator
带电气体风力发电机　electrogaseous dynamic wind driven generator
带浮力箱的冷水管　cold-water pipe with buoyancy tank
带冠涡轮　shrouded turbine
带硅　silicon ribbon
带宽，频带宽度　bandwidth

带式干燥机　belt dryer
带透明盖板集热器　glazed collector
带隙材料　bandgap material
带隙构造　bandgap structure
带状电池　ribbon cell
带状硅太阳电池　ribbon silicon solar cell
带状热储　zoned reservoir
袋式除尘器　baghouse
袋式过滤器　bag filter
戴维南等效回路　Thevenin equivalent circuit
戴维南等值参数　Thevenin equivalent parameter
戴维南等效电路　Thevenin's equivalent circuit
戴维斯水力涡轮机　Davis hydro turbine

dan

丹倍效应　Dembet effect
丹麦风力机概念　Danish wind turbine concept
（风电的）丹麦概念　Danish concept
单 p-n 结恒能带宽度太阳能电池　single p-n junction constant energy gap solar cell
单 p-n 结均匀能带宽度太阳能电池　single p-n junction form energy gap solar cell
单板计算机　single board computer
单层玻璃盖板　single glass cover
单层玻璃盖板平板集热器　single glass covered flat plate collector
单层玻璃盖板平面集热器　single layer glass cover flat plate collector
单层玻璃集热器　single-glazed collector
单层反射玻璃　single glazing reflecting

单层盖板集热器　single-cover collector
单层透明玻璃　single glazing clear glass
单层透明盖板黑吸收表面集热器　single glazed black painted collector
单层透明盖板集热器　single glazed collector
单层透明盖板空气型集热器　single glazed flat plate air collector
单层吸热玻璃　single glazing heat-absorbing glass
单朝向风轮　single weather rotor
单程热交换器　single pass heat exchanger
单窗格玻璃　single-pane glass
单导叶接力器　individual guide vane servomotor
单点系泊系统　single point mooring system
单电子理论　one electron theory
单定日镜太阳能炉　single heliostat furnace
单独系统　stand-alone system
单阀滤池　single-valve filter
单个操作　one-to-one control
单管集热器　single tubular collector
单管蒸汽系统　one-pipe steam system
单罐太阳能热水器　single-tank solar-gas water heater
单轨系统　monorail system
单轨运输系统　monorail system
单回路　single loop
单回路系统　single loop system
单机容量　unit capacity
单机系统　stand alone system
单级变速箱　single-stage gearbox
单级行星齿轮系　single planetary gear train
单级燃烧　single-stage combustion
单级太阳能蒸馏器　single effect solar still
单级消化池　single stage digester
单级蒸发器　single effect evaporator, one effect evaporator
单结（太阳能）电池　unijunction solar cell
单结太阳能电池　single-junction solar cell
单晶电池　single crystal cell
单晶电池片　monocrystalline cell
单晶硅　single crystalline silicon, monocrystal silicon, monocrystalline silicon
单晶硅棒　monocrystalline ingot
单晶硅方棒　monocrystalline block
单晶硅片　monocrystalline wafer
单晶硅太阳能电池　single crystalline silicon solar cell, mono-crystalline solar cell
单井筒技术　single bore-hold technology
单井眼（同心套管）技术　single bore-hole (concentric pipe) technology
单壳单程热交换器　one-one exchanger
单壳双程热交换器　one-two exchanger
单孔　single bore hole
单库　single pool
单库潮汐能工程　single-pool tidal power project
单库单向潮汐工程设施　single-pool one way tide working installation
单库单向潮汐能生产方案　single-basin unidirectional power production scheme
单库单向方案　single-basin single-effect scheme

单库单向式潮汐电站　single basin/single effect tidal power plant
单库单作用方案　single basin/single effect scheme
单库方案　single basin project, single-basin scheme
单库—加泵水双向方案　single basin-double effect scheme with pumping
单库双向潮汐功能设施　single-pool two way tide working installation
单库双向系统　single-pool two-way system
单库双作用方案　single basin/double effect scheme
单库退潮发电系统　single-pool ebb-generation system
单库系统　single-pool system
单面表面磁铁　single-sided with surface magnets
单面的　single-side
单色波　monochromatic wave
单色发射率　monochromatic emissivity
单色发射能力　monochromatic emissive power
单色辐射力　monochromatic emissive power
单色光辐射　monochromatic radiation
单色光响应　monochromatic response
单色吸收率　monochromatic absorptivity
单绳　single rope
单鼠笼式转子　single squirrel cage rotor
单速泵　single-speed pump
单台阶式地表起伏　single step of surface relief
单体功率　unit wattage
单体太阳电池　single solar cell

单体太阳电池的有效光照面积　active area of a solar cell
单位　unit
单位产出率　specific production index
单位产量　specific capacity
单位出力　specific duty
单位电抗　perunit reactance
单位电能投资　cost per kilowatt-hour
单位跨度　per unit span
单位流量　specific discharge
单位千瓦投资　cost per kilowatt
单位深度地温差　geothermic step
单位生产率　specific production index
单位矢量　unit vector
单位蒸发量　specific evaporation duty
单相电流　single-phase current
单相发电机　single-phase generator
单相流　single-phase flow
单相流动　single phase flow
单相位值　single-phase value
单相系统　single-phase system
单相压力　single-phase pressure
单箱系统　one-tank system
单箱蓄热　single tank storage
单向瓣阀　uni-directional flapper
单向潮汐电站　single-effect tidal plant
单向传输　simplex transmission
单向单库方案　single-effect single-basin scheme
单向的　unidirectional
单向发电（该潮汐电站只能在涨潮或落潮时发电）　one-way operation
单向分解　single direction decomposition
单向海流能源　uni-directional ocean-current energy source
单向拍动板　uni-directional flapper
单向透平　uni-directional turbine

单向涡轮机　uni-directional turbine
单向系统　one-way system
单向装置　single-effect unit
单效贯流涡轮机　single-effect straflo turbine
单效吸收式制冷机　single effect absorption chiller
单效蒸发器　single effect evaporator, one effect evaporator
单循环燃气轮发电机组　single-cycle combustion turbine unit
单叶片风车　single bladed windmill
单叶片风力机　single-bladed wind turbine
单叶片风轮　single bladed rotor
单叶片螺旋桨　single-bladed propeller
单液电池　single fluid cell
单元　location
单元机组保护　boiler-turbine-generator unit protection
单元集中控制　unit centralized control
单元控制　unit control
单元控制室　unit control room
单质结　homojunction
单轴　single shaft
单轴负载　uniaxial loading
单轴跟踪聚光器　concentrator with single axis
单轴跟踪太阳能聚光器　one axis tracking solar concentrator
单轴太阳跟踪器　one-axis tracker
（海水）淡化方法　desalting process
（海水）淡化工厂（装置）　desalting plant
淡水　sweet water, saline-free water
淡水渗透　freshwater bleed
淡水生浮萍　freshwater duckweed
淡阴影　light shading
弹式热量计　bomb calorimeter

弹丸凿岩　projectile drilling
蛋白石　opal
蛋白质降解　protein degradation
氮保障系统　nitrogen support system
氮化硅　silicon nitride
氮化硼奈米管　boron nitride nanotubes
氮磷钾比例　nitrogen to phosphorus to potassium ratio
氮维持系统　nitrogen support system
氮氧化物　nitrogen oxide
氮源　nitrogen source

dang

当潮闸　tide gate
当地地质　local geology
当地实度　local solidity
当地速比　local speed ratio
当地太阳能强度　local solar intensity
当量比　equivalence ratio
当量长度　equivalent length
挡板　wall barrier
挡板阀　flap valve
挡板式洗涤器　baffle scrubber
挡板调节　damper control
挡潮堤　tidal embankment
挡潮闸　tidal lock, tidal sluice, tide-gate
挡风板　wind shield
挡风玻璃　windscreen
挡风墙　windshielding
挡泥板　fender
挡热层　temperature barrier
挡水板　breakwater

dao

刀具　cutter
刀闸（隔离开关）　isolator
（螺旋桨叶片的）导边　leading edge

导磁系数　permeability
导带　conduction band
导电环　conducting ring
导电排　conducting bar
导电性　electrical conductivity
导电性　conductivity
导电氧化物薄膜　transparent conductive oxide
导电黏合剂　conductive adhesive
导风板　air deflector
导风轮叶片　inlet bucket
导风器式风力机　deflector wind machine
导风筒　ventilating duct
导管　conduit
导管接线盒　conduit box
导管配件　conduit fitting
导管式风轮　ducted rotor
导管式风轮风能发电机　ducted rotor wind machine
导管引入装置　conduit entry
导辊　guide roller
导流　abstraction of river
导流板　deflector, air flow guide, air-flow guide, turning vane
导流墙　guide wall
导流叶片　guide vane, guide blade
导流罩　air flow guide, air-flow guide
导轮　guide wheel, idler pulley
导纳　admittance
导频激励器　pilot exciter
导热材料　heat conducting material
导热体　thermal conductor
导热系数　thermal conductivity
导热性　thermal conduction, temperature conductivity
导入装置　gatherer

导水沟　headrace
导水机构　water deflection part, water guide mechanism
导水率　hydraulic conductivity
导水墙　guide wall
导水装置　distributor
导体　conductor
导线　harness
导线厚度　conductor thickness
导线间电压　circuit voltage
导向板　guide plate, guiding fin
导向槽　guideway
导向杆　guide rod
导向隔板　finger baffle
导向工具　guide tool, steering tool
导向管　guide tube
导向棍　guide bar
导向护套　guide sheath
导向环　guide ring
导向块　guide block
导向叶片　turning vane, guide vane, guide wheel bucket
导向支架　guide support
导向轴　guiding shaft
导叶　guide vane
导叶接力器　guide vane servomotor
导叶轴　guide vane trunnion
导叶轴颈　guide vane trunnion
导则　guide rule
岛式运行　island operation
倒角机　chamfer machine
倒流　flow reversal
倒数　inverse
倒烟　down draft
倒焰窑　down-draught kiln
倒易密度　reciprocal density
稻壳　rice husk

de

(德国)《2005年可再生能源法》 Germany's 2005 Renewable Energy Sources Act
德国风力能源研究所 German Wind Energy Institute
德国风力涡轮机 German wind turbines
德国劳埃德规范 GL rules = Germanischer Lloyd's Regulation for the Certification of Wind Energy Conversion
(德国)Valorga反应器 Valorga reactor

deng

灯 light
灯泡式发电装置 bulb-type generating unit
灯泡式水轮机 bulb-turbine
灯泡型外壳 bulb casing
灯丝 filament
灯丝电压 heating voltage, heat voltage
灯座 lamp holder
等边三角形 equilateral triangle, isosceles triangle
等边体 isometric projection
等地温线 isogeotherm
等电位连接带 bonding bar
等电位连接导体 bonding conductor
等电位联结 equipotential bonding
等动力采样 isokinetic sampling
等动能采样头 isokinetic probe
等高线 contour line
等高线图 contour map
等厚图 isopach data
等候工况 waiting situation
等截面叶片 constant chord blade
等离子电弧气化 plasma-arc gasification
等离子焚化炉 plasma incinerator
等离子腐蚀 plasma etching
等离子流 plasma jet
等离子室 plasma chamber
等离子体 plasma
等离子体磁流体发电 plasma magnetohydrodynamics power generation
等离子体反应器 plasma reactor
等离子体焚烧炉 plasma incinerator
等离子体射流 plasma jet
等离子体焰炬 plasma torch
等离子体柱 plasma column
等曝辐量线 isorad
等日照线 isohel
等熵变化 isentropic change
等熵的 isentropic
等熵分析 isentropic analysis
等熵过程 isentropic process
等熵混合 isentropic mixing
等熵面 isentropic surface
等熵条件 isentropic condition
等熵图 isentropic chart
等熵线 isentrope
等熵线图 isentropic chart
等熵效率 isentropic efficiency
等熵压缩 isentropic compression
等熵运动 isentropic motion
等势图 equipotential map
等梯度线 isogradient
等通量层 constant flux layer
等温层 isothermal layer
等温的 isothermic
等湿度线 isohume
等温线 isothermal line
等温压缩率 isothermal compressibility
等温压缩性 isothermal compressibility

等弦宽叶片　blade of constant chord width
等效电感　equivalent inductance
等效电路　equivalent circuit, equivalent electrical circuit
等效电路参数　equivalent circuit parameter
等效电容　equivalent capacitance
等效负载小时　equivalent full load hour
等效孔径　effective aperture
等压线　isobar
等盐度线　isohaline
等值曲线　contour curve
等轴晶区　equiaxed crystal zone

di

低/中固分解系统　low/medium solid digestion system
低层大气　lower atmosphere
低潮　low tide
低潮位　low tide level, low water level, low water stand
低成本太阳能阵列　low-cost solar array
低发射率涂料　low emissivity coating
低沸点　low boiling point
低分量　harmonic component
低分子量气体　low molecular weight gas
低辐射　low insolation
低辐射玻璃　low emission glass
低辐射率涂料　low emissivity coating
低负荷效率　part load efficiency
低负荷性能　part load behaviour
低—高结理论　low-high junction theory
低功率太阳能电池　low-power solar cell

低共熔材料　eutectic material
低共熔混合物　low melting eutectic mixture
低共熔温度　eutectic temperature
低谷期　trough
低固体含量　low-solids content
低耗电活塞环　low-leakage piston ring
低级热能回收　low level recovery
低级夜间喷嘴　low-level nighttime jet, low-level nocturnal jet
低集热效率　long collection efficiency
低集中度　low concentration
低焦油燃料　low-tar fuel
低聚光集热器　low concentration collector
低库　low pool, low reservoir
低库库容　low-level reservoir capacity
低库库址　low-level reservoir site
低库区　low-level reservoir basin
低库水面　low pool level
低摩擦损耗（失）　low friction loss
低浓度　low concentration
低浓度的　low-concentrate solvent
低品位能　low-grade energy
低品位太阳能热　solar low grade heat
低品位余热　low-grade waste heat
低气压　low pressure, trough
低热导率　lower temperature conductivity
低热焓　low enthalpy
低热量　net calorific value
低热源温度，低温热源　lower source temperature
低热值　lower heating value
低日照率　low insolation
低溶盐（太阳能储热器的一种储热介质）　eutectic salt
低渗透压　low osmotic pressure
低实度风轮　low-solidity rotor

低水头（数米至十余米的发电水头） low-head
低水头波能水轮机 low-head wave energy water turbine
低水位水库 low-level reservoir
低速齿轮比 low-gear ratio
低速多叶片风力涡轮机 low-speed multi-blade wind turbine
低速风轮 slow speed rotor
低速软弱夹层 low-velocity layer
低速运行风轮 slow running rotor
低速轴 low-speed shaft
低碳钢焊接 mild-steel welding
低头双向涡轮机 low-head bi-directional turbine
低洼地 low-lying flat
低位发电量 net calorific value
低位法 low-order method
低位能 low-grade energy
低位热回收 low-level heat recovery
低位热能 low-grade thermal energy, low-grade heat energy
低位热值 low heat value
低位势热能 low potential thermal energy
低温储热 lower temperature storage
低温地热流体 low-temperature geothermal fluid
低温发射率 low-temperature emissivity
低温干馏气 low-temperature carbonization gas
低温集热器 low-temperature collector
低温集热系统 low-temperature collector system
低温集热装置 low-temperature collector device
低温平板型太阳能集热器 low temperatureflat-plate solar collector

低温气化炉 low temperature gasifier
低温区 low-temperature area
低温燃料电池 low temperature fuel cell
低温热 low temperature heat
低温热能 low temperature heat energy
低温热源 source of low temperature heat
低温太阳能转换系统 low-temperature solar conversion system
低温太阳能转换装置 low-temperature solar conversion system
低温碳化气 low-temperature carbonization gas
低温梯度测量 geothermal gradient survey
低温脱硝装置系统 low-temperature denitration equipment system
低温限 lower temperature limit
低温蓄热库 low-temperature reservoir
低温溢出 cryogenic spill
低效技术法（合理利用太阳能的一种方法） low impact technology solution
低压侧 step-down side
低压电器 low voltage
低压端 step-down side
低压段（级） low pressure stage
低压空气 low-pressure air
低压凝汽式汽轮机 low-pressure condensing turbine
低压排出回路 lower-pressure escape route
低压闪蒸容器 low pressure flash vessel
低压闪蒸室 low pressure flash evaporator chamber
低压水蒸气 low-pressure steam

低压头水电站　low-head water-power plant
低压头水电装置　low-head hydro-electric unit
低压线圈　low voltage coil
低压蒸汽　low-pressure steam, low pressure steam
低压蒸汽室　low-pressure steam chest
低叶尖速度比风轮　rotor of low tip speed ratio
低中温太阳能集热器　low-moderate-temperature solar collector
低中温太阳能热电站　low-moderate temperature solar thermal power plant
堤坝　dike, dyke
堤坝式水电站　power station with reservoir
堤道　causeway
滴给器　drip feeder
迪尔姆希尔恩—绍贝勒尔总日射计　Dirmhirn-Sauberer pyranometer
笛卡尔坐标系　Cartesian coordinates
底部冷水　cold bottom water
底部热损失　bottom heat loss
底层　substrate
底层冷水　colder deep water
底床　sea floor
底负荷　part load
底格里斯河　Tigris
底灰　incinerator bottom ash
底焦　bed coke
底片观察用光源　film viewer
底涂层　first coat
底循环　bottom cycle
底座　base plate
底座底　base bottom
底座面　base top
地表比热　surface specific heat
地表反照率　surface albedo
地表风　surface wind
地表观察　surface observation
地表摩擦系数　ground surface friction coefficient
地表起伏　surface relief
地表浅层　shallow subsurface
地表浅层温度　shallow subsurface temperature
地表浅层温度测量　shallow surface thermometry
地表倾斜　dip of surface
地表热活动　surface thermal activity
地表热显示　geothermal manifestation
地表水补给区　surface water intake area
地表水排泄区　surface water discharge area
地表水热泵　surface water heat pump
地表水热泵系统　surface-water heat pumps
地表温度　temperature at earth's surface
地表温度异常　surface temperature anomaly
地表显示　surface manifestation
地槽　geosyncline
地层　geological strata
地层电阻率因素　formation factor
地层厚度　stratigraphic thickness
地层结构　stratum structure
地层剖面　stratigraphic section
地层温度　formation temperature
地层学　stratigraphy
地层压力　formation pressure
地层因素　formation factor
地层中的岩石组成　stratigraphy
地电法　（勘探地热资源的一种地球物理方法）　geoelectric method

地电流　terrestrial current
地电效应　geoelectric effect
地方公用电网　local utility grid
地方性　locality
地核　core of earth
地基　foundation
地壳结构　crustal structure
地壳深处　deep crust
地壳岩石　crustal rock
地理信息系统　geographic information system
地貌　land configuration, landscape, physiognomy
地貌面　bedding plane
地面沉降　land subsidence
地面抽水储能　aboveground pumped hydrostorage
地面粗糙度　surface roughness
地面大型储存装置　above-ground large-scale storage unit
地面发电机　ground-mounted generator
地面反射　ground reflection
地面反射光　ground reflected light
地面方阵　terrestrial array
地面风　surface wind, ground wind
地面辐射　ground radiation
地面辐射光谱　terrestrial irradiance spectra
地面辐射强度计　pyrheliometer
地面辐照度　terrestrial irradiance
地面覆盖率　ground cover ratio
地面高度　ground elevation
地面跟踪　ground track
地面环境　terrestrial environment
地面极端风　extreme surface wind
地面集热器阵列　ground mounted collector array
地面技术　surface technology
地面接收站　ground station
地面聚光器方阵　terrestrial concentrator array
地面控制器　ground controller
地面零点　hypocenter
地面逻辑　ground logic
地面排水　land drainage
地面平板方阵　terrestrial flat plate array
地面日辐射量　daily terrestrial radiation
地面日照　terrestrial insolation
地面设备站　ground equipment station
地面式塑料太阳能热水器　ground-level plastic solar water heater
地面太阳电池方阵　terrestrial solar cell array
地面太阳能电池系统　terrestrial photo-voltaic system
地面太阳能辐射　terrestrial insolation
地面太阳能供电设备　terrestrial solar-powered device
地面套管　surface casing
地面现场　surface site
地面折射　terrestrial refraction
地面直接辐射量　terrestrial beam radiation
地能系统　earth energy system
地耦合热泵　earth-coupled heat pump
地平经圈　vertical circle
地平面　horizontal plane
地平圈　horizontal circle
地平式跟踪器　altazimuth mount, altazimuth tracker
地平线的　horizontal
地壳　crust
地球大气层外的太阳辐照度　extra-rrestrial irradiance
地球潮汐　earth tide

地球辐射 terrestrial radiation, earth radiation
地球辐射表 pyrgeometer
地球辐照度 terrestrial irradiance
地球化学 geochemistry
地球化学成分 geochemical composition
地球化学和物理调查 geochemical and geophysical survey
地球化学勘探 geochemical survey
地球化学勘探方法 geochemical method
地球化学取样 geochemical sample
地球内部能源地热能 energy from earth's interior
地球物理勘探 geophysical survey
地球物理方法 geophysical method
地球物理取样 geophysical sample
地球物理探测 geophysical survey
地球物理学技术 geophysical technique
地球质心 center of mass of the earth
地球自转 earth's rotation
地区风力资源 regional wind resource
地热 geothermal heat, geothermy, terrestrial heat
地热（供热）站 geothermal heat plant
地热饱和蒸汽 geothermal saturated steam
地热泵 geothermal heat pump
地热表 geothermometer
地热测量 geothermal measurement
地热产业 geothermal industry
地热成分试验设施 geothermal component test facility
地热成因 origin of geothermal heat
地热抽提技术 geothermal extraction technology
地热储层 geothermal reservoir, thermal reservoir
地热储层分析 geothermal reservoir analysis
地热储层工程 geothermal reservoir engineering
地热储层估计 geothermal reservoir evaluation
地热储存 terrestrial heat storage
地热创新 geothermal innovation
地热带 geothermal girdle
地热的 geothermal, geothermic
地热等温线 geoisotherm
地热地表显示（区） geothermal surface manifestation
地热地带 geothermal belt
地热电厂 geothermal power plant
地热电厂热能利用率 heat rate of geothermal power plant
地热电站 terrestrial heat power station, geothermal-electric power plant
地热度 geothermic degree
地热发电 geothermal electricity, geothermal electricity generation, geothermal generation, geothermal power generation
地热发电厂 geothermal power station, geothermal-electric power plant, terrestrial heat power station
地热发电能力 geothermal electricity generating capacity
地热发电容量 geothermal electricity generating capacity
地热发电设施 geothermal power facility, geothermal electricity generation facility
地热发电市场 geothermal power market
地热发电站 geothermal electric power station, geothermal generating station, geothermal electric power plant,

geothermal plant
地热法 geothermal method
地热房 geothermal vault
地热放射 terrestrial heat emission
地热分布 distribution of geothermal heat
地热分布网 geothermal distribution system
地热孵化 geothermal incubation
地热辐射采暖 floor radioactive space heating
地热干燥 geothermal drying
地热干蒸汽发电 geothermal dry steam power
地热干蒸汽发电站 geothermal dry steam power plant
地热工程 geothermal engineering, geothermal project
地热供暖 geothermal space heating
地热供热 geothermal heating
地热供热成本 geothermal heating cost
地热构造带 geothermal tectonic zone
地热锅炉 geothermal boiler
地热含量 terrestrial heat content
地热换热器 geothermal heat exchanger
地热回灌 geothermal reinjection
地热混合凝气器 geothermal mixture condenser
地热活动 geothermal activity, geothermal event
地热机制 mechanism of heat transfer
地热机组 geothermal set
地热集中供热系统 geothermal district heating system
地热计 geothermometer
地热加热费用 geothermal heating cost

地热间接供暖系统 geothermal indirect space heating system
地热介质 geothermal media
地热经济学 geothermal economics
地热井 geothermal well
地热井测井 geothermal well logging
地热井孔 geothermal bore
地热井口发电机组 geothermal wellhead generating unit
地热井流体测量 geothermal well fluid measurement
地热井设计 geothermal well design
地热井水 geothermal bore water
地热井水回路 geothermal bore water circuit
地热井响应 geothermal well response
地热井性能 geothermal well performance
地热井与井口装置 geothermal well and wellhead installation
地热开发 geothermal development, geothermal exploitation
地热开发对环境的影响 environmental impacts associated with geothermal development
地热开发环境保护 environmental protection of geothermal development
地热开发商 geothermal developer
地热勘探 geothermal exploration, geothermal prospect, geothermal prospecting, geothermal reconnaissance survey
地热勘探程序 geothermal prospecting procedure
地热控制室 geothermal vault
地热库 geothermal vault
地热快速勘探 fast geothermal reconnaissance

地热矿藏　geothermal deposit
地热类型　kinds of earth heat
地热利用率　geothermal efficiency, rate of geothermal utilization
地热流　geothermal stream
地热流测量　measurement of terrestrial heat flow
地热流量　geothermal flux, geothermal heat flow, terrestrial heat flux
地热流量计　terrestrial heat flow gauge, terrestrial heat flow meter
地热流量控制　terrestrial heat flow control
地热流量密度　geothermal flux density
地热流量数据　terrestrial heat flow data
地热流体　geothermal fluid, geofluid
地热流体回灌策略　geothermal fluid reinjection strategy
地热流体取样　sampling of geothermal fluid
地热流体温度　temperature of geothermal fluid
地热流体消耗　specific geofluid consumption
地热流体质量　quality of geothermal fluid
地热卤水　geothermal brine
地热卤水矿床　geothermal brine deposit
地热民用取暖　geothermal domestic heating
地热能　geothermal energy, geothermal power, terrestrial heat energy
地热能采量　geothermal energy recovery
地热能电站　geothermal energy plant
地热能发电　electricity from geothermal energy
地热能计划　geothermal power project

地热能勘探　exploration of geothermal power
地热能冷却　geothermal cooling
地热能利用　use of geothermal energy, utilization for geothermal energy
地热能热源　heat source for geothermal power
地热能容量　geothermal capacity
地热能应用　application of geothermal energy
地热能源　source of geothermal energy
地热能源激励法　stimulation of geothermal energy resource
地热能直接利用　direct utilization of geothermal energy
地热能贮量　geothermal energy reserve
地热能资源　geothermal energy resource
地热能综合利用　comprehensive utilization of geothermal energy
地热暖房　geothermal greenhouse
地热暖水储层　geothermal warm water reservoir
地热平衡　terrestrial heat balance
地热剖面　geothermal profile
地热普查　geothermal reconaissance
地热起源　origin of thermal spring
地热气　geothermal gas
地热汽轮机　geothermal steam turbine
地热清洗剂　thermal cleaning bath
地热区　geothermal area, geothermal region
地热区地质条件　geological condition in geothermal area
地热区域供热　geothermal district heating
地热泉　geothermal spring
地热扰动　geothermal disturbance
地热扰动带　geothermal disturbed zone

地热热泵模块　geothermal heat pump module
地热热泵系统　geothermal heat pump system
地热热抽提　geothermal-heat extraction
地热热风供暖　geothermal hot-air space heating
地热热水储层　geothermal hot water reservoir
地热热水田　geothermal hot water field
地热热水源　geothermal hot water source
地热容量　terrestrial heat capacity
地热设施　geothermal application, geothermal facility
地热深度　geothermal depth
地热深度梯度　geothermal depth gradient
地热生产层　geothermal producing zone
地热蚀变　geothermal alteration
地热势　geothermal potential
地热试验电站　geothermal experimental power station
地热数据　geothermal data
地热水　geopressured water, geothermal water
地热水产养殖　geothermal aquaculture
地热水除砂　remove sand from geothermal water
地热水除铁　remove iron from geothermal water
地热水防腐　geothermal water anticorrosion
地热水防垢（处理）　geothermal water scale prevention
地热水回灌　geothermal water reinjection
地热特性　geothermal characteristic
地热特征　geothermal feature
地热梯度　geothermal gradient

地热梯度变换　geothermal gradient change
地热梯度钻井　geothermal gradient drilling
地热梯级综合利用　geothermal stair comprehensive utilization
地热田　geothermal field
地热田储层　reservoir of geothermal field
地热田监测　geothermal field monitoring
地热田开发　geothermal field development
地热田开发效应　effect of exploration on geothermal field
地热田流量　geothermal field output
地热田模型　geothermal field model
地热田能量潜力　energy potential of geothermal field
地热田排水系统　field drainage system
地热田寿命　geothermal field life
地热田压力　geothermal pressure
地热条件　condition of geothermal heat, geothermal condition
地热调查　geothermal investigation
地热图　geothermal map
地热脱盐　geothermal desalination
地热温度　geotemperature
地热温度记录仪　geothermograph
地热温度梯度　geothermal temperature gradient
地热温室　geothermal greenhouse
地热问题　geothermal problem
地热吸收制冷　geothermal absorption refrigeration
地热系统　geothermal system
地热系统激励　stimulation of geothermal system
地热现象　geothermal manifestation

地热效应　geothermic efficiency
地热选项　geothermal option
地热学　geometrics
地热岩心钻探　geothermal core drilling
地热研究　geothermal research
地热盐水　geothermal brine
地热盐水矿床　geothermal brine deposit
地热盐水利用工厂　geothermal brine utilization plant
地热异常　geothermal abnormity, geothermal anomaly
地热异常区　geothermal anomaly area
地热异常现象　geothermal anomaly phenomenon
地热余热　geothermal waste heat
地热原动机　geothermal prime mover
地热源　terrestrial heat source, heat source in earth's interior
地热源温度　geothermal source temperature
地热噪声　geothermal noise
地热增温率　geothermal gradient, geothermal temperature increment ratio
地热增温钻井　geothermal gradient drilling
地热增压区　geopressured zone
地热增压装置　geopressurised system
地热增长　geothermal growth
地热站　geothermal heat station
地热蒸汽　geothermal steam
地热蒸汽抽提装置　geothermal steam extraction plant
地热蒸汽储层　geothermal steam reservoir
地热蒸汽发电　electricity generation with geothermal steam, geothermal steam power
地热蒸汽发电厂　geothermal steam plant
地热蒸汽轮机　geothermal turbine
地热蒸汽喷射　geothermal steam ejection, geothermal steam injection
地热蒸汽汽井　geothermal steam well
地热蒸汽收集器　geothermal steam collector
地热蒸汽田　geothermal steam field
地热蒸汽消耗量　geothermal steam consumption
地热蒸汽压力　geothermal steam pressure
地热蒸汽钻井　drilling for geothermal steam
地热正常区　geothermal normal area, geothermal normal region
地热直接供暖系统　geothermal direct space heating system
地热直接利用　geothermal direct use
地热指标　geothermal indicator
地热治疗　geothermal therapy
地热转换技术　geothermal conversion technology
地热装置　geothermal plant
地热咨询　geothermal consultant
地热资料　geothermal data
地热资源　geothermal resource
地热资源估算　evaluation of geothermal resource
地热资源勘查　geothermal resources exploration
地热资源勘探　exploration of geothermal resource, prospect for geothermal resource
地热资源面积　geothermal resource area
地热资源评价　assessment for geothermal

resource, geothermal resource assessment
地热资源评价参数　assessment parameter of geothermal resource
地热资源评价方法　assessment methods of geothermal resource
地热资源特征　character for geothermal resource, characteristic of geothermal resource
地热租赁处理　geothermal lease processing
地热钻井（探）　geothermal drilling
地渗出位置图　location map of thermal seepage
地台区域　platform region
地外（日射）辐照度　extraterrestrial irradiance
地外太阳辐射　extraterrestrial solar radiation
地温　geothermal heat
地温表　ground thermometer, earth thermometer
地温测量　geothermometry
地温计　geothermometer
地温梯度　thermal gradient
地文特征　physiographic feature
地下抽水储能　underground pumped hydrostorage
地下地热水　geothermal underground water
地下发电站　underground power station
地下构造　subsurface structure
地下井　groundwater well
地下耦合热泵系统　ground-couple heat pump
地下热储　underground reservoir
地下热交换器　ground heat exchanger
地下热能储存系统　underground thermal energy storage system
地下熔岩　subterranean molten rock
地下水　groundwater
地下水储层　groundwater reservoir
地下水储量　groundwater amount
地下水径流　groundwater runoff
地下水库　groundwater reservoir
地下水热泵　ground water heat pump
地下水位　water table, ground water level
地下水位表　ground water table
地下水污染　groundwater pollution
地下水源热泵　ground-water source heat pump
地下水运动　groundwater movement
地下温度　subsurface temperature
地下增温深度　geothermic depth
地下蒸汽　underground steam
地下蒸汽田　subterranean steam field
地下储能　underground storage
地心引力　gravitational force
地形　land configuration
地形粗糙度　terrain roughness
地形分类　terrain category
地形高度　terrain height
地形评价　assessment of landscape
地形特征　topographic feature
地形条件　topographic condition
地形效应　topographic effect
地形性上升风　orographic upward wind
地形性中尺度系统　terrain induced mesoscale system
地形学　topography
地形因素　topographical factor
地形诱导下洗　terrain downwash
地学工具　geoscientific tool

地压地热系统　geopressured geothermal system
地压力资源　geopressured resource
地压区域　geopressurized zone
地压砂岩层　geopressured sand
地压水　geopressured water
地压系统　geo-pressurized system
地压型地热资源　geopressured geothermal resource
地压型热储　trap reservoir
地音探测器（仪）　geophone
地源热泵　geothermal heat pump, ground source heat pump
地源热泵系统　ground heat exchanger
地源系统　ground-source system
地噪声　ground noise
地噪声测量（用于地热勘探的一种地球物理方法）　ground noise survey
地震波反射　seismic reflection
地震波曲线　seismogram
地震反射　seismic reflection
地震海浪　seismic sea wave
地震活动　seismic activity
地震活动性　seismicity
地震检波器　geophone
地震勘探法　seismic method
地震烈度　earthquake intensity
地震脉动　microseisms
地震速度　seismic velocity
地震折射　seismic refraction
地震折射技术　seismic refraction technique
地震震级　earthquake magnitude
地址因素　geological factor
地质背景　geological setting
地质不整合面　geological unconformity
地质储量　oil-in-place

地质地层　geologic formation
地质构造　geologic formation
地质环境　geological setting
地质绘图　geologic mapping
地质交换系统　geo-exchange system
地质控制　geologic control
地质流体　geofluid
地质数据　geological data
地质条件　geologic condition
地质温标　geological thermometer
地质温度计　geological thermometer
地质相似度　geological similarity
地质信息　geological information
地质制图　geologic mapping
地转风　geostrophic wind
地转风场　geostrophic wind field
地转风高度　geostrophic wind level, geostrophic wind height
地转风气流　geostrophic current
地转风矢量　geostrophic wind vector
地转风速　geostrophic wind velocity
地转流　geostrophic current
地转拽力定律　geostrophic drag law
递变波导器　tapered waveguide
递变截面波导　tapered waveguide
递减率　lapse rate
递增效应　incremental effect
第一调查阶段　first stage survey
第二代装置　second generation device
第二级提水风车　second water-pumping windmill
第三代装置　third generation device
第三级热泵　tertiary pump
第四调查阶段　fourth-stage survey
第四纪火山中心（它与地热田有密切关系）　quaternary volcanic center

第一表面反射镜　first-surface mirror
第一级空气预热器　primary air heater
第一扩散方程　first diffusion equation
碲化镉薄膜太阳能电池　CdTe thin-film solar cell
碲化镉太阳能电池　cadmium telluride solar cell

dian

点聚焦聚光器　point-focus concentrator
（索尔特）点头鸭波能转换装置　Salter nodding duck wave energy conversion device (Salter Duck)
电池阵列的总效率　gross array efficiency
电力汇集系统　power collection system

die

迭代最大值似然法　iterated maximum likelihood method
碟式—斯特林太阳能热发电　solar dish-Stirling engine

ding

定焦聚光器　fixed target concentrator
定日镜　heliostat
定日镜场　heliostat field
定日镜阵列控制器　heliostat array controller

dong

动焦聚光器　moving target concentrator

du

镀银聚合物反射镜　silvered polymer mirror

duo

多级式吸热器　multistage receiver

e

额定功率　rated power

fa

发电机组年发电量　annual electric output of generating unit
发电机组年利用小时数　annual utilization hours of generator unit
发电机组年运行小时数　annual service hours of generating unit

fan

反射镜表面变形　deformation of reflective mirror

fang

方解石　calcite
方头螺栓集热器　lag-bolt collector
方位　orientation
方位—俯仰跟踪　azimuth-elevation tracking
方位角　azimuth, azimuth angle
方向　orientation
方向比较保护　directional comparison protection
方向比较保护系统　directional comparison protection system
方向波谱　directional wave spectrum
方向舵　direction vane
方向控制，定向控制　directional control
方阵场　array field
方阵的面积利用率　area utilization of a solar array
方阵联结开关系统　array switching system
芳香性　aromaticity

芳香族化合物 aromatic compound
芳族聚合物 aromatic polymer
防爆罐 blow-out can
防爆膜 bursting disc
防冰器 deicer
防波堤 seawall, wave breaking revetment
防潮 moistureproof
防潮堤 tide embankment
防潮性 moisture resistance
防潮堰堤 tidal barrage
防潮闸 tide gate
防潮闸门 tidal gate
防尘 dust protected
防冲板 impingement plate
防冻剂 antifreeze solution
防冻加热器 heat tracing
防冻溶液 antifreeze solution
防冻液 antifreeze
防反射涂层 anti-reflective coating
防风加强环 wind guarder
防风篱 wind break fence
防风林 wind-break forest
防风林带 shelter belt
防风林网 wind-break network
防风林系统 wind break system
防风墙 wind break barrier
防风土埂 wind-break earth ridge
防风雨 weather protection
防风雨的 weatherproof
防腐 corrosion resistance
防海浪的 protected against heavy seas
防洪库容 flood control storage
防洪限制水位 beginning water level for flood control
防护 shielding
防护板 fender
防护林带 shelter belt

防回流阀 backflow preventer
防火间隔 fire barriers
防浪板 breakwater
防浪堤 breakwater
防雷区 lightning protection zone
防雷系统 lightning protection system
防喷器 blowout preventer
防热的 heat proof
防砂堤 groin
防失速的 stall proof
防湿 moistureproof
防水 waterproof
防水壁 bulkhead
防水层 waterproof layer
防水接线盒 waterproof junction box
防水式电动机 water proof machine
防松螺母 locknut
防通阀 bypass valve
防污涂层 anti-fouling coating
防污涂料 antifouling paint
防雨 rain protection
防振锤 damper
防止海洋污染 marine pollution prevention
防止日光照射的 antisun
房顶池太阳能装置 solar roof pond system
房顶太阳能集热器 solar roof collector
仿生材料 bionics material
仿生计算机 bionic computer
仿生学 bionics
仿真（等效）载荷 dummy load
放（出）能（量）的 exothermic (-mal)
放出电流 releasing current
放出热 abstracted heat, released heat
放大器盘 amplifier panel
放电 electric discharge, electrical discharge

放电电极　discharge electrode
放电电流　discharge current
放电电压　discharge voltage, spark-over voltage, sparking voltage
放电管　discharge tube
放电路径　discharge path
放电效率　discharging efficiency
放空　emptying
放能的　exothermal
放能反应　exergonic reaction
放气　air release
放气阀　bleed valve
放热　heat release, heat liberation, thermal discharge
放热的　exothermal, heat producing
放热反应　exothermic reaction
放热化学反应　exothermal chemical reaction, exothermic chemical reaction
放热化学反应机制　exothermic chemical reaction mechanism
放热率　heat release rate
放热衰变　exothermic disintegration
放热速率　rate of heat liberation
放热吸收器　exothermic absorber
放热吸收体　exothermic absorber
放热转化（放出能量）　exothermic disintegration
放射　emittance
放射率　emissivity
放射器　emitter
放射性尘埃　active dust
放射性废物　radioactive waste
放射性气体　radgas
放射性衰变　radioactive decay
放完电的电池　exhausted cell
放泄阀　bleed valve
放渣口　slag tap

fei

飞灰处理　fly ash disposal
飞灰过滤器　fly ash filter
飞灰含碳值　carbon-in-ash value
飞灰稳定化　fly ash stabilify
飞机发动机　aeromotor
飞机辅助电源装置　aircraft auxiliary power unit
飞溅点　flashpoint
飞轮［机械］　flywheel
飞轮储能　flywheel energy storage
飞球式调速器　fly ball governor
飞行高度　flight level
非（无）跟踪太阳能聚光器　non-tracking photovoltaic concentrator
非本征半导体　extrinsic semiconductor
非本征光电导率　extrinsic photoconductivity
非本征寿命　extrinsic lifetime
非本征吸收　extrinsic absorption
非标准电压　off standard voltage
非成像集热器　non-imaging collector
非成像聚光器　non-imaging concentrator
非磁性材料　non-magnetic material
非电子太阳能跟踪机　nonelectronic solar tracker
非定常空气动力学　unsteady aerodynamics
非对称太阳能聚光器　asymmetrical concentrator
非高峰荷载　off peak load
非高峰用电　off peak electricity
非跟踪聚光器　non-tracking concentrator
非跟踪式复合抛物面集热器　nontracking CPC
非跟踪式聚光型集热器　nontracking

concentrator
非工作齿面　non-working flank
非共格界面　noncoherent boundary
非化学计量化合物　nonstoichiometric compound
非机械分离工艺　nonmechanical separation process
非甲烷挥发性有机化合物　non-methane volatile organic compound
非尖峰的　off-peak
非搅拌燃料堆　non-agitated fuel pile
非晶硅　amorphous silicon, non-crystalline silicon
非晶硅太阳电池　non-crystalline silicon solar cell
非晶硅太阳能电池　amorphous silicon solar cell, amorphous solid silicon solar cell
非晶硅太阳能电池　amorphous silicon PV cell
非晶态半导体材料　noncrystalline semiconductor material
非晶态材料　amorphous material
非晶态磁性材料　non-crystal magnetic material
非晶态太阳电池　non-crystal solar cell
非晶体膜　amorphous film
非镜面导体　non-specular conductor
非聚光型集热器　nonconcentrating collector
非聚焦型聚能器　nonimaging concentrator
非均匀地下水　non-uniform ground fluid
非均匀形核　heterogeneous nucleation
非可燃物　noncombustible
非矿物燃料　nonfossil fuel
非冷凝气体效应　effect of non-condensible gas
非连续能源　non-continuous source
非连续相　discontinuous phase
非零升力系数　nonzero lift coefficient
非能动控制调节系统　passive system control
非凝性气体　noncondensable gas
非牛顿流体　non-Newtonian fluid
非燃性废料　non-combustible waste
非燃性垃圾　non-combustible waste
非热能　non-thermal energy
非商品能源　non-commcrical energy
非设计工况操作　off-design operation
非设计性能　off-design performance
非时变流　time-invariant flow
非食物生物质　non-food biomass
非水能可再生能源　non-hydro renewable energy
非太阳能　nonsolar energy
非太阳能成本　nonsolar energy cost
非太阳能辅助设备　nonsolar auxiliary
非太阳能辅助设备系统　nonsolar auxiliary system
非太阳能加热子系统　nonsolar heating subsystem
非太阳能系统　nonsolar system
非太阳能源　nonsolar energy source
非太阳能子系统　nonsolar energy subsystem
非微晶叠层　micromorph
非稳定热界面　unstable thermal interface
非稳态系数　transient coefficient
非稳态系数方程　equation of transient coefficient
非线性波相互作用　nonlinear wave interaction
非线性传播　nonlinear propagation

非线性空气动力学　non-linear aerodynamics
非选择性表面　nonselective surface
非有机垃圾　non-biogenic waste
非原位　ex situ
非载荷期　off-load period
非折射吸收　nondeviated absorption
非整比化合物　nonstoichiometric compound
非正弦交流电压　non-sinusoidal alternating voltage
非转移弧　non-transferred arc
菲克第二定律　Fick's second law
菲克第一定律　Fick's first law
菲涅尔透镜　Fresnel lens
菲涅耳带　Fresnel zone
菲涅耳反射镜集热器　Fresnel mirror collector
菲涅耳方程　Fresnel's equations
菲涅耳集热器　Fresnel collector
菲涅耳聚光器　Fresnel concentrator
菲涅耳透镜集热器　Fresnel lens collector
废活性污泥　waste activated sludge, surplus activated sludge, excess sludge
废料能　waste material energy
废木材　waste wood
废气　effluent, exhaust gas, flue gas, off-gas
废气净化处理过程　flue gas cleaning process
废气排出控制　emission control
废气燃烧器　gas flare
废气损失　waste-gas loss/ stack loss
废气洗涤器　gas scrubber
废弃燃烧装置　after burner
废弃物管理　waste management
废弃物衍生燃料　residue derived fuel
废汽　dump steam
废燃料　waste fuel
废热　waste heat, exhaust heat
废热发电　cogeneration
废热锅炉　waste heat boiler
废热回收　waste-heat recovery
废热回收锅炉　heat-recovery boiler
废热回收蒸汽发生器　heat recovery steam generator
废热利用　waste heat recovery
废水处理　effluent treatment
废水生物处理　biological wastewater treatment
废物处理　waste disposal
废物分级　waste hierarchy
废物焚化　waste incineration
废物管理　waste management
废物减量化算法　waste reduction algorithm
（欧盟）废物框架指令　Waste Framework Directive
废物利用　salvage
废物排放标准　waste discharge standard
废物原料　waste feedstock
废物源流　waste stream
废物再生能源　waste-to-energy, energy-from-waste
废渣处理　waste residue treatment
沸点　vaporizing point
沸点—比重常数　boiling point-gravity constant
沸点温度表　boiling point thermometer
沸泥塘　boiling mud pool
沸水反应堆　boiling water reactor
沸腾传热系数　boiling coefficient
沸腾泥浆坑　boiling mud pit
费密能级　Fermi level

费密能量　Fermi energy
费托柴油　Fischer-Tropsch diesel
费托合成　FT synthesis
费托合成反应器　Fischer-Tropsch reactor
费托合成过程　Fischer-Tropsch process

fen

分贝　decibel (dB)
分辨率　resolution
分布　distribution
分布板　distributor plate
分布电压　spread voltage
分布电容　distributed capacitance
分布绕组　distributed winding
分布式电源客户侧模型　distributed energy resources customer adoption model
分布式发电　distributed generation
分布式供电技术　distributed power supply technology
分布式集热器　distributed collector
分布式计算　distributed computing
分布式控制系统　distributed control system
分布式能源　distributed energy resources
分布式系统　distributed system
分布图　scatter diagram
分布系数　distribution coefficient
分布系统　distributted system
分布型线集热器系统　distributed linear collector
分层床气化炉　stratified-bed gasifier
分层流体　stratified fluid
分层试验　stage test
分层下吸式气化炉　stratified downdraft gasifier
分叉（指令）　fork

分潮　tidal constituent, tidal component
分潮岭　tidal divide
分带性　zonation
分点大潮高潮位　high water equinoctial spring tide
分动箱　power take-off
分段多项式　piecewise polynomials
分段试验　stage test
分断　breaking
分光光度法　spectrophotometric method
分光镜检查　spectroscopy
分光谱电池　split spectrum cell
分光总日射表　spectral pyranometer
分级机　grader
分级绝缘线圈　winding with nonuniform insulation
分级能带宽度太阳能电池　graded energy gap solar cell
分级燃烧　staged combustion
分接头位置信息　tap position information
分解　dissolution
分解代谢　catabolism, destructive metabolism
分解能　decomposition energy
分解蒸馏　destructive distillation
分类仓　sorting cabin
分类机　grader
分类收集　separate collection
分离器　separator
分离器容器　separator vessel
分离器疏水冷却热交换器　cooling separator drain heat exchanger
分离式集热器　separate collector
分离式沼气池　separate gasholder digester
分离压力　separation pressure
分立式太阳能集热器　separated collector
分裂能　break-up energy

分裂凿岩　spallation drilling
分流　side stream
分流器　shunt
分路电容器　shunt capacitor
分路阻抗　shunt impedance
分配系数　distribution coefficient
分配系统　delivery system
分批干燥器　batch dryer
分批燃烧　batch combustion
分散　dissipation
分散关系　dispersion relation
分散控制系统　distributed control system
分散率　diversity factor
分散气溶胶　dispersion aerosol
分散热　heat of decomposition
分散式电容　distributed capacitance
分散式发电　dispersed generation
分散式发电装置　distributed power generation facility
分散式风能系统　decentralized wind energy system
分散式太阳能发电系统　distributed solar power production system
分散式太阳能热发电　solar thermal distributed power system
分散式系统　distributed system
分散式制冷负荷　separate refrigeration load
分散系数　diversity factor
分散型集热器　distributed collector
分散型太阳能电站　decentralized solar power station
分散型系统　decentralized system
分时电价　time-of-use power price
分体式(太阳能热水)系统　remote-storage (solar water heating) system

分析器　analyser
分线盒　junction box
分项安全系数　partial safety factor
分压器　voltage divider
分样　sub-sample
分支　bifurcation, fork
分支接续　branch connection
分子　molecule
分子泵　molecular pump
分子晶体　molecular crystal
分子量　molecule weight
分子能　molecular energy
分子热　molecular heat
分子散射　molecular scattering
分子筛　molecular sieve
分子束质谱技术　molecular beam mass spectrometry
分子重量　molecular weight
分组方法　method of bins
酚醛塑料　phenolics
焚风　foehn wind
焚化　incineration
焚化场　incineration plant
焚化炉残渣　incinerator residue
焚烧　incineration
焚烧炉　incinerator
焚烧炉上限垃圾处理量　upper limit waste treatment capacity for incinerator
焚烧炉上限垃圾低位热值　upper limit LHV of waste for incinerator
焚烧炉下限垃圾处理量　lower limit waste treatment capacity for incinerator
焚烧炉下限垃圾低位热值　lower limit LHV of waste for incinerator
焚烧线　incineration line
粉尘层　dust cake
粉尘排放　dust emission

粉化　pulverization
粉煤灰　fly ash
粉末燃料　pulverized fuel
粉末燃料燃烧　pulverized fuel combustion
粉末燃料燃烧系统　pulverized fuel combustion system
粉碎　comminution
粉碎的生物质燃料　pulverized biofuel
粉状　trituration
粪便　manure, dung
粪便收集　dung collection

feng

风（堆）积　aeolian accumulation
风（机驱）动发电机　fan-driven generator
风暴　windstorm, storm
风暴潮　wind tide, storm tide
风暴潮预报　storm surge forecast
风暴风量筒　storm drum
风暴警报　storm warning
风暴浪　storm wave
风暴期　storm duration
风泵　wind pump
风变差　wind variation
风标　wind cone
风柴电池混合系统　wind-diesel-battery hybrid system
风柴电压　wind-pv-diesel
风柴系统　wind-diesel system
风场　wind site, wind field
风场电气设备　site electrical facility
风场评价　wind assessment
风车　windmill
风车（产生的）噪声，风力发电机（产生的）噪声　windmill-produced noise
风车泵　windmill pump
风车尺寸　windmill size
风车风速　wind velocity for windmill
风车式风速表　windmill anemometer
风车式风速计　anemometer of windmill type
风车式水轮机　windmill turbine
风车式制动装置　windmill brake
风车型生热器　windmill heat generator
风车型式　windmill type
风车制动状态　windmill-braking condition
风成波浪　wind ripple mark
风成沉积　aeolian deposit, wind-laid deposit
风成沉积物　wind borne sediment
风成的　aeolian
风成海浪　wind-generated sea wave
风成浪　wind-induced wave
风成平原　aeolian plain
风程　wind run, run of the wind
风程表　wind-run anemometer
风程风速计　wind-run anemometer
风冲子　air ram (mer)
风吹损失　windage, windage loss
风带　wind belt, wind streak
风挡　damper plate
风挡板　air damper
风挡刮水器　windscreen wiper
风挡雨雪刷　windscreen wiper
风道　air course
风道湍流　wind tunnel turbulence
风的测定　wind measurement
风的切变　shear of wind
风的状况　wind regime
风电变流器　wind converter
风电并网　wind power grid integration
风电并网系统　grid-connected wind power system

风电产能 wind generating capacity	风电系统 wind system, wind power system
风电厂 wind-power plant	
风电厂并网发电 grid-connected wind plant	风电装机容量 wind-power capacity, installed wind capacity
风电场 wind power station, wind farm	风动冲击凿岩机 air hammer drill
（风电场）电网连接点 grid connection point for wind farm	风动锤 air ram (mer)
	风动海洋波能转换 pneumatic ocean wave energy conversion
风电场并网点 point of interconnection of wind farm	
	风动履带式凿岩机 air-track drill
风电场布局 wind farm layout	风动伸缩式气腿 air leg
风电场部件 wind power plant component	风动涡旋 wind spun vortex
风电场管理系统 wind power plant management system	风动系统 pneumatic system
	风动凿岩机 airleg rock drill
风电场规模 wind farm sizing	风洞 wind tunnel
风电场模拟信息 wind power plant analog (ue) information	风对流系数 wind convection coefficient
	风阀 air valve
风电场筛选图表 wind farm screening chart	风帆助航 sail assisting
	风干重 air dry weight
风电场无功功率 reactive power of wind farm	风镐 air pick, pneumatic pick, air ram (mer)
风电场有功功率 active power of wind farm	风功率密度 wind power density
	风管 air pipe
风电场噪声 wind-farm noise, noise of wind power plant	风—光伏—柴油混合系统 wind-pv-diesel hybrid power
风电场噪声评估 assessment of wind-farm noise	风—光伏电池混合系统 wind-pv-battery hybrid system
风电机端口 wind turbine terminal	风光互补 wind-photovoltaic complementary
风电机排列布置 array of wind turbine generator	风—光互补电站 wind-PV hybrid power system, wind photovoltaic power system
风电机组低电压穿越 low voltage ride through of wind turbine	
	风—光互补路灯 wind and PV hybrid streetlights
风电机组塔架 tower for wind turbine generator system	
	风荷载 wind load, wind loading
风电机组现场试验 field testing of wind energy conversion system	风化 weathering
	风化度 degree of weathering
风电机组效率 efficiency of WECS, efficiency of WTGS	风化壳 weathering crust
	风化裂缝 weathering crack

风化裂隙　weathering fissure
风化煤　weathered coal, windblown coal
风化岩　weathered rock
风化作用　weathering
风环流　wind circulation
风机　fan
风机齿轮箱　wind turbine gearbox
风机出口　fan outlet
风机导流装置　fan casing
风机导叶驱动控制　fan pitch drive control
风机对流器　fan convector
风机负压　fan suction
风机盘管　fan coil
风机盘管机组　fan coil unit
风机盘管系统　fan coil system
风机驱动部件　fan drive assembly
风机入口　fan inlet
风机升降台　wind turbine pad
风机送风量　fan delivery
风机塔架　wind turbine tower
风机特性　fan performance
风机特性曲线　fan performance curve
风机性能　fan performance
风机叶片　fan blade
风机展弦比　fan aspect ratio
风积的　aeolian
风积土　aeolian soil
风积物　aeolian deposit, aeolian material
风激加速度　wind excited oscillation
风级　wind scale, scale of wind-force
风迹　wind path, wind trajectory
风剪力　wind shear
风角　wind angle
风接近角　wind-approach angle
风井　air shaft

风井口　air end
风镜　goggle
风可利用的总功率　total power available in the wind
风口　tuyere, air opening, wind gap
风况　wind regime
风廓线　wind profile
风廓线风切变律　wind profile wind shear law
风廓线指数　wind speed profile coefficients
风廊　wind porch
（风浪）平息区　decay area
风浪　wind-induced wave
风浪负荷　wind-wave load
风浪模型　wind-wave model
风浪水力学　wind-wave hydraulics
风浪载荷　wind-wave load
风冷的　air cooled
风冷冷凝器　air condenser, air-cooled condenser
风冷指数　wind chill factor
风力　wind strength, wind force
风力泵　wind pump
风力补偿　wind compensation
风力测试站　wind test station
风力—柴油发电系统　wind-diesel photovoltaic system
风力场　wind field
风力充电机　wind charger
风力充电器　wind charger
风力出口　wind-power out
风力传感器　wind wandler
风力萃取系统　wind-power extraction system
风力的　wind-powered
风力等级　wind force scale, wind power class

风力地图 wind map
风力—电力水泵 wind-electric water pump
风力电力转换系统 wind-electric system
风力电站 wind-power plant
风力电站布置图 wind-power plant layout
风力发电 wind-generated electricity, wind power generation, wind electric power generation
风力发电厂 wind power plant
风力发电抽水装置 wind-eletric water pumping system
风力发电储能系统 stored energy system of wind power, energy storage system of wind power
风力发电机 windmill, windmill generator, fan driven generator, wind generator, wind genny, wind-driven electrical generator, wind-driven dynamo, wind-driven generator, wind-powered generator
风力发电机互联 interconnection of aerogenerators
风力发电机塔 wind generator tower
风力发电机组 wind turbine generator system (WTGS)
风力发电机组度电成本 cost per kilowatt hour of the electricity generated by WTGS
风力发电机组精度 accuracy for WTGS
风力发电机组输出端 wind turbine terminals
风力发电机组输出特性 output characteristic of WTGS
风力发电设备 wind-power generation equipment
风力发电设施 wind generator facility
风力发电试验场 pilot wind power plant
风力发电网络 wind-power network
风力发电系统 wind generating system
风力发电站 wind-power station
风力发电阵列 wind generator array
风力发电装置 wind-power electric generating unit
风力发动机 wind motor
风力发动机工作轮叶 windmill sail
风力分级 air classification, wind classification
风力分级机 air classifier
风力工程 wind power engineering
风力—光伏发电系统 wind-solar photovoltaic system
风力机 wind turbine
风力机模拟 wind turbine simulation
风力机潜在场址 potential site for wind machine
风力机输出特性 output characteristic of WTGS
风力机塔架 turbine tower
风力机停机 parked wind turbine
风力机尾流 turbine wake, wind turbine wake
风力机效率 wind turbine efficiency
风力机械 wind-power machine, wind-powered machine
风力机械设计 wind machine design
风力机性能曲线 performance curve of wind turbine
风力机叶片 wind turbine blade
风力机噪声 wind-turbine noise
风力机最大功率 maximum power of wind turbine

风力记录仪　anemograph
风力加热　wind heating
风力监察站　wind monitoring station
风力减弱　abatement of wind, retardation of wind
风力降温　windchill
风力偏流修正表　wind and drift chart
风力偏移　wind deflection
风力曝气机　wind powered aerator
风力曝气系统　wind powered aeration system
风力侵蚀　wind abrasion, wind corrosion, wind erosion
风力驱动装置　wind-driven device
风力热泵　wind driven heat pump, wind powered heat pump
风力筛分试验　air analysis (test)
风力设备　wind power installation
风力输送　pneumatic conveying
风力输送干燥　pneumatic conveying drying
风力输送机　wind conveyer
风力输送系统　pneumatic conveying system
风力水泵系统　wind-powered water pumping system
风力损失　windage loss
风力提水　wind power water pumping
风力提水机　wind-driven water pump
风力提水机组　wind water-lifting set, wind power draw water machine
风力调速器　windmill governor
风力透平　wind turbine
风力图　wind-force diagram, wind force diagram
风力涡轮发电机　wind turbine generator
风力涡轮发电机开发　wind turbine generator development
风力涡轮发电机模型　wind turbine generator model
风力涡轮机　wind turbine spin, wind turbine
风力涡轮机舱　wind turbine nacelle
风力涡轮机动力学　wind turbine dynamics
风力涡轮机发电机组　wind turbine generator set
风力涡轮机基本参数　wind turbine foundation parameter
风力涡轮机声标准　wind turbine acoustic standard
风力涡轮机数据采集系统　wind turbine data acquisition system
风力涡轮机效率　wind turbine efficiency
风力涡轮机性能　wind turbine performance
风力涡轮机噪声　wind turbine noise
风力涡轮机转鼓　wind turbine drum
风力涡轮机塔架　wind turbine tower
风力涡轮转矩　wind turbine torque
风力系统　wind power system
风力修正　wind correction
风力修正器　wind corrector
风力运输　air transport
风力站（风力发电站）　wind station
风力指示器　wind indicator
风力制热　wind power heating
风力装置　wind power system
风力资源　wind resource, wind power resource
风力资源地图　Wind Resource Maps
风力资源评估项目　Wind Resource Assessment Program
风力作用　wind action
风量　air quantity, air volume, air flow

rate, blowing rate, airflow, air output, quantity of air flow
风量分配 air distribution
风量估算 air quantity estimation
风量机翼测量装置 air foil
风量计 air volume meter
风量计算法 air volume calculation method
风量调节 air conditioning, air quantity control, air regulation
风量自然分配 flow natural distribution, natural distribution of air flow
风临度 air velocity
风流 air stream, air circulation, air flow
风流测量站 air-measuring station
风流反向 inverted air current
风流分支 air split ventilation parting
（风轮）摆振方向 lead-lag
（风轮）挥舞方向的 flapwise
风轮 wind wheel, wind turbine rotor, wind rotor
风轮闭锁装置 rotor locking device
风轮波力电动机 air turbine wave motor
风轮额定转速 rated turning speed of rotor
风轮风速计 wind wheel anemometer
风轮功率系数 rotor power efficient
风轮机 wind machine, aeroturbine
风轮机外壳 wind turbine housing
风轮基准面 reference area of the rotor
风轮界面 rotor interface
风轮空气动力特性 aerodynamic characteristics of rotor
风轮力矩 rotor torque
风轮扭矩标准差 rotor torque standard deviation
风轮扭矩波动 rotor torque fluctuation

风轮偏侧式调速机构 regulating mechanism of turning wind rotor out of the wind sideward
风轮偏航角 yawing angle of rotor shaft
风轮偏角 yaw angle of rotor
风轮平面内速度 velocity in the rotor plane
风轮区域 rotor area
风轮扫掠面积 swept area of rotor, rotor thrust, rotor swept area
风轮实度 rotor solidity
风轮推力 rotor thrust
风轮推力功率谱 power spectrum of rotor thrust, rotor thrust power spectrum
风轮尾流 rotor wake
风轮涡旋状态 rotor whirl modes
风轮斜置角度 rotor tilt
风轮叶片理论 rotor blade theory
风轮叶片扭曲 rotor blade twist
风轮迎风 orientation of the rotor
风轮迎风面积 rotor frontal area, rotor thrust
风轮预锥角 rotor pre-cone
风轮直径 blade diameter
风轮轴 rotator axis, rotor hub, rotor shaft
风轮轴仰角 tilt angle of rotor shaft
风轮转动方向 turning direction of rotor
风轮转动圆盘面 rotor disc
风轮转矩 rotor rotation moment
风轮转速 rotor speed
风轮锥角 cone angle of rotor
风轮最高转速 maximum turning speed of rotor
风玫瑰 windrose
风煤气 air gas
风门 air door
风敏感度 degree of wind sensitivity

风模拟　wind simulation
风摩擦　wind friction
风能　aeolian energy, aeolic energy, wind power, wind energy
风能的收集　collection of wind energy
风能电力　wind electricity
风能发电机螺旋桨　aerogenerator propeller
风能分布图　wind atlas
风能工业　wind-power industry, wind power industry
风能换热器　wind-power heat exchanger
风能机械　wind-power machine
风能技术　wind technology
风能监测器　wind energy monitor
风能经济学　wind power economics
风能开发　wind energy development
风能开发导则　guidelines for wind energy development
风能利用　wind energy utilization, wind energy application
风能利用率　utilization rate of wind energy
风能利用系数　wind power utilization coefficient, wind energy utilization coefficient
风能玫瑰图　wind energy rose
风能密度　wind energy density, wind energy flux
风能频谱图　wind power duration curve, wind energy distribution curve
（风能资源评估）代表年　representative year for wind energy resource assessment
风能潜力　potential of wind energy
风能潜势　wind power potential
风能热交换器　wind power heat exchanger

风能盛行方向　prevailing wind power direction
风能时域分析仪　wind energy time-domain analyzer
风能收集器　collector of wind energy, wind energy collector
风能系统　wind system
风能系统时域　wind energy system time domain
风能与能源安全　wind energy and energy security
风能转换　wind energy conversion
风能转换器　wind energy converter
风能转换系统　wind turbine generator system (WTGS), wind-driven generating system, wind-driven generating unit, wind-driven generator group, wind-driven generator set, wind-generator set, wind-turbine generator set, wind generator group
风能转换装置　wind conversion device, wind driven power-conversion device
风能装置　wind-power plant
风能资源　wind resources, wind energy resources
风能资源评估　wind resource assessment
风偏角　angle of wind deflection
风频率分布　wind frequency distribution
风墙　air stopping
风桥　air crossing
风切变　wind shear
风切变律　wind shear law
风切变幂律　power law for wind shear
风切变影响　influence by the wind shear
风切变指数　wind shear exponent
风琴式风速计　eolian anemometer
风区　wind field

风驱波 wind driven wave
风驱流 wind driven current
风扰动 wind disturbance
风沙 winddrift sand
风沙分选法 air-sand process
风扇 fan, fanner
风扇电机 fan motor
风扇加热器 fan heater
风扇轮毂 fan hub
风扇螺管装置 fan-coil unit
风扇皮带 fan belt
风扇叶轮 fan propeller
风扇叶片 fan blade
风扇罩 fan cover
风扇转子 fan rotor
风生（波）浪 wind generated wave
风生电 wind-generated electricity, wind power generation, wind power generation, wind electric power generation
风生海流 wind induced current
风生湍流 wind induced turbulence
风生运动 wind induced motion
风生重力波 wind-generated gravity wave
风时 wind duration
风蚀 wind abrasion, wind corrosion, wind erosion, eolation
风蚀（作用） aeolian erosion
风矢 wind arrow
风矢杆 wind-direction shaft, wind shaft
风矢量 wind vector, wind velocity
风数据记录器 wind data logger
风速 wind speed, wind velocity
风速比 wind speed ratio
风速变化特性 wind change characteristic
风速标准差 standard deviation of wind velocity

风速表 wind gauge
风速不足 wind speed deficit
风速测量 wind speed measurement
风速测量系统 wind measuring system
风速测量站 wind speed measurement station
风速传感器 air velocity transducer
风速的概率分布 probability distribution of wind speed
风速的高度换算 wind velocity on elevation basis
风速分布 wind speed distribution
风速分布曲线 wind speed profile, wind speed distribution curve
风速分布图 wind profile
风速风压记录器 anemobigraph
风速积分器 air speed integrator
风速计 air meter, wind gage
风速记录图 anemogram
风速廓线 wind shear law, wind speed profile
风速历时曲线 wind speed duration curve, wind speed duration curve
风速脉动 wind velocity fluctuation
风速频率 frequency of wind speed, wind speed frequency
风速频率曲线 wind speed frequency curve
风速气压计 anemobarometer
风速器 wind gauge
风速时间历程 time history of wind speed
风速特性 wind speed characteristic
风速线图 isovent map
风速压力 velocity pressure
风速预测 wind speed prediction
风速直方图 wind speed histogram

风速值　wind speed value
风速最大值　maximum value of wind speed
风速最小值　minimum value of wind speed
风损　windage loss, loss of wind
风塔　wind tower
风特性　wind characteristic
风筒　air shaft
风图　wind map
风图谱　wind atlas
风湍流　wind turbulence
风涡　wind eddy
风系统　wind system
风箱　windbox
风响应热性　wind response characteristic
风向　wind flow, wind direction
风向变化　wind variation
风向标　wind cock, wind direction vane, wind vane, vane, weather vane, weathercock
风向波动　wind direction fluctuation
风向测定　wind-direction measurement
风向测量　wind-direction measurement
风向传感器　wind direction sensor
风向袋　sock, wind sleeve, wind sock
风向风速表　anemorumbometer
风向风速计　wind vane and anemometer
风向风速仪　aerovane, anemoclinograph
风向计　registering weather vane
风向玫瑰图　wind diagram
风向频率　wind direction frequency
风向频率图　wind diagram
风向图　wind rose
风向仪　anemoscope
风向指示器　wind direction indicator
风向自记器　wind direction recorder

风向最小值　minimum value of wind direction
风巷　air way
风斜表　anemoclinometer
风信标　wind indicator
风信旗　vane
风性波动　wind fluctuation
风选　air cleaning
风穴　wind scoop
风学　anemology
风压　wind pressure
风压表　draftmeter
风压测量孔　wind pressure tap
风压浮标（一种波浪能装置）　air pressure ring buoy
风压混合系统　wind-PV hybrid system
风压径间　wind span
风压力　wind pressure
风压应力　wind stress
风眼　air way
风翼板　windsail
风翼升力　aerofoil lift
风翼阻力　aerofoil drag
风迎角　wind angle
风影热　wind shadow thermal
风壅水幅度　amplitude of wind tide
风与生物质共燃　wind and biomass co-firing
风载　air load
风载尘　wind borne dust
风载挡距　wind span
风载能力　wind load capacity
风载物质　wind borne material
风载雪　wind borne snow
风增水　wind onset
风障　wind break
风筝式风速计　kite anemometer

风致荷载　wind induced load
风致加热器系统　wind induced heater system
风致热损失　wind induced heat loss
风致热系统　wind induced heat system
风致热装置　wind induced heat system
风致压力　wind induced pressure
风致噪声　wind generated noise
风致振荡　wind induced oscillation
风致振动　wind induced vibration
风轴风力机　wind-axis wind machine
风轴系　wind axis system
风状态启动条件　starting under wind-milling condition
风阻　windage resistance, wind drag
风阻荷载　wind drag load, drag wind load
风阻力　wind resistance
风阻面积　drag area
风阻试验　drag test
风阻损失　windage loss
风阻特性曲线　air way characteristic curve
风阻影响　windage effect
风钻　air hammer drill
封闭　lock
封闭孔隙度　storage porosity
封闭热循环　closed thermal cycle
封闭式　closed-cycle
封闭式循环发动机　closed-cycle engine
封闭式循环系统　closed-cycle system
封闭式液体太阳能集热器　closed liquid solar collector
封闭式蒸发器　shell evaporator
封闭系统　closed system
封闭油循环　closed oil cycle
封场后渗滤液　closed landfill leachate
封隔器　packer
封锁　latching

封装材料　encapsulant material
封装太阳能电池板的薄膜　ethylen-venyl-acetat
峰点电流　peak point current
峰点电压　peak point voltage
峰谷比　peak to valley ratio
峰荷电厂　peak load plant
峰荷电能　on-peak energy
峰荷发电机组　peaking unit
峰荷水电站　peak load hydropower plant
峰间电压　peak to peak voltage
峰期　peak period
峰期功率跟踪方案　peak power tracking scheme
峰容量　peak capacity
峰数量　peak magnitude
峰瓦　watts peak
峰压　peak voltage
峰值点　peak point
峰值电流　peak current
峰值反向电压　peak reverse voltage
峰值负荷　peak load
峰值负荷发电机组　peak-load generating unit
峰值负荷功率　peak load power
峰值功率持续时间率　peak load effective duration factor
峰值功率点　peak power point
峰值功率跟踪器　peak power tracker
峰值间电流　peak to peak current
峰值力　peak force
峰值频率　peak frequency
峰值日照时数　peak sun hour
峰值温度　peak temperature
峰值正流向阳极电压　peak forward anode voltage
蜂窝复合结构（置于透明盖板和集热板

之间类似蜂窝状的隔热材料） honeycomb compound material
蜂窝状（太阳能）集热器 honeycomb collector
蜂窝状固定太阳池 stationary honeycomb solar pond
蜂窝状太阳能集热器 cellular solar collector
疯狗浪 rogue wave

fu

呋喃 furan
弗朗萨式太阳能炉 Francia type furnace
伏角 pitch
伏特/赫兹比 volts/hertz ratio
伏特（安培） volt-amperes (VA)
氟 fluoride
氟掺杂氧化锡 fluorine-doped tin oxide
氟化物 fluoride
氟利昂 Freon
氟碳聚合物 fluorocarbon polymer
浮标 buoy
浮充电池组 floating battery
浮动的 floating
浮动开关 float switch
浮动平板 floating platform
浮动气罩沼气池 floating cover digester
浮动式球阀 ball float valve
浮阀 float valve
浮垢 scum
浮接充电 trickle charge, floating charge
浮控阀 float valve
浮控开关 float switch
浮力 buoyancy, buoyancy force
浮力效应 buoyancy effect
浮球阀 float valve
浮球开关 float switch
浮区 float zone, float-zone
浮区熔法 float zone, float-zone
浮式潮汐电站 floating tidal plant
浮式海洋热能转换站 floating ocean thermal energy conversion plant
浮式机械设备 floating plant
浮式开式循环（海水温差）电站 floating plant with an open cycle
浮式太阳能箱 floating solar pond
浮式太阳能蒸馏器 floating solar still
浮水植物 floating plant
浮筒 floating buoy
浮筒系统 float system
浮箱波能发电机 raft energy generator
浮箱单元 raft unit
浮选 float
浮游的 pelagic
浮游生物 plankton
浮游生物网 plankton net
浮游生物消长 plankton pulse
浮游藻类 errant algae
浮渣层 scum layer
浮罩式沼气池 floating holder digester
浮舟 pontoon
浮子 float device, float chamber
浮子臂 float arm
浮子式液面计 float gauge
浮子系统 float system
幅—频响应 amplitude-frequency response
辐（射）出（射）度 radiant exitance
辐（射）功率 radiant power
辐（射）照度的光谱密集度 spectral concentration of irradiance
辐（射能）通量 radiant energy flux, radiant flux, flux of radiation
辐射 emittance, radiation
辐射安全 radiation safety
辐射板 radiant panel

辐射保藏	radiation preservation
辐射本底	radiation background
辐射本领	radiant capacity, radiantting capacity, radiantting power
辐射表	radiometer
辐射表面	radiantting surface
辐射波浪力	radiation wave force
辐射不稳定性	radiation instability
辐射测量技术	radiometric technique
辐射产额	radiation yield
辐射产物	radiation product
辐射常数	radiation constant
辐射场	radiation field
辐射处理堆	radiation processing reactor
辐射传热	radiant heat transfer, heat transfer by radiation
辐射窗	radiation window
辐射电离	radiation ionization
辐射电离室	radiation ionization chamber
辐射电热器	radiant electric heater
辐射发射率	radiant emissivity
辐射反应	radiation reaction
辐射防护	radiation protection
辐射防护装置	radiation protection device
辐射放热系数	radiant heat transfer coefficient
辐射分解速率	radiation decom position rate
辐射俘获	radiation capture
辐射俘获率	radiative capture rate
辐射腐蚀	radiation corrosion
辐射感生蠕变	radiation induced creep
辐射干扰	radiation interference
辐射功率	radiation power
辐射管	radiant tube
辐射管理	radiation control
辐射管理区域	radiation control area
辐射光谱	spectrum of a radiation
辐射化学	radiation chemistry
辐射化学产额	radiation chemical yield
辐射化学反应	radiation chemical reaction
辐射化学合成	radiation chemical synthesis
辐射换热	radiation heat transfer
辐射激发	radiation excitation
辐射计	radiationmeter
辐射剂量计	radiation dosimeter
辐射加热	radiative heating
辐射监测系统	radiation monitoring system
辐射交换系数	radiation exchange factor
辐射角	radianted angle, radiation angle
辐射角系数	radiative angle factor
辐射截面	radiation cross section
辐射警报系统	radiation alarm system
辐射聚合	radiation polymerization
辐射控制	radiation control
辐射宽度	radiation width
辐射冷却	radiation cooling, radiative cooling
辐射力	emissive power, radiation force
辐射量	radiation quantity
辐射灵敏度	radiation sensitivity
辐射率	emissivity, emittance
辐射面	emitting surface
辐射面积	emitting area
辐射敏感器	radiation sensor
辐射耐受能力	radiation tolerance status
辐射能	radianted energy, radiant energy, radiating power
辐射能力	emissivity
辐射能量	radiation energy
辐射能量交换	radiant energy exchange
辐射能量流	radiant energy flow
辐射能量密度	radiant energy density

辐射能吸收 absorption of radiant energy
辐射能性质 quantity of radiant energy
辐射逆反应 radiation back reaction
辐射碰撞 radiative collision
辐射平衡通量聚光度 radiation balance flux concentration
辐射屏蔽 radiation shield, radiation shielding
辐射破坏 radiation destruction
辐射曝露量 radiant exposure
辐射强度 radiation intensity, intensity of radiation, radiant intensity
辐射强度计网络 pyranometer network
辐射区 radiation zone, radiation area
辐射区段 radiant section
辐射燃料 irradiated fuel
辐射热 radiating heat, heat of radiation
辐射热交换 radiant interchange, radiation heat exchange
辐射热流 radiant heat flow
辐射热能交换 radiant heat exchange
辐射热损失 radiation heat loss, radiative heat loss
辐射热通量 radiant heat flux
辐射散射 radiation scattering
辐射射线 radiant ray
辐射生物化学 radiation biochemistry
辐射矢量 radiation vector
辐射式再热器 radiant reheater
辐射事故 radiation accident
辐射受热面 radiation heating surface
辐射衰减 radiation attenuation
辐射衰减因数 radiation decrement factor
辐射水平 radiation level
辐射松 radiate pine
辐射损伤 radiation breakdown, radiation injury
辐射损伤反应 radiation damage reaction
辐射损伤敏感区 radiation damage susceptibility
辐射损失 radiation loss, radiative loss
辐射太阳能 radiant solar energy
辐射态 radiant state
辐射探测 radiation detection
辐射体 radiant body, radiating body, radiation emitter
辐射通量 radiant flux
辐射通量密度 radiant flux density, radiation flux density
辐射稳定 radiation stability
辐射稳定聚合物 radiation stable polymer
辐射问题 radiation problem
辐射吸收 radiant absorption, radiation absorption
辐射吸收剂量 radiant absorbed dose
辐射吸收介质 radiation absorption medium
辐射吸收率 radiation absorptivity
辐射系数 radiation coefficient, emissivity
辐射效率 radiant efficiency
辐射效应研究（反应）堆 radiation effect reactor
辐射泄漏 radiation leakage
辐射型密度计 radiation-type density gauge
辐射型气体加热器 radiant-type heater
辐射选择性 selectivity of radiation
辐射压力 radiation pressure
辐射仪 radiac
辐射仪表 radiac instrument
辐射影响 radiation alarm system
辐射诱发遗传效应 radiation induced genetic effect

辐射阈能	radiation threshold energy
辐射元件	radiating element
辐射原子	radianted atom
辐射源	radiation source
辐射阵面	radiation front
辐射植物学	radiation botany
辐射指示器	radiation indictor
辐射中心	radiation center
辐射状色谱法	radial chromatography
辐射自分解	radiation selfdecomposition
辐射阻抗	radiation impedance
辐射阻力	radiation resistance
辐射阻尼	radiation damping
辐射作业	radiation action
辐照脆化	irradiation embrittlement
辐照度	irradiance
辐照度计	irradiance meter
辐照二氧化铀燃料元	irradiated uranium dioxide fuel element
辐照防护	radiation protection
辐照感生生长	irradiation induced growth
辐照畸变	irradiation distortion
辐照剂量	irradiation dose
辐照监督管	irradiation surveillance capsule
辐照量	irradiation, insolation
辐照率	irradiance ratio
辐照燃料后处理	irradiated fuel reprocessing
辐照燃料检验	irradiated fuel inspection
辐照蠕变	irradiation creep
辐照试验台	irradiation rig
辐照损伤	irradiation damage
辐照危险性	irradiation hazard
辐照系数	radiation factor
辐照效应	effect of irradiation
辐照影响	irradiation effect
俯仰	pitching
俯仰反应	pitch response
俯仰力矩	pitching moment
辅酶	coferment
辅以太阳能的气体能量系统	solar assisted gas energy system
辅助电动机	auxiliary motor
辅助电路	auxiliary circuit
辅助电气设备	ancillary electrical equipment
辅助发电机	auxiliary generator
辅助服务	ancillary service
辅助加热器	auxiliary heater, booster heater, back up heater
辅助加热系统	auxiliary heating system
辅助接点	auxiliary switch
辅助进给阀	auxiliary feed valve
辅助冷却设备	supplementary cooling equipment
辅助励磁机	pilot exciter
辅助能源	auxiliary energy source, backup power source
辅助燃烧器	auxiliary burner
辅助热源	auxiliary heat source
辅助生产区	auxiliary item
辅助受热面	auxiliary heating surface
辅助系统	ancillary system, accessory system
辅助装置	auxiliary device
腐蚀	corrosion
腐蚀防护层	corrosion protective layer
腐蚀机理	corrosion mechanism
腐蚀控制	corrosion control
腐蚀性	corrosiveness
腐蚀性测定计探测技术	corrosometer probe technique

腐蚀性卤水　corrosive brine
腐蚀性岩层　abrasive formation
腐殖土　humus soil
负电荷　negative charge
负电极板　negative electrode plate
负电流　negative current
负电压　negative voltage
负感现象　solarization
负荷变化　load variation
负荷波动　load variation
负荷的增长　load growth
负荷角　load angle
负荷历时　load duration
负荷历时曲线　load-duration curve
负荷曲线　load profile
负荷调节　load control
负荷暂态　load transient
负极　negative electrode
负极性终端　negative polarity terminal
负离子　anion
负能量　negative energy
负扭矩　negative torque
负排放　negative emission
负压　negative pressure
负载　load
负载饱和曲线　load-saturation curve
负载比　duty ratio, duty factor
负载比　duty ratio
负载变化　load variation
负载成长　load growth
负载递增　load growth
负载电流　load current
负载电压　load voltage
负载分配　load sharing, load distribution
负载分配策略　load sharing strategy
负载功率　load power
负载共享　load sharing, load distribution
负载换相变频器　load-communicated inverter
负载简档　load profile
负载角　load angle
负载配置文件　load profile
负载特性　load characteristic
负载系数　load factor
负载因数　load factor
附壁效应　coanda effect
附加电池　end cell
附加荷载　extraneous loading
附加冷却器　aftercooler
附加能量　parasitic energy
附加热　added heat
附加热量　parasitic energy
附加热损失　parasitic heat loss
附加热泄露　parasitic heat leak
附加受热面　auxiliary heating surface
附加质量系数　added mass coefficient
附加压降　parastic pressure drop
附加阳光间式　attached sunspace
附加阳光间式被动太阳房　attached sunspace passive solar house
附加载荷　parasitic load
附加质量　added mass
附着点　point of attachment
附着流动区　attached flow regime
复潮　double tide
复幅值　complex amplitude
复共轭控制　complex conjugate control
复合材料　matrix material
复合磁场　resultant field
复合电流　recombination current
复合抛物面集热器　compound parabolic concentrator collector
复合抛物面聚光器　composite parabolic condenser

复合抛物面型集光器 compound parabolic concentrator
复合曲率集热器 compound-curvature collector
复合式供暖系统 complexed heating system
复合填料 jointing compound
复合透镜 composite lens
复合循环厂 combined cycle plant
复合循环锅炉 combined circulation boiler
复合循环燃气轮机 combined cycle combustion turbine
复励 compounded
复励发电机 compound generator
复热器 recuperator
复式旋风器 multi-cyclone
复数 complex number
复数阻抗 complex impedance
复现 reproduction
复相燃烧 heterogeneous combustion
复员 demobilization
复原系统 recuperated system
复杂地形 complex terrain
复杂地形带 complex terrain
副励磁机 pilot exciter
副热带无风带 horse latitude
副翼 aileron
副翼操纵 aileron control
傅里叶变换 Fourier transform method
傅里叶分析 Fourier analysis
傅里叶级数 Fourier series
傅里叶系数 Fourier coefficient
傅里叶转换方法 Fourier transform method
傅立叶红外分光光谱仪 Fourier transform infrared spectroscopy
傅立叶序列 Fourier series
富含甲烷气 methane rich gas, methane-rich gas
富含植物纤维质残渣 cellulose rich residue
富氧 excess oxygen, oxygen enrichment
覆盖 coverage
覆盖层结构 capping structure
覆盖层深度 cased depth

ga

伽伐尼电池（组） galvanic battery
伽马函数 gamma function

gai

改变钻井位置 changing drilling location
改进燃料 reformed fuel
改进型发电设施 second generation device
改良西门子法 modified siemens process
改性发酵剂 modified yeast
钙华 travertine
钙化合物 calcium compound
盖板温度 envelope temperature
盖格计数器 Geiger counter
（丹麦）盖瑟风力机 Gedser wind turbine
盖瑟斯地热电站（美国） Geysers geothermal power plant (USA)
盖岩 cap rock
概率分布函数 probability distribution function
概率密度函数 probability density function
概率模型 probabilistic model
概念模型 conceptual model

gan

干饱和蒸汽 dry saturated steam

干沉降　dry deposition
干电池　dry battery, dry cell
干发酵　dry fermentation
干法分解　dry digestion
干法刻蚀　dry etching
干固体垃圾　dry solid waste
干涸盐湖盐差能动力装置　power plant of energy from salinity gradients in the wizened salt lake
干基　dry basis
干净能源　clean power
干绝热递减率　dry adiabatic lapse rate
干馏　destructive distillation
干膨胀式蒸发器　dry expansion evaporator
干片式光电池　dry disk photoelectric cell
干球节能控制　dry-bulb economizer control
干球温度　dry bulb temperature
干燃料　dry fuel
干扰电压　disturbing voltage
干扰试验　interference test
干热岩　hot dry rock
干热岩地热系统　dry-hot-rock geothermal system
干热岩发电　hot dry rock geothermal power
干涉　intervene
干涉滤光片　interference filter
干式变压器　dry type transformer
干式除尘器　dry scrubber
干式开挖技术　dry excavation technique
干式冷却塔　dry cooling tower
干式洗涤器　dry scrubber
干物质　dry matter
干物质含量　dry matter content
干吸收剂喷射系统　dry sorbent injection system
干性　drying
干厌氧堆肥过程　dry anaerobic composting process
干预　intervene
干燥　torrefaction, drying
干燥床　drying bed
干燥段　drying section
干燥固体废物　dry solid waste
干燥剂　desiccant, drying medium
干燥器　dehumidifier
干燥设备　drying machinery
干燥室　drying kiln
干燥塔　dry tower
干燥无灰基　dry ash-free basis
干燥窑　drying kiln
干蒸汽　dry steam
干蒸汽电厂　dry steam power plant
干蒸汽动力厂　dry steam plant
干蒸汽发电站　dry steam power plant
干蒸汽能源系统　dry steam energy system
干蒸汽田　dry steam field
干蒸汽型（地热电厂）　dry steam type
干蒸汽资源　dry-steam source
甘醇　ethylene glycol
甘油　glycerin
甘蔗乙醇　ethanol-cane
甘蔗渣　bagasse, sugar cane bagasse
坩埚　crucible
柑桔废水　citrus waste water
杆木　stemwood
杆木片　stemwood chip
杆塔　pole tower
杆状浮标　spar buoy
感潮河段　tidal reach
感潮河口　tidal estuary

感潮界限　tide head
感潮水道　tidal waterway, tidal channel
感潮水域　tidal waters
感光计　actinometer
感抗　inductive reactance
感生电流　induced current
感性（无功）分量　inductive component
感性无功功率　lagging reactive power
感应的　inductive
感应电动机　induction motor
感应电流　electromagnetic induction, induced current
感应电压　induced voltage
感应发电机　induction generator
感应分类系统　induction sorting system
感应结光电池　induced junction photovoltaic cell
感应结光伏电池　induced junction photovoltaic cell
感应力　induced force
感应器　inductor
感应速度场　induced velocity field
感应系数　inductance
橄榄残渣　olive residue
橄榄石　olivine

gang

刚度　stiffness
刚接式旋翼　rigid rotor
刚体　rigid body
刚性齿轮　rigidity gear
刚性联轴节　solid coupling
刚性联轴器　rigid coupling
刚性梁　rigid beam
刚性支承　rigid support
刚性转子　stiff rotor
钢管　steel tube
钢棍　steel rod
钢化玻璃　toughened glass
钢结构塔筒　tubular steel tower
钢筋混凝土　reinforced concrete
钢锯　hacksaw
钢丝锯　wire saw
钢索　wire rope
钢套　steel bushing
钢条　steel rod
钢温　billet temperature
钢制塔筒　tubular steel tower
缸体　cylinder, cylinder block
缸头　cylinder head
缸头垫片　cylinder-head gasket
港湾沉积　estuary deposit
杠杆定律　lever rule

gao

高比速透平　high specific speed turbine
高侧窗　clerestory window
高掺杂浓度　high dopant concentration
高产断裂系统　high yield fracture system
高潮　high tide, high water
高潮标　high water mark
高潮位　high water level, high water stand, high-tide level
高纯度硅（光伏发电）　solar silicon
高低潮位　higher low water
高地势　high elevation
高电压光电转换器　high-voltage photovoltaic converter
高度一方位定日镜　altitude-azimuth heliostat
高度一高度定日镜　altitude-altitude heliostat
高度角　altitude angle, elevation angle
高度真空　high vacuum

高发热值　high heating value (HHV)
高发射率涂料　high emissivity coating
高反射表面　highly reflective surface
高反射金属薄片　highly reflective foil
高沸点　high boiling point
高分子聚合物　polymer
高分子绝缘材料　polymer insulating material
高风带　high wind zone
高风速　high wind speed
高峰能耗　peak power consumption
高峰容量　peak capacity
高伏特光电转换器　high-voltage photovoltaic converter
高辐射率涂料　high emissivity coating
高负荷（功率）　high duty
高高潮位　higher high water
高固分解系统　high solid digestion system
高固体含量　high-solids content
高光谱测算　hyperspectral survey
高光谱数据　hyperspectral data
高光谱仪器　hyperspectral instrument
高含量有机废物　high-strength organic waste
高挥发性煤　high-volatility coal (hv-coal)
高级能　high grade energy
高级汽轮机系统　advanced turbine system
高级热交换器　sophisticated heat exchanger
高级往复式内燃机　advanced reciprocating internal combustion engine
高集聚集热器　high concentration collector
高集热率　high collection efficiency

高架电线　overhead wire
高架起重机　overhead crane
高架输电线路　overhead transmission line
高精度聚焦型集热器　focusing collector of high precision
高空　high altitude
高空风　upper air wind, wind aloft
高空生物学　aerobiology
高库　high basin
高库水面　high pool level
高宽比　aspect ratio
高亮度光束　high brightness beam
高亮度聚光度　high brightness concentration factor
高亮度聚光因子　high brightness concentration factor
高岭石　kaolinite
高炉　blast furnace
高炉燃烧器　furnace burner
高密度聚乙烯管　high-density polyethylene pipe
高密度聚乙烯管道　high-density polyethylene piping
高密度聚乙烯换热器　high-density polyethylene heat exchanger
高密度塑料管道　high density plastic pipe
高能辐射　high energy radiation
高能激励源　high energy excitation
高能级太阳辐射　high-level of solar radiation
高频电压　high frequency voltage
高频发电机　high frequency generator
高频加热　high frequency heating
高频谐波　high-frequency harmonics
高频信号注入　frequency signal injection

高品位能源　high-grade energy
高气压　anticyclone
高强度　high strength, hightensile
高强度辐射源　high level radiation source
高清晰度　high definition
高日照率　high insolation
高日照区　high insolation area
高渗透压　high osmotic pressure
高升力　high lift
高实度风轮　high-solidity rotor
高斯分布　Gauss distribution
高速车床　high-speed lath
高速齿轮比　high-gear ratio
高速海水侵蚀　high-velocity seawater erosion
高速活性污泥废水处理　high-rate activated sludge waste treatment
高速连续旋转运动　high speed-continuous rotary motion
高速水流掺气　air entrainment of high velocity flow
高速湍流　high turbulence
高速污泥消化　high-rate sludge digestion
高速消化　high-rate digestion
高速旋转叶轮轴扭应力　high cyclical rotor shaft torque stress
高速轴　high-speed shaft
高塔　tall tower
高通量低温太阳能集热器　high-flux low-temperature solar collector
高透射率玻璃罩　high transmissivity coverglass
高位法　high-order method
高位热值　higher heating value, gross calorific value
高温　high temperature

高温储热器　higher temperature storage
高温地热田　high temperature geothermal field
高温发酵池　thermophilic digester
高温反应堆　high temperature reactor
高温集热器　high temperature collector
高温井　high-temperature well
高温绝热分离器　high temperature insulating separator
高温裂化　pyrolytic cracking
高温密封剂　high temperature sealant
高温强度　high temperature strength
高温驱动吸收式冷冻机　heat-driven absorption chiller
高温缺氧气化　pyrolytic gasification
高温上升液流面积　high-temperature upflow area
高温太阳能转换系统　high-temperature solar conversion system
高温特性　high temperature property
高温梯度　high-temperature gradient
高温涂层　high temperature coating
高温稳定性　high temperature stability
高温性能　high temperature property
高温蓄热　high temperature storage
高温蓄热箱　high temperature storage tank
高温岩　hot dry rock
高温煮解　high-temperature digestion
高吸收剂表面　highly absorbent surface
高效光电池　efficient photovoltaic cell
高效光电微型组件　high-efficiency photovoltaic module
高效率　high efficiency, high performance
高效率太阳能电池　high efficiency solar cell

高性能　high performance
高性能的　high-performance
高压变配电系统　HV transformer system
高压冲洗泵　high-pressure flushing pump
高压地热能　geopressured geothermal activity
高压电源　high voltage power supply
高压感应电机　high voltage induction motor
高压给水　high pressure feed water
高压含水层　geopressured aquifer
高压级　high pressure stage
高压交流电　high voltage alternating current
高压井　geopressured well
高压钠灯　high pressure sodium lamp
高压盆地　geopressured basin
高压气体　geopressured gas
高压线圈　high voltage coil
高压型沉积　geopressured deposit
高压型储层　geopressured reservoir
高压蒸汽　high pressure steam
高压直流电　high voltage direct current (HVDC)
高压直流输电　high voltage direct current transmission (HVDC)
高黏度（性）　high viscosity
高蒸汽参数　elevated steam condition

ge

搁栅　joist
割缝衬管　slotted liner
割谷机　swather
格构式杆架　lattice mast
格栅　lattice
格栅风口　air grill (e)
格子形杆架　lattice tower
隔板　diaphragm
隔断阀　isolating valve
隔离二极管　blocking diode
隔离阀　isolation valve
隔离开关　disconnect switch
隔离膜　separator
隔膜泵　diaphragm pump
隔膜电池　diaphragm cell
隔膜密封　diaphragm seal
隔热　thermal insulation
隔热百叶窗　insulating shutter
隔热材料　heat shield material
隔热的　adiabatic
隔热浇注料　insulating castable
隔热体　insulator, insulation
隔热性　ablative insulating quality
隔热砖　insulating brick
隔水层　confining layer
隔音　deadening
隔音作用　deadening
葛劳渥经验关系　Glauert empirical relation
个人废弃物（垃圾）燃料　individual waste fuel
各向同性体　isotropic body
各向异性等能面　anisotropic energy surface
各向异性　anisotropy
各向异性的　aeolotropic
各向异性样本　anisotropic sample
各向异性应力　anisotropic stress
铬　chrome
铬黑太阳能选择性膜　black chrome selective coating
铬矿质耐火材料　chromite refractory

gei

给料　feedstock
给水泵　feed pump
给水补给　feed water supply
给水处理　feedwater treatment, feed water treatment
给水管路　feedwater line
给水井　feed well
给水流量　feedwater flow
给水软化　feedwater softening
给水调节器　feedwater regulator
给水系统　feedwater cycle

gen

跟踪　tracking, padding
跟踪反射镜　tracking mirror
跟踪反射器　tracking reflector
跟踪机构（实现对太阳跟踪的装置）　tracking system
跟踪集热器　tracking collector
跟踪精度　tracking precision
跟踪聚光器　solar tracking concentrator
跟踪式定日镜　tracking heliostat
跟踪式接收器　tracking receiver
跟踪式聚光器　tracking concentrator
跟踪式聚焦集热器　movable focusing collector
跟踪式太阳能集热器　tracking solar collector
跟踪式太阳能热水器　solar tracking water heater
跟踪式太阳能装置　solar tracking device
跟踪式折射器　tracking refractor
跟踪太阳光发电　sun-tracking photovoltaic power generation
跟踪误差　tracking error
跟踪吸收器式固定球面反射太阳能集热器　stationary spherical reflector tracking absorber solar collector
跟踪吸收器式固定球面反射镜集热器　stationary reflector tracking-absorber spherical mirror collector
跟踪系统　follow-up system
跟踪准确度　tracking accuracy
跟踪准确度控制器　tracking accuracy controller

geng

更换砂浆　slurry exchange
耿贝尔分布　Gumbel distribution

gong

工厂规模　plant size
工程财务评价　project financial assessment
工程分析　engineering analysis
工程管理信息系统　project management information system
工程国民经济评价　project national economic assessment
工程海洋学　engineering oceanography
工程计划管理　project planning management
工程技术管理　project technical management
工程决算　project final account
工程竣工验收　project final acceptance
工程判断　engineering judgement
工程师用便携式温度计　engineer's pocket thermometer
工程投资管理　project investment management

工程延期　project delay
工程移交生产验收　project handover for operation
工程预算　project cost budget
工程质量管理　project quality management
工程资料　engineering information
工程总投资　total project cost
工地工作　field work
工地监测　site monitoring
工地制作的　field fabricated
工具集　toolset
工具箱　kit
工况　operating condition
工频磁场　power frequency magnetic field
工频电场　power frequency electricalfield
工业分析　proximate analysis
工业硅　industrial silicon
工业计时器　industrial timer
工业计算机　industrial computer
工业天然蒸汽田　commercial natural steam field
工业用电力　utility electricity
工业有机废物　industrial organic waste
工业沼气　industrial biogas
工业装置　commercial plant
工艺废料　process steam
工艺流程图　process flowsheet
工艺设备　producer gas
工艺用热　process heat
工艺蒸汽　process steam
工艺装置　producer unit
工质出口温度　fluid outlet temperature
工质储热　heat storage of material
工质回路　working-fluid loop
工质进口温度　fluid inlet temperature
工质平均温度　mean fluid temperature
工质蒸汽　working-fluid vapor
工作标准太阳电池　working standard solar cell
工作齿面　working flank
工作点　operating point
工作电流　operating current
工作电压　operating voltage
工作风速范围　range of effective wind speed
工作辐射表　field radiometer
工作负载　service load
工作接地　working earthing
工作流体　working fluid
工作面电压　face voltage
工作频率　working frequency
工作气体　working gas
工作情况　service condition, work condition
工作条件　run condition
工作温度　service temperature, running temperature, operating temperature
工作误差　operating bias
工作液体　driving fluid
工作原理　working principle, principle of operation
工作状态　in-service condition, operating state
弓形片弹簧圆规　bow-spring compass
公称横截面　nominal cross-section
公吨　metric ton
公共连接点　point of common coupling
公共耦合点　point of common coupling
公共事业委员会　Public Utility Commission
公海　open ocean
公式　expression
公用电缆　utility line
公用电力　utility electricity

公用电网　utility grid
公用工程费用　utility cost
公用事业电站　utility power plant
公用天然气　utility natural gas
功角　power angle
功率　power
功率半导体　power semiconductor
功率变换器　power transformer
功率变流器　power converter
（风电）功率波动　power fluctuation
功率采集系统　power collection system
功率电流　power current
功率方程　power equation
功率放大器　power amplifier
功率—风速曲线　power versus wind speed curve
功率峰值　power spike
功率计　dynamometer
功率角　power angle
功率晶体管　power transistor
功率矩阵　power matrix
功率控制　power control
功率控制算法　power control algorithm
功率流　power flow
功率流计　power flow meter
功率密度　power density
功率密度分布　power density distribution
功率密度频谱　power density spectrum
功率密度图　power density diagram
功率疲劳　power fatigue
功率平衡　power balance
功率谱　power spectrum
功率谱密度　power spectral density
功率谱密度函数　power spectral density function
功率曲线　power curve
功率输出　power delivery

功率输出电缆　power delivery cable
功率调节　power conditioning
功率调节和存储单元　power regulation and storage unit
功率调节器　power conditioner, power conditioning unit
功率通量　power flux
功率系数　power performance (coefficient)
功率因素　power factor
功率因素矫正电容器　power factor correction capacitor
功率因素控制电容器　PF control capacitor
功率圆图　power circle diagram
功率转换效率　power conversion efficiency
功能部件　functional component
功能器件　functional device
功能调节　setup control
功能元件　functional component
功能组级控制　function group control
拱度控制　camber control
共轭线　conjugate line
共格界面　coherent boundary
共格孪晶界　coherent twin boundary
共价键　covalent bond, electron-pair bond
共晶反应　eutectic reaction
共晶温度　eutectic temperature
共晶组织　eutectic structure
共模电压　common-mode-voltage
共栖　syntrophy
共析体　eutectoid
共用接地系统　common earthing system
共用样品　common sample
共振电压　resonance voltage
共振滤波器　resonance filter

共振能　delocalization energy
共振散射表面　resonance scattering surface
共振系统　resonant system
共振响应　resonant response
共振叶尖响应　resonant tip response
共轴的　coaxial
共轴反转式螺旋桨　co-axial contra rotating propeller
贡贝尔分布　Gumbel distribution
供电阀　electric supply valve
供电可靠性　power supply reliability, service reliability
供电质量　quality of supply
供风管道　air supply line pipe
供给链　supply chain
供给量调节器　delivery regulator
供料系统　feed system
供暖用太阳能　solar energy for heating
供气管道　air supply line pipe
供气喷嘴　air-feed nozzle
供气装置　air charging system
供热　heat input
供热（反应）堆　heat reactor
供热水　heating water
供水槽　feed water tank
供水井　supply well
供水流量　water delivery rate
供水温度　water delivery temperature
供需链　supply chain
供应　furnish
供应集管　supply header
供应链　supply chain
供应线　supply line

gou

沟渠基底　channel base

钩（连）接　hook's joint
钩结合　hook's joint
构造　formation
构造板块　tectonic plate
构造背景　tectonic setting
构造破碎带　fracture zone
构造相图　tectofacies map
构造应力　tectonic stress
购电协议　power purchase agreement
购买力评价　purchasing power parity
垢的成因　cause of scale formation

gu

孤岛系统　island system
孤岛运行　island operation
孤立运行　isolated operation
孤立阵风统计方法　statistical discrete gust method
古地磁　paleomagnetics
谷底　valley floor
谷类植物能　energy from cereal grains
谷物秸秆　cereal straw
谷物燃烧　cereal combustion
谷酰胺　cobamide, glutamine
骨架　beam trammel
鼓风　airblast, windage
鼓风机　forced draft fan
鼓风口　tuyere
鼓风炉　up-draught stove, blast furnace
鼓泡流化床　bubbling fluidized bed
鼓泡器，起泡器　bubbler
鼓式干燥机　drum dryer
鼓式加料机　drum feeder
鼓式削片机　drum chipper
鼓形转子　cylindrical rotor
鼓状刹车　drum brake
毂孔　hub bore

固氮细菌活动	activity of nitrogen fixing bacteria
固定	fixing
固定表面水	stationary surface water
固定传感器	stationary sensor
固定床气化	gasification of fixed-bed
固定床气化炉	fixed bed gasifier
固定床燃烧	fixed bed combustion
固定床系统	fixed-bed system
固定电池组	stationary battery
固定电枢	stationary armature
固定反射镜动吸收器系统	fixed-mirror moving absorber system
固定反射镜跟踪吸收器类槽式集热器	fixed-mirror tracking receiver through like collector
固定反射镜跟踪吸收器式集热器	fixed-mirror tracking receiver collector
固定反射镜配焦	fixed mirror/ distributed focus
固定反射镜配焦太阳能—电能转换	fixed mirror/distributed focus solar-to-electrical conversion
固定反射镜线聚焦系统	fixed mirror line focus system
固定反射器跟踪吸收器式太阳能集热器	stationary reflector tracking absorber collector
固定集热管	stationary receiver tube
固定集热器	stationary collector
固定桨毂	fixed hub
固定接受管	stationary receiver tube
固定聚光集热器	fixed concentrator
固定聚光器	fixed concentrator
固定立式集热器	stationary vertical collector
固定连接	dead joint
固定炉排	fixed grate
固定抛物线形反射镜	fixed parabolic mirror
固定倾斜阵列	fixed tilt array
固定式半球面聚光（太阳能）集热器	stationary compound parobolic concentrator
固定式波能吸收装置	fixed system for absorbing wave energy
固定式槽型聚光器	fixed trough type collector
固定式动力装置	land-based power plant
固定式复合抛物面聚光器	fixed compound parabolic concentrator
固定式集热器	fixed collector
固定式桨叶	fixed blade
固定式结构物	fixed structure
固定式聚光器	stationary concentrator
固定式倾斜式太阳能集热器	fixed tilt solar collector
固定式球面聚光型集热器	fixed spherical concentrator
固定式太阳能收集器	fixed solar collector
固定式圆形槽聚光器	fixed circular trough concentrator
固定碳	fixed carbon
固定污染源	stationary pollution source
固定吸尘器	station absorber
固定吸热器式旋转反射板集热器	fixed receiver rotating-slat reflector collector
固定弦	constant chord
固定斜面	fixed sloped surface
固定斜置集热器	fixed tilt collector
固定性悬浮固体	fixed suspended solid
固—固扩散	solid-solid diffusion
固化	solidification

固化成形　solid forming
固化剂　hardener
固化时间　cure time
固井水泥浆　cement slurry
固一气相变储热　solid-gas phase change heat storage
固溶度（性）　solid solubility
固态反应　solid state reaction
固态扩散　solid diffusion
固态粒子　solid-state particle
固态逆变器　solid-state inverter
固态排渣鲁奇气化炉　dry-ash Lurgi gasifier, dry-ash-removal Lurgi gasifier
固态燃料电池　solid fuel cell
固态岩浆　magma solid
固态岩石　solid rock
固态氧化物燃料电池　solid-oxide fuel cell
固态整流器　solid state rectifier
固体材料蓄热单元　solid material storage unit
固体潮　earth tide
固体材料蓄热单元　solid material storage unit
固体沉降槽　settling tank
固体电路　solid state circuit
固体废物　solid waste
固体化石燃料　solid fossil fuel
固体回收燃料　solid recovered fuel
固体燃料　solid fuel, dry fuel
固体燃料动力学　solid fuel kinetics
固体热力学　thermodynamics of solid
固体生物质燃料　solid biofuel, solid biomass fuel
固体体积　solid volume
固体停留时间　solids retention time
固体氧化物燃料电池　solid oxide fuel cell
固体杂质　solid impurity
固体再生燃料　solid recovered fuel
固相扩散　solid phase diffusion
固相蓄热　solid phase heat storage, solid phase thermal storage
固相蓄热介质　solid phase storage medium
固相蓄热物质　solid storage material
固相储能　solid phase thermal storage, solid phase heat storage
固相储能介质　solid storage medium
固相储能物质　solid storage material
固性比　solidity ratio
固液平衡　solid-liquid equilibrium
固液相变蓄热　solid-liquid phase change storage
固液储能系统　solid-liquid storage system
固有灰分　inherent ash, natural ash
固有频率　eigen frequency, natural frequency
固有自旋　intrinsic spin
固有阻尼　inherent damping
故障　fault, hazard
故障持续时间　duration of the fault
故障代码　error code
故障电流　fault current
故障接地　fault earthing
故障前电压　V pre-fault
故障停机时间　downtime
故障预测　failure predication
故障诊断　fault diagnosis
故障状态　fault condition

guai

拐点　knee point

拐角流　corner flow

guan

关闭阀　shut off valve
关闭时间　switched-off time
关合触头　make contact, on-off contact
关机　shutdown
关联—测量—预测方法　correlate-measure-predict method
关态漏电流　off leakage current
观测船　observation boat
观测风速　observed wind speed
观测孔　observation door, peephole
观测台　observation desk
观测用浮标系统　observation buoy system
观察时数据　recent data
管板　tube plate, tube sheet
管板液式集热器　liquid-type collector with tube and plate
管壁厚度　tube wall thickness
管道　duct
管道安装工　pipe fitter
管道泵　pipeline pump
管道补压泵　jockey pump
管道布置　piping layout
管道风机　channel style fan
管道工程　pipework
管道工程系统　pipework system
管道摄像检查系统　pipe camera inspection system
管道式水轮机　tube turbine, tube-type turbine
管道损失　duct loss
管道阻力　duct loss
管夹　hose clip
管径　pipe size
管距　tube spacing
管距布置　tube spacing
管壳式热水器　tube-in-sheet water heater
管壳式蒸发器　tube-in-sheet evaporator
管内清垢　tube cleaner
管排配置　tube spacing
管钳子　box spanner
管式集热器　tube-type collector
管式接受器　tubular receiver
管式炉　tube furnace
管式球磨机　tube ball mill
管式热交换器　tube-type heat exchange
管式太阳能集热器　tubular solar energy collector
管式吸热器　pipe receiver
管式蒸发器　tubular evaporator
管束干燥机　tube bundle dryer
管束烘干机　tube bundle dryer
管塔　tube tower
管网　pipeline network
管下扩眼器　underreamer
管形直线感应电动机　tubular linear induction motor
管眼扩大器　underreamer
管状风速表　tube anemometer
管状集热器　tubular collector
管状吸收器　tubular absorber
管子管板之间的密封　tube plate packing
管子鳍片节距　tube fin pitch
贯入度仪　penetrometer
贯轴振荡器　transverse oscillator
冠岩　cap rock
惯性　inertia
惯性除尘器　inertial dust separator
惯性矩　moment of inertia
惯性开关　inertia switch
惯性离心力　inertial centrifugal force

惯性力 inertia force
惯性轮 flywheel
惯性中心 center of inertia
灌浆材料 grouting material
灌浆法 grouting procedure
灌浆工序 grouting procedure

guang

光 light
光程 optical distance
光磁耦合 photomagnetic coupling
光弹性应力可视化 photoelastic stress visualization
光导式传感器 photoconductive sensor, photoconductive transducer
光导体 photoconductor, lighting conductor
光的增透 light trapping
光电 photoelectricity, photovoltaic power
光电板 photovoltaic panel
光电倍增管 photomultipliler tube
光电倍增器电池 photomultiplier cell
光电变换器件 photoactor
光电材料 photoelectric material
光电常数 photoelectric constant
光电池 photo cell, photovoltaic cell, photronic cell
光电池阵 photovoltaic array, photovoltaic collector
光电池组件 photovoltaic collector
光电磁效应 photomagnetoelectric effect
光电导 photoconduction
光电导的光谱响应 spectral response of photoconductivity
光电导电流 photoconductive current
光电导率 photoconductivity
光电导器件 photoconductive device
光电导响应 photoconductive response
光电导效应 photo-conductive effect
光电导性 photoconductivity
光电的 photoelectric, photovoltaic
光电电池 photovoltaic cell
光电电动势 photoelectromotive force
光电电站 photovoltaic power plant
光电动力泵 photovoltaic powered pump
光电二极管暗电流 photodiode dark current
光电二极管 photodiode
光电发电 photovoltaic electric power generation, photovoltaic power generation
光电发电装置 photovoltaic power generating unit
光电发射 photoelectric emission
光电发射管 photo emissive cell
光电发生器 photovoltaic generator
光电阀值能量 photoelectric threshold energy
光电分光光度计 photoelectric spectrophotometer
光电伏打 photovoltaic
光电辐射探测器 photovoltaic radiation detector
光电功率转换 photovoltaic power conversion
光电管 photo cell, phototube
光电光度计 photoelectric photometer
光电过程 photogalvanic process, photovoltaic process, helioelectrical process
光电合成（作用） photoelectrosynthesis
光电合成反应 photoelectrosynthetic reaction
光电化学 photoelectrochemistry
光电化学电池 photoelectrochemical cell

光电化学能转换　photoelectrochemical energy conversion
光电化学效应　photoelectrochemical effect
光电换能器　photoelectric transducer, photoelectric power converter
光电极　photoelectrode
光电结电池　photojunction battery
光电解（作用）　photoelectrolysis
光电聚光器系统　photovoltaic concentrator system
光电开关　photoelectric switch
光电离　photoionization
光电灵敏度　photoelectric sensitivity
光电流　photoelectric current
光电能　photovoltaic energy
光电能系统　photovoltaic energy system
光电耦合器　optoelectronic coulper
光电器件　photoelectric device
光电热能（综合利用）系统　photovoltaic thermal system
光电三极管　phototransistor
光电太阳能转换装置　photovoltaic solar energy conversion device
光电温度计　photoelectric thermometer
光电吸收　photoelectric absorption
光电吸收系数　photoelectric absorption coefficient
光电系统技术　photovoltaic system technology
光电现象　photoelectric phenomenon, photoelectricity
光电效应　photoeffect, photoelectric effect
光电效应常数　photoelectric constant
光电效应的　photoelectric
光电性质　optoelectronic property

光电学　photoelectricity
光电阴极材料　photoelectric cathode material
光电元件　photocell
光电照明控制　photoelectric lighting controlling
光电阵列支持结构　PV array support structure
光电转换　photo-electricity conversion, photovoltaic conversion
光电转换器　Photovoltaic Converter
光电转换系统　photovoltaic conversion system
光电转换效率　photoconversion efficiency, photoelectric conversion efficiency
光电转换装置　ovonic device
光电装置　photovoltaic device
光电装置技术　photovoltaic device technology
光电子　photoelectron
光电子产额　photoelectric yield
光电子化学光电池板　photo electronics chemical solar cell
光电子学　optoelectronics
光电组合机　photocomposing machine
光电作用　photovoltaic action
光帆　solar sail
光反应性叶绿素　photoreactive chlorophyll
光伏板　photovoltaic panel
光伏并网发电系统　grid-connected photovoltaic system
光伏充电器　photovoltaic charger
光伏打装置技术　photovoltaic device technology
光伏电池板　photovoltaic panel
光伏电池组件　photovoltaic array
光伏发电系统　photovoltaic system

光伏发电最大功率峰值跟踪系统　peak power tracking photovoltaic power system
光伏方阵　photovoltaic array
光伏光热混合 CPC 组件　PV-thermal hybrid CPC module
光伏光热系统　photovoltaic-thermal system
光伏焊带　pv ribbon
光伏建筑材料　photovoltaic building material
光伏建筑一体化　building integrated photovoltaics
光伏模块　solar module
光伏幕墙　photovoltaic panel wall
光伏器件　photovoltaic device
光伏实验（示范）电厂　photovoltaic demonstration plant
光伏式传感器　photovoltaic sensor, photovoltaic transducer
光伏水泵　photovoltaic water pump
光伏太阳能转换　photovoltaic solar energy conversion
光伏太阳能转换装置　photovoltaic solar energy conversion device
光伏探测器　photovoltaic detector
光伏微功率系统　photovoltaic micropower system
光伏系统　photovoltaic system
光伏系统最大功率点跟踪　maximum power point tracker
光伏效应　photovoltaic effect
光伏阴极保护　photovoltaic cathodic protection
光伏阵列　photovoltaic array
光伏转换　photovoltaic conversion
光伏组件　photovoltaic module
光伏组列　pv string
光伏作用　photovoltaic action
光功率计　optical power meter
光合作用　photosynthesis
光滑面　glaze
光滑涂料　gloss paint
光化（学）反应　photochemical reaction
光化玻璃　actinic glass
光化辐射　actinic radiation
光化吸收　actinic absorption, photochemical absorption
光化学　photochemistry
光化学电池　photochemical cell
光化学电池能量转换装置　photochemical cell energy conversion unit
光化学二极管　photochemical diode
光化学发电　photochemical electric power
光化学能量转换　photochemical energy conversion
光化学气相沉积　photo chemical vapor deposition
光化学氧化潜势　photochemical oxidation potential
光化学转换　photochemical conversion
光化学转换效率　photochemical conversion efficiency
光化转化　photochemical inversion
光化作用　photo actinic action
光接触　optical contact
光金属水平仪　light metal spirit level
光控开关　photoswitch
光控系统　light control system
光阑　diaphragm
光量计　quantometer
光流密度　light current density
光密度　optical density

光密度计	optical densitometer
光敏材料	photosensitive material
光敏电阻	photoconductive resistance, photoresistance, photovaristor
光敏电阻和光电导探测器材料	photoresistances and photoconductive detector materials
光敏电阻器	photoresistor
光敏晶体管	photistor
光敏探测器	light sensitive detector
光能	photoenergy, optical energy, light energy
光能隙损失	optical energy gap loss
光能测定计	actinoscope
光谱	sun spectrum
光谱带宽	spectral bandwidth
光谱发射	spectral emittance
光谱发射率	spectral emittance
光谱分辨率	spectral resolution
光谱分析	spectral analysis, spectrum analysis
光谱辐照度	spectral irradiance
光谱辐照度分布	spectral irradiance distribution
光谱感光度	spectral sensitivity
光谱光子辐照度	spectral photon irradiance
光谱灵敏度	spectral sensitivity
光谱灵敏度特性	spectral-sensitivity characteristic
光谱能量分布曲线	spectral energy distribution curve
光谱失配	spectral mismatch
光谱碳棒	spectroscopic carbon rod
光谱线	spectral line
光谱响应	spectral response
光谱响应率	spectral responsivity
光谱学	spectroscopy
光强	intensity of light, lighting intensity
光球（层）	photosphere
光热（直接）转换	photothermal conversion
光热（直接）转换器	photothermal converter
光热（转换）装置	photothermal device
光热电效应	photothermoelectric effect
光热能量转换	photo thermal energy conversion
光散射	light scatter
光渗燃料	irradiated fuel
光生电流	light-generated current, photo-generated current
光生电流（光电流）	photocurrent
光生电压	photovoltage
光速	velocity of light
光探测和测距	light detection and ranging
光通量	luminous flux
光吸收	photoabsorption, light absorption
光吸收系数	photoabsorption coefficient
光纤	fiber optic
光纤束	cable bundle
光线追踪	ray-tracing
光效率	optical efficiency
光学	optics
光学（量）传感器	optical (quantity) sensor, optical (quantity) transducer
光学测量	optical measurement
光学常数	optical constant
光学传感器	optical sensor
光学的	optical
光学电子	optical electron
光学镀层玻璃	optically coated glass
光学法测量	optical measurement

光学风速传感器	optical wind speed sensor
光学高温计	optical pyrometer
光学恒量	optical constant
光学厚度	optical thickness
光学耦合	optical coupling
光学容量	optical capacity
光学扫描技术	optical scanning technique
光学涂料	optical coating
光学吸收	optical absorption
光学吸收常数	optical absorption constant
光学吸收谱	optical absorption spectrum
光学吸收限	optical absorption edge
光学系统	optics
光学效率	optical efficiency
光学增温器	optical booster
光源	light source
光源种类	LED type
光增强屏	intensifying screen
光栅	optical grid
光栅型（光电电池）	grating type
光照分布	irradiation distribution
光致扩散电压	photodiffusion voltage
光致效应	Staebler-Wronski effect
光轴	optic axis
光转化效率	photoconversion efficiency
光子	photon
光子效应	photo effect
光子增强型热离子辐射	photon enhanced thermionic emission

gui

规定电流	rated electrical output
规格	specification
规模沉积	Scale Deposition
规模效应	scale effect
规则波	regular wave
规准尺	gauge board
硅带太阳电池	silicon ribbon solar cell
硅锭	silicon ingot
硅光电池	silicon photocell
硅光电池等能量响应	silicon photocell equi-energy response
硅光电池直接日射计	photovoltaic silicon pyrheliometer
硅光电装置	silicon photovoltaic device
硅光伏电池	silicon photovoltaic cell
硅光伏装置	silicon photovoltaic device
硅化钙	calcium silicide
硅基薄膜太阳电池	thin-film silicon solar cell
硅胶黏合	silicon rubber adhesive
硅晶体	silicon crystal
硅控整流器	thyristor
硅块	silicon block
硅铝钙合金	calcium-aluminium-silicon
硅铝裂化催化剂	silica-alumina cracking catalyst
硅片	silicon wafer
硅片剥离清洗	strip-cleaning of silicon wafer
硅石	silica brick/dinas
硅石耐火砖	dinas firebrick
硅酸钙	calcium silicate
硅酸铝耐火材料	alumina silicate refractory
硅酸盐化合物	silicate compound
硅太阳能电池	silicon solar cell
硅太阳能电池参数	silicon-solar-cell parameter
硅太阳能转换器	silicon solar energy converter
硅铁	ferrosilicon
硅烷法	silane method

硅烷热分解　decomposition of silane
硅油　silicon oil
硅有机树脂　silicone
硅藻土　diatomite/kiesel guhr, diatomaceous earth, diatomaceous silica
硅遮雨缘　silicone weather shed
硅整流元件　thyristor element
硅质岩沉积　siliceous sediment
硅砖　silica brick/dinas
轨道太阳能发电　on-orbit power generation
轨道太阳能发电厂　solar-powered orbiting generating station
轨道太阳能站　orbiting solar power station
轨迹　locus
轨迹长度　path length
柜门　cabinet door

gun

辊式炉条　roller grate
滚筒（柱）式输送机　roller conveyor
滚筒分离筛　trommel screen sieve
滚筒筛　trommel screen
滚压机　roller mill
滚珠导螺杆　ball screw
滚珠螺杆　ball screw
滚珠丝杆　ball screw
滚珠丝杠　ball screw (shaft)
滚珠旋致动器　ball screw adctuator
滚珠支撑　ball saddle
滚珠轴承　ball bearing
滚柱轴承　roller bearing

guo

锅炉房　conventional boiler
锅炉给水泵　boiler feed water pump
锅炉管　boiler tube
锅炉灰　boiler ash
锅炉机组　steam generator block
锅炉间　boiler plant
锅炉漏风试验　air leakage test for boiler furnace
锅炉炉身温度　stack temperature
锅炉燃料　boiler fuel
锅炉热效率　boiler thermal efficiency
锅炉容量　boiler capacity
锅炉设备　boiler plant
锅炉效率　boiler efficiency
锅炉压力　boiler pressure
锅炉蒸汽压　boiler steam pressure
国际标准单位　SI unit
国际大电网会议　International Conference on Large High voltage Electric System
国际风电产业化现状　status of wind power in the world
国际可持续发展和碳认证　International Sustainability and Carbon Certification
国际可持续性和碳认证体系　International Sustainability and Carbon Certification System
国际能源机构光伏电力系统项目　International Energy Agency's Photovoltaic Power Systems Programme
国家电气安全代码　National Electrical Safety Code
国家电网　national grid
国家风力资源　national wind resource
国家能源建模系统　National Energy Modeling System
果实生物质　fruit biomass
过饱和蒸汽　supersaturated steam
过程加热器　process heater

过程加热系统　process heat system
过程控制系统　process control system
过程柔性　process flexibility
过充电电流　recharge current
过电流　over circuit
过电流保护　over current protection
过电压　over-voltage
过电压保护　over voltage protection
过度激发　overexcitation
过渡性岩石区　transitional rock site
过负载止向电流　overload forward current
过激励　overexcitation
过冷度　degree of supercooling
过冷液体　subcooled liquid, supercooling melt
过励磁　overexcitation
过梁　lintel
过量空气控制　excess air control
过量空气系数　excess air coefficient
过零周期　zero crossing period
过滤　percolation
过滤设备　filtration equipment
过热　abnormal temperature rise
过热的　superheated
过热度　degree of superheat
过热降温器　desuperheater
过热炉　superheater
过热器　superheater
过热蛇形管　superheater coil
过热设备　superheater
过热水　superheated water
过热蒸汽　superheated steam, superheated
过热蒸汽干燥机　superheated steam dryer
过热蒸汽机车　superheated steam locomotive
过热蒸汽降温器　desuperheater
过热状态　superheated state
过生空气量　excess air quantity
过剩空气系数　excess air ratio
过压　excess voltage
过载电流　excess current
过载度　ratio of over load
过载功率　over power

hai

海岸保护　coastal protection
海岸潮　coastal tide
海岸带综合管理　coastal zone management
海岸工程　marinetime engineering
海岸沙丘　coastal sand dune
海岸线　shoreline
海岸线地带　shoreline
海拔　altitude
海拔高度　sea level elevation
海边浪　edge wave
海波计　wave recorder
海草　seaweed
海曾—威廉系数　Hazen Williams coefficient
海潮　coastal tide
海床　sea floor, seabed
海床安装系统　sea-bed mounted system
海道测量　marine hydrographic (al) survey
海德里环形防喷器　hydril annular preventer
海的　maritime
海底　seafloor, seabed, ocean floor, sea bed
海底电缆　marine cable, submarine power cable, subsea cable
海底电线　marine cable, submarine power cable, subsea cable
海底—海洋界面　ocean floor-ocean

interface
海底扩展 sea-floor spreading
海底流 submarine river
海底摩擦力 bottom friction
海底热能源 thermal subsea power source
海底山 seamount
海底太阳能电站 ocean-based solar plant
海底压力差动式（多安装在近岸，固定于海底，由波浪造成的海水起伏带动差动器，进行发电）submerged pressure differential
海底钻井 ocean floor drilling
海风 off-sea wind, sea breeze, off-shore breeze
海风锋 sea breeze front
海港盐度差能转换装置 estuarine salinity gradient energy converter
海谷底线 thalweg
海谷深泓线 thalweg
海脊 searidge
海况 sea state, sea condition
海浪 wind sea, wave condition, wind swell, ocean surface waves, ocean wave
海浪发电系统 sea wave electric power system
海浪能 wave power
海浪谱 sea spectra, wave spectrum
海浪要素 wave parameter
海浪预报 wave forecast
海浪装置 sea wave device
海里 nautical mile
海流 low-lying flat, ocean current
海流发电 ocean current generation
海流计 oceanographic (al) current meter
海流能 marine current power, ocean current energy
海流能量利用率 energy efficiency in ocean current
海流能密度 density of ocean current energy
海流速度剖面仪 marine current velocity profiler
海流涡轮机 marine current turbine
海陆风 sea land wind, sea-land breeze
海漫 pitching
海面波能 energy from ocean surface wave
海面粗糙度 sea surface roughness
海面热图 thermal map of sea surface
海面温度 sea surface temperature
海面有效回辐射 efficient back radiation of sea surface
海面杂乱回波 wave clutter
海面状况 state of sea
海明波浪发电装置 Kaimei wave power facility
海明波浪发电装置试验（日本）Kaimei experiment
海能发电厂 wave-power park
海平面气压 sea level pressure
海上保险 marine insurance
海上变电站（平台）offshore substation platform
海上的 off-shore/offshore
海上定位 marine positioning
海上发电站 ocean sited power plant
海上风力发电 offshore wind power
海上风力发电工程 offshore wind power project
海上风力发电系统 offshore wind power

system
海上风能 offshore wind energy
海上风能系统 ocean wind power system
海上浮标 marine buoy
海上管道开沟 marine pipeline trenching
海上禁捕区 marine sanctuary
海上漏油 marine oil spill
海上平台 off-shore oil rig, offshore platform
海上设备 offshore device
海上试航 sea trials, sea trial, field trial
海上应用 offshore application
海上自然保护区 marine sanctuary
海生植物丛 marine growth
海蚀 marine erosion
海蚀平原 marine plain
海蚀崖 sea cliff
海事的 maritime
海水 ocean water, saline water
海水泵 sea water pump
海水抽提联氨 hydrazine from sea water
海水淡化 sea water desalination, desalination
海水淡化反应器 desalting reactor
海水淡化装置 sea water desalting plant
海水电池 sea water battery, saltwater battery, saltwater-activated battery
海水电阻率 electrical resistivity of sea water
海水腐蚀 sea water corrosion, marine corrosion
海水混合 ocean wave mixing
海水浸入作用 saltwater intrusion
海水空调 seawater air conditioning
海水冷凝器 sea water condenser
海水浓度差发电 electric power of ocean energy from concentration gradients
海水入侵 saltwater intrusion, saline intrusion
海水渗透压能 sea water osmotic pressure energy
海水提铀 uranium extraction from seawater
海水温差电站 sea temperature gradient power station
海水温差发电 ocean thermal energy conversion
海水温差发电系统 ocean thermal energy conversion power system
海水温度表 sea-water thermometer
海水盐差发电 seawater salinity gradient power generation
海水盐差能 osmotic power
海水元素提取 extraction of elements from seawater
海水原电池 sea water primary battery
海水蒸馏 sea water by distillation
海水自净 sea water self purification
海塘 seawall
海图 sea chart, sea map, marine map
海相 marine facies
海相层 marine bed
海相沉积 marine deposit
海啸 tidal bore, tsunami
海崖 sea cliff
海洋保护 marine protection
海洋保护区 marine protected area
海洋表面 ocean surface
海洋表面波 ocean surface wave
海洋波浪 ocean wave
海洋波浪发电机 ocean wave electric

generator
海洋波浪能　ocean wave energy, ocean wave power
海洋波浪能利用　ocean wave energy utilization
海洋波浪能资源　ocean wave energy resource
海洋波浪能吸收结构物　ocean wave energy absorbing structure
海洋波浪能转换　ocean wave energy conversion
海洋波浪能转换系统　marine wave energy conversion system
海洋波透镜　ocean wave lens
海洋测深学　bathymetry
海洋潮汐　ocean tide, oceanic tide
海洋潮汐动量　ocean tidal movement
海洋潮汐能　ocean tidal power
海洋沉积物　marine sediment, ocean sediment
海洋处置（废物）　marine disposal, ocean disposal
海洋船　oceanic vessel
海洋大气　marine atmosphere
海洋的　naval
海洋地壳　oceanic crust, oceanic lithosphere
海洋地貌　marine geomorphy
海洋地质调查　marine geological survey
海洋电化学　marine electrochemistry
海洋动力学　ocean dynamics
海洋动植物　marine plant and animal
海洋度　oceanity
海洋对流层　oceanic troposphere
海洋防污法　marine pollution prevention law
海洋风况　marine wind regime
海洋风力　oceanic wind power
海洋风力资源　oceanic wind power resource
海洋负荷应变潮汐　ocean loading strain tide
海洋工程　marinetime engineering, ocean engineering, oceaneering
海洋工程地质调查　marine engineering geology investigation
海洋工程系统开发　ocean engineering system development
海洋工程系统研制　ocean engineering system development
海洋观察站　hydrographic (al) station
海洋管道　marine pipeline
海洋光合作用带/区　epipelagic zone
海洋光谱　sea spectra
海洋化学资源　marine chemical resource
海洋环境　ocean environment, marine environment
海洋环境能源　ocean environmental energy source
海洋环境质量　marine environment quality
海洋环流　oceanic gyre, ocean circulation
海洋环流模式　oceanic general circulation model
海洋环流圈　ocean gyre
海洋洄游　oceanodromous migration
海洋开发　exploration of the ocean, ocean exploitation
海洋开发产业　ocean development industry
海洋开发技术　oceanology
海洋勘探设备　marine exploration equipment

海洋科学 marine science
海洋空间规划 maritime spatial planning
海洋空间约束 sea space constraint
海洋空气 marine air, marine air (mass)
海洋矿物资源 marine mineral resource
海洋冷水团 ocean cold-water-mass
海洋流速仪 oceanographic (al) current meter
海洋内波 internal wave of ocean
海洋能 energy from ocean, marine power, ocean energy
海洋能地段 ocean energy farm
海洋能风系统 ocean-wind system
海洋能开发 ocean energy development
海洋能联合行动 co-ordinated action on ocean energy
海洋能系统 ocean energy system
海洋能应用 marine energy application
海洋能源 marine energy source, ocean energy source
海洋能源开发 ocean energy exploitation
海洋能转换 marine energy conversion
海洋能转换换热器 OTEC heat exchanger
海洋能转换技术 ocean energy conversion technique
海洋能转换器 ocean energy converter
海洋能转换因子 ocean energy conversion factor
海洋农场 ocean farm
海洋盆地 ocean basin
海洋气流 marine current
海洋气团 marine air
海洋气象站 ocean weather station
海洋倾（倒）废（物） ocean dumping
海洋热力学 marine thermodynamics
海洋热流 ocean heat flow
海洋热能 thalassothermal energy, thermal energy from sea, thermal energy of ocean
海洋热能的总能量 total energy of ocean thermal
海洋热能发电 sea thermal power generation
海洋热能发电厂 ocean thermal power plant
海洋热能远程利用 remote harnessing of ocean thermal energy
海洋热能转换 ocean thermal energy conversion (OTEC)
海洋热能转换板式换热器 OTEC plate-type heat exchanger
海洋热能转换泵 OTEC pump
海洋热能转换闭式循环系统 OTEC closed-cycle system
海洋热能转换传热 OTEC heat transfer
海洋热能转换电功率集成化 OTEC power integration
海洋热能转换电站 OTEC station, OTEC plant, OTEC power plant
海洋热能转换动力循环 OTEC power cycle
海洋热能转换发电 OTEC power generation
海洋热能转换发电系统 OTEC power system
海洋热能转换放牧场 OTEC gazing plant
海洋热能转换工场 OTEC workshop
海洋热能转换计划 OTEC program

海洋热能转换技术　OTEC technology
海洋热能转换甲醇电站船　OTEC methanol plantship
海洋热能转换经济学　OTEC economics
海洋热能转换开发　OTEC development
海洋热能转换开式循环系统　OTEC open-cycle system
海洋热能转换壳—管汽化器　OTEC shell and tube evaporator
海洋热能转换控制功能　OTEC control function
海洋热能转换冷水管计算机辅助设计法　OTEC-CWP computer-aided design methodology
海洋热能转换冷水管设计　OTEC-CWP design
海洋热能转换冷水管现场试验　OTEC-CWP field experiment
海洋热能转换平台　OTEC platform
海洋热能转换汽化器　OTEC evaporator
海洋热能转换热交换器　OTEC heat exchanger
海洋热能转换热能　OTEC thermal energy
海洋热能转换设计方法　OTEC design process
海洋热能转换试验平台　OTEC test platform
海洋热能转换塔　OTEC tower
海洋热能转换提升循环　OTEC lift cycle
海洋热能转换应用　OTEC application
海洋热能转换用管式换热器　OTEC tube heat exchanger
海洋热能转换蒸汽—提升循环　OTEC steam-lift cycle
海洋热能转换致密换热器　OTEC compact heat exchanger
海洋热能转换中试管道系统　OTEC pilot plant pipe system
海洋热能转换装置　OTEC apparatus
海洋热能转换资源　OTEC resource
海洋热能转换组件　OTEC module
海洋热能资源　ocean thermal resource
海洋热梯度　ocean thermal gradient
海洋热液系统　oceanic hydrothermal system
海洋热转换　ocean thermal conversion, ocean thermal electric conversion
海洋三角洲　oceanic delta
海洋生态系统　marine ecosystem
海洋生态学　marine ecology
海洋生物　marine growth, marine biota, marine life, marine organism
海洋生物腐蚀　marine organism corrosion
海洋生物量　marine biomass
海洋生物污染　marine biological pollution
海洋生物学　marine biology
海洋生物养殖　mariculture
海洋生物质　marine biomass
海洋生物质能转化　marine biomass conversion
海洋石油开发　marine petroleum exploitation
海洋水位　ocean level
海洋水文测量　marine hydrographic (al) survey
海洋水文站　hydrographic (al) station
海洋水下环境　ocean underwater environment
海洋水域　ocean volume
海洋太阳能　sea solar power
海洋太阳能电站　solar sea power plant
海洋替代能源　ocean alternative energy

海洋调查船 oceanographic research ship
海洋同温层 oceanic stratosphere
海洋土力学 marine soil mechanics
海洋湍流 oceanic turbulence
海洋微生物污染 marine microbial pollution
海洋温差 ocean temperature difference, ocean thermal difference, ocean thermal gradient
海洋温差电站 ocean thermal difference power plant
海洋温差发电 ocean thermal gradient power generation, ocean thermal power generation
海洋温差发电系统 ocean thermal gradient electricity generating system
海洋温差发电站 ocean thermal different power plant
海洋温差能 ocean thermal energy, ocean thermal gradient energy
海洋温差系统 ocean temperature gradient system, ocean thermal gradient system
海洋温差装置 ocean thermal gradient system
海洋温度 ocean temperature
海洋涡轮系统 ocean turbine system
海洋污垢 marine fouling
海洋污染 marine pollution
海洋污染监测 marine pollution monitoring
海洋污染物 marine pollutant
海洋物理学 physical oceanography, oceanophysics
海洋斜温层 ocean thermocline
海洋新能源 ocean alternative energy

海洋性空气 maritime air
海洋性气候 marine climate, ocean climate, oceanic climate
海洋悬崖 ocean scarp
海洋学 oceanography, oceanology
海洋学国际总有机碳分析计 oceanographic international total organic carbon analyzer
海洋学家 oceanographer
海洋循环 oceanic cycle
海洋研究浮标 oceanographic research buoy
海洋盐差能 ocean energy from salinity gradients
海洋盐浓度差发电 salinity difference power generation
海洋养殖 marine culture, offshore aquaculture
海洋涌 ocean swell
海洋油库 ocean depot
海洋有机物 marine organism
海洋预测中心 Ocean Prediction Center
海洋温跃层 ocean thermocline
海洋中心 midocean
海洋主温跃层 sea main thermocline, permanent thermocline
海洋资料发送器 ocean data transmitter
海洋资料站 ocean data station
海洋资料站浮标 ocean data station buoy
海洋资源 marine resource, ocean resource
海洋资源开发 ocean resource development
海洋综合设施 marine complex
海洋组件测试设备 marine component testing facility

海涌　swell
海用引擎　marine engine
海域结构物　offshore structure
海藻　seaweed
氦氖激光器　helium neon laser

han

含氚量　tritium content
含灰量　ash content
含空气煤气　air gas
含硫量　sulphur content
含能涡旋　energy containing eddy
含能作物　energy crop
含汽量　steam content
含氢的非结晶性硅　hydrogenated amorphous silicon
含沙风　sand bearing wind
含湿量　moisture capacity
含水层　water bearing layer
含水量　water content, moisture content
含水煤气　moisture-laden gas
含水重量　wet weight
含碳物质　carbon-containing material
含铁金属　ferrous metal
含盐的　saline
含盐度　salineness
含盐浓度　salt concentration
含油气盆地　petroliferous basin
焓　enthalpy, heat content
焓轮　enthalpy wheel
焓—熵图　enthalpy entropy diagram
涵道螺旋桨　ducted propeller
韩国始华湖潮汐电厂　Sihwa Lake Tidal Power Station
焊工　welder
焊机　welder
焊剂　flux, fluxing agent

焊井地热测量　downhole geothermal measurement
焊料/焊丝　filler metal
焊条　filler rod

hang

行对行电压　line to line voltage (VLL)
航磁测量　aeromagnetic survey
航道　waterway
航空灯　aviation light
航空地磁学　aeromagnetics
航空发电机　aero generator
航空发动机　aeromotor
航空衍生型燃气轮机　aero-derivative gas turbine
航空翼型　aviation airfoil
航空用汽轮机　aero derivative turbine
航向稳定性　yaw stability
航油　jet fuel

hao

毫秒　millisecond
好氧堆肥　aerobic composting
好氧菌　aerobic bacteria
好氧曝气器　aerobic aeration bin
耗能工质　energy-consumed medium
耗能装置　energy dissipation device
耗气量　air consumption
耗散　dissipation

he

合并　absorption
合成柴油　synthetic diesel
合成风向　resultant wind direction
合成孔径雷达　Synthetic Aperture Radar (SAR)
合成路线　synthesis route

合成气　syngas, synthesis gas
合成汽油　synthetic gasoline
合成射流　synthetic jet
合成橡胶密封　elastomeric seal
合成样品　combined sample
合环　ring closing
合计　aggregate
合金　alloy
合金钢　alloy steel
合理使用能源　rational use energy
合力　resultant force
合适位置特征　appropriate site characteristics
和风　moderate breeze
河口　estuary
河流能量　river energy
河流侵蚀　stream erosion
河湾　winding bend
荷电状态　state of charge
荷兰风车　Holland windmill, Dutch windmill
荷兰四臂型风车　Dutch four arm type mill
核安全标准　nuclear safety criterion
核安全限额　nuclear safety limit
核安全准则　nuclear safety criterion
核材料　nuclear material
核材料控制　nuclear material control
核材料控制系统　nuclear material control system
核厂房及设备　nuclear island
核磁共振温度计　nuclear magnetic resonance thermometer
核岛　nuclear island
核电　nuclear electricity
核电站　nuclear power plant, nuclear power station, nuclear steam electric plant, nuclear energy plant
核电站选址方针　nuclear siting policy
核电子　nuclear electron
核动力　nuclear power
核动力站　nuclear power plant, nuclear power station
核动力装置　nuclear power unit, nuclear power complex
核发动机　nuclear engine
核反应堆安全性　nuclear reactor safety
核反应能　nuclear reaction energy
核工程　nuclear engineering
核光电效应　nuclear photoeffect, nuclear photoelectric effect
核锅炉　nuclear fuelled boiler
核激发　nuclear encitation
核激发能　nuclear encitation energy
核级　nuclear grade
核力　nuclear force
核粒子　nuclear particle
核裂变　nuclear fission
核裂变链式反应　nuclear fission chain reaction
核能　nuclear energy
核能工业　nuclear power industry
核能级　nuclear level
核能级间距　nuclear level spacing
核能级密度　nuclear level density
核能极　nuclear energy level
核能面　nuclear energy surface
核能生产　nuclear power production
核能隙　nuclear energy gap
核能装置　nuclear energy plant
核燃料　nuclear fuel
核燃料循环　nuclear fuel cycle
核燃料资源　nuclear fuel resource

核热量　nuclear heat
核—热流体动力不稳定性　nuclear thermohydrodynamic instability
核设施　nuclear installation
核势（位）能　nuclear potential energy
核物质　nuclear matter
核转动能　nuclear rotational energy
核转动能带　nuclear rotation band
核转动能级　nuclear rotational level
核子成分　nucleonic component
核子的　nucleonic
核子分离能　nucleonic separation energy
核子分裂　nuclear fission
核子力　nucleon-nucleon force
核子碰撞　nucleon-nucleon collision
核子散射　nucleon-nucleon scattering
核子激发　nucleonic excitation
核子位形　nucleonic configuration
核子运动　nucleonic motion
核子组成　nucleonic component
荷载工况　load case
褐煤　lignite
褐色砂质泥土　brown sandy clay
褐色石灰岩　brown limestone

hei

黑（色）电池　black cell
黑底水池　water pond with black bottom
黑度　degree of blackness
黑化铝　blackened aluminum
黑金属基座　black metal base
黑金属片　black metal sheet
黑金属吸热器　black metal receiver
黑沥青　abbertite
黑色多孔吸热顶盖　black porous absorbing head
黑色多孔蒸发器　black porous evaporator
黑色金属　ferrous metal
黑色金属网　black gauze
黑色金属网式太阳能聚热器　black gauze solar collector
黑色聚乙烯基塑料　black polyvinyl plastic
黑色硫化铅颗粒　black lead sulfide particle
黑色铝板　blackened aluminum sheet
黑色水箱　black tray of water
黑色太阳能吸收器　black solar absorber
黑色涂层　black coating
黑色吸热板　black absorbing sheet
黑色吸热材料　black absorbing material
黑色吸热面积　black absorbing area
黑色吸热体　black absorber
黑色吸热体表面　black absorber surface
黑色叶片　black blade
黑体　black body
黑体（间）辐射换热　black body radiant exchange
黑体辐射　black body radiation, black radiation, blackbody radiation
黑体辐射功率　black body emissive power
黑体辐射计　black body radiator
黑体辐射强度　intensity of black-body radiation
黑体火焰温度　black body flame temperature
黑体机械吸收体　black mechanical absorber
黑体平板表面　plane black surface
黑体吸热器　black body receiver
黑小麦　triticale
黑液　black liquor

heng

恒定风　permanent wind
恒定负载　steady load
恒定光照　constant illumination
恒定流量　constant rate
恒定体积　constant volume
恒定压力　constant pressure
恒定叶尖速比方案　constant tip-speed ratio scheme
恒定值　steady state value
恒流　constant flow
恒流热线风速计　constant current wire anemometer
恒能带宽度太阳能电池　constant energy gap solar cell
恒频发电机　constant frequency generator
恒频输出　constant frequency output
恒速　constant speed, constant rate
恒速恒频发电系统　constant speed-constant frequency system
恒速恒频风电机组　constant speed constant frequency wind turbine generator system
恒速区域　constant speed region
恒速运行　constant-rpm operation, fixed-speed operation
恒温器　thermostat, calorstat
恒压　constant pressure
恒载　dead load
恒转矩　constant torque
横波　shear wave
横断面　cross-section
横断面视图　cross-sectional view
横风驰振　cross wind galloping
横风抖振　cross wind buffeting
横风桨板式风力机　crosswind paddles wind machine
横风扩散　cross wind diffusion
横风弥散参数　cross wind dispersion parameter
横风尾流力　cross wind wake force
横风位移　cross wind displacement
横风稳定性　cross wind stability
横风向　cross wind direction
横风载荷　cross wind loading
横风振动　cross wind viberation
横风轴风力机　cross wind axis wind machine
横风装置　cross wind installation
横管　horizontal pipe
横截面积　cross-sectional area
横流　cross flow
横流角　cross flow angle
横流式风轮机　cross-flow wind turbine
横流式热交换器　cross flow heat exchanger
横流永磁　longitudinal flux permanent magnet (LFPM)
横切头　cross-cut end
横头锤　cross-peen hammer
横卧式涡轮机　horizontal axis turbine
横向层压（薄板）　transverse lamination
横向磁场电机　transverse flux machine
横向磁场永磁　transverse flux permanent magnet (TFPM)
横向磁场永磁电机　transverse flux permanent magnet machine (TFPMM)
横向磁场永磁电机技术　TFPMM technology
横向磁通　transversal flux
横向风　across-wind
横向风试验　cross wind test
横向负荷　side load

横向扩散　lateral diffusion
横向流　crossflow
横向弥散　lateral dispersion
横向密度变化　lateral density variation
横向阵风　cross wind gust, lateral gust
横轴磁化电抗　quadrature-axis magnetizing reactance
横轴电抗　q-axis reactance
横轴风车　horizontal axis windmill
横轴结构　horizontal axis configuration
横坐标轴　abscissa axis

hong

烘焙　torrefaction
烘干的木材　oven dry wood
烘干窑　drying kiln
红泥塑料　red-clay plastics
红外成像探测　infrared imagery survey
红外辐射　infrared radiation
红外辐射加热量　infrared radiant heater
红外辐射探测器　infrared radiation detector
红外辐射温度计　infrared radiation thermometer
红外高温测量　infrared pyrometry
红外高温计　infrared pyrometer
红外光谱法　infrared spectrometry
红外加热　infrared heating
红外气体分析仪　infrared gas analyser
红外吸收光谱　infrared absorption spectrum
红外线反射涂层　IR reflection coating
红外线辐射　infrared radiation (IR)
红外线区　infrared region
红外线热处理　infrared heating
虹吸回排系统　siphon drainback system
虹吸吸管太阳能集热器　thermosiphoning solar collector
虹吸吸管太阳能热水系统　thermosiphon solar water heating system
虹吸吸管太阳能热水装置　thermosiphon solar hot-water device

hou

后报　hindcast
后触点　back contact
后行桨叶　retreating blade
后掠桨叶　sweptback blade
后掠面积　swept area
后掠叶片　sweptback vane
后掠翼　sweptback wing
后期渗滤液　anaphase leachate
后燃　post-combustion
后燃催化控制系统　post-combustion catalytic control system
后燃烧室　secondary combustion chamber
后向散射　back-scattering
后缘　tailing edge
厚度　thickness
厚度变化　thickness variation
厚度测量　thickness measurement
厚度函数　thickness function
厚度偏差　thickness variation
厚度误差　thickness variation
候选场址　candidate site

hu

弧口凿　gouge
弧线　camber line
胡克定理　Hooke's Law
湖面波动　seiche
湖震　seiche
蝴蝶型隔离阀　butterfly-type isolating valve

互动率谱密度 cross-spectral density
互感磁通 mutual magnetic flux
互换能量 exchange energy
互连（风力发电机组） interconnection (for WTGS)
互连系统 interconnected system
互养 syntrophy
户外加速寿命试验 accelerated outdoor exposure testing
户外实时寿命试验 real-time outdoor exposure testing
户用光伏电源 family photovoltaic power system
护板 guard plate
护壁板散热器 baseboard radiator
护壁板式加热器 baseboard heater
护目镜 goggle
护坡 pitching
护墙 parapet
护圈 retainer
护罩 shroud
斗式（冲击式）水轮机 impulse turbine

hua

花岗闪长岩 granodiorite
花岗岩 granite
花岗岩层 granitic layer
花键连接 splined coupling
花键液压泵 spline hydraulic pump
花砖火炉 tiled stove
华氏度数 degrees Fahrenheit
华氏温标 Fahrenheit, Fathrenheit temperature scale
华氏温度计 Fahrenheit
滑差功率 slip power
滑差频率 slip frequency
滑动制动器 sliding shoe
滑阀 slide valve
滑环 slip ring
滑环转子 slip ring rotor
滑块 moving slider
滑块连接 Oldham coupling
滑流 slipstream, slip flow
滑流收缩 slipstream contraction
滑流速度 slipstream velocity
滑流旋转 race rotation, racing
滑轮 chain wheel
滑轮组 block and tackle
滑坡涌浪 land slide surge
滑石 soapstone
滑脱带 slip band
滑移带 slip band
滑移系 slip system
滑移线 slip line
滑脂枪 grease gun
化肥原料 fertilizer feedstock
化粪池 septic tank
化工分离技术 chemical fractionation technique
化合热 heat of combination
化合物 compound
化合物半导体太阳能电池 compound semiconductor solar cell
化灰机 ash melter
化学爆炸激励法 method of chemical explosive stimulation
化学池沉积 chemical bath deposition
化学动力学 chemical kinetics
化学电解质 chemical electrolyte
化学镀 electroless plating
化学反应热储存 chemical reaction heat energy storage
化学腐蚀 chemical corrosion
化学激光器能量 chemical lasing energy

化学兼容性　chemical compatibility
化学键　chemical bond
化学均衡　chemical equilibrium
化学能　chemical energy
化学平衡　chemical equilibrium
化学气相沉积　chemical vapor deposition
化学热泵　chemical heat pump
化学蚀变　chemical alteration
化学势　chemical potential
化学相互作用　chemical interaction
化学需氧量　chemical oxygen demand (COD)
化学凿岩　chemical drilling
化学转换技术　chemistry conversion technology
画面　frame
画图板　drawing board
划定装机容量评估　delineated installed capacity estimate
划定资源评估　delineated resource estimate
桦木胶合板　birch plywood

huan

环保冲击　environmental impact
环保意识　environmental concern
环规　ring gauge
环轨　circular track
环行接地体　ring earth electrode
环礁湖　lagoon
环境传感器　environmental sensor
环境大气　ambient atomosphere
环境递减率　atmospheric lapse rate
环境风　environmental wind
环境风速　surrounding air speed, ambient wind speed, ambient wind velocity
环境感应器　environmental sensor
环境管理法　environmental management act
环境环流　surrounding flow
环境监测　environmental monitoring
环境空气　ambient air
环境空气质量　ambient air quality
环境空气质量标准　ambient air quality standard
环境控制技术　environmental control technology
环境绿化工程　landscape engineering
环境美化　landscaping
环境能源　ambient energy source
环境容量　environmental capacity
环境条件　environment condition
环境温度　ambient temperature
环境限制　environmental restriction
环境压力　ambient pressure
环境影响　environmental impact
环境影响评估　environmental impact assessment
环境影响潜值　potential environmental impact
环境噪声　ambient noise
环境质量模式　environmental quality pattern
环境质量指数　environmental quality index
环链式输送机　chain trough conveyor
环梁壳　mantle
环量控制型风轮　circulation control rotor
环流　circulating current
环面　annulus
环模制粒机　ring die pelletizer
环日辐射　circumsolar radiation
环生成元　ring generator
环形齿轮　ring gear

环形焊缝　girth weld
环形锯　band saw
环形流体　annular fluid
环形气隙　annular air gap
环形太阳能电池　circular solar cell
环形太阳能集热器　circular solar collector
环形凸轮液压泵　ring-cam hydraulic pump
环氧复合材料　epoxy composite
环氧胶　epoxy glue
环氧树脂　epoxies, epoxy
环氧树脂钢　epoxy lined steel
环氧涂层　epoxy coating
环氧黏合剂　epoxy adhesive
环状裂缝　ring fracture
环状面积　annular area
缓变率　ramp rate
缓冲槽　surge tank, buffer tank
缓冲储热器　buffer storage
缓冲罐　surge tank
缓冲剂　buffer solution
缓冲力　dampening force
缓冲器　shock absorber
缓冲装置　fender
缓启动　soft start
换接　change-over switching
换能器　energy conversion device, transducer
换气　air change
换气次数　air change rate, air circulation rate
换气负荷　ventilation load
换气量　air displacement
换气率　air change rate
换气扇　ventilator
换气设备　air regenerating device
换热　heat exchanging
换热管　heat exchange pipe
换热机理　heat transfer mechanism
换热介质　heat transfer medium
换热面积　heat exchange area
换热器　heat exchanger
换热性能　heat transfer property
换算器　scaler
换向　commutation
换向齿轮　idler gear
换向阀　reversing valve
换向片　commutator segment
换向器　commutator
换向器　commutator-brush combination
换向器节距　commutator pitch
换向状况　commutation condition

huang

荒料　quarry stone
黄赤交角　obliquity of the ecliptic
黄铜　brass

hui

灰白色白云岩（石）　gray dolomite
灰沉积　ash deposition
灰半球温度　ash hemisphere temperature
灰变形温度　ash deformation temperature
灰流动温度　ash flow temperature
灰软化温度　ash softening temperature
灰斗　ash hopper
灰分　ash content
灰分分析　ash analysis
灰分含量　ash content
灰回收　ash recycling
灰烬　ember
灰聚结　ash coalescence
灰泥　mortar
灰熔点　ash melting point

灰熔聚 ash agglomeration	回风道 air return way
灰熔聚炉 ash agglomeration gasifier	回灌井 injection well
灰熔融温度 ash fusion temperature	回归潮 tropic tide
灰熔融性 ash fusibility/ash melting behaviour	回归线无风带 horse latitude
灰熔温度 ash melting temperature	回馈线性化 feedback linearization
灰色砂岩 gray sandstone	回流 return flow
灰色砂质黏土 gray sticky clay with sand	回流防止器 backflow preventer
灰色石灰岩 gray limestone	回流管 return pipe
灰烧结 ash sintering	回流管路 return line
灰造粒 ash granulation	回流极地气团 returning polar air
灰渣输送机 ash conveyor	回流集管 return header
灰渣形成 ash formation	回流区 flow recirculation zone, recirculation region, reclaimed stock
挥发 volatilization	回流软管 return hose
挥发度 volatility	回流系统 drainback system
挥发分 volatile matter, volatile	回流线 return line
挥发物含量 volatile content	回流箱 return tank
挥发性 volatility	回路 loop
挥发性固体 volatile solid	回路电流 loop current
挥发性化合物 volatile compound	回气系统 air return system
挥发性馏分 volatile fraction	回热循环 regenerative cycle, feedwater cycle
挥发性燃料 volatile fuel	回热再热循环 regenerative reheat cycle
挥发性燃烧 volatile combustion	回收波能装置 device for recovering wave energy
挥发性有机化合物 volatile organic compound	回收槽 accumulator
挥发性蒸发 volatile evaporation	回收的建筑木材 recovered construction wood
挥发性脂肪酸 volatile fatty acid	回收锅炉 recovery boiler
挥发性组分 volatile component	回收浆料 reclaimed stock
挥发油 naphtha	回收率 recycling rate
辉光放电 glow discharge	回收能量 recovered energy
辉石 pyroxene	回收效率 recovery efficiency
辉长岩 gabbro	回水 return water
回采 winning	回水效应 backwater effect
回弹 resilience	回旋射束 rotating beam
回动阀 reversing valve	回旋吸收 cyclotron absorption
回返对流 return convection	

回旋装置　gyroscope
回压　back pressure
回用水　reclaimed water
回游鱼类　migratory fish
回转包装机　rotary packer
回转框架　swing frame
回转力　gyroscopic force
回转力矩　gyroscopic moment
回转式焚烧炉　rotary kiln incinerator, rotary kiln
回转式烟气换热器　rotary gas heat exchanger
回转速度　rotating speed
回转台　rotary table
回转效应　gyroscopic effect
回转窑　rotary kiln
回转叶片，风标　swivelling vane
回转仪　gyroscope
汇集系统　collector system
汇流条　bus bar, bus
汇水面积　catchment area
绘图点　drawing point

hun

混合　mixing
混合槽　mixing tank
混合层　mixing layer
混合潮　mixed tide
混合潮流　mixed current
混合动力汽车　hybrid electric vehicle
混合动力系统　hybrid power system, hybrid system
混合阀　mixing valve
混合价格　lump sum price
混合冷却　hybrid cooling
混合冷却系统　hybrid cooling system
混合流体　mixed fluid
混合器　shaker
混合燃料　dual fuel
混合热　heat of mixing
混合设计　hybrid design
混合式　hybrid
混合式光电系统　hybrid photovoltaic system
混合式光伏打系统　hybrid photovoltaic system
混合式凝汽器　direct contact condenser
混合式热交换器　direct-contact heat exchanger
混合式太阳能系统　hybrid solar energy system, hybrid solar system
混合式循环系统　hybrid cycle system
混合室　mixing chamber
混合太阳能热发电　hybrid solar thermal power
混合位错　mixed dislocation
混合物　admixture
混合循环系统　mixed cycle system
混合液挥发性悬浮固体　mixed liquor volatile suspended solid
混合液悬浮固体　mixed liquor suspended solid
混合永磁电机　hybrid permanent magnet machine
混合整数线性规划　mixed integer linear programming
混流式蓄能泵　mixed-flow storage pump
混凝土　concrete
混凝土导管　tremie pipe
混凝土顶盖　concrete cap
混凝土风塔　tubular concrete
混凝土基座　concrete pad
混凝土块太阳能储热壁　concrete block solar thermal storage wall

混凝土墙　concrete wall
混凝土圆坯　circular platform of concrete
混凝土制太阳能集热器　concrete solar collector
混凝土柱　concrete column
混凝土柱帽　concrete cap
混凝土钻　concrete drill
混气液　aerated fluid
混燃　co-combustion
混烧　co-firing

huo

活动扳手　adjustable spanner
活动导叶　guide vane
活动底板　false bottom
活动断层　active fault
活动化作用　mobilization
活动积温　active accumulated temperature
活动梁　walking beam
活动炉底　removable bottom
活动炉排加煤机　traveling grate stoker
活动炉栅　moving grate
活动水槽　tilting flume
活动性八叠菌　sarcina mobilis
活度系数　activity coefficient
活化　activation
活火山　active volcano
活桨叶　feathering paddle
活塞　piston
活塞泵　piston pump
活塞阀蒸汽机　piston valve steam engine
活塞环　piston ring
活塞流　plug flow
活塞流连续消解器　plug flow digester
活塞式波浪力发电装置　piston wave power device
活塞式发动机　reciprocating engine
活塞式蒸汽机　steam piston engine, piston steam engine
活塞造波机　piston wavemaker
活表面积　active surface area
活性参数　activity parameter
活性断层　active fault
活性能　activition energy
活性燃料　reactive fuel
活性炭　activated carbon, active carbon, absorbent carbon
活性炭储仓　active carbon storage silo
活性炭过滤器　active carbon filter, carbon filter
活性炭吸附　sorption with activated carbon
活性炭吸附器　activated-charcoal absorber
活性微生物　active bacteria
活性污泥　activated sludge
活性污泥法（污水处理）　activated sludge process
活性污泥系统　activated sludge system
活性污泥氧化　activated sludge oxidation
活性细菌　active bacteria
活叶螺旋桨　reversible propeller
活翼式泵　wing pump
火成岩　igneous rock
火床　fire bed
火管锅炉　fire-tube boiler
火花点火　spark ignition
火花放电电压　flash over voltage
火花隙　spark gap
火炬管线　flare line
火炬气　gas flare
火力　thermal power
火力发电　thermal power

火山活动 volcanism
火山裂缝 volcanic rent
火山岩 lava, volcanic rock
火室燃烧 suspension combustion
火焰冲击 flame impingement
火焰带 flame zone
火焰辐射 flame radiation
火焰监视器 flame scanner
火焰离子化检测器 flame ionization detector (FID)
火焰亮度 flame luminance
火焰炉 over-fire boiler
火焰通口 flame port
火焰原子吸光谱法 flame atomic absorption spectrometry
火焰长度 flame length
火用效率 exergetic efficiency
获能 energy extraction, harvesting energy, power extraction
获能系数 power extraction coefficient
霍尔式传感器 Hall sensor
霍尔效应 Hall effect

ji

击穿 breakdown
击穿电压 shorting voltage, flash over voltage, sparking voltage
机舱 nacelle
机舱变压器柜 nacelle transformer cabinet
机舱底架 bedplate
机舱机柜 nacelle cabinet
机舱偏航 nacelle yaw
机舱罩 nacelle cover
机电换能器 electromechanical transducer
机电能量变换 electromechanical power conversion, electromechanical energy conversion
机电偏航机制 electromechanical yaw mechanism
机电暂态性能 electromechanical transient performance
机电装置 electromechanical device
机端 generator terminal
机柜 frame
机架 frame
机器额定载荷 machine rated load
机器人气象浮标 robot weather buoy
机械齿轮 mechanical gear
机械除尘器 mechanical collector
机械房布局 mechnical room layout
机械分类城市生活垃圾 mechanically sorted municipal solid waste
机械风速计 mechanical anemometer
机械负载 mechanical load
机械功率 mechanical power
机械共振 mechanical resonance
机械故障 mechanical failure
机械弧度 mechanical radian
机械换向器 mechanical commutator
机械集尘器 mechanical collector
机械加工 machining, mechanical processing
机械控制 mechanical control
机械控制系统 mechanical control system
机械力 mechanical force
机械零件 machine element
机械率 mechanical power
机械能 mechanical energy
机械抛光 mechanical polishing
机械强度 mechannical strength, mechanical durability
机械韧性 mechanical resilience

机械生物处理　mechanical biological treatment
机械师　mechanic
机械寿命　mechanical endurance
机械损失　mechanical loss
机械通风　mechanical draft
机械湍流　mechanical turbulence
机械系统　mechanical system
机械效率　machine efficiency
机械性不稳定　mechanical instability
机械应力　mechanical stress
机械运动　mechanical motion
机械爪　robotic claw
机械振动布袋清灰　mechanical shake cleaning
机械制动　mechnical brake
机械制动器　mechanical brake
机械制动系　mechanical braking system
机械转动式浓度差能发电　mechanical rotating salinity gradient energy conversion
机械转换器　mechanical commutator
机械转矩　mechanical turbine torque
（机翼的）前缘　leading edge
（机翼型）轴流风机　aerofoil fan
机翼　aerofoil
机翼横截面　aerofoil cross-section
机翼叶片　airfoil-shaped blade
机翼振荡式风能转换系统　oscillating wing WECS
机油冷却器　oil cooler
机载电子测量控制　airborne electronic survey control
机载勘测　airborne survey
机载热红外成像　airborne thermal infrared imagery

机组　water-turbine generator set
机组效率　unit efficiency
机组性能试验　unit performance test
奇函数　odd function
积淀　sedimentary deposit
积分球（内壁镀有漫反射涂层的空腔圆球。球壁上任一面受照后，其反射辐照度在整个空腔圆球内将是均匀的，利用积分球的这一特性可测定材料的反射率）　integrating sphere
积分下限　integration
积复励电动机　cumulatively compounded motor
积灰　ash deposition
积累层　accumulation layer
积温　accumulated temperature
积盐　salification
基板　substrate
基本反应速率　elementary reaction rate
基本风速　basic wind speed
基本负荷　base load
基本负荷动力　base load power
基本负荷功率　base load power
基本负荷机组　base load unit
基本绝缘等级　basic insulation level
基本原理　principal theorem
基波振幅　amplitude of first harmonic
基础　foundation
基础接地体　foundation earth electrode
基础密度　basic density
基础温度　basic temperature
基础阻抗　base impedance
基底材料　base material
基尔霍夫电压定律　Kirchhoff's voltage law
基尔霍夫定律　Kirchhoff's law

基荷电厂 base load plant
基荷动力 base load power
基荷机组 base load unit
基荷容量 base load capacity
基极 base
基片 substrate
基态 ground state
基体开裂 matrix crack
基岩 basic rock, bed rock
基岩井 bedrock well
基元反应速率 elementary reaction rate
基载电力 baseload power
基址探测 site exploration
基质 substrate
基质降解 substrate degradation
基准粗糙长度 reference roughness length
基准发射面 principal radiating surface
基准高度 reference height
基准距离 reference distance
基准器件（是一种以标准太阳光谱分布为依据，用来测量辐照度或校准太阳模拟器辐射度的光伏器件） reference device
基准水 standard water
基准温度 fiducial temperature
基准误差 basic error
基准站 reference station
基座 foundation
畸变 distortion
畸形波 freak wave
激潮 tidal rip, tide blow
激磁电流 magnetizing current, excitation current
激磁线圈 field coil
激发场 excitation field

激发光谱 excitation spectrum
激发极化 induced polarization
激发力 exciting force, excitation force
激发能 excitation energy
激发能级 excitation level
激光多普勒风速计 laser Doppler anemometer
激光风速计 laser anemometer
激光划片 laser scribing
激光划线 laser scribing
激光溅射沉积 pulsed laser deposition
激光刻蚀 laser etching
激光轮廓测量技术 laser profilometry
激光能量转换 laser energy conversion
激光切割 laser cutting
激光束 laser beam
激光凿岩 laser drilling
激光钻孔 laser drilling
激活 activation
激活能 activition energy
激活作用 wave action
激冷区 chill zone
激励整流器 exciter rectifier
激碎波 surging breaker
激振力 excitation force
吉布斯函数 Gibbs function
吉布斯熵 Gibbs entropy
吉布斯相律 Gibbs phase rule
吉布斯自由能 Gibbs free energy
吉吨 gigatonne (Gt)
级联多结电池 cascade multijunction cell
极大风 extreme wind
极大风速 extreme wind speed
极大值 maximum value
极端梯度风 extreme gradient wind
极端英里风速 extreme mile wind

speed
极端最高　extreme maximum
极锋气流　polar front jet
极化电流　polarization current
极化电压　polarizing voltage
极矩　polar moment
极限荷载　ultimate load
极限量规　limit gauge
极限强度　ultimate strength
极限数　limiting value
极限限制状态　ultimate limit state
极限真空　vacumm limit
极限转差率　breakdown slip
极限转矩　breakdown torque
极限转速　run away speed
极限状态　limit state
极性分子　polar molecule
极轴全跟踪装置　polar axis fully tracking mounting
极轴太阳跟踪　polar solar tracking
极轴装置（全跟踪集热器的）　polar mounting
急潮流　tide race
急冷技术　splat cooling
急冷热交换器　quenching heat exchanger
急冷塔　quench tower
急流　jet stream
疾风　moderate gale, near gale
集成门极换流晶闸管　integrated gate commutated thyristor
集成气化组合循环技术　integrated gasification combined cycle
集成型非晶硅太阳电池　integrated α-Si solar cell
集电环　slip ring, collector, collector ring
集电极　collector electrode
集电极反向电流　collector reserve

current
集电器　collector
集风罩式风力机　shrouded wind turbine
集肤效应　skin effect
集管　header
集气管　air header
集气罐　air collector
集气器　gas trap
集气系统　gas gathering system
集热—蓄热墙式太阳房　Trombe wall solar house
集热（蓄热）墙式　heat collection (storage) wall
集热（量）　heat collection
集热板　collecting plate
集热场　collector field
集热场年效率　annual efficiency of collector field
集热场效率　efficiency of collector field
集热管　heat collecting tube
集热介质　collection medium
集热开口面积　collecting aperture area
集热器　heat collection device, thermal collector
集热器泵　collector pump
集热器表面　collector surface
集热器表面的波长选择性　wavelength selectivity of collector surface
集热器产能速率　collector energy production rate
集热器尺寸　collector size
集热器出口　collector outlet
集热器出口温度　outlet temperature of collector
集热器—储罐太阳能热水器（太阳能热水器两种基本类型之一）　collector-and-storage-tank solar water heater

集热器单位水平表面上接受直接太阳能的速率 collector insulation
集热器的朝向 orientation of collector
集热器的热交换器 collector heat exchanger
集热器定向角 collector orientation angle
集热器发电机 collector generator
集热器反射镜 collector mirror
集热器方程式 collector equation
集热器方位 collector orientation
集热器方位角 collector azimuth angle
集热器负载 collector load
集热器盖板 collector cover plate
集热器盖层 collector cover
集热器跟踪器 collector tracker
集热器管件布置 collector plumbing layout
集热器光学效率 collector optical efficiency
集热器黑化吸热板 black collector plate
集热器回路 collector loop
集热器集热板流量分布 collector plate flow distribution
集热器进口 collector inlet
集热器净有效面积 net effective area of collector
集热器空气流率 collector air flow rate
集热器空气流速 collector air flow rate
集热器框架 collector frame
集热器理论 collector theory
集热器流动因子 collector flow factor
集热器流体 collector fluid
集热器面积 collector area
集热器面积要求 collector area requirement
集热器倾角 tilt angle of collector, angle of inclination for collector, collector tilt
集热器倾斜度 collector slope

集热器—热交换器效率 collector-heat exchanger efficiency
集热器—热交换器修正系数 collector-heat exchanger correction factor
集热器热交换效率 collector heat removal efficiency
集热器热容量 collector thermal capacity
集热器热损失 thermal loss from collector
集热器热损失传导率 collector heat loss conductance
集热器热转移因子 collector heat removal factor
集热器设置方式 collector configuration
集热器施釉 collector glazing
集热器试验 collector testing
集热器输出（能量） output of collector
集热器—水箱热交换器 collector-tank heat exchanger
集热器瞬时效率 collector instantaneous efficiency
集热器外壳 collector casing
集热器吸热板 collector absorber plate
集热器吸收器表面 collector absorber surface
集热器效率 collector efficiency
集热器效率方程 collector efficiency equation
集热器效率因子 collector efficiency factor
集热器性能 collector performance
集热器造价 cost of collector
集热器长期性能 long-term performance of collector
集热器阵结构 collector array structure
集热器阵列 collector array
集热器阵列面积 solar field size
集热器阵列总面积 gross collector

array area
集热器中流体种类 type of collector fluid
集热器储存系统 collector storage system
集热器子系统 collector subsystem
集热器总面积 gross collector area
集热器总能量损失系数 collector overall energy loss coefficient
集热器总热损系数 collector overall heat loss coefficient
集热器最佳方向 optimum collector orientation
集热器最佳温度 optimum collector temperature
集热墙 wall collector
集热透镜 collector lens
集热系统 collecting system
集热系统构件 part of collecting system
集热效率 collecting efficiency
集热装置 heat collection device
（地热流体的）集输系统 collection and transmission system
集水区 catchment area
集雪器 snow trap
集中控制 centralized control
集中式焚烧 mass burn system
集中式风能转换系统 centralized wind energy conversion system
集中式太阳能发电 concentrated solar power
集中式太阳能热发电系统 solar thermal central receiver system
集中式制冷负荷 integrated refrigeration load
几何半径 geometric radius
几何变换 geometrical transformation
几何聚光率 geometrical concentrator ratio
几何位置 geometrical position
几何弦长 geometric chord
几何相似 geometric similarity
挤压（成形） extrusion
挤压叶片 extruded blade
挤制铝 extruded aluminum
脊线撞击 ridges impinge
计测器框架 instrument rack
计测器支架 instrument rack
计划中的潮汐发电方案 proposed tidal power scheme
计数器 scaler
计算机仿真 computer simulation
计算流体动力学 computational fluid dynamics
计算流体力学模型 computational fluid dynamic model
记录带度计（带有遥感球） recording thermometer
记录风速计 recording anemometer
记忆函数 memory function
技术参数 technical parameter
技术规范 techinical specification
技术可行性 technical feasibility
技术潜势 technical potential
技术套管 intermediate casing
技术特性 technical characteristic
技术资源 techinical resource
季风 seasonal wind
季风环流 monsoon circulation
季风降水（量） monsoon precipitation
季风气候 monsoon climate, monsoon weather pattern
季风转换 reversal of the monsoon
季节风 monsoon
季节控制 seasonal control

季节效率　seasonal efficiency
季节性变化　seasonal variation
季节性供热负荷　seasonal heating load
季节性能系数　seasonal performance factor
季节性气候类型　seasonal weather pattern
季节性性能系数　seasonal performance factor
季节性蓄能　seasonal storage
季节性资源（指风力资源）　seasonal resource
季平均法线太阳能日通量　seasonally averaged daily direct-normal solar flux
季运行方式　seasonal operating plan
剂量　dosage
迹线长度　path length
继电保护　relay protection
继电保护装置　relay protection device
继电器　relay
继电器逻辑　relay logic
继电器输出　relay output
寄生电流　parasitic current
寄生电压　parasitic voltage
寄生电阻　parasitic resistance
寄生功率　parasitic power
寄生功率损耗　parasitic power loss
寄生能源　parasitic power
寄生热损失　parasitic heat loss
寄生热泄漏　parasitic heat leak
寄生阻力　parasitic resistance
绩效指标　performance indicator

jia

加工　fabrication
加工单位　process unit
加工废料　process residues
加工液　working fluid
加固板　gusset plate
加力式涡轮　augmented turbine
加料速度　feed rate
加料推杆　feed ram
加料斜槽　inlet chute
加强的玻璃纤维塑料阀　fibre-glass reinforced plastic valve
加氢脱氧　hydrodeoxygenation
加热　heating up
加热伴热　heat tracing
加热电压　heat voltage
加热范围　heating range
加热负载　heating load
加热构型　heating configuration
加热过程　heating process
加热结构　heating configuration
加热盘管　heater coil
加热期　heating up period
加热器　boiler, heater, temperature booster
加热器蒸汽盘管　heat steam coil
加热区段　heating section
加热设备　heating plant
加热时间　heating up time
加热室　heating box
加热速度　heating rate
加热线圈　heater coil
加热叶片　heated blade
加热元件　heating element
加热站　heater plant
加热主导型垂直闭合环路　heating dominated vertical closed loop
加热组件　heating element
加速度传感器　acceleration sensor, acceleration transducer

| 加速度幅值 | acceleration amplitude
| 加速计 | accelerometer
| 加速磨耗 | accelerated wear
| 加速试验 | accelerated test
| 加速寿命测试 | accelerated exposure testing
| 加锁 | lock
| 加压酸溶 | pressurized acid digestion
| 加重料 | weighting material
| 加州分布式能源指南 | California Distributed Energy Resources Guide
| 挟沙风 | sand driving wind
| 家电 | household appliance
| 家庭（太阳能）热水器 | domestic hot water heater
| 家庭废物 | domestic waste
| 家庭危险废弃物 | household hazardous waste
| 家庭有害垃圾 | household hazardous waste
| 家用锅炉 | domestic boiler
| 家用水加热系统 | domestic water heating system
| 家用太阳（能）热水器 | domestic solar water heater
| 家用太阳能供暖 | domestic solar heating
| 家用太阳能热水器 | solar domestic hot water heater
| 家用太阳能系统 | domestic solar energy system
| 夹带 | entrainment
| 夹紧力 | clamping force
| 夹具 | fixture, jig
| 夹具系统 | clamping
| 夹钳 | clamp
| 夹线器 | conductor holder
| 夹杂 | entrainment

夹杂空气　air entrainment
甲醇　methanol
甲醇—空气电池　methanol-air battery
甲基叔丁基醚　methyl tertiary butyl ether
甲烷　methane
甲烷测定器　methane tester
甲烷产菌　methane-producing bacteria
甲烷发酵　methane fermentation
甲烷放出，沼气释放　methane liberation
甲烷分层　methane layering
甲烷化　mathanization
甲烷化物　methanides
甲烷记录器　methane recorder
甲烷监控系统　methane monitoring system
甲烷菌　methane bacterium
甲烷离子化室　methane ionization chamber
甲烷喷出　methane outburst
甲烷气　methane gas
甲烷气柜　methane gas reservoir
甲烷生成　methanogenesis
甲烷势　methane potential
甲烷水合物　methane hydrate
甲烷厌氧菌　methane-oxidizing bacteria
甲烷运输船　methane tanker
甲烷指示器　methane indicator
甲烷煮解器　methane-producing digester
钾　potassium
假潮　seiche
假顶　intermediate ceiling
假负载　dummy load
假荷载　fictitious load
价带　valence band
价电子　valence electron
架　frame

架空输电线　overhead transmission line
架空线　air wire

jian

尖顶形集热器　cusp collector
尖端制动　tip brake
尖峰负荷发电厂　peak load power plant, standby station
尖峰负荷运行　peaking service
尖峰能力　peaking capability
尖峰周期　peak period
尖头叶片　sharp-nosed blade
间壁冷凝器　surface condenser
间断发电　intermittent generation
间隔　interval
间隔环　distance ring
间隔角　angular interval
间接混烧　indirect co-firing
间接集热（作用）　remote collection
间接膨胀式低温太阳能热发电　indirect expansion low temperature solar thermal power
间接气动波能转换器　indirect-pneumatic wave energy converter
间接气化　indirect gasification
间接燃烧炉　indirect fired oven
间接式太阳辐射吸热器　indirectly irradiated solar receiver
间接太阳能　indirect solar energy
间接太阳能转换系统　indirect solar conversion system
间接土地用途变更　indirect land-use change
间接系统　indirect system
间接增益（系统）　indirect gain
间隙固溶体　interstitial solid solution
间隙扩散　interstitial diffusion
间隙位置　interstitial site
间隙原子　interstitial atom
间歇　intermittent
间歇发电　intermittent power generation
间歇负载　intermittent load
间歇供热　batch heating
间歇加热　intermittent heating
间歇井喷　geyser
间歇流　intermittent flow
间歇泉　geyser
间歇式反应器　batch reactor
间歇式炉　batch furnace, intermittent furnace
间歇式太阳能热水器　batch-type solar water heater
间歇性阳光直射　intermittent direct sunlight
间歇运行　intermittent operation
间歇运行控制　intermittent operation control
监测井　monitoring well
监控点　monitoring node
监控光电管　monitor photoelectric cell
监控器　monitor, supervisory controller
监控系统　supervisory control system
监视器　monitor
监视信息　monitored information
兼性菌　facultative bacteria
兼性厌氧微生物　facultative anaerobe
检波　detection
检波电流　demodulation current
检测　detection
检测井　monitoring well
检测器　detector
检漏　gas leakage detecting
检修接地　inspection earthing

检修孔　manhole
检修门　access door
检验　verification
减反射　antireflection
减废算法　waste reduction algorithm
减风效率　wind reduction efficiency
减幅交流　damped alternating current
减排信用额度　emission reduction credit
减热器　desuperheater
减少盐分　desalination
减湿器　dehumidifier, attemperation
减税　tax abatement
减温器　desuperheater
减温器盘管　desuperheating coil
减压　de-pressurise
减压阀　pressure reducing valve, reducing valve
减压井　bleeder well
减压室　decompression chamber
减载　load shedding
减震　vibration suppression
减震阀　damped valve
减震技术　mitigation technique
减震力　dampening force
减震器　shock absorber, vibration isolator, dashpot, vibration absorber
减震造波机　absorbing wavemaker
减阻　drag reduction
减阻装置　drag reducing device
剪切波　shear wave
剪切荷载　shear loading
剪切角　angle of shearing strength
剪切力　shear stress
剪切流　shear flow
剪切应力　shear stress
剪应力　shear stress

简单回路　simple circuit
简单回路微型燃气轮机　simple cycle microturbine
简易太阳房采暖器　simple solar room-heater
简易太阳灶　simple solar cooker
碱含量　alkali content
碱化合物　alkali compound
碱金属氯化物　alkali chloride
碱金属盐　alkali metal salt
碱性感测　alkali-sensitive
碱性空气电池　alkaline air cell
碱性硫酸盐　alkali sulphate
碱性燃料　alkali fuel
碱性生物质燃料　alkali biomass fuel
碱性蓄电池　alkaline accumulator, alkaline storage battery, steel alkaline battery
碱蓄电池　edison battery
建模和分析方法　methods for modelling and analysis
建筑安装工程费　civil and erection cost
建筑空气动力学　architectural aerodynamics
建筑垃圾　construction debris
建筑密集城市　densely built-up city
建筑物　erection
建筑物加热负荷　building heating load
建筑物加热负荷系数　building heat load coefficient
建筑物加热设计负荷　design building heating load
建筑物热量损失　building heat loss
建筑物热（量）增益　building heat gain
建筑物太阳能供暖　heating of building with solar energy
建筑物太阳能供热　solar heating building

建筑物外体 building envelope
建筑物尾流 building wake
渐进加速势方法 asymptotic acceleration potential method
渐进性技术 incremental technology
键槽 keyway
键孔锯 keyhole saw
键能 bonding energy
键盘操作 key-operate

jiang

浆材料 grouting material
浆叶梢部 blade tip
浆液泵 slurry pump
桨 paddle
桨距角 pitch angle
桨距控制 pitch control
桨距调节 pitch regulated
桨叶节距角变化率 blade pitching rate
桨叶扭旋 blade twist
桨叶有效面积 effective blade area
降低压力阀 pressure reducing valve
降膜式吸热器 falling film receiver
降水 precipitation
降膜式吸热器 falling film receiver
降雨量 precipitation level
降雨型 rainfall regime
降噪技术 muffling technique

jiao

交变电压 alternating voltage
交变荷载 alternating loading
交变压力 alternating pressure
交变应力 alternating stress
交叉地 crosswise
交叉角 crossing angle
交叉流 crossflow
交叉流动 cross flow
交叉流式冷却塔 cross flow cooling tower
交叉流式热交换器 cross flow heat exchanger
交叉通风炉 cross-draught stove
交错浪 cross sea
交换场 exchange field
交换剂 exchanger
交换率 exchange rate
交换能 exchange energy
交换速度 exchange rate
交角 inclination
交流变频器 AC frequency converter
交流变速传动 AC-adjustable speed drive
交流电 alternating current (AC)
交流电机 alternating current machine
交流电压 alternating voltage
交流断流板 AC breaker panel
交流断路器 AC breaker
交流发电机 AC machine, alternator
交流负载 AC load
交流光伏组件 AC-pv module
交流电动机 AC motor
交流换热 regenerative heating
交流换热器 regenerative heat exchanger
交流回路 AC circuit
交流机 AC machine
交流励磁电流 AC excitation current
交流励磁系统 AC excitation system
交流伺服电机 AC servo motor
交流整流器 AC rectifier
郊区太阳能草坪灯工程 solar lawn light for suburb
胶带 adhesive tape
胶合 glue

胶合板残渣　plywood residue
胶皮管　hose
胶体充电系统　colloid charging system
焦耳　Joule
焦耳定律　Joule law
焦耳加热　Joule heating
焦耳热　Joule heat
焦化加煤机　coking stoker
焦炭床层　coke bed
焦线　focal line
焦油裂解气化炉　tarcracking gasifier
焦油热裂解　thermal tar cracking
角撑板　gusset plate
角锉　angle grinder
角顶　vertex
角动量　angular momentum
角动量方程　angular momentum equation
角度　viewing angle, beam angle
角干扰系数　angular induction factor
角焊　fillet weld
角砾岩区　breccia zone
角落流　corner flow
角盘　angle plate
角频　angular frequency
角闪岩　amphibolite
角速度　angular speed, angular velocity
角速度矢　angular velocity vector
角位置　angular position
角向运动　angular motion
角运动　angular motion
绞车　drawwork
绞盘车床　capstan lathe
铰接桨叶　articulated blade
铰链弹簧转子　hinge-spring rotor
铰链接合　knuckle joint
脚泵　foot pump
搅拌器　stirrer, agitator, pulsator

酵母　yeast
酵素　enzyme

jie

阶梯置放蝶式太阳能蒸馏器　tilted tray solar still
阶跃电流　step current
阶跃响应　step response
接触电压　touch voltage
接触器　contactor
接触时间　contact time
接触式电压　contact voltage
接触式风杯风速计　contact cup anemometer
接触式风速计　contact anemometer
接地　earthing, grounding
接地棒　ground rod
接地导体　grounding conductor
接地电流　earth current
接地电路　earthed circuit, resistance of an earthed conductor
接地回路　ground loop
接地回路热交换器　ground loop heat exchanger
接地基准点　earthing reference point
接地开关　earthing switch
接地体　earth electrode
接地铜网格　underground copper grid
接地线　earth conductor
接地线圈　ground coil
接地装置　earth termination system
接电装置　switchgear
接箍定位器　collar locator
接合的　articulated
接合垫　gasket
接合线　bonding wire
接户线　service main

接口　interface
接口电路　interface circuit
接力泵　booster pump
接力抽出器　booster ejector
接力器　servomotor
接力器行程　servomotor stroke
接力器基础板　servomotor foundation plate
接入风速　cut-in wind velocity
接闪器　air-termination system
接收面　receiving surface
接收器采光口　receiver aperture
接受能量　received energy
接通　cutting in
接通时间　connect time
接线端　terminal
接线盒　junction box
接线器　connector
秸秆　straw
秸秆燃烧　straw combustion
节点　pitch point
节点电价　locational marginal price
节点温差　pinch point temperature difference
节距控制　pitch controlling
节距圈　pitch circle
节理　joint
节流阀　throttle, throttle valve, shut off valve, restrictor
节流损失　throttling loss
节面角（锥齿轮的）　pitch angle
节能　energy conservation
节能灯　energy saving lamp
节能技术　energy efficiency technology
节能绩效保证合约　energy savings performance contracting
节能率　fractional energy savings

节能新技术　new energy conservation technology
节气门　gasoline throttle
节热器　economiser
节约装置　economizer
节制闸　control gate
节锥锥进　pitch coning
洁净能源　clean energy
结冰传感器　ice sensor
结构动力学　structural dynamics
结构刚度　structural stiffness
结构载荷　structural load
结构支撑　structural support
结垢沉积物　fouling deposit
结垢损失　scaling loss
结合　combination
结合电容　coupling capacitor
结合混合料　binder matrix
结合键　binding bond
结合水　bound water
结晶　crystallization
结晶变质岩　crystalline metamorphic rocks
结晶粒　close grain
结晶学　crystallography
结晶岩石　crystalline rock
结渣　slag deposit
结渣性　slagging propensity
截口损失　kerf loss
截流　closure
截面阻塞影响　blockage effect
截铁器　iron cutter
截位傅里叶序列分解法　truncated Fourier series decomposition method
截止电压　non-conducting voltage
截止阀　cut out valve, on-off valve, stop valve, check valve

解环　ring opening
解聚作用　depolymerization
解缆　untwist
解列　splitting
解列点　splitting point
解散　dissolution
解调器　demodulator
解析公示　analytical equation
介电层　dielectric layer
介电常数　dielectric constant
介入　intervene
介稳态过渡相　metastable phase
介质试验　dielectric test
介质损耗　dielectric loss
界面波　capillary wave
界面能　interface energy
界面缺陷　interface defect

jin

金刚砂　emery
金刚砂布　emery cloth
金刚砂旋转磨石　emery wheel
金刚石晶体结构　diamond crystal structure
金刚石钻进　diamond drilling
金刚石钻井　diamond drilling
金属部件　metallic component
金属惰性气体焊接　metal-inert gas welding
金属废料　metallic scrap
金属腐蚀　corrosion of metal
金属盒通道吸热板　metal box duct absorber plate
金属间化合物　intermetallic
金属键　metallic bond
金属绝缘硅　metal-insulator-silicon
金属—绝缘体—半导体太阳电池　metal-insulator-semiconductor solar cell
金属锂　metallic lithium
金属卤化物灯　metal halogen lamp
金属疲劳　metal fatigue
金属氢化物　metal hydride
金属纤维　metal fibre
金属盐水合物　metallic salt-hydrate
金属氧化物半导体场效应晶体管　metal oxide semiconductor field effect transistor
金属轴　metal shaft
金丝雀草　canary grass
襟翼　flap
襟翼空气制动　flap air brake
襟翼偏转角　flap angle
襟翼造波机　flap wavemaker
紧凑式热交换器　compact heat exchanger
紧凑式太阳(能)热水器　close-coupled solar water heater
紧凑型潮力发电站　compact tidal power station
紧固件　fastener
紧急关机　emergency shutdown
紧急停车按钮　emergency stop push button
紧急停运　emergency outage
紧急停止　emergency stop
紧急烟道出口　emergency stack outlet
紧急止流阀　emergency valve
紧急制动系统　emergency braking system
进潮口　tidal creek, tidal inlet
进潮量　tidal influx tidal prism
进动轴　precession axis
进风　air supply, air intake, inflow wind
进风过滤器　air-intake filter
进风口　air inlet, air intake, air scoop
进风口开度　air intake opening

进给阀　feed valve, charging valve
进给速率　feed rate
进口导叶　inlet bucket
进口冷盐罐　molten salt receiver inlet vessel
进口能源　imported energy
进口压力　entrance pressure
进料槽　feed chute
进料斗　feed hopper
进料混合器　feed mixer
进料搅拌机　feed mixer
进气槽　intake pit
进气端　admission end
进气阀　intake valve
进气管　intake pipe
进气管道　air intake duct
进气加热器　air intake heater
进气口开度　air intake opening
进气喇叭口　air scoop
进气量　air input
进气面积　capture area
进气歧管　intake manifold
进气温度　inlet air temperature
进气支管　air intake branch
进汽端　admission end, upstream end
进水　water intrusion
进水道　intake pipe
进水口　water inlet, water intake
进水前池　forebay
近岸海洋热能电站　nearshore ocean thermal power plant
近岸环流　nearshore circulation
近岸流　coastal current, nearshore current
近岸水域　nearshore waters
近岸水域装置　nearshore device
近滨带　nearshore zone
近场　near field
近地表地下水　near surface groundwater
近地表风　near surface wind
近地表岩石　near surface rock
近地点　perigee
近地点潮　（指月球运行到近地点附近时潮差随之增大的潮汐）perigean tide
近地点大潮　perigean spring tide
近地风　near surface wind
近海风电场　offshore wind farm
近海风力　offshore wind power
近海风力发电站　offshore wind station
近海风力发电装置　offshore wind power plant
近海风能系统　offshore wind power system
近海结构　offshore structure
近海排放　offshore discharge
近海区　offshore area
近海石油平台　off shore oil rig
近海钻井　offshore drilling
近环境压力操作　near-ambient pressure operation
近间距　closer spacing
近区　near field
近日点　perihelion
近似分析　proximity effect
近尾流　near wake
劲风　fresh breeze
浸出　leaching
浸出法　leaching process
浸灰剂　liming agent
浸入罐体式加热器　tank immersion heater
浸水水线面　flooded waterplane
浸提致滞后断裂　Infusion-induced delayed fracture
浸透　percolation

浸渍　bucking
禁带宽度　energy gap

jing

经济可行性　economic feasibility, economic viability
经济能力　economic viability
经济潜势　economic potential
经济性分析　economic analysis
经济性评价　economic assessment, economic evaluation
经济蕴藏能量　economic energy potential
经向流动　diametral flow
经验测试　empirical test
经验关系　empirical relationship
经验系数　empirical coefficient
经营成本　operating cost
晶胞　unit cell
晶格　crystal lattice
晶格参数　lattice parameter
晶格空位　vacant lattice site
晶硅电池　crystalline silicon cell
晶硅太阳能电池　crystalline-silicon photovoltaics
晶核　nucleus
晶核生长　nucleus growth
晶界　grain boundary
晶界迁移　grain boundary migration
晶界形核　boundary nucleation
晶粒　crystalline grain, grail
晶粒大小　grain size
晶粒取向电用硅钢片　grain-oriented electrical steel
晶粒细化　refining
晶片　wafer
晶片清洗　wafer cleaning
晶体　crystal

晶体半导体　crystalline semiconductor
晶体表面　crystal surface
晶体管　transistor
晶体硅　crystal silicon, crystalline silicon
晶体硅太阳电池　crystalline silicon solar cell
晶体结构　crystal structure
晶体缺陷　crystal defect, crystal imperfection
晶体生长　crystal growth
晶体生长提拉法　Czochralski method
晶体学取向关系　crystallographic orientation
晶体闸流管逆变器　thyristor rectifier
晶系　crystal system
晶圆切割机　dicing saw
晶圆清洗　wafer cleaning
晶闸管　silicon controlled rectifier, thyratron
晶闸管保护电路　thyristor protective circuit
晶闸管触发电路　thyristor trigger circuit
晶闸管阀　thyristor valve
晶闸管换流阀　thyristor valve
晶闸管级　thyristor level
晶闸管可控串补偿　thyristor controlled series compensation
晶闸管控制电感　thyristor controlled inductance
晶闸管控制电抗器　thyristor controlled reactor
晶闸管软启动装置　thyristor soft-start unit
晶闸管整流器　silicon controlled rectifier, thyristor rectifier
晶闸管组件　thyristor module
晶族　crystal group
精炼　reflectance

精米工厂	rice milling plant
精密计时表	chronometer
精细材料	fine material
精细粉碎机	fine grinding mill
精研机	lapping machine
井壁	borehole wall
井壁基环	crib
井壁坍塌	wellbore collapse
井产量	well production
井底记录温度仪	downhole recording temperature gauge
井底压力	bottom hole pressure
井管	well casing
井管滤网	well screen
井距	well spacing
井孔	borehole
井孔寿命	life of well
井口钢管	steel pipe at wellhead
井口设备	wellhead equipment
井口压力	wellhead pressure
井漏带	loss of circulation zone, lost-circulation
井漏区	lost-circulation zone
井内流体	well fluid
井内压力恢复	pressure build-up in well
井喷	well blowout
井喷防护	blowout preventer
井群	group of wells
井水源热泵	well water heat pump
(地热井)井下操作	downhole operation
井下测量	downhole measurement
井下点温	spot downhole temperature
井下热泵	downhole heat pump
井下热交换技术	downhole heat-exchanger technology
井下温度	downhole temperature, subsurface temperature
井下资料	subsurface information
井斜	well deviation, borehole deviation
井斜测量	inclination measurement
井眼尺寸	borehole size
井眼清洗过程	wellbore cleaning process
井眼直径	wellbore diameter
井之间的干扰	interference between wells
景观	landscape
景观工程	landscape engineering
景观评价（估）	landscape assessment
径流	runoff
径流泵	radial flow pump
径流强度	runoff intensity
径流式发电厂	run-of-river power station
径流式系统	runoff system
径流式压气机	radial flow compressor
径流涡轮机	radial flow turbine
径向波函数	radial wave function
径向场	radial field
径向分布	radial distribution
径向峰值热点因子	radial peaking hot spot factor
径向干扰	radial interference
径向荷载	radial load
径向混合	radial mixing
径向加速度	radial acceleration
径向间隙感应机	radial gap induction machine
径向聚焦	radial focusing
径向力	radial force
径向量子数	radial quantum
径向流动	radial flow
径向曲率	radial buckling
径向热管	radial heat pipe
径向速度	radial velocity

径向通量 radial flux	净热交换器 net heat exchanger
径向无反射层的 radially bare	净热效率 net thermal efficiency
径向销连接 radial pin coupling	净热增益 net heat gain
径向叶片 radial vane	净热值 net calorific value
径向叶片风机（扇） radial blade fan	净输出电量 net energy output
径向引出 radial extraction	净水头 net head
径向应力 radial stress	净损失 dead loss
径向再生区 radial blanket	净太阳辐射 net solar radiation
径向增殖区 radial breeder	净现值 net present value
径向振荡 radial oscillation	净值 clean value, net value
径向轴 radial axis	净重 net weight
径向坐标 radial coordinate	静触头 fixed contact
净长波辐射 net longwave radiation	静电 static electricity
净参考热负荷 net reference thermal load	静电除尘器 electrostatic precipitator
净电荷 net charge	静电功率输出 net electric power output
净电量计量法 net metering	静电过滤器 electrostatic filter
净短波辐射 net shortwave radiation	静电透镜 electrostatic lens
净吨 net ton	静电学 electrostatics
净发电量 net generation	静风 calm wind
净（全）辐射 net total radiation	静负荷 dead load
净（全）辐射表 net pyrradiometer	静强度 static strength
净辐射通量 net radiation flux	静区 shadow
净负荷 payload	静水 static water
净功率 net power	静水面 still-water level
净换热器 net heat exchanger	静水模型 hydrostatic model
净集热效率 net collection efficiency	静水位 hydrostatic level
净化 distillation	静水压力 hydrostatic force, hydrostatic pressure
净空间电荷 net space charge	静水压试验 hydrostatic pressure test
净力矩 net torque	静态电流 quiescent current
净能比 net energy ratio	静态工程投资概算 static project cost estimate
净能计量 net energy metering	
净能量产出 net energy output	静态集光器 static concentrator
净能量增益 net energy gain	静态聚合物钻液井 static polymeric drilling fluid
净气器 gas scrubber	
净热 net heat	静态水体 static water
净热耗率 net heat rate	静态稳定极限 steady state stability

limit	
静态稳定限度	steady state stability limit
静态无功补偿器	static var compensator
静态无功控制器	static var controller
静态型同步补偿器	static synchronous compensator
静态总线阻抗	static bus impedance
静压孔	static pressure hole
静压力	static pressure
静叶持环	blade carrier
静叶片	stator blade
静载	dead load
静载荷试验	static load testing
静止空气	air at rest
静止能量	rest energy
镜面导管	specular conductor
镜面反射比	specular reflectance
镜面清洁度	mirror cleanliness
镜子	mirror

jiu

九级风	strong gale
酒精	ethyl alcohol, fuel bioethanol
酒精发酵	alcohol fermentation
酒精燃料	alcohol fuel
就地储能	on-site energy storage
就地控制	local control

ju

居间带太阳能电池	Intermediate band solar cell
居里温度	Curie point
居住建筑	domestic building
居住区风—光互补系统	wind and PV hybrid power system for residential areas
局部供热	district heating
局部供热装置	space heater
局部环流	local circulation
局部平衡	localized equilibrium
局部热处理	local heat treatment
局部温度扰动	local temperature disturbance
局部效应	local effect
局部阻力	local resistance
局地风	local wind
局间的	interoffice
矩鞍形填料	intalox saddle
矩形管	rectangular tube
矩形波脉冲	square wave pulse
矩形堰	rectangular weir
矩阵解	matrix solution
矩阵式变换器	matrix converter
矩阵运算	matrix operation
举力	lift force
巨大海浪	breaking sea
巨浪	heavy sea
巨石	boulder
巨型风车	jumbo windmill
巨型风力发电机	jumbo windmill generator
巨涌	surge
剧刃	blade for iron saw
距离常数	distance constant
飓风	hurricane, hurricane force
飓风塔	tornado tower
锯木厂	sawmill
锯木厂废弃物	sawmill waste
锯木厂垃圾	sawmill waste
锯屑	sawdust
聚变反应	fusion reaction
聚波	wave focusing
聚风器	wind concentrator

聚光板　solar panel
聚光场　concentrator field
聚光场采光面积　concentrator field aperture area
聚光场年效率　annual efficiency of concentrator field
聚光场效率　efficiency of concentrator field
聚光电池　concentrated cell, concentrator cell
聚光光学系统　concentrating optics
聚光镜面　concentrating mirror surface
聚光率　concentration ratio
聚光器　concentrator
聚光器表面轮廓误差　concentrator surface counter error
聚光器采光面积　concentrator aperture area
聚光器跟踪精确度　tracking precision
聚光器跟踪误差　tracking error
聚光器跟踪准确度　tracking accuracy
聚光太阳电池方阵　photovoltaic concentrator array
聚光太阳电池方阵场　photovoltaic concentrator array field
聚光太阳电池组件　photovoltaic concentrator module
聚光太阳能电池　concentrator solar cell, concentrating solar cell
聚光太阳能电池阵列　array of concentrating solar cell
聚光太阳能发电　concentrating solar power
聚光太阳能热电联供系统　concentrating solar cogeneration system
聚光型集热器的吸热体面积　absorber area of concentrating collector
聚光型太阳能集热器　concentrating solar collector
聚光型太阳灶　concentrating solar cooker
聚合波道　converging wave channel
聚合燃料电池　polymer fuel cell
聚合物半导体电池　polymer semiconductor cell
聚合物半导体太阳电池　polymer semiconductor solar cell
聚合物薄膜太阳能集热器　polymer film solar collector
聚合物苯基丙烷　polymer phenyl propane
聚合物电解质　polymer electrolyte
聚焦反射镜　focusing mirror
聚焦光学器件　focusing optics
聚焦集热器　concentrating (focusing) collector, focusing collector
聚焦式太阳能发电　concentrating solar power
聚焦系统　focusing system
聚焦效率　focusing efficiency
聚焦型塑料集热器　focusing collector of plastics
聚焦型（性）太阳能集热器　focusing solar energy collector, solar concentrating collector
聚晶金刚石复合片　polycrystalline diamond compact
聚硫化溴电池　polysulfide bromide battery (PSB)
聚氯乙烯　polyvinyl chloride
聚马来酸　polymaleic acid
聚乳酸塑料　polylactic acid plastic
聚四氟乙烯薄膜　Teflon film
亚胺薄膜　polyimide film
聚酰亚胺　polyimide

聚乙烯管（PE管）　polyethylene pipe
聚酯　polyester
聚酯树脂　polyester resin

juan

涓流集热器　trickle collector
卷包式太阳电池　wrap-around type solar cell
卷边工具　crimping tool
卷积积分　convolution integral
卷浪　plunging breaker
卷绕机　winder
卷碎波　plunging breaker
卷筒　winding drum
卷扬机　winder
卷云　cirrus cloud

jue

决定性变量　deterministic variable
绝对辐射表　absolute radiometer
绝对光谱灵敏度　absolute spectral sensitivity
绝对光谱灵敏度特性　absolute spectral sensitivity characteristic
绝对光谱响应　absolute spectral response
绝对黑体　absolute black body
绝对零度　absolute zero
绝对零值　absolute zero
绝对日射测量　absolute pyrheliometer measurement
绝对日射测量计　absolute pyrheliometer
绝对日温计　absolute pyrheliometer
绝对湿度　absolute humidity
绝对式编码器　absolute encoder
绝对温标　absolute (temperature) scale
绝热（性）　heat insulation
绝热百叶窗　thermal insulation shutter
绝热板　insulating plate
绝热层　thermal barrier
绝热大气　adiabatic atmosphere
绝热的　heat insulated
绝热底座　insulation bed
绝热反应温度　adiabatic reaction temperature
绝热隔板　thermal baffle
绝热工况　adiabatic condition
绝热过程　adiabatic process
绝热井筒流动　adiabatic wellbore flow
绝热膨胀　adiabatic expansion
绝热曲线　adiabatic curve
绝热燃烧温度　adiabatic combustion temperature
绝热交换器　adiabatic exchanger
绝热热力系统　adiabatic thermodynamic system
绝热式热量计　adiabatic calorimeter
绝热炭气化　adiabatic char gasification
绝热条件　adiabatic condition
绝热温度　adiabatic temperature
绝热效率　adiabatic efficiency
绝热性能　heat insulating property
绝热增温　adiabatic heating, adiabatic warming
绝热指数　adiabatic exponent
绝热储能　adiabatic energy storage
绝热状态　adiabatic condition
绝缘板　insulating board, insulating plate, laminate
绝缘材料　insulating material
绝缘等级　insulation level
绝缘电阻　insulation resistance
绝缘隔板　insulating separator
绝缘功率因数　insulation power factor
绝缘硅钢片　insulated steel lamination

绝缘漆 insulating varnish
绝缘手套 insulating glove
绝缘套 insulating sleeving
绝缘套管 insulated sleeve, insulating bushing
绝缘体 electrical insulator, insulator, insulation
绝缘填料 filled insulation
绝缘物 insulant
绝缘靴 insulating boot
绝缘栅双极晶体管 insulated gate bipolar transistor
绝缘砖 insulating brick
绝缘子 insulator
攫取 grab

jun

均方根值 root mean square value
均分负载 load sharing, load distribution
均衡 equilibrium
均衡阀 balancing valve
均化 homogenization
均化作用 homogenization
均立方风速 root mean cube wind speed
均热段 soaking zone
均热区 soaking zone
均性化 homogenization
均压变压器 balancer transformer
均压环 grading ring
均匀的核化 homogeneous nucleation
均匀地形 homogeneous terrain
均匀度 degree of consistency
均匀化热处理 homogenization heat treatment
均匀搅拌反应器 perfectly stirred reactor
均匀流动 flow uniformity
均匀热源 homogeneous heat source
均匀态 homogeneous state
均质体 isotropic body
均质土壤 homogeneous soil

ka

卡口 bayonet
卡琳娜低温太阳能热发电 low temperature solar thermal power with Kalina cycle
卡内基—米隆系统（一种利用氨作工作流体的闭环海洋温差发电系统）CMU system
卡诺效率 Carnot efficiency
卡诺循环 Carnot cycle
卡诺循环效率 Carnot cycle efficiency
卡普兰水轮机 Kaplan turbine
卡普曼燃料 K-fuel
卡钳 caliper
卡线钳 conductor clamp

kai

开槽式太阳能热水器 open trough solar water heater
开敞角 opening angle
开断触头 break contact, b-contact
开断电流 drop out current
开尔文 Kelvin
开尔文波 Kelvin wave
开发方式 development scheme
开放焚烧 open burning
开放式（海洋发电厂） open cycle
开放式集热器 open loop collector
开放式空气预热系统 open-loop air-preheating system
开放式循环动力机 open cycle engine
开放式循环发动机 open cycle engine

开放式循环系统　open cycle system
开放式循环周期　open loop cycle
开关　switching, switch
开关磁阻　switched reluctance
开关磁阻电机　switched reluctance machine
开关磁阻发电机　switched reluctance generator
开关电路　switching circuit
开关设备　switchgear
开关象限　switching quadrant
开管淀积　open tube deposition
开环控制系统　open loop control system
开环应用程序　open loop application
开环主动式太阳能系统　open loop active solar system
开火角　firing angle
开口　hatch
开口管　open-ended pipe
开口铆钉　bifurcated rivet
开口式空气吸热器　open loop air receiver
开口式液体单板（太阳能）集热器　open flow flat plate collector
开口系统　vented system
开口销　cotter pin
开阔地带　open site
开阔水面　open water
开裂式萨沃纽斯形风轮机　split Savonius wind turbine
开路　open circuit
开路电抗　open circuit reactance
开路电流　open current
开路电压　open circuit voltage
开路费　right-of-way fee
开路光生电压　open circuit photovoltage
开路式风洞　open ended wind tunnel
开路特性　open circuit characteristic

开路循环海洋热能转换装置　open cycle ocean thermal energy conversion unit
开路运行　open circuit operation
开启电压　firing voltage
开始　commencement
开氏度　Kelvin
开氏温标　degree Kelvin
开式防滴型　open drip proof
开式海洋温差电站　open circuit OTEC plant
开式回热循环　regenerative open cycle
开式循环　open cycle
开式循环海洋热能转换　open cycle ocean thermal energy conversion, open cycle OTEC
开式循环海洋热能转换装置　open cycle ocean thermal energy conversion unit
开式循环海洋温差电站　open cycle ocean thermal difference power plant
开式循环涡轮机　open cycle turbine
开式循环系统　open loop system, open cycle system
开式盐差能动力装置　open power plant ennergy from salinity gradient
开态电流　on-state current

kan

勘测工具　reconnaissance tool
勘探策略　exploration strategy
勘探评估　prospect evaluation
勘探钻井　exploratory drilling
坎贝尔图　Campbell diagram

kang

抗（耐）湿电池　moisture resistant cell

抗扯力	tear resistance
抗冲击	impact resistance
抗断强度	breaking strength
抗反射膜	anti-reflection coating
抗风构架	wind frame
抗风桁架	wind resisting truss
抗风肋	wind rib
抗风梁	wind beam
抗风柱	end panel column
抗辐射的	radiation hardened
抗辐照损伤	resistance to irradiation damage
抗腐蚀	corrosion resistant
抗腐蚀性	corrosion resistance
抗机械强度高	strong mechanical resistance
抗拉试验	tensile test
抗磨的	abrasion-resistant
抗磨损性	abrasion resistance
抗挠刚度	flexural rigidity
抗扭强度	torsional strength
抗疲劳极限	fatigue endurance limit
抗疲劳强度	fatigue resistance
抗疲劳性	fatigue resistance
抗扰性	immunity
抗热的	heat proof
抗热冲击阻力	thermal shock resistance
抗热震阻力	thermal shock resistance
抗散裂强度	spalling resistance
抗渣性	slag resistance
抗张强度	tensile strength

ke

苛性溶液	caustic chemical solution
柯肯达尔效应	Kirkendall effect
科克雷尔筏波能转换器	Cockerel raft wave energy conversion device
科里奥利	Coriolis
科里奥利力	Coriolis force
科氏	Coriolis
科氏力	Coriolis force
科氏偏转	Coriolis deflection
科氏修正	Coriolis correction
颗粒	grail
颗粒壁炉	pellet stove
颗粒床储热	particle bed storage
颗粒分离	particle separation
颗粒含水层	granular aquifer
颗粒间蒸发（作用）	intergranular vaporization
颗粒粒度	particle size
颗粒密度	particle density
颗粒排放	particulate emission
颗粒燃烧机	pellet burner
颗粒筒仓	pellet silo
颗粒维数	particle dimension
颗粒物排放	particulate emission
颗石藻	coccolithophre
壳管热交换器	shell and tube heat exchanger
壳管式	shell and tube
壳体吨位	casing tonnage
壳体式太阳电池阵	body-mounted type solar (cell) array
壳体下入深度	casing setting depth
壳体直径	casing diameter
壳体种类	casing type
可编程逻辑控制器	programmable logic controller
可编程序控制	programmable control
可变比热	variable specific heat
可变反射层	variable reflective layer, variable reflective coating
可变反射膜	variable reflective coating

可变几何翼型风力机 variable geometry type wind turbine
可变频率驱动式起动机 variable frequency drive starter
可捕获风能 capturable wind power
可采资源 recoverable resource
可拆桨叶 detachable blade
可沉入水中的 submersible
可持续容量 sustainable capacity
可持续生物燃料圆桌论坛 roundtable on sustainable biofuels
可储存的 storable
可达深度 attainable depth
可锻铸铁 casting, malleable iron
可感热 sensible heat
可忽略排放 negligible emission
可回圈寄生能源 parasitic power
可回收波能资源 recoverable wave energy resource
可回收资源 recoverable resource
可获能量 extractable energy
可见辐射 visible radiation
可见光 visible light
可见光传感器 visible light transducer
可见光谱 visible spectrum
可交变温度 interchangeable temperature
可锯金属的弓形锯 hacksaw
可开采量 exploitable reserve
可开发（热）储层 exploitable reservoir
可靠峰值输出功率 dependable peaking capacity
可靠能源指数 energy index of reliability
可靠性 reliability
可靠性测定试验 reliability determination
可控桨叶长度 controllable length blade
可控闪蒸过程 controlled flash evaporation process
可控源音频大地电磁法 controlled source audio-frequency magnetotellurics
可利用率 availability
可磨性指数 Hardgrove grindability index
可能最大风暴潮 probable maximum storm surge
可能最大飓风 probable maximum hurricane
可能最大热带气旋 probable maximum tropical cyclone
可能最大温带风暴 probable maximum extratropical storm
可逆电池 reversible cell
可逆阀 reversing valve
可逆过程 reversible process
可逆化学反应储热 reversible chemical reaction heat storage
可逆绝热压缩 reversible adiabatic compression
可逆循环 reversible cycle
可潜式排水泵 submersible type sump pump
可燃废物 combustible waste
可燃废物燃气焚烧炉 gas-fired waste combustible incinerator
可燃垃圾 combustible waste
可燃气体 burnable gas, flammable gas
可燃气体，可燃性气体 combustible gas
可燃物负荷 fuel load
可燃物负荷量 fuel load
可燃物类型 fuel type
可燃吸收体 burnable absorber
可燃性废料处理 treatment of combustible waste
可溶矿物质 soluble mineral
可溶性单体 soluble monomer
可溶性硅 reactive silica

可溶性盐　soluble salt
可溶盐　soluble salt
可渗透岩层　permeable rock
可湿性生物质　wettable biomass
可提取电力　extractable power
可替代能源　alternative energy
可调桨叶角度的水轮机　variable pitch turbine
可调角度　adjustable angle
可调聚焦集热器　adjustable focusing collector
可调螺距　adjustable pitch
可调螺距螺旋桨　variable pitch propeller
可调螺距水轮机　variable pitch turbine
可调式节流阀　adjustable choke
可听噪声　audible noise
可消化性　digestibility
可行性试验　feasibility test
可行性研究　feasibility study
可延长体　elongated body
可用发电　available power generation
可用功率　useful power
可用集热　usable heat collection
可用能　available energy
可用能量的损失　available power loss
可用热　available heat
可用装机容量评估　possible installed capacity estimate
可用资源评估　possible resource estimate
可再生电　renewable electricity
可再生海洋能源　renewable ocean energy source
可再生技术　renewable technology
可再生能源　renewable energy, renewable energy resources
可再生能源电力　renewable energy source electricity
可再生能源发电　renewable generation
可再生能源法　renewable energy law
可再生能源技术　renewable energy technology
可再生能源配额制　renewable portfolio standard
可再生能源信贷　renewable energy credits
可再生能源指令　renewable energy directive
可再生能源组合　renewable energy portfolio
可再生取暖油　renewable heating oil
可再生热　recoverable heat
可再生生物能源　renewable biological source
可再生太阳能　renewable solar energy
可再生一次能源　renewable primary energy
可再生自然资源　renewable natural resource
可照时数　duration of possible sunshine
克莱姆法则　Cramer's Rule

keng

坑木　creosoted timber
坑蚀　pitting corrosion

kong

空间变化　spatial variation
空间电荷　space charge
空间电荷层　space charge layer
空间电荷密度　space charge density
空间电荷区　space charge region
空间电荷效应　space charge effect
空间分布　spatial distribution

空间环境效应　space environment effect
空间加热器　space heater
空间冷却　space cooling
空间能源系统　space energy system, space power system
空间矢量　space vector
空间太阳能　space solar energy
空间太阳能电池　space solar cell
空间太阳能发电站　solar-powered orbiting generating station, space solar energy station, space solar power station
空间太阳能系统　space solar power system
空间谐波　space harmonics
空间用多节太阳能电池　multijunction space solar cell
空间用太阳能激光动力站　solar laser power station for space use
空间用太阳能源　solar source in space
空间制冷　space cooling
空壳层　vacant shell
空冷凝汽器　air-cooled condensor
空冷式冷凝器　air condenser
空冷系统　air-cooled system
空泡流　cavity flow
空气饱和值　air saturation value
空气比　air ratio
空气边界层　air boundary layer
空气波　air wave
空气测量设备　air metering device
空气测微计　air micrometer
空气抽吸装置　air-suction system
空气储备　air-storage
空气储气瓶　air accumulator
空气处理　air handling
空气处理单元　air handling unit (AHU)
空气处理机　air handler
空气处理机组　air handling unit
空气吹洗　air purging
空气锤　air hammering
空气磁力断路器　air magnetic power circuit breaker
空气弹簧悬架　air spring suspension
空气弹性变形的不稳定（性）　aeroelastic instability
空气导阀　air pilot valve
空气电池　air cell
空气电极　air electrode
空气垫　air cushion
空气动力　aerodynamic force
空气动力刹车　aerodynamic brake
空气动力刹车系统　aerodynamic braking system
空气动力的　aerodynamic
空气动力计算　aerodynamic calculation
空气动力力矩　aerodynamic moment
空气动力设计　aerodynamic design
空气动力损失　aerodynamic loss
空气动力特性　aerodynamic behavior
空气动力调节　aerodynamic regulation
空气动力性能　aerodynamic performance
空气动力学效率　aerodynamics efficiency
空气动力噪声　aerodynamic noise, air dynamic noise
空气动力阻尼　aerodynamic damping
空气断路器　air circuit breaker
空气对流　air convection
空气发动机　air engine
空气发生炉煤气　air producer gas
空气分布　air distribution
空气分布板　air distribution plate
空气分布器　air distributor
空气分级　air staging

空气分级燃烧 staged-air combustion
空气分离器 air separator
空气浮选机 air cell
空气浮选装置（污水处理） air floatation unit
空气辐射温度计 aerial radiation thermometer
空气附面层 air boundary layer
空气干燥机 air drying machine
空气干燥基 air dry basis
空气干燥基水分 moisture in the air dry basis
空气干燥器 air dryer
空气更新率 air renewal rate
空气共振 air resonance
空气供暖装置 air heating system
空气供热 air heating
空气供热系统 air heating system
空气鼓风机 air blower
空气鼓风型曝气槽 air blow system aeration tank
空气鼓风悬浮分离装置 air blow floatation unit
空气管道 air pipe
空气管道系统 air circuit
空气过滤机组 air filter unit
空气过滤器 air filter, air strainer
空气过滤设备 air filtration unit
空气过滤吸附机组 air filter and absorber unit
空气环境参数 air parameter
空气缓冲器 air damper
空气换热器 air heat exchanger
空气混合箱 air mixing chamber
空气集热器 air (heating) collector
空气剂量 air dose
空气加热 air heating

空气加热器 air heater
空气加热式平板集热器 air-heating flat plate collector
空气加热式太阳能集热器 air-heating solar collector
空气加热系统 air-heating system
空气加热型集热器效率因数 air heating collector efficiency factor
空气夹层 air blanketing
空气监测器 air monitor
空气减湿 dehumidification
空气接触器 air contactor
空气节能器 air economizer
空气净化 air cleaning
空气净化器 air cleaner, air purifier
空气净化设备 air cleaning facility
空气净化系统 air cleaning system
空气净化装置 air purification equipment, air cleaning equipment
空气—空气换热器 air-to-air heat exchanger
空气—空气热泵 air-to-air heat pump
空气扩散曝气仪 air diffusion aerator
空气扩散器 air diffuser
空气冷却（反应）堆 air-cooled reactor
空气冷却剂 air coolant
空气冷却器 air cooler
空气冷却蛇管 air coil
空气粒子 air particle
空气粒子检测器 air particle monitor
空气量 air capacity
空气量平衡 air quantity balance
空气流动试验 air flow test
空气流动吸热器 absorber with air flow
空气流量 air flow rate, air mass flow
空气流量计 air flowmeter
空气流泄 air drain
空气滤清装置 air filtration unit

空气马力　air horsepower
空气密度　air density
空气密度修正因数　air density correction factor
空气摩擦阻力　air friction
空气幕　air curtain, air screen
空气凝汽器　vent condenser
空气排放　air emission
空气喷射　air jet
空气品质　air quality
空气屏障　air curtain
空气气化　air gasification
空气强化脱硫过程　air solutizer process
空气燃料比　air fuel ratio, air-fuel ratio
空气热泵　air heat pump
空气蛇管　air coil
空气射流　air jet
空气渗透，漏风　air infiltration
空气湿度　air humidity
空气式热泵　air cycle heat pump
空气室　air cell, air chamber
空气输入系统　air inlet system
空气—水界面　air-water interface
空气—水冷却　air-water cooling
空气—水毛细压力　air-water capillary pressure
空气—水相互作用　air-water interaction
空气调节　air conditioning
空气调节机组　air conditioning unit
空气调节器　air governor, air register, air conditioner
空气调节系统　air consitioning system
空气调节装置　air conditioning plant
空气调温器　air attemperator
空气通风　air ventilation
空气涡轮　air turbine
空气涡轮波力电动机　air turbine wave motor
空气涡轮波力发电机　wave-activated air turbine generator
空气涡轮机型波能吸收器　air turbine type wave power absorber
空气涡旋　air eddy
空气污染　air contamination
空气污染分析　air pollutant analysis
空气污染监测器　air contamination monitor
空气污染检测仪　air monitoring instrument
空气污染控制　air pollution control
空气污染控制设备　air pollution control equipment
空气污染浓度　air pollution concentration
空气污染物　air contaminant
空气污染指数　air pollution index
空气雾化喷燃器　air atomizing burner
空气吸热器　air receiver
空气洗涤法　air cleaning
空气洗涤器　air scrubber
空气型单层玻璃盖板平板集热器　air-type single-layer glass cover flat plate collector
空气型平板集热器　air flat-plate collector
空气型真空管集热器　air-type evacuated tube collector
空气蓄电池　air accumulator, air battery
空气蓄压器　air accumulator
空气循环设备　air regenerating device
空气循环效率　air cycle efficiency
空气压缩机　air compressor
空气压缩机房　air compressor station
空气—岩石太阳能供暖系统　air to rock solar heating system
空气引射　air injection

空气预热　air preheating
空气预热器　air preheater
空气造冷循环　air refrigeration cycle
空气增湿器　air humidifier
空气制动　air brake
空气制动系统　air braking system
空气质量　air mass
空气质量和排放标准　air quality and emission standard
空气质量减少　air mass attenuation
空气中凝固　air set
空气轴承　air bearing
空气阻力　air drag, air resistance
空气阻尼　air damping
空气钻井　air drilling
空腔辐射　cavity radiation
空腔共振波能转换装置　cavity resonator
空腔加热器　cavity heater
空晒　exposure
空晒（集热器内只有空气而不充入其他载热工质时接受太阳辐射的工作状态）　stagnation test
空晒温度　exposure temperature
空速管　airspeed head, air speed tube
空调场所（A级）　air conditioned area (class A)
空调机组　air conditioning unit
空调区域　air conditioned location
空调装置　air conditioning apparatus
空位机理　vacancy mechanism
空位扩散　vacancy diffusion
空位浓度　vacancy concentration
空位缺陷　vacant defect
空心管　hollow pipe
空心纤维束　hollow fiber bundle
空心砖石墙　hollow-core masonry wall
空穴边界　void boundary

空穴锯　hole saw
空穴系数　void coefficient
空穴效应　void effect
空穴型半导体　positive-type semiconductor
空压机　air compressor
空压站　air compressed system
空载　no-load, free load, idle load
空载电流　non-load current
空载功耗　no-load power consumption
空载功率　idle capacity
空载时间　dead time
空载损失　no-load loss
空载消耗　no-load consumption
空载运行　no-load run, no-load operation
空载值　quiescent value
空载转速　no-load speed
空载最大加速度　maximum bare table acceleration
空转齿轮　idler gear
空转轮　idler pulley
空转时间　race time
空转消耗　no-load consumption
孔板量计　orifice meter
孔板流量计　orifice meter
孔壁坍塌　collapse of hole well
孔底试样采集　collection of bottom hole sample
孔径光阑　aperture diaphragm
孔口　orifice
孔流/闸门出流　sluice flow
孔蚀　pitting corrosion
孔隙　pore, pore space
孔隙比　void ratio
孔隙度　porosity
孔隙结构系数　coefficient of pore

structure
孔隙介质 porous media
孔隙率 void ratio, porosity
孔压 pore pressure
控制参数 governing parameter
控制齿轮 control gear
控制触头 control contact
控制电极 control electrode
控制电缆 control cable
控制电路 control circuitry, control circuit
控制电器 control apparatus
控制电压 control voltage
控制电子设备 control electronics
控制断层的泉水系统 fault-controlled spring system
控制阀 control valve
控制方案 control scheme
控制方程 governing equation
控制柜 control cabinet
控制孔 controllable orifice
控制面板 control panel
控制盘台 control panel and console
控制棚室 control shed
控制平台 control platform
控制器 controller, governor
控制器作用 controller action
控制强光 controlling glare
控制输入 control input
控制算法 control algorithm
控制塔 control tower
控制台 console
控制体积 control volume
控制系统 control system
控制线路 control wiring
控制再循环 controlling recirculation
控制中心 control center

控制装置 control device
（瑞士）厌氧分解装置 Kompogas AD facility

ku

枯竭井 dead well
枯柳 dry willow
库存泵 spare pump
库仑力 coulombic force
库容 capacity of reservoir
库容曲线 capacity curve, curve of reservoir capacity

kua

跨（短）接片 jumper
跨步电压 step voltage
跨季节（能量）储存 interseasonal storage
跨接 jumper
跨接线 jumper
跨向流动 spanwise flow

kuai

快热锅炉 flash boiler
快速裂解 fast pyrolysis
快速启动 quick startup
快速调谐 fast tuning
快速退火 rapid thermal annealing
快速响应仪表 fast response instrument
快速蒸汽发生器 flash steam generator

kuan

宽能带隙 wide band gap
宽频带辐射 broadband irradiance

kuang

狂暴 turbulence

狂风　whole gale, full gale, storm
矿产回收　mineral recovery
矿化垃圾　aged-refuse
矿化垃圾生物反应床　aged-refuse-based bioreactor
矿坑　stall
矿区信息　field information
矿石筛　trommel
矿物成分　mineral composition
矿物分析　mineralogical analysis
矿物共生　mineral assemblage
矿物颗粒　mineral particle
矿物燃料能量源　fossil-fuel-based energy source
矿物溶解　mineral dissolution
矿物提取　mineral extraction
矿物杂质　mineral impurity
矿物质沉积　mineral deposition
矿物质析出　mineral precipitation
矿物组成　mineralogical composition
矿物组合　mineral assemblage
框架　frame
框图　block diagram

kui

窥视孔　peephole
馈电电路　feed circuit
馈线　feeder, feeder line
溃坝　dam failure

kun

捆束机　bundler
捆扎生物质燃料　bundled biofue

kuo

扩充器　expander
扩大器　expander
扩风器增力型风托　diffuser-argumented rotor
扩开螺栓　expansion bolt
扩孔器　underreamer
扩散　dispersion
扩散泵　diffusion pump
扩散电容　diffusion capacity
扩散段损失　diffuse loss
扩散方程　diffusion equation
扩散工艺　diffusion process
扩散光电压　diffusion photo-voltage
扩散机理　diffusion mechanism
扩散结　diffused junction
扩散流量　diffusion flow rate
扩散率　diffusion rate
扩散面积　scattering area
扩散器　diffuser
扩散区　diffusion region
扩散速度　diffusion velocity
扩散特性　diffusion property
扩散体增强型风力机　diffuser augmented wind turbine
扩散通量　diffusion flux
扩散透射　diffuse transmission
扩散温度　diffusion temperature
扩散误差　diffuseness error
扩散系数　diffusion coefficient
扩展器　expander
扩张极大似然法　extended maximum likelihood method
扩张器　expander
扩张最大熵值法　extended maximum entropy principle
阔叶树　hardwood

la

垃圾场另类覆盖层　alternative daily

cover
垃圾车　garbage truck
垃圾处理　waste disposal
垃圾处理场　garbage disposal plant
垃圾处理设施　waste disposal infrastructure
垃圾处置　garbage disposal
垃圾储料仓　trash storage bunker
垃圾电厂　garbage power plant
垃圾堆　landfill
垃圾堆填区　landfill site
垃圾发电　waste to energy, municipal solid waste power generation
垃圾分类　waste classification
垃圾分类和管理　waste classification and management
垃圾分选　waste seperation
垃圾焚化　waste incineration
垃圾焚化炉　garbage furnace
垃圾焚烧　waste burning
垃圾焚烧处理设备　garbage incineration disposal device
垃圾焚烧电站　garbage incineration power station
垃圾焚烧发电　waste-to-energy, energy-from-waste, waste burning power generation
垃圾焚烧发电厂　waste-to-energy plant, waste-to-energy power plant
垃圾焚烧指令（欧盟）　waste incineration directive
垃圾焚烧炉　incinerator stoker, refuse incinerator
垃圾焚烧余热锅炉　waste incineration boiler
垃圾管理　waste management
垃圾锅炉　garbage-fired boiler

垃圾接收工艺　garbage collection technology
垃圾坑　dump pit
垃圾料斗　feed hooper
垃圾磨碎机　garbage grinder
垃圾能　waste energy, garbage power
垃圾起重机　waste crane
垃圾气化　waste gasification
垃圾倾倒费　gate fee, tipping fee
垃圾收集　garbage collection
垃圾填埋场　landfill site
垃圾填埋场沼气　landfill methane gas
垃圾填埋地　landfill
垃圾填埋气　landfill gas
垃圾填埋气回收　landfill gas recovery
垃圾喂入槽　waste feed chute
垃圾喂入机　refuse feeder
垃圾箱　garbage bin, garbage can
垃圾衍生燃料　refuse derived fuel
（欧盟）《垃圾掩埋法令》　Landfill Directive
垃圾预处理　waste pretreatment
拉德瑞罗地热电站（意大利）　Larderello geothermal power plant (Italy)
拉挤玻璃钢　pultruded FRP
拉挤成型　pultrusion
拉缆　guy cable
拉力　tensile force
拉力试验　tensile test
拉姆达控制探针　lambda control probe
拉普拉斯变换　Laplace transform
拉塞尔整流波能转换装置　Ressel rectifier wave energy conversion device
拉伸断裂　tensile failure
拉伸应变　tensile stress
拉绳　guy wire
拉索式格栅　guyed lattice
拉索式塔架　guyed tower

拉西环　Raschig ring
拉线杆　guyed pole
拉应力　expansion stress

lai

来料控制　incoming quality control
来料螺旋输送机　feed screw conveyor
来流风迎角　wind-approach angle
来流湍流　oncoming turbulence

lan

拦潮坝　tidal barrage
拦潮堤　tidal barrier
拦潮闸门　tidal gate
拦河坝　barrage
拦浪发电站　wave power breakwater
拦污栅　trashrack
拦蓄　impoundment
蓝灰色白云石　blue-gray dolomite
蓝色能量　blue energy
蓝藻　cyanobacteria
缆索连接系统　cable connection system

lang

榔头　club hammer
朗肯循环　Rankine cycle
朗肯循环发动机　Rankine cycle engine
朗缪尔环流　Langmuir circulation
浪成三角洲　wave delta
浪费　dissipation
浪高仪　wave height gauge
浪基面　wave base
浪轮机　wave turbine
浪蚀　marine erosion
浪蚀阶地　wave-cut terrance
浪蚀台　wave-cut platform
浪蚀台地　abrasion (abraded) platform
浪向　wave direction
浪涌　swell
浪涌保护器　surge protective device
浪涌电流　surge current

lao

捞沙滚筒　sand reel
劳加奈斯地热田　Laguarnes filed
老化试验　ageing tests

lei

雷暴　thunderstorm
雷电流　lightning current
雷诺数　Reynolds number (Re)
肋片管　extended surface tube
肋状管吸收器　fin tube absorbor
类槽形聚光式太阳能集热器　troughlike concentrator
类低共熔体　eutectoid
累积容量　cumulative capacity
累计损伤　cumulative damage
累计值　integrated value

ling

棱波　edge wave
棱镜窗　prismatic window
棱镜聚光器　prism concentrator
棱镜太阳能聚光器　prism solar concentrator
棱锥光学聚光镜　pyramidal optical condenser

leng

楞次定律　Lenz law
冷备用　cold standby reserve

冷槽　heat sink
冷脆（性）　cold shortness
冷的大气水　cold meteoric water
冷冻水　chilled water
冷风机　fan cooler
冷负荷　cooling load
冷空气　cool air
冷凝　condensation
冷凝槽　condensate tank
冷凝器　condenser
冷凝器—热交换机　condenser-heat exchanger
冷凝器温度　condenser temperature
冷凝器压力　condenser pressure
冷凝水　condensate water
冷凝液罐　condesate drum
冷气羽流　cold gas plume
冷却　cooling
冷却负载　cooling load
冷却功率式风速计　cooling-power anemometer
冷却机　cooler
冷却剂　coolant
冷却结构　cooling configuration
冷却介质　cooling medium
冷却介质损耗　cooling media loss
冷却能力　chiller capacity
冷却盘管　cooling coil
冷却片　cooling fin
冷却曲线　cooling curve
冷却设计负荷　design cooling load
冷却水　cooling water, chilled water
冷却水生物污染　biological pollution in cooling system
冷却塔　cooling tower
冷却塔系统　cooling tower system
冷却线圈　chilled-water coil
冷却需求　cooling requirement
冷却旋管　cooling coil
冷却循环　cooling cycle, refrigeration cycle
冷却主导型垂直闭合环路　cooling dominated vertical closed loop
冷却装置　chiller
冷热电联产　combined cold heat and power
冷水泵　cold water pump
冷水层　cold water layer
冷水池　cold water pond
冷水出口　cold water outlet
冷水管　cold water pipe
冷水机组容量　chiller capacity
冷水进水管　cold water intake pipe
冷水流量　cold water flow rate
冷水排放　cold water discharge
冷水团　cold water mass
冷水系统　chiller system
冷水再灌　cold water recharge
冷温水机组　chiller-heater
冷盐储槽　cold salt storage tank
冷营养海水　cold nutrient sea water
冷源　heat sink

li

离岸的　offshore
离岸风　off shore breeze, offshore wind
离岸风能转换系统　offshore wind-energy conversion system
离岸风切变　offshore wind shear
离岸工程结构物　offshore structure
离岸流　rip current
离海风　off-sea wind

离合器	clutch
离合器制动器	clutch brake
离散化	discretization
离散控制器	discrete controller
离散脉冲调制	discrete pulse modulation
离散时间	discrete time
离散图形	scatter diagram
离散系统	discrete system
离散阵风模型	discrete gust model
离体剂量	exit dose
离网电力	off-grid electricity
离网发电	off-grid electricity generation
离网风电系统	off-grid wind-electric system
离网型发电系统	stand-alone power system
离网型风电厂	stand-alone wind farm
离网型风力发电机	off-grid wind power generator
离网型风力发电系统	wind stand-alone power system
离心	centrifugal
离心泵	centrifugal pump
离心单元	centrifugal unit
离心力	centrifugal force
离心力矩	centrifugal moment
离心式分离器	centrifugal separator
离心式鼓风机	centrifugal blower
离心式蓄能泵	centrifugal storage pump
离异共晶	divorsed eutectic
离子传导	ionic conduction
离子电导率	ion-conductivity
离子键	ionic bonding
离子迁移率	ionic mobility
理论聚光	theoretical concentration
理论聚光比	theoretical concentration ratio
理论效率	theoretical efficiency
理论资源	theoretical resource
理论最大值	theoretical maximum value
理想的功率系数	ideal power coefficient
理想电源	ideal source
理想反射	perfect reflection
理想反射镜面	perfect mirror surface
理想跟踪反射镜	perfect tracking mirror
理想光学系统	perfect optical system
理想黑体	ideal blackbody, perfectly black body
理想黑体表面	ideal black surface
理想集热器	perfect collector
理想燃烧	perfect combustion
理想同步	ideal synchronizing
锂电池	lithium-ion battery
锂聚合物电池	lithium-polymer battery
力（扭）矩系数	moment (torque) coefficient
力觉反馈	force feedback
力矩方程	moment equation
力偶矩	moment of couple
力系数	power coefficient
力学机制	mechanical mechanism
历年气象要素	meteorological elements over the years
立方金刚石晶格	diamond cubic lattice
立方米	cubic meter
立管蒸发器	vertical-tube evaporator
立环式生物反应器	vertical loop biological reactor
立式槽管型换热器	vertical flute-tube heat exchanger
立式车床	vertical lathe
立式集热器	vertical collector
立式离心化工泵	vertical centrifugal

chemical pump
立轴风力机 vertical type wind turbine
立柱式太阳能支架 pole-mounted panel
励磁 excitation
励磁不足 underexcitation
励磁电流 excitation current, field current
励磁电压 exciting voltage
励磁机 exciter
励磁频率 excitation frequency
励磁绕组 field winding
励磁损失 excitation loss
励磁系统 excitation system
励磁线圈 excitation winding
励磁响应 excitation response
励磁涌流 inrush current
利用 harness
沥滤 leaching
例图 instance
砾石充填 gravel pack
砾石过滤层 gravel pack
砾石过滤器 gravel filter
粒度 grail
粒度分布 particle size distribution
粒度分析样品 size analysis sample
粒度破碎 particle size reduction
粒化 pelletisation
粒间孔隙度 intergranular porosity
粒状含水层 granular aquifer
粒状载热体加热器 pebble heater
粒子 particulate
粒子分离器 particle separator
粒子辐射 particulate radiation
粒子通量 particle flux
粒子吸热器 particle receiver
粒子蓄热床 particle storage bed

lian

连杆 rod
连杆环 tie-rod ring
连接半导体 connection semiconductor
连接分支 branch of joint
连接杆 connecting rod
连接热泵 connected heat pump
连结 link
连结物 link
连通度 connectivity
连通库方案 linked-basin scheme
连续超松弛 successive overrelaxation
连续冲模 dies-progressive
连续发电 continuous power generation
连续焚烧方式 continuous incineration
连续跟踪集热器 continuous tracking collector
连续跟踪式聚光能器 continuous tracking concentrator
连续搅拌反应釜 continuously stirred tank reactor
连续介质 continuum
连续可控设备 continuously controllable apparatus
连续雷电 continuous thunder and lightning
连续流反应器 continuous flow reactor
连续排污扩容器 continuous blow-down flash tank
连续燃烧 continuous combustion
连续溶解保温炉 continuous melting & holding furnace
连续时间 run on time
连续式辐射管式炉 continuous radiant-tube furnace
连续式加热炉 continuous heating furnace
连续式太阳能干燥机 continuous solar drier

连续调节　stepless control
连续系统　continuous system
连轴器　junction box
怜床反应堆　fluidized bed reactor
联邦能源管理计划　Federal Energy Management Program
联动装置轴承　linkage bearing
联合　conjunction
联合电力系统　power pool
联合发酵　co-fermentation
联合工艺　process integration
联合几率　joint probability
联合循环　combined cycle
联合循环发电　combined cycle power generation
联合循环发电厂　combined cycle power plant
联集管　manifold
联接螺旋夹　jubilee clip
联结　connection
联生藻类　adnate algae
联锁装置　interlocker
联网（太阳能发电装置）　net coupling
联轴器　coupling
炼炉　forge
链板式输送机　chain drag
链传动　chain drive
链环　link
链式钳　chain vice
链条炉排加煤机　chain-grate stoker
链条输送机　chain conveyor

liang

粮食能源一体化系统　integrated food and energy system
两级循环　dual circulation
两级循环锅炉　dual circulation boiler
两级真空泵　two-stage vacuum pump
两极性　dipolar nature
两面凹镜　biconcave lens
两面凸镜　biconvex lens
两系统同步　synchronization of two systems
两相　double phase
两相运行　two phase operation
两液电池　double fluid cell
亮度　luminous flux
量表　gauge
量纲分析　dimentional analysis
量规　gauge, dial gauge
量热箱　calorimeter box
量子点　quantum dots
量子点太阳能电池　quantum dot solar cell
量子开采　quantum exploit
量子能级　quantum level
量子能量　quantum energy
量子数　quantum number
量子效应　quantum effect

lie

烈风　strong gale
裂缝　flaw
裂缝半径　fracture radius
裂缝带　fracture zone
裂缝间距　fracture spacing
裂缝开度　fracture aperture
裂缝密度　fracture density
裂缝面　fracture face
裂缝网络　fracture network
裂缝系统　fracture system
裂缝性油藏　reservoir fracture

裂解　pyrolysis
裂解气　pyrolysis gas
裂解油　pyrolysis oil
裂流　ripple current
裂隙间距　fracture spacing
裂隙开度　fracture aperture
裂隙网络　fracture network
裂隙岩层　fractured formation
裂隙岩体　fractured rock

lin

邻避效应主义　not in my backyard
邻近地区　vicinity
邻近位　juxtaposition
邻近效应　proximity effect
林茂次涡轮机　Lynmouth turbine
林业废弃物　forest residue
临界安全限额　nuclear safety limit
临界功率　activation power
临界荷载　critical loads
临界晶核　critical nucleus
临界晶核半径　critical nucleus radius
临界能量　threshold energy
临界压力　critical pressure
临界压强　critical pressure
临界值　critical value
临界转差率　breakdown slip
临界转速　activation rotational speed
临界阻尼　critically damped, critical damping
临时紧急措施　stop-gap emergency
磷　phosphorus
磷光体和荧光材料　phosphors and fluorescent materials
磷化处理　phosphate treatment
磷酸燃料电池　phosphoric acid fuel cell

磷酸盐　phosphate
磷酸盐处理　phosphate treatment

ling

灵活交流输电系统　flexible AC transmission system
灵活燃料车辆　flex-fuel vehicle
灵活燃料汽车　flexible fuel vehicle
灵敏度　sensitivity
灵敏度分析　sensitivity analysis
灵敏值　figure of merit
灵敏转杯风速表　fast response cup anemometer
菱镁矿　magnesite
零电阻　zero resistance
零方位角　zero azimuth
零级风　calm
零间隙壁炉　zero clearance fireplace
零交叉波周期　zero crossing wave period
零能房屋　zero energy house
零速度测试　zero speed test
零损失集热器效率　zero loss collector efficiency
零线　zero conductor
零相位　zero phase
领海　marginal seas

liu

流槽　sluice
流场　flow field
流程图　flow chart, flow sheet, flow diagram
流出　flux
流出速度　flow velocity
流出物　effluent
流道参数　flow parameter

流动	flux
流动参数	flow parameter
流动反应器	flow reactor
流动回路	flow circuit
流动均匀性	flow uniformity
流动可视化	flow visualization
流动空气	moving air
流动摩阻	flow friction
流动温度	flow temperature
流动显示	flow visualization
流动相	flowing phase
流动性	flowability
流动状态	flow condition
流沟	flow ditch
流管	stream tube
流化	liquefaction
流化床	fluidized bed
流化床反应器	fluidized bed reactor
流化床焚烧炉	fluidized bed incinerator
流化床加热炉	fluidized bed furnace
流化床气化	cycling fluidized bed
流化床气化反应器	fluidized bed gasification reactor
流化床燃烧法	fluidized bed combustion
流化床燃烧锅炉	fluidized bed boiler
流化床燃烧室	fluidized bed combustor
流化床吸热器	fluidized bed receiver
流化容器	fluidization vessel
流量	flow volume
流量测定	flow measurement
流量测量	flow measurement
流量传感器	flow sensor
流量特性	discharge characteristic
流量调节	flow control
流量调节器	flow regulator
流量系数	efflux coefficient, discharge ratio, flow coefficient
流量需求	flow rate requirement
流率	flow rate
流谱	airflow pattern
流沙	heaving sand
流砂	winddrift sand
流速	velocity, flow velocity, flow rate, flow speed
流速分布	velocity distribution
流态	flow regime
流态床反应法	fluidized bed method
流体	fluid
流体的黏滞性	fluid viscosity
流体动力技术	hydrokinetic technology
流体动力学	fluid dynamics
流体动力学的	hydrokinetic
流体动力阻尼	hydrodynamic damping
流体工作介质	working fluid
流体供应温度	fluid supply temperature
流体构成	fluid composition
流体行为	fluid behavior
流体化学	fluid chemistry
流体静力	hydrostatic force
流体静力学试验	hydrostatic test
流体力学	fluid mechanics
流体流动	fluid flow
流体流动能量转化	fluid flow energy conversion
流体流量	fluid flow
流体面	fluid surface
流体摩擦	fluid friction
流体黏度	fluid viscosity
流体黏滞性效应	viscous effect
流体品质	fluid quality
流体属性	fluid property
流体损失	fluid loss
流体温度	fluid temperature
流体性质	fluid property, fluid behavior

流体循环　fluid circulation
流体压力　fluid pressure
流体质点　fluid particle
流体质量　fluid quality
流体阻力　fluid resistance
流纹岩　rhyolite
流线　streamline
流线密集效应　effect of steamline squeezing
流线型　streamline
流泄风　drainage wind
流域　river basin
流阻　resistance of flow
硫醇　mercaptan
硫含量　sulphur content
硫化层　fluidized bed
硫化镉基薄膜太阳能电池　cadmium sulfide based thin film solar cell
硫化镉太阳能电池　cadmium sulphide solar cell
硫化镉陶瓷太阳电池　cadmium sulphide ceramic solar cell
硫化氢　hydrogen sulfide
硫化铊光电装置　thallium-sulfide photovoltaic device
硫化铊光伏打装置　thallium-sulfide photovoltaic device
硫化物　sulfide
硫化物分解（量）　decomposition of sulphide
硫化物化学变化　chemical transformation of sulfides
硫化物矿石　sulfidic ore, sulphide ore
硫化物应力破裂　sulfide stress cracking
硫酸铜试验柱　copper sulfate test column
硫酸盐水　acid sulfate water
柳属　salt accumulator
柳枝稷　switchgrass
六角扳手　hexagon spanner
六方密排结构　hexagonal close-packed
六级风　strong breeze
六角板牙　hexagon screw die
六角车床　turret lathe
六角螺母　hexagonal nut
六角螺丝钢板　hexagon screw die
六相绕组　six-phase winding

long

龙波　Wave Dragon
龙卷风　wind spout
隆起　heaving
隆胀　heaving

lou

漏磁场　leakage field
漏磁电抗　leakage reactance
漏磁通　leakage flux
漏地电阻　ground leakage resistance
漏电断路保护器　leakage circuit breaker
漏电感　leakage inductance
漏电系数　coefficient of losses
漏斗　funnel
漏风　air leakage
漏风系数　air leakage factor
漏失通道　porous media
漏水　water oozing
漏损系数　scattering coefficient
漏泄场　leakage field
漏泄试验　leak test
漏渣　grate siftings, fall slag

lu

炉　kiln
炉箅加热　combustion fixed bed
炉底渣　bottom ash
炉负荷　hearth load
炉拱　furnace arch
炉灰　furnace dust
炉排　grate
炉排机械负荷　mass load of grate
炉排炉　grate furnace
炉排燃烧　grate combustion, stoker combustion
炉排燃烧器（室）　stoker combustor
炉排热负荷　grate heat release rate
炉排系统　grate system
炉腔　furnace chamber
炉墙　furnace wall
炉箅　grate surface
炉膛　furnace chamber
炉膛负压　furnace draft, furnace hearth pressure
炉膛截面　furnace cross-section
炉膛容积　furnace volume
炉膛容积热负荷　combustion chamber volume heat release rate
炉膛压力　furnace hearth pressure
炉条　grate bar
炉温　furnace temperature
炉渣　slag, cinder, furnace slag
炉渣热灼减率　loss of ignition
炉栅　grate bar
卤化有机化合物　halogenated organic compound
卤素灯　halogen lamp
卤钨灯　tungsten halogen lamp
鲁滨逊风速表　Robinson's anemometer
鲁奇气化炉　Lurgi gasifier
鲁瑟尔整流器　Russel rectifier
鲁瑟尔整流器型波能电站　Russel rectifier wave power plant
陆地闭（式循）环（海水温差）电站　land-based plant with a closed cycle
陆地沉积　terrestrial deposit
陆地形状　land configuration
陆风（自陆地吹向海上的微风）　land breeze
陆内熔融异象　intracontinental melting anomalies
陆棚外域　outer continental shelf
陆上动力装置　land-based power plant
陆上装置　onshore device
陆相地热流体　terrestrial geothermal fluid
路程长度　path length
露点差　dew point spread
露点湿度表　dew point hygrometer
露点温度　dew point temperature
露天焚烧　open burning
露天锅炉　outdoor boiler
露天矿　strip mine, surface mine
露天气候　open-air climate
露天气温　outdoor air temperature
露天装置　outdoor equipment

luan

卵石床　pebble bed
卵石加热器　pebble heater

lun

轮齿　gear tooth
轮齿啮合　teeth mesh
轮辐　spoke
轮毂　hub
轮毂刚度　hub rigidity

轮毂高度　hub height
轮毂罩　hubcap
轮机排汽　turbine exhaust
轮式装载机　wheel loader
轮胎面　toroidal surface
轮胎衍生燃料　tire derived fuel
轮叶泵　wing pump
轮缘　flange
轮缘涡轮机　rim turbine
轮轴　axle

luo

罗比茨型双金属日射计　bimetallic actinograph of Robitzsh
罗斯比波　Rossby wave
罗斯比重力波　Rossby-gravity wave
罗伊德能源系统　Lloyd energy system
逻辑集成电路　logic chip
逻辑片　logic chip
逻辑图　logic diagram
螺齿铣刀　helical cutter
螺桨式风速计　propeller type anemometer
螺距力　pitch power
螺距调节　pitch control
螺栓　bolt
螺丝攻　diestock
螺丝眼　eye screw
螺丝锥　gimlet
螺线管磁场　solenoidal magnetic field
螺旋　helix
螺旋泵　helical pump
螺旋齿　helical gearing, helical teeth
螺旋动力泵　spiral dynamic pump
螺旋加料器　screw feeder
螺旋桨　propeller
螺旋桨滑流　airscrew slip
螺旋桨式风速表　helicoidal anemometer
螺旋桨式风速计　prepeller anemometer
螺旋桨式涡轮机　propeller turbine
螺旋桨涡轮机　propeller turbine
螺旋桨型风轮（转子）　propeller type rotor
螺旋桨叶片　propeller blade
螺旋角　pitch angle
螺旋聚光(太阳能)灶　spiral concentrator cooker
螺旋聚光器　spiral concentrator
螺旋式风机　propeller fan
螺旋输送机　screw conveyor
螺旋送料器　corkscrew feeder
螺旋推运器　auger drive
螺旋型点聚焦聚光器　spiral snaped point focusing concentrator
螺旋型风力机　propeller type wind machine
螺旋轧机　screw rolling mill
螺旋状物　helix
螺旋钻　auger
螺旋钻法　auguring
螺旋钻探采矿　auger drill mining
裸电池　bare cell
裸管　bare tube
裸管换热器　borehole heat exchanger
裸管热存储　borehole heat storage
裸露接收器　exposed receiver
洛夫太阳房（一种主动式太阳能房）　Lof house
洛克数　Lock number
络合多糖碳水化合物　complex polysaccharide carbohydrate
落差　head
落潮　tidal ebb, tidal fall, ebb tide
落潮潮量　volume of ebb
落潮发电　ebb generation

落潮历时 duration of ebb
落潮流 ebb current, ebb tide current
落潮时 duration of ebb
落潮水道 ebb channel
落锤 drop hammer
落在井内的工具 lost tool

lv

铝 aluminum
铝（吸收）片 aluminum sheet
铝箔太阳能集热器 aluminum-foil solar collector
铝电池 aluminum cell
铝电解槽 aluminum cell
铝反射镜 aluminized mirror
铝反射翼平板太阳能集热器 flat plate collector with reflecting aluminum
铝黄铜 aluminum brass
铝桥 aluminum bridge
铝青铜 aluminum bronze
铝酸硅 alumino-silicate
铝阳极氧化选择性吸收涂层（在铝材上采用电解着色阳极氧化方法制备的选择性吸收涂层，它具有良好的选择性吸收性能） anodized aluminum selective coating
铝制玻璃支架 aluminum glass support
铝制吸收器翅片 aluminum absorber fin
铝制叶片 aluminum blade
铝质金属外壳 aluminum alloy housing
绿潮 green tide
绿地项目 greenfield project
绿化 landscaping
绿色发电 green power generation
绿色火电 green thermal power
绿色建筑 green architecture
绿色节能 green power
绿色煤电 green coal-based power
绿色能源 green energy source
绿色能源机器 green energy machine
氯 chlorine
氯含量 chlorine content
氯化钠 sodium chloride
氯化钠水 sodium chloride water
氯化物 chloride
氯化物含量 chloride content
氯蓄电池 chloride storage battery
滤饼 filter cake, press cake
滤布 filter cloth
滤层 filter layer
滤尘器 dust filter
滤床工艺 leach bed process
滤光片 filter
滤光器 filter, optical filter
滤膜电池 filter diaphragm cell
滤气器 air cleaner
滤失量 fluid loss
滤水池 water filter
滤水器 water filter
滤网 impounding
滤液回收 leachate recycling

ma

麻风树 jatropha
马格努斯效应 Magnus effect
马格努斯效应式风力机 Magnus effect type wind turbine
马赫数 Mach number
马力 horsepower
马氏体 martensite
马氏体相变 martensite phase transformation
马蹄形磁铁 horseshoe magnet

马尾藻　gulfweed
马纬度　horse latitude

mai

埋地换热器　underground heat exchanger
埋入式接点太阳电池　buried-contact solar cell
迈克尔森双金属直接日射计　Michelson bimetallic pyrheliometer
麦考密克涡轮机　Mc Cormick turbine
麦斯威尔方程　Maxwell's equation
脉波调制　pulse wavelength modulation
脉冲　impulse
脉冲布袋清灰　pulse jet cleaning
脉冲带宽　pulse bandwidth
脉冲电流　impulse current
脉冲电压　pulse voltage
脉冲阈能　pulse threshold energy
脉冲峰压　peak pulse voltage
脉冲浮标　impulse buoy
脉冲宽度调制　pulse width modulation
脉冲宽度调制转换器　pulse width modulated converter
脉冲宽度调制晶体闸流管　pulse width modulated thyristor
脉冲量输入　counter input
脉冲流　pulse stream
脉冲能量　pulse energy
脉冲喷吹袋式除尘器　pulsejet bag filter
脉冲频率调制　pulse frequency modulation
脉冲燃烧技术　pulse firing technology
脉冲热线风速计　pulsed wire anemometer
脉冲式太阳模拟器　pulse solar simulator
脉冲输入　impulsive input

脉冲响应　impulse response
脉冲信号　pulse signal
脉动　pulsation
脉动电流　pulsatory current, pulsating direct current, pulsating current
脉动电压　ripple voltage, pulsating voltage
脉动风　turbulence wind
脉动风速谱　fluctuation wind speed spectrum
脉动流　oscillatory flow
脉动压力　fluctuating pressure
脉幅调制　pulse amplitude modulation
脉宽调制　pulse amplitude modulation

man

满潮　high tide
满潮时距　lunitidal interval
满载　full load
满载转矩　full-load torque
曼螺　Ekman spiral
漫反射　scattered reflection
漫辐射　diffuse radiation
漫辐射形态　diffuse radiation form
漫射材料　diffusing material
漫射常数　diffusion constant
漫射传输　diffuse transmission
漫射功率　diffusing power
漫射光　diffuse light, diffused light
漫射光照　diffuse illumination
漫射率　diffuse reflectance
漫射太阳光　diffuse sunlight
漫射源　diffusing source
漫溢　overtopping
漫溢设备　terminator device
慢滤池　slow sand filter
慢性滑行　freewheeling

mang

芒草	miscanthus
芒刺	burr
芒属	miscanthus
芒属植物	miscanthus
盲板	blind ram, blinding plate
盲区	shadow

mao

毛密度	gross density
毛水头	gross head
毛细管能量	capillary energy
锚泊浮体	moored vessel
锚泊系统	mooring system
锚定螺栓	anchor bolt
锚具	anchorage
锚链滑环	cable slip-ring
冒气地面	fumarolic area
冒气区域	fumarolic area
冒汽地面	steaming ground

mei

媒介载体	carrier medium
楣	lintel
煤仓	coal bunker
煤尘沉降	dust precipitation
煤斗	fuel hopper
煤粉锅炉	pulverized coal boiler
煤粉炉膛	pulverized fuel combustor
煤焦燃烧	char combustion
煤焦氧化	char oxidation
煤气化	gas gasification
煤气化联合循环发电厂	coal gasification combined cycle power plant
煤气火焰	gas burner
煤气净化	gas cleaning
煤气冷却器	gas cooler
煤气炉	gas-fired furnace
煤气喷嘴	gas jet
煤气设备	gas installation
煤气压缩机	gas compressor
煤气灶	gas burner
煤气储柜	gas holder
煤气总管	gas main
煤炭气化（煤气化）	coal gasification
煤油	kerosene
煤砖	coal briquette
酶	enzyme
酶催化分解	enzymatic breakdown
酶的催化作用	catalytic intervention of enzyme
酶的诱导	enzyme induction
酶电池	enzyme cell
酶电极	enzyme electrode
酶活性	enzyme activity
酶水解	enzyme hydrolysis
酶抑制	enzyme inhibition
每分钟转数	revolution-per-minute (rpm)
每小时炉油损耗	hourly furnace oil consumption
每小时销毁容量	hourly destruction capacity
每月日热负荷图	daily variation graph of heat consumption in one month
每转负荷	per rev load
美吨	short/net ton
美国"百万太阳能屋顶计划"	u.s. "Million Solar Roof Initiative"
美国风车	American windmill
美国国家电气规程	National Electrical Code
美国国家可再生能源实验室	National Renewable Energy Laboratory

美国国家太阳辐射数据库　National Solar Radiation Data Base
美国海岸工程手册　Coastal Engineering Manual
美国经济恢复和再投资法案　American Recovery and Reinvestment Act
美国联邦能源管理项目　Federal Energy Management Program
美国能源部可再生能源实验室　Department of Energy National Renewable Energy Laboratory
美国能源部西北太平洋国家实验室　Pacific Northwest National Laboratory
美国应用海洋能管理系统　Practical Ocean Energy Management Systems

men

门阀　gate valve
门极　gate pole
门极可关断晶闸管　gate turn off thyristor
闷晒　stagnation
闷晒式太阳（能）热水器　integral collector-storage solar water heater
闷晒温度　stagnation temperature

mi

弥散度　degree of dispersion
弥散燃料　dispersion fuel
米/秒　metres per second (m/s)
密度　density
密度比　density ratio
密度差　density difference
密度分选　density separation
密度剖面　density profiles
密度梯度　density gradient
密封管　gland

密封系统　sealed system
密封储藏　airtight storage
密勒指数　miller indice
密网格　fine grid
幂定律　power law
幂律廓线　power law profile
幂律指数　power law exponent

mian

棉花纤维　cotton fiber
棉籽壳　cottonseed hull
面板保护二极管　panel protection diode
面包箱式间歇加热器　breadbox hatch heater
面包箱式热水器　breadbox-type water heater
面包箱式太阳能热水器　breadbox solar water heater
面波频散　surface wave dispersion
面式减温器　convection type desuperheater
面式冷凝器　surface condenser
面下流　subsurface current

min

民用闭环太阳能热水器　closed-loop domestic water heater
民用锅炉　domestic boiler
民用过电流保护断路器（小型断路器）　civilian over-current protection circuit breaker
民用开环太阳能热水器　open-loop domestic water heater
民用燃气　town gas
民用太阳电池方阵　residential solar array
民用太阳能供暖　domestic solar

heating
民用太阳能热水器 domestic solar water heater
民用住宅 domestic dwelling
敏感负荷 sensitive load
敏感元件 pick off

ming

明暗 shading
命令 command

mo

模块化 modularity
模块化两级燃烧 modular two-stage combustion
模拟光合（作用） mimic photosynthesis
模拟荷载 fictitious load
模拟控制 analogue control
模拟量输入端子 analog input terminal
模拟盘 analogue board
模拟示波器 analog oscilloscope
模拟输出 analog output
模拟输入 analog input
模拟 数字转换器 analog-to-digital converter
模拟信号 analog signal
模拟型 analog type
模绕 form-wound
模式控制器 mode controller
模式速度 mode speed
模数 module
模型开发 model development
模型线性化 model linearization
模型预测控制 model predictive control
膜传热系数 film heat transfer coefficient
膜片 diaphragm
膜片压力表 diaphragm manometer

膜生化反应器 membrane biochemical reactor
膜生物反应器 biomass membrane bioreactor
膜系数 film coefficient
膜组件 membrane module
摩擦 friction
摩擦辐合 frictional convergence
摩擦辐散 frictional divergence
摩擦力 friction
摩擦密封 friction seal
摩擦盘 friction grip
摩擦水头 friction head
摩擦损耗 friction loss, frictional dissipation
摩擦损失 frictional loss, friction head
摩擦系数 friction coefficient, friction factor
摩擦消散 frictional dissipation
摩擦压力 friction pressure
摩擦压力损失 frictional pressure loss
摩擦阻力 friction drag, friction resistance, drag friction
摩擦作用 rubbing action
摩尔比率 molar ratio
摩尔表面能 molar surface energy
摩尔浓度 molar concentration
摩尔旋光（度） molar rotation
摩尔折射（度） molar refraction
摩尔自由能 molar free energy
磨床 grinder, grinding machine
磨床工作台 grinder bench
磨掉 grind off
磨耗 abrasion
磨料盘 abrasive disc
磨木 ground wood
磨蚀 abrasion

磨碎　trituration
磨碎机　grinding drum
磨损　abrasion
磨损故障　wear-out failure
磨损检查　abrasion inspection
磨损率　wear rate
磨损指数　abrasive index
磨斜棱　beveling
磨屑　grinding dust
磨削工具　grinding tool
末端朝前或向上的　endwise
末端挡板　end stop
末端荷载　end load
末级除氧器　final stage deaerator
末级过热器　final superheater
末级冷却器　aftercooler
末级阳极电压　final anode voltage
末级再热器　final reheater
末级蒸发器　final evaporator
莫尔 戈尔钦斯基直接日射强度计　Moll-Gorchinski pyrheliometer
莫尔 戈尔钦斯基总日射表　Moll-Gorchinski pyrheliometer
莫克来斯塔德法　Myklestad method

mu

模板　gauge board
模具　jig
母线　bus bar
母线槽　bus duct
母线电压　bus voltage
母线间隙垫　bus bar separator
母线伸缩节　bus bar expansion joint
木本能源作物　woody energy crops
木材废料　lumber scrap
木材含水率　wood moisture content
木材加工厂　wood processing plant
木材加工业副产品和残渣　wood processing industry by-product and residue
木柴燃料　wood fuel
木柴转换为液体燃料　conversion of wood to liquid fuel
木槌　mallet
木钉　dowel
木灰　wood ash
木料加工残留物　wood processing residue
木馏油　creosote
木刨花　wood shaving, cutter shaving
木片　wood chip
木薯　cassava
木炭　charcoal
木炭气化炉　charcoal gasifier
木炭燃烧　charcoal combustion
木糖　xylose
木屑颗粒　wood pellet
木屑颗粒燃料　wood pellet fuel
木屑炉　wood pellet stove
木质颗粒　wood pellet
木质燃料　wood fuel, wood based fuel, wood-derived biofuel
木质生物质　woody biomass
木质生物质燃料　woody biomass fuel
木质素　lignin
木质素残渣　lignin residue
木质素复合物　lignin complex
木质纤维生物乙醇　lignocellulosic bioethanol
木质纤维素　lignocellulose
木质纤维素能源作物　lignocellulosic energy crop
木质纤维素原料　lignocellulosic feedstock
木质纤维无烟煤　lignocellulosic anthracite

目标 定位跟踪 target-aligned tracking
目标函数值 objective function value
牧草作物 pasture crop
钼 molybdenum

na

纳潮量 tidal prism
纳缔克式太阳能炉 Natick furnace
纳滤 nanofiltration
纳米技术 nanotechnology
纳米粒子 nano-particle
钠 sodium
钠离子 sodium ion
钠 汽热交换器 Na-to-steam heat exchanger

nai

耐风设计 wind proof design
耐风性能 wind proof performance
耐辐射 radiation resistant
耐辐射性 radiation resistance
耐腐蚀材料 corrosion-resistant material
耐腐蚀性 corrosion resistance
耐腐试验 corrosion resistance test
耐高温漆 heat resistance paint, heat resistant paint
耐高温氧化物 refractory oxide
耐火材料 refractory
耐火衬砌 refractory oxide
耐火内衬 refractory lining
耐火燃料 slow-burning fuel
耐火涂料 fire retardant paint
耐火砖 firebrick
耐久试验 endurance test
耐久性 durability
耐磨性 abrasion resistance
耐热 heat resisting

耐热薄膜 heat resistant film
耐热玻璃 heat resistant glass
耐热瓷 heat resisting porcelain
耐热度 heat resistance
耐热合金 high temperature alloy, heat resistant alloy
耐热漆 heat resistance paint, heat resistant paint
耐热性 heat resistance, temperature resistance
耐晒塑料 sun-resistant plastic
耐晒透明塑料 transparent sun-resistant plastic
耐湿 moistureproof
耐湿性 moisture resistant
耐蚀材料 corrosion resistant material
耐蚀性 corrosion resistance
耐受性 survivability
耐酸保护层 acid proof coating
耐用性 durability

nan

南北向跟踪集热器 NS tracking collector
南赤道洋流 south equatorial current
南方"猪沼果"模式 south swine-biogas-fruit pattern
南极绕极流 antarctic circumpolar current

nao

挠曲 flexure
挠性管 flexible pipe
挠性卷缩太阳能阵列 flexible rolled-up solar array
挠性联轴节 flexible coupling
挠性轴 flexible hinge
挠性转子 flexible rotor

nei

内混式盐差能动力装置　power plant of energy from salinity gradients of mixer in the inside
内半轴　inner shaft
内包装　inner packing
内部报酬率　internal rate of return
内部防雷系统　internal lighting protection system
内部脉动风载　fluctuating internal wind loading
内部摩擦　internal wear
内部收益　internal gain
内部收益率　internal rate of return
内部水运动　internal water motion
内潮　internal tide
内衬　lining
内齿轮　internal gear
内齿轮副　internal gear pair
内电压　internal voltage
内电阻　internal resistance, internal impedance
内防喷器　inside blow-out preventer
内耗　internal friction
内环　inner ring
内建冗余　built-in redundancy
内聚光式真空集热管　internal concentrative vacuum tube
内卡钳　internal caliper
内冷凝蒸馏器　internally condensing still
内摩擦　internal friction
内摩擦角　angle of internal friction
内能　internal energy
内燃　internal combustion
内燃机　internal combustion engine
内燃机车　diesel locomotive
内燃循环　internal combustion cycle
内散射　scattering-in
内损　internal loss
内透射比　internal transmittance
内吸收比　internal absorptance
内循环　internal circulation
内置测温器　embedded temperature detector
内轴　inner shaft
内装式集热器　site-built collector
内阻　internal resistance
内阻抗　internal resistance, internal impedance

neng

能带　energy band
能带结构　energy band structure
能带宽度　energy band gap
能带移动　band shift
能当量　energy equivalent
能动系统　active system
能含量　energy content
能耗　energy consumption
能级　energy level
能级图　energy level diagram
能量不足　energy deficit
能量草　energy grass
能量储存　energy storage
能量传递　energy transmission, energy transfer
能量传送系统　energy transport system
能量代谢　energy metabolism
能量单位　energy unit
能量的当量值　energy calorific value
能量的等价值　energy equivalent value
能量的季节储存　seasonal storage of

energy
能量等级 energy grade
能量动用 energy mobilization
能量分布 energy distribution, energy spectrum
能量浮标 power buoy
能量耗散 energy dissipation
能量厚度 energy thickness
能量回收 energy recovery
能量回收系数 energy recovery factor
能量获取 energy capture
能量计量 energy meter
能量计算 energy calculation
能量交换 energy exchange
能量金字塔 energy pyramid
能量均衡 energy balance
能量开采设备 energy extraction device
能量密度 energy density
能量密度峰值 energy density peak
能量模式 energy mode
能量木材 energy wood
能量平衡 energy balance
能量平衡分析 energy balance analysis
能量强度 energy intensity
能量容量 energy capacity
能量容限 energy capacity
能量色散X射线光谱法 energy-dispersive X-ray spectroscopy
能量森林树 energy forest tree
能量生产还本周期 energy return period
能量释放 energy liberation
能量释放率 energy resease rate
能量收益 energy gain
能量收支 energy budget
能量守恒 energy conservation
能量输出 energy output

能量输入 energy input
能量输入比 energy input ratio
能量损耗 energy loss
能量梯度 energy gradient
能量提取 energy extraction
能量通量 energy flow
能量位垒 energy barrier
能量吸收 energy absorption
能量效率方程 energy efficiency equation
能量直接转换 direct energy conversion
能量种植树 energy plantation tree
能量储存池 energy storage pond
能量转化机 energy conversion machine
能量转换 energy conversion, energy transformation
能量转换器 energy converter
能量转换设备 energy conversion device
能量转换系统 energy conversion system
能量转换效率 energy conversion efficiency, power conversion efficiency
能量转换装置 power conversion unit
能量资源 energy resource
能流 energy flux, energy current, energy flow
能谱 energy spectrum
能通量 energy flux
能通量密度 energy flux density
能隙 band gap, energy gap
能隙梯度 graded band-gap
能源 energy source
能源安全 energy security
能源标准 energy standard
能源偿还时间 energy payback time
能源当量 energy equivalent
能源分布 energy distribution
能源管理器 energy manager
能源管理与控制系统 energy management

and control system
能源回收期　energy payback time
能源回收时间　energy payback time
能源经济　economy of energy
能源开发　energy development
能源利用　energy utilization
能源利用的潜在污染　potential pollution from use of energy resource
能源农业　energy agriculture
能源农作物　energy farming
能源输送因子　energy delivery factor
能源损耗　energy penalty
能源替代　energy alternative
能源效率　energy efficiency
能源效率比值　energy efficiency ratio
能源研究和开发　energy research and development
能源载体　energy carrier
（美国）能源政策法案　Energy Policy Act
能源植物　energy plant
能源种植　energy cropping
能源种植场　energy farm, energy plant
（为用作合成燃料或发电的原料而种植的）能源植物丛　energy plantation
能源总量　energy capacity
能源作物　energy crop

ni

泥浆泵　slurry pump, mud pump
泥浆测井仪　mudlogger
泥浆池　suction pit, slurry pit
泥浆管线　mud line
泥浆罐　mud tank
泥浆搅拌器　mud agitator
泥浆循环　mud circulation
泥浆钻井　mud drilling

泥浆钻头喷嘴　nozzle of mud drilling bit
泥煤　peat
泥气分离器　mud-gas separator
泥沙　silt
泥沙泵　sand pumping
泥沙输送　sediment transport
泥沙输移　sediment transport
泥炭块　peat
泥炭煤球　peat briquette
逆变器控制　inverter control
逆潮　reversed tide
逆冲断层　overthrust, thrust fault
逆地　adversely
逆断层　thrust fault
逆恶浪　cross sea
逆风　adverse wind, head wind, dead wind
逆风螺旋架　upwind propeller
逆浪　cross sea
逆流　upstream, backwash, reverse flow, reverse current, opposite flow
逆流而上　upstream
逆流喷雾室　counter-current spray chamber
逆流燃烧　counter-current combustion
逆流热交换器　counterflow heat exchanger, reverse current heat exchanger
逆流式系统　contraflow system
逆流蒸发器　counterflow evaprator
逆渗透淡化　reverse osmosis desalination, reverse osmosis
逆时针方向　counter clockwise
逆温　thermal inversion
逆温层　thermal inversion layer
逆温层高度　inversion height

逆向渗透 reverse osmosis desalination
逆循环 reverse cycle
逆掩断层 overthrust
溺湾 rhyolitic dome

nian

年变幅 annual amplitude
年变化（风速或风功率密度） annual variation
年度热值 annual heating value
年度废物流 annual waste flow
年度温度变化 annual temperature change
年发电量 annual energy output, annual energy production
年风况 annual wind regime
年风速频率分布 annual wind speed frequency distribution
年际变化 inter-annual variation
年均发电利用率 annual capacity factor
年均风能潜势 mean annual wind power potential
年均风速 mean annual wind speed
年均理论产值 mean annual theoretical production
年利用小时 annual operation hour, annual utility hour
年利用小时数 annual available hour
年轮 annual zone
年能量供给量 annual energy delivery
年能量交换量 annual energy displacement
年平均 annual average
年平均风能密度 annual average wind power density
年平均风速 annual average wind speed, annual mean wind speed
年平均负荷 annual average load
年平均功率 annual average power
年平均雷暴日 mean annual thunderstorm day
年平均日照量 annual average insolation
年平均日照小时 annual average sunshine hour
年太阳辐射量 annual solar radiation
年太阳能节能量（太阳能建筑相对于非太阳能建筑每年所节省的能量） annual solar saving
年太阳热量 annual solar heat
年运行方式 yearly operating plan
年照射量百分率 annual load fraction
年最高 annual maximum
年最高日平均温度 annual extreme daily mean of temperature
黏度 viscosity
黏度计 viscosimeter
黏度效应 viscosity effect
黏度指数 viscosity index
黏胶树脂 adhesive resin
黏结剂 adding binder
黏土衬垫 clay liner
黏土矿物 clay mineral
黏土质耐火材料 fireclay refractory
黏性 viscosity
黏性层 viscous layer
黏性耗散 viscous dissipation
黏性力 viscous force
黏性流 viscous flow
黏性流动烧结机制 viscous flow sintering mechanism
黏性流体 viscous fluid
黏性尾流 viscous wake
黏性系数 viscosity coefficient
黏性液体 viscous liquid

黏性阻力　viscous drag, viscosity resistance
黏液残渣　viscose residue
黏质熔岩　viscid lava
黏滞阀　sticking valve
黏滞耗散　viscous dissipation
黏滞扩散　viscous dissipation
黏滞摩擦　viscous friction
黏滞效应　viscous effect
捻度　twist
捻度分布　twist distribution
碾压　laminate

nie

啮合　engagement mesh
啮合干涉　meshing interference
镍镉电池　NiCd battery (nickel-cadmium battery)
镍铬　镍铝热电偶　chromel-alumel thermocouple
镍合金　nickel alloy
镍氢电池　nickel-metal hydride (NiMH) battery

ning

凝固　solidification
凝固精炼　solidification refining
凝固热　heat of solidification
凝灰岩　tufa
凝结气胶　condensation aerosol
凝结 融化温度　freeze-thaw temperature
凝结水泵　condensate pump
凝结水循环　condensate return
凝结损失　condensation loss
凝结液　condensed fluid
凝结蒸汽 液体热交换器　condensing vapor-to-liquid heat exchanger
凝聚物　agglomerant

凝汽器　condenser
凝汽器热交换面　condenser cooling surface
凝汽器压力　condenser pressure
凝汽装置　condensing plant
凝析带　condensation zone

niu

牛顿第二定律　Newton's second law
牛粪浆　cattle slurry
牛脂　tallow
扭矩　torque, moment of torsion
扭矩传感器　torque transducer
扭矩计　torque meter
扭矩减震器　torque damper
扭矩水平　torque level
扭缆　cable twist
扭力　torque force
扭力计　torque pickup, torsion indicator
扭应力　torsional stress, twisting stress
扭转变形　torsion deflection
扭转常数　torsion constant
扭转弹性　torsion elasticity
扭转分布　twist distribution
扭转刚度　torsional rigidity
扭转刚度系数　coefficient of torsional rigidity
扭转荷载　torsional load
扭转晶界　twist grain boundary
扭转强度　torsional strength
扭转失速颤振　torsional stall flutter
扭转系统　torsional system
扭转叶片　twist balde
扭转应变　torsional strain
扭转应力　torsional stress
扭转振动　torsional vibration

nong

农场型风车　farm-type windmill
农村电气化　village electrification
农副产品　agricultural by-product
农工副产品　agro-industrial by-product
农化产品　agri-chemical
农田残茬　crop residue
农业废料　agricultural waste
农业废（弃）物　agricultural residues, agricultural waste
农业燃料　agrofuel
农业生态分区　agro-ecological zoning
农作物副产品　crop by-product
农作物废料　crop waste
农作物废弃物　crop waste
农作物生产残渣　crop production residue
浓差电位　electric potential by concentration
浓淡电池　salinity cell
浓淡电池式浓度差能发电　salinity cell energy conversion
浓淡电位差　electric potential by concentration
浓度差能转换　salinity gradient energy conversion
浓度单位　concentration unit
浓度范围　concentration range
浓度分布曲线　concentration profile
浓度三角形　concentration triangle
浓度梯度　concentration gradient
浓缩材料　enriched material
浓缩汽体　enriched gas
浓缩物　condensate, enriched material
浓阴影　dark shading

nuan

暖风机　unit heater
暖空气　warm air
暖空气散热片　air-warmed radiant panel
暖气管　flue
暖气锅炉　heating boiler
暖水管　warm-water pipe
暖水进口　warm-water inlet
暖水进水管　warm-water intake pipe
暖水排放　warm-water discharge
暖水入口　warm-water intake
暖通空调（暖气、通风及空气调节）　heating, ventilation and air conditioning (HVAC)

nuo

诺顿等效电路　Norton's equivalent circuit

ou

欧拉方程　Euler equation
欧姆　ohm
欧姆加热　ohmic heating
欧姆接触　ohmic contact
欧姆损耗　ohm loss
欧洲波浪和潮汐能源会议　European Wave and Tidal Energy Conference
耦合器　electric coupling
耦极电阻率法（用于地热勘探的一种地球物理方法）　dipole resistivity
耦合　coupling
耦合等离子体原子发射光谱法　coupled plasma-atomic emission spectrometry technique
耦合性　coupling bolt

pa

耙　harrow
帕斯卡（压力单位）　pascal
帕斯卡定律　Pascal's law

pai

拍动板系列　flap system
拍动板型波力发动机　flap-type wave motor
排出阀　release valve
排出管　drip pipe
排出孔　discharge hole
排出口　delivery end
排出压力　exhaust pressure
排除　eliminate
排除故障　clearance
排放涤气系统　emission scrubber system
排放阀　bleed valve
排放管　discharge pipe
排放极限　emission limit
排放减排系统　emission abatement system
排放控制　emission control
排放控制设备　emission control equipment
排放速度　emission rate
排放速率　emission rate
排放系数　emission factor
排放系统　draindown system
排放限度　emission limit
排放限值　emission limit, emission limit value
排放因子　emission factor
排放值限定　emission limit value
排风过滤　air-exhaust filter
排风机　air exhaust fan, air exhauster
排风量　air output
排风扇　draft fan
排管　tube bank
排灰斗　ash discharge hopper
排气　exhaust, air exhaust
排气道　air chute, air chimney, air passage
排气阀　vent valve, drain tap, exhaust valve
排气风机　fan exhauster
排气管　air vent pipe, exhaust pipe
排气管线　exhaust line
排气加热热交换器　exhaust heat exchanger
排气孔　gas vent
排气口　air vent
排气量　air displacement
排气湿度　final moisture content
排气竖井　air shaft
排气损失　exhaust loss
排气调节阀　exhaust regulator
排气通道　vent stack
排气系统　exhaust system
排气线　exhaust line
排气消音器　exhaust silencer
排气循环　exhaust cycle
排气压力　exhaust pressure
排气烟羽　air plume
排气装置　air exhausting device
排气总管　exhaust manifold
排汽　steam dumping, dump steam
排热　heat rejection
排热系统　heat extraction system
排水　drain
排水泵　effluent pump, drip pump
排水阀　vent valve
排水沟　waterway
排水管　effluent pipe
排污阀　delivery cock, drip cock
排污许可证　emission permit
排泄阀　vent valve
排烟　flue gas

排渣　slagging
迫击炮　mortar
派生（指令）　fork

pan

盘管加热器　coil heater
盘簧减震器　coil spring damper
盘区开采法　panel method
盘式离合器　plate clutch
盘式太阳能电站　solar parabolic dish power generation
盘式削片机　disc chipper
盘式制动器　disk brake
盘形聚热器　dish collector

pang

旁流　side stream
旁流阀　spill valve
旁漏电流　bleeder current
旁路电容　bypass capacity
旁路电容器　shunt capacitor
旁路二极管　bypass diode, bypass (shunt) diode
旁滤器　by-pass filter
旁通阀　bypass damper, bypass valve
旁通活门　bypass valve

pao

抛光　lap
抛光分离器　polishing separator
抛光化合物　buffing compound
抛光轮　buffing wheel
抛光洗涤器　polishing scrubber
抛煤机锅炉　spreader stoker boiler
抛煤机炉排　spreader stoker
抛物槽　parabolic trough
抛物槽场　parabolic trough field
抛物槽式表面　parabolic trough surface
抛物面槽式聚光器　parabolic trough concentrator
抛物槽式太阳能电厂　parabolic trough power plant
抛物面槽式太阳能热发电　parabolic trough solar power
抛物面槽式吸热管　parabolic trough receiver tube
抛物面碟式聚光器　parabolic dish concentrator
抛物碟形太阳能聚光器系统　parabolic dish solar concentrator system
抛物镜　parabolic mirror
抛物面（聚光）槽　paraboloidal trough
抛物面槽技术　parabolic trough technology
抛物面槽式聚光器　parabolic trough concentrator
抛物面槽式吸热管　parabolic trough receiver tube
抛物面光电化学系统　paraboloidal photoelectrochemical system
抛物面碟技术　parabolic dish technology
抛物面碟式聚光器　parabolic dish concentrator
抛物面碟形集热器　parabolic dish collector
抛物面定日镜　parabolic heliostat
抛物面反射镜　parabolic dish, parabolic reflector
抛物面反射器　parabolic dish
抛物面菲涅耳反射镜　parabolic Fresnel mirror
抛物面集热器　paraboloid collector
抛物面聚光器　parabolic concentrator, paraboloidal concentrator

抛物面盘状太阳能集热器　paraboloidal dish solar collector
抛物面太阳能聚光集热器　paraboloidal concentrating solar collector
抛物面太阳能聚光器　paraboloidal solar concentrator
抛物面形集热器　parabolic shaped collector
抛物面柱面形聚光集热器　concentrating collector of parabolic cylindrical type
抛物盘热器　parabolic dish collector
抛物体凿岩　projectile drilling
抛物线　parabola
抛物线方程　parabolic equation
抛物线形槽反射器　parabolic trough reflector
抛物线形反射镜　parabolic reflector
抛物线形支架　parabolic support
抛物线形聚焦集热器　parabolic focusing collector
抛物线状点聚集集热器　parabolic point focus collector
抛物形反射面　parabolic reflecting surface
抛物形集热器　parabolic collector
抛物形铝模　parabolic aluminum mold
抛物形渗铝反射器　parabolic aluminized reflector
抛物柱面　parabolic cylinder
抛物柱面反射镜集热器　parabolic cylinder mirror collector
抛物柱面反射器　parabolic dish
抛物柱面集热器　parabolic cylinder collector
抛物柱面镜　parabolic mirror
抛物柱面镜集热器　parabolic mirror collector
抛物柱面镜聚焦太阳能集热器　parabolic mirror focusing solar collector
抛物柱面聚焦集热器　cylindrical parabolic focusing collector
抛物柱面太阳能反射器　parabolic cylinder solar reflector
刨削槽　gouging
泡壳包装　clam shell
泡利不相容原理　Pauli principle
泡流（地热蒸汽泡沿管道的上部流动，其速度与液体的速度相近）　bubble flow
泡沫水泥　foamed cement
泡沫芯层　foam core
泡式涡轮机　bulb turbine
炮耳轴　trunnion pin
炮眼　borehole
炮眼钻井　blasthole drilling

pei

佩尔顿水轮机　Pelton wheel (turbine)
配电板　distribution board, distributor plate, electrical panel
配电变压器　distribution transformer
配电电器　distributing apparatus
配电公司　distribution company
配电盘　distribution board, distributor plate, electrical panel, switch board
配电盘控制　panel control
配电屏　distribution board
配电系统　distribution system
配电线路　distribution feeder, distribution line
配电箱　distribution board
配电小间　kiosk substation
配对齿轮　mating gear
配料　dosage

| 配料仓　mixing tank
| 配煤　coal blend
| 配气公司　distribution company
| 配水系统　distribution system
| 配位多面体　coordination polyhedron
| 配位数　coordination number
| 配置　disposal
| 配置性基础设施　distribution infrastructure

pen

喷灯　blowlamp
喷溅涂覆法　sputtering
喷淋塔　spray tower
喷抛清理　blast cleaning
喷漆枪　spray gun
喷气发动机　air breathing engine
喷枪　spray gun
喷泉　geyser, spout
喷射泵　injection pump
喷射气流　jet stream
喷射器　ejector
喷射式浮选机　air jet floatation machine
喷射式热交换器　spray type heat exchanger
喷射式涡轮　jet turbine
喷涂流体　spray flow
喷涂枪　spray gun
喷雾干燥机　spray dryer
喷雾干燥减震器　spray dryer absorber
喷雾干燥器　spray dryer
喷雾干燥吸收器　spray dryer absorber
喷雾器　aerosol
喷雾式加湿器　spray humidifier
喷雾室　spray chamber
喷雾增湿器　spray humidifier
喷雾嘴　spray nozzle
喷油器集箱（母管）　injector manifold

喷油器歧管　injector manifold
喷嘴　ejection nozzle, nozzle, jet
喷嘴速度　nozzle velocity
盆地充填沉积　basin-fill sediment
盆形太阳能蒸馏器　basin type solar still

peng

蓬松的东西　flue
硼硅玻璃3.3（又称特硬玻璃）borosilicate glass 3.3
硼硅酸玻璃　pyrex
硼漫射背面场电池　boron diffused BSF cell
膨胀　dilation
膨胀阀　expansion valve
膨胀金属　expanded metal
膨胀能量　expansion energy
膨胀水箱　expansion tank
膨胀系数　expansion coefficient
膨胀应力　expansion stress
碰撞率　collision rate

pi

批　lot
劈管　split spoon
皮带传动　belt drive
皮带运输机　belt conveyor
皮尔逊莫斯科维茨谱　Pierson-Moskowitz (PM) spectrum
皮托管　pitot tube
皮托管式风速计　pitot-tube anemometer
皮瓦　picowatt
疲劳荷载　fatigue load
疲劳极限　endurance limit
疲劳加载　fatigue loading
疲劳螺栓　fag bolt

疲劳强度　fatigue stress
疲劳试验　fatigue test
疲劳寿命曲线　fatigue life curve
疲劳损伤　fatigue damage
疲劳载荷　fatigue loading
p-n 结硅光电池　silicon p-n junction photocell

pian

偏差　deviation, drift
偏航　yaw, yawing
偏航齿轮　yaw gear
偏航电机　yaw motor
偏航环　yaw ring
偏航机构　yawing mechanism
偏航基座　yaw base
偏航角　yaw angle
偏航角速度　yaw rate
偏航角误差　yaw misalignment
偏航控制　yaw control, yaw controlling
偏航控制系统　yaw control system
偏航力矩　yaw moment
偏航力矩系数　yawing moment coefficient
偏航盘　yaw base
偏航偏移　yaw offset
偏航气流　yawed flow
偏航驱动　yaw drive, yawing driven
偏航失调　yaw misalignment
偏航误差　yaw error
偏航系统　yaw system, yawing system
偏航制动器　yaw brake
偏航轴　yaw axis
偏航轴承　yaw bearing
偏航装置　yaw mechanism
偏航阻尼器　yaw damper
偏角　deviation angle
偏向　deflection

偏斜分布　skew distribution
偏心环流　eccentric circulation vortex
偏心转子发动机　eccentric rotor engine
偏轴风　off-axis wind
偏轴聚光太阳灶　yaw axis concentration solar cooker
偏转　yaw
偏转板　deflector
偏转风速表　deflection anemometer
偏转力　defleting force, deviating force
片麻岩　gneiss
片式散热器　fin type radiator

piao

漂浮风轮机　floating wind turbine
漂浮装置　float device
漂移　drift
漂移场　drift field
漂移场电池　drift field cell
漂移电流　drift current
漂移电压　drift voltage
漂移力　drift force
漂移率　drift mobility
漂移速度　drift velocity
漂移效应　drift effect
漂移型光伏器件　drift type photovoltaic device

pin

频率　frequency, periodicity
频率表式风速计　frequency-meter anemometer
频率范围　frequency domain
频率谱　frequency spectrum
频率响应　frequency response
频率域　frequency domain
频谱　cross-spectra, frequency spectrum

频谱分裂　spectrum splitting
频谱分析　spectral analysis
频谱灵敏度　spectral sensitivity
频谱形状　spectral shape
频散关系　dispersion relation
频闪　stroboflash
频数比　frequency ratio
频数分布　frequency distribution
频数估计　frequency estimation
频域　frequency domain
频域测量　frequency domain measure
频域法　frequency domain method
频域分析　frequency domain analysis
频域分析器　frequency domain analyser
频域模型　frequency domain modelling
频域偏移　frequency domain migration
品质因数　figure of merit

ping

平（太阳）时　mean solar time
平凹镜　plano-concave lens
平板光电池组件　photovoltaic flat plate module
平板集热器　flat collector, flat-plate collector
平板接受器　flat receiver
平板卡车　flat bed truck
平板炉　flat plate hearth
平板平均温度　average plate temperature
平板热转换器（装置）　flat plate converter
平面　planar
平板　flat plate
平板式换热器　flat plate heat exchanger
平板式集热器　flat plate collector
平板式吸热器　billboard receiver
平板式组件　flat plate module

平板吸收器　flat absorber
平板形集热器　flat plate collector
平板形集热器元件　components of flat plate collector
平板形太阳能集热器　solar flat plate collector, solar panel
平板形太阳能集热器热容量　heat capacity of flat plate collector
平板形真空管（太阳能）集热器　flat plate evacuated tube collector
平板阻力　flat plate drag
平潮　slack water, tidal stand
平底容器　flat-bottomed vessel
平方　立方定律　square-cube law
平方律特性电压　square law voltmeter
平管热交换器　flat tube heat exchanger
平光电池板　photovoltaic flat plate
平行共燃　parallel co-firing
平行混燃　parallel co-firing
平行流　parallel flow, parallel stream
平行流动热交换器　parallel flow heat exchanger
平行燃烧　parallel combustion
平行轴　parallel shaft
平行轴齿轮副　gear pair with parallel axes
平衡　equilibrium
平衡阀　balancing valve
平衡分布系数　equilibrium distribution coefficient
平衡浮子　displacer
平衡环　gimbal
平衡架　gimbal
平衡凝固　equilibrium solidification
平衡器　equalizer
平衡设备　balancing equipment
平衡位置　equilibrium position

平衡压力　equilibrium pressure
平衡压强　equilibrium pressure
平衡载流子　equilibrium carrier
平滑风　smooth wind
平滑流　smooth flow
平均波浪周期　mean wave period
平均潮差　mean range, mean range of tide, mean tide range
平均潮面　mean tide level
平均大潮低潮面　mean low water spring
平均大潮升　mean spring rise
平均低潮面　mean low water
平均低低潮面　mean lower low water
平均电池电压　average cell voltage
平均风　average wind
平均风速　average wind speed, average wind velocity, mean wind speed
平均风向　mean wind direction
平均辐(射)照度　average irradiance
平均高潮位　mean high water
平均高高潮面　mean higher high water
平均功率　mean power
平均海平面，平均海面　mean sea level
平均河水位　mean river level
平均荷载　mean load
平均几何弦长　mean geometric chord length
平均流量　mean discharge, mean flow
平均能量输出　mean power output
平均热损系数　average heat loss coefficient
平均日效率（在有太阳辐照的一天内，太阳热水器储水所获得的热量与照射到太阳集热器采光表面上的太阳辐射能量之比）　average daily efficiency
平均时间　mean time
平均寿命　mean life

平均速度　average velocity, mean speed
平均停留　mean stay
平均温度　average temperature
平均小潮低潮面　mean low water neap
平均有机加载率　average organic loading rate
平均月潮低潮间隙　mean low water lunitidal interval
平均月潮高潮间隙　mean high water lunitidal interval
平均噪声　average noise level
平均涨潮间隙　mean rise interval
平均蒸汽产量　average steam production
平均值　mean value
平均最大风速　maximum average wind velocity
平炉钢　open hearth steel
平螺母　flat nut
平面　plane
平面波　plane wave
平面反射式平板集热器　plane receiver with plane reflector
平面加强反射镜　planar booster reflector
平面镜　planar mirror
平面性　planar
平面旋转　planar rotation
平模制粒机　flat die pelletizer
平太阳　mean sun
平太阳日　mean solar day
平坦地区　flat terrain
平头铆钉　flat-head rivet
平凸镜　plano-convex lens
平推流　plug flow
平稳随机负荷　stationary random load

平叶片 paddle blade
平正午 mean noon
平直管线 slick line
平准化电力成本 levelized cost of electricity
平准价格 levelized cost
屏蔽 shielding
屏极（阳极）电池组 anode battery
屏幕显示 screen display
屏栅极 screen grid (SG)

po

坡板 ramp
坡风 slope wind
迫近 loom
破袋机 bag-breaking machine
破坏性试验 destructive test
破坏性振动 destructive viberation
破浪波 breaking wave
破裂角 angle of rupture
破裂系 fracture system
破裂型式 fracture pattern
破裂压力 breakdown pressure
破碎 size reduction
破碎波 breaking wave
破碎的生物质燃料 shredded biofuel
破碎机 crusher
破碎压力 breakdown pressure
破碎岩层 fracture zone
破碎凿岩 spallation drilling

pu

葡糖糖发酵 fermentation of glucose
蒲氏风力等级 wind of Beaufort force
普朗克常量 Planck constant
普朗克常数 Planck constant
普朗克定律 Planck's law

普通二级管（整流二级管） general purpose diode
谱 spectrum
谱的能量分布 spectral energy distribution
谱分布 spectral distribution
谱矩 spectral moment
谱空缺 spectral gap
谱轮廓参量 shape parameter
谱隙 spectral gap
谱信号 spectral signal
谱形 spectral shape
谱硬化 spectral hardening
蹼轮 paddle-wheel

qi

七级风 moderate gale, near gale
栖息地丧失 habitat loss
期望风速 expected wind speed
期望输出值 desired output
鳍管热交换器 fin heat exchanger
鳍片管 extended surface tube
鳍片受热面 finned surface
鳍形接受器 fin type receiver
起动 start-up
起步阻力 breakaway force
起潮力 tide reproducing force
起动变压器 starting transformer
起动冲击电流 starting inrush current
起动齿轮 starter gear
起动电动机 starting motor
起动电流 starting current
起动电压 starting voltage
起动风速 starting wind speed, starting wind velocity, start-up wind speed, starting wind speed, threshold wind speed
起动继电器 start relay

起动力矩　starting torque
起动力矩系数　starting torque coefficient
起动扭矩　starting torque
起动燃烧器　start-up burner
起动瞬变量　starting transient
起动损失　starting loss
起动系统　start-up cycle
起动信号　starting signal
起动运行　start up operation
起动周期　start-up cycle
起动装置　start-up device, starter gear
起动阻力　starting resistance
起灰机　ash hoist
起火焰　pilot flame
起货钩　crowbar
起裂压力　initial pressure
起沙风　sand moving wind
起始瞬态电抗　subtransient reactance
起重扒杆　gin pole
起重机　hoist
气泵房　air pump room
气布比　air cloth ratio, air-to-cloth ratio
气锤　air ram (mer)
气道　air way
气电量计　oxyhydrogen voltameter
气垫　air cushion
气垫钻井平台　air cushion drilling platform
气顶　gas cap
气动动力　pneumatic power
气动给进架式风钻　air-feed drifter
气动给进钻机　air-feed drill
气动给进钻岩机　air-feed drill
气动功率　pneumatic power
气动互制　aerodynamic interaction
气动记录调节器　air operated recording regulator
气动架式凿岩机　air drifter (drill)
气动力下洗　aerodynamic downwash
气动扭矩　aerodynamic torque
气动伸缩式凿岩机　air stopper
气动失速　aerodynamic stall
气动式波能转换器　pneumatic wave-energy converter
气动试验泵　air operated test pump
气动输送　pneumatic conveying
气动输送系统　pneumatic conveying system
气动往复式压气机　pneumatic-reciprocating compressor
气动系统　pneumatic system
气动效率　aerodynamic efficiency
气动运输机　pneumatic conveyor
气动载荷　air load
气动凿岩机　air sinker
气动阻力　air drag, air resistance
气动钻井　air drilling
气—沸腾液体热交换器　gas-to-boiling-liquid heat exchanger
气氛　gas atmosphere
气割　gas cutting
气冠　gas cap
气罐　gas tank
气候　climate, weather
气候极限载荷　climatic limit loading
气候极值　climatic extreme
气候条件　climate condition
气候图集　climatic atlas
气候学　climatology
气候影响　climate effect
气候志　climatography
气化　gasification
气化带　gasification zone
气化反应　gasification reaction

气化反应器　gasification reactor
气化机组　gasifier set
气化剂　gas agent
气化炉　gasifier
气化炉反应堆容器　gasifier reactor vessel
气化区　gasification zone
气化燃料电池发电厂　gasification fuel cell power plant
气化效率　gasification efficiency, gasified efficiency, vaporization efficiency
气解电量计　gas voltameter
气井　gas well
气井钻台　gas platform
气举　gas lift
气孔　pore
气块　air parcel
气浪　air wave
气冷蛇管　air coil
气冷蛇形盘管　air coil
气冷式移动炉排炉　air-cooled moving grate furnace
气力发电机　aero generator
气力输送　airslide
气量　air flow rate
气量表　gasometer
气量计　gasometer
气流　air current, air draft, air flow, airstream, wind stream, moving air
气流参数　flow parameter
气流床气化反应器　entrained-flow gasification reactor
气流分离　flow separation
气流分配　air distribution
气流畸变　flow distortion
气流角　flow angle
气流结构　air flow structure

气流脉动　airflow pulsation
气流速度　air velocity, gas velocity
气流提升　gas lift
气流调节器　barometric damper
气流通道　air duct
气流瓦特计　air flow wattmeter
气流噪声　airflow noise
气流坐标系　wind axis system
气密　airseal, air-tight seal, pressurized acid digestion
气密层，内衬层　air barriers
气密性　airtightness
气密铸件　airtight cast
气膜　gas film
气膜冷却　air film cooling
气囊　air-bag
气囊支架　air-bag support
气喷净法　air blast
气一气换热　gas-gas heat exchange
气一气热交换器　gas-to-gas heat exchanger
气侵　air cutting
气热式（太阳能）集热器　gas heating collector
气溶胶　aerosol
气溶胶电（学）　aerosol electricity
气溶胶静电过滤器　aerosol electrostatic filter
气溶胶粒子　aerosol particle
气升法　airlift method
气升式浮选机　air-lift (floatation) machine
气升式搅拌机　air-lift type agitator
气式集热器　air-type collector
气式平板集热器　air-type flat plate collector
气式太阳能系统　air-type solar system
气水热泵（一种利用户外空气作热源或

冷源，通过第二介质——水，向需要空调的户内空间供热的热泵） air-to-water heat pump
气水热交换器　air-to-water heat exchanger
气速　gas speed
气锁　air lock
气态　gas phase
气态离子　gas ion
气体　gaseous fluid
气体捕集器　gas trap
气体产生装置　gas-generating device
气体常数　gas constant
气体电池组　gas battery
气体定律　gas law
气体动力学　aerodynamics
气体发生器　gas generator
气体分布器　gas distributor
气体分离器　gas trap
气体分配器　gas distributor
气体分析器　gas analyzer
气体分析仪　gas analyzer
气体分压　gas partial pressure
气体分油器　gas trap
气体辐射　gaseous radiation
气体混合器　gas mixer
气体混合物　gaseous mixture
气体混合仪　gas mixer
气体接受器　gas receiver
气体净化　gas cleaning
气体净化处理　gas conditioning
气体净化装置　gas cleaning unit
气体静力学　aerostatics
气体扩散　gaseous diffusion
气体扩散厌氧消化　anaerobic digestion with gas diffusion
气体冷却器　gas cooler
气体离子　gas ion
气体流程　flow path
气体排放　gaseous emission
气体排放法规　air emissions regulation
气体喷流　gas jet
气体燃料　gaseous fuel
气体生物燃料　gaseous biofuel
气体收集器　gas collector
气体调节　gas conditioning
气体微粒净化系统　gas-particulate cleanup system
气体温度表　gas thermometer
气体污染物　gaseous pollutant
气体吸附　gas absorption
气体洗涤器　gas scrubber
气体旋流器　air cyclone
气体压缩机　gas compressor
气体转化器　gas reformer
气体装置　gas installation
气团分析　ar mass analysis
气团密度　air mass density
气腿凿岩机　airleg rock drill
气温日较差　daily range of air temperature
气洗　air purging
气隙　air gap
气隙磁场　gap magnet
气隙磁化线　air-gap line
气隙磁通　air-gap flux
气隙磁通分布　air-gap flux distribution
气隙磁阻　air gap reluctance
气相　gas phase, vapour phase
气相沉积　vapor phase deposition
气相动力学　gas phase kinetics
气相扩散　vapor phase diffusion
气相燃烧　gas phase combustion
气相色谱检测　gas chromatographic detection

气相色谱仪　gas chromatograph (GC)
气相色谱仪鉴定器　GC detector
气相渗透率　gas permeability
气相生长　vapor growth, vapor phase growth
气相外延　vapor phase epitaxy
气象　weather
气象参数　weather parameter
气象潮　meteorological tide
气象观测塔　instrumentation tower
气象局　weather bureau
气象气球　weather balloon
气象潜势　meteorological potential
气象视距　meteorological optical range
气象数据　meteorological data
气象塔　meteorological tower, meteorological mast
气象条件　weather conditions, climatic conditions, meteorological conditions, meteorology
气象系统　meteorological system
气象学　meteorology
气象要素　weather element, meteorological element, meteorological essential
气象预报　meteorological forecast
气象站　meteorological observatory, meteorological institution, meteorological station
气旋　cyclone
气旋波　cyclonic wave
气旋床反应器　cyclonic bed reactor
气旋风　cyclonic wind
气旋粒子　cyclonic particle
气旋生成　cyclogenesis
气旋涡度　cyclonic vorticity
气旋涡量　cyclonic vorticity
气旋消失　cyclolysis

气旋型涡流　cyclonic whirl
气穴　air sink
气穴流涡空流　cavity flow
气压　air pressure
气压表　air gauge
气压差异　air pressure difference
气压缸　pneumatic reservoir
气压计　barometer
气压探头　air pressure probe
气压梯度　pressure gradient
气压梯度力　pressure gradient force
气压调节器　air pressure regulator
气压响应　barometric response
气液热交换器　air-to-liquid heat exchanger, gas-to-liquid heat exchanger
气油冷却器　air-to-oil cooler
气穴现象　cavitation
气源　air supply, gas source
气闸　air lock, air brake
气震钻井　air-hammer drilling
气柱　gas column
气阻　air lock
汽车发动机罩　bonnet
汽袋　steam pocket
汽电共生系统　cogeneration system
汽鼓　steam drum
汽锅　steam kettle
汽耗　steam consumption
汽化（潜）热　vaporization heat
汽化面　evaporating surface
汽化能　vaporization energy
汽化器　carburetor, vaporizer
汽化潜热　latent heat of vaporization
汽孔　steam vent
汽轮发电机　steam turbine generator
汽轮发电机组　turbogenerator set
汽轮发电机组额定发电功率　rated

electric power of turbine-generator unit
汽轮发电机组启动　start up of turbine generator unit
汽轮机定压运行　constant pressure operation of turbine
汽轮机滑压运行　sliding pressure operation of turbine
汽轮机技术　turbine technology
汽轮机排气　turbine exhaust
汽轮机装置　steam turbine plant
汽轮机组　turboset
汽煤比　evaporation factor
汽—汽热交换器　steam-to-steam heat exchanger
汽水分离　steam water separation
汽水分离器　steam separator
汽水混合物　steam-water mixture
汽水循环　steam water cycle
汽温　steam temperature
汽相　steam phase
汽压　steam pressure, vapor pressure
汽油添加剂　gasoline additive
器件连接　device connection
憩流　slack tide
憩流憩潮　slack water

qian

千吨　kilotonne (kton)
千伏安　kilovolt-ampere (kVA)
千斤顶　jack
千帕　kilopascal
千瓦　kilowatt (kW)
千瓦热能　kilowatt thermal (kWth)
千瓦时　kilowatt hour (kWh)
千瓦时计　kilowatt-hour meter
千兆瓦　gigawatt (GW)
千兆瓦时　gigawatt hour (GWhr)
牵出同步　pulling out of synchronism
牵出转矩　pull-out torque
牵索　guy wire
牵引环　clevis drawbar
牵引机　towed machine
铅箔　lead foil
铅管　lead pipe
铅酸蓄电池　lead acid battery
铅直风速表　vertical anemometer
前壁薄膜太阳能电池　frontwall thin film solar cell
前池　forebay
前触点　front contact
前导喷射　pilot injection
前港谐振效应　front port resonant efficiency
前寒武纪岩床　Precambrian bedrock
前行叶片　advancing blade
前后　pitching
前膜光电池　front boundary layer cell
前膜光电管（池）　front boundary cell
前缘半径　leading edge radius
前缘缝翼　leading edge
前置泵　booster pump
前置回路　primary circuit
前置机　front end processor
钳位（电路）　cramp
钳形表　clamp ammeter
潜伏故障　latent fault dormant failure
潜流　underrun
潜热　latent heat, potential heat
潜热储存　latent heat energy storage, latent heat storage
潜热量热器　latent heat calorimeter
潜入式电动机　submersible motor
潜水泵　submersible pump

qiang

潜水面　water table
潜水喷发　phreatic eruption
潜水位　ground water level
潜油涡轮泵　submersible turbine pump
潜在风能场址　potential wind power site
潜在结垢　scaling potential
潜在可用风能　potentially available wind power
潜在污染物　potential pollutant
浅表层断裂　subsurface fracturing
浅表面含水层　subsurface aquifer
浅层地壳构造　shallow crustal structure
浅层地热能　shallow geothermal energy
浅层地震　shallow earthquake
浅层电阻率测量　shallow resistivity survey
浅层温度测量　shallow temperature surveying
浅沟　shallow ditch
浅海　shallow water
浅海沉积　shallow sea deposit
浅灰色砂岩　light gray sandstone
浅井　shallower well
浅孔　shallow bore hole
浅扩散电池　shallow diffused cell
浅埋结构　subsurface structure
浅水　shallow water, shoaling
浅水波　shallow water wave
浅水域　shallow water
浅滩　shallow water, shoaling
欠电压　under-voltage
欠励磁　underexcitation
欠阻尼振荡　underdamped oscillation
嵌段共聚物　block copolymer
嵌入式pc　embedded pc

腔式吸热器　cavity receiver
腔体吸收器　cavity absorber
腔体式绝对辐射表　absolute cavity radiometer
强潮流　tidal race, tidal rapid
强电控制　strong current control
强风　strong breeze, energetic wind, blast
强辐射　hard radiation
强化地热系统　enhanced geothermal system
强加函数　forcing function
强迫风冷式变压器　air-blast transformer
强迫循环太阳能热水器　forced-circulation solar water heater
强迫振动　forced vibration
强压通风扇　forced draft fan
强阵风　heavy gust
强制函数　forcing function
强制通风　forced draught, blast draft, mechanical draft
强制循环　forced circulation
强制循环水加热系统　forced circulation water heater system
强制循环太阳热水器　forced cycle solar water heater
强制循环系统　forced circulation system
强阻尼　hard damping
墙式锅炉　wall fired boiler
墙式集热　wall collection
墙式集热器　wall collector
抢夺　grab

qiao

桥联性　bridging, arching

桥式起重机　overhead travelling crane, electric double girder travelling crane
翘板式风轮　seesaw rotor
翘板型轮毂　teetered hub

qie

切（纸）　guillotine
切变风　shear wind
切出（断）风速　cut-out wind speed
切断机　cutting-off machine
切割进给速度　cutting feed rate
切割木片　cutter chip
切割盘　cutting disk
切割生物质燃料　cut biofuel
切换开关　changeover switch
切换运行　switching operation
切克劳斯基晶体生长装置　Czochralski artefaction system
切孔　cutting opening
（硅等的）切片　wafering
切片　slab, wafer cutting
切片理论　strip theory
切入风速　cut-in wind speed
切碎机　straw shredder
切铁刀　iron cutter
切线力　tangential force
切线应力　tangential stress
切向风流诱导因子　tangential flow induction factor
切向力　tangential force
切向燃烧锅炉　tangentially fired boiler
切向速度　tangential velocity
切削性　machinability

qin

侵略地　aggressively
侵入体　intrusive body

qing

青铜　bronze
青储饲料　ensilage
青储玉米　maize silage
轻便钻机　mobile rig
轻柴油　light diesel oil
轻风　light air, light breeze
轻骨料　lightweight aggregate
轻砂石　lightweight aggregate
轻质水泥　lightweight cement
氢化物电池　hydride battery
氢化植物油　hydrotreated vegetable oil
氢键　hydrogen bond
氢能发电　hydrogen power generation
氢气浸提　hydrogen infusion
氢溴电池　hydrogen bromine battery
氢氧电量计　oxyhydrogen voltameter
氢氧化钙　hydrated lime
氢氧化钾电解质　potassium hydroxide electrolyte
氢氧化镍　nickel hydroxide
氢氧燃料电池　hydrogen oxygen fuel cell
倾覆风速　overturning wind speed
倾覆力矩　overturning moment
倾角　tilt angle
倾斜　obliquity, pitch
倾斜衬里　tilted wick
倾斜度测定仪　gradienter
倾斜光线　oblique ray
倾斜角　angle of inclination
倾斜平板型集热器　tilted flat-plate collector
倾斜式集热器　tilted collector
倾斜式热虹吸太阳能集热器　tilted thermosiphoning solar collector

倾斜式热水槽 tilted water-heating tray
倾斜式太阳能系统 tilted solar system
倾斜式往复炉排 inclined reciprocating grate
倾斜式振动筛 inclined vibrating screen
倾斜式蒸馏器 tilted still
倾斜装置 tilted device
倾斜状管式集热器 tilted tubular collector
清灰 ash removal
清洁车 garbage truck
清洁发展机制 clean development mechannism
清洁剂 detergent
清洁可再生能源 carbon-free renewable energy
清洁能源 cleaner energy, clean source of power, green power
清洁箱 garbage bin
清晰度 clearness index
清晰度系数 clearness factor
情况 instance
晴空指数 clearness index
晴天接受到的太阳辐射，晴天单位水平表面上接受直接太阳能的速率 clear day insolation
晴天太阳脉冲曲线 clear day sun pulse curve
晴天突起的（热带）暴风 white squall

qiu

丘陵区 rolling terrain
球床吸热器 pebble-bed receiver
球阀 ball valve
球棍 mallet
球面聚光器 spherical concentrator
球磨机 bowl mill
球墨铸铁管 ductile iron pipe
球跑轨磨 ball-race mill
球泡灯 bulb light
球头定位器 insulator ball guide
球头挂钩 ball-hook
球头挂环 ball-eye
球形阀 globe valve
球形集热器 spherical collector
球形面板 spheral panel
球形散流器 ball diffuser
球型硅太阳电池 spherical silicon solar cell

qu

区划 zonation
区间 interval
区时 zone time
区域电网 area grid
区域风 regional wind
区域供冷 district cooling
区域供暖 district heating
区域供暖系统 district heating system
区域供暖站 district heating plant
区域供热系统 district heating system
区域供热站 district heating station, district heating plant
区域开发 district exploration
区域热交换器 district heat exchanger
区域输电组织 regional transmission organization
区域太阳能供暖 solar space heating
区域性普查 regional reconnaissance
区域选择控制器 zone-selected controller
驱动齿轮 driving gear
驱动发动机的发电机 engine-driven generator
驱动机构 drive mechanism

驱动力 driving force
驱动器 driver
驱动拓扑 drive topology
屈服强度 yield strength
屈服信息 yielding information
趋势分析 trend analysis
曲柄 bell crank
曲柄轴箱 crankcase
曲弧度 camber
曲头钉 brad
曲线槽 curvilinear groove
曲线图 graph
曲线弯曲点 knee point
曲线修正系数 curve correction coefficient
曲线因数 curve factor
曲折连接绕组 interconnected star winding
曲轴 crankshaft
取暖油 heating oil
取水流量 water draw-off rate
取水温度 draw-off temperature
取芯钻进 core drilling
取芯钻井装置 core drilling assembly
取芯钻头 coring bit
取样电路 sample circuit
取样电压 sampling voltage
取样系统 sampled system
去除效率 removal efficiency
去灰 deashing
去矿物质 demineralization

quan

全波 full wave
全玻璃集热器 all-glass collector
全玻璃真空管集热器 all-glass evacuated tube collector
全玻璃真空集热管 all-glass evacuated collector tube
全程合成 de novo synthesis
全电路定律 all-circuits law
全额电能转换 full converter
全钒液流电池 all vanadium redox flow battery
全封闭风扇冷却 totally enclosed fan cooled
全辐射 total incident radiation
全辐射表 pyrradiometer
全辐射体 full radiator, full emitter
全功耗 total power dissipation
全贯流式水轮机 straight flow turbine, rim-generator turbine
全活叶螺旋桨 full feathering propeller
全浸式蒸发器 flooded evaporator
全绝缘线圈 winding with uniform insulation
全量焚烧系统 mass burn system
全流式地热系统 geothermal total flow system
全流式地热发电系统 total flow system for electric power production from geothermal energy
全硫 total sulfur
全年太阳辐射总量 total annual solar radiation
全桥驱动电路 full bridge circuit
全球变暖潜值 global warming potential
全球标定法 global calibrating method
全球地热能 global geothermal power
全球风 global wind
全球风环流模式 global wind circulation pattern
全球风力格局 global wind pattern

全球环流　global circulation
全球暖化潜势　global warming potential
全球气压　global pressure
全球热污染　global thermal pollution
全球日照率　global insolation
全球生物能伙伴关系　global bioenergy partnership
全球所受到的太阳能辐射　global insolation
全球增温潜势　global warming potential
全日潮　single day tide, diurnal tide
全树　complete tree
全水分　total moisture, total insolation
全速测试　full-speed test
全太阳能光电能系统　total solar photovoltaic energy system
全太阳能家庭　all-solar energy home
全天候通道　all-weather access
全天候运行　all-weather operation
全屋顶集热器　all-roof collector
全线电压　line-to-line voltage
全向风　omnidirectional wind
全叶片变桨　full-span pitch control
全藻类水热液化　whole algae hydrothermal liquefaction
全自动太阳能电池组件层压机　automatic solar module laminator
铨孔锯　keyhole saw

que

缺陷　defect imperfection, flaw
确认　acknowledgement
确认装机容量评估　confirmed installed capacity estimate
确认资源评估　confirmed resource estimate

qun

群速　group velocity
群体供暖系统　group heating system
群体供热系统　group heating system
群形古风车　smoke mill

ran

燃点　ignition temperature
燃尽　burnout
燃尽风　overfire air
燃烬段　burning section
燃料比　fuel ratio
燃料草　fuel grass
燃料层　fuel bed
燃料成本　fuel cost
燃料成本节约　fuel cost saving
燃料成分　fuel composition
燃料尺寸　fuel size
燃料处理　fuel handling
燃料床　fuel bed
燃料床温　fuel bed temperature
燃料存储　fuel storage
燃料电池　fuel cell
燃料电池车　fuel cell vehicle
燃料电池催化剂　fuel cell catalyst
燃料电池的燃料　fuel cell fuel
燃料电池电解质　fuel cell electrolyte
燃料电池汽车　fuel cell powered vehicle
燃料电极　fuel electrode
燃料堆　fuel bed
燃料费　fuel cost
燃料分配　fuel distribution
燃料供给　fuel feed
燃料供给器　fuel feeder
燃料供应　fuel supply
燃料加料器　fuel feeder

燃料进给和处理系统　fuel feeding and handling system
燃料进料斗　fuel feed hopper
燃料径向倒换　radial shuffling of fuel
燃料颗粒　fuel particle
燃料 空气比例调节　fuel-air ratio control
燃料木材　fuel wood
燃料能 电能换能效率　fuel energy to electrical conversion efficiency
燃料喷射　fuel injection
燃料品质　fuel quality
燃料气　fuel gas
燃料球芯块　fuel pellet
燃料燃烧层　active fuel bed
燃料烧尽　burnout
燃料适应性　fuel flexibility
燃料属性　fuel property
燃料特性　fuel property
燃料替代率　fuel substitution rate
燃料同质性　fuel homogeneity
燃料消耗率　fuel consumption rate
燃料效率　fuel efficiency
燃料芯块　fuel pellet
燃料乙醇　fuel bioethanol, fuel ethanol
燃料预处理　fuel pretreatment
燃料预处理技术　fuel pretreatment technology
燃料预处理系统　fuel pretreatment system
燃料原料　fertile element
燃料渣　waste fuel
燃料值　fuel value
燃料制备过程　fuel preparation process
燃料质量　fuel quality
燃料转化厂　conversion plant
燃料装卸　fuel handling
燃料组成　fuel composition

燃料组件　fuel assembly
燃料作物　fuel crop
燃煤电厂（站）　coal-fired power station
燃煤炉　coal-fired furnace
燃木供热设备　wood-fired heating plant
燃气　fuel gas, gas vapor
燃气发电厂　gas-fired power plant
燃气发电机组　gas-fueled generator set (genset)
燃气锅炉　gas fired boiler
燃气机　gas-fired engine
燃气净化器　gas cleaner
燃气轮发电机　combustion turbine generator
燃气轮机　combustion turbine, gas turbine
燃气轮机联合循环发电厂　gas turbine combined cycle power plant
燃气燃烧器　gas-fired burner
燃气涡轮　gas turbine
燃烧　combustion
燃烧车间　combustion plant
燃烧段　combustion section
燃烧发电　power generation by combustion
燃烧反应　combustion reaction
燃烧过程　combustion process
燃烧过程控制　combustion process control
燃烧过程控制器　combustion process controller
燃烧技术　combustion technique, combustion technology
燃烧空气　combustion air
燃烧炉　combustion furnace
燃烧率　combustion rate
燃烧气体　combustion gas
燃烧器，燃烧室　combustor

燃烧强度 intensity of combustion
燃烧区 combustion zone
燃烧设施 combustion plant
燃烧室 combustion chamber
燃烧温度 combustion temperature
燃烧系统 combustion system
燃烧效率 combustion efficiency
燃烧性能 combustion performance, combustion behaviour
燃烧值 fuel value
燃用城市垃圾电厂 city garbage-burning power station
燃油供给 fuel feed
燃油供应 fuel supply
燃油锅炉 oil-fired furnace
燃油喷射 fuel injection
染料敏化太阳能电池 dye-sensitized solar cell
染料渗透试验法 dye penetrant examination

rao

扰动电流 disturbing current
扰动加速度 disturbing acceleration
扰动力 disturbing force
扰流板 spoiler
扰流器 spoiler
绕南极流 Antarctic Circumpolar Current
绕射力 diffraction force
绕射问题 diffraction problem
绕线电枢 wound armature
绕线器 winder
绕线式发电机 wound rotor generator
绕线式感应（异步）发电机 wound rotor induction generator
绕线式异步电动机 wound rotor asynchronous moter
绕线柱 winding cylinder
绕线转子 wound rotor
绕线转子同步发电机 wound rotor synchronous generator
绕组单元 winding section
绕组电阻测量 winding resistance measurement
绕组端部 winding overhang
绕组节距 winding pitch
绕组温度 winding temperature
绕组系数 winding factor
绕组线 winding wire

re

热备用 hot reserve, spinning reserve
热泵 heat pump, thermal pump
热泵或地下耦合热泵 ground coupled heat pump
热泵模块 heat pump module
热泵年度使用效率 annual heat pump coefficient of performance
热泵蒸发器 heat pump evaporator
热泵制冷剂 heat pump refrigerant
热泵装置年度使用效率 annual heat pump plant coefficient of performance
热泵装置年度使用期 annual heat pump plant utilisation period
热泵装置年度有效供热 annual useful heat supplied by heat pump
热边界层 thermal boundary layer
热变电阻 thermal resistance
热变量热计 fission calorimeter
热变形 thermal strain
热变质 thermal metamorphism
热剥落 thermal spalling
热补偿 thermal compensation
热采原油 thermal oil

热槽 hot tub
热测绘图 thermal mapping
热测量法 calorimetric procedure
热产生 heat production
热场 hot zone
热成层作用 thermal stratification
热成型玻璃反射镜 hot-molded glass mirror
热弛豫 thermal relaxation
热冲击 thermal shock, temperature shock
热冲击系数 thermal shock parameter
热出力 heat producing capability
热储 hydrothermal reservoir
热储存 thermal storage
热储存点损失系数 overall storage heat loss factor
热储工程 geothermal resorvoir engineering
热储管理 reservoir management
热储化学 reservoir chemistry
热储模拟 reservoir simulation
热储模型 reservoir modeling
热储容积变动 reservoir volumetric change
热储衰减 reservoir decay
热储压力衰减 reservoir pressure decline
热储岩体 reservoir rock
热处理 thermal treatment, heat treating
热触点 thermal contact
热传导 thermal conduction, heat exchanging, heat conduction
热传导迟缓 thermal lag
热传递 heat transfer, heat transmission, thermal transmission
热传递系数 thermal transmittance
热传输效率 heat transfer effectiveness
热磁 thermomagnetic
热存储 heat storage
热带海洋区域 tropical ocean region
热带水 tropical water
热当量 thermal equivalent
热导电涂层 thermally-conductive coating
热导换热 transfer by conduction
热导率 thermal conductivity, temperature conductivity, heat conductivity
热导率探测器 thermal conductivity detector
热导系数 thermal conductivity
热岛 heat island, thermal island
热的耗损 heat rejection
热等离子体 thermal plasma
热电 hot spot
热电发电机 thermoelectric generator
热电材料 thermoelectric material
热电厂 cogeneration plant, power and heating plant
热电堆 thermopile
热电发电机 thermoelectric generator
热电共生厂 combined heat and power station
热电合供 electrical-thermal cogeneration
热电金属 thermpelectric metal, thermoelectric metal
热电晶体 pyroelectric crystal
热电冷三联产 trigeneration
热电联产 cogeneration, electricalthermal cogeneration
热电联产电厂 co-generation power plant
热电联产机组 combined heating power, cogeneration system
热电联产能力 cogeneration capacity

热电联供　combined heat and power, electrical-thermal cogeneration
热电联供系统　combined heat and power system
热电联供装置　cogenerator
热电流　thermal current
热电偶　thermocouple
热电偶材料　materials for thermocouple
热电日射表　thermoelectric actinometer
热电式感温元件　thermoelectric cell
热电效应　pyroelectric effect
热电性　pyroelectricity
热电压　thermal voltage
热电站　steam power plant, cogeneration plant, thermal power station, power and heating plant
热电直接日射计　thermoelectric pyrheliometer
热电转换(作用)　thermoelectric conversion
热(光)电转换器件　ovonic device
热电转换系统　thermoelectric conversion system
热动力　thermodynamic force
热动式调节阀　thermal valve
热对流　thermal convection, thermal convection current
热二极管太阳板　thermodiode solar plate
热发电厂　thermal power generation plant
热发射(率)　thermal emittance (emissivity)
热发射电极　hot emitted electrode
热发生速率　rate of heat generaion
热反射薄膜　heat reflective coating
热反射玻璃　heat reflecting glass
热反射涂层　heat reflective coating
热反应　thermal response
热防护层　thermal shielding

热防护能力　heat shield capability
热分布　heat distribution
热分解　thermal decomposition
热分解法　thermal decomposition
热分离器　heat seperator
热分配　heat distribution
热分析　thermal analysis
热风速仪　thermal anemometer
热风筒　thermal chimney
热封套　thermal envelope
热肤效应　thermal skin effect
热辐射　thermal radiation
热辐射计　kampometer
热辐射率　thermal emissivity
热辐射器　thermal radiator
热辐射源　thermal source
热负荷　heat load
热干化　thermal drying
热工保护　thermal process protection, thermal equipment protection
热工况　thermal condition
热工水力建模方法　thermal hydraulic modeling approach
热功率　thermal power, heat output, heat rating, thermal output, thermal rating
热管　heat tube
热管(太阳电池等传热元件)　heat pipe
热管式吸热器　heat pipe receiver
热管式真空管集热器　heat pipe evacuated tube collector
热管真空集热管　heat pipe vacuum tube
热惯性　thermal lag, thermal inertia
热光电太阳能集热器　thermal-photovoltaic solar collector
热光电转换器　thermophotovoltaic

converter
热光子转换器　thermophotonic converter
热辊型　thermal profile
热害截面　thermal poison cross section
热含量　heat content
热函　enthalpy
热耗率　heat rate
热红外技术　thermal infrared technique
热虹吸管　thermosiphon
热虹吸管冷却　thermosiphon cooling
热虹吸空气集热器　thermosiphoning air collector
热虹吸水收集器　thermosiphon water collector
热虹吸系统　thermosiphon system
热虹吸作用水泵　thermosiphoning action pump
热（能）化　thermalization
热化学生物质转换技术　thermochemical biomass conversion technology
热化学吸热反应器　thermochemistry receiver reactor
热化学转化　thermochemical conversion
热化学转换技术　thermochemical conversion technology
热回收装置　heat recovery unit
热混合　thermal mixing
热机　thermal engine
热机械效应　thermomechanical effect
热机循环　engine cycle
热机转换效率　thermomechanical conversion efficiency
热激发能量　thermal excitation energy
热（中子）激活截面　thermal activation crosssection
热激励热泵　heat-actuated heat pump
热激作用　heat activation

热继电器　thermal relay
热降　thermal drop
热降解　thermal degradation
热交换　thermal exchange
热交换法　hot exchange method
热交换器　heat exchanger, heat extractor, heat interchanger, exchanger
热交换器回路　heat exchanger circuit
热交换器清洗系统　heat exchanger cleaning system
热交换器效率　heat exchanger effectiveness
热交换系统　heat exchange system
热交换因子　heat removal factor
热接受器　thermal acceptor, thermal receiver
热解　thermal decomposition, pyrolysis
热解产物　pyrolysis procuct
热解改质　hydrothermal upgrading
热解聚　thermal depolymerization
热解气化　pyrolytic gasification
热解气化炉　pyrolitic gasifier
热解油　pyrolytic oil
热介质循环　circulation of hot medium
热镜　heat mirror
热镜薄膜　heat mirror coating
热镜涂层　heat mirror coating
热绝缘　thermal insulation
热开采　heat mining
热抗计　thermal rising
热抗散　thermal diffusion
热空气出口　hot air outlet
热空气干燥箱　air oven
热空气气轮机　hot air turbine
热控制　thermal steering
热库　heat store, heat reservoir
热扩散　thermal diffusion

热扩散率　thermal diffusivity
热扩散系数　coefficient of thermal diffusion, thermal diffusion coefficient, thermodynamic coefficient
热浪　hot wave, heat wave
热力参数　thermal parameter
热力阀　thermal valve
热力过程　thermodynamic process
热力经济性　thermal effectiveness, thermodynamic effectiveness
热力面　thermodynamic surface
热力平衡　thermodynamic equilibrium
热力势　thermodynamic potential
热力双向流反应器　thermal flow-rever-sal reactor
热力损失　thermodynamic loss
热力特性　thermal performance
热力学　thermodynamics
热力学不可逆性　thermodynamic irreversability
热力学参数　thermodynamic parameter
热力学第二定律　second law of thermodynsmics
热力学定律　law of thermodynamics
热力学计算　thermodynamic calculation
热力学温(度)标　thermodynamic scale of temperature
热力学温标　thermodynamic scale
热力学效率　thermodynamic efficiency
热力学性质　thermodynamic property
热力学循环效率　thermodynamic cycle efficiency
热力循环　thermodynamic cycle
热力循环效率　thermodynamic cycle efficiency
热力指数　thermal parameter
热利用　thermal utilization

热量　heat quantity
热量传感器　heat sensor
热量单位　thermal unit
热量放出　heat rejection
热量废弃　heat rejection
热量分布　heat distribution
热量回收　heat reclaim, heat return
热量计　calorimeter
热量利用　heat reclaim
热量平衡　calorific balance
热量输出　heat output
热量输入　heat input
热量输送　heat transport
热量提取设备　heat extraction equipment
热量吸取　heat absorption
热量吸收　thermal absorption
热量资源　thermal resource
热裂　pyrolysis
热裂（作用）　pyrolytic cracking
热裂化　thermal cracking
热裂解　thermal cracking
热灵敏性　thermal sensitivity
热流　thermal current, heat flow, heat flux
热流测量技术　thermal flow measurement technique
热流方向　direction of heat flow
热流计　heat flow meter
热流密度　heat flux, heat flux density
热流强度　rate of heat flow
热流数据　heat flow data
热流体　hydrothermal fluid
热流体比热　specific heat of hot fluid
热流体等高线　heat flow contour
热流体密度　density of hot fluid
热流图　heat flow diagram
热漏泄系数　thermal leakage coefficient

热卤水　hot brine
热录像仪　thermograph
热率　heat rate
热敏电阻　thermistor, thermal resistance, thermal resistor
热模拟　thermal simulation
热模型　thermal model
热膜风速计　hot film anemometer
热能　heat energy
热能储存　thermal energy storage
热能储存系统　thermal energy storage system
热能废物　thermal waste
热能梯度　thermal gradient
热能通量　thermal energy flux
热能转换效率　thermodynamic conversion efficiency
热泡抗计　thermal rising
热喷涂　thermal spraying
热膨胀　thermal expansion
热膨胀系数　thermal expansion coefficient
热平衡　thermal equilibrium
热平衡计算　heat account
热屏　heat shield, thermal shielding
热屏蔽　thermal shield
热气化　thermal gasification
热迁移　thermophoresis
热强度　thermal intensity
热侵蚀　thermal etching
热泉　hot spring, thermal spring
热泉保护区　area for protection of hot springs
热泉变化　hot spring fluctuation
热泉换热　heat transfer by hot spring
热泉流量　hot spring flow
热泉温度　temperature of hot spring
热缺陷　thermal defect

热容　heat capacity, caloricity
热容量　calorific heat, heat capacity, thermal capacity
热容率　thermal capacitivity
热散射　thermal scattering
热渗出位置图　location map of thermal seepage
热声斯特林热机　thermoacoustic-Stirling heat engine
热失配　thermal mismatch
热时间常数　thermal time constant
热示踪物　thermal tracer
热势　thermal potential
热室　hot cell
热释电材料　pyroelectric material
热受主　thermal acceptor
热衰减　thermal decline
热水　heating water
热水储热箱　hot water heat store tank
热水地热系统　hot water geothermal system
热水供给　hot water supply
热水井　hot water well, hot well
热水渗出　hot water seepage
热水田　hot water field
热水箱　hot water tank
热水钻井　drilling for hot water
热丝化学气相沉积　hot wire chemical vapor deposition
热丝式风速计　hot wire anemometer
热速度　thermal velocity
热损传导率　heat loss conductance
热损伤　heat transfer impairment
热损失　heat loss, thermal loss, temperature loss
热损失百分数　heat loss percent
热损失系数　heat loss factor

热损失转储负载	heat loss dump load
热锁	thermal lock
热塔	hot tower, thermal tower
热探测器	thermal detector, thermal defector
热探针	thermal probe
热探针法	hot probe method
热套	hot jacket
热特性	thermal property
热梯度	thermal gradient
热田地质史	geological history of field
热田面积	area of field
热调节器	thermistor
热通量	heat/thermal flux
热通量峰值	peak heat flux
热退火	thermal annealing
热网络	thermal network
热稳定剂	heat stabilizer
热稳定性	thermal stability
热污染	heat pollution, thermal pollution
热污染效应	effect on thermal pollution
热吸收管	heat absorbing pipe
热系数	thermal coefficient
热显示区	thermally active area, thermally active zone
热线	heat ray
热线风速计	hot wire anemometer
热响应	thermal response
热像仪	thermal imaging system
热效比	thermal efficiency ratio
热效率	thermal efficiency, heat efficiency, heat rate
热效应	thermal effect
热行为	thermal behaviour
热性能	thermal property, thermal behavior
热选技术（日本）	thermoselect technology
热选择性表面	selsective heat surface
热循环	hot run, thermal cycle
热压封	thermal pressure sealing
热烟气	hot flue gas
热烟气管	thermal chimney
热延时	thermal delay
热延时继电器	thermal time delay relay
热岩	hot rock
热岩层	hot rock reservoir
热岩构造	hot rock formation
热岩技术	hot rock technology
热岩浆体	hot magma body
热盐	hot brine
热盐环流	thermohaline circulation
热盐水池	hot brine pool
热盐水管	hot brine pipe
热盐水田	hot brine field
热氧化	thermal oxidation
热氧化作用	thermal oxidation
热液产能	hydrothermal capacity
热液活动	hydrothermal activity
热液流体成分	composition of hydrothermal fluid
热液流体腐蚀性	corrosivity of hydrothermal fluid
热液能（地下蒸气产生的）	hydrothermal energy
热液容量	hydrothermal capacity
热液系统	hydrothermal system
热液循环	hydrothermal circulation
热液异常	hydrothermal anomaly
热液资源	hydrothermal resource
热液作用	hydrothermal action
热溢流	heat outflow
热应变	thermal strain
热应力	thermal load, thermal stress

热泳　thermophoresis
热源　heat source, thermal source
热源富集　concentration of heat source
热源深度　heat source strength
热源温度　source temperature, temperature of hot source
热跃迁　thermal transition
热运动　thermal motion
热噪声　thermal noise
热增益　heat gain
热沾　hot-dip
热障　thermal barrier
热真空　thermal vacuum
热真空循环　thermal vacuum cycling
热振动　thermal vibration
热震参数　thermal shock parameter
热蒸发　thermal evaporation
热值　calorific value, heating capacity, caloricity
热值传感器　calorific value sensor
热质量　thermal mass
热致发光　thermoluminescene
热滞后　thermal hystersis
热中心　thermal centre
热（能）中子　thermal energy neutron
热中子剂量　thermal dose
热中子吸收　thermal absorption
热中子转换反应堆　thermal converter
热重分析　thermogravimetric analysis
热柱　thermal column
热转换　thermal conversion
热转换装置　thermal conversion device
热资源　thermal resource, heat source
热阻　thermal resistance, heat transfer resistance, heat resistance
热阻率　thermal resistivity

ren

人工地热储层　artificial geothermal reservoir
人工分选台　band separator
人工举升　artificial lift
人工裂缝　artificial fracture
人工上升流　artificial upwelling
人工引喷　artificially induced blow
人孔　manhole
人体毒性潜值　human toxicity potential
人为失误　human error
人因诱发沉降　anthropogenic-induced subsidence
人造地热能　enhanced geothermal system
人造纤维　synthetic fiber
人造橡胶密封　elastomeric seal
人字（齿）轮　herringbone gear
人字形齿轮　double helical gear
仁川潮汐电站　Incheon Tidal Power Station
刃型位错　edge dislocation
任选测试条件　optional test condition

ri

日（地）距　solar distance
日斑　solar spot
日变动　daily variation
日变风　diurnal wind
日变幅　daily amplitude, daily range
日变化　diurnal variation
日产出量（吨）　tons per day (tpd)
日产量　daily output
日常检查　daily observation, daily surveillance
日潮　solar tide

日潮不等 diurnal inequality
日潮差 diurnal range
日潮汐分量 diurnal tidal component
日成风 heliotropic wind
日出 sunrise
日磁周期 solar magnetic cycle
日地距离 sun-earth distance
日地空间 solar terrestrial space
日地平均距离 mean earth-sun distance
日地物理学 solar terrestrial physics
日分潮 diurnal tidal component
日辐射比 daily radiation ratio
日负荷变动运行 daily load fluctuating operation
日负荷曲线 daily load curve
日负荷系数 daily load factor
日光 light
日光层 heliosphere
日光反射装置 heliostat
日光疗法 heliotherapy
日光能量测定器 actinometer
日光燃料 sunfuel
日光效应 sun effect
日光浴室 solarium
日光源 daylight source
日光照明 daylight illumination
日光作用 daylight effect
日耗水量 daily water consumption
日集热时间 collection length of day
日降水量 daily precipitation amount
日轮亮度 brightness of solar disk
日轮亮度分布 brightness distribution of solar disk
日没 sunset
日冕 solar corona
日面 solar disk, sun's disk
日面经度 solar longitude

日平均温度 daily mean temperature
日平均值 daily mean, daily mean value
日晴朗系数（地面水平日辐射总量与大气上界对应平面辐射总量之比） daily clearness index
日热负荷 daily heat load
日晒盐 solar salt
日晒盐度差能 solar salt power
日晒作用 solarization
日射 insolation
日射表 solarimeter, actinometer
日射测量学 actinometry
日射观测 solar radiation observation
日射率 insolation, insolation rate
日射强度计 solarimeter, pyrheliometer
日射微粒流 solar wind
日射温度表 solar radiation thermometer
日射总量表 solarimeter
日射总量计 solarigraph
日射总量曲线 solarigram
日食 solar eclipse
日太阳辐射量 daily solar radiation amount
日太阳辐射总量 total daily solar radiation
日温计 pyrheliometry
日下点 subsolar point
日效率 daily efficiency
日循环 diurnal circle
日影仪 heliodon
日用油箱 daily oil tank
日有效集热时数 hours of useful collection per day
日月（之间）的 luni-solar
日晕 solar halo
日照 sunshine
日照百分率 percentage of sunshine
日照峰值 peak insolation
日照计 heliograph

日照记录仪　sunshine recorder
日照角　sun angle
日照面积　sun-illuminated area
日照区　insolation area
日照区域　sun-illuminated area
日照时间　insolation duration
日照时数　sunshine duration, hour of insolation, number of sunshine hour
日照型风力机　sunlight wind machine
日照值　daily insolation, insolation value
日转风　heliotropic wind
日最低温度　daily minimum temperature
日最高温度　daily maximum temperature
日最小　daily minimum

rong

绒面电池（无反射电池，黑电池）　textured cell (non-reflective cell, black cell)
绒面硅太阳电池　textured silicon solar cell
绒面太阳电池　textured solar cell
容（体）积密度　bulk density
容积　bulk volume
容积流量　volumetric flow rate
容积流率　volumetric flow rate
容积式斯特林吸热器　volumetric Stirling receiver
容积式吸热器　volumetric receiver
容积压缩系数　volumetric compressibility factor
容抗　capacitive reactance
容量系数　capacity factor
容量限度　capacity limit
容器式热水器　storage water heater
容许电压　allowable voltage
容许负荷　permissible (allowable) load
容许压力　allowable stress

容易　facilitate
溶剂　fluxing agent, solvent
溶解　liquefaction
溶解成分　dissolved constituent
溶解固形物　dissolved solid (matter)
溶解矿物质　dissolved mineral
溶解气体　dissolved gas
溶解曲线　melting curve
溶解物　dissolved matter
溶解箱　dissolving tank
溶解性固体　dissolved solid
溶解性有机碳　dissolved organic carbon
溶解氧　dissolved oxygen
溶解氧分析仪　dissolved oxygen analyzer
溶解有机碳　dissolved organic carbon
溶解质　dissolved solids
溶解装置　dissolver
溶气浮选　dissolved floatation
溶液黏度　solution viscosity
溶液培养学　hydroponics
溶液栽培法　hydroponics method
溶质　solute
熔池　molten bath
熔点　fusion/melting point
熔断　fused disconnect
熔断器　fuse
熔化　fusion
熔化层　melting zone
熔化带　melting zone
熔化炉　melting furnace
熔化潜热　melting latent heat
熔化热　heat of fusion
熔化温度　melting temperature
熔解炉　melting furnace
熔解热　heat of liquefaction, heat of solution
熔炉　furnace, forge

熔凝石英　fused quartz
熔热处理炉　heating treatment funace
熔融　fusion
熔融硅　molten silicon
熔融曲线　melting curve
熔融石英　fused quartz
熔融碳酸盐燃料电池　molten carbonate fuel cell
熔融温度　melting temperature
熔融盐　molten salt
熔融盐吸热器　molten salt receiver
熔体结构　structure of melt
熔盐　molten salt
熔盐储槽　hot salt storage tank
熔盐储罐　hot salt storage tank
熔盐—传热流体换热器　salt-heat transfer fluid exchanger
熔盐换热器　salt-heat transfer exchanger
熔盐接收器　molten salt receiver
熔盐容量　molten salt capacity
熔盐蓄热箱　molten salt heat storage tank
熔盐转换堆　molten salt convertor reactor
熔窑　melting furnace
熔渣　slag
融雪摄影　snow-melt photography
冗余　redundancy
柔性齿轮　flexible gear
柔性钢丝绳　flexible wire rope
柔性滚动轴承　flexible rolling bearing
柔性梁　flexbeam
柔性转子　flexible rotor
蠕动泵　hose pump

ru

入海河口　ocean outfall
入孔门　access door
入口阀　inlet valve
入口管　inlet pipe
入口空气　inlet air
入口压头　inlet head
入射波　incident wave
入射光　incident light
入射光强度　intensity of incident light, incident intensity
入射角　angle of incidence, incidence angle
入射角修正系数　incidence angle modifier
入射强度　incident intensity
入射太阳辐射　incoming solar radiation
入射太阳能　incident solar energy
入射阳光　incident sunlight
入射余角　grazing angle

ruan

软百叶帘集热器　venetian blind collector
软风　slight air
软辐射　soft radiation
软管　flexible conduit, hose, flexible pipe
软管泵　hose pump
软管夹　hose clamp
软管支架　hose clamp
软化　softening
软化点　softening point
软化剂　emollient
软化器　demineralizer
软件逻辑　software logic
软件平台　software platform
软模式　soft mode
软木（材）　softwood
软木残渣　cork residue
软木废料　cork waste
软能源　soft energy source

软启动电路　soft-start circuit
软启动器　soft starter

rui

瑞利分布　Rayleigh distribution
瑞利散射　Rayleigh scaffering

run

润滑　lubricate
润滑泵　lubric ant pump
润滑剂　antifriction
润滑系统　lubrication system
润滑油　lubricant, grease

ruo

弱潮河口　microtidal estuary
弱电控制　weak current control
弱电网　weak grid
弱对流　slope convection
弱气驱　weak gas drive

sa

萨伏纽斯风车　Savonius windmill
萨伏纽斯式　Savonius
萨伏纽斯转子　Savonius rotor

sai

塞流反应器　plug flow reactor
塞流式沼气池　plug flow digester
塞流厌氧发酵罐　plug flow anaerobic fermenter

san

三层玻璃　triple glass
三代发电装置　third generation device
三蝶烯基　triptyl
三分子层太阳能转换器　three-layer solar energy converter
三级风　gentle breeze
三级风口　tertiary tuyere
三级环流　tertiary circulation
三角浪　windlop
三角形机翼　delta wing
三角形联结　delta connection
三角形启动器　star-delta starter
三角洲沉积物　river delta sediment
三脚架结构　tripod piled structure
三结复合太阳电池　triple junction compound solar cell
三结合沼气池　three-in-one digester
三结太阳电池　triple junction solar cell
三库系统　three-pool system
三联产　trigeneration
三氯氟代甲烷(太阳热间接转换为电能的一种工质)　trichloromonofluoromethane
三通阀　triple valve
三维边界层效应　three-dimensional boundary layer effect
三维地震探测　3D seismic survey
三维矩形栅格　rectangular three-dimensional grid
三维失速效应　three dimension stalling effect
三相并网　three phase grid system
三相电机　three phase machine
三相电流　three phase current
三相负载　three phase load
三相感应电动机　three phase induction motor
三相功率转换　three phase power conversion
三相换向电动机　three phase commutator

machine
三相交流电 three phase AC power
三相交流系统 three phase AC system
三相绕组 three phase winding
三相凸极同步电机 three phase synchronous machine
三相隐极同步电机 three phase non-salient synchronous machine
三氧化钼 molybdenum trioxde
三叶风力发电机 three-blade wind generator
三叶风力涡轮机 three-blade wind turbine
三叶风轮 three-blade wind wheel
三叶逆风涡轮机 three-blade upwind turbine
三叶片风车 three-blade windmill
三元催化剂系统 three-way catalyst system
三圆计算作图法 three circles calculative design method
散布图 scatter diagram
散辐射 scattered radiation
散裂 spalling
散热剂 heat releasing fluid
散热片 radiantting fin, radiantting rib, radiantting gill
散热片式采暖系统 radiant heating system
散热片凸缘 radiating flange
散热片效率 fin efficiency
散热器 heat sink, exchanger, radiator
散热损失 external heat loss, heat leakage
散射 diffusion, scattering
散射（日射）辐照度 scattering solar irradiance, diffuse solar irradiance

散射波 scattered wave
散射参数 scattering parameter
散射场 scattered field
散射电子 scattered electron
散射定律 scattering law
散射阈 scattering threshold
散射反应 scattering reaction
散射幅度 scattering amplitude
散射辐射 scattered radiation, diffuse insolation, diffuse radiation
散射辐照度 diffuse irradiance
散射干涉 scattering interference
散射光 scattered light
散射函数 scattering function
散射机理 scattering mechanism
散射角 scattering angle
散射截面 scattering cross section, scattering cross-section
散射介质 emitting medium, scattering medium
散射力 scattering force
散射粒子 scattered particle, scattering particle
散射媒质 diffusing media
散射面积 scattering area
散射面积系数 scattering area coefficient
散射能力 diffusing power
散射能量 diffuse energy
散射碰撞 scattering collision
散射日射 scattering solar radiation
散射日总量 daily total scattered radiation
散射矢量 scattering vector
散射束 scattered beam
散射衰减系数 scattering attenuation coefficient
散射损失 scattering loss

散射体 scatterer
散射条纹 scattered pattern
散射图 scatter diagram
散射吸收系数 scattering absorption coefficient
散射系数 scattering coefficient
散射线 scattered radiation, scattered ray
散射效应 scattering effect
散射因子 scattering factor
散射源 diffuse source
散射振幅 scattered amplitude
散射—直射比 diffuse to beam radiation ratio
散射状态 scattering state
散失热 abstracted heat
散失热量 rejected heat
散体积 loose volume
散纤维 bulk fiber
散装材料 bulk material
散装废料 bulk waste
散装垃圾 bulk waste

sao

扫掠面积 area swept
扫描电流 sweep current
扫描电压 scanning voltage
扫描电子显微镜 scanning electron microscopy

se

色彩 color
色球爆发 chromosphere eruption
色球层 chromosphere
色散关系 dispersion relation
色散系数 dispersion coefficient

色温 color temperature

sen

森林残留物 forest residue
森林废弃物 forest arising/waste
森林和人造林木材 forest and plantation wood
森林垃圾 forestry waste
森林木片 forest chip
森林内残留物 in-forest residue
森林培育学 silviculture
森林燃料 forest fuel
森林认可认证计划 programme for the endorsement of forest certification
森林生产率 forest productivity
森林生物量 forest biomass
森林资源 forest resource
森斯托太阳能热水器 Sunstor solar water heater
森特拉普平板型集热器 Sunstrap flat plate collector

sha

杀菌剂 biocide
沙克 pg (picograms, 10-12 grams)
沙砾 grit
沙漏原理（集热器跟踪太阳的一种原理） principle of hourglass
沙丘 sandbank, coastal sand dune
沙滩 sandbank, coastal sand dune
沙洲 sandbank, coastal sand dune
刹车盘 brake disc
刹车油 brake fluid
砂泵 sand pumping
砂布 emery cloth
砂袋式塑料太阳能蒸馏器 plastic solar

still with sandbag
砂轮　emery wheel, grinding wheel
砂泥岩　siltstone

shai

筛板　sieve plate
筛分机　screen
筛上物颗粒　over size particle

shan

山顶　crest
山风　mountain wind
山谷风　mountain valley breeze
删去　eliminate
栅电池　grid cell
栅格单元　grid cell
栅极接线　grid connection
栅式太阳电池　grating solar cell
栅—阴极电压　cathode-grid voltage
闪变　flicker
闪变阶跃系数　flicker step factor
闪电流　lightning current
闪电密度　lightning flash density
闪发式蒸发器　flash evaporator
闪光玻璃　actinic glass
闪光灯　flashlight
闪光灯电池　flashlight battery
闪光焊　flash welding
闪急蒸发器　flash evaporator
闪路电压　shorting voltage
闪络　flashover
闪熔镀层　flash plate
闪岩　amphibolite
闪蒸　flash
闪蒸/两相联合循环　flash/binary combined cycle
闪蒸地热发电系统　geothermal power system using steam flashed from hot brine
闪蒸电厂　flash power plant
闪蒸发　flash evaporation
闪蒸工发器　flash evaporator
闪蒸罐　flash tank
闪蒸锅炉　flash boiler
闪蒸器　flash vessel, flash evaporator
闪蒸设施　flash facility
闪蒸式电厂　flash steam power plant
闪蒸系统　flash steam system
闪蒸蒸发器　flash evaporator
闪蒸蒸汽发电站　flash steam power plant
扇形地背斜轴　salient
扇形片　fanning strip
扇形尾（一种靠风力本身来操纵的自动调向机构）　fantail

shang

商品性太阳能建筑　commercial solar building
商业性潮汐电站　commercial tidal power plant
熵产生　entropy production
上层土　topsoil
上冲断层　overthrust
上风面（的）　windward
上风式风轮　upwind rotor
上风向　upwind, upwind direction
上风向布置　upwind configuration
上风向风力机　upwind wind turbine
上风向式风电机组　upwind wind turbine generator system
上光　glaze

上光材料　glazing material
上流区　upflow zone
上流式厌氧污泥床　upflow anaerobic sludge blanket
上坡风　slope up wind
上汽包　steam drum
上升风　rising gust
上升流　rising flow
上升流带　upwelling zone
上升暖气流　thermal
上升气流　updraft, rising air, anaflow
上网电价　feed in tariff
上涌破浪　surging breaker
上釉　glaze
上釉集热器　glazed collector

shao

烧变岩　metamorphic rock
烧结法　sintering process
烧结温度　sintering temperature
烧尽　incineration
烧失量　loss on ignition
烧蚀　burn through
烧嘴砖　burner/quarl block
(桨叶)梢速　tip speed
少数载流子　minority charge carrier, minority carrier
少子寿命　minority carrier lifetime

she

蛇管　hose
舍弃式刀头　disposable tool holder bit
设备故障　equipment failure
设备故障信息　equipment failure information
设备利用率　plant availability
设备连接　device connection
设定点　set point
设定压力　setting pressure
设定值　set point valve
设计参数　design parameter
设计低位热值　low heat value for design
设计点　design point
设计点功率　design point power
设计风速　design wind speed
设计风载　design wind load
设计工况　design situation
设计供热负荷　design heating load
设计基准风暴　design basis storm
设计极限　design limit
设计叶尖速比　design tip speed ratio
设计气象条件　meteorology condition for design
设计任务书　design assignment
设计容量　design capacity
设计温差　design temperature difference
设计性能　specified performance
设计压力　design pressure
设计要素　design consideration
设计折衷　design tradeoff
设施规划　facility design
设施设计　facility design
设置控制　setup control
社会环境　social environment
射流　jet stream, jet
射流式涡流发生器　air-jet vortex generator
射汽泵　steam jet pump
射束能量　beam energy
射水法　water jetting
射线发射　gamma ray emission
涉禽　wading bird

涉水鸟　wading bird
摄取能量　capture energy
摄取能量系统　capture energy system
摄氏度　degree Celsius

shen

伸缩　flex
伸缩管　extension tube
伸缩接头　expansion joint
砷化镓半导体　gallium arsenide semi-conductor
砷化镓电池　gallium arsenide cell
砷化镓太阳能电池　gallium arsenide solar cell
深凹球面反射镜　deep spherical mirror
深部地幔　deep mantle
深部地幔热源　heat source in deep mantle
深层地热能　deep geothermal energy
深层电法勘探　deep electrical exploration
深成岩体　pluton
深度沸点　boiling point for depth
深度计　depth gauge
深度控制仪　depth control equipment
深度温度检测　deep temperature monitoring
深海　abyssal sea
深海沉积　abyssal deposit
深海沉积物　abyssal sediment
深海的　pelagic
深海换能器　sea depth transducer
深海锚定浮标　deep ocean moored buoy
深海水应用　deep ocean water application
深海同位素海流分析仪　deep-water isotopic current analyzer
深井回灌式水源热泵系统　ground water heat pumps
深能级　deep level
深水　deep water
深水波　deep water wave
深水港　deep water port
深水排放系统　deep water discharge system
深水盐度差能转换装置　deep water salinity-gradient energy converter
深水钻井　deep water drilling
深钻井　deep drilling
渗透流　osmotic flow
渗漏　oozing
渗滤层　filter layer
渗滤空气　infiltration air
渗滤液处理　leachate disposal
渗滤液收集管　leachate collection pipe
渗水　water oozing
渗水岩　permeable rock
渗碳处理　carburizing
渗碳体　cementite
渗透（现象）　seepage
渗透泵　osmotic pump
渗透电流　penetration current
渗透断层　permeable fault
渗透联系　permeable linking
渗透膜　permeable membrane
渗透能力　ability to penetrate
渗透能量转换　osmotic energy conversion
渗透热管　osmotic heat pipe
渗透势　osmotic potential
渗透系数　hydraulic conductivity
渗透性　permeability
渗透性增强　enhancement of permeability
渗透压　osmotic pressure
渗透压差　osmotic pressure difference
渗透压力差　osmotic pressure differential

渗透压能　osmotic pressure energy
渗透压式盐浓度差能发电　osmotic pressure salinity gradient power conversion
渗透压梯度　osmotic pressure gradient
渗透原理　osmosis principle
渗透作用　osmosis
渗析膜　dialyzer
渗析器　dialyzer
渗盐　salt penetration

sheng

升柴油当量（能量含量为36.1MJ/L）　liters of diesel equivalent
升级冶金级硅　upgraded metallurgical silicon
升降机井　elevator shaft, elevator well
升降型　lift-type
升举式机械　lift-type machine
升力　lift force
升力系数　lift coefficient
升力型风机　lift-type machine
升力型叶轮　lift-type rotor
升流管　riser pipe
升流式厌氧污泥床反应器　up-flow anaerobic sludge blanket reactor
升起　hoist
升汽油当量（能量含量为33.5MJ/L）　liters of gasoline equivalent
升温　heating up
升温时间　heating up period, heating up time
升温速率　heating rate
升压泵　booster pump
升压变压器　step-up transformer
升压器　step-up transformer
升压去磁　boost buck

升压压风机　booster compressor
升致阻力　drag from lift
升阻比　lift-drag ratio
升阻型　lift-drag type
生（物）化（学）作用　biochemical action
生产　fabrication
生产策略　production strategy
生产层　producing zone
生产费用　operating cost
生产规模深孔钻井　production scale deep drilling
生产井　field well, producing well
生产力指数下降　productivity index decline
生产税抵免　production tax credit
生产套管　production casing
生产套管坍塌　collapse of production casing
生产性风洞　production type wind tunnel
生产压差　production pressure difference
生产用蒸汽　process steam
生产者责任延伸　extended producer responsibility
生潮力　tide-generating force
生成阶段　generation phase
生成率　generation rate
生成压力　build-up pressure
生化的　biochemical
生化反应池　biochemical reaction basin
生化反应池曝气器　aerator for biochemical reaction basin
生化废料　organic waste
生化过程，生物过程　biological process
生化能量循环　biochemical energy cycle
生化燃料电池　biochemical fuel cell
生化氧化　biochemical oxidation, bioche-

mical process
生化指示剂　biochemical indicator
生化转换技术　biochemical conversion technology，biochemistry conversion technology
生命周期　life cycle
生命周期成本　life cycle cost
生命周期能耗　life cycle energy
生命周期评估　life cycle assessment
生命周期清单　life cycle inventory
生命周期影响评价　life cycle impact assessment
生石灰　quicklime
生态保护　ecosystem conservation
生态过程　ecological process
生态评估　ecological assessment
生态评价　ecological assessment
生态系统　ecosystem
生态系统破坏　ecosystem damage
生态系统研究　ecosystem research
生态系统中的能量　energy in ecosystem
生态相互作用　ecological interaction
生态效率　ecological efficiency
生物法废物处理　biological waste treatment
生物去污　biological depollution
生物半衰期　biological half-life
生物材料　biomaterial
生物层　biosphere
生物柴油　biodiesel
生物柴油厂　biodiesel plant
生物柴油处理设施　biodiesel processing facility
生物柴油燃料　biodiesel fuel
生物沉积物　biological deposit
生物传感器　biosensor
生物催化剂　biocatalyst

生物催化作用　biocatalysis
生物带（包括陆地、海洋和淡水域）　biocycle
生物的　biotic
生物电　bioelectricity
生物电池　biobattery
生物电流　bioelectricity
生物电性　bioelectricity
生物电源　biopower
生物丁醇　bio-butanol
生物动力学　biokinetics
生物多样性　biological diversity
生物发酵　biochemical fermentation
生物反应器　bioreactor
生物防护　biological protection
生物废料　bio-waste
生物废弃物　biowaste
生物分解　biological breakdown, biological decomposition
生物风化　biological weathering
生物腐蚀　biological corrosion
生物工程　bioengineering
生物过滤　biological filtration
生物过滤膜　biological filtering membrane
生物航油　bio-jet fuel
生物化学沉积　biochemical deposit
生物化学甲烷势　biochemical methane potential
生物活性　biological activity
生物活性絮体　biologically active floc
生物积累因子　biological accumulation factor
生物基化学品　biobased bulk chemical
生物剂量　biological dose
生物剂量计　biological dosimeter
生物甲烷　biomethane

生物降解　biodegradation
生物接触氧化　biological contact oxidation
生物净化　biological purification
生物静力学　biostatics
生物控制　biocontrol
生物力学　biomechanics
生物量传感器　biological sensor, biological transducer
生物量全利用　biomass total utilization
生物流化床　biological fluidized bed
生物滤池　bacteria filter, bio-filter, biological filter
生物膜反应器　biofilm reactor
生物膜流化床反应器　biofilm fluidized bed reactor
生物能源　biological energy
生物能源生产　bioenergy production
生物能源与碳捕获与储存　bio-energy with carbon capture and storage
生物能源作物　bioenergy crop
生物能转换技术　biomass conversion technology
生物黏膜　biological slime
生物浓集因子　biological concentration factor
生物屏蔽　biological shield
生物曝气　bio-aeration
生物气象学　biometeorology
生物亲和性　biological affinity
生物取样　biological sampling
生物群　biota
生物燃料　biofuel
生物燃料电池　biological fuel cell
生物燃料电池电厂　biomass fuel cell power plant
生物燃料中间体　biofuel intermediate
生物染色剂　biological stain
生物实验孔道（反应堆内）　biological tunnel
生物衰变　biological decay
生物衰变常数　biological decay constant
生物损伤　biological damage
生物探头　biological probe
生物碳　biocarbon, biochar
生物体　organism
生物通道　biological hole
生物统计学　biostatistics
生物污染　biological fouling
生物系统　biological system
生物效应　biological effectiveness
生物芯片　biochip
生物需氧量　biological oxygen demand
生物絮凝　bioflocculation
生物学　biology
生物学的　biological
生物氧化　biological oxidation
生物液体燃料　biology liquid fuel
生物医学辐射计数系统　biomedical radiation-counting system
生物医学剂量学　biomedical dosimetry
生物医学射线照相　biomedical radiography
生物乙醇　bioethanol
生物乙醇的经济性　bioethanol economy
生物乙醇燃料　bioethanol fuel
生物淤积（指液体流经的管道等物体的表面上黏附细菌或水下船底贝属动物的沉积或黏附）　biofouling
生物预处理　biological pretreatment
生物质　biomass material
生物质残留物　biomass residue
生物质柴油　biomass-to-liquids diesel
生物质成本　biomass cost
生物质成型燃料　biomass moulding

fuel
生物质储备　biomass reserve
生物质的燃烧特性　combustion characteristics of biomass
生物质电厂　biomass-based power plant
生物质电厂烟气脱硝技术装备　denitration equipment of biomass-fired power plant
生物质发电产业　biomass power generation industry
生物质发电机　biomass generator
生物质发电技术　biomass power generation technology
生物质发电量　biomass generating capacity
生物质发电容量　biomass generating capacity
生物质发电系统　electricity generating system of biomass
生物质废弃物　biomass waste
生物质共燃　biomass co-firing, co-firing of biomass
生物质共燃技术　biomass co-firing technology
生物质固定床气化炉　fixed bed biomass gasifier
生物质合成天然气　bio-synthetic natural gas
生物质灰　biomass ash
生物质混燃　biomass co-firing
生物质火力发电厂　biomass-fired power station
生物质火力发电站　biomass-fired power station
生物质浆　biosludge
生物质颗粒燃料　biomass moulding fuel
生物质利用　biomass utilization
生物质内燃涡轮机　biomass internal combustion gas turbine
生物质能　biomass energy, bioenergy, energy from biomass
生物质能发电　bio-energy electricity, biomass electric power generation
生物质能发电厂　biomass power plant
生物质能燃料　biomass energy fuel
生物质能源产业　biomass energy industry
生物质能源技术　biomass energy technology
生物质能源作物　biomass energy crop
生物质能转化　biomass conversion
生物质能转换　biomass energy conversion
生物质能转换系统　biomass energy conversion system
生物质能资源　biomass energy resource
生物质气化发电　biomass gasification power generation
生物质气化发电厂　biomass gasification power plant
生物质气化技术　biomass gasification technology
生物质气化炉　biomass gasification set, biomass gasifier
生物质气化装置　biomass gasification plant
生物质气体　biogas
生物质气化　biomass gasification
生物质气化的热化学特性　thermalchemical property of biogasification
生物质燃料　biomass fuel
生物质燃料掺合物　biofuel blend
生物质燃料成本　biomass fuel cost

生物质燃料处理系统　biomass fuel handling system
生物质燃料混合物　biofuel mixture
生物质燃料块　biofuel briquette
生物质燃料沫　fuel powder, fuel flour
生物质燃料燃烧　combustion of biomass fuel
生物质燃料丸　biofuel pellet
生物质燃料消耗　biomass fuel consumption
生物质燃料屑　biomass fuel dust
生物质燃料预处理　biofuel preprocessing
生物质燃料资源　biomass fuel resource
生物质燃烧　biomass combustion
生物质燃烧技术　biomass combustion technology
生物质燃烧炉　biomass furnace, biomass burner
生物质燃烧器　biomass burner
生物质热分解　biomass pyrolysis
生物质热干化　thermal drying of biomass
生物质热转换过程　biomass therm-oconversion
生物质设备　biomass facility
生物质生产　biomass production
生物质提炼　biorefinery
生物质压块　biomass briquette
生物质衍生燃料　biomass-derived fuel
生物质液化　biomass liquefaction
生物质与煤共燃　co-firing of biomass with coal
生物质原料　biomass feedstock
生物质整体气化联合循环发电　biomass internal gasification combined cycle
生物质直燃　direct fired biomass
生物质直燃锅炉　direct fired biomass boiler
生物质中间体　biomass intermediate
生物质种植园　biomass plantation
生物质资源　biomass resource
生物质资源利用　biomass energy utilization
生物质作物　biomass crop
生物转化　bioconversion
生物转化之物理程序　physical process in bio-conversion technology
生物资源　biological resource
生物组织　tissue
生长激素　growth hormone
生长结光电池　grown junction photocell
生质油　bio-oil
生质原油　bio-crude
生质转化技术　conversion technology for advanced biofuels
声波吹灰器　sonic soot-blower
声波反向散射　acoustic backscattering
声波风速计　sonic anemometer
声波探测和测距　sonic detection and ranging
声的基准风速　acoustic reference wind speed
声级　weighted sound pressure level, sound level
声脉冲　acoustic pulse
声频发射　acoustic emission
声学多普勒感应系统　acoustic Doppler sensor system
声学多普勒海流剖面仪　acoustic Doppler current profiler
声压级　sound pressure level
牲畜残留物　livestock residue
省柴灶　fuel saving stove
盛行风　prevailing wind
盛行流　prevailing current
剩余储藏量　remaining reserve

剩余（活性）污泥 waste activated sludge, surplus activated sludge, excess sludge
剩余溶液 residual liquid
剩余油含量 residual oil content

shi

失速 stalling
失速点 stall point
失速后翼型 post-stall airfoil
失速控制 stall control, stall regulation
失速控制风轮 stall control rotor
失速气流 stalled flow
失速区 stall region, stalled area
失速式限速风轮 stall regulated rotor
失速调节 stall regulated
失速叶片 stalled blade
失速转矩 stall torque
失速状态 stalled condition
失调电流 offset current
失调电压 offset voltage
失相 loss of phase
失效—安全 fail-safe
失效时间 dead time
失载负荷 run out load
施工安全管理 construction safety management
施工工程 construction work
施工机械管理 construction machine management
施工控制性进度 construction critical path schedule
施工力能供应 energy and gases supply for construction
施工临时设施 temporary facility for construction
施工物资供应计划 material supply plan for construction
施工专业化 construction specialization
施工总布置 construction general layout
施工总进度 construction general schedule
施工组织设计 construction organization design
施釉保温 glazing insulation
施釉绝缘 glazing insulation
施照度 illuminance
施主原子 donor atom
湿沉降 wet deposition
湿处理废物 wet process waste
湿垫蒸发冷却器 wet-pad evaporative cooler
湿度 humidity
湿度表（计） moisture meter
湿度指示器 moisture indicator
湿吨 moisture ton
湿发酵 wet fermentation
湿法分解 wet digestion
湿法腐蚀 wet etching
湿法刻蚀 wet etching
湿负荷 humidity load
湿基 wet basis
湿绝热过程 saturation adiabatic process
湿绝热曲线 saturation adiabatics
湿绝热线 saturation adiabat
湿绝热直减率 saturation adiabatic lapse rate
湿空气 damp air
湿量基准 wet basis
湿气 humidity
湿球温度 wet bulb temperature
湿球温降 wet bulb depression
湿润因子 moisture factor
湿润指数 moisture index
湿式除尘器 wet scrubber
湿式机械预处理 wet mechanical pretr-

eatment
湿式烟气处理　wet flue gas treatment
湿树皮　wet bark
湿蒸气储藏　wet steam reservoir
湿蒸汽　wet steam, damp steam
湿蒸汽田　wet steam field
湿重　wet weight
十二分规则　Rule of Twelfths
十进的　decimal
十亿吨　gigatonne (Gt)
十亿瓦特　gigawatt (GW)
十字标记　cross mark
十字长孔　cross slotted screw
十字头　crosshead
石化厂　petroleum plant
石灰藻　calcareous algae
石蜡储热　paraffin wax heat storage
石棉　asbestos
石墨　graphite
石墨电极　graphite electrode
石脑油　naphtha
石英　quartz, quartz crystal
石英坩埚　quartz crucible
石英晶体　quartz crystal
石英晶体谐振器　quartz crystal resonator
石英溶解度　quartz solubilities
石英砂　quartz sand
石英温度计　quartz thermometer
石英岩　quartzite
石油焦　petroleum coke
石钻　masonry drill
时变系统　time varying system
时差　equation of time
时间常数　time constant
时间间隔　interval
时间平均（海流）能密度　mean current energy density of time

时间同步　time synchronization
时间响应　time response
时间域　time domain
时间域模型　time domain modelling
时角　hour angle
时平均风速　mean hourly speed
时平均接受到的太阳辐射　hourly average insolation
时圈　hour circle
时域电磁学　time-domain electromagnetics
时域分析　time domain analysis
时域与频域　time domain and frequency domain
识别标志　identification mark
实测流　observed current
实测最大风速　observed maximum wind velocity
实船实验　protype testing
实地测试（试验）　field trial
实度　solidity
实度损失　solidity loss
实际尺寸　physical dimension
实际风　actual wind
实际风能　actual wind power
实际风速频率分布　actual wind speed frequency distribution
实践资源　practical resource
实例　instance
实施潜势　implementation potential
实时　real time
实时系统　real time system
实验建模　experimental modelling
实验模型　experimental modelling
实验室样品　laboratory sample
实验条件　laboratory condition
实验用光电池　experimental photovoltaic cell

实验用光伏打电池	experimental photovoltaic cell
实用载荷	service load
食品加工业残渣	food processing industry residue
食物链能量损失	food chain energy loss
蚀变带	alteration zone
蚀变岩	altered rock
蚀刻机	etching machine
史密斯—普特南（实验风力发电）机	Smith-Putnam machine
史密斯—普特南大型风轮机	Smith-Putnam wind turbine
史密斯—普特南风力发电机	Smith-Putnam wind generator, Smith-Putnam wind turbine
矢量	vector
矢量和	vector sum
矢量控制调速系统	vector control speed system
矢量图	phasor diagram
使便利	facilitate
使潮湿，使沮丧	dampen
使磁化	magnetize
使活跃	energizing
使容易	facilitate
使用范围	effective range
使用极限状态	serviceability limit state
使用期限	service life
使用情况	service condition
使用时段	time of use
使用寿命	service life, useful life
使用温度	service temperature
示波测试	oscillographic testing
示波器	oscillograph
示波图	oscillograph trace
示差日射表	differential actinometer
示差温度表	differential thermometer
示潮器	tidal indicator
示范厂	demonstration plant
示踪气体	trace gas, tracer gas
示踪试验	tracer test
世界波	world wave
世界辐射测量基准	world radiation reference
世界时	universal time
世界主要地热带	principal geothermal zone in the world
市场预测	demand forecast
市电同价	grid parity
似稳电流	quasi stationary current
势垒	potential barrier
势垒层电容	barrier layer capacity
势垒能	barrier energy
势能	potential energy
势能储能系统	potential energy storage system
事故	hazard
事故备用泵	emergency pump
事故备用电源	emergency auxiliary power
事故工况	emergency condition
事故后运行方式	operating plan after accident
事故排气	emergency dump steam
事故消耗	accidental consumption
事故信号灯	emergency light
事故运行方式	emergency operating plan
事故状态	emergency condition
事件信息	event information
试航	sea trial
试剂	reagent

试井 well testing
试验场地 test site
试验数据 test data
试验台 observation desk, test bed
试验性地热发电站 experimental geothermal electric power plant
试验装置 test rig
试样量 test portion
试运行阶段 trial phase
试运转 green test
视电导率 apparent conductivity
视电阻 apparent resistance
视荷电率 apparent chargeability
视觉冲击 visual impact
视觉的 optical
视口门 access door
视太阳日 apparent solar day
视在功率 apparent power
视在声功率级 apparent sound power level
适温性生物 mesophilic organism
适温厌氧污泥消化 mesophilic anaerobic sludge digestion
适温厌氧消化 mesophilic anaerobic digestion
适应控制 adaptive control
适应气候 climatize
室内空气质量 indoor air quality
室内模型试验 laboratory modelling
室内气候 indoor climate
室内设计条件 indoor design condition
室内小气候 cryptoclimate
室外设计条件 outdoor design condition
室外设计温度 outdoor design temperature
室温 room temperature
室温变化 room temperature change
释放 lease

释放热量 heat release
释荷阀 unloading valve
释能反应 exergonic reaction
释能转换 exergonic conversion
释热率 heat release rate
嗜热发酵池 thermophilic digester
嗜热厌氧细菌 thermophilic anaerobic bacteria

shou

收尘电极 dust collection electrode
收到基 as received basis
收集器 accumulator
收集系统 gathering system
收集效率 collection efficiency
收气 gettering
收缩波道式 contracting wave channel
收缩水道波流式电站 tapered channel wave power station
收叶风速（风力机械在风速比灾害性风速低得多时的停机风速） furling windspeed
收叶速度（风力发电机必须关机的最低风速） furling speed
收支平衡点 break-even point
手持工具 handheld instrument
手持式风速计 handheld anemometer
手电筒 flashlight
手动控制 manual control
手工具 hand tool
手钻 gimlet
首次猜想 first guess
首次闪蒸 initial flash
寿命 life span, longevity
寿命极端风速 extreme lifetime wind speed
受电端电压 receiving end voltage
受风暴破坏的 storm-swept

受控急速蒸发	controlled flash evaporation
受控温区	controlled temperature zone
受热变形	temperature distortion
受热集热器	heated collector
受热量	received heat
受热面	heat receiving surface, heat absorbing surface
受热面面积	heating surface area
受油器	oil head
兽脂	tallow

shu

疏水泵	drain water pump
疏水箱	drain water tank
疏松沉积	unconsolidated sediment
输出电流	output current
输出电压	output voltage
输出端电流	output-terminal current
输出端损失	exit loss
输出功率	output power
输出功率预测	power output prediction
输出力矩	output torque
输出联接	output coupling
输出扭矩	output torque
输出温度	output temperature
输出轴	output shaft
输出转矩	output torque
输出阻抗	output impedance
输导通道	pathway
输电干线	power main
输电和配电系统	transmission and distribution system
输电网	transmission grid
输电网络	electrical grid
输电线	power line
输电线路	power transmission line
输煤设备	coal handling plant
输配电设备	power transmission and distribution equipment
输配子系统	distribution subsystem
输入	input
输入功率	input power
输入管	inlet pipe
输入能量峰值	energy input peak
输入扭矩	input torque
输入压力	input pressure
输入轴	input shaft
输水隧道	water tunnel
输送泵	delivery pump
输送管	transportation pipe
输送管道	transmission pipeline
输送管线上电流	line current
输送链	conveying chain
输油管道	oleo-duct
输油管线	oleo-duct
蔬菜废弃物	vegetable waste
熟料形成	clinker formation
熟石灰	hydrated lime
属性值	attribute value
蜀黍	sorghum
鼠笼感应（异步）电机	squirrel cage induction generator
鼠笼式	squirrel-cage
鼠笼式推斥电动机	squirrel cage repulsion motor
鼠笼式转子	squirrel cage rotor (cage rotor)
束能量	beam energy
树段	tree section
树胶分泌	gumming
树木生长年轮	growth ring of tree
竖井开采矿	shaft mine
竖炉	shaft furnace
竖起	erection

竖式炉　shaft furnace
竖式叶轮泵　vertical turbine pump
竖向应力　vertical stress
竖直地埋管换热器　vertical ground heat exchanger
竖轴风车　vertical-axis windmill
竖轴风机（风车）　vertical axis wind turbine
竖轴风力涡轮机　vertical-axis turbine
竖轴风轮　vertical axis totor
竖轴结构　vertical axis configuration
竖轴式风车　vertical-shaft windmill
竖轴式风机　vertical-axis machine
竖轴式风机叶轮　vertical axis wind rotor
竖轴式风机转子　vertical axis wind rotor
竖轴式风力涡轮机空气动力性能　vertical axis wind turbine aerodynamic performance
竖轴式风力涡轮机叶轮　vertical axis wind turbine rotor
竖轴式风力涡轮机转子　vertical axis wind turbine rotor
竖轴式螺旋桨型风车　vertical-shaft-propeller-type windmill
竖轴式水轮机　vertical axis turbine
竖轴式叶轮型风力涡轮机　vertical-axis rotor-type wind turbine
数据采集系统　data acquisition system
数据采集与监控　supervisory control and data acquisition
数据采集与监控系统　supervisory control and data acquisition system
数据电路　data circuit
数据记录器　data logger
数据记录系统　data recording system
数据库　data base
数据终端设备　data terminal equipment

数据组（测试功率特性）　date set (for power performance measurement)
数据组功率特性测试　data set for power performance measurement
数位微波传输设备　digital microwave radio
数学规划　mathematical programming
数值模拟　numerical simulation
数字传输　digital transmission
数字高程模式　digital elevation model
数字故障记录仪　digital fault recorder
数字划线地图　digital line graph
数字计数器　digital counter
数字建模　numerical modelling
数字控制　digital control
数字量输出端子　digital output terminal
数字量输入端子　digital input terminal
数字面板　digitizing tablet
数字—模拟转换器　digital-to-analog converter
数字输出　digital output
数字输入　digital input
数字数据记录器　digital data logger
数字位移　digital displacement
数字型　digital type
数字指示器　digital indicator
数组　array

shuai

衰减　attenuation
衰减参数　decay parameter
衰减常数　decay constant
衰减函数　decay function
衰减交流　damped alternating current
衰减量　decrement
衰减器　attenuator
甩负荷　loss of load

shuan

栓行浮标 spar buoy
栓孔锯 keyhole saw

shuang

双(多)井眼(钻井)技术 dual (multi) bore-hole technology
双壁热交换器 double wall heat exchanger
双层壁换热器 double-walled heat exchanger
双层玻璃 double glass
双层玻璃窗 double glazing, double-glazed window
双层玻璃盖 double glass cover
双层玻璃盖板 double cover glass
双层玻璃盖板集热器 double glass cover collector
双层玻璃盖板平板集热器 double glass flat plate collector, double glazed flat plate collector
双层玻璃系统 double glass system
双层地下系统 two-layered ground system
双层盖板集热器 double cover collector
双层减反射涂层 double layer anti-reflection coating
双层炉 dual-zone stove
双层上釉反射玻璃 double glazing reflecting glass
双层上釉透明玻璃 double glazing clear glass
双层上釉吸热玻璃 double glazing heat-absorbing glass
双层式线圈 double layer winding
双层塑料薄膜反射镜 double layer cover plastic film mirror
双层透明盖板集热器 double glazed cover collector
双层外壳 double casing
双层泄水闸 double sluice gate, double-layer sluice
双潮 double tide
双池式布置(潮汐发电) two-pool plan
双窗格玻璃 double-pane glass
双低潮 tidal double ebb
双电层电容器 electrical double-layer capacitor
双动力潮汐电磨机 dual-powered tidal current mill
双动力电磨机 dual-powered mill
双动片电容器 twin rotor capacitor
双对数曲线图 log-log plot
双对数图 bi-logarithmic graph
双峰分布 double humped distribution
双峰曲线 double humped curve
双高潮 tidal double flood
双工传输 duplex transmission
双罐太阳能热水系统 dual-tank solar hot water system
双回路测深法 two-loop sounding method
双回路剖面法 two-loop profiling method
双回路系统 double loop system
双击式水轮机 cross flow turbines
双级泵 double stage pump
双级式燃气轮机 two-stage gas turbine
双极结型晶体管 bipolar junction transistor
双介质理论 double medium theory
双金属的 bimetallic
双金属杆拨温度计 bimetallic-stem dial thermometer
双金属直接日射计 bimetallic phrheliometer
双进汽涡轮机 dual admission turbine

双卡头　double clamp
双可调发电装置　double-regulated generating unit
双库　two-pool
双库一单作用方案　double-basin-single effect scheme
双库方案　paired-basin scheme
双库贯连方案　double-basin transconnected scheme
双库间断　two pool intermittent
双库连续　two pool continuous
双库三向潮汐电站　double-basin three way tidal power plant
双库系统　two pool system, double-pool system
双馈　doubly-fed
双馈发电机　doubly-fed induction generator
双馈感应电机　doubly-fed induction machine
双馈感应发电机　double-fed induction generator
双馈异步发电机　double-fed asynchronous generator
双流体喷嘴　dual-fluid nozzle
双路太阳能空气加热器　two-pass solar air heater
双螺旋齿轮　herringbone gear
双螺旋曝气头　double helical aerator
双马达的　bimotored
双面(太阳能)电池板　bifacial flat panel
双面光电地板　bifacial photovoltaic flat panel
双面光伏模块　bifacial combined photovoltaic module
双面焊接　both sides welding
双面太阳能电池　bifacial solar cell
双面太阳墙　solar wall with two sides
双面吸热式平板集热器　double exposure flat plate collector
双谱模型　bispectral model
双汽循环　binary-vapor cycle
双曲率聚光集热器　double curvature concentrating collector
双曲率聚光器　double curvature concentrator
双曲率聚光装置　double curvature concentrating device
双曲率装置　double curvature device
双燃循环　dual combustion cycle
双石英变频器　digiquartz transducer
双石英压力　digiquartz pressure
双鼠笼式转子　double squirrel cage rotor
双体半潜式钢船(温差电站的一种装置)　twin semi-submersible steel vessel
双头螺栓　stud-bolt
双透镜聚光器　double lens concentrator
双透镜聚光型太阳能集热器　double lens concentrator
双凸极永磁发电机　doubly salient permanent magnet generator
双位阀　on-off valve
双吸式泵　double suction pump
双吸式的叶轮　double suction impeller
双箱储能系统　two-tank storage system
双相的　biphase
双相含水层　two-phase aquifer
双相混合物　two-phase mixture
双相冷却　two-phase cooling
双相流　two-phase flow
双相流体　two-phase fluid
双相热储存　two-phase thermal storage
双相位值　two-phase value

双相蓄热 two-phase thermal storage
双向电流 bidirectional current, duo-directional current
双向电路 bilateral circuit
双向发电（该潮汐电站在涨潮、落潮时均能发电） double-way operation, ebb and flood operation, reversible operation
双向分解 double direction decomposition
双向风向标 bivane
双向隔离阀 two-way isolating valve
双向井用涡轮机 bi-directional wells turbine
双向涡轮机 bi-directional turbine
双向泄水闸 double-way sluice
双效介质 dual purpose medium
双效冷却装置 double effect chiller
双效吸收式冷水机 double-effect absorption chiller
双效吸收式制冷机 double-effect absorption chiller
双效制冷机 double effect chiller
双旋翼 twin rotor
双旋翼系统 twin rotor system
双循环 binary cycle
双循环地热发电厂 binary cycle geothermal power plant
双循环地热发电系统 geothermal power systems using binary cycle
双循环发电厂（站） binary cycle power plant
双循环系统 binary cycle system
双压沸水堆 dual pressure boiling water reactor
双叶风力发电站 two-bladed wind station
双叶风力涡轮机 two-bladed wind turbine
双叶螺旋桨 two-bladed propeller
双叶片风车 double bladed windmill
双叶片农用风车 Fales-Stuart windmill
双叶水平风轮 two-blade horizontal rotor
双叶水平转子 two-blade horizontal rotor
双叶转子 two-blade rotor
双元示踪技术 double tracer technique
双源蒸发 two-source evaporation
双增速齿轮 double step-up gear
双真空管集热器 double vacuum tube collector
双值电容式电动机 two-value capacitor motor
双指向性管道卧式涡轮机 bi-directional ducted horizontal turbine
双重壁管 dual-wall pipe
双重曝光平板 double exposure flat-plate
双重效应 double effect
双重影响 double effect
双轴定向方阵 two-axis oriented array
双轴跟踪碟形抛物面反射器 two-axis tracking dish
双轴跟踪聚光器 concentrator with two axes
双轴跟踪太阳能聚光器 two-axis tracking solar concentrator
双轴太阳跟踪器 two-axis tracker, dual-axis suntracker
双轴太阳追踪器 dual-axis suntracker
双轴阳光传感器 two-axis sun sensor
双转子 double rotor
双转子涡轮机 dual rotor turbine
双作用潮汐电站 double-effect tidal plant
双作用一单库方案 double-effect single-basin scheme

双作用装置　double-effect unit
霜凇　rime

shui

水坝—环礁　dam-atoll
水波动力脱盐　wave-powered desalination
水槽　water tunnel, water tank
水槽温度　sink temperature
水层　water column
水产养殖　aquaculture
水车　waterwheel, hydraulic wheel
水尺　tide staff, tide pole
水冲法　water jetting
水冲击　water impinge
水处理系统　water disposal system
水锤（现象）　water hammer
水袋式集热器　water bag collector
水袋式太阳房　water bag house
水的反向流　counterflow of water
水电　hydroelectric power, hydropower
水电厂　hydroelectric power plant
水电开发　hydropower development
水电站　hydropower station
水电站总效率　overall hydropower plant efficiency
水动力特性　hydrodynamic performance
水动力系数　hydrodynamic coefficient
水动力效率　hydrodynamic efficiency
水动力性能　hydrodynamic performance
水动力学模型　hydrodynamic model
水动力学性能　hydrodynamic performance
水洞　water tunnel
水斗式水轮机　pelton turbine
水分分离器　moisture catcher
水分分析样品　moisture analysis sample
水分损失　moisture loss
水分有效性　water availability

水封式沼气池　hydraulic sealing digester
水垢　scale deposits
水垢热阻系数　scale coefficient
水管式锅炉　water tube boiler
水合熔盐　hydrous salt
水合盐　salt hydrate
水合作用　aquation
水化学　water chemistry
水化盐物储热　hydration salt heat storage
水击（现象）　water hammer
水加热系统　water-heating system
水夹套　water chamber
水解反应器　hydrolysis reactor
水解酶　hydrolytic enzyme
水解细菌　hydrolytic bacteria
水解作用　hydrolysis
水介质蓄热器　water type heat storage
水介质旋流器　hydro-cyclone
水晶基岩　crystalline bedrock
水库　reservoir
水库面积　area of reservoir
水冷　water cooling
水冷壁　water wall
水冷炉排　water cooled grate
水冷却法　water cooling
水冷式燃烧室　water cooled combustion chamber
水冷式太阳能热电联产系统　water cooled photovoltaic-thermal cogeneration system
水冷系统　water cooled system
水力　hydraulic power
水力参数　hydraulic parameter
水力传导率　hydraulic conductivity
水力电气　hydroelectricity
水力发电　hydroelectric power
水力发电机　hydroelectric generator
水力发电站　hydroelectric power station

水力发展　hydropower development
水力仿真分析　computational fluid dynamics analys
水力计算　hydraulic calculation
水力能　hydraulic energy
水力能量　hydraulic energy
水力能源　hydraulic energy
水力破碎　hydrofracture
水力渗透性　hydraulic permeability
水力损失　hydraulic losses
水力特性　hydraulic property
水力停留时间　hydraulic retention time
水力拖拽力　hydraulic drag
水力旋流器　hydrocyclone
水力学　hydraulics
水力研究站整流装置　hydraulic research station rectifier
水力压裂　hydraulic fracturing, hydraulic stimulation
水力阻力　hydraulic drag
水利发展　hydropower development
水利化浆机　hydropulper
水流　chute of water
水流参数　flow parameter
水流冲击（潮轮）上部而转动的　overshot
水流冲击（潮轮）下部而转动的　undershot
水流冲击（潮轮）中部而转动的　midshot
水流速度　water velocity
水路　waterway
水轮　waterwheel
水轮发电机　water-turbine-driven generator
水轮发电机组　water-turbine generator set, water turbine generator set
水轮机　water turbine, hydraulic turbine
水轮机泵　water turbine pump
水轮机测功机　turbine dynamometer
水轮机顶盖　turbine head cover
水轮机进水弯管　turbine inlet bend
水轮机气蚀系数　turbine cavitation factor
水轮机叶片　water turbine blade
水轮机闸门　turbine gate
水轮机轴　turbine shaft
水密度　waterdensity, water density
水面标高　surface elevation
水泥　cement
水泥材料　cementing material
水泥沉箱　concrete caisson
水泥衬里　cement lined piping
水泥护层　cement sheath
水泥环　cement sheath
水泥浆　cement grout
水泥浆液　cement grout
水泥壳　cement sheath
水泥塞　cement plug
水泥窑　cement kiln
水泥渣　cement slurry
水平闭合回路　horizontal closed loop
水平变位　horizontal displacement
水平大地耦合系统　horizontal earth coupling system
水平的　horizontal
水平地埋管热换器　horizontal ground heat exchanger
水平多结太阳电池　horizontal multijunctions solar cell
水平放水渠　horizontal tailrace
水平放泄阀　horizontal discharge valve
水平风场　horizontal wind field
水平风切　horizontal wind shear
水平风矢量　horizontal wind vector
水平辐射日总量　daily total horizontal radiation
水平管　horizontal pipe

水平横楣梁 lintel beam
水平衡 water balancing
水平轮 horizontal wheel
水平埋管地源热泵系统 horizontal ground-coupled heat pump
水平面 horizontal area
水平面积 horizontal area
水平清晰度 horizontal resolution
水平燃料床 horizontal fuel bed
水平燃烧室 horizontal combustion chamber
水平渗透系数 horizontal hydraulic conductivity
水平尾水渠 horizontal tailrace
水平位移 horizontal displacement
水平线 horizontal line
水平压力梯度 horizontal pressure gradient
水平仪 level instrument
水平移位 horizontal displacement
水平应力 horizontal stress
水平运动 horizontal motion
水平震动 horizontal motion
水平支架定日镜 horizontal yoke helicostat
水平轴潮汐涡轮机 horizontal axis tidal turbine
水平轴灯泡型装置 horizontal shaft bulb unit
水平轴风车 horizontal axis wind turbine
水平轴风力发电机 horizontal axis aerogenerator, horizontal axis wind turbine
水平轴流式灯泡型水轮机 horizontal axis bulb turbine
水平轴涡轮机 horizontal axis turbine
水气并动波浪发电装置 hydropneumatic wave power device
水气并动波能装置 hydropneumatic wave power device
水气式波能装置 hydropneumatic wave power device
水汽含量 water vapour content
水汽密度 water vapour density
水汽吸收 water vapour absorption
水汽压 water vapour pressure
水墙 waterwall
水渠 penstock
水热（作用）的 hydrothermal
水热爆炸 hydrothermal explosion
水热处理 hydrothermal processing
水热发电 hydrothermal electricity production
水热喷发 hydrothermal eruption
水热蚀变 hydrothermal alteration
水热系统 hydrothermal system
水溶性焊剂 water flux
水乳胶 water emulsion
水深测量 bathymetry
水生的 aquatic
水生生态系统 aquatic ecosystem
水生食物链 aquatic food chain
水蚀 water erosion
水势 water potential
水室 water chamber
水—水热泵 water-to-water heat pumps
水塔 water tower
水塔储能 water tower storage
水体结构 water column structure
水通量 water flux
水筒 water tunnel
水头 head
水头损失 head loss
水位 water potential, river stage
水位线 water line
水温自动调节仪 aquastat
水文地球化学 hydrogeochemistry
水文地质流 hydrogeological flow

水文地质调查 hydrogeological investigation
水文地质学 hydrogeology
水文气象学 hydrometeorology
水文气象资料 hydrometeorology information
水文效应 hydrologic effect
水文学家 hydrologist
水文张力 hydrologic tension
水文状况 water regime
水雾化系统 water misting system
水下操作器 underwater manipulator
水下风车 underwater turbine
水下海流发电装置 underwater ocean current plant
水下机械手 underwater manipulator
水下建筑 underwater construction
水下冷水泉 submarine fresh water spring
水下热电站 underwater thermal power-plant
水下施工 underwater construction
水险 marine insurance
水线 water line
水线（平）面 waterplane area
水线面力矩 waterplane moment
水锈 incrustation
水旋转接头 water swivel
水循环 water cycle
水循环系统 water circulating system
水压 hydraulic pressure, water pressure
水压参数 hydraulic parameter
水压的 hydraulic
水压试验 hydrostatic test
水压致裂 hydrofracture
水一岩石储能仓 water-rock storage bin
水岩相互作用 rock-water interaction

水养动力驳船 aqua power barge
水翼船 hovercraft
水印横线 waterline
水源 water supply, headwater
水源井 source well
水源热泵 water-source heat pump, water source heat pump
水栽法 hydroponics
水闸 sluice, dam, water gate, barrage
水蒸发器 water evaporator
水蒸气 water vapour
水蒸汽 steam vapor
水蒸汽表 steam table
水质稠密度 waterdensity, water density
水质污染 aquatic pollution
水中生物群 aquatic biota
水中水车 underwater mill
水中野生动物 aquatic wildlife
水柱 water column
水柱位能 potential energy of column of water
水准仪 gradienter

shun

顺风 downstream wind (downwind)
顺风的 downwind
顺风风力机 downwind turbine
顺风间距 downwind spacing
顺风螺旋桨 downwind propeller
顺风面 downwind side
顺桨 feathering
顺桨螺桨 feathering airscrew
顺桨位置 feather position
顺流 downstream, parallel flow
顺流燃烧 co-current combustion
顺流热交换器 parallel flow heat exchanger
顺流位置 feathered position

顺水　downstream
顺向低泵工作　consequent low pumping effort
顺压差　favorable pressure difference
顺压梯度　favorable pressure gradient
顺转风　veering wind
瞬变电流　transient, transient current
瞬变现象　transient
瞬间的　instantaneous
瞬间热解　flash pyrolysis
瞬间蒸发技术　flash evaporation technique
瞬时　instantaneous
瞬时变化　instantaneous variation
瞬时测量　instantaneous measurement
瞬时电功率　instantaneous electric power
瞬时电压　instaneous voltage, transient voltage
瞬时风　gust, gust wind
瞬时风速　instaneous wind speed, instantaneous wind speed
瞬时负荷　momentary load, momentary duty
瞬时功率　momentary power, instantaneous power
瞬时功率输出　instantaneous power output
瞬时荷载　transient load
瞬时机械功率　instantaneous mechanical power
瞬时力　transient force
瞬时效率　instantaneous efficiency
瞬时效率特性　characteristic of instantaneous efficiency
瞬时值　instantaneous value
瞬时转矩　transient torque
瞬时最大风速　momentary maximum wind velocity

瞬态电抗　transient reactance
瞬态机电扭矩　transient electromechanical torque
瞬态稳定边界　transient stability limit
瞬态性能　transient performance
瞬态性能测试　transient electrical performance test

su

朔望潮　syzygy tide
朔望月　synodic month

si

丝网分离器　wire mesh separator
丝网印刷工艺　screen printing process
丝网印刷术　screen printing
斯蒂芬—波尔兹曼常数　Stefan-Boltzmann constant
斯蒂芬—波尔兹曼定律　Stefan-Boltzmann's law
斯涅尔定律　Snell's law
斯特兰福特湾海洋发电站　Strangford Lough SeaGen
斯特林电动机　Stirling thermal motor
斯特林发动机　Stirling engine
斯特林机回热器　Stirling regenerator
斯特林机热头　Stirling heater head
斯特林太阳能热空气发动机　Stirling solar hot air engine
斯特林吸热器　Stirling receiver
斯特林循环　Stirling cycle
斯特林循环引擎　Stirling cycle engine
斯托克斯波　Stokes wave
斯托克斯漂移　Stokes drift
死谱带　deadband
死水区　dead area
四冲程　four-stroke

四冲程发动机 four-stroke engine
四冲程循环发动机 Otto cycle engine
四级风 moderate breeze
四探针法 four probe method
四象限变频器 four-quadrant frequency converter
四象限运行 four-quadrant operation
四溴化钼 molybdenum tetrabromide
四爪卡盘 four-jaw chuck
伺服机构 servo actuator
饲料切碎机 forage chopper
饲料玉米 field corn

song

松姆沃克太阳房 Zomeworks solar house
松散沉积物 unconsolidated deposit
松散含水层材料 unconsolidated aquifer material
送风口 air outlet
送风量 air output
送风调节器 air governor, air register, air conditioner
送料阀 feed valve
送射能量 irradiate energy

sou

搜索算法 search algorithm

su

速度 velocity, speed
速度比例 velocity scale
速度不足 velocity deficit
速度传感器 speed sensor
速度幅值 velocity amplitude
速度控制 speed control
速度落差 velocity head

速度模式 speed mode
速度压力 velocity pressure
速率 velocity
速率控制 rate control
速率信号 velocity signal
速燃燃料 fast-burning fuel
速位差 velocity head
塑晶 plastic crystal
塑料薄膜 plastic film
塑料薄膜顶盖 plastic film cover
塑料薄膜吸收体 plastic film absorber
塑料一玻璃叠片 plastic-glass laminate
塑料袋式热水器 plastic bag solar water heater
塑料菲涅尔透镜 plastic Fresnel lens
塑料换热器 plastic heat exchanger
塑料管路 plastic piping
塑料集热器 plastic collector
塑料晶体 plastic crystal
塑料抛物面反射镜 plastic parabolic reflector
塑料太阳能集热器 plastic solar energy heat collector
塑料太阳能蒸馏器 plastic solar still
塑料太阳能热水器 plastic solar water heater
塑料太阳灶 plastic solar cooker
塑料沼气池 plastic digester
塑料枕形热水器 plastic pillow water heater
溯流而上的 upstream

suan

酸沉降 acid deposition
酸化 acidize
酸化反应 acidification reaction
酸化潜势 acidification potential

酸化实验反应器　acidogenesis experimental reactor
酸性火山岩　acid volcanic
酸性降水　acid precipitation
酸性金属氧化物　acidic metal oxide
酸性煤尘　acid smut
酸性蓄电池　acid battery
酸性岩　acid rock
酸性转炉钢　bessemer steel
酸雨　acid precipitation

sui

随机反应　stochastic response
随机风场模拟　stochastic wind field simulation
随机风轮推力波动　stochastic rotor thrust fluctuation
随机风载　stochastic wind load
随机故障　random failure
随机海浪中的波能　wave energy in random seas
随机荷载　stochastic load
随机塔架弯矩　stochastic tower bending moment
随机响应　stochastic response
随机性法　stochastic method
随机振动　random vibration
随机状态　random manner
随机最优控制　stochastic optimal control
随钻测量　measurement while drilling
碎波　breaking wave
碎波带　surf zone
碎波区　surf zone
碎浪　breaker
碎木块　chunk wood
碎片　debris
碎芯机　core breaker
碎渣机　grinder/crusher
隧道窑　tunnel kiln

sun

损耗　dissipation
损耗系数　coefficient of losses
损失比　damage ratio
损失指数　damage index

suo

缩放系数　zoom factor
索尔特点头鸭式波浪发电装置　Salter nodding duck wave power device
索尔特鸭式波浪发电机　Salter duck water power generator
索拉布莱克涂料　Solarblak coating
索拉雷托集热器板　Solarator collector panel
锁（紧，定）　lock
锁存器　latch
锁定（风力机）　blocking (for wind turbine)
锁定转子　locked-rotor
锁定转子保护　locked rotor protection
锁定转子转矩　locked-rotor torque
锁定装置　locking device, locking set
锁紧（止动，防松）垫圈　lock washer

ta

塔顶法　rainstorm
塔顶重量　tower top weight
塔架　tower
塔架底部前后弯矩　tower base fore-aft bending moment
塔架刚度　tower stiffness
塔架高度　tower height
塔架荷载　tower loading
塔架基础临界荷载工况　critical load

case for tower base
塔架运动　tower motion
塔架阵列　tower array
塔架自振频率　tower natural frequency
塔聚焦式发电厂　tower focus power plant
塔逻辑　tower logic
塔式太阳能电站　solar tower power generation
塔式太阳热发电系统　solar thermal power generation system with tower and heliostat plant
塔式太阳热能发电厂　heliostat power plant
塔式蒸发器　tower evaporator
塔弯曲　tower bending
塔影　tower shadow
塔影荷载　tower shadow loading
塔影效应　influence by the tower shadow
塔振动　tower vibration

tai

台车　dolly
台风　typhoon
台风转向点　vertex
台架试验　test on bed
台钳　combination pliers
太（拉）瓦时　Terawatt-hour
太空单结太阳能电池　single-junction space solar cell
太瓦（兆兆瓦）　terawatt (TW)
太瓦时　terawatt-hours (TWh)
太阳倍数　sloar multiple
太阳（辐照度）模拟器　solar (irradiance) simulator
太阳（能）电池　solar battery, solar battery cell
太阳（能）反射率　solar reflectivity
太阳（能）透射率　solar transmissivity, solar transmission rate, solar transmittance
太阳（能）吸收率　solar absorptivity, solar absorptance
太阳（能）炉　solar furnace
太阳（能）物理学　solar physics
太阳（能）物理学家　solar physicist
太阳（能）灶　solar furnace
太阳（射电）噪声　solar noise
太阳半径　solar radius
太阳暴　solar burst
太阳爆发　solar burst
太阳背点　solar antapex
太阳倍数　solar multiple
太阳闭环控制系统　solar close loop control system
太阳表面能量辐射　energy radiation from surface of sun
太阳测量仪表　solar measuring instrumentation
太阳常数　solar constant
太阳潮汐　solar tide
太阳城　solar town
太阳池　solar pond
太阳池的盐分扩散　salt diffusion in solar pond
太阳池发电　solar pool power generation
太阳池发电厂　solar pond plant
太阳池发电站　solar pond power generation
太阳池发电装置　power unit using solar pond
太阳池集热效率　solar pond collection efficiency

太阳池热能　solar pond thermal energy
太阳池热效率　solar pond thermal efficiency
太阳池效率　solar pond efficiency
太阳赤纬　solar declination
太阳磁　solar magnetism
太阳磁像仪　solar magnetograph
太阳磁学　solar magnetism
太阳村　solar village
太阳大气潮　solar atmospheric tide
太阳大气光学质量　solar air mass
太阳的　solar
太阳低能量等离子体　low-energy solar plasma
太阳—地球距离　sun-earth distance
太阳电池材料　solar cell material
太阳电池的串联电阻　series resistance of solar cell
太阳电池的等效电路　equivalent circuit of a solar cell
太阳电池的电压温度系数　voltage temperature coefficient of a solar cell
太阳电池的工作温度　operating temperature of solar cell
太阳电池等效电路　equivalent circuit of solar cell
太阳电池底板　solar cell basic plate
太阳电池电极　contact of solar cell
太阳电池帆板　solar panel
太阳电池方阵的重量—功率比　weight to power ratio of a solar array
太阳电池伏安特性曲线　I-V characteristic curve of solar cell
太阳电池面积　solar cell area
太阳电池子方阵　solar cell subarray
太阳电池组件表面温度　solar cell module surface temperature
太阳电池组件面积　solar cell module area
太阳定位角　solar position angle
太阳发射微粒流　solar steam
太阳帆　solar sail
太阳方位　solar azimuth angle
太阳方位角　solar azimuth angle, solar azimuth
太阳房　solar house
太阳房采暖器　solar room-heater
太阳房尺寸　solar house size
太阳房供暖系统　house heating system
太阳房供热水系统　house hot water supply system
太阳风　solar wind
太阳风暴　solar storm
太阳辐射　solar radiation
太阳辐射测量　solar radiation measurement, solar radiometry
太阳辐射测量仪器　solar radiation instrument
太阳辐射发电　electricity from solar radiation
太阳辐射仿真　solar radiation simulation
太阳辐射光谱　solar radiation spectrum
太阳辐射计　solar radiation meter
太阳辐射计标准化　standardization of solar radiometer
太阳辐射记录　solar radiation record
太阳辐射聚光器　solar radiation concentrator
太阳辐射量　solar radiation
太阳辐射能　radiant energy from sun
太阳辐射匹配　solar radiation matching
太阳辐射强度　intensity of solar radiation
太阳辐射强度计　solarimeter
太阳辐射热电转换　thermoelectric

conversion of solar radiation
太阳辐射数据　solar radiation data, solar insolation data
太阳辐射衰减　attenuation of solar radiation
太阳辐射速率　solar radiation rate, rate of solar radiation
太阳辐射通量　solar flux, solar radiation flux
太阳辐射照度　solar irradiance
太阳辐射最大值　maximum solar radiation
太阳辐照　solar irradiation
太阳辐照误差估计　solar irradiance error estimate
太阳赋能冷却　solar-energized cooling
太阳赋能模式　solar-energized mode
太阳干涉边界　solar interference boundary
太阳干涉等值线　solar interference contour
太阳干燥器　solar dryer
太阳高度　solar altitude, solar height
太阳高度角　solar altitude angle, solar elevation angle, solar altitude, solar elevation, sun angle
太阳跟踪表面　sun-following surface
太阳跟踪控制器　sun-tracking controller
太阳跟踪器　solar tracker, sun tracker
太阳跟踪伺服系统　sun-following servosystem
太阳跟踪系统结构　sun-tracking structure
太阳跟踪仪　sun tracker
太阳光电　photovoltaics
太阳光度　solar luminosity
太阳光发电　sunlight power generation
太阳光反射比　solar reflectance
太阳光反射系数　solar reflectance
太阳光伏能源系统　solar photovoltaic energy system
太阳光合（作用）　solar photosynthesis
太阳光谱　solar spectrum
太阳光谱辐照度　spectral solar irradiance
太阳光谱强度　solar spectral intensity
太阳光散　solar beam spread
太阳光通路　solar access
太阳光纤照明　solar optical fiber lighting
太阳行程图　sun-path diagram
太阳黑点　solar spot
太阳红外辐射　solar infrared radiation
太阳活动　solar activity
太阳活动高峰期　solar maximum
太阳活动极小期　solar minimum
太阳活动周期　solar cycle
太阳角模拟器　solar angle simulator
太阳镜象亮度　brightness of solar image
太阳聚光器　solar concentrator, solar dish
太阳聚焦辐射　focused solar radiation
太阳炉　solar furnace
太阳炉光学系统　optical system of furnace
太阳轮　sun gear
太阳罗经　sun compass
太阳模拟器　sun simulator
太阳膜　solar film
太阳能　solar energy
太阳能（玻璃隔板）集热墙　solar mass wall
太阳能（通量）集热器（收集器）　solar flux collector
太阳能（蒸馏器的）蒸馏量　limit of solar distillation
太阳能板　solar thermal panel
太阳能薄膜　solar film
太阳能保存　solar saving
太阳能泵　solar powered pump

太阳能闭环控制系统 solar close loop control system
太阳能冰箱 solar energy fridge, solar refrigerator
太阳能材料 solar energy material, solar material
太阳能材料老化 solar degradation
太阳能采集板 solar collector panel
太阳能采暖暖房 solar heated green house
太阳能草坪灯 solar energy lawn light, solar lawn lamp, solar lawn light
太阳能测量设备 solar energy measuring equipment
太阳能产业 solar industry
太阳能车 solar powered vehicle
太阳能成本 solar energy cost
太阳能充电包 solar rechargeable pack
太阳能充电灯 solar energy charging light
太阳能充电控制器 solar charge controller, PV charge controller
太阳能充电器 solar power charger
太阳能抽运装置 solar pumping device
太阳能除湿 solar dehumidification
太阳能储存 solar energy storage
太阳能储存器 solar energy storage
太阳能储存（水）箱 solar tank
太阳能储存效率 efficiency of storing solar energy
太阳能储藏池 solar pond
太阳能储罐 solar storage tank
太阳能储热（器） solar heat storage
太阳能储热池 solar heat storage pond
太阳能传感器 solar sensor
太阳能传输 solar energy transmission
太阳能传输损失 solar attenuation loss
太阳能窗口 solar window
太阳能磁流体发电 solar magnetic fluid power generation
太阳能带辅助热源系统 solar-plus-supplementary system
太阳能单独系统 solar-only system
太阳能单轴跟踪器 solar one-axis tracker
太阳能的等价小时 sun equivalent hours
太阳能的工业应用 commercial use of solar energy
太阳能的光热转换器 solar energy photothermal conversion unit
太阳能的间歇现象 intermittency of solar energy
太阳能的间歇性 intermittency of solar energy
太阳能的可利用性 solar feasibility
太阳能的能量储存 storage of energy from solar energy
太阳能的生物转换 bioconversion of solar energy
太阳能灯标 solar powered light beacon
太阳能灯具 solar lamp
太阳能等通量图 solar isoflux map
太阳能等通量线 solar isoflux line
太阳能点火 solar firing
太阳能电池 solar cell
太阳能电池板 solar cell panel, solar panel
太阳能电池板组件 solar power flat plate module
太阳能电池背膜 backsheet
太阳能电池材料 solar cell material
太阳能电池的并联电阻 shunt resistance

of solar cell
太阳能电池发电系统 solar cell power generation system
太阳能电池方阵 solar cell array
太阳能电池结构 solar cell structure
太阳能电池经济学 economics of solar cell
太阳能电池理论 solar cell theory
太阳能电池模块 solar module
太阳能电池石英钟 quartz clock powered by solar cell
太阳能电池吸收效率 solar cell absorption efficiency
太阳能电池效率 solar cell efficiency
太阳能电池叶片 solar paddle
太阳能电池有效面积 active area of solar cell
太阳能电池阵 solar array
太阳能电池组 solar cell battery
太阳能电池组件 solar cell module, photovoltaic module
太阳能电池组件输出功率 solar power module's power output
太阳能电话 solar telephone
太阳能电力 solar electricity
太阳能电力卫星 solar power satellite
太阳能电力系统 solar electric system
太阳能电能系统 solar electric energy system
太阳能—电能转换 solar electric conversion
太阳能电源 solar electric source, sun-generated electric power
太阳能电—热式平板集热器 solar electric-thermal flat-plate collector
太阳能定日镜 sun-tracking mirror
太阳能动力泵 solar power pump

太阳能动力灌溉 solar powered irrigation
太阳能动力机 solar power machine
太阳能动力系统 solar energy power system
太阳能二极管 solar diode
太阳能发电 solar electric power generation, solar-generated power, solar-produced electricity, solar electricity, solar power generation
太阳能发电厂 solar power plant, solar farm plant
太阳能发电机 solar generator
太阳能发电机（装置） solar driven power generator
太阳能发电设备 solar electric equipment
太阳能发电系统 solar energy generating system
太阳能发电站 solar generating station, solar-powered electric-generating station
太阳能发动机 solar energy engine, solar engine, solar powered engine
太阳能帆船 solar sailer
太阳能反光材料 solar reflector material
太阳能反射表面 solar reflecting surface
太阳能反射指数 solar reflectance index
太阳能反应堆 solar breeder
太阳能方位角 solar azimuth angle
太阳能方阵的实际效率 practical efficiency of a solar array
太阳能房供暖 solar house heating
太阳能仿真器 solar simulator
太阳能飞机 solar plane, solar powered aeroplane
太阳能非晶硅组件，非晶硅太阳能面板 amorphous silicon solar panel

太阳能分布　solar energy distribution, solar distribution
太阳能分散收集系统　solar decentralized collection system
太阳能风力发电站　wind-power station utilizing solar energy
太阳能—风能发电装置　solar-wind power generator
太阳能风能混合系统　solar-wind energy hybrid system
太阳能—风能联合转换系统　combined solar-wind energy conversion system
太阳能—风能系统　solar-wind energy energy system
太阳能峰值辐射（量）　peak solar radiation
太阳能服装　solar clothes
太阳能辐射观测　solar radiation observation
太阳能辐射集热器　collector of solar radiation
太阳能辐射能　solar radiation energy
太阳能辐射误差　solar radiation error
太阳能辐射吸收器　solar radiation absorber
太阳能辐射资源地图　solar radiation map
太阳能辅助冷却　solar assisted cooling
太阳能辅助热泵　solar-assisted heat pump
太阳能辅助热泵系统　solar assisted heat pump system
太阳能辅助吸收冷却　solar assisted absorption cooling
太阳能辅助吸收冷却系统　solar assisted absorption cooling system
太阳能负荷比法　solar load ratio method
太阳能负载部分　solar load fraction
太阳能感生电流　solar-induced current
太阳能干燥泥炭　sun drying peat
太阳能干燥器　solar drier
太阳能干燥系统　solar drying system
太阳能干燥效率　efficiency of solar drying
太阳能干燥装置　solar air-heating system
太阳能跟踪聚光型集热器　solar tracking concentrator
太阳能跟踪装置　solar tracking device
太阳能工程　solar energy engineering, solar engineering
太阳能工程师　solar engineer
太阳能工业　solar industry, solar energy industry
太阳能工业加热　solar industrial process heat
太阳能工艺学　solar energy technology
太阳能公文包　solar document bag
太阳能共享　solar energy sharing
太阳能供电设备　solar powered facility
太阳能供电系统　solar power system
太阳能供电装置　solar powered facility
太阳能供给量　solar energy delivery
太阳能供暖　solar heating
太阳能供暖—冷却系统　solar heating-cooling system
太阳能供暖系统　solar heating system
太阳能供暖效率　efficiency of solar heating
太阳能供暖制冷联合装置　combined solar heating-cooling installation
太阳能供热　solar heating

太阳能供热和冷却(系统) solar heating and cooling
太阳能供热和冷却示范法律 solar heating and cooling demonstration act
太阳能供热和冷却系统 solar heating and cooling system
太阳能供热和热水系统 solar heating and hot water system
太阳能供热技术 solar heating technology
太阳能供热经济学 solar heating economics
太阳能供热量 solar contribution
太阳能供热设备 solar heating equipment
太阳能供热试验 solar heating experiment
太阳能供热系统 solar heating system
太阳能供热装置 solar space heater
太阳能供应(量) solar energy supply
太阳能光电电池 photovoltaic solar cell
太阳能灌溉泵 solar irrigation pump
太阳能灌溉系统 solar irrigation pumping system
太阳能光电板 photovoltaics
太阳能光电池组件 solar photovoltaic module
太阳能光电收集器 solar photoelectric collector
太阳能充电围栏 solar photovoltaic fence
太阳能光电转换 solar photovoltaic conversion, photovoltaic conversion of solar energy
太阳能光电转换技术 solar photovoltaic conversion technology
太阳能光电转换效率 solar photovoltaic conversion efficiency
太阳能光电装置 solar-operated photovoltaic unit
太阳能光发电 solar electric direct conversion
太阳能光伏板 pv string
太阳能光伏打装置 solar-operated photovoltaic unit
太阳能光伏电池 solar photovoltaic cell
太阳能光化学过程 solar photochemical process
太阳能光化学技术 solar photochemical technology
太阳能光化学系统 solar photochemical system
太阳能光化学转换 solar photochemical conversion
太阳能光谱分布 spectral distribution of sun's energy
太阳能光热发电系统 solar thermal power system
太阳能光热发电站 solar thermal power station
太阳能光学材料 solar optical material
太阳能光子 solar photon
太阳能规划 solar energy planning
太阳能轨道反射器系统 orbiting solar reflector system
太阳能锅炉 solar energy boiler
太阳能锅炉辅助锅炉系统 solar boiler auxiliary boiler
太阳能锅炉—蒸汽轮机系统 solar boiler-steam turbine system
太阳能海水淡化 solar desalination
太阳能海水淡化系统 solar desalination system

太阳能海水温差发电　solar sea water temperature difference power generation
太阳能海水蒸馏器　solar sea water still
太阳能海洋热电站　solar sea thermal power plant
太阳能和风能的综合利用　hybrid use of solar and wind energy
太阳能黑电池　black solar cell
太阳能烘干过程　solar drying process
太阳能烘干试验　solar drying experiment
太阳能烘干系统　solar drying system
太阳能—化学能转换装置　solar chemical energy conversion unit
太阳能化学热泵　solar energy chemical heat pump
太阳能换热器　solar heat exchanger
太阳能活动低潮　solar ebb
太阳能火箭　solar rocket
太阳能获得量　solar gain, solar heat gain
太阳能机械　solar machine
太阳能机械能（转换）过程　solar mechanical process
太阳能激光器　solar laser, solar-excited laser
太阳能级硅　solar grade silicon
太阳能级硅电池　solar grade silicon cell
太阳能集热面积　solar collection area
太阳能集热器　solar energy collector, solar energy concentrator, solar thermal collector, solar energy thermal collector
太阳能集热器储热水系统　solar collector hot water storage system
太阳能集热器的窗口（进光口）面积　aperture area of a solar collector
太阳能集热器方向　solar collector orientation
太阳能集热器分类　classification of solar collector
太阳能集热器流体流速　solar collector flow rate
太阳能集热器坡度　solar collector slope
太阳能集热器热力发电系统　solar collector thermal power system
太阳能集热器设计　solar thermal collector design
太阳能集热器吸收管　solar collector tube
太阳能集热器斜度　solar collector tilt
太阳能集热器追踪装置　solar collector tracking device
太阳能集热器组件　solar collector assembly
太阳能集热墙　solar wall
太阳能集热系统　solar heat collection system, solar thermal collection system
太阳能集热效率　solar collection efficiency
太阳能集热形式　form of solar energy collection
太阳能计算机模型　solar energy computer model
太阳能计算器　solar calculator
太阳能记录器　solar recorder
太阳能技术　solar technology, heliotechnics
太阳能加（供）热系统　solar heat system
太阳能加热　solar heating
太阳能加热方向（位）　solar heating orientation
太阳能加热集热器　solar heating collector

太阳能加热技术 solar heating technology
太阳能加热取向 solar heating orientation
太阳能加热热离子转化器 solar heated thermionic converter
太阳能加热系统 solar heating system
太阳能加热效率 efficiency of solar heating
太阳能加热游泳池 solar heated swimming pool
太阳能加热装置 solar heating installation, solar space heater
太阳能家用发电系统 solar home system
太阳能家用热水器 solar domestic water-heater
太阳能家用用水加热 solar domestic water heating
太阳能间接形态 indirect form of solar energy
太阳能间歇式加热器 intermittent heater with solar energy
太阳能建设新农村工程 solar project for new villages
太阳能建筑 solar architecture, solar building
太阳能建筑物中的日光 daylighting in solar building
太阳能降解 solar degradation
太阳能交通灯 solar energy traffic light, solar energy traffic signal
太阳能交通信号灯 solar traffic signal
太阳能接收器 solar receiver, sun-heat absorber
太阳能经济 solar economy
太阳能经济分析 solar economic analysis
太阳能经济学 solar economics, economics of solar energy
太阳能净水器 solar water distiller
太阳能聚光 solar concentration
太阳能聚光倍数 solar concentration factor
太阳能聚光比 solar concentration ratio
太阳能聚光灯 solar spot light
太阳能聚光电池阵 photovoltaic concentrating array
太阳能聚光器 solar concentrator, solar dish
太阳能科学 solar science
太阳能科学家 solar scientist
太阳能空间电站 solar space power plant
太阳能空间供热 solar space heating
太阳能空间供热设备 solar space heating facility
太阳能空间加热 space heating with solar energy
太阳能空气集热器 solar air collector
太阳能空气加热集热器 solar air heating collector
太阳能空气加热器 solar air heater
太阳能空气加热系统 solar air heating system, solar air-heating system, air based solar-heating system
太阳能空气气球 solar ballon
太阳能空气系统 solar air system
太阳能空调 solar air conditioning
太阳能空调机 solar air conditioner
太阳能空调循环 solar air conditioning cycle
太阳能控制器 solar controller, solar regulator
太阳能控制系统 solar control system

太阳能雷达　solar radar
太阳能冷冻机　solar refrigerator
太阳能冷却（制冷）系统　solar cooling technology
太阳能冷却（系统）　solar powered cooling
太阳能冷却机房　solar cooling house
太阳能冷却—加热（取暖）系统　solar cooling-heating system
太阳能冷却应用　solar cooling application
太阳能理疗　solar physiotherapy
太阳能利用　solar energy application, solar energy use, solar energy utilization, solar utilizaiton, utilization of solar energy, application of solar energy
太阳能利用技术　solar utilization technique
太阳能利用率　soalr utilization rate
太阳能利用装置　solar energy utilizing device, helioplant
太阳能量　quantity of solar energy
太阳能量密度　solar power density, solar flux density
太阳能量输出　energy output of sun
太阳能流密度　solar energy flux
太阳能炉　solar stove
太阳能路灯　solar energy street lamp, solar street lamp
太阳能灭虫器　solar pest killer
太阳能模拟器　solar simulator
太阳能农场　solar farm
太阳能暖房　solar heated green house
太阳能抛物面集热器　solar parabolic collector
太阳能喷射式制冷　solar ejector cooling

太阳能烹饪　solar cooking
太阳能平板集热器　flat plate solar collector, flat plate solar heat collector, flat plate solar energy collector
太阳能谱　energy spectrum from sun
太阳能气体接受器　solar gas receiver
太阳能汽车　solar automobile, solar car
太阳能—氢能系统　solar-hydrogen energy system
太阳能区域供暖　space heating with solar energy
太阳能区域供暖设备　solar space heating facility
太阳能驱动的风机　solar-power-driven fan
太阳能驱动的风扇　solar-power-driven fan
太阳能驱动的热机系统　solar power heat engine system
太阳能取暖设备　solar heating installation
太阳能去毒　solar detoxification
太阳能全能量系统　solar total energy system
太阳能全能系统试验设施　solar total-energy-system test facility
太阳能燃料　solar fuel
太阳能—燃气联合循环槽式热发电系统　integrated solar combined cycle system
太阳能热（收集）系统　solar thermal system
太阳能热泵　solar heat pump, solar-assisted heat pump, solar energy heat pump

太阳能热泵系统　solar heat pump system, solar energy heat pump system
太阳能热储存　solar thermal energy storage
太阳能热电发电机　solar thermoelectric generator
太阳能热电系统　solar thermoelectric system
太阳能热电站　solar energy heat power station
太阳能热电转换　thermoelectric conversion of solar energy
太阳能热—电转换　solar thermal-electric conversion
太阳能热电转换器　solar thermoelectric converter
太阳能热动力　solar power, solar thermal power
太阳能热发电　solar thermal electric power generation, solar thermal electricity
太阳能热发电厂　solar thermal power station
太阳能热发电厂单位发电耗水率　water consumption rate per kWh of solar thermal power plant
太阳能热发电厂容因子　capacity factor of solar thermal power plant
太阳能热发电厂辅助用电量　auxiliary power quantity of solar thermal power plant
太阳能热发电厂年发电量　annual electric output of solar thermal power plant
太阳能热发电厂年效率　annual efficiency of solar thermal power plant
太阳能热发电厂容量　solar thermal power plant capacity
太阳能热发电厂厂用电　auxiliary power quantity of solar thermal power plant
太阳能热发电厂厂用电率　auxilia power consumption rate of solar thermal power plant
太阳能热发电基本负荷电厂　base load solar thermal power plant
太阳能热发电调峰电厂　peak load solar thermal power plant
太阳能热发电系统　solar thermal power generation system
太阳能热过程　solar energy system thermal process
太阳能热机　solar heat engine, solar power heat engine
太阳能热交换器　solar heat exchanger
太阳能热离子发电　solar thermionic power generation
太阳能热离子发电机　solar thermionic generator
太阳能热离子系统　solar-thermionic system
太阳能热离子转换系统　solar energy thermionic conversion system
太阳能热力泵　solar thermal power pump
太阳能热力发电　solar thermal power generation
太阳能热力发电设备　solar thermal power facility
太阳能热力发电系统　solar thermal power system
太阳能热力发电站　solar thermal power plant
太阳能热力发电装置　solar thermal power installation
太阳能热量收集器　solar heat

collector, solar energy thermal collector
太阳能热率 solar heat rate
太阳能 热能的电能转换 solar-thermal electric conversion
太阳能 热能电能转换系统 solar-thermal electric energy conversion system
太阳能 热能分散式发电系统 distributed solar thermal power system
太阳能 热能分散收集系统 solar-thermal distributed receiver system
太阳能 热能分散收集系统发电厂 solar-thermal distributed power station
太阳能热能集热器（收集器） solar-thermal collector
太阳能 热能集中收集系统 solar-thermal central receiver system
太阳能 热能系统的环境影响 environmental impact of solar thermal system
太阳能 热能转换 solar thermal conversion
太阳能热能转换过程 solar energy system thermal process
太阳能 热能转换设备 solar-thermal conversion facility
太阳能 热能转换系统 solar-thermal conversion system
太阳能热气机 solar hot-air engine, solar-operated hot-air engine
太阳能热容量 solar thermal capacity
太阳能热水 solar hot water
太阳能热水锅炉 solar water boiler
太阳能热水器 solar energy water heater, solar hot water heater, solar water collector, solar water heater
太阳能热水器预热系统 solar water preheat system
太阳能热水系统 solar water heating system, solar hot water system
太阳能热水浴室 solar hot water bath house
太阳能热箱 solar hot box
太阳能热增湿（淡化）法 solar humidification
太阳能热转换系统 solar power thermal conversion system
太阳能容量 solar capacity
太阳能入射角 solar incidence angle
太阳能闪光测量系统 solar flare measuring system
太阳能设备 solar energy equipment
太阳能生物技术 solar biotechonology
太阳能 生物能（转换）过程 solar biological process
太阳能生物能转换过程 solar bioconversion process
太阳能生物能转换效率 solar bioconversion efficiency
太阳能 生物质转换效率 solar-to-biomass conversion efficiency
太阳能时代 solar era
太阳能实验室 solar energy laboratory
太阳能使用权 solar right
太阳能收集 solar energy collection
太阳能收集板 solar panel, solar energy panel
太阳能收集板实验架 solar panel test stand
太阳能收集器（集热器） sun collector
太阳能收集塔发电厂 solar power tower plant

太阳能收集塔系统 solar power tower system
太阳能手电筒 solar flashlight
太阳能输出 solar energy sharing
太阳能输出（量） solar energy output
太阳能输入 solar input
太阳能输送系统 solar transport system
太阳能树脂工艺品 solar energy polyresin craft
太阳能双轴跟踪器 two-axis solar tracker
太阳能水泵 solar pump, solar water pump, solar power water pump
太阳能水泵及喷泉 solar water pump and fountain
太阳能水池集热器 solar pond collector
太阳能水力发动机 solar hydraulic engine
太阳能水压式沼气池 solar assisted liquid pressurized biogas digester
太阳能水预热器 solar water preheater
太阳能斯特林热空气发动机 solar Stirling hot-air engine
太阳能塑料 solar plastic
太阳能塔 solar tower
太阳能探测器 solar probe
太阳能条型聚光器 solar strip concentrator
太阳能庭院灯 solar courtyard light, solar energy courtyard light, solar yard lamp
太阳能通量 solar energy flux
太阳能通量放大率（倍数） solar flux amplification
太阳能通量年平均值 annual mean solar flux
太阳能通量图 solar flux map
太阳能投资 solar investment, solar capital
太阳能涂层 solar coating
太阳能推进 solar propulsion
太阳能脱水器 solar dehydrator
太阳能脱水设备 solar dehydration facility
太阳能脱盐 solar desalting
太阳能望远镜 solar telescope
太阳能微波传输 microwave transmission of solar energy
太阳能微波中继站 solar microwave relay station
太阳能卫星 solar power satellite, solar satellite
太阳能卫星站 solar power satellite station, solar powered satellite station
太阳能温差发电 solar temperature difference power generation
太阳能温室 solar greenhouse
太阳能涡轮发电机 solar turbo power plant
太阳能涡轮发电驱动 solar turboelectric drive
太阳能涡轮机 turbine with solar energy, solar-operated turbine
太阳能屋顶 solar roof
太阳能吸附式制冷 solar sorption cooling
太阳能吸热片 solar absorber plate
太阳能吸热器 solar receiver
太阳能吸热器支撑塔 solar receiver support tower
太阳能吸入量 inflow of solar energy
太阳能吸收表面 solar absorbing surface
太阳能吸收冷却器 solar absorption cooler

太阳能吸收率　solar absorbtivity
太阳能吸收去湿系统　solar absorption and dehumidification system
太阳能吸收式制冷　solar absorption cooling
太阳能吸收涂层　solar coating
太阳能吸收因数　solar absorption factor
太阳能吸收指数　solar absorption index
太阳能系统　solar system
太阳能系统费用（成本）方程　solar system cost equation
太阳能系统基本投资　solar system capital
太阳能系统控制器（装置）　solar system controller
太阳能系统设计　solar energy system design
太阳能系统效率　solar system efficiency
太阳能系统总效率　overall efficiency of solar energy system
太阳能系统最佳经济成本（费用）　economic optimal solar system cost
太阳能系统最优经济性选择　solar system economic optimization
太阳能咸水淡化　solar salt-water conversion
太阳能咸水转换　solar salt-water conversion
太阳能效率保证（太阳能热水系统产权方和太阳能热水工程实施责任方之间的一种合同）　solar energy benefit guarant
太阳能信号灯　solar signal lamp
太阳能蓄热温度　solar storage temperature

太阳能选择（性）涂层　solar selective coating
太阳能选择性薄膜　solar selective film
太阳能选择性表面　solar selective surface
太阳能选择性接收器　solar selective receiver
太阳能选择性吸收表面　solar selective absorbing surface
太阳能选择性吸收器　solar selective absorber
太阳能学会　solar energy society
太阳能压缩式制冷　solar compression cooling
太阳能牙刷　solar toothbrush
太阳能烟囱　solar chimney
太阳能液态金属磁流体发电系统　solar liquid metal magnetohy-drodynamic system
太阳能液体集热器　power generation solar liquid collector
太阳能液体系统　solar liquid system
太阳能液体制冷系统　solar powered liquid cooling system
太阳能应用　solar application
太阳能应用进展　advance in applied solar energy
太阳能游泳池加温装置　solar swimming pool heater
太阳能预热　solar preheating
太阳能预热系统　solar preheating system
太阳能源　solar power source, solar energy source
太阳能源装置示范产品　solar power unit demonstrator
太阳能再生系统　solar regenerative

system
太阳能灶 solar stove
太阳能增益 solar gain
太阳能站 solar power station
太阳能蒸发（作用） solar evaporation
太阳能蒸馏 solar distillation
太阳能蒸馏厂 solar distillation plant
太阳能蒸馏淡化（法） desalination by solar distillation
太阳能蒸馏器 solar still, solar distillator
太阳能蒸馏水 solar-distilled water
太阳能蒸馏作业 solar distilling operation
太阳能蒸汽发电厂 solar steam plant
太阳能蒸汽发电站 solar steam generating station
太阳能蒸汽发动机 solar steam engine
太阳能蒸汽发生器 solar vapor generator
太阳能蒸汽机 solar-operated steam engine, solar-powered steam engine
太阳能直接利用 direct use of solar energy
太阳能直接入射通量 direct incident solar flux
太阳能直接转换系统 solar power direct conversion system
太阳能指数 solar fraction
太阳能制冷 solar refrigeration, solar cooling
太阳能制冷（系统） solar powered cooling
太阳能制冷设备（装置） solar driven refrigeration unit
太阳能制冷系统 solar cooling system, solar refrigeration system
太阳能制氢 solar hydrogen production, solar production of hydrogen
太阳能制氢经济 solar-hydrogen economy
太阳能制氧 solar prepared hydrogen
太阳能制冷 solar refrigeration
太阳能中心接受器发电系统 solar central receiver electric generation system
太阳能煮水器 solar water boiler
太阳能住宅系统 solar residential system
太阳能转换 conversion of solar energy, solar energy conversion
太阳能转换成化学能 converting solar to chemical energy
太阳能转换法 solar conversion method
太阳能转换技术 solar conversion technology, solar energy conversion technology
太阳能转换率 solar transformity
太阳能转换器 solar converter
太阳能转换设备 solar energy conversion facility
太阳能转换为热能 conversion of solar energy into heat
太阳能转换系统 solar energy conversion system
太阳能转换效率 solar conversion efficiency
太阳能转换装置 solar-energy device, solar energy conversion device
太阳能装置 solar device, solar installation, solar-operated device, solar energy device
太阳能装置阵列 solar array
太阳能资源 solar energy resource, solar resource
太阳能自动跟踪装置 solar tracking device
太阳能总反射比 total solar reflectance
太阳能总辐射量 total solar radiation
太阳能总量 total amount of solar

energy, solar crude
太阳能总入射量　total solar energy incidence
太阳偏角　solar declination
太阳强度　solar intensity
太阳—氢能系统　solar hydrogen energy system
太阳倾斜角　solar declination
太阳热发电　solar heat power generation
太阳热过程　heliothermal process
太阳热量计　pyrheliometry, pyroheliometer
太阳热量记录仪　recording pyrheliometer
太阳热能　solar heat, solar-generated thermal energy, solar thermal energy
太阳热能分散收集发电系统　DSTP system (distributed solar thermal power system)
太阳热能控制窗　solar heat control window
太阳热能驱动（装置）　solar heat exchanger drive
太阳热能与蓄热系统　solar thermal and heat storage system
太阳热直接转换为电能　conversion of solar heat directly into electricity
太阳日照数据　solar insolation data
太阳入射角　solar angle of incidence
太阳伞　solar umbrella
太阳散射辐射　diffuse solar radiation
太阳色散　solar dispersion
太阳闪射　solar flare
太阳射电波　solar radio wave
太阳射电噪声　solar radio noise
太阳射能爆发　solar radio burst
太阳射线电子　solar radio electron
太阳摄谱仪　solar spectrograph

太阳时　solar time
太阳时角　solar hour angle
太阳矢量　solar vector
太阳视差　solar parallax
太阳天顶角　solar zenith angle
太阳同步轨道　sun synchronous orbit
太阳微粒　solar corpuscular
太阳微粒流　solar stream
太阳温度　solar temperature
太阳系　solar system
太阳向点　solar apex
太阳小时　sun hour
太阳型反应　solar type reaction
太阳烟囱发电　power generation by a solar chimney
太阳盐度差能　solar salt power
太阳仰角　elevation angle of sun
太阳翼　solar panel
太阳因子潮　solar component tide
太阳映像　solar image
太阳宇宙射线　solar cosmic ray
太阳源　solar source
太阳灶　solar cooker
太阳灶采光面积　aperture area of solar cooker
太阳灶操作高度　operating height of solar cooker
太阳灶操作距离　operating distance of solar cooker
太阳灶煮水热效率　water-boiling thermal efficiency of solar cooker
太阳张角　sun field angle
太阳照辐射量　solar insolation
太阳照射强度　solar radiation intensity
太阳正午　solar noon
太阳直射辐射　beam radiation
太阳直射通量　direct normal radiation

太阳质量　solar mass
太阳中微子　solar neutrino
太阳周年潮　solar annual tide
太阳紫外线辐射　solar ultraviolet radiation
太阳自东向西的轨迹　sun path from east to west
太阳总辐射　solar total irradiance, global solar radiation
太阴潮　lunar tide
太阴出差潮　lunar evectional tide
太阴全日潮　lunar diurnal tide
太阴日　lunar day
太阴日潮　lunar diurnal tide
太阴日周潮　lunar diurnal tide
钛铁矿　ilmenite
泰勒填料　Tellerette

tan

滩海　shallow water
弹簧　spring
弹簧常数　spring constant
弹簧回位　spring return
弹簧铰链　hinge spring
弹簧压力　spring pressure
弹力　elastic force
弹力系统　elastic system
弹力效应　elastic force effect
弹圈　coil spring
弹性变形　elastic deformation
弹性袋（一种波能装置）　flexible bag
弹性挡圈　circlip
弹性结构　elastic structure
弹性联接　elastic coupling
弹性模量　elastic modulus
弹性模数　elastic modulus, elastic moduli
弹性能　elastic energy

弹性碰撞　elastic collision
弹性燃料汽车　flexible fuel vehicle
弹性软管泵　elastomeric hose pump
弹性释水系数　elasticity releasable factor
弹性阻力　elastic resistance
探测　detection
探测覆盖　detection coverage
探测器　detector
探测站　sounding station
探勘井　test well
碳捕获与封存　carbon capture and storage
碳氮比　C/N ratio, carbon-to-nitrogen ratio (C/N)
碳电极　carbon electrode
碳含量　carbon content
碳化硅衬套　SiC sleeve
碳化钨　tungsten carbide
碳化物刀具　carbide cutter
碳粒　char particle
碳排放量　carbon footprint
碳氢化合物　hydrocarbon
碳氢化合物沉积　deposit of hydrocarbon, hydrocarbon deposit
碳氢燃料　hydrocarbon fuel
碳热还原　carbothermic reduction
碳刷　carbon brush
碳刷磨损　brush wastage
碳水化合物　carbohydrate
碳丝　carbon filament
碳丝灯泡　carbon-filament lamp
碳素材料　carbonaceous material
碳酸钙　calcium carbonate
碳酸盐型地热水　carbonate-type geothermal water
碳纤维　carbon fiber

碳循环过程　carbon cycle
"碳中性"排放　"carbon neutral" emission

tang

汤姆逊太阳能房　Thomason solar house
（华盛顿区）汤姆逊太阳能房　Thomason house
汤姆逊型滴流式集热器　Thomason-type trickle collector
搪孔头　boring head
搪磨机　honing machine
镗床　boring machine
镗孔　borehole
糖化　saccharification
糖基柴油　sugar-based diesel
糖基乙醇　sugar-based ethanol
糖类发酵　carbohydrate fermentation
糖类作物　sugar crop

tao

陶瓷　ceramics
陶瓷坩埚　ceramic crucible
陶瓷硫化镉太阳能电池　cadmium sulfide ceramic solar cell
陶瓷卵石　ceramic pebble
陶瓷卵石床　ceramic pebble bed
陶瓷密封　ceramic seal
陶瓷球　ceramic ball
陶瓷太阳能电池　ceramic solar cell
陶瓷涂层　ceramic coating
陶瓷纤维　ceramic fiber
陶瓷小珠　ceramic bead
陶瓷制品　ceramic
套管　junction box
套管板配置　tube-in-plate configuration
套管板式吸收器　tube-in-plate absorber
套管串　casing string
套管导向　conductor guide
套管底环　casing shoe
套管法兰盘　casing flange
套管换热器　double pipe exchanger
套管挤坏　casing collapse
套管接箍　casing coupling
套管接箍位置　casing collar location
套管接头　casing joint
套管冷却器　double pipe cooler
套管侵蚀　casing erosion
套管射孔　casing perforating
套管损坏　casing collapse
套管头　casing head
套管靴　casing shoe
套管柱　casing string

te

特定场地地质　site-specific geology
特级耐火材料　super-duty refractory
特罗姆勃被动式太阳能墙　Trombe wall
特殊废弃物条例　Storage Tank Regulations
特殊污泥（指腐败的粪便）　infrequently-led sludge
特性曲线　characteristic curve
特征频率　eigenfrequency
特征水位　characteristic water level
特征向量　eigenvector
特征值　eigenvalue
特种聚合体　special polymer

ti

梯度风　gradient wind
梯度风速　gradient velocity
梯度吸收膜　graded absorbing film
梯度指数吸收器　graded index absorber

梯形逻辑　ladder logic
提纯　refinement
提纯硅　purified silicon
提供　furnish
提前期　lead time
提取器　extractor, extraction apparatus
提取塔　extraction tower, extraction column
提取系统　extraction system
提取效率　extraction efficiency
提取蒸馏　extraction distillation
提升（在矿井中提升煤、矿物、人员或材料等的作业）　winding
提升装置　lift device
提水风车　windmill for water lifting
体积流量　volumetric flow rate
体积收缩　volume shrinkage
体积弹性模量　bulk modulus
体积稳定性　volume stability
体积压缩率　compressibility
体内平衡　homeostasis
体缺陷　volume defect
体心立方　body-centered cubic
体心立方结构　body-centered cubic
替代操作模式　alternative operating mode
替代能源　alternative energy sources, alternative energy resource, alternative energy

tian

天赤道　celestial equator
天底　nadir
天顶　zenith
天顶角　zenith angle
天顶距　zenith distance
天极　celestial pole
天空（有效）温度　(effective) sky temperature
天空辐射　sky radiation
天空漫射辐射　diffuse sky radiation
天气　weather
天气电码　weather code
天气图　weather chart
天气资料　weather data
天球　celestial sphere
天球子午圈　celestial meridian
天然地热条件　natural geothermal condition
天然放热量　natural heat release
天然裂缝　natural fracture, nature fissure
天然气车辆　natural gas vehicle
天然气处理　natural gas conditioning
天然气联合循环　natural gas fired combined cycle
天然气热电联供系统　natural-gas cogeneration
天然气水合物　natural gas hydrate
天然气往复式发动机　natural-gas-fired reciprocating engine
天然热流出　natural outflow of heat
天然热流量法（估计地热田能量潜力的一种方法）　natural heat flow
天然热泉　natural hot spring
天然热源　natural heat source
天然热输出　natural heat output
天然热液储层　natural hydrothermal reservoir
天然热资源　natural heat resource
天然水　natural water
天然水位　natural water level
天然蒸汽　natural steam
天然蒸汽田　natural steam field

天然状态外　ex situ
天体运动　celestial movement
天文潮　astronomical tide
天文单位　astronomical unit
天线　aerial
天轴　celestial axis
田间试验　field trial
甜菜　sugar beet
甜高粱　sweet sorghum
填充层　packed bed
填充床　packed bed
填充床储热　packed bed storage
填充床洗涤器　packed-bed scrubber
填充铅块　explanatory quad
填充塑料　filled plastics
填充塔　packed distillation column
填充物　fill
填充因数　fill factor
填充因子　fill factor
填充蒸馏塔　packed distillation column
填缝　flushing
填角焊　fillet weld
填砾　grail
填料　padding
填埋密度　landfill density
填埋气体　landfill gas
填隙合金（材料）　caulking metal
填隙式固溶体　interstitial solid solution
填隙式扩散　interstitial diffusion

tiao

调和成分　harmonic component
调和分析　harmonic analysis
调速轮　flywheel

tou

透光层　photic zone

透光区　photic zone

tong

通量聚光比　flux concentration ratio

wai

外置式吸热器　external receiver

wan

弯角扳手　hook spanner

wei

尾流干扰　wake interaction
位垒　potential barrier
位置反馈　position feedback
卫星太阳能发电站　satellite solar power station

wen

温差　temperature difference
温差能　energy by thermal gradient
温水出口　hot water outlet

wu

污水储留池　lagoon
无线电干扰　radio inference
物理模型试验　physical model test

xi

吸热板　absorber panel
吸收率　absorption rate
吸热体　absorber
吸热器（或接收器）　receiver
吸收器采光口　receiver aperture
吸热器额定热功率　design point thermal power of receiver
吸热器峰值辐射通密度　receiver peak

flux density
吸热器截断因子（或溢出因子） intercept factor (spillage factor) of receiver
吸热器净热功率 net thermal power of receiver
吸热器疏水放气系统 receiver vent and drain system
吸热器效率 receiver efficiency
吸热器系统 receiver system
吸热器溢出 receiver spillage

xia

瑕疵 flaw
下部加料锅炉 underfeed stoker
下层逆流（海面下的） undertow
下顶点 bottom dead-centre
下风（的） leeward
下风螺旋桨 downwind propeller
下风弦 lee side
下风向 down wind, downwind
下风向布置 downwind configuration
下风向风力机 downwind wind turbine
下风向风轮 downwind rotor
下风向扇形区 downwind sector
下风向式风电机组 downwind wind turbine generator system
（海洋在地壳板块压力下的）下降，沉降流 downwelling
下开口（太阳能）腔体接受器 downward facing cavity receiver
下坡风 downslope wind, slope down wind
下坡扩散 downhill diffusion
下射式燃烧 down-shot firing
下水道 drain
下吸式生物质气化炉 biomass downdraft gasifier
下行 downstream
下行通风 down draft
下限 lower limit
下引风 down draft
下游 downstream
下游阀 downstream valve
夏季热量增益 summer heat gain
夏季用通风口 vent for summer use
夏季遮荫 summer shade
夏季制冷系统 summer cooling system
夏季最大日照率 maximum summer insolation

xian

先进生物柴油 advanced biodiesel
先进先出 first-in first-out
纤维板残渣 fibreboard residue
纤维缠绕 filament winding
纤维过滤器 fibrous filter
纤维浆 fibre sludge
纤维缆绳 fibre rope
纤维缆索 fibre rope
纤维热绝缘层 fibrous thermal insulation
纤维蔬菜废弃物 fibrous vegetable waste
纤维素 cellulose
纤维素废弃物 cellulose waste
纤维素复合物 cellulose complex
纤维素类生物质 cellulosic biomass
纤维素乙醇 cellulosic ethanol
纤维素乙醇工厂 cellulosic ethanol plant
纤维增强混凝土 fiberre inforced concrete
咸的 saline
咸水侵入 saltwater intrusion
咸水入口 salt water inlet

嫌气分解　anaerobic decomposition
显热　sensible heat
显热储存　sensible heat energy storage
显热储存　sensible heat storage
显色指数　color rendering
显微结构　microstructure
现场安装　field installation
现场测量　field measurement
现场发电　on-site generation
现场可靠性试验　field reliability test
现场控制　site control
现场控制点　site control point
现场能量转换　on-site energy conversion
现场配电盘　local panel
现场评估　site assessment
现场试验　field testing
现场数据　field data
现场调查　field survey
现场信息　field information
现场装配　field fabrication
现场总线　field bus
现场作业　field work
现代风机　modern wind turbine
现金流分析　cash flow analysis
限波器　line trap
限度　limitation
限幅器　limiter
限流电路　limited current circuit
限速开关　limit speed switch
限速器　governor
限位开关　limit switch
限制量计　limit gauge
限制器　restrictor
线电压　line voltage
线动量　linear momentum
线痕　wire mark

线锯　fret saw
线锯床　wire saw
线聚焦集热器　line-focus collector
线聚焦聚光器　line-focus concentrator
线聚焦系统　line focus system
线卡子　guy clip
线路电流　line current
线圈　coil
线圈节距　coil pitch
线圈绕组　coil winding
线圈温度　winding temperature
线圈形状　winding configuration
线圈压紧结构　winding pressure arrangements
线缺陷　line defect
线速度　linear velocity
线速度，线速　linear speed
线性波动理论　linear wave theory
线性传动装置　linear actuator
线性定理　linear theory
线性动量　linear momentum
线性规划　linear programming
线性发电机　linear electrical generator, linear generator
线性菲涅耳式太阳能热发电　linear Fresnel reflector solar power
线性化法　linearized method
线性聚焦集热器　line focus collector, linear focused collector
线性聚焦型式　line focus configuration
线性理论　linear theory
线性区　linear zone
线性驱动器　linear actuator
线性热源　linear heat source
线性水力泵　linear hydraulic pump
线性太阳能集热器　linear solar collector
线性同步发电机　linear synchronous

	generator
线性系统	linear system
线性液压泵	linear hydraulic pump
线性运动效应	linear kinematic effect
线性振动	linear vibration
线性致动器	linear actuator
线与中性点间的	line-to-neutral
线组电阻	winding resistance

xiang

乡镇公路太阳能路灯的应用	solar streetlight for rural road
乡镇太阳能庭院灯	solar garden light for towns
相变	phase transition, phase change
相变（储热）材料	phase change material
相变材料储存	phase change material storage
相变储存系统	phase change storage system
相变储能	phase change energy storage
相变储热	latent heat storage
相变储热元件	phase change heat storage element
相变介质	phase change medium
相变区	phase change section
相变热	heat of transformation
相变太阳能热水器	phase change solar water heater
相变太阳能系统	phase change solar system
相变物质	phase change material
相变蓄能	phase change storage
相变蓄热	phase change thermal storage
相变蓄热容器	phase change storage container
相变蓄热物质	phase change storage substance
相变盐	phase-changing salt
相变转化	phase-change transition
相变转换	phase-change transition
相称性	proportionality
相电压	star voltage, phase voltage
相对捕获宽度	relative capture width
相对方向	relative direction
相对风	relative wind
相对风速	relative wind velocity
相对风向	relative wind direction
相对辐射表	relative radiometer
相对光谱灵敏度	relative spectral sensitivity
相对光谱灵敏度特性	relative spectral sensitivity characteristic
相对光谱响应	relative spectral response
相对渗透率	relative permeability
相对生物效应	relative biological effectiveness
相对湿度	relative humidity
相负荷	phase load
相关法律	affinity law
相关危险	hazard
相互扩散	mutual diffusion
相互连接的筏串	series of inter-connected rafts
相互连接的浮箱	inter-connected raft
相间短路保护	phase fault protection
相角	phase angle
相界	phase boundary
相界反应	phase boundary reaction

相界交联　phase boundary crosslinking
相界线　phase boundary line
相控整流器　phase controlled rectifier
相量电压　phasor voltage
相量图　phasor diagram
相匹配（文件）　assorted file
相平衡　phase equilibrium
相速度　phase velocity
相位补偿器　phase controller
相位超前　phase advance
相位角　phase angle
相位控制　phase control
相位控制率　phase control factor
相位控制设备　phase control device
相位耦合　phase coupling
相位线圈　phase coil
相位滞后　phase lag
相消干涉　destructive interference
相序　sequential order of the phase
相移角　phase angle
相转变　phase conversion
相长干涉　constructive interference
箱式太阳灶　box solar cooker
箱型吸收器　box absorber
箱状分隔（整流器）　box-like compartment
镶齿钻头　insert bit
镶嵌物　incrustation
镶嵌钻头　insert bit
镶式钻头　insert bit
响应时间　response time
响应阵风因子　response gust factor
向（或在）上游　upstream
向岸风　on shore wind onshore wind
向风面　windward side
向风面（的）　windward
向风漂流　windward drift
向海坡面　seaward slope
向极的　poleward
向量方程　vector equation
向量风场　vector wind field
向量和　vector sum
向量积　cross product
向前的　endwise
向阳玻璃窗　south-facing glazed window
向阳立式太阳能集热器　south-facing vertical solar collector
向洋　oceanward
向洋侧　oceanward
向洋方　oceanward
项目法人责任制　responsibility system of project legal person
橡胶带输送机　rubber belt conveyor
橡胶垫　rubber mat
橡胶缓冲器　rubber bumper
橡皮管　rubber tube

xiao

肖特基二极管　Schottky diode
肖特基势垒电池　Schottky barrier cell
肖特基太阳电池　Schottky solar cell
消波岸坡　absorbing beach
消波堤　wave breaking revetment
消除　eliminate
消除有害影响的系统　abatement system
消防斧　fireman's axe
消防水泵　fire fighting water pump
消耗　dissipation, drain
消耗热　chargeable heat
消化处理　degestion process
消化后处理机组　post-digestion processing unit
消化后处理器　post-digestion processing unit

消化剂　digestant
消化器性能　digester performance
消化污泥　digested sludge
消气剂（吸附真空器件中残留气体的材料，吸气剂可分为蒸散型和非蒸散型两大类）　getter
消融气体　dissolved gas
消融物质　ablating material
消声技术　muffling technique
消声设备　noise muffling equipment
消石灰　hydrated lime
消音器　silencer
硝酸钠　sodium nitrate
销子　dowel
小边界角跟踪式集热器　small rim angle tracking collector
小潮　neap tide
小潮潮差　neap range
小潮潮流　neap tidal current
小潮平均高潮位　mean high water neap
小潮升　neap rise
小齿轮　pinion, pinion gear
小电抗　small reactance
小风　light wind
小沟　shallow ditch
小弧度集热器　collector with less curvature
小角度晶界　low angle grain boundary
小井　slimhole
小卵石　pebble
小木头　small wood
小片淡水域　lagoon
小区太阳能发电系统　residential area PV power system
小曲率长聚焦集热器　long-focus collector with less curvature
小水电　small hydro

小水电发展的政策　policy for small hydropower development
小型并网光伏电站　small on-grid PV power station
小型潮汐电站　mini-tidal power station
小型风力发电系统　small wind system
小型风力机　small-size wind machine
小型风能工程　small wind energy system
小型风能转换装置　small-scale WECS
小型浮游生物　microplankton
小型高效风力发电站　windworks
小型海洋热能转换电站　mini-OTEC plant
小型相变盐蓄能器　small container of phase-changing salt
小样本连续研究法　panel method
小叶　leaflet
小翼　winglet
小圆石　pebble
小直径井　slimhole
小锥　bradawl
校正系数　correction factor
校准因子　calibration factor
效率　efficiency
效率比　efficiency ratio
效能评估　efficiency assessment
效能数据　efficiency data
效能指数　efficiency index
效益成本　utility cost
效用成本　utility cost

xie

协方差函数　covariance function
协同作用　synergy
斜棒栅水膜除尘器　inclined rod grid water film scrubber

斜不锈钢槽 inclined stainless steel chute
斜齿轮 helical gear, bevel gear
斜齿圆柱齿轮 helical gear, single-helical gear
斜度 obliquity
斜风 skew wind
斜击式水轮机 inclined jet turbine, Turgo impulse turbine
斜流 oblique flow
斜面反射板增温器 tilted reflector booster
斜温层 thermocline
斜温层蓄热系统 thermocline thermal storage system
斜向风 yawed wind, oblique wind
斜缘薄钢板 bevel-edge steel
斜支柱 inclined supporting strut
斜置绒布式太阳能蒸馏器 tilted wick type solar still
斜置双层玻璃盖板 double sloped glass cover
斜轴风力机 inclined-axis-rotor WECS
谐波 harmonic wave, harmonics
谐波变化 harmonic variation
谐波的 harmonic
谐波分量 harmonic component
谐波电压源 harmonic voltage source
谐波分析 harmonic analysis
谐波失真 harmonic distortion
谐波抑制 harmonic elimination
谐振 resonance
谐振变换器 resonant converter
谐振频率 harmonic frequency
谐振曲线 resonance curve
携带式仪表 field instrument
携带水分能力 moisture carrying capacity
泄放阀 drip cock

泄放管 drip pipe
泄洪道 spillway
泄漏电流 leakage current
泄漏管道 leaking pipeline
泄漏围堵系统 spill containment system
泄漏检测系统 leak detection system
泄水道 sluiceway
泄水台 water table
泄水锥 runner cone
泄压阀 bleed off valve
泄油 drain
泻湖 lagoon
卸荷 off-loading
卸荷期 off-load period
卸料 off-loading
卸料端 delivery end
卸料间 tipping floor
卸料斜槽 fuel chute
卸载 load shedding
谢尔比斯变速传动 Scherbius variable-speed drive
谢尔比斯电机 Scherbius machine

xin

心轴 mandrel
辛烷值 octane rating
锌空电池 zinc-air battery
锌碳粉 zinc carbon powder
锌溴电池 zinc bromine battery
新概念型风能转换系统 innovative wind energy conversion system
新建项目 greenfield project
新能源 new energy resource
新能源开发 new energy development
新能源资源 new energy resource
新生代火山活动 Cenozoic volcanism
新添加的燃料 green fuel

新鲜固体消化　fresh solid digestion
新鲜基　green basis
新鲜木片　green chip
新型波能转换装置　new wave-energy converter
薪材　fuelwood
薪柴　fire wood
薪炭林　fire forest
信风　trade wind
信号电路　signal circuit
信号器　annunciator
信号调节器　signal modifier
信号调整器　signal conditioner
信噪比　signal-to-noise ratio

xing

星形阀　star valve
星型联结　star connection
星型绕组　star winding
星载测量　spaceborne measurement
行波热声热机　traveling-wave thermoacoustic heat engine
行程长度　path length
行频　line frequency
行星边界层　planetary boundary layer
行星齿轮　epicyclic gear, planet gear
行星齿轮传动机构　planetary gear drive mechanism
行星齿轮系　planetary gear train
行星齿轮箱　planetary gearbox
行星架　planet carrier
行星减速箱　planetary gearbox
形成　formation
形成热　heat of formation
形核功　nucleation energy
形核率　nucleation rate
形式　formation

形状参数　shape parameter
型线浮箱（一种利用波浪能的设备，由一列或一排浮箱组成，能将浮箱运动中的能量转换为流体中的高压），流面筏　contouring raft
型芯沙　core sand
性能　performance
性能参数　performance parameter
性能分析　performance analysis
性能系数　coefficient of performance
性能指数　performance index

xiong

胸压手摇钻　breast drill

xiu

修复时间　repair time
修整机　finishing machine
修琢　chipping

xu

虚部　imaginary part
虚负载　dummy load
需求预测　demand forecast
需氧发酵　aerobic fermentation
需氧微生物　aerobiont
需氧细菌　aerobic bacteria
需氧消化　aerobic digestion
许用负荷　permissible load
序列相关　serial correlation
序批式厌氧堆肥　sequential batch anaerobic composting
絮凝剂　flocculating agent
蓄潮池　tide pool
蓄电池　storage cell, secondary battery, secondary cell, rechargeable battery
蓄电池板　storage battery plate

蓄电池充电电路　battery charging circuit
蓄电池充电调节器　battery charge regulator
蓄电池充电转换器　battery charge converter
蓄电池放电调节器　battery discharge regulator
蓄电池放电转换器　battery discharge converter
蓄电池组　battery bank, accumulator battery
畜牧业残渣　animal husbandry residue
蓄能　energy storage
蓄能床　storage bed
蓄能电站　storage power station
蓄能反应　endergonic reaction
蓄能光化学反应　endergonic photochemical reaction
蓄能器　accumulator
蓄能容量　energy storage capacity
蓄能损失　storage loss
蓄能系统　storage system
蓄能站　storage station
畜禽副产品　animal by-product
蓄热　heat storage, thermal storage
蓄热储存　thermal storage of energy
蓄热床　storage bed
蓄热单罐　single thermal storage tank
蓄热的　heat retaining
蓄热地板　storage floor, heat floor slab
蓄热供热系统　thermal storage heat supply system
蓄热锅炉　heat storage boiler
蓄热极限　storage limit
蓄热粒子床　storage particle bed
蓄热量　thermal storage capacity
蓄热流体流速　storage fluid flow rate
蓄热流体热容量　storage fluid heat capacity
蓄热炉　heat-storing stove
蓄热卵石床　pebble bed storage
蓄热密度　storage density
蓄热能力　thermal storage capacity, heat retaining capacity, heat storage capacity
蓄热器　storage water heater, thermal storage device, container for storing heat
蓄热器组　thermal storage battery
蓄热墙　thermal storage wall
蓄热容　storage capacity
蓄热式加热器　thermal storage heater
蓄热体　heat retainer, heat retaining mass
蓄热屋顶　storage roof
蓄热岩石床　storage rock bed
蓄热装置　thermal storage unit
蓄水　impoundment
蓄水层　aquifer, aquifer storage
蓄水池　storage basin, tidal basin, storage ponds, reservoir
蓄水库　storage reservoir
畜业废弃物　animal husbandry waste

xuan

玄武岩层　basaltic layer
悬臂梁　cantilever, cantilevered beam
悬浮固体　suspended solid
悬浮颗粒气化炉　suspended particle gasifier
悬浮区熔法　float-zone-procedure
悬浮燃烧　suspension combustion
悬浮微粒　particulate
悬挂　hanging

悬空燃烧　suspension firing
旋风　cyclone
旋风层　Ekman spiral
旋风除尘器　cyclone dust collector
旋风分离器　cyclone separator, cyclonic separator
旋风过滤器　cyclone filter
旋风灰　cyclone ash
旋风喷雾室　cyclone spray chamber
旋风汽水分离器　cyclone steam separator
旋风器　air cyclone
旋风燃烧锅炉　cyclone fired boiler
旋风式风力涡轮机　tornado-type wind turbine
旋风水膜式除尘器　cyclone gaswasher
旋环　water swivel
旋桨型风车　propeller type windmill
旋流　cyclone flow, rotary current
旋流燃烧器　cyclone burner
旋涡　eddy
旋涡聚集式风能装置　vortex concentrator device
旋涡衰减　vortex decay
旋涡增力型装置　vortex augmentor
旋液分离器　hydrocyclone
旋翼涡轮机　turbine with rotary wing
旋翼制动器　rotor brake
旋转　swivel, revolution, rotate, revolution-per-minute
旋转臂式隔离开关　rotary arm disconnector
旋转变流机　rotary converter
旋转波　wave of oscillation
旋转采样　rotational sampling
旋转采样风矢量　rotationally sampled wind velocity
旋转磁场　rotating magnetic field
旋转挡板调节　grid valve control

旋转的　rotary
旋转电机　electrical rotating machine
旋转阀　rotary valve
旋转分量　rotational component
旋转风速计　rotating anemometer
旋转鼓轮　rotating drum
旋转加料器　feed auger
旋转角　rotation angle
旋转接头　rotating union
旋转流化床　revolving fluidised bed
旋转抛物面集热器　parabolic-dish collector
旋转喷雾器　rotary atomiser
旋转频率　rotational frequency
旋转失速　rotating stall
旋转式风速计　rotary anemometer
旋转体　rotator
旋转温度表　sling thermometer
旋转雾化器　rotary atomizer
旋转液压泵　rotary hydraulic pump
旋转栅　rotating grate
旋转周期　rotational period
旋转锥形钻头　rotary cone bit
旋转钻孔　rotary drilling
旋转钻头　roller bit
漩涡　vortex
漩涡脱离　vortex shedding
漩涡脱离转换器　vortex shedding converter
漩涡脱体　detachment of vortice
漩涡泄离　vortex shedding
选择反射　selective reflection
选择辐射　selective radiation
选择散射　selective scattering
选择散射系数　selective scattering coefficient
选择性薄膜　selective film
选择性表面　selective surface

选择性催化还原　selective catalytic reduction
选择性反射板　selective reflection plate
选择性反射镜　selective mirror
选择性反射器（板）　selective reflector
选择性非催化还原　selective non-catalytic reduction
选择性黑色涂料　selective black paint
选择性接收器　selective receiver
选择性扩散　selective diffusion
选择性面吸收器　selective surface absorber
选择性渗透　selective permeability
选择性透射膜　selective transmission film
选择性透明膜　selective transparent film
选择性透射　selsective transmission
选择性涂层　selective coating
选择性温室　selective greenhouse
选择性吸收　selective absorption
选址　sitting

xue

削磨残渣　thinning residue
削片机　chipper
雪载　snow load

xun

巡航速度　cruise speed
循环　circulation, circular
循环泵　circulation pump
循环操作温度　loop operating temperature
循环供给系统　circulation supply system
循环荷载　cyclic load
循环换流器　cycloconverter
循环回路　flow circuit
循环加压设备　circulation pressure equipment
循环经济性　cyclic economy
循环冷却水　circulating cooling water
循环冷却水系统　circulating cooling water system
循环流化床　circulating fluidized bed
循环流率　loop flow rate
循环流体　circulating fluid
循环流体　cycle fluid
循环能源效率　round trip energy efficiency
循环区　recirculation zone
循环驱动力　cyclic driving force
循环热耗率　cycle heat rate, cycle thermal efficiency, thermodynamic cycle efficiency
循环水泵　circulating pump
循环水泵房　circulating water pump house
循环水系统　circulating water system
循环索系　moving cable
循环系统　circulating system
循环效率　cycle efficiency
循环液漏失区　lost-circulation zone
循环蒸汽发生器　circulating steam generator
训练　drill
迅速　velocity

ya

压（力）头　pressure head
压板　pressure plate
压板风速表　plate anemometer
压差　differential pressure, pressure differential
压差式水平轴风机　Enfield-Andreau wind machine
压差阻力　drag from pressure

压电（现象）	piezoelectricity
压电传感器	piezoelectric pickup
压电电阻率	piezoresistivity
压电高分子	piezoelectric polymer
压电晶体	piezocrystal
压电效应	piezoelectric effect
压风落煤	air-shooting
压降	pressure drop
压降计算	pressure drop calculation
压紧顶帽	guide tube support plate
压块	briquetting, lamination, wafering, pressing piece
压力表	pressure gauge
压力波动	pressure oscillation
压力波能转换装置	pressure wave energy conversion device
压力舱	pressure chamber
压力传感器	pressure sensor, pressure transducer
压力范围	pressure range
压力风速表	pressure anemometer
压力封闭酸溶	pressurized acid digestion
压力管风速表	pressure tube anemometer
压力计	piezometer, manometer
压力降	pressure drawdown
压力交换器	pressure exchanger
压力角	pressure angle
压力进给系统	force feed system
压力控制阀	pressure control valve
压力脉动	pressure fluctuation
压力面	pressure side
压力室	pressure chamber, vacuum chamber
压力瞬变	pressure transient
压力瞬时分解	pressure transient analysis
压力梯度	pressure gradient
压力通风机	forced draft fan
压力温差安全器	pressure-temperature safety device
压力稳定作用	pressure stabilization
压力延迟渗透	pressure retarded osmosis
压力元件	pressure element
压力振荡	pressure oscillation
压粒机	pellet press
压裂工艺	fracturing technology, fracturing technique
压裂技术	fracturing technology, fracturing technique
压裂液罐	frac tank
压裂液效率	fluid efficiency
压滤机	filter press
压滤器	filter press
压滤式电池	filter press cell
压煤砖机	briquette press
压敏电阻	piezoresistor
压气冲击钻井	air hammer drilling
压强计	piezometer
压燃式发动机	compression ignition engine
压入式通风系统	blowing system of ventilation
压缩比	compression ratio
压缩波	compressional wave
压缩冲程	compression stroke
压缩打包机	baling press, baler
压缩点火	compression ignition
压缩辅料	pressing aid/additive
压缩机	compressor
压缩机控制模块	compressor control module
压缩空气	compressed air, pressurized air
压缩空气储能	compressed air energy

storage
压缩空气储能系统　compressed air energy storage
压缩空气存储　compressed air storage
压缩空气钻井　air drilling
压缩生物质燃料　compressed biofuel
压缩液体　compressed liquid
压缩制冷循环　compression refrigerating cycle
压载水　ballast water
压铸冲模　die casting dies
压铸机　die casting machine
鸭式波浪能抽提装置　duck wave energy extractor
鸭式波浪能发电机　duck wave power generator
牙轮钻头　cone bit, roller bit
牙嵌式联接　castellated coupling
亚临界制冷循环　subcritical refrigerating cycle
亚麻荠属　camelina

yan

烟尘排放　dust emission
烟道　flue
烟道采样　stack sampling
烟道除尘系统　flue dust removal system
烟道灰　flue ash
烟道气　flue gas
烟道气出口　flue gas outlet
烟道气体化合物　flue gas compound
烟道取样　stack sampling
烟道损失　waste-gas loss, stack loss
烟道阻力　draft loss
烟洞　flue
烟灰　soot
烟煤　soft coal, bituminous coal
烟气　flue gas
烟气侧腐蚀　flue gas-side corrosion
烟气挡板　flue gas baffle
烟气换热器　flue gas heat exchanger
烟气加热器　flue gas heater
烟气净化　flue gas cleaning
烟气净化区　flue gas treatment hall
烟气净化系统　flue gas cleaning system
烟气冷凝　flue gas condensation
烟气冷凝装置　flue gas condensation unit
烟气流速　flue gas velocity
烟气排放连续监测　continuous emissions monitoring
烟气脱硫　flue gas desulfurization
烟气循环　flue gas recirculation
淹没灰输送机　submerged ash conveyor
淹溢模式　inundation model
延迟分离　delaying separation
延性　tensility
延展性　ductility
严重故障　catastrophic failure
严重失速　deep stall
岩层　rock strata
岩基　batholith, batholite
岩浆　magma
岩浆（源）流体　magma solid
岩浆房　magma chamber
岩浆库　magma chamber
岩浆热液系统　magmatic hydrothermal system
岩浆室　magma chamber
岩浆收集器　magma trap
岩浆岩　igneous rock
岩浆岩侵入体　igneous intrusion

岩块弛豫时间	block relaxation time
岩脉	dykes
岩石床储热装置	rock bed heat storage unit
岩石萃热	heat extraction from rock
岩石分析	petrographic analysis
岩石孔隙中的各种流体	geofluid
岩石破碎技术	rock fracture technique
岩石热导体	thermal conductivity of rock
岩石特征构造	rock characteristic formation
岩石性质	rock property
岩石压力	rock pressure
岩石压裂水力技术	hydraulic technique of rock fracturing
岩体风化	weathering of rock mass
岩体绘图	lithologic mapping
岩系	rock formation
岩屑	rock cuttings
岩心提取器	core breaker
岩心钻进	core drilling
岩心钻探	core drilling
岩性	lithology
岩盐	rock salt
沿岸带	tidal land
沿岸流	coastal current
沿岸漂砂	longshore drift
沿岸漂移	longshore drift
沿岸区替代能源	coastal zone alternative energy
沿岸物质流	longshore drift
沿海的	maritime
沿海地区，沿海的，海滨的	littoral
沿海风能转换器	coastal wind energy converter
沿翼展分布	spanwise airfoil distribution
研磨机	grinder, crusher
研磨性岩层	abrasive formation
研碎	trituration
盐（度）差	salinity gradient
盐（度）差（度）能转换装置	salinity gradient energy converter
盐（度）差能	salinity energy
盐差	osmotic pressure differential
盐差能	energy from salinity gradients
盐差能动力装置	power plant of energy from salinity gradients
盐场风车	salt windmill
盐沉积	salt deposit
盐电池	salt battery
盐动力	salinity power
盐度	salinity
盐度差电站	salinity power plant
盐度差发电装置	salinity power device
盐度差膜	salinity membrane
盐度差能	salinity-gradient energy
盐度差能转换	conversion of salinity gradient energy
盐度电桥	salinity bridge
盐度梯度	salinity gradient energy, salinity gradient
盐分	saline matter, salt content
盐分探测器	salt detector
盐风	salt wind
盐海	saline sea
盐化作用	salinification
盐结皮	salt crust
盐量计	salinity indicator
盐浓度	salt concentration
盐浓度梯度	salt concentration gradient
盐侵	salt intrusion
盐丘	salt dome
盐溶液	salting liquid
盐渗透能	saline energy

盐生植物　halophyte
盐水　brine, salting liquor, saline
盐水腐蚀　salt water corrosion
盐水浸入作用　salt water intrusion
盐水浓度　brine concentration
盐水入侵　saline intrusion
盐水转化　saline water conversion
盐水组分　brine composition
盐酸溶液　hydrochloric acid solution
盐梯度太阳池　salt gradient pond
盐田　salt pan
盐雾　salt fog, salt spray
盐液比重计　salinometer
盐沼　salt-marsh
盐汁　salting liquor
衍射　diffraction
衍射角　diffraction angle
眼睛睁视　goggle
眼式太阳能集热器　solar eyeball collector
厌氧处理　anaerobic treatment
厌氧处理方法　anaerobic treatment process
厌氧发酵　anaerobic digestion, anaerobic fermentation, anaerobic decomposition
厌氧发酵过程　anaerobic fermentation process
厌氧废物处理　anaerobic waste treatment
厌氧分解　anaerobic breakdown
（厌氧分解产生的）沼渣沼液　digestate
厌氧腐蚀　anaerobic corrosion
厌氧过程　anaerobic process
厌氧过程设计　design of anaerobic process
厌氧降解力　anaerobic biodegradability
厌氧接触　anaerobic contact

厌氧接触法　anaerobic contact process
厌氧菌　anaerobic bacteria, anaerobic microbe
厌氧滤器　anaerobic filter
厌氧培养　anaerobic culture
厌氧培养皿　anaerobic petri dish
厌氧生物处理过程　anaerobic biological treatment process
厌氧生物分解　anaerobic biomass decomposition
厌氧酸　anaerobic acid
厌氧塘　anaerobic pond
厌氧微生物　anaerobic microorganism, anaerobiont
厌氧污泥消化　anaerobic sludge digestion
厌氧污泥消化池　anaerobic sludge digester
厌氧污水处理　anaerobic sewage treatment
厌氧细菌　anaerobic bacterial
厌氧消化（作用）　anaerobic digestion
厌氧消化器　anaerobic digester
厌氧消化污泥　anaerobically digested sludge
厌氧性的　anaerobic
厌氧性生物　anaerobe
厌氧氧化池　anaerobic lagoon
厌氧植物　anaerophyte
验潮标　tide staff, tide pole
验潮杆　tide-pole
验潮井　tide gauge well, tide well
验潮水准点　tidal benchmark
验潮仪　tidal ball, tidal gauge, tide gauge, tidal meter
验潮站　tidal gaging station, tidal station, tide station

堰缺口　weir notch
堰上水头　weir head

yang

扬尘量　dust emission
扬程　delivery head, delivery lift
扬料板　lifting flight
羊八井地热电站（中国）Yangbajing geothermal power plant (China)
羊皮纸　membrane
阳光百分率　percent possible sunshine
阳光板（双层透明聚碳酸酯板，它是一种具有透光、高强、轻质、隔热、阻燃、耐候等优良性能的新型建筑材料，可以用作太阳能装置的透明盖板）sunlight panel
阳光反射率　reflectivity for sunlight
阳光防护窗　solar shield
阳光辐射能　radiant energy in sunlight
阳光辐射作用　solarization
阳光计划（日本的一项全国性计划，其目标是开发除原子能以外的所有新能源）Sun-Light Plan
阳光季节性变化　seasonal variation of sunlight
阳光聚焦理疗仪　sunlight focusing instrument for physiotherapy
阳光吸收率　solar absorptivity
阳光消毒法　solar disinfection
阳光（照射）型式　sunlight pattern
阳光直射　direct sunlight
阳光直射温度　direct sun temperature
阳光转换原理　principle of converting sunlight
阳极　anode
阳极斑点　anode spot
阳极板　anode plate
阳极辐射热　radiant heat of anode
阳极静电流　anode rest current
阳极射线电流　anode ray current, positive ray current
阳极氧化　anodic oxidation, anodization
阳离子　cation
阳离子交换器　cation exchanger
阳坡　adret, adretto
阳转子　male rotor
杨树　populus
洋床　ocean bed
洋底　ocean bed, ocean floor
洋风　ocean wind
洋浪场　ocean wave field
洋流　seaflow, marine current, oceanic current, ocean current
洋面海流　ocean surface current
洋面温度　ocean surface temperature
洋盆　ocean basin
仰角　tilt angle
养分含量　nutrient content
养蜂业　apiculture
氧传感器　lambda sensor
氧含量　oxygen content
氧化　combustion
氧化层　oxide layer
氧化催化剂　oxidation catalyst
氧化带　oxidation zone
氧化氮　nitrogen oxide
氧化的气化（作用）　oxygen gasification
氧化钒　vanadium oxide
氧化反应　oxidation reaction
氧化钙　quicklime
氧化锆分析仪　zirconium oxide analyzer
氧化锆　zirconia

氧化硅气凝胶　silica aerogel
氧化还原电位　oxidation-reduction potential
氧化还原体系　oxidation-reduction system, redox system
氧化还原液流电池　Redox-flow battery
氧化剂　oxidizing agent
氧化精炼　oxidation refining
氧化磷酸化作用　oxidative phosphorylation
氧化膜　oxide layer
氧化钼　molybdenum oxide
氧化气化器　air-blown gasifier
氧化区　oxidation zone
氧化物燃料电池电力发电机　oxide fuel cell power generator
氧化物蒸汽压　oxide vapor pressure
氧化硒光电池　selenium oxide photovoltaic cell
氧化硒光伏打电池　selenium oxide photovoltaic cell
氧化锡铟　tin doped indium oxide
氧化压力气化　air-blown pressure gasification
氧化亚铜光伏打电池　cuprous oxide photovoltage cell
氧化作用　oxidation reaction
氧气气化炉　oxygen gasifier
氧乙炔炬　oxyacetylene torch
氧杂茂　furan
氧—蒸汽（反应）系统　oxygen-steam system
样板　gauge board
样机试验　prototyping experiment
样品　instance, sample
样品破碎　sample reduction
样品缩分　sample division
样品制备　sample preparation

yao

窑　kiln
摇动炉排，摇杆炉排　rocking grate
遥测度数刻度盘式温度计　remote-reading dial thermometer
遥测验潮仪　tide gauge telemetry
要求　instance
耀斑　solar flare

ye

椰子壳　coconut shell
冶金　metallurgy
冶金级硅　metallurgical grade silicon
冶金焦　metallurgical coke (met coke)
冶金术　metallurgy
野外试验　field testing
业绩指标　performance indicator
叶（片）弦　blade chord
叶（片）顶（部）漏泄　tip leakage
叶顶漏泄损失　tip leakage loss
叶端损失　tip loss
叶根　root of blade
叶根荷载　blade root load
叶根损耗系数　blade root loss factor
叶根弯矩　blade root bending moment
叶根涡流　blade root vortex
叶尖（叶片）　tip of blade
叶尖半径　tip radius
叶尖段桨距　tip section pitch
叶尖舵片增力型风轮机　tip vane augmented wind turbine
叶尖接收器　blade tip receptor
叶片剖面半径　radius of blade section
叶尖失速　tip stall

叶尖顺桨 tip feathering
叶尖速比 tip speed ratio, tip speed ratio
叶尖速度 blade tip speed
叶尖速度比 tip velocity ratio
叶尖损失 tip loss
叶尖损失因子 tip loss factor
叶尖脱流 tip stall
叶尖涡旋 tip vortice
叶轮 blade wheel, impeller
叶轮变速 variable rotor speed
叶轮的鼓风作用 windmilling action of rotor
叶轮风机 paddle wheel fan
叶轮风速计 vane anemometer
叶轮干扰速度 induced velocity at the rotor
叶绿素 chlorophy
叶绿素电池 chlorophyll solar cell
叶绿体膜 chloroplast membrane
叶面残片 surface debris
叶片 wing, wind cock, wind direction vane, wind vane
叶片安装环 blade mounting ring
叶片安装角 setting angle of blade, wing setting angle
叶片半径 blade radius
叶片材料 blade composition, blade material
叶片颤振 blade flutter
叶片颤振参数 blade flutter parameter
叶片弹性轴 blade elastic axis
叶片顶部 blade top
叶片根部 blade root
叶片根部组件 blade root fixing
叶片根梢比 ratio of tip-section chord to root-section chord
叶片共振 blade resonance

叶片固定点 blade anchor point
叶片后缘 trailing edge
叶片几何参数 blade geometry parameter
叶片几何攻角 angle of attack of blade
叶片几何形状 blade geometry
叶片几何形状优化 blade geometry optimum
叶片计算 blade calculation
叶片坚固性 blade solidity
叶片桨距变化 blade pitch change
叶片桨距角 blade pitch angle
叶片桨距角设置角度 blade pitch set angle
叶片接力器 blade servomotor
叶片节距 blade pitch
叶片结构 blade configuration, blade construction, blade structure
叶片结构设计 blade structural design
叶片连接 blade connection
叶片轮毂 blade hub, blade contour
叶片内部 inner portion of blade
叶片扭角 twist angle of blade, twist of blade
叶片扭曲分布 blade twist distribution
叶片疲劳 blade fatigue
叶片偏差 blade deflection
叶片频率试验 blade frequency test
叶片剖面 blade section
叶片前缘 leading edge of blade
叶片强度 blade strength
叶片倾角 blade angle
叶片式燃烧器 blade type burner
叶片数 number of blades
叶片顺桨 blade feathering
叶片损失 blade loss
叶片—塔架的净空 blade-tower clearance
叶片通过频率 balde passing frequency,

blade passage frequency
叶片通流 blade path
叶片投影面积 projected area of blade
叶片弯曲 blade bending
叶片弯曲载荷 blade bending load
叶片微元动量理论 blade element momentum theory
叶片涡流发生器 vane vortex generator
叶片形状 blade shape
叶片压力 blade pressure
叶片翼型 blade profilc
叶片优化设计 optimal blade design
叶片元素理论 blade element theory
叶片元速度 blade element velocity
叶片运动 blade motion
叶片噪声 blade noise
叶片长度 blade length
叶片振动 blade vibration
叶片重量 weight of blade, blade weight
叶片轴承 blade bearing
叶片轴线 blade axis
叶片装配工 blader
叶片自然频率 blade natural frequency
叶片组装 blade assembly
叶梢上翘 tip deflection
叶梢上翘功率谱 power spectrum of tip deflection
叶素动量 blade element momentum
叶素一动量理论 blade element-momentum theory
叶素方法 blade element method
叶栅 blade cascade
(叶轮)叶梢速率 tip-speed ratio
夜间风速 nighttime wind speed
夜间有效辐射 effective nocturnal radiation
液舱试验 tank testing

液动 hydraulic drive
液动阀 oil controlled valve
液缸 hydraulic cylinder
液固比 liquid-solid ratio
液化 liquefaction, liquefy
液化热 heat of liquefaction
液化石油气 liquefied petroleum gas
液化作用 devolatilization
液基太阳能加热系统 liquid-based solar heating system
液力耦合器 fluid coupling
液力涡轮 hydraulic turbine
液流电池 flow battery
液面线 flux line
液式太阳能集热器 liquid type solar collector
液态 liquefaction
液态丙烷再循环 liquid propane recirculation
液态地核 liquid core
液态排渣鲁奇气化炉 liquid slagging Lurgi gasifier
液态岩浆 magma liquid
液体表面 fluid surface
液体池 liquid pool
液体动力 hydrokinetic power
液体肥料 liquid fertilizer
液体供暖系统 liquid heating system
液体活塞 liquid piston
液体集热器 liquid (heating) collector
液体加热集热器 liquid heating collector
液体加热平板集热器 liquid heating flat plate collector
液体加热系统 liquid heating system
液体力学 fluid mechanics
液体喷射焚烧炉 liquid injection incinerator
液体品质 fluid quality

液体平板型集热器 liquid flat plate collector
液体气动水下电源转换 hydropneumatic subsurface power converter
液体球面聚光集热器 fluid spherical concentrator
液体燃料 wet fuel, liquid fuel
液体渗透探伤 liquid penetrant examination
液体生物燃料 liquid biofuel
液体生物质燃料 liquid biofuel
液体松香 tall oil
液体速度 liquid velocity
液体损失 hydraulic loss
液体温度计 liquid-in-glass thermometer
液体效率 fluid efficiency
液体循环加热（冷却）系统 hydronic system
液体占优势的储层 liquid-dominated reservoir
液体占优势的地热田 liquid-dominated field
液体质量 fluid quality
液体组合式太阳能电池 liquid combined solar cell
液位玻璃管 gage glass
液位开关 level switch
液相 liquid phase
液相沉淀法 liquid deposition
液相流速 liquid velocity
液压 hydraulic pressure
液压泵 hydraulic pumping
液压臂 hydraulic arm
液压参数 hydraulic parameter
液压储能器 hydraulic accumulator
液压储油柜 hydraulic reservoir
液压单元 hydraulic unit
液压动力 hydraulic power
液压动力元件 hydraulic power unit

液压阀 hydraulic valve
液压缸 hydraulic cylinder
液压给料机 hydraulic feeder
液压工具 hydraulic power tool
液压过滤器 hydraulic filter
液压发动机 hydraulic motor
液压回转缸 hydraulic rotary cylinder
液压活塞 hydraulic piston
液压机液体 hydraulic fluid
液压技术 hydrokinetic technology
液压加料推杆 hydraulic feed ram
液压接头 hydraulic joint
液压块 hydraulic block
液压马达 hydraulic motor
液压驱动 hydraulic drive
液压系统 hydraulic system, hydraulic circuit
液压油缸 hydraulic ram
液压油箱 hydraulic reservoir
液压元件 hydraulic component
液压制动系统 hydraulic braking system
液柱 fluid column
曳力 drag force

yi

一般分析试样 general analysis sample
一般分析试样水分 moisture in the general analysis sample
一次电流 primary current
一次电压 primary voltage
一次风 primary air
一次风比率 primary air ratio
一次风分布 primary air distribution
一次风控制 primary air control
一次风扩口 primary air outlet
一次风率 primary air ratio

一次风压力控制系统 primary air pressure control system
一次回路 primary circuit
一次空气 primary air, underfire air
一次空气供应 primary air supply
一次能源 primary energy, primary energy source, primary power source
一次能源供应量 amount of primary energy supply
一次能源构成 primary energy mix
一次燃烧 primary combustion
一次绕组 primary winding
一次热媒 primary fluid
一次生物质能供应量 primary bioenergy supply
一次再热循环 single reheat cycle
一等标准直接日射表 primary standard pyrheliometer
一点谱 one-point spectrum
一对一控制方式 one-to-one control mode
一个太阳值（自然太阳辐照的最大值） on sun
一级（工作）全辐射表 first class pyrradiometer
一级（工作）直接日射表 first class pyrheliometer
一级（工作）总日射表 first class pyranometer
一级标准太阳电池 primary standard solar cell
一级风 light air
一阶响应 first-order response
一氯代苯（太阳热间接转换为电能的一种工质） monochlorobenzene
一维动量理论 one-dimensional momentum theory
一氧化碳 carbon monoxide
医疗垃圾 medical waste
仪表气源 instrument air
移动表面水 moving surface water
移动波 wave of translation
移动车 dolly
移动床气化反应器 moving-bed gasification reactor
移动的 floating
移动电荷 moving charge
移动滑车 traveling block
移动炉排 travelling grate
移动炉排焚烧炉 moving grate incinerator
移动炉排炉 travelling grate furnace
移动平均数 moving average
移动摄影车 dolly
移动式电机组 mobile generator set
移动式发电装置 mobile electro-generating unit
移动式海水淡化装置 mobile desalination plant
移动式聚焦集热器 mobile concentrating collector
移动式太阳能集热器 movable solar collector
移动式太阳能药浴设备 movable solar medication bath equipment
移动式应急电源 mobile emergency power supply
移动通信基站太阳能电源系统 PV power system for GSM base station
移动通信信号站太阳能发电系统 solar PV power system for mobile communication signal station
移频键控 frequency shift keying
疑惑 doubt
乙醇 ethanol, ethyl alcohol

乙醇混合物	ethanol blend
乙醇降解菌	ethanol degrading bacteria
乙二醇	ethylene glycol
乙基叔丁基醚	ethyl tertiary butyl ether
乙酸	acetic acid
乙酸铵盐	ammonium acetate solution
乙酸钾	potassium acetate
乙烯基酯	vinyl ester
已开采过的井	dead well
以气为动力的	pneumatic
以生态系为本管理机制	ecosystem-based management
以液体为主的热液资源	liquid-dominated hydrothermal resource
以沼气为纽带的农村能源生态模式	rural energy-ecology pattern by biogas
异步电机	asynchronous generator
异步发电机	asynchronous generator
异常暴（风）雨状况	unusual storm condition
异常波	freak wave
异常波型	abnormal wave
异常高热流量	anomalous high heat flow
异常高压地层	abnormally high-pressure formation
异常性热流量	anomalous heat flow
异常压力地层	abnormally pressured formation
异相电流	out of phase current
异相电压	out of phase voltage
异相界面	heterophase boundary
异形流动	variable flow
异形漩流	maelstrom
异养藻类	heterotrophic algae
异质结电池	heterojunction cell
异质结太阳能电池	heterojunction solar cell
译码	decode
易怒的	inflammable
易燃的	inflammable
易溶盐	soluble salt
易散性排放	fugitive emission
逸出电流	outgoing current
意外消耗	accidental consumption
溢洪道	spillway
溢洪道排水系统	spillway draining system
溢流阀	spill valve, bleeder valve
溢流管	upflow tube
翼动方程式	flapping equation
翼根弦	wing root chord
翼根翼型	root airfoil
翼尖刹车	tip brake
翼尖翼型	tip airfoil
翼截面	wing section
翼肋	rib
翼梁	spar
翼面	plane
翼面方向的	flapwise
翼片后缘	trailing edge
翼剖面	wing section
翼梢旋涡	tip vortex
翼弦	chord, wing chord
翼弦线	chord line
翼型	airfoil
翼型板风力机	airfoil wind machine
翼型测风装置	aerofoil flow measuring element
翼型厚度	thickness of airfoil
翼型零升力线	aerofoil zero lift line
翼型剖面	airfoil section
翼型剖面特性	airfoil section characteristic
翼型剖面阻力	airfoil section drag
翼型气动弦线	aerodynamic chord of

airfoil
翼型升力　airfoil lift
翼型相对厚度　relative thickness of airfoil
翼型族　the family of airfoil
翼型阻力　airfoil drag
翼型阻力系数　airfoil drag coefficient, section drag coefficient
翼缘　wing, wind cock, wind direction vane, wind vane
翼阻力　wing resistance

yin

阴极　cathode, negative electrode
阴极斑点　cathode spot
阴极板　cathode plate
阴极保护系统　cathodic protection system
阴极电压　cathodic voltage
阴极射线管　cathode ray tube
阴离子交换器　anions exchanger
阴影　shading
阴影法　hill shading
阴影损失　shading loss
阴影效应　shadow effect
阴影因子　shading factor
阴转子　female rotor
荫蔽效应　shadow effect
荫影效应　shadow effect
音频大地电磁法　audio-frequency magnetotelluric method
音值　tonality
铟锌氧化物　indium zinc oxid
银盘式直接日射计　silver-disk pyrheliometer
引潮力　tide force, tide producing force, tide reproducing force, tide-generating force, tide-generation force, tidal generation force
引出端剂量　exit dose
引导气流　steering current
引风机　induced draft fan
引力波　gravity wave
引力场　gravitational field
引力质量　gravitational mass
引擎机床　nacelle machine bed
引燃火焰　pilot flame
引燃喷射　pilot injection
引入管　inlet pipe
引示激磁机　pilot exciter
引水渠　approach channel
引汐力　tidal raising force
引下线　down conductor
饮用水　potable water
隐现　loom
印版　forming
印记　imprint
印刷电极　print electrode

ying

英国热量单位　British Thermal Unit (BTU)
迎波面坡　apron slope
迎风壁　windward wall
迎风侧　windward
迎风潮　windward flood
迎风舵　windward rudder
迎风风力机　upwind turbine
迎风风速　face velocity
迎风机构　orientation mechanism
迎风面（的）　windward
迎风面积　wind area, frontal area
迎风坡　windward slope
迎风气流　wind stream
迎风驱动装置　orientation drive

迎风装置　wind ward rudder
迎风阻力　frontal drag
迎角　attack angle, incidence angle
迎面风　oncoming wind, opposing wind
迎面流　oncoming flow
迎面阻力　drag force
荧光灯　fluorescent lamp
荧光粉检漏　fluorescence detection of leak, fluorescent leak detection
荧光光谱　fluorescence spectroscopy
荧光聚光器　fluorescent solar concentrator
盈利筛选图表　profitability screening chart
营养成分　nutrient content
营运成本　operating cost
营运费　operating cost
影响评估　impact assessment
影响评价　impact assessment
应变　strain
应变计　strain gauge
应变片　strain gauge
应急电池　emergency cell
应急冷却系统　emergency core cooling system
应急燃料　emergency back-up fuel
应急蓄水池　emergency holding pond
应力测量　stress measurement
应力场　stress field
应力幅度　amplitude of stress
应力腐蚀　stress corrosion
应力剖面　stress profile
应力周期　stress cycle
应用图　application drawing
硬（软）板（片）材及自由发泡板机组　hard/soft and free expansion sheet making plant
硬地层　hard formation
硬化剂　hardener
硬件　hardware
硬件逻辑　hardware logic
硬件平台　hardware platform
硬壳　incrustation
硬木　hardwood
硬木材　hardwood
硬能源　hard energy source
硬水　hard water
硬岩　hard rock
硬脂酸钙　calcium stearate
硬脂酸锂　lithium stearate
硬质合金　tungsten carbide
硬质合金刀具　carbide cutter

yong

壅高（海侧流通口处的水位与实际潮位的差值，是正向发电时的水头损失）　obstruction height
永磁发电机　permanent magnet alternator, permanent magnet generator
永磁方波电机　permanent magnet square wave motor
永磁极　permanent magnet pole
永磁同步发电机　permanent magnet synchronous generator
永磁直流电机　permanent magnet DC machine
永久磁铁　permanent magnet
永久负荷　permanent load
永久系泊（系统）　permanent mooring
涌潮　tidal bore, bore, tidal surge
涌浪能量转换器　surging wave energy

convertor
用暗销接合 dowel
用锄耕地 hoe
用电质量 quality of consumption
用过的木材 used wood
用户电力系统 customer power system
用卡盘夹住 chuck
用能单位 energy consumption unit
用水力发的电 hydropower
用酸处理 acidize

you

油菜饼 colza cake
油菜籽 rapeseed
油藏工程师 reservoir engineer
油藏模拟 reservoir simulation
油藏增产措施 reservoir stimulation
油层动态 reservoir behavior
油滴盘 drip pan
油动阀 oil controlled valve
油封 oil seal
油膏 grease
油管 grease nipple
油耗量 oil consumption
油浸式变压器 oil-immersed type transformer
油—空气分离器 air-oil separator
油冷却器 oil cooler
油料提取 oil extraction
油料作物 oil crop, oilseed crop
油炉 oil furnace
油路 hydraulic circuit
油轮 tanker
油门 gasoline throttle
油喷枪 oil lance
油气分离罐 catch pot

油气罐 air oil reservoir
油气混合物 fuel-air mixture
油气井增产措施 well stimulation
油气联产 oil and gas co-production
油燃料跳闸 oil fuel trip
油燃烧器 oil burner
油乳胶 oil emulsion
油水分离器 decanter
油田卤水 oilfield brine
油纹 oil streak
油压缸 hydraulic cylinder
油棕榈 oil palm
游标尺 vernier gauge
游标混合永磁电机 vernier hybrid permanent magnet machine
游动滑车 traveling block
游离水 free water
游丝 hairspring
游隙 windage
游泳池用太阳能系统 swimming pool solar system
有（无）源电路元件 active (passive) circuit elements
有槽凸圆柱头螺钉 cheese head screw
有潮港 tidal harbour
有潮河 tidal river
有潮河口 tidal estuary
有潮界限 tide head
有潮区 tidal compartment
有潮水域 tidal water
有磁性的 magnetic
有功电流 active current
有功分量 active component
有功功率 active power
有光泽的 glossy
有害空气污染物 hazardous air pollutant
有弧度翼型 cambered airfoil

有机半导体太阳电池 organic semiconductor solar cell
有机废物 organic waste
有机废物能 energy from organic waste
有机干物质 organic dry matter
有机金属化学气相沉积法 metal organic chemical vapor deposition
有机垃圾 biogenic waste
有机朗肯循环 organic Rankine cycle
有机朗肯循环技术 organic Rankine cycle technology
有机流体 organic fluid
有机生活垃圾 organic household waste
有机生物降解物质 organic biodegradable matter
有机食品农场 organic food farm
有机太阳能电池 organic photovoltaic cell
有机碳总量 total organic carbon
有机体 organism
有机物 organic matter, organism
有机物含量 organic content
有机物质发酵 fermentation of organic substance
有机质 organic matter
有机质负荷率 organic loading rate
有机质含量 organic content
有缺陷的 defective
有涂层的玻璃 coated glass
有涂层的集热器 coated collector
有限差 finite difference
有限差分 finite difference
有限尺度 finite size
有限深度 finite depth
有限元技术 finite element method
有效安装空间尺寸 active dimension of mounted space
有效安装面积尺寸 active dimension of mounted area
有效半衰期 effective half life
有效波高 significant wave height, effective wave height
有效出力 effective out-put
有效地球辐射 effective terrestrial radiation
有效电流 effective current
有效分布系数 effective distribution coefficient
有效范围 effective range
有效风速 effective wind speed
有效浮力 effective buoyancy
有效辐射功率 effective radiated power
有效负荷 payload
有效高度 effective height
有效功率 available power, useful power, effective capacity, effective horsepower, effective power
有效光阑 aperture diaphragm
有效光强 effective luminous intensity
有效集热 usable heat collection
有效集热面积 effective collection area
有效集温 effective accumulated temperature
有效接触面积 effective contact area
有效镜面辐射 effect specular radiation
有效聚光比 effective concentration ratio
有效均匀温度 effective uniform temperature
有效空隙率 effective porosity
有效孔径 effective aperture
有效孔隙率 effective porosity
有效控制 active control
有效冷却表面 active cooling surface

有效马力 effective horsepower
有效面积 effective area
有效能 useful energy
有效能的损失 available power loss
有效气动下洗 effective aerodynamic downwash
有效倾卸因子 effective tilt factor
有效热阻 effective thermal resistance
有效容量 effective capacity
有效渗透率 effective permeability
有效渗透性 effective permeability
有效受光角 effective acceptance angle
有效受热面 effective heating surface
有效太阳能辐射 effective solar radiation
有效太阳能蒸馏器 efficient solar still
有效太阳热能 useful solar heat
有效推力 effective thrust
有效弯度 effective camber
有效温度 effective temperature
有效下洗 effective downwash
有效性 availability
有效压力 effective pressure
有效迎角 effective angle of attack
有效源面积 effective source area
有效再能 effective carrying capacity
有效增值系数 effective multiplication factor
有效栅电压 effective gate voltage
有效阵风速度 effective gust velocity
有效值 effective values
有效阻抗 effective impedance
有序无序转变 disordered order transition
有旋流 rotational flow
有焰燃烧 flaming combustion
有用功率 useful power
有用能 useful energy
有用太阳热能 useful solar heat

有源电力滤波器 active power filter
有源空腔辐射计 active cavity radiometer
有源谐波滤波器 active harmonic filter
有载运行 on-load operation
有载指示灯 on-load indicator
有罩风轮 shrouded rotor
有阻挡层的光电池 photoelectric barrier layer cell
诱导的 inductive
诱导通风 induced draught
诱发辐照蠕变 irradiation-induced creep
釉料 glaze
釉面 glaze

yu

淤积 siltation
淤泥 silt
余烬 ember
余流 residual current
余能 complementary energy
余热 waste heat, residual heat, excess heat, residue derived fuel
余热锅炉 waste heat boiler, heat recovery steam generator, heat recovery boiler
余热锅炉热效率 thermal efficiency of waste incineration boiler
余热回收 waste heat recovery, heat recovery
余热回收锅炉 heat recovery boiler
余热利用 residual heat utilization
余弦损失 cosine loss
余弦因子 cosine factor
鱼道 fish pass, fish way
鱼类洄游 fish migration
鱼类通道 fish passage
渔场 fishing ground
与太阳能电池并联的二极管 bypass-

diode	
宇宙能源系统	space energy system, space power system
宇宙日辐射量	daily extraterrestrial radiation
宇宙太阳能	space solar energy
宇宙太阳能同量	extraterrestrial solar flux
雨量型	rainfall regime
雨流计数法	rainflow cycle counting
雨棚	weather shed
玉米棒子	corncob, maize cob
玉米秆	corn stalk
玉米壳	corn husk
玉米湿磨机	corn wet mill
玉髓	chalcedony
预定极限	preset limit
预定义操作	predefined operation
预混火焰	premixed flame
预计发电量	predicted energy production
预燃室	mixing chamber
预燃室式柴油机	antechamber diesel engine
预热	preheating
预热炉	preheater
预热器	preheater
预热序列	preheat sequence
预涂层	pre-coating
预先勘察	preliminary reconnaissance
预消化处理器	pre-digestion processing unit
预压系统	pressurized system
阈限值	threshold limit value

yuan

元素分析	ultimate analysis/elementary analysis, elemental analysis
园林管理残渣	landscape management residue
园林建筑学	landscape architecture
园艺残渣	horticultural residue
原材料	primary material
原电池（组）	galvanic battery
原动机	prime mover
原光电流	primary photoelectric current
原理	theorem
原理图	schematic
原料	feedstock
原木干材	stemwood log
原绕组	prime mover
原生燃料	primary fuel
原生水	connate water
原生纸浆	virgin pulp
原始化学产物	initial chemical product
原始热状态	original thermal conditioin
原始设备制造商	original equipment manufacturer
原始岩石温度	original rock temperature
原位	in situ
原型风速	prototype wind speed
原子	atom
原子电池	atomic battery
原子堆积因数	atomic packing factor
原子辐射	atomic radiation
原子光谱	atomic spectra, atom spectrum
原子光谱线的精细结构	fine structure of atomic spectra line
原子轨道	atomic orbit
原子键	atomic bond
原子结构	atomic structure
原子结合能	atomic binding energy
原子晶体	atomic crystal
原子量	atomic weight
原子面密度	atomic planar density

原子能　atomic energy
原子能（动力）　atomic power
原子能电站　atomic power station
原子能发电装置　atomic power station
原子能工业　nuclear power industry
原子能蓄存器　atomic energy storage battery
原子吸收谱　atom absorption spectroscopy
原子序数　atomic number
原子质量单位　atomic mass unit
圆底形白土容器　round-bottomed clay vessel
圆底形金属容器　round-bottomed metal vessel
圆顶螺母　domed nut
圆顶熔炉　cupola furnace
圆规　divider
圆环发电机　torus generator
圆锯　circular saw
圆木　log
圆筒分离筛　trommel screen sieve
圆筒筛　trommel screen
圆筒式塔架　tubular tower
圆筒形反射镜　cylindrical reflector
圆筒形热管　cylindrical geometry heat pipe
圆筒形同步发电机　tubular synchronous machine
圆筒形直线永磁同步电机　tubular permanent magnetic linear synchronous motor
圆头锤　ball-peen hammer
圆图　circle diagram
圆形菲涅尔透镜　circular Fresnel lens
圆形色谱法　radial chromatography
圆周　circumference
圆周侧隙　circumferential backlash
圆周速度　circumferential speed
圆柱齿轮　cylindrical gear
圆柱滚子轴承　cylindrical roller bearing
圆柱形　cylindrical
圆柱形菲涅尔透镜　cylindrical Fresnel lens
圆柱形集热器　cylindrical collector
圆柱形接收器　cylindrical receiver
圆柱形金属条反射镜　cylindrical slat mirror
圆柱形聚光镜　cylindrical focusing mirror
圆柱形聚焦集热器　cylindrical focusing collector
圆锥的　conical
圆锥滚子轴承　tapered roller bearing
源电压　source voltage
源分类城市生活垃圾　source separated municipal solid waste
源分类垃圾　source-separated waste
源岩流体界面　source-rock-fluid interface
远场　far field
远场压力—时间历程　far-field pressure-time history
远程发电　remote power generation
远程监视　telemonitoring
远动终端　remote terminal unit
远方动力　remote power
远红外成像（确定热显示区的一种方法）　infrared imagery
远红外辐射　far-infrared radiation
远距离监控　remote monitoring
远日点　aphelion
远尾流　far wake
远洋垂直浮标　ocean-going heaving buoy
远洋的　pelagic

远源场　far field
远月潮　apogean tide
远月潮流　apogean tidal current
远震　teleseism
远震P波滞后　teleseismic P-wave delay
远震滞后　teleseismic delay
远紫外辐射　far-ultraviolet radiation

yue

约束涡　confined vortex
月潮　lunar tide
月潮间隙　lunitidal interval
月亮质心　center of mass of the moon
月平均加热负荷　average monthly heating load
月平均温度　mean monthly temperature
月球上升力　lunar-raising force
月球延滞　lunar retardation
越波型　overtopping

yun

晕渲法　hill shading
云母　mica
云杉　spruce
云水　cloud water
云状空化　cloud cavitation
允许的峰值瞬变电压　peak allowable transient voltage
允许辐射量　radiation tolerance level
允许能级　allowed energy level
允许能量　allowed energy
孕镶金刚石钻头　impregnated diamond bit
孕镶钻头　impregnated diamond bit
运动波　kinematic wave
运动发酵单细胞菌　zymomonas mobilis
运动稳定性　stability of motion

运行程序　running program
运行方式　mode of operation
运行风速　operational wind speed
运行工况　run condition, mode of operation
运行管理　operation management
运行期　period of duty
运行情况　serviceable condition
运行时间　period of duty
运行特性　operational characteristic
运行调节　operational modulation
运行与维护　operation and maintenance
运行周期　duty cycle
运输线　supply line
运营维护　operation & maintenance
运油飞机　tanker aircraft
运载系统　delivery system
运转时间　run on time

za

匝数比　turns ratio
杂交高粱　hybrid sorghum
杂散电感　stray inductance
杂散响应抑制比　spurious response rejection ratio
杂砂岩　graywacke
杂质的去除过程　impurity elimination process
杂质　impurities, inclusion, impurity
杂质含量　impurity content
杂质浓度　impurity concentration
杂质原子　impurity atom

zai

载波　carrier
载波电流　carrier current
载重水线面　load waterplane
再（加）热　reheat

再充电电流 recharge current
再次环流 tertiary circulation
再辐射损失 reradiation loss
再冷却液 subcooled liquid
再汽化 revaporization
再燃烧 reburning
再热锅炉 reheat boiler
再热循环 reheat cycle
再热蒸汽 reheat steam
再生柴油 renewable diesel
再生电池 regenerative cell
再生技术 renewable technology
再生喷气燃料 renewable jet fuel
再生汽油 renewable gasoline
再生式燃料电池 regenerative fuel cell
再生性 renewability
再生循环 regenerative cycle
再循环电流 recirculation current
再循环燃料 recycled fuel
在半途 halfway
在下游地 downstream
在原地 in situ
在中途 halfway

zan

暂存区 staging area
暂态稳定边界 transient stability limit
暂停模式 pause mode

zao

凿子 chisel, firmer chisel
藻类 algae
藻类高发 algae bloom
藻类能量转换 algae energy conversion
藻类燃料 algae fuel
藻类生物柴油 algae oil based biodiesel
藻类生长潜力 algae growth potential
藻类消化池 algae digester pond
藻类悬浮物 algal suspension
藻脂改质 algal lipid upgrading
皂石 soapstone
造波机 wavemaker
造波水池 wave tank, wave basin
造链机 chain making tool
造林学 silviculture
造山带 orogenic zone
造线机 cable making tool
造渣 slag formation
造纸厂 paper processing plant
噪声电流 noise current
噪声电平 noise level
噪声电压 noise voltage
噪声级 noise level
噪声控制 noise control
噪声能量 noise energy
噪声屏蔽 noise shield
噪声抑制 noise reduction

zeng

增稠剂 thickener
增光屏 intensifying screen
增量式编码器 incremental encoder
增量式输出 incremental output
增强式平板集热器 boost flat plate collector, flat plate collector with planar reflector
增热器 temperature booster
增湿塔 conditioning tower
增速比 speed increasing ratio
增速齿轮 step-up gear
增速齿轮副 speed increasing gear pair
增速齿轮系 speed increasing gear train
增速传动比 gear-up ratio
增速传动装置 step-up gear

增速器　speed increaser
增效　synergy
增压　boost, pressurization
增压泵　booster pump
增压锅炉　supercharged boiler
增压空气　pressurized air
增压器　supercharger
增压热水　hot pressurized water

zha

渣粒　slag particle
轧断的禾草　chopped straw
轧机炉　rolling mill furnace
闸板　damper
闸衬片　brake lining
闸刀　guillotine
闸垫　brake pad
闸阀　gate valve
闸沟　sluiceway
闸瓦　brake shoe
闸式阀　sluice valve

zhai

窄带隙半导电体　narrow band gap semiconductor, narrow gap semiconductor

zhan

斩波电路　chopper circuit
展开　deploy
展向流动　spanwise flow
辗转法　iterative approach
占空比　duty ratio, dutyfactor
栈测试　stack testing
战略环境绩效指标　strategic environmental performance indicator
战略环境评价　strategic environmental assessment
站间脉冲响应法　interstation impulse response method
站列井　standing column well
站柱　standing column

zhang

张拉结构　tensioned structure
张力　tensile force
张力荷载　tension load
张力结构　tensile structure
张力轴　tension axis
张量描述　tensor description
长大率　growth rate
涨波　bulge wave
涨潮　tidal flood, tidal lift, flood tide
涨潮潮量　volume of flood
涨潮发电　flood generation
涨潮界　tidal limit
涨潮力　tidal raising force
涨潮历时　duration of flood
涨潮流　flood current, flood tide current
涨潮时　duration of flood
涨潮线　tideline
涨潮线，逐浪　rising tide
涨落潮流构造　ebb-and-flow structure
涨溢箱　expansion tank
障　hindrance
障碍物　obstacle

zhao

招投标制　bidding and tendering system
着火点　ignition temperature
沼气　biogas (methane gas/sewage gas/swamp gas)
沼产气率　biogas production rate
沼气　methane

沼气池　biogas digester, biogas generator, biogas plant
沼气池涂料　paint for biogas digester
沼气池污泥　biogas muck
沼气灯　biogas lamp
沼气电站　biogas power station
沼气发电　biogas generation, biogas power generation
沼气发酵　biogas fermenting
沼气工程　biogas engineering
沼气供应系统　biogas supply system
沼气浆肥效　fertilizer effect of biogas slurry
沼气脱硫　biogas desulphurizing
沼气脱水　biogas dewatering
沼气微生物　biogas microorganism
沼气厌氧消化　biogas anaerobic digestion
沼气灶　biogas cooker
沼气综合利用　biogas comprehensive utilization
兆比率　parts per million (ppm)
兆焦　megajoule (MJ)
兆瓦　megawatt (MW)
兆瓦级风电机组　megawatt wind turbine
兆兆瓦（特）时　terawatt-hour (TWh)
照度　luminous intensity
照明电池　illuminated cell
照明度　illuminance
照明剂　illuminant
照明器材　lighting fixture
照明用气　illuminating gas
照明装置　illuminator
照射方向自动跟踪　setpoint tracing
照射评价　assessment of exposure
罩温　envelope temperature

zhe

遮蔽　shading, shadow
遮蔽区　sheltered area
遮挡损失　blocking loss
遮挡因子　blocking factor
遮风屏　wind shield
遮光带　shadow band
遮光环　shade ring
遮光片　shade disk
遮光系数　blocking factor
遮阳板　sun shield
遮阳的　adumbral
遮雨缘　weather shed
折流板风力机　shrouded windmill
折流板—风轮面积比　shrouded-to-rotor area ratio
折流板风能发电机　shroud aerogenerator
折流板螺旋桨　shrouded propeller
折流力　shroud force
折曲　flex
折射波　refracted wave
折射波法勘探　refraction survey
折射角　refraction
折射聚光器　refracting concentrator
折射率　refractive index
折射率温度系数　thermal refraction index coefficient
折射面　refracting surface
折射式太阳能聚光器　refracting solar concentrator
折射系数　refraction coefficient
折射指数　refractive index
折尾　furling
褶升区　culmination
蔗秆　bagasse straw
蔗渣　sugarcane bagasse

zhen

针入度仪　penetrometer

针形探头	needle probe
针叶树	softwood
帧	frame
真空泵	vacuum pump
真空玻璃管	glass cylinder with vacuum
真空玻璃管（圆柱面）盖板	evacuated glass tube cover
真空玻璃管集热器	evacuated glass tube collector
真空镀膜	sputtering
真空断路器	vacuum circuit breaker
真空管（状）太阳能集热器	evacuated tubular solar collector
真空管集热器	evacuated tube collector, evacuated tubular collector, vacuum tube collector
真空管式太阳能集热器	vacuum tube solar collector, evacuated-tube solar collector
真空管吸收器	evacuated tube absorber
真空过滤	vacuum filtration
真空混合器	vacuum mixer
真空集热管	evacuated collector tube
真空集热器	evacuated collector
真空搅拌机	vacuum mixer
真空接收器	evacuated receiver
真空精制	vacuum refining
真空熔炼法	vacuum melting method
真空室	vacuum chamber
真空脱气剂	vaccum degasifier
真空蒸镀薄膜	vacuum evaporated film
真内摩擦角	angle of true internal friction
真太阳日	solar day
真太阳时	apparent (solar) time, true solar time
阵点空位	vacant lattice site
阵风	gust
阵风的	gusty
阵风发生器	gust generator
阵风风速	gust speed
阵风风速计	gust anemometer
阵风荷载（工况）	gust loading
阵风极值	extreme gust
阵风记录仪	gust recorder
阵风切片	gust slicing
阵风切片效应	gust slicing effect
阵风速度	gust speed
阵风探测仪	gustsonde
阵风探空仪	gustsonde
阵风系数	gust factor, gustiness factor
阵风响应系数	gust response factor
阵风性	gustiness
阵风影响	gust influence
阵风影响系数	gust factor
阵风载荷	gust load
阵风最大风速	gust peak speed
阵列	array
阵列损失	array loss
振荡波	oscillatory wave
振荡的	oscillatory
振荡电流	oscillating current, oscillatory current
振荡电压	oscillatory voltage
振荡浪涌转换器	oscillating wave surge converter
振荡力	oscillating force
振荡流动	oscillatory flow
振荡器	oscillator
振荡式风能转换系统	oscillating WECS
振荡水柱	oscillating water column
振荡水柱式波能转换装置	（由波浪运动驱动固定在岸边或半潜在海面的腔体内的水柱上下振荡，压迫空腔内的空

气，产生往复气流，推动空气涡轮机发电的装置）oscillating water column wave energy converter
振荡水柱型 oscillating water column
振荡体 oscillating body
振荡运动 oscillatory motion
振动 oscillatory motion, vibratory motion, oscillation
振动的 oscillatory
振动给料机 vibratory feeder
振动级别 vibration level
振动力 oscillating force
振动炉排 vibrating grate
振动模态分析 modal vibration analysis
振动泥浆筛 shale shaker
振动器 shaker
振动窑 oscillating kiln
振动翼 oscillating airfoil
振动源 vibration source
振动阻尼 vibration damping
振幅 oscillation, amplitude
振幅比 amplitude ratio
振幅磁导率 amplitude permeability
振幅控制 amplitude control
振型 mode shape
震波 seismic wave
震波折射测勘 seismic refraction survey
震荡能 vibrational energy
震荡水柱 oscillating water column
震荡水柱系统 oscillating water column system
震动水平 vibration level
震动水翼 oscillating hydrofoil
震源机构 focal mechanism
震源机制 focal mechanism
震中 epicenter

zheng

征地问题 land acquisition issue
蒸镀减反射膜 evaporated anti-reflection coating
蒸发 evaporation
蒸发比率 evaporation ratio
蒸发池 evaporation pond, evaporation tank
蒸发回路 evaporating circuit
蒸发计 evaporogragh
蒸发可能率 evaporation opportunity
蒸发空间 evaporative space
蒸发冷却器 evaporative cooler
蒸发冷却系统 evaporative cooling system
蒸发量 evaporative capacity, steam capacity, steam out-put, steam relieving capacity, rate of evaporation
蒸发率 evaporation rate, evaporative rate, steaming rate
蒸发面 evaporating surface
蒸发凝汽器 evaporative condenser
蒸发器 evaporator, evaporated dish
蒸发器盘管 evaporator coil
蒸发器蛇形管 evaporator coil
蒸发器室 evaporator chamber
蒸发强度 rate of flow
蒸发区 evaporation zone, steaming zone
蒸发热 evaporation heat
蒸发式冷却器 evaporative cooler
蒸发式冷却塔 evaporative cooling tower
蒸发受热面 evaporating heating surface, generating surface
蒸发损失 evaporation loss
蒸发温度 vaporizing point
蒸发系数 evaporation factor

蒸发氧化薄膜	evaporated oxide film
蒸馏	distillation
蒸馏法	distillation
蒸馏气体	still gas
蒸馏燃料油	distillate fuel oil
蒸馏水	distilled water
蒸馏物	distillation
蒸馏过程	distillation process
蒸气压差式盐度差能发电	salinity gradient vapor pressure power conversion
蒸汽包	steam pocket
蒸汽泵	steam pump
蒸汽表	steam table
蒸汽采集系统	steam gathering system
蒸汽参数	steam parameter
蒸汽产量	steam production, output of steam
蒸汽储层	steam reservoir, steam-only reserve
蒸汽储量	steam deposit
蒸汽处理	steam processing
蒸汽纯度	steam purity
蒸汽带	steam zone
蒸汽导电度	steam conductivity
蒸汽的	steamy
蒸汽的溶解固体总含量	total dissolved solid content of steam
蒸汽地热田	vapor-dominated geothermal field
蒸汽电站	steam power station, steam electric generating station
蒸汽动力装置	steam power plant
蒸汽发电厂	steam power plant
蒸汽发生器	steam generator
蒸汽发生系统	steam generation system
蒸汽发生装置	steam generation unit
蒸汽反向流	counterflow of steam
蒸汽分离器	steam separator
蒸汽负荷	steam load
蒸汽改质（过程）	steam reforming
蒸汽改质过程	steam reforming process
蒸汽干度	steam quality
蒸汽干燥	steam drying
蒸汽鼓筒	steam drum
蒸汽管道	steam line
蒸汽锅炉	steam boiler
蒸汽核心流	core stream of vapor
蒸汽回热器	steam reheater
蒸汽加热空气预热器	steam type airheater
蒸汽加热系统	steam heating system
蒸汽加热蒸发器	vapor heated evaporator
蒸汽井	steam well
蒸汽净化	steam purification
蒸汽量	vapor quantity
蒸汽裂口	steam vent
蒸汽流量	steam throughput, steam flow
蒸汽流量表	steam flow meter
蒸汽流量率	steam flow rate
蒸汽轮机	steam turbine
蒸汽喷流	steam jet
蒸汽喷射	steam jet
蒸汽喷射器	steam jet ejector
蒸汽喷嘴	steam jet
蒸汽品质	steam quality
蒸汽驱动	steam drive
蒸汽驱动的	steam driven
蒸汽—燃气联合循环	combined steam and gas turbine cycle
蒸汽散热器	steam radiator
蒸汽闪蒸系统	steam flashing system
蒸汽射流	steam jet
蒸汽生产井	steam production well
蒸汽生产率	steam production rate
蒸汽似的	steamy

蒸汽提升泵　steam lift pump
蒸汽提升器　steam lift
蒸汽田　steam field
蒸汽温度　steam temperature
蒸汽吸热器　steam receiver
蒸汽型地热资源　vapor dominated geothermal resource
蒸汽蓄热器　steam accumulator
蒸汽循环　steam cycle
蒸汽压　vapor pressure
蒸汽压力　steam pressure, vapor pressure
蒸汽压力表　steam gauge
蒸汽压缩　vapour compression
蒸汽压缩装置　vapor compression equipment
蒸汽优势储层的干燥　dry-out of vapor-dominated reservoir
蒸汽羽烟　vapor plume
蒸汽质量　steam quality, vapor quality
蒸汽终湿度　final moisture content
蒸汽转化（过程）　steam reforming
蒸汽转化过程　steam reforming process
蒸汽转化炉　steam reformer
蒸汽转化器　steam reformer
蒸汽总管　steam main
蒸汽钻井　drilling for steam
整流电流　commutating current
整流器　converter, flow straightener
整流器谐波　rectifier harmonics
整流型波能发电站　rectifier wave power plant
整流罩　nose cone, radome
整树片　whole-tree chip
整数　integer
整套启动试运　unit start-up and commissioning
整体二极管太阳电池　integral diode solar cell
整体煤气化　integrated coal gasification
整体煤气化联合循环发电系统　integrated gasification combined cycle
整体转筒　rigid rotor
整圆转子　round rotor
正常大地热流量　normal terrestrial heat flow
正常地面水　normal ground water
正常地热热流量　normal geothermal heatflow
正常电流　normal current
正常风　normal wind
正常负荷　normal duty
正常工作电流　running current
正常关机　normal shutdown
正常检修运行方式　normal maintenance operating plan
正常地热梯度　normal geothermal gradient
正常运行　normal operation
正常制动系　normal braking system
正常状态　normal condition
正齿轮　spur gear
正电极板　positive electrode plate
正电流　positive current
正电压　positive voltage
正电子　positron
正丁烷　n-butane
正断层带　normal fault zone
正多边形　regular polygon
正极　positive electrode, anode
正极电压　positive polarity voltage
正极输出端　positive output terminal
正极输出线接线柱螺帽　positive output terminal nut
正极输出线接线柱扭矩　positive output terminal torque

正极性直流电压　positive polarity dc voltage
正交电流　quadrature current
正交力　perpendicular force
正交流动　crossflow
正交轴线　quadrature axis
正离子反向电流　positive-ion back current
正排量泵　positive displacement pump
正切液压压缩机　tangential hydraulic compressor
正射投影　orthographic projection
正渗透　pressure-retarded osmosis, forward osmosis
正态概率密度函数　normal probability density function
正态分布　normal (Gaussian) distribution
正午太阳入射角　solar noon incidence angle
正弦变化　sinusoidal variation
正弦波　sinewave, sinusoidal wave
正弦波发生器　sinusoidal wave generator
正弦波形　sinusoidal waveform
正弦电流　sinusoidal current
正弦电路　sinusoidal circuit
正弦电压　sinusoidal voltage
正弦输入　sinusoidal input
正弦响应　sine response
正弦信号　sinusoidal signal
正向抽水　direct pumping
正向传递函数　forward transfer function
正向发电　direct generation
正向力　normal force
正向渗透　forward osmosis
正原始光电流　positive primary photoelectric current
正则载荷　regular load
政府间气候变化专门委员会　intergovernmental panel on climate change

zhi

支撑管　standpipe
支撑剂　proppant
支撑结构　support structure
支撑轮毂　supporting hub
支持系统　holding system
支持物　holderiprop
支架　bearer
支流　tributary
支柱　brace
枝晶偏析　dendritic segregation
枝力　principal stress
织布机　loom
织物　fabric, textile, tissue
织物过滤器　fabric filter
脂肪燃料发电厂　fat-fuelled power station
脂肪酸　fatty acid
脂肪酸甲酯　fatty acid methyl ester
脂肪酸氧化　fatty acid oxidation
脂酰脱氢酶　acyl dehydrogenase
脂质　lipid
直齿圆柱齿轮　spur gear
直读式仪表　direct reading instrument
直横 (d-q) 轴模型　d-q axis model
直交流转换器　inverter
直角　right angle
直角投影　orthographic projection
直接传动螺旋桨　direct drive propeller
直接肥料　direct fertilizer
直接辐射　direct radiation, beam insolation
直接辐射计　normal incidence pyrheliometer (NIP)
直接辐射季总量　seasonal total beam radiation

直接辐射日总量　daily total beam radiation, daily total diffuse radiation
直接负载光伏发电　direct-to-load photovoltaic power generation
直接供液蒸发器　direct feed evaporator
直接光束辐射　direct beam radiation
直接还原　direct reduction
直接回路管道　direct return piping
直接混烧　direct co-firing
直接火焰加热　direct firing
直接集热　direct collection
直接加热回转　direct rotary
直接加热式回转干燥器　direct heat rotary dryer
直接接触凝结器　direct contact condenser
直接接触式换热器　direct contact heat exchanger
直接冷却系统　direct cooling system
直接能量强度　direct energy density
直接能隙半导体　direct gap semiconductor
直接能隙材料　direct gap material
直接耦合井水　direct coupling well water
直接膨胀　direct expansion
直接膨胀式低温太阳能热发电　direct expansion low temperature solar thermal power
直接气动波能转换器　direct pneumatic wave energy converter
直接驱动　direct drive
直接驱动泵　direct acting pump
直接燃烧　direct combustion
直接燃烧炉　direct fired oven
直接日射　beam solar radiation
直接日射表　pyrheliometer
（直接日射表）视场角　field of view angle (of pyrheliometer)
直接日射辐照度　direct solar irradiance
直接日射学　pyrheliometry
直接熔炼　direct smelting
直接使用　direct use
直接式太阳辐射吸热器　directly irradiated receiver
直接式蒸发冷却器　direct-type evaporative cooler
直接受益式　direct gain
直接受益型太阳房　direct-gain solar house
直接太阳辐射　direct solar radiation
直接太阳能　direct solar energy
直接太阳能增益　direct solar gain
直接太阳能转换　direct solar conversion
直接吸收接收器　direct absorption receiver
直接系统　direct system
直接旋转钻机　direct rotary rig
直接增益被动式（太阳能）系统　direct-gain passive system
直接增益被动式加热系统　direct-gain passive heating system
直接增益采光面积　direct gain aperture
直接增益建筑　direct gain building
直接增益墙　direct gain wall
直接增益天窗　direct gain skylight
直接增益屋顶　direct gain roof
直接增益系统　direct gain system
直接蒸发　direct expansion
直接转换太阳能电池　direct transition solar cell
直径　diameter
直立　erection
直列翅片太阳能空气加热器　straight finned solar air heater
直列肋片太阳能空气加热器　straight finned solar air heater
直流/交流电压上（下）变换器　DC/AC

	up (down) converter
直流/直流电压上（下）变换器	DC/DC up (down) converter
直流潮流	DC load flow
直流电	direct current
直流电动机	DC motor
直流电机	direct current machine
直流电流	DC current
直流电压	direct voltage
直流电源	DC power supply
直流电整流	DC rectifier
直流电阻率测深	DC resistivity sounding
直流发电机	direct current generator
直流负载	DC load
直流锅炉	once through boiler
直流—交流逆变器	DC/AC inverter
直流冷却	once through cooling
直流流动	once through circulation
直流母线	DC bus
直流式水轮机	straight-flow turbine
直流式太阳能热水器	direct flow solar water heater
直喷分层充气（发动机）	direct injection stratified charge
直驱多极	direct drive multi-pole
直驱式发电机	direct-driven generator
直驱式风电机组	gearless wind turbine generator system, gearless WTGS
直驱式风力发电厂	direct-driven wind turbine generator
直驱无电池风电系统	direct-driven batteryless wind-electric system
直取式热水	direct-piped hot water
直燃式加热炉	direct fired heater
直燃式生物质发电厂	direct-fired biomass power station, direct-fired biomass power plant
直燃式天然气冷冻机	direct-fired natural gas chiller
直烧蒸发器	direct fired evaporator
直射辐照	direct irradiation
直射辐照度	direct irradiance
直射光特性校准系统	beam characterization system
直射太阳辐照	direct insolation
直通式真空集热管	direct flow vacuum tube
直线波动说	linear wave theory
直线发电机	linear generator
直线水力泵	linear hydraulic pump
直线图	straight-line plot
直线液压泵	linear hydraulic pump
直线翼垂直轴风力机	straight-bladed vertical axis wind turbine
直轴	direct axis
直轴磁化电抗	direct axis magnetizing reactance
直轴电抗	d-axis reactance
直轴瞬变时间常数	direct axis transient time constant
植被重建	revegetation
植被指数	vegetation indicator
植冠密度	canopy density
植物物质	vegetal matter
止回阀	check valve
止推轴承	thrust bearing
纸浆	pulp
纸浆厂	pulp mill
纸浆缓冲罐	pulp buffering tank
指令输入	command input
指示灯	display lamp
指数分布	exponential distribution
指向性	directivity
酯化	esterification

酯化作用　esterification
酯基转移　transesterification
酯交换（是用于改变油和脂肪理化特性的处理过程之一，是一种甘油分子上的酰基进行重排的反应）　interesterification
酯交换反应　transesterification
酯类　ester
至点潮　solsticial tide
制程控制　input process quality control
制动机构　braking mechanism
制动力　retarding force
制动扭矩　braking torque
制动器　brake
制动器闭合　brake setting
制动器杆　actuator rod
制动器释放　braking releasing
制动系统　braking system
制动转子　blocked rotor
制冷负荷　refrigeration load
制冷机系统　chiller system
制冷剂　refrigerant
制冷剂管　refrigerant piping
制冷剂蒸汽　refrigerant vapor
制冷设备　chiller plant
制冷系统　chiller system
制冷循环　refrigeration cycle
制氢　hydrogen production
制图板　drawing board
制造　fabrication
制造容差　fabrication tolerance
制作（加工）图　fabrication drawing
质量矩　mass moment
质量流　mass flow
质量流量　mass flow
质量流率　mass flow rate
质量中心　center of mass

质心能量　centre-of-mass energy
质子交换膜燃料电池　proton exchange membrane fuel cell
致冷设备　chiller plant
致密度　consistency
致密化　densification
致密生物质燃料　densified biofuel
智能控制　smart control
智能旋翼桨叶　smart rotor blade
滞后功率因数　lagging power factor
滞后回线　hysteresis loop
滞后时间　dead time
滞后无功电流　lagging reactive current
滞后作用　hysteresis
滞水　stagnant water
滞止层　stagnant layer
滞止尾流　stagnant wake
滞止状态　stagnation condition
置换型固溶体　substitutional solid solution

zhong

中和电压　neutralizing voltage
中和剂　neutralizing agent
中弧线　mean camber line, mean line
中间泵站　intermediate pumping station
中间负荷发电机组　intermediate load generating unit
中间流体海洋温差电站　intermediated fluid OTEC plant
中间套管　intermediate casing
中密度纤维板　medium density fibreboard
中期渗滤液　medium-term leachate
中深波　intermediate depth wave
中途　halfway
中温发酵池　mesophilic digester
中温消化池　mesophilic digester

中线电流　neutral current
中性导体　neutral conductor
中心冲　center puncher
中心等轴晶区　center equiaxial crystal zone
中心电力站　central power station
中心接收集热器　central receiver collector
中心接收器电源装置　central receiver power system
中心接收器定日镜阵列　central receiver heliostat
中心接收器光学系统　central receiver optical system
中心距　center distance
中心控制器　hub controller
中心轮　center gear
中心式卵石床　central pebble rock pile
中心塔太阳能—电能转换系统　central tower solar-electric system
中心塔太阳热电动力站　central tower solar thermal electric powerplant
中心位　center bit
中性（导）线　neutral conductor
中性点　neutral point
中性点有效接地系统　system with effectively earthed neutral
中性空气电池　neutral air cell
中性湍流流动　neutral turbulent flow
中性原子　neutral atom
中性长石　andesite
中性轴　neutral axis
中压　medium-voltage
中央处理器　computer processing unit (CPU)
中央接受式集热器　central reveiver collector
中央空调热泵冷水机组　central air conditioner heat pump water chiller
中央空调系统　central air system
中央立管　central rising pipe
中央收集系统太阳能发电站　central receiver solar-thermal power plant
中央收集系统太阳能热力发电站　central receiver solar thermal power plant
中央主控制箱　central main control cabinet
中叶尖速度比风轮　rotor of medium tip speed ratio
中子　neutron
中子俘获　neutron capture
中子结合能　neutron binding energy
中子流　neutron current
中子吸收　neutron absorption
中子吸收材料　neutron absorbing material
中子吸收剂　neutron absorber
中子衍射　neutron diffraction
终板　endplate
终点挡板　end stop
终点能量　end point energy
终点止动装置　end stop
终端电池　end cell
终端电压　terminal voltage
终端能源　end-use energy
终端设备　end device
终端设施　end-user facility
终端用户　ultimate consumer, end user
终端载荷　end use load
终结　wind-up
终碛　terminal moraine
终止子　terminator
钟形阀　bell valve
种植作物　cropping
重点波　gravity wave

重晶石　baryte
重空穴　heavy hole
重空穴带　heavy hole band
重矿石　heavy mineral
重矿物　heavy mineral
重力波　gravity wave
重力测量　gravity survey
重力场　gravitational field
重力场测量　gravitational field measurement
重力沉降　gravitational settling
重力分选机　gravity separator
重力荷载　gravity load
重力加速度　gravitational acceleration
重力能　gravitational energy
重力热管　gravimetric heat pipe
重力式过滤器　gravity filter
重力式基础　gravity base gravity foundation
重力势能　gravitational potential energy
重力调查　gravity survey
重力位势　gravitational potential
重力循环　gravity circulation
重力油箱　gravity tank
重力铸造机　gravity casting machine
重量百分数　weight percent
重量比　weight ratio
重水冷却反应堆　heavy water cooling reactor
重碳酸盐水　bicarbonate water
重型车床　heavy duty lath
重型汽轮机　heavy frame turbine
重油　heavy oil
重质燃料　heavy fuel

zhou

周变间距　cyclic pitch
周界风　surrounding air
周期不规则性　cyclic irregularity
周期剪应变振幅　cyclic shear strain amplitude
周期性　periodicity
周期性变化　cyclic variation
周期性传热　periodic heat transfer
周期性负荷　periodic duty
周期性工作　periodic duty
周期性脉动风速　periodic fluctuating windspeed
周期性应力极限　cyclic stress limit
周期性源测试　periodic source testing
周期性运行　cyclic service
周期振动　periodic vibration
周圈接缝　circumferential joint
周围风能密度　ambient wind power density
周围空气温度, 环境空气温度　ambient air temperature
周向速率　circumferential speed
周游实验　round robin experiment
周转齿轮　epicyclic gear
周转率　velocity
轴　shaft
轴衬　bushing
轴承　bearing
轴承加工机　bearing processing equipment
轴承密封　bearing seal
轴承磨损　bearing wear
轴承配件　bearing fittings
轴负载故障　axial load failure
轴功率　shaft power
轴流　axial flow
轴流式发动机　axial flow engine
轴流式风机　propeller type fan

轴流式汽轮机　axial flow steam turbine
轴流式水轮机　axial flow water turbine
轴流式水轮机组　axial flow hydroelectric unit
轴流式涡轮机　axial flow turbine, axial turbine
轴扭矩　shaft torque
轴向齿距　axial pitch
轴向磁通电动机　axial flux motor
轴向磁通电机　axial flux machine
轴向磁通感应电机　axial flux induction machine
轴向电流密度　axial current density
轴向负载　axial load
轴向干扰系数　axial induction factor
轴向干涉因子　axial interference factor
轴向荷载　axial load
轴向间隙　axial clearance
轴向间隙感应机　axial gap induction machine
轴向流动推进器　axial flow propeller machine
轴向推力　axial thrust
轴向尾流诱导（扰动）气流　wake axial induced flow
轴座　shaft seat
肘杆泵　toggle pump
肘型尾水管　elbow draft tube
昼夜流　diurnal flow
昼夜平分点　equinox

zhu

逐波调优　wave-by-wave tuning
逐步控制装置　stepwise controllable apparatus
逐层负载分析　zone-by-zone load analysis
逐级控制　step-by-step control

烛式过滤器　candle filter
主厂房　main power house
主触头　main contact
主导风向　dominant wind direction, prevailing wind direction
主电路　main circuit
主动齿轮　driving gear
主动存储系统　active storage system
主动阀　active valve
主动控制　active control
主动偏航　active yaw
主动失速　active stall
主动式（蓄热）墙　active wall
主动式闭环太阳能热水器　closed-loop active solar water heater
主动式集热器　active collector
主动式空气加热集热器　active air-heating collector
主动式冷却　active cooling
主动式太阳房　active solar house
主动式太阳能供暖系统　active solar heating system
主动式太阳能供热　active solar heating
主动式太阳能供热系统　active solar heating system
主动式太阳能加热系统　active solar heating system
主动式太阳能空间加热　active solar space heating
主动式太阳能冷却系统　active solar cooling
主动式太阳能系统（一种由集热器、储热装置和将太阳能转换为热能的传热流体构成的系统）　active solar system
主动式太阳能住宅供暖系统　active residential solar heating system
主动式太阳热设计　active solar thermal

design
主动式太阳热收集装置　active solar thermal collecting device
主动太阳能加热　active solar heating
主动探测器　active sensor
主动调向风轮　yaw active rotor
主动轴　main drive shaft
主断路器　main circuit breaker
主发电机　primary generator
主发动机　main engine
主阀　master valve
主风道　air main
主风向　predominant wind direction
主管道　main pipe
主机　main engine
主接触器　main contactor
主控制系统　master control system
主励磁机　main exciter
主梁　main beam, main spar
主脉动风速　deterministic fluctuating wind speed
主燃区　primary combustion zone
主燃烧器　primary burner
主要功能　main function
主要燃烧室　primary combustion chamber
主要施工技术方案　major construction technical scheme
主要图表　master schedule
主要作业表　master schedule
主叶　master blade
主液压缸　master cylinder
主应力　principal stress
主阵风　deterministic gust
主直流断开　main DC disconnect
主轴　main shaft
主轴接口　spindle interface
主轴螺杆　head screw

主轴密封　main shaft seal
主轴箱　headstock
煮解能　digester energy
煮解器槽　digester tank
助力器　servo actuator
助滤剂　filter aid
助滤器　filter aid
助溶剂　fluxing agent
助溶剂氧化脱硫法　air solutizer-process
住宅供暖系统　house heating system
住宅规模电力系统　residential scale power system
住宅热水系统　house hot water system
住宅用太阳能电池　solar cell module for residential house
住宅用太阳能发电装置　residential solar power device
注出量　flow volume
注浆泵　injection pump
注孔　orifice
注气搅拌机　air-lift type agitator
注射泵　injection pump
注水井　injection well
注油机　lubricator
注油枪　grease gun
驻波　clapotis, standing wave
驻流　standing current
柱面光学系统　cylindrical optical system
柱面光学装置　cylindrical optical mounting
柱面镜　cylindrical mirror
柱面聚光型集热器　cylindrical concentrator
柱销　pin
柱销套　roller
铸锭　ingot

铸钢　cat steel
铸灰口铁　cast gray iron
铸件　casting
铸铝　cast aluminium
铸铝转子　cast-aluminum rotor
铸坯　ingot
铸铜　cast copper
铸造　foundry
铸造厂　foundry
铸造设备　foundry equipment

zhua

爪形（式）离合器　dog clutch

zhuan

专属系统　dedicated system
专用机床　dedicated machine
专用控制系统　specialized control system
专用能源作物　dedicated energy crop, purpose-grown energy crop
专用系统　dedicated system
专用液体循环　dedicated fluid loop
砖石墙　masonry wall
转杯风速表　cup-cross anemometer
转臂起重机　derrick, jib crane
转变热　heat of tansformation
转变熵　entropy of transition
转差功率　slip power
转差率　slip
转潮　change of tide
转潮点　amphidromos
转底炉　rotary hearth furnace
转动不稳定性　rotational instability
转动挡板　flap shutter
转动导流叶片　rotating guide vane
转动动能　rotational kinetic energy
转动刚度　rotational stiffness
转动面　plane of rotation
转动能　rotational energy, rotation energy
转动能级　rotational level
转动频率　rotation frequency
转动叶片　rotating blade
转动叶片式挡板　flap type damper
转动质量　rotating mass
转动轴　rotation axis
转动坐标系　rotating coordinate system
转废为能　energy from waste
转风点　amphidromos
转化程序　conversion process
转化技术　conversion technology
转换电路　switching circuit
转换开关　changeover switch
转换器　translator
转换设备　conversion device
转基因酵母菌株　transgenic yeast strain
转接开关　transfer switch
转矩　torque moment of rotation
转矩变换器　torque converter
转矩波动　torque fluctuation
转矩控制　torque control
转矩脉动　torque pulsation
转矩—速度特性　Torque-Speed Characteristic
转轮式发动机　rotary engine
转轮叶片操作接力器　blade operating servomotor
转轮叶片接力器　runner blade servomotor
转能惯量　rotor inertia
转盘　rotary table, swash-plate
转速　rotational speed
转速控制　shaft speed control, speed control
转速频率控制　speed control with frequency signal

转筒筛　trommel screen
转向齿轮　steering gear
转向机械　steering mechanism
转向架式退火炉　bogie type annealing furnace
转向蜗杆　steering worm
转向系统　steering system
转向轴　steering shaft
转向装置　deflector, steering gear
转叶（旋转叶片）　rotating vane
转移弧　transferred arc
转轴　rotating shaft
转子　rotator
转子半径　rotor radius
转子尺寸　rotor geometry
转子冲片　rotor punching
转子磁场　rotor field
转子磁场线圈　rotor field coil
转子磁极　rotor pole
转子导体　rotor conductor
转子导体电阻　rotor conductor resistance
转子导条　rotor bar
转子电流　rotor current
转子电流控制结构　OptiSlip
转子电阻　rotor resistance
转子动力学　rotor dynamics
转子方位角　rotor azimuth angle
转子惯量　rotor inertia
转子机械速度　rotor mechanical speed
转子角　rotor angle, rotator
转子临界转速　rotor critical speed
转子笼　rotor cage
转子漏磁电抗　rotor leakage reactance
转子轮盘　rotor disk
转子偏心度监视器　rotor eccentricity monitor
转子起动器　rotor spindle

转子绕组　rotor winding
转子式测速仪　rotameter
转子损耗　rotor loss
转子铁芯片　rotor lamination
转子系统　rotor system
转子线圈　rotor winding
转子效率　rotor efficiency
转子芯　rotor core
转子性能模型　rotor performance modelling
转子叶片　rotor blade
转子应力场　rotor stress field
转子制动器　rotor brake
转子轴　rotator axis
转子轴承　rotor bearing
转子轴向推力　rotor axial thrust

zhuang

桩　pier
装备　furnish, harness
装机容量　installed capacity
装料传送机　feeder conveyor
装料斗　feed hopper
装有极轴的集热器　polar mounted collector
装置功率　plant capacity
状态观测器　state observer
状态空间　state space
状态信息　state information
撞击　impinge

zhui

追算　hindcast
追算模型　hindcast model
追踪阵列　tracking array
锥度　taper
锥形波导　tapered waveguide

锥形阀　cone valve
锥形孔道　tapered channel
锥形物　cone
锥形钻头　cone bit
锥子　drill, awl

zhun

准备运行　ready for operation
准定态　stationary state
准静态风载　quasi static wind load
准可再生能源形态　quasi renewable energy form
准抛物槽聚光器　quasi paraboloidal trough concentrator
准抛物槽聚光式集热器　quasi paraboloidal trough concentrator
准稳电流　quasi stationary current
准稳态　quasi steady state
准则　guide rule
准直标定法　normal incidence calibrating method

zhuo

桌型水轮机　table-shaped turbine

zi

资本金制　capital fund rule
资用功（量）　available work
资源不确定性　resource uncertainty
资源潜力　resource potential
(美国)《资源保护与回收法案》　Resource Conservation and Recovery Act
子晶　matted crystal
子样　increment
紫电池　violet cell
紫光太阳电池　violet solar cell
紫外辐射　ultraviolet radiation, UV radiation
紫外光　ultraviolet light
紫外光传感器　ultraviolet light sensor, ultraviolet light transducer
紫外激光器　ultraviolet laser
紫外线　ultraviolet ray
紫外线灯　ultraviolet lamp
紫外线辐射　ultraviolet radiation, ultraviolet
紫外线光谱法　ultraviolet spectroscopy
紫外线能量　ultraviolet energy
紫外线谱带　ultraviolet band
紫外线吸收　ultraviolet absorption
紫外线吸收光谱　ultraviolet absorption spectrum
紫外线照射　ultraviolet radiation
紫外总日射表　ultraviolet pyranometer
自动补水系统　automatic watering system
自动操作　automatic operation
自动车床　automatic lathe
自动电压调节(整)器　automatic voltage regulator
自动调度系统　automatic dispatch system (ADS)
自动对风　self-orientating
自动发电控制　automatic generation control (AGC)
自动防故障系统　fail-safe system
自动防故障装置　fail-safe device
自动防止故障运转　fail-safe operation
自动风速表　anemometrograph
自动给料盘　automatic feed tray
自动跟踪　autotrack
自动关联　auto-correlation
自动海退岩系　auto-regressive series
自动回归　auto-regression

自动接通周期	auto-reclosing cycle
自动净化	self purging
自动滤净膜	self cleaning membrane
自动启动能力	self starting capability
自动倾卸车	tipper truck
自动湿度记录计	hygrograph
自动太阳能跟踪装置	automatic solar tracker
自动调节风车	self regulating windmill
自动调平	self leveling
自动调温器	thermostat
自动调向	self orientating
自动同期系统	automatic synchronized system(ASS)
自动烟气抽风机	automatic flue gas suction fan
自动预热烧嘴	self recuperative burner
自动找平	self leveling
自动遮光装置	solar tracker with shade disk kit
自放电	self-discharge
自辐射	self irradiation
自攻螺丝	expansion bolt, self tapping screw
自行控制	self-acting control
自换相逆变器	self-commutated inverter
自回归序列	auto-regressive sery
自激	self-excitation
自激式洗涤器	impingement scrubber
自给电池	self-contained cell
自记测波仪	wave recorder
自记测风器	wind recorder
自记潮位仪	tide recorder
自记风力表	self-registering anemometer
自记风向计	recording wind vane
自记气压表	self-registering barometer
自记湿度表	self-recording hygrometer
自记温度表	self-registering thermometer
自记验潮仪	tide recorder
自拉	self spring
自来水缓冲器	water buffer
自力推进压捆机	self-propelled baler
自力推进制粒机	self-propelled pelletizer
自励	self-excitation
自励电容	self-excitation capacitance
自励发电机	self-excited generator
自流压	artesian pressure
自能	self-energy
自耦变压器	auto transformer
自喷阶段	flowing phase
自喷井	unloading well
自平衡热线风速计	self-balancing hot wire anemometer
自起动	self-startup
自然变热的地下水	natural heated groundwater
自然电场法	self-potential method
自然电位	self-potential
自然电位法	self-potential method
自然风	natural wind
自然风边界层	natural wind boundary layer
自然风场	natural wind field
自然风环境	natural wind environment
自然干燥	natural drying
自然环境	natural environment
自然界边界层风	natural boundary layer wind
自然冷却	natural cooling
自然频率	natural frequency
自然箝位变换器	natural clamped converter
自然通风	natural draught

自然通风冷却塔　natural draft cooling tower
自然循环　natural circulation, gravity circulation
自然循环太阳热水器　thermosyphon solar water heater
自然循环系统　natural circulation system
自然阵风　natural wind gust
自燃　self-ignition
自燃(着火)点　self-ignition point
自身放电　self-discharge
自锁螺母　locking nut
自调整　self regulation
自同步　self-synchronization, motor synchronizing
自相关函数　autocorrelation function
自养生物　autotrophic organism
自迎风　self orientation
自由(气)流静压力　free stream static pressure
自由表面　free surface
自由表面能　free surface energy
自由电子　free electron
自由度　degree of freedom
自由对流　free convection
自由对流传热　free convection heat transfer
自由风　free stream wind, free wind
自由风轮　free rotor
自由浮碇浮标　free floating moored buoy
自由浮动式波能浮标　free floating wave power buoy
自由管长度　length of free pipe
自由基　radical
自由流　free flow
自由流动涡轮　free flow turbine
自由流风速　free stream wind speed
自由流速度　free stream velocity
自由面　free surface
自由能　free energy
自由膨胀　free expansion
自由偏航系统　free yaw system
自由曲面聚光器　free form concentrator
自由水分　free water
自由体　free body
自由脱扣　trip-free
自由液面　free surface
自由振动　free oscillation
自由转子　free rotor
自有震荡　free oscillation
自走式捡拾压捆机　self-propelled baler
自走式制粒机　self-propelled pelletizer
字节　byte
字模　matrix, typehead

zong

综合废物管理　integrated waste management
综合规划模型　integrated planning model
综合海洋养殖与能源生产　integrating mariculture and energy production
综合耗能　comprehensive energy consumption
综合楼　administration multiple building
综合气象因数　synthetic weather factor
综合图表　master schedule
(欧盟)《综合污染防治指令》　Integrated Pollution Prevention and Control Directive
综合资源规划　integrated resource planning

棕褐色砂岩　tan sandstone
棕黄色砂岩　tan sandstone
棕色砂质泥土　brown sandy clay
棕丝　palm fibre
总变电站　main substation
总出力　gross capability
总传热　overall heat transfer
总传热系数　overall heat transfer coefficient
总动力密度　total power density
总发电量　gross generation
总发电容量　total generating capacity, total power generating capacity
总阀　master valve
总反力　resultant force
总辐射　global radiation, total radiation
总辐射平均强度表指数　helioelectric index
总辐照度　total irradiance, global irradiance
总功率　gross power, gross capacity
总功率损耗　total power loss
总功率消耗　total power consumption
总固体含量　total solid content
总固体量　total solid
总光合效率　overall photosynthetic efficiency
总能量　overall energy
总燃料跳闸　master fuel trip
总热量　gross calorific value
总热效率　gross thermal efficiency
总热阻　entire thermal resistance, overall thermal resistance
总日射表　solarimeter, pyranometer
总日射计　pyranograph
总日射量　solar total radiation
总容量　gross capacity
总溶解固体量　total dissolved solid
总太阳能通量　total solar flux
总太阳热能增益　overall solar heat gain
总线耦合器　bus coupler
总谐波失真　total harmonic distortion
总悬浮固体　total suspended solid
总有机碳　total organic carbon
总闸门　main block valve
总轴　main shaft
总装机容量　total installed capacity
纵的　vertical
纵舵调整器　gyroscope
纵向层压（薄板）　longitudinal lamination
纵向的　longitudinal
纵向涡流　vertical eddy
纵向长的　lengthwise
纵轴　direct axis

zou

走行系统　running gear

zu

族太阳电池　group solar cell
阻垢剂　scaling inhibitor
阻火器　flame arrestor
阻抗　impedance
阻抗电压　impedance voltage
阻力　drag, drag force
阻力板　drag spoiler
阻力损失　resistance head
阻力系数　drag coefficient
阻力型　drag type
阻力型风杯风速计　drag cup anemometer
阻力型风力机　drag type wind machine
阻力型风轮　drag type rotor
阻尼　damping
阻尼板　spoiling flap
阻尼常数　damping constant
阻尼缓冲结构　damped structure

阻尼率 damping ratio
阻尼损失 damper loss
阻尼系数 damping coefficient
阻尼振荡 damped oscillation
阻尼振动 damped vibration
阻气阀 choke valve
阻气门 choke
阻塞效应 blockage effect, blocking effect
阻塞修正 blockage correction
阻性负载 resistive load
阻焰器 flame arrestor
组成材料 constituent material
组合集热器/存储系统 integral collector/storage system
组合式被动太阳房 combination passive solar house
组合式空调机组 air handling unit
组合式浪涌保护器 surge protective device assembly
组合式燃烧器锅炉 combination burner-boiler
组合式太阳能系统 combination solar energy system
组合损失 assembling loss, combined loss
组件表面温度 module surface temperature
组件额定电压 nominal module voltage
组件实际效率 practical module efficiency
组件效率 component efficiency, module efficiency
组元 component
组装式的 modular

zuan

钻 drill
钻床 drill, drilling machine
钻床工作台 drilling machine bench
钻铤 drill collar
钻杆 drill pipe, drill stem
钻规 drill gauge
钻机 drill rig
钻进深度 drilling depth
钻进深度自动记录仪 penetrometer
钻进速度 rate of penetration
（为了了解地质水文而挖的）钻井 borehole
钻井 well
钻井测试 well logging
钻井成本 drilling cost
钻井承包商 drilling contractor
钻井队 drill team
钻井过程 drilling process
钻井记录 well log
钻井技术 drilling technique
钻井绞车 drawwork
钻井经验 drilling experience
钻井孔 wellbore
钻井临时因素 drilling contingency factor
钻井能力 drilling capacity
钻井泥浆 drilling mud
钻井泥浆添加剂 drilling mud additive
钻井偏差 borehole deviation
钻井平台 drilling platform
钻井曲线 drilling curve
钻井权变因素 drilling contingency factor
钻井日志 well log
钻井信息 drilling information
钻井液 drilling fluid
钻井液循环系统 drilling fluid circulating system
钻井意外因素 drilling contingency factor
钻井装置透视图 perspective view of drilling assembly

钻具　drill stem
钻孔　geothermal bore hole, drill, borehole, drill hole
钻孔机　drill, driller
钻孔热阻　borehole thermal resistance
钻孔容量　drilling capacity
钻孔深度　drilling depth
钻孔坍塌　borehole collapse
钻孔套管　well casing
钻孔弯曲　well deviation
钻孔液体　drilling fluid
钻孔者　driller
钻孔直径　well diameter
钻孔直径记录图　caliper log
钻模　jig
钻取法　auger method
钻石刀具　diamond cutter
钻速　drilling rate, drilling velocity
钻台　rig floor
钻探　exploratory drilling
钻探技术　drilling technique
钻探与完井　drilling and completion
钻头　drill
钻头磨损　bit wear
钻头设计　bit design
钻头寿命　bit life
钻芯　drilled rock core
钻液　drilling fluid
钻柱　drill string, drill stem

zui

最初成本　initial cost
最初的　initial
最大波高　maximum wave height
最大波形高度　extreme wave height
最大波周期　maximum wave period
最大测量功率　maximum measured power
最大出功风速　cutout wind speed
最大反射集热器　maximum reflector collector
最大风速　maximum wind speed, maximum wind velocity
最大负荷　peak load
最大负荷功率　peak load power
最大功率　maximum power
最大功率点　maximum power point
最大功率电流　maximum power current
最大功率电压　maximum power voltage
最大功率跟踪　maximum power tracking
最大功率跟踪法　maximum power tracking
最大极限状态　ultimate limit state
最大聚光比　maximum concentration ratio
最大可实现控制技术　maximum achievable control technology
最大连续蒸发量　maximum continuous rating
最大亮度集聚　maximum brightness concentration
最大能量　maximum energy
最大能量密度　maximum energy density
最大逆电流　peak inverse current
最大屏极电流　peak plate current
最大前向电压　peak forward voltage
最大日供应量　maximum daily output
最大设计风速　maximum designed wind speed
最大升力系数　maximum lift coefficient
最大输出功率　maximum power output
最大瞬时风速　peak wind speed
最大退潮（流速）　maximum ebb strength
最大弯曲力矩　maximum bending moment

最大系泊力　maximum mooring force
最大系统电压　maximum system voltage
最大需求量　maximum end-use demand
最大阳极电流　peak plate current
最大英里风　fastest mile wind
最大允许功率　maximum permitted power
最大运行方式　maximum operating plan
最大运行库　maximum operating pool
最大涨潮流速　tidal flood strength
最大阵风　peak gust
最大蒸发量　peak evaporation capacity
最大正向电压　peak forward voltage
最大值　maximum value
最大转速　maximum rotational speed
最低潮　dead tide
最低潮位　extreme low tide
最低过量空气　minimum excess air
最高工作温度　maximum operating temperature, maximum service temperature
最高能量级　maximum energy level
最高平均温度　maximum average temperature
最高日射温度　solar maximum temperature
最高温度　maximum temperature
最高温度表　maximum thermometer
最高效率　peak efficiency
最高阵风风速　gust peak speed
最佳朝向　optimal orientation, optimum facing
最佳出力　optimum output
最佳负荷　optimum load
最佳负载　optimum load
最佳工作电流　optimum operating current
最佳工作电压　optimum operating voltage

最佳集热器面积　optimum collector area
最佳聚光比　optimum concentration ratio
最佳可行控制技术　best available control technology
最佳流速　optimum flow rate
最佳倾角　optimum angle of inclination
最佳倾角（一定时间内固定式太阳能装置获得太阳辐射能最大时与地面的倾角）　optimum tilt angle
最佳使用年限　optimum useful life
最佳嗜热温度　mesophilic temperature optimum
最佳输出　optimum output
最佳效率　optimum efficiency
最佳叶尖速比　optimal tip speed ratio
最佳迎风速度　optimum speed to windward
最佳折射系数　optimum refrative index
最佳蒸汽温度　optimal steam temperature
最佳中温　thermophilic temperature optimum
最佳状态　groove
最深谷底线　thalweg
最先进的发电厂　state-of-the-art power plant
最小触发角控制　minimum α control
最小负荷　minimum load
最小关断角控制　minimum γ control
最小落潮流　minimum ebb
最小入射角　minimum angle of incidence
最小运行方式　minimum operating plan
最小运行库　minimum operating pool
最小涨潮流　minimum flood
最小阻力点　minimal resistance
最优额定风速　optimum rated wind speed
最优功率　optimal power

最优化准则　optimization criterion
最优控制　optimal control
最优叶尖　optimum tip
最优叶尖速度比　optimum tip speed ratio
最终（风能）系统设计　final system design
最终工作温度　final operating temperature
最终需求量　end-use demand
最终压力　resulting pressure

zuo

左手定则　left-hand rule
左旋螺纹　left-hand thread
左右摇摆　yawing
作物残茬　crop residue
作物残渣　crop residue
作业维护　operation & maintenance
作用力矩　applied moment
坐标系　coordinate system
座环　stay ring

附录1 电力工程常用缩略语

a=absolute 绝对的
a=acceleration 加速度
A=accumulator 累加器
A=alarm 报警
A=amplitude 振幅
A=area 面积
A/D=analog to digital 模/数
A/M=automatic/manual 自动/手动
AAC=amplitude absorption coefficient 振幅吸收系数
AAC=automatic amplitude control 自动幅度控制
AAP=associative array processor 相联阵列处理机
AAS=advanced administrative system 先进管理系统
AAS=atomic absorption spectrometry 原子吸收光谱法
AAS=atomic absorption spectroscopy 原子吸收光谱
AAS=automatic addressing system 自动访问系统
AB=address bus 地址总线
ABRO=air bump rinse operation 空气擦洗
ABS=acrylonitrile-butadiene-styrene 丙烯腈—丁二烯—苯乙烯共聚物
abt=about 大约
AC=alternating current 交流（电）
ACC=accepted 已承兑
ACC=accessories 辅助设备，附件
ACC=air cooled condenser 空气冷却器（直接空冷系统）
ACC=automatic combustion control 自动燃烧控制
ACE=automatic computing equipment 自动计算装置
ACOE=automatic checkout equipment 自动检测装置
ACP=asbestos cement pipe 石棉水泥管
ACP=auxiliary control panel 辅助控制（仪表）盘
ACS=auxiliary cooling system 辅助冷却系统
ACSR=aluminum conductor steel reinforced 钢芯铝绞线
ACSS=analog computer subsystem 模拟计算机子系统
ADC=analogue-digital converter 模—数转换器
ADR=American depositary receipt 美国存托凭证
ADS=automatic dispatch system 自动调度系统
AE=air entraining 加气，掺气
AE company=architectural engineering company 工程承包公司，AE公司
af=as fired (basis) 应用基
AF=automatic following 自动跟踪
AF=available factor 可用系数
AFC=automatic following control 自动跟踪控制
AFC=automatic frequency control 自动频率控制
AFC=automatic fuel control 自动燃料控制
AFD=anode-to-film distance 焦距
AFWC=automatic feed water control 自

动给水控制
AGC=automatic gain control 自动增益控制
AGC=automatic generating control 自动发电控制
AH=air preheater 空气预热器
Ah=ampere-hour=amp-hour 安培—小时，安—时
AH hopper=air preheater hopper 空气预热器灰斗
AI=analogue input 模拟量输入，遥测量输入
AID=application industry data 应用工业资料
AIS=alarm indication signal 报警指示信号
alt=altitude 标高，海拔，高程
AM=air mass 空气质量
AMDES=automatic meter data exchange system 自动仪表数据交换系统
amp=ampere 安培
AO=analogue output 模拟量输出
AOL=aircraft obstruction lamp 航空障碍灯
AOP=automatic operation panel 自动操作（仪表）盘
AOP=auxiliary oil pump 辅助油泵
AP=American Patent 美国专利
APC=automatic phase control 自动相位控制
APC=automatic power control 自动功率控制
APC=automatic program control 自动程序控制
APEC=Asia-Pacific Economic Cooperation 亚太经济合作组织
app=apparatus 机器，仪器，设备，仪表，装置
approx=approximately 近似地
APR=automatic reactive power regulator 自动无功功率调整器
ARC=automatic remote control 自动遥控
ARD=application reference data 应用参考资料（数据）
ARS=asbestos roof shingle 石棉屋顶瓦
AS=American Standard 美国标准
AS=auxiliary specifications 辅助规范
AS=auxiliary steam (system) 辅助蒸汽（系统）
asb=asbestos 石棉
ASD=allowable stress design 许用应力设计
ASES=American Solar Energy Society 美国太阳能协会
ASL=above sea level 海拔高度
ASR=asynchronous send/receive 异步发送/接收（装置）
ASS=automatic synchronizing system 自动同步系统
ASSEM=assembly=assemble 组（合）件，部件
ass'y=assembly 组装，装配
ASTM=American Standard of Testing Materials 美国材料试验标准
ATC=acoustical tile ceiling 吸声砖吊顶
ATC=automatic turbine control 汽轮机自动控制
ATF=asphalt tile floor 沥青砖地面
atm=atmosphere 大气
atm=atmospheric pressure （标准）大气压
ATM=Automatic Teller Machine 自动

柜员机
aux=auxiliary 辅助的
avg=average 平均,平均的
AVT=all volatile treatment 全挥发性处理
AWU=air washer unit 新风机组
B&S=beam and stringer 横梁与纵梁
B/C=benefit-cost ratio 效益费用比,益本比
B/L=bill of lading 提货单,正本提单
balun=balanced-unbalanced transformer 平衡—不平衡变压器
BBS=British standard sieve 英国标准筛
BBS=bulletin board system 公告牌系统,电子公告板
BCD=binary-coded decimal 二进制编码的十进制
BCS=boiler control system 锅炉控制系统
BCS=burner control system 燃烧器控制系统
BCT=bushing current transformer 套管式电流互感器
BCU=base controller unit 基本控制器单元
BCU=buffer control unit 缓冲控制器
Be=Baume (degree) 波美度(测量液体相对密度用)
BER=bit error rate 误码率
BF=boiler feedwater (system) 锅炉给水(系统)
BF=boiler follow mode 锅炉跟踪方式
BFBP=boiler feedwater booster pump 锅炉给水前置泵
BFE=boiler front equipment 炉前点火控制设备

BFP=boiler feedwater pump 锅炉给水泵
BFPT=boiler feedwater pump turbine 驱动锅炉给水泵的汽轮机
BFSP=boiler feedwater startup (standby) pump 锅炉给水启动(备用)泵
BFV=butterfly valve 蝶阀
BFW=boiler feed water 锅炉给水
BHN=Brinell hardness number 布氏硬度数
BI=boiler island 锅炉岛
BIL=basic insulation level 绝缘基本冲击耐压水平
BIPV=building integrated photovoltaics 光伏建筑一体化
BIS=bank for international settlement 国际清算银行
BL=boundary line 边界线
BLT=build-lease-transfer 建设—租赁—移交
BMCR=boiler maximum continuous rating (condition) 锅炉最大连续蒸发量(工况)
BMLR=boiler minimum combustion stable load rate 锅炉最低稳燃负荷
BMS=burner management system 燃烧器管理系统
BOD=biochemical oxygen demand 生化耗氧量
BOD=biological oxygen demand 生物需氧量
BOO=build-own-operate 建设—拥有—运营
BOOS=build-own-operate-sell 建设—拥有—运营—出售
BOOT=build-own-operate-transfer 建设—拥有—运营—移交

BOP=balance of plant　电厂配套设施
BOQ=bill of quantity　数量清单
BOT=build-operate-transfer　建设—运营—移交
BOTP=build-operate-transfer project　建设—运营—移交项目
BPS=bit per second　每秒传送位数
BRL=boiler rated load　锅炉额定出力
BSD=British standard dimension　英国度量标准
BSI=British Standards Institution　英国标准协会
BSS=base station system　基站系统
BTG=boiler, turbine and generator　锅炉、汽轮机和发电机
BTO=build-transfer-operate　建设—移交—运营
BTU=British thermal units　英制热单位
BV=ball valve　球阀
BW=butt weld　对接焊
C=carbon　碳
C=centigrade　摄氏
c=channel　槽钢
c=coefficient　系数
c=cold　冷的
C=condensate (system)　凝结水（系统）
c/c=center-to-center　中心距
C/C=center-to-center　中心距
C/C=code converter　代码转换器
C/d=Charge/discharge of the battery　电池充放电
c/f=carried forward　转入下页
c/o=care of　转交
c/s=cycles per second　转/秒
cab=cellulose acetate butyrate　醋酸丁酸纤维素

CAD=computer aided design　计算机辅助设计
CADS=computer aided design system　计算机辅助设计系统
CAF=cost and freight　成本加运费价格
cal=calorie　卡
CAP=computer aided production　计算机辅助生产
CATV=cable television　有线电视
CB=concentration basin　浓缩池
CBA=cost-benefit analysis　费用—收益分析，成本—效益分析
CBD=Central Business District　中央商务区
CBR=California bearing ratio　加州承载比
CCCW=closed cycle cooling water　闭式循环冷却水
CCCW=closed cycle cooling water (system)　闭式循环水冷却（系统）
CCR=capacity continuous rating　能力连续工况
CCR=counter current regeneration　逆流再生
CCS=closed loop control system　闭环控制系统
CCS=coordinated control system　协调控制系统
CCS=unit coordinated control system　单元机组协调控制系统
CCW=counter-clockwise　逆时针
CD-test=consolidated-drained triaxial compression test　固结排水三轴压缩试验
CEA=cost-effectiveness analysis　费用—效果分析
CEMS=continuous emission monitoring

system of flue gas 烟气连续排放监测系统

CEMS=continuously emission monitoring system 连续排放监测系统
CEP=condensate extraction pump 凝结水抽水泵
CES=cement enveloped sand 水泥裹砂（法）（一种喷混凝土的施工工艺）
CES=corporate engineering standard 公司技术标准
CFB=circulating fluidized bed 循环流化床
CFVV=constant flux voltage variation 恒磁通调压
chap=chapter 章，篇
CI=cast iron 铸铁
CIA=cash in advance 预付货款
CIF=cost, insurance and freight 到岸价
CISPR=International Special Committee on Radio Interference 国际无线电干扰特别委员会
CL=center line 中心线
CLE=cycle life expenditure 寿命损耗率
CLK=clock 时钟
CL-test=continual loading test 连续加荷固结试验
cm=centimeter 厘米
CMR=continuous maximum rating 连续最大功率
CNT=carbon nanotube 碳纳米管
co=care of 转交
co=carried over 转入
co=certificate of origin 原产地证明书
co=checkout 检查，调整，测试
COD=cash on delivery 到货付款
COD=chemical oxygen demand 化学需氧量
COD=commercial operation date 商业运营日
COE=cost of electricity 发电成本
const=constant 常数
Corp=corporation 公司
COS=change-over switch 切换开关
CP=candle power 烛光
CP=condensate pump 凝结水泵
CPE=chlorinated polyether 氯化聚醚
CPM=critical path method 关键路径法
CPP=condensate polishing plant 凝结水精处理站
CPS=corporate product specifications 公司产品规范
CPT=cone penetration test 圆锥触探试验
CPU=central processing unit 中央处理单元，中央处理器
CPU=condensate polishing unit 凝结水精处理设备
CPVC=chlorinated polyvinyl chlorite 氯化聚氯乙烯
CQC=complete quadratic combination 完全二次项平方根组合（法）
CR=cold reheat 冷再热，低温再热
CRC=cyclic redundancy check 循环冗余码校验
CRP-test=constant rate of penetration test 等速贯入试验
CRT=cathode ray tube 显示器，阴极射线管
CRU-test=constant rate of uplift test 等速上拔试验
cs=carbon steel 碳钢
cs=cast steel 铸钢

CS=certification specification 认可规范

CS=configuration specification 组态说明书

CS=control switch 控制开关

CSP=concentrating solar power 聚焦式太阳能发电

CST=centistoke 厘沲

CST=condensate storage tank 凝结水箱

CT=current transformer 电流互感器

CT=current transmitter 电流变送器

CTC=competition transition charge 竞争过网收费

CU=coefficient of utilization 利用系数

cu=cubic 立方

CU-test=consolidated undrained triaxial compression test 固结不排水三轴压缩试验

CV=curriculum vitae 履历，简历

CVT=capacitor voltage transformer 电容式电压互感器

CVV=combined voltage variation 混合调压

CW=circulating water 循环水

CW=clockwise 顺时针

CWD=control wiring diagram 控制接线图

CWT=combined water treatment 联合水处理

d=degree 度

D=derivative 微分

D=diameter 直径

D controller=derivative controller 微分调节（控制）器

D/A=digital-to-analog 数/模

D/AC=digital analogue converter 数—模转换器

D/O=delivery order 交货单，提货单

d/p=differential pressure 差压

d/p transmitter=differential pressure transmitter 差压变送器

D-action=derivative action 微分作用

daf=dry ash-free basis 无灰干燥基

DAR=Doppler acoustic radar 多普勒声雷达

DAS=data acquisition, process and supervisory system 数据采集系统

DAS=distribution automation system 配电自动化

dB (a)=A-weighted decibel A加权分贝

DB=data base 数据库

dB=decibel(s) 分贝

DBE=design basis earthquake 设计基准地震

DBF=design basis failure 设计基准故障

DBR=design basis report 基本设计进度报告

DC=direct current 直流（电）

DCC=drag chain conveyer 板捞渣机，刮板式输送机

DCE=data circuit terminating equipment 数据电路终接设备

DCS=distributed control system 分散控制系统

DDC=downhole dynamic compaction 孔内动力压实法

DDF=digital distribution frame 数字配线架

deg=degree 度

DEH=digital electro-hydraulic system 数字电液控制系统

DI=digital input 开关量输入
DM=degraded minute 降级分
DMB=distributed mixing burner 分布式混合燃烧器
DMS=distribution management system 配电管理系统
DN=domain name 域名
DNI=direct normal insolation 直射太阳辐射通量（强度）
DO=digital output 开关量输出，数字量输出
DOE=Department of Energy 美国能源部
DP=drip-proof 防滴型
DPC=damp proof course 防潮层
DPDM=digital pulse duration modulation 数字脉冲宽度调制
DPDT=double-pole double-throw 双刀双掷（开关）
DPE=data processing equipment 数据处理装置
DPG=digital pattern generator 数字模式发生器
DPM=data processing machine 数据处理机
DPO=delayed pulse oscillator 延迟脉冲振荡器
DR=data receiver 数据接收器
DR=data recorder 数据记录器
DR=data register 数据寄存器
DR=digital resolver 数字分解器
DTE=data terminal equipment 数据终端设备
DTS=dispatcher training system 调度员培训系统
dwg=drawing 图纸
e.g.=exempli gratia 例如

EA=environmental assessment 环境评价
EAR=engineering action request 技术工作请求书
EAT=earnings after tax 税后收益额
EBIT=earnings before interest and taxes 息税前利润
ECR=economical continuous rating 经济连续出力
ECU=European Currency Unit 欧洲货币单位
ED=economic dispatching 经济调度
ED=electrodialysis 电渗析
ED=engineering description 技术说明书
EDC=economic dispatch control 经济调度控制
EDF=energy delivery factor 能源输送系数
EDI=electrodeionization 电除盐
EDR=electrodialysis reversal 倒极电渗析
EDR=engineering documentation release 技术文件发布
ee=errors excepted 允许误差
EEC=European Economic Community 欧洲经济共同体
eff=efficiency 效率
EHC=electrohydraulic control 电液控制
EHP=effective horsepower 有效功率
EHV=extra-high voltage 超高压
EHWL=extreme high water level 极高水位
EHX=external heat exchanger 外部热交换器
EIA=environmental impact assessment

环境影响评价
EIS=environmental impact statement 环境影响报告书
EL=elevation 标高
ELCB=earth leakage circuit breaker 对地漏电断路器，接地保护断路器
EMF=electromagnetic field 电磁场
EMF=electromotive force 电动势
EMI=electric magnetic interference 电磁干扰
EMS=energy management system 能量管理系统
EMS=express mail service 邮政特快专递
ENE=east-northeast 东北偏东
EOT=extension of time 工期延长
EPACT=Energy Policy Act 能源政策法
EPC=engineering procurement and construction 设计—采购—建造总承包，工程总承包
EPROM=electrically programmable read only memory 电可编程只读存储器
EPS=earnings per share 每股盈利
EQ=environmental quality 环境质量
ERP=Electric Reliability Panel 可靠性小组（隶属北美电力可靠性协会）
ERS=earth resources satellite 地球资源卫星
ES=extraction steam (system) 抽汽（系统）
ESD=electrostatic discharge 静电放电
ESDS=electrostatic discharge sensitivity 静电放电敏感度
ESE=east-southeast 东南偏东

ESP=electrostatic precipitator 静电除尘器，电除尘器
ESS=environmental survey satellite 环境勘测卫星
ESV=emergency stop valve 危急关闭阀，紧急切断阀
etc.=et cetera 等等
ETP=effluent treatment plant 污水处理设备
ETS=emergency trip system 事故跳闸系统
EWD=elemental wiring diagram 原理接线图
EWS=engineer working station 工程师工作站
exp=exponential function 指数函数
F=Fahrenheit 华氏（温度）
f=foot 英尺
f=force 力
f=frequency 频率
FAS=free alongside ship 船边交货价格
FAT=factory acceptance testing 工厂验收测试
FB=fluidized bed 流化床
FBHE=fluid bed heat exchanger 流化床换热器
FC=fail close 故障关
FC=foot-candle 英尺—烛光
FC=high-voltage current limiting fuse and vacuum contactor 高压限流熔断器及真空接触器
FCAN=full capacity above normal 高于正常值的全容量
FCB=fast cut back 速切负荷，快速切回，快速减负荷
FCBN=full capacity below normal 低

于正常值的全容量
FCS=field bus control system 现场总线控制系统
FDF=forced draft fan 送风机
FF=Field Bus Foundation=Foundation Field Bus 现场总线基金会（国际性组织）
FF=flat flange 平（焊）法兰
FGD=flue gas desulfurization 烟气脱硫
FIB=free into barge 驳船上交货价格
FIDIC=International Federation of Consulting Engineers 国际咨询工程师联合会（简称"菲迪克"）
FIFO=first in first out 先进先出
fig=figure 图，数值，形状，外形，轮廓
FIRR=financial internal rate of return 财务内部收益率
FL=floor 层
FM=frequency modulation 调频
FNPV=financial net present value 财务净现值
FO=fail open 故障开
FOA=forced oil air cooled 强油风冷
FOA=forced oil cooled with forced air cooler system 强油风冷系统
FOB=free on board 离岸价格，船上交货价格
FOR=free on rail 火车上交货价格
FOS=free on ship 船上交货价格
FOW=free on wagon 火车（货车）上交货价格
FRP=fiber-glass reinforced plastic 玻璃钢
FS=feasibility study 可行性研究
FSK=frequency shift keying 移频键控

FSS=fuel safety system 燃料安全系统
FSS=furnace safeguard system 炉膛安全系统
FSSS=furnace safeguard supervisory system 炉膛安全监控系统
FST=fuel-sodium reaction 燃料—钠反应
FSU=final signal unit 最终信号单元
FTA=fault tree analysis 故障树分析
FTC=fast time constant 快速时间常数
FTC=frequency time control 频率时间控制
FTL=fault-tree-linking method 故障树连结法
FTM=frequency time modulation 频率时间调制
FTP=file transfer protocol 文件传输协议
FTS=fuel transfer system 燃料输送系统
FW=feed water 给水
G=generator 发电机
g=gram 克
g=gravity 重力
GAC=gap automatic control (system) 间隙自动控制（系统）
GATT=general agreement on tariffs and trade 关税及贸易总协定
GDC=general design criteria 总设计准则
GDP=gross domestic product 国内生产总值
GIS=gas insulated switchgear 气体绝缘组合电器设备，气体绝缘开关设备
GIS=geography information system 地理信息系统

GLP=general layout plan 总布置平面图
GNE=gross national expenditure 国民总支出
GNI=gross national income 国民总收入
GNP=gross national product 国民生产总值，国民总产值
GP=generalized programming 通用程序设计
GPAC=general purpose analog computer 通用模拟计算机
GPC=general peripheral control 通用外围控制
GPC=general purpose computer 通用计算机
GPDC=general purpose digital computer 通用数字计算机
GPIB=general purpose interface bus 通用接口总线
GPKD=general purpose keyboard and display control 通用键盘和显示控制器
GPL=general purpose language 通用语言
GPL=generalized programming language 通用程序设计语言
GPL=graphic programming library 图形程序设计库
GPM=general purpose macrogenerator 通用宏生成程序
GPM=general purpose macroprocessor 通用宏处理程序
GPP=general purpose processor 通用处理机
GPS=global position system 全球定位系统

grad=gradient 梯度
GSH=gland steam heater 汽（轴）封蒸汽加热器
GW=gigawatt (=1000000 kilowatts) 千兆瓦，十亿瓦特
GW=gross weight 毛重
H=enthalpy 焓
H=head 压头
h=hour 小时
H=hydrogen 氢，氢气
HBR=Brinell hardness 布氏硬度
HD=heater drain and vent (system) 加热器疏水及放气（系统）
HH=high-high 高高（极高）
HH nut=hexagon-headed nut 六角螺母
HMI=human machine interface 人机界面
HMS=human machine system 人机系统
HP=high pressure 高压
HPC=high pressure cylinder 高压缸
HPRS=high pressure recirculation system 高压再循环系统
HR=harmonic ratio 谐波含有率
hr=hour 小时
HRSG=heat recovery steam generator 余热锅炉
HSR=historical data storage and retrieval 历史数据存储和检索
HTR=heater 加热器
HTTP=hypertext transfer protocol 超文本传送协议
HV=high voltage 高压
HVAC=heating, ventilation and air conditioning 采暖，通风和空气调节
HVDC=high-voltage direct-current tran-

smission 高压直流输电
Hz=hertz 赫兹，周/秒
I controller=integral controller 积分控制（调节）器
I&C=instrumentation and control 仪表和控制
I&C island=instrumentation and control island 仪表控制岛，I&C 岛
i.e.=id est 即
I/O=input/output 输入/输出
I/P=electric to pneumatic 电/气
IA=instrument compressed air (system) 仪用压缩空气（系统）
IBR=Indian Boiler Regulations 印度锅炉规程
ICB=international competitive bidding 国际竞争性招标
ICC=International Chamber of Commerce 国际商会
ICP=international comparison project 国际比较项目
ID=identification 标志
ID=inside diameter, internal diameter, inner diameter 内径
id.=idem 同上，同前
IDC=interest during construction 建设期利息
IDF=induced draft fan 引风机
IF=intermediate frequency 中频
IGCC=integrated gasification combined cycle 整体煤气化联合循环
IOU=investor owned utility 投资者所有电力公司
IP=intermediate pressure 中压
IP=international protection 国际防护
IP code=international protection code 国际防护代码

IPB=isolated phase bus duct 离相封闭母线
IPC=integrated pollution control 污染综合治理
IPP=independent power producer 独立发电商，独立电力生产者
IPR=intellectual property right 知识产权
IPTS—68=International Practical Temperature Scale of 1968 国际实用温标（IPTS—1968）
IQC=incoming quality control 进料品质控制
IR=internal report 内部报告
IRR=internal rate of return 内部回收率，内部效益率
ISCC=integrated solar combine cycle 整体太阳能联合循环
ISO=independent system operator 独立系统调度机构
ISP=internet service provider 因特网服务提供者，因特网服务提供机构
ISU=idle signal unit 闲置信号单元
ISU=initial signal unit 初级信号单元
IT=information technology 信息技术
ITC=international tendering company 国际招（投）标公司
ITO=Indium-Zinn-Oxid=indium tin oxide 铟锌氧化物
J=joule 焦尔
K=Kelvin 开尔文
kcal=kilocalorie 千卡
kg=kilogram 千克
KGV=knife gate valve 刀型闸阀
kJ=kilojoule 千焦
km=kilometer 千米
KSR=keyboard send/receive 键盘发

送/接收装置
kV=kilovolt 千伏
kVA=kilovolt ampere 千伏安
kvar=kilovar 千乏
kW=kilowatt 千瓦
kWh=kilowatt-hour 千瓦时
l=length 长度
l=liter 升
L/C=letter of credit 信用证
L/G=liquid to gas ratio 液气比
LAD=laboratory application data 实验室应用资料
LAN=local area network 本地局域网
lb=pound 磅
LCB=local competitive bidding 国内竞争性招标
LCD=liquid-crystal display 液晶显示（器）
LCL=local 就地
LCOE=levelized cost of energy 投资费用平衡
LDD=luminaire dirt depreciation 照明器污秽减光系数
LED=light emitting diode 发光二极管
LG=letter of guarantee 保证书，保函，信用保险证书
LH=left hand 左手
LIB=limited international bidding 有限国际招标
LIBOR=London inter-bank offered rate 伦敦银行拆借利率
LIFO=last in, first out 后进先出
LILIS=international and local inquiry shopping 国际国内询价采购
LL=low-low 低低，极低
LLD=lamp lumen depreciation 光通量衰减系数

LOCA=loss-of-coolant accident 冷却剂丧失事故
LP=low pressure 低压
LPG=liquefied petroleum gas 液化石油气
LRFD=load and resistance factor design 荷载和抗力系数设计
LSB=last-stage bucket 末级叶片
LSU=lone signal unit 单一信号单元
Ltd=limited 有限的
LTR=loop transfer recovery 回路传输恢复，环路传递复现，回路传递恢复，回路转移函数回复
LVRT=low voltage ride through 低电压穿越
LX=lux 勒克司
M=mass 质量
M=mega 兆
m=meter 米
M=module 模数（件）
M=mole 摩尔
M=moment 力矩
M/A=manual/automatic 手/自动
m/s=meters per second 米/秒
mA=milliampere, milliamp 毫安
MAE=mean absolute error 平均绝对误差
magamp=magnetic amplifier 磁放大器
max=maximum 最大
M-BFP=motor driven boiler feedwater pump 电动锅炉给水泵
MCB=miniature circuit breaker 微型断路器
MCC=maintenance control center 维护控制中心
MCC=motor control center 电动机控制中心

MCCB=moulded case circuit breaker 塑壳式断路器
MCD=manual control device 人工控制设备
MCDR=maintenance control data register 维护控制数据寄存器
MCES=main condenser evacuation system 主冷凝器抽真空系统
MCHF=maximum critical heat flux 最大临界热流密度
MCHFR=minimum critical heat flux ratio 最小临界热流密度比
MCK=maintenance check 维护检查
MCP=measure-correlate-predict 测量—相关性分析—预测
MCR=maximum continuous rating 最大连续出力
MCR=maximum continuous revolution 最高连续转速
MCS=modulation control system=modulation control system 模拟量控制系统
MDBFP=motor driven boiler feedwater pump 电动锅炉给水泵
MEH=modular electro-hydraulic control system=DEH system for boiler feedwater pump turbine 小汽机数字电液控制系统
MF=maintenance factor 维护系数
MFT=main fuel trip 总燃料跳闸
MFT=master fuel trip 总燃料跳闸
mg=milligram 毫克
MHC=mechanical hydraulic control 机械液压式控制
MHF=mounting height above the floor 距地面安装高度
MHWP=mounting height above the work-place 距工作面安装高度
MI=master instruction 使用说明书
min=minimum 最小
min=minute 分钟
MIS=management information system （厂级）管理信息系统
MIS=metal-insulator-silicon 金属绝缘硅
misc=miscellaneous 杂项，其他
ml=milliliter 毫升
mm=millimeter 毫米
MM=modified Mercalli scale 修订麦加利地震烈度，修订麦加利地震烈度表
MMI=man-machine interface 人机交换界面（接口）
MOA=metal oxide arrester 金属氧化物避雷器
MOC=mechanism operated cell switch 机构操作式开关
MOV=motor operated valve 电动（阀）门
MPa=megapascal 兆帕［压强单位］
MPF=maximum probable flood 最大可能洪水
MPP=max power point 最大功率点
MS=main, reheat and by-pass steam (system) 主蒸汽,再热及旁路（系统）
MS=manufacturing specifications 制造规范
MS=mild steel 低碳钢
msl=mean sea level 平均海拔
MSV=main stop valve 主汽门
MTBF=mean time between failure （两次间隔）平均无故障工作时间
MTTR=mean time to repair 平均故障修复时间
MTTR=mean time to restore 平均修复

时间
mV=millivolt 毫伏
MW=Megawatt (=1000 kilowatts) 兆瓦
MWP=maximum working pressure 最大操作（工作）压力
NA=Avogadro's constant 阿伏伽德罗常数
N=negative 负的
N=Newton 牛顿
N/A=not available, not applicable 不供的，不用的，不适用的
NC=normally closed 常闭
ND=not detected 未检出
NDE=normal de-energized 正常失电
NDI=non-destructive inspection 无损检验
NDT=non-destructive testing 无损探伤
NE=normal energized 正常带电
NETA=new electricity trading arrangement 电力交易新模式
NGR=neutral grounding resistor 中性点接地电阻
NMV=normal mode voltage 串模电压
NO=normally open 常开
No=number 数字，编号
NPB=nonsegregated phase bus duct 共箱母线
NPSH=net positive suction head 净正吸入头
NPV=net present value 净现值
NRC=National Research Council 美国国家研究委员会
NREL=National Renewable Energy Laboratory 国家再生能源实验室
NRV=non-return valve 止回阀，单向阀

NTS=not to scale 不按比例
Nu=Nusselt number 努谢尔特数
NW=net weight 净重
O&M=Operation and maintenance 运行维护
o/d=on demand 即付
O/D=over draft 透支
OA=office automation 办公自动化
OCCW=open cycle cooling water system 开式循环冷却水系统
OCCWP=open cycle cooling water pump 开式循环冷却水泵
OCO=open-close-open 开闭开
OCO=open-close-open contact 开闭开接点
OCS=optical control system 光控制系统
OD=operational directive 执行导则
OD=outside diameter, outer diameter 外径
OD=outside dimension 外部尺寸
ODF=optical distribution frame 光配线架
OFA=overfire air 燃尽风
OFT=oil fuel trip 燃料油跳闸
OHM=ohmmeter 欧姆表
OLTC=on-load tap-changer 有载分接开关
OPC=overspeed protection control 超速保护控制
OPEC=Organization of Petroleum Exporting Countries 石油输出国组织
OPF=optimal power flow 最优的电力潮流
OPGW=composite optical fiber groundwire 地线复合光纤
OPR=organizational peer review 组织

同行评审
OPT=overspeed protection trip 超速跳闸保护
ORP=oxidation-reduction potential 氧化—还原电位
ORP transmitter=oxidation-reduction potential transmitter 氧化—还原电位变送器
OSART=operational safety review team （核电厂）运行安全评议（检查）组
OSI=open system interconnect 开放系统互连
P=per 每
P=poise 泊［黏度单位］
P=power 功率
P=pressure 压力
P=pump 泵
P&ID=piping and instrument diagram 管道及仪表流程图
P&ID=process and instrument diagram 工艺及仪表流程图
Pa=pascal 帕
PA=polyamide 聚酰胺
PAD=product application data sheet 产品应用资料
PAF=primary air fan 一次风机
PAYE=pay as you earn 预扣所得税
PB=polybutylene 聚丁烯
PBT=pay back time 投资回收期
pc=pitch circle 节圆
PC=polycarbonate 聚碳酸酯
PC=power center 动力中心
PC=project controlling 项目总控制
PC=pulverized coal 煤粉
PCB=polychlorinated biphenyl 多氯联苯
PCI=pulse counter input 脉冲输入
PCV=pressure control valve 压力控制阀
PD=proportional plus derivative 比例微分
PDM=plant design memorandum 电厂设计备忘录
PDR=post disturbance review 事故追忆
Pe=Peclet number 贝克列数（雷诺数×普朗特数）
PE=polyethylene 聚乙烯
PE=power exchange 电力交易（所）
PERT=program evaluation and review technique 计划评审技术
PFBC=pressured fluidized bed combustion 增压流化床燃烧
PG=performance guarantee 性能保证
PHE=plate heat exchanger 板式热交换器
PI=proportional plus integral 比例积分
PI=pulse input 脉冲量输入，电能量输入
PID=proportional plus integral plus derivative 比例积分微分
PL=parts list 零部件清单
PLC=power line carrier 电力线路载波机
PLC=power-line carrier channel 电力线载波通道
PLC=programmable logic controller 可编程控制器
PLC=public limited company 股份有限公司
PM=particulate matter 颗粒物
PM=project management 项目管理
PMC=process management and control 过程管理和控制
PMC=project management contractor

项目管理承包商
PMF=probable maximum flood 可能最大洪水
PMMA=polymethylmethacrylate 聚甲基丙烯酸甲酯，有机玻璃
PMP=probable maximum precipitation 可能最大降水（量）
PMR=post mortality review 事故追忆
PMU=project management unit 项目管理机构
PO=polyolefin 聚烯烃
PO=post office 邮政局
PO=postal order 邮政汇票
POQAP=plant operational quality assurance procedure 电厂运行质量保证程序
PORC=plant operation review committee 电厂运行审评委员会
PORT=photo-optical recorder tracker 光电记录跟踪装置
PP=polypropylene 聚丙烯
PPA=power purchase agreement 购电合同
PPL=polypropylene lined 衬塑
PPS=plant protection system 电厂保护系统
PROM=programmable read-only memory 可编程序只读存储器
PS=polystyrene 聚苯乙烯
PS=price sheet 价目单
PSK=phase shift keying 移相键控
PSS=power system stabilizer 电力系统稳定器
PSS=product specification sheet 产品说明书
PT=potential transformer 电压互感器
PTFE=polytetrafluoroethylene 聚四氟乙烯
PUV=per unit value 标幺值
PV=photovoltaic 光电池的，光致电压的，光生伏打的
PVC=polyvinyl chloride 聚氯乙烯
PVDF=polyvinylidene fluoride 聚偏二氟乙烯
PVF=polyvinyl fluoride 聚氟乙烯
PWM=pulse width modulation 脉冲宽度调制
Q=quantity 量
QA=quality assurance 质量保证
QBS=quality-based selection 根据质量选择
QC=quality control 质量控制
QF=qualifying facility 资格认证手段
QFs=qualified facilities 限定设备
QI=quality index 质量指数，质量指标
QMS=quality management system 质量管理体系
R=Rankine degree 朗肯温度
R=Reaumer degree 列氏温度
R&D=research and development 研究与开发
R/A=revision appendix 修改附件
RA=reliability analysis 可靠性分析
rad=radian 弧度
RAM=random access memory 随机存取存储器
RAS=reliability, availability, serviceability 可靠性，可用性，可维护性
RB=runback 快速减负荷
RC=reinforced concrete 钢筋混凝土
RC=Rockwell hardness C 洛氏硬度C
RD=rundown 迫降（负荷）
rebar=reinforcing bar 钢筋
Ref=reference 参考，标准

REV No=revision number 修改版号
RFI=radio frequency interference 射频干扰
RH=reheat 再热
RH=right hand 右手
RH=Rockwell hardness 洛氏硬度
RMS=root mean square 有效值，均根
RN=Reynolds number 雷诺数
RO=reverse osmosis 反渗透
ROE=return on equity 资本收益率
ROM=read only memory 只读存储器
rpm=revolutions per minute 转/分
RSPL=recommended spare parts list 推荐的备件清单
RTD=resistance temperature detector=resistive thermal detector 热电阻
RTG=regional transmission group 地区性输电集团
RTU=remote terminal unit 远方终端机
RU=run up 迫升（负荷）
RV=relief valve 减压阀
S=entropy 熵
s=second 秒
S=south 南
S/MH=spacing-to-mounting height ratio 距高比
SA=security analysis 安全分析
SAFP=shop assembled fabricated piece 车间组装的制作件
SAMA=Scientific Apparatus Manufacturer Association 科学仪器仪表制造商协会
SAT=site acceptance testing 工地验收测试
SAT=site availability test 现场可利用率试验

SCADA=supervisory control and data acquisition 监视控制和数据采集系统
SCR=selective catalytic reduction 触媒式脱硝工艺
SCS=sequence control system 顺序控制系统
SDD=system design description 系统设计说明
SDI=silt density index 污染密度指数
SDP=standard depth of penetration 标准穿透深度
SDR=special drawing right 特别提款权
SE=state estimate 状态估计
SE by E=southeast by east 东南偏东
sec=second 秒
SEGS=solar energy generating system 太阳能发电系统
SER=sequence event recorder 事故顺序记录仪
SI=special information 专用信息
SI=special instruction 专用说明书
SI=System International 国际单位制，公制
SI unit=standard international unit 标准国际单位制
SIL=switching impulse level 操作冲击水平
SOC=state of charge of the battery 电池充电状况
SOE=sequence of event 事故顺序记录
SPC=stored-program control 存储程序控制
SPDT=single pole double throw 单刀双掷

SPEC No=specification number 规范号
SPL=sound pressure level 声压级
SPST=single pole single throw 单刀单掷
SPT=standard penetration test 标准贯入试验
SR=standard rating 额定容量
SR=styrene-rubber 苯乙烯橡胶
SRSS=square root of the sum of squares 平方和开平方组合法
SS=select switch 选择开关
SS=stainless steel 不锈钢
SS=suspended solid 悬浮物
SSC=submerged scraper conveyer 水浸式刮板捞渣机
SSU=subsequent signal unit 后续信号单元
SSU=synchronization signal unit 同步信号单元
SV=safety valve 安全阀
SV=slide valve 滑阀
SV=sluice valve 闸阀
SV=solenoid valve 电磁阀
SV=stop valve 主汽阀
SVP=service processor 服务器
SVP=stop-valve pressure 主汽门压力
SWAS=steam and water analysis system 汽水分析系统
SWR=standing-wave ratio 驻波比
t=temperature 温度
t=time 时间
t=ton 吨
T&D=transmission and distribution 输电和配电
T/T=telegraphic transfer 电汇
TAC=Tariff Advisory Committee 关税咨询委员会
T-BFBP=turbine driven boiler feedwater booster pump 汽动锅炉给水前置泵
T-BFP=turbine driven boiler feedwater pump 汽动锅炉给水泵
TC=thermocouple 热电偶
TCC=transmission congestion contract 输电阻塞合同
TCP=transmission control protocol 传输控制协议，传送控制协议
TDH=total dynamic head 总动压头
TDM=transient data management 瞬态数据管理系统
TDM=turbine transient data management system 汽轮机瞬态数据管理系统
TDS=total dissolved salt 总含盐量
TDS=total dissolved solid 总溶解固体物
TE=temperature element 温度元件，温感元件
TEFC=totally enclosed with fan cooler 全封闭风冷型
temp=temperature 温度
TEWAC=totally enclosed with water cooled air cooler 全封闭水空冷型
TF=turbine follow mode 汽轮机跟踪方式
TG=turbo-generator 透平发电机
TI=technical information 技术资料
TI=turbine island 汽机岛
TMCR=turbine maximum continuous rating 汽机最大连续出力
TOC=truck operated cell switch 小车操作式开关
TOT=transfer-operate-transfer 转让—经营—转让
TQC=total quality control 全面质量管理
TSI=turbine supervisory instrument 汽轮机监视仪表

TSI=turbine supervisory instrumentation 汽轮机（本体）监视仪表
TSP=total suspended particulates 总悬浮颗粒物
TSS=total suspended solid 总悬浮物
UAT=unit auxiliary transformer 厂用变压器
UBC=Uniform Building Code 统一建筑法规（美国）
UCE=unit cost of energy 能量单位成本
UCR=unit control room 单元控制室
UDC=utility distribution company 配电公司
UD-test=unconsolidated-drained test 不固结排水剪切试验
UHV=ultra-high voltage 特高压
UN=United Nations 联合国
UNIDO=United Nations Industrial Development Organization 联合国工业及发展组织
UPC=universal product code 条形码
UPS=uninterrupted power supply 不间断电源
USCS=unified soil classification system 土的统一分类
UU-test=unconsolidated-undrained test 不固结不排水剪切试验
V=vacuum extraction 抽真空
V=velocity 速度
V=volt 伏特
V=voltage 电压
V=volume 容积
VA=volt ampere 伏安
VAT=value-added tax 增值税
VCB=vacuum circuit breaker 真空断路器

VDF=voice distribution frame 音频配线架
VFVV=variable flux voltage variation 变磁通调压
VHN=Vickers hardness number 威氏硬度
VIP=very important person 贵宾，重要人物
Vol=volume 卷
VPH=Vickers pyramid hardness 威氏角锥硬度
VWO=valve wide open 阀门全开
VWO=valve wide open capacity 汽阀全开容量
W=watt 瓦特
W=work 功
WACC=weighted average cost of capital 资本的加权平均成本
WAIS=wide area information server 广域信息服务系统，广域信息服务器
WC=water closet 盥洗室，（冲水）厕所
WECS=wind energy conversion system 风能转换系统，风力发电机组
WHO=World Health Organization 世界卫生组织
WL=water level 水位
WTGS=wind turbine generator system 风电机组
WTO=World Trade Organization 世界贸易组织
WWW=world wide web 环球网，万维网，全球浏览系统
XLPE=cross linked polyethylene 交联聚乙烯
Yd=yard 码
ZT=position transmitter 位置变送器

附录2 电力工程常用计量单位

量的名称	量的符号	计量单位		应废除的单位	
		名称	符号	名称	符号
[平面]角 angle(plane angle)	$\alpha,\beta,$ $\gamma,\theta,$ Φ	弧度	rad		
		度 [角]分 [角]秒	° ′ ″		
立体角 solid angle	Ω	球面度	sr		
长度 length	l, L	千米 米 厘米 毫米	km m cm mm	埃[①] 费密 尺	Å
		海里	n mile		
面积 area	A, S	平方米 平方厘米 平方毫米	m^2 cm^2 mm^2	靶恩[①]	b
		公顷	hm^2	亩	
体积 volume	V	立方米 立方厘米	m^3 cm^3		
		升	l, L		
时间 time	t	秒	s		
		分 [小]时 日,(天)	min h d		
速度 velocity	v c u, v, w	厘米每秒 米每秒	cm/s m/s		
		千米每[小]时 节	km/h kn		
加速度 acceleration	α	米每二次方秒	m/s^2	伽[①]	Gal
频率 frequency	f, ν	赫[兹]	Hz		

续表

量的名称	量的符号	计量单位		应废除的单位	
		名称	符号	名称	符号
旋转频率 rotational frequency	n	每秒	s^{-1}		
		转每分 转每秒	r/min r/s		
角频率 angular frequency, pulsatance	ω	弧度每秒 每秒	rad/s s^{-1}		
场[量]级 level of a field quantity	L_F	分贝 奈培	dB Np		
噪声 noise		分贝	dB		
质量 mass	m	千克,(公斤)	kg	克拉[米制]	
		吨 原子质量单位	t u		
体积质量 volumetric mass [质量]密度 mass density, density	ρ	千克每立方米	kg/m³		
		吨每立方米 千克每升	t/m³ kg/L		
线质量 linear mass 线密度 linear density	ρ_l	千克每米	kg/m		
		特[克斯]	tex		
转动惯量,(惯性矩) moment of inertia	$I(J)$	千克平方米	kg·m²		
动量 momentum	P	千克米每秒	kg·m/s		
力 force	F	牛[顿]	N	达因 千克力	dyn kgf
动量矩 moment of momentum 角动量 angular momentum	L	千克二次方米每秒	kg·m²/s		
力矩 moment of force	M	牛[顿]米	N·m	千克力米	kgf·m

续表

量的名称	量的符号	计量单位 名称	计量单位 符号	应废除的单位 名称	应废除的单位 符号
压力(强度,应力,压头) pressure (strength, stress, head)	P	帕[斯卡] 千帕 兆帕	Pa kPa MPa	巴① 标准大气压 千克力每平方米 托 工程大气压 约定毫米水柱 约定毫米汞柱	bar atm kgf/m^2 Torr at mmH_2O mmHg
[动力]黏度 viscosity, dynamic viscosity	$\eta,(\mu)$	帕[斯卡]秒	Pa·s	泊	P
运动黏度 kinematic viscosity	v	二次方米每秒	m^2/s	斯[托克斯]	St
能(量) energy 功 work	E W, A	焦[耳] 瓦[特][小]时 电子伏	J W·h eV	千克力米 尔格	kgf·m erg
功率 power	P	瓦[特] 千瓦[特] 兆瓦[特]	W kW MW	千克力米每秒 [米制]马力	kgf·m/s
热力学温度 thermodynamic temperature	$T,(\Theta)$	开[尔文]	K		
摄氏温度 celsius temperature	t, Θ	摄氏度	℃		
热 heat 热量 quantity of heat	Q	焦[耳] 千焦[耳]	J kJ	国际蒸汽表卡 热化学卡 15℃卡	cal_{IT} cal_{th} cal_{15}
热导率,(导热系数) thermal conductivity	$\lambda,(k)$	瓦[特]每米开[尔文]	W/(m·K)		
热容 heat capacity	C	焦[耳]每开[尔文]	J/K		
电流 electric current	I	毫安[培] 安[培]	mA A		

续表

量的名称	量的符号	计量单位		应废除的单位	
		名称	符号	名称	符号
电荷(量) electric charge, quantity of electricity	Q	库[仑] 安[培][小]时 千安培	C A·h kA		
电场强度 electric field strength	E	伏[特]每米	V/m		
电位,电势 electric potential	V, ϕ	伏[特]	V		
电位差(电势差),电压 potential difference, tension	$U, (V)$				
电动势 electromotive force	E				
电容 capacitance	C	法[拉] 微法[拉]	F μF		
介电常数,(电容率) permittivity	ε	法[拉]每米	F/m		
磁场强度 magnetic field strength	H	安[培]每米	A/m		
磁通[量]密度 magnetic flux density 磁感应强度 magnetic induction	B	特[斯拉]	T		
磁通[量] magnetic flux	ϕ	韦[伯]	Wb		
自感 self inductance	L	亨[利]	H		
互感 mutual inductance	M, L_{12}				
电阻 resistance	R	欧[姆]	Ω		
电导 conductance	G	西[门子]	S		
发光强度 luminous intensity	$I, (I_v)$	坎[德拉]	cd		

续表

量的名称	量的符号	计量单位 名称	计量单位 符号	应废除的单位 名称	应废除的单位 符号
光通量 luminous flux	$\Phi, (\Phi_v)$	流[明]	lm		
[光]照度 illuminance	$E, (E_v)$	勒[克斯]	lx		
物质的量 amount of substance	$n, (v)$	摩[尔]	mol		
摩尔质量 molar mass	M	千克每摩[尔]	kg/mol		
B 的质量分数 mass fraction of B	W_B	—	1		
B 的浓度 concentration of B B 的物质的量的浓度 amount-of-substance concentration of B	c_B	摩[尔]每立方米 摩[尔]每升	mol/m³ mol/L		
溶质 B 的质量摩尔浓度 molality of solute B	B_b, B_m	摩[尔]每千克	mol/kg		
[放射性]活度 activity	A	贝可[勒尔]	Bq	居里[1]	Ci
吸收剂量 absorbed dose	D	戈[瑞]	Gy	拉德[1]	rad
剂量当量 dose equivalent	H	希[沃特]	Sv	雷姆[1]	rem
照射量 exposure	X	库[仑]每千克	C/kg	伦琴[1]	R
电压 voltage	V	伏[特] 千伏[特]	V kV		
焓 enthalpy	h	千焦[耳]每千克	kJ/kcal		
熵 entropy	s	千焦[耳]每千克开	kJ/(kcal·K)		

续表

量的名称	量的符号	计量单位 名称	计量单位 符号	应废除的单位 名称	应废除的单位 符号
㶲 exergy		千焦[耳]每千卡	kJ/kcal		
比容 specific volume	V	立方米每千克	m^3/kg		
传热系数 heat transfer rate		瓦[特]每平方米开[尔文]	$W/(m^2 \cdot K)$		
膨胀系数 expansion rate		米每米摄氏度	$m/(m \cdot ℃)$		
比热容 specific heat capacity		千焦[耳]每千克开	$kJ/(kg \cdot K)$		
低发热量 net calorific value		千焦[耳]每千克	kJ/kg		
高发热量 gross calorific value		千焦[耳]每千克	kJ/kg		
荷载 load	p_α	牛每平方米	N/m^2		
容量 capacity		伏安 千伏安 兆伏安	$V \cdot A$ $kV \cdot A$ $MV \cdot A$		
CPU 存储容量 CPU storage capacity		字节 千字节 兆字节	Byte kilo Byte mega Byte		
浓度 concentration		毫克每升 克每升	mg/l g/l		
电导 conductivity		微西门子 西门子	μS S		
电能 electrical energy		千瓦[小]时	$kW \cdot h$		
细度 fineness		微米	μm		
流量 flow		千克每小时 立方米每小时 立方米每秒	kg/h m^3/h m^3/s		

续表

量的名称	量的符号	计量单位		应废除的单位	
		名称	符号	名称	符号
重量 weight		克 千克 公吨	g kg t		
电感 inductance		亨利	H		
力矩 moment of force 力偶矩 moment of a couple 转矩 torque	M M M, T	牛[顿]米	N·m		
电阻率 resistivity	ρ	欧·米	Ω·m		
阀门接头额定压力 valve fittings rating		帕斯卡 千帕 兆帕	Pa kPa MPa		

注：1. 圆括号中的名称，是它前面的名称的同义词。
  2. 方括号中的字可以省略。
  3. 计量单位中，一般只给出 SI 单位；非 SI 的单位列于 SI 单位之下，并用虚线与相应的 SI 单位隔开。
① 应废除的单位名称，为国际上暂时还允许使用的单位。

附录3　电力工程常见的国外、国内组织(机构)或公司名称

国外部分(按英文名称排序)

Advanced Research Projects Agency	美国远景研究规划局
African Development Bank	非洲开发银行
Agence Pour L'Energie Européenne Nucléaire	欧洲核能局(巴黎)
Agency for International Development	国际开发署(美国)
AGRA Earth & Environmental Limited	地球与环境有限公司(加拿大)
Allgemeine Elektrizitats-Gesellschaft	德国通用电气公司
Allmanna Svenska Electriska Aktiebolaget	瑞典通用电气公司
American Academy of Arts and Science	美国艺术与科学学院
American Association of Engineers	美国工程师协会
American Association of Scientific Workers	美国科学工作者协会
American Automatic Control Council	美国自动控制委员会
American Chemical Society	美国化学学会
American Coal Ash Association	美国煤灰协会
American Concrete Institute	美国混凝土学会
American Council for an Energy-Efficient Economy	美国能源有效经济利用理事会
American Electric Power System	美国电力公司
American Electronics Incorporated	美国电子工程公司
American Institute of Steel Construction	美国钢结构学会
American National Standards Institute	美国国家标准学会
American Nuclear Energy Council	美国核能理事会
American Nuclear Society	美国核学会
American Power Conference	美国电力会议
American Public Power Association	美国公用电力协会
American Society for Industrial Security	美国工业安全学会
American Society for Testing and Materials	美国试验与材料学会
American Society of Civil Engineers	美国土木工程师学会

续表

American Society of Heating, Refrigerating, and Air conditioning Engineers	美国供暖、制冷与空气调节工程师学会
American Society of Mechanical Engineers	美国机械工程师学会
American Society of Nondestructive Testing	美国无损检验学会
American Solar Energy Society	美国太阳能学会
American Standard Association	美国标准协会；美国标准委员会
American Water Works Association	美国自来水厂协会
American Wind Energy Association	美国风能协会
Argonne Code Center	阿贡法规中心（美国）
Asian Development Bank	亚洲开发银行
Asian Development Centre	亚洲开发中心
Asian Development Fund	亚洲开发基金
Asian Information Center for Geotechnical Engineering	亚洲岩土工程情报中心（泰国）
Associated Electrical Industries	联合电气工业公司（英国）
Associated General Contractors of American	美国承包商总会
Atomic Energy Authority	原子能管理局（英国）
Atomic Energy Commission	原子能委员会（美国）
Atomic Energy Establishment Winfirth	原子能中心（英国）
Atomic Fuel Corporation	原子能燃料公司（日本）
Atomic Power Construction Ltd	英国原子能建设有限公司
Atomic Power Development Associates Inc.	原子能发电研究联合公司（美国）
Atomic Safety & Licensing Appeal Board	原子安全和审批上诉委员会（美国）
Atomic Safety & Licensing Boards	原子核安全和注册申请委员会（美国）
Australian Nuclear Science and Technology Organization	澳大利亚核科学与技术机构
Babcock & Wilcox Company	巴布科克·威尔科克斯公司（美国）
Bank for International Settlements	国际清算银行
Bechtel Power Corporation	贝克特电力公司（美国）

续表

Beckwith Electric Company, Inc.	贝克维斯电气公司(美国)
BHP Steel Building Products Ltd	BHP 建筑钢品有限公司(澳大利亚)
BIS = Bureau of Indian Standards	印度标准局
Bituminous Coal Research Inc.	烟煤研究公司(美国)
Black & Veatch International, Inc.	博莱克威奇国际公司(美国)
British Atomic Energy Corporation	英国原子能公司
British Electricity Authority	英国电力管理局
British Insulated Callender's Cables Ltd	英国卡伦德绝缘电缆公司
British Motor Corporation Ltd	英国电动机有限公司
British Motor Trade Association	英国电动机贸易公司
Building Research Institute	建筑研究院(美国)
Building Research Station	建筑研究所(英国)
Bureau of Mines	矿务局(美国)
Bureau of Ocean Energy Management	美国海洋能源管理局
Bureau of Safety and Environmental Enforcement	美国安全和环境执法局
CAE Electronics Ltd	CAE 电子有限公司(加拿大)
Canada China Power Inc.	加华电力公司(加拿大)
Canadian General Electric Co., Ltd	加拿大通用电气公司
Canadian International Development Agency	加拿大国际开发署
CCI-Sulzer Valves	CCI 苏尔寿阀门公司(美国)
CEA Asia, Inc.	CEA 亚洲有限公司
Center for International Research of Environment and Development	国际环境与发展研究中心
Central Electrical Authority	英国电气管理局
Central Electricity Generation Board	中央发电局(英国)
Central Electricity Research Laboratories, UK	英国电力研究所
Ceske Energeticke Zavody	捷克电力公司
Chemetron Fire Systems, Inc.	科美全消防系统工程公司(美国)
Chubb National Foam, Inc.	Chubb 泡沫消防公司(美国)
Chubu Electric Power Co. Inc.	日本中部电力公司

续表

Chugoku Electric Power Co. Inc.	日本中国电力公司
Coastal Power Company	美国高达发电总公司
Combustion Engineering (CE)	燃烧工程公司(美国)
Combustion Equipment Association	燃烧设备联合公司(英国)
Commission for Hydrometeorology	水文气象委员会(世界气象组织)
Commission for the Geological Map of the World	世界地质图委员会
Committee on International Geophysics	国际地球物理学委员会
Consortium for Electric Reliability Technology Solutions	美国电气可靠性技术解决方案联合会
Consultation Committee of International Telegraph and Telephony	国际电话电报咨询委员会
Council for Mutual Economic Assistance	经济互助委员会
Department of Energy	能源部(美国)
Department of Trade and Industry (DTI)	英国贸易工业部
Deutsche Babcock AG	德国巴高克股份公司(集团)
Dimosia Epiheirisis Ilektrismou	希腊电力公司
DKK Corporation	电气化学计器株式会社(日本)
Dresser Industries, Inc.	德莱赛工业公司(美国)
Dresser Valve & Controls Division	德莱赛工业阀门部(美国)
Ecology & Environment Inc.	生态与环境公司(美国)
Economic and Social Commission for Asia and the Pacific	亚洲及太平洋地区经济和社会委员会
Economic Commission for Europe	欧洲经济委员会
Economic Development Institute	世界银行经济发展研究所
Edward Valves, Inc.	爱德华阀门公司(美国)
Egyptian Electricity Authority	埃及电管局
Electrabel	比利时电力局
Electric Power Development Co., Ltd	电源开发株式会社(日本)
Electric Power Research Institute	美国电力研究院
Electricidade de Portugal	葡萄牙电力公司
Electricitè de France	法国电力公司

续表

Electricity Corporation of New Zealand Ltd	新西兰电力公司
Electricity Generating Authority of Thailand	泰国发电局
Electro Tec, Corporation	电技术公司(美国)
Electromagnetic Research Corporation	电磁研究公司(美国)
Electronics Corporation of America	美国电子有限公司
ELIN Energieversorgung GmbH	伊林电气公司(奥地利)
Elsag-Bailey Process Automation	萨格-贝利过程自动化集团(美国)
Enctech Engineering Systems Ltd	英德工程系统公司(新加坡)
Energy Information Administration	能源信息管理局
English Electric Co., Ltd	英国电气有限公司
Ente Nazionale Per l'Energia Elettrica	意大利电力公司
Environmental Protection Agency	环境保护委员会(美国)
Erisson	爱立信有限公司(瑞典)
Eskom	南非供电局
European Energy Equation	欧洲能量方程
European Marine Energy Centre (EMEC)	欧洲海洋能源中心
European Investment Bank	欧洲投资银行
European Union's Commission	欧盟能源委员会
European Seismological Commission	欧洲地震委员会
European Wind Energy Association	欧洲风能协会
Falk Corporation, USA	美国福克公司
Federal Energy Regulation Commission	联邦能源管理委员会(美国)
Federal Hydro-Electric Board	联邦水电局(美国)
Federation International Des Ingenieus-Conseils/ International Federation of Consulting Engineers (FIDIC)	国际咨询工程师联合会(简称"菲迪克")
Felten & Guilleaume Energietechnik AG	F&G电气设备公司(德国)
Fisher-Rosemount Inc.	希尔-罗斯蒙特公司(美国)
Flender German	德国弗兰德公司
Forest Stewardship Couoncil	森林管理委员会

Foster Wheeler International, Inc.	福斯特惠勒能源国际公司(美国)
Foxboro Company	福克斯波罗公司(美国)
Framatome	法马通公司(法国)
Fraunhofer Institute for Solar Energy System	弗劳恩霍夫太阳能系统研究所
Fuller-Kovako Corporation	福乐公司(美国)
GEA Thermal & Energy Technology Division	GEA 热能技术部(德国)
GEC Alstom	通用电气阿尔斯通公司(法国)
GEC Alstom Protection & Control Limited Far East Office	GEC 阿尔斯通保护及控制有限公司远东办事处
GEC Alstom Protection & Control Limited St Leonards Works	GEC 阿尔斯通保护及控制有限公司圣罗纳兹工厂
General Cable Corporation	通用电缆公司(美国)
General Controls Corporation	通用控制(设备)公司(美国)
General Electric Co., Ltd	通用电力公司(英国)
General Electric Company	通用电气公司(美国)
General Electric Power Systems (GE)	美国通用电气公司动力系统集团
Geothermal Energy Association	地热能协会
German Wind Energy Institute	德国风力能源研究所
Group of European Manufacturers for the Advancement of Turbine Technology	欧洲透平技术集团
HAMON Group	哈蒙集团公司(比利时)
Hitach, Ltd	株式会社日立制作所(日本)
Hokkaido Electric Power Co. Inc.	日本北海道电力公司
Hokuriku Electric Power Co.	日本北陆电力公司
Howden Group Plc.	豪顿集团(英国)
Hydro Quebec	加拿大魁北克水电局
Ingersoll-Dresser Pumps	英格索兰-德莱赛泵浦公司(美国)
Ingersoll-Rand Far East U. S. A.	美国英格索兰远东公司
Institute for the Development of Renewable Natural Resources	再生资源发展研究所

Institute of Electrical and Electronics Engineers	电气和电子工程师协会(美国)
Inter-American Development Bank	泛美开发银行
Intergovernmental Panel on Climate Change	政府间气候变化专门委员会
International Atomic Energy Agency	国际原子能机构(奥地利维也纳)
International Bank for Economic Co-operation	国际经济合作银行
International Bank for Reconstruction and Development	国际复兴开发银行(简称"世界银行")
International Business Machines Corporation (IBM)	国际商业机器公司(美国)
International Cogeneration Society	国际热电联产学会
International Combustion Engineering Ltd	国际燃料工程公司(英国)
International Commission of Scientific Management	国际(企业)科学管理委员会
International Commission on Groundwater	国际地下水委员会
International Commission on Large Dams	国际大坝委员会
International Commission on Surface Water	国际地表水委员会
International Commission on Water Quality	国际水质委员会
International Commission on Water Resources Systems	国际水资源系统委员会
International Conference on Large High Voltage Electric System	国际大电网会议
International Consultants Ltd (International Consultancy)	国际咨询公司
International Electronics Manufacturing Co.	国际电子仪器制造公司
International Electrotechnical Commission	国际电工委员会
International Energy Agency (IEA)	国际能源署
International Energy Agency's Photovoltaic Power Systems Programme (IEA-PVPS)	国际能源机构光伏电力系统项目
International Financial Corporation	国际金融公司
International Flame Research Foundation (IFRF)	国际火焰研究基金会
International Fund for Agricultural Development	国际农业发展基金
International Fusion Superconducting Magnet Test Facility	国际聚变超导磁铁试验设施(美国)

International Ground Source Heat Pump Association (IGSPHA)	国际地源热泵协会
International Hydrographic Bureau	国际水文局
International Hydrological Programme	国际水文计划组织
International Institute for Applied Systems Analysis (IIASA)	国际应用系统分析研究所
International Institute of Seismology and Earthquake Engineering	国际地震学与地震工程研究所
International Investment Bank	国际投资银行
International Monetary Fund	国际货币基金组织
International Nuclear Cooperation Center	国际核合作中心(日本)
International Organization for Standardization	国际标准化组织
International Research and Development Corporation	国际研究与发展公司
International Seismological Centre	国际地震中心
International Solar Energy Society (ISES)	国际太阳能学会
International Solid Waste Association (ISWA)	国际固体废物协会
International Standard Electric Corporation	国际标准电气公司
International Telephone and Telegraph Corporation	国际电话电报公司(美国)
International Union of Geodesy & Geophysics	国际大地测量学及地球物理学联合会
International Union of Producers and Distributors of Electrical Energy	国际发供电联盟
Ishikawajima-Harima Heavy Industries Co., Ltd	石川岛播磨重工业株式会社(日本)
Itochu Corporation	伊藤忠商事株式会社(日本)
Japan Electronic Computer Company	日本电子计算机公司
Japan Nuclear Fuel Limited	日本核燃料有限公司
Japan Steel Works Ltd	日本钢铁公司
Japanese Meteorological Agency	日本气象厅
Johannes Möller Hamburg Engineering GM	穆勒工程公司(德国)
Joint Committee on Seismic Safety	地震安全联合会
Joint Stock Company "SIBENERGOMASH"	西伯利亚电力设备联合股份公司
Kansai Electric Power Co. Inc.	日本关西电力公司

续表

Kenticott Water Systems Ltd	肯尼特水处理有限公司(英国)
Keystone International Inc. U.S.A.	美国基士敦国际有限公司
Kobe Steel, Ltd	株式会社神户制钢所(日本)
Korea Electric Power Corporation (KEPCO)	韩国电力公社
Korea Heavy Industries & Construction Co., Ltd	韩国重工业株式会社
Korean Nuclear Fuel Company	韩国核燃料公司
Kraftwerk Union AG	电站联盟公司(德国)
Kyushu Electric Power Co. Inc.	日本九洲电力公司
LG Cable & Machinery Ltd	LG电线公司(韩国)
Livermore Pool-Type Reactor	利弗莫尔池式反应堆(美国)
Low Power Test Facility	低功率试验装置(美国)
Minerals Management Service	美国矿产管理局
Mitsubishi Corporation	三菱商事株式会社(日本)
Mitsubishi Electric Corporation	三菱电机株式会社(日本)
Mitsubishi Heavy Industries, Ltd	三菱重工业株式会社(日本)
Mitsui & Co., Ltd	三井物产株式会社(日本)
Mitsui Babcock Energy Ltd	三井巴布科克能源有限公司(日本)
Motorola	摩托罗拉公司(美国)
Motorola Information Systems Group	摩托罗拉公司信息系统部(美国)
Nagangkeiki Co, Ltd	日本长野计器株式会社
Nash Engineering Company	纳氏工程公司(美国)
National Academy of Science	国家科学院(美国)
National Air Surveillance Network	全国大气监视网点(美国)
National Center for Air Pollution Control	国家空气污染控制中心(美国)
National Coal Board	国家煤炭部(英国)
National Earthquake Information Center	国家地震情报中心(美国)
National Electrical Manufacturing Association	国家电气制造协会(美国)
National Fire Protection Association	全国消防协会(美国)
National Geodetic Survey	国家大地测量局(美国)

续表

National Ground Water Association	美国地下水协会
National Institute of Standards and Technology	国家标准与技术局(美国)
National Oceanic and Atmospheric Administration (NOAA)	美国国家海洋和大气管理局
National Power Corporation	菲律宾国家电力公司
National Power PLC	英国国家电力公司
National Reactor Testing Station	全国反应堆测试站(美国)
National Renewable Energy Laboratory	美国国家可再生能源实验室
National Research Council	国家研究理事会(美国)
National Thermal Power Corporation	印度火电公司
National Weather Service	美国国家气象局
Netherlands Organisation for Applied Scientific Research	荷兰应用科学研究组织
Northern American Electric Reliability Council (NERC)	北美电力可靠性协会
North America Electric Reliability Organization (NERO)	北美电力可靠性机构
NGK Insulators Ltd	日本 NGK 公司
Nippon Kokan Kaisha Ltd	日本钢管公司
Nissho Iwai Corporation	日商岩井株式会社(日本)
Nohmi Fire Protection System	能美消防公司(美国)
Nuclear Fuel Complex	核燃料综合体(印度)
Nuclear Fuel Industries, Ltd	核燃料工业有限公司(日本)
Nuclear Management and Resources Council	核管理与资源理事会(美国)
Nuclear Power Corporation of India Limited	印度核电有限公司
Nuclear Power Plant Co.	核动力设备公司(英国)
Nuclear Power Plants Equipment Manufacturing Company	罗马尼亚核电厂设备制造公司
Nuclear Regulatory Commission	核管理委员会(美国)
Nuclear Safety Center	核安全中心(美国)
Oak Ridge National Laboratory	橡树岭国家实验室(美国)
Office of Saline Water	咸水淡化局(美国)

续表

Okano Valve MFG. Co.	冈野阀门制造株式会社(日本)
Organization for Economic Co-operation and Development	经济合作与发展组织
Organization for European Economic Cooperation	欧洲经济合作组织
Organization of Arab Petroleum Exporting Countries	阿拉伯石油输出国组织
Pacific Northwest National Laboratory	美国能源部西北太平洋国家实验室
Power Gen PLC	英国电能公司
Power Reactor and Nuclear Fuel Development Corporation	原子能发电与核燃料开发公司(日本)
Power System Resources	力源公司(美国)
Public Works Administration	公用(事业管理)局(美国)
Radio Corporation of America	美国无线电公司
Raytheon Engineers & Constructors	雷神工程与建筑公司(美国)
Rockwell Automation-Dodge	罗克韦尔自动化—道奇公司(美国)
Rolls-Royce PLC	罗尔斯·罗伊斯公共有限公司(英国)
Rotork Actuation	罗托克执行器公司
Sargent & Lundy Engineers, Ltd	萨金兰迪工程有限公司(美国)
Schneider Electric SA	施耐德电气公司(法国)
Shikoku Electric Power Co. Inc.	日本四国电力公司
Shinkawa Electric Co., Ltd	新川电机株式会社(日本)
Sihwa lake tidal power station	(韩国)始华湖潮汐电厂
Siemens AG	西门子股份公司(德国)
Siemens Energy & Automation, Inc.	西门子能源与自动化公司
Siemens Vacuum Pump & Compressor Co., Ltd	西门子真空泵压缩机有限公司
Singapore Valve & Fitting Pte., Ltd	新加坡阀门及配件有限公司
SKODA a. s.	斯科达公司(捷克)
Soil and Water Conservation Research Division	水土保持研究部(美国)
Soil Conservation Service	土壤保持局(美国)
Solid Smokeless Fuels Federation	固体无烟燃料联合会(英国)

续表

Statkraft	挪威国家电力局
Sulzer Technology Corporation	苏尔寿公司(瑞士)
Sumitomo Corporation	住友商事株式会社(日本)
Sumitomo Electric Industries, Ltd	住友电气工业株式会社(日本)
Sumitomo Metal Industries, Ltd	住友金属工业株式会社(日本)
Super Link Australia Limited	捷丰集团有限公司(澳大利亚)
Svedala Industry AB	斯维达拉工业集团(瑞典)
Swedish State Power Board	瑞典国家动力局
Tavanir Co.	伊朗电力局
Thorn Security Protection Ltd	科艺保安消防公司(英国)
Tohoku Electric Power Co. Inc.	日本东北电力公司
Toko Trading Co., Ltd	东红贸易株式会社(日本)
Tokyo Electric Power Services Co., Ltd (TEPSCO)	东电设计株式会社(日本)
Toshiba, Corporation	东芝公司(日本)
Trane Corporation USA	美国特灵公司
Turkiye Elektrik Kurumu (TEK)	土耳其电力局
U.S. Department of Energy	美国能源部
U.S. Department of Energy government research laboratory	美国能源部可再生能源实验室
U.S. Electric Power Institute	美国电力学会
U.S. Government Patents Board	美国政府专利局
U.S. National Oceanic and Atmospheric Administration	美国国家海洋和大气局
Union Electric Co.	美国联合电力局
United Conveyor Corporation	联合输送公司(美国)
United Nations Development Programme	联合国开发计划署(联合国机构,美国纽约)
United Nations Economic Development Administration	联合国经济发展局
United Nations Environment Programme	联合国环境规划署(联合国机构,肯尼亚内罗毕)
United Nations Industrial Development Organization	联合国工业发展组织

续表

United Nations Technical Assistance Administration	联合国技术援助局
United States Agency for International Development	美国国际开发署
United States Atomic Energy Commission	美国原子能委员会
United States Committee on Large Dams	美国大坝委员会
United States Filter Corporation	美国过滤器公司
United States Patent Office	美国专利局
United States Weather Bureau	美国气象局
Universal Training Reactor	通用训练(反应)堆(美国)
University of Teheran Research Reactor	德黑兰大学研究堆(伊朗)
University of Virginia Reactor	弗吉尼亚大学(反应)堆(美国)
Vattenfall	瑞典国家电力公司
Vereinigte Enegiewerke AG	德国联合电力公司
Vinking Corporation	威景消防公司(美国)
VO "Technopromexport"	俄外经国家"技术工业出口联合公司"
Water Pollution Research Laboratory	水污染研究所(英国)
Water Resources Scientific Information Center (USA)	美国水资源科学情报中心
Weir Pumps Ltd	威尔泵有限公司(英国)
Wescon Technology, Inc.	威斯康技术公司
Westinghouse Electric Corporation	西屋电气公司(美国)
World Bank	世界银行
World Energy Council	世界能源委员会
World Environment and Resources Council	世界环境和资源委员会
World Health Organization	世界卫生组织
World Intellectual Property Organization	世界知识产权组织
World Meteorological Organization	世界气象组织
World Trade Organization	世界贸易组织
Wormald Fire Systems	威武消防系统公司(澳大利亚)
Yankee Atomic Electric Company	美国原子发电公司
Yarway Corporation	亚威公司(美国)

国内部分(按中文汉语拼音排序)

北京北重汽轮电机有限责任公司	Beijing Beizhong Steam Turbine Generator Co., Ltd
电建所杆塔试验站	Transmission Tower Test Station of Electric Power Construction Research Institute
电力规划设计总院	China Electric Power Planning & Engineering General Institute
电力建设工程咨询公司	Power Construction Engineering Consulting Corporation
电子工程咨询公司	Electronic Engineering Consultants (Electronic Engineering Consulting Company)
东方电机有限公司	Dongfang Electric Machinery Co., Ltd
东方锅炉股份有限公司	Dongfang Boiler Group Co., Ltd
东方汽轮机有限公司	Dongfang Turbine Co., Ltd
对外业务部	International Cooperation Department (International Operations Department)
工程顾问协会	Institute of Consulting Engineers
工商行政管理总局	State Administration for Industry & Commerce
公路工程咨询公司	Highway Engineering Consultants Inc.
公路局	Highway Administration
广州美罗钢格板有限公司	Guangzhou Metro Grating Co., Ltd
规划设计院	Planning and Design Institute
国际商业机器(IBM)中国有限公司	IBM China Company Limited
国家电力公司(前电力部)	The State Power Corporation (Former MOEP)
国家电力公司产品质量标准研究所	Standard & Quality Control Research Institute, SPC
国家电力公司热工研究院	Thermal Power Research Institute, SPC
国家电网有限公司	State Grid Corporation of China
国家科学基金会	National Science Foundation
国家绿化委员会	National Afforestation Committee
国家气象局	National Weather Service
哈尔滨电机厂有限责任公司	Harbin Electric Machinery Co., Ltd

续表

哈尔滨电气集团有限公司	Harbin Electric Corporation
哈尔滨锅炉厂有限责任公司	Harbin Boiler Co., Ltd (HBC)
哈尔滨汽轮机厂有限责任公司	Harbin Steam Turbine Co., Ltd
海洋资源工程委员会	Engineering Committee on Oceanic Resources (ECOR)
杭州机械设计研究所	Hangzhou Machinery Design and Research Institute
湖北—戴蒙德机械有限公司	Diamond Power-Hubei Machine Co., Ltd
华能国际电力股份有限公司	Huaneng Power International, Inc.
华能国际电力开发公司	Huaneng International Power Development Co.
华能核电开发有限公司	Huaneng Nuclear Power Development Co., Ltd
华能新能源股份有限公司	Huaneng Renewables Corporation Limited
机械工程咨询公司	Machinery Engineering Consultation Corporation (Consultants)
联合开利(上海)工程服务有限公司	United Carried (Shanghai) Engineering & Service Co., Ltd
绿色煤电有限公司	GreenGen Co., Ltd
南京电力自动化设备总厂	Nanjing Electric Power Automation Equipment General Factory
南京汽轮发电机(集团)有限责任公司	Nanjing Turbine & Electric Machinery (Group) Co., Ltd
南京自动化研究院	Nanjing Automation Research Institute
轻工业工程咨询公司	Light Industry Engineering Consultants (Consulting Corporation)
上海凯士比泵有限公司	KSB Shanghai Pump Co., Ltd
上海电气集团股份有限公司	Shanghai Electric Group Co., Ltd
上海福伊特西门子水电设备有限公司	Shanghai Voith Siemens Hydro Power Generation Co., Ltd
上海锅炉厂有限公司	Shanghai Boiler Works Co., Ltd
上海汽轮机有限公司	Shanghai Turbine Company Ltd
上海西门子线路保护系统有限公司	Siemens Circuit Protection System Ltd, Shanghai (SPC)
神华国华能源投资有限公司	Shenhua Guohua Energy Investment Co., Ltd
神华集团有限责任公司	Shenhua Group Corporation Limited

续表

水利水电规划设计总院	China Renewable Energy Engineering Institute
水利水电建设工程咨询公司	Water Resources and Hydropower Engineering Consulting Corporation
天津天威有限公司	Tianjin VEGA Co., Ltd
天津威津安全系统有限公司	Tianjin Vijin Security System Co., Ltd
武汉锅炉股份有限公司	Wuhan Boiler Company Limited
西安超滤净化工程有限公司	Xi'an Unifilter Engineering Co., Ltd
西北电力集团公司	Northwest Electric Power Group Corporation
西北电力设计院	Northwest Electric Power Design Institute
西北勘测设计研究院有限公司	Northwest Engineering Corporation Limited
中国成套设备进出口集团有限公司	China National Complete Plant Import and Export Group Corporation Limited
中国出版对外贸易总公司	China National Publishing Industry Trading Corporation
中国出口商品包装公司	China National Export Commodities Packaging Corporation
中国出口商品基地建设总公司	China National Export Bases Development Corporation
中国船舶工业总公司	China State Shipbuilding Corporation
中国电机工程学会	Chinese Society for Electrical Engineering
中国电力工程顾问集团公司	China Power Engineering Consulting (Group) Corporation
中国电力技术进出口公司	China Electric Power Technology Import & Export Corporation
中国电力建设集团有限公司	Power Construction Corporation of China
中国电力科学研究院	China Electric Power Research Institute (CEPRI)
中国电力企业联合会	China Electricity Council (CEC)
中国电力投资集团公司	China Power Investment Corporation (CPI)
中国电子技术进出口公司	China Electronics Technology Import and Export Corporation
中国东方电气集团公司	Dongfang Electric Corporation
中国对外贸易咨询与技术服务公司	China Foreign Trade Consultation and Technical Service Corporation

续表

中国葛洲坝集团公司	China Gezhouba Group Corporation (CGGC)
中国工商银行	Industrial and Commercial Bank of China
中国国电集团公司	China Guodian Corporation
中国国际工程咨询公司	China International Engineering Consulting Corporation
中国国际信托投资公司	China International Trust and Investment Corporation
中国核工业集团有限公司	China National Nuclear Corporation
中国华电电站装备工程公司	China Huadian Power Equipment Engineering Corporation
中国华电集团公司	China Huadian Corporation
中国华能集团公司	China Huaneng Group
中国环境标志产品认证委员会	China Certification Committee for Environmental Labeling Products (CCEL)
中国环境监测总站	China National Environmental Monitoring Centre
中国机械对外经济技术合作总公司	China Machine-Building International Corporation (CMIC)
中国机械进出口(集团)有限公司	China National Machinery Import and Export Corporation (CMC)
中国机械设备进出口总公司	China National Machinery & Equipment Import & Export Corporation (CMEC)
中国技术进出口总公司	China National Technical Import & Export Corporation
中国技术进口总公司	China National Technical Import Corporation
中国建材集团有限公司	China National Building Material Group Corporation
中国建筑工程总公司	China State Construction Engineering Corporation
中国进出口公司	Chinese National Import and Export Corporation
中国科学院	Chinese Academy of Sciences
中国能源建设集团有限公司	China Energy Engineering Group Co., Ltd
中国神华能源股份有限公司	China Shenhua Energy Co., Ltd (CSEC)
中国石化总公司	China Petrochemical Corporation
中国水电建设集团新能源开发有限责任公司	Sinohydro Renewable Energy Co., Ltd
中国水利工程公司	China Water Conservancy Engineering Corporation

续表

中国太平保险集团公司	China Taiping Insurance Group Co
中国通用石化机械工程总公司	China National Petroleum Chemical & General Machinery Engineering Corporation
中国土木工程公司	China Civil Engineering Corporation
中国五金矿产进出口公司	China National Metals and Minerals Import and Export Corporation
中国冶金进出口公司	China National Metallurgical Products Import and Export Corporation
中国仪器进出口(集团)公司	China National Instruments Import & Export (Group) Corporation (INSTRIMPEX)
中国有色金属总公司	China National Nonferrous Metals Industry Corporation
中国原子能工业公司	China Atomic Energy Industrial Corporation
中国远洋运输公司	China Ocean Shipping Company
中国再保险(集团)股份有限公司	China Reinsurance (Group) Corporation
中国质量保证协会	China Society Quality Assurance (CSQA)
中国中化集团公司	Sinochem Group

附录4 电力工程常用的国外和国内标准

国外部分(按英文缩写或编号列字母顺序排列)

英文缩写或编号	英文全称	中文
美国及国际标准		
AA	Aluminum Association	美国铝业联合会
AAA	American Arbitration Association	美国仲裁协会
AABC	Associated Air Balance Council	空气平衡联合会
AABC-Vol. 1	Vol. 1 National Standards for Field Measurements and Instrumentation, Testing & Balancing	第1卷:现场测定和仪器仪表试验与平衡的国家标准
AAMA	Architectural Aluminum Manufacturers Association	美国建筑用铝制造商协会
AASHTO	American Association of State Highway and Transportation Officials	美国州级公路和运输官员联合会
ABAI	American Boiler & Affiliated Industries	美国锅炉和附属设备制造厂商协会
ABMA	American Boiler Manufacturers Association	美国锅炉制造商协会
ABS	American Bureau of Standards	美国标准局
ACI	American Concrete Institute	美国混凝土学会
ACME	Association of Consulting Management Engineers	管理顾问工程师协会
ACRI	Air Conditioning and Refrigeration Institute	空调和制冷学会
ADC	Air Diffusion Council	大气扩散委员会
AEC	The United States Atomic Energy Commission	美国原子能委员会
AEIC-CS6	Specification for EPR Insulated Shielded Power Cables Rated 5 Through 69kV	额定电压5~69kV的乙丙烯橡胶绝缘护套动力电缆规范
AEPC	American Electric Power Co.	美国电力公司
AESC	American Engineering Standards Committee	美国工程标准委员会
AFBMA	Anti-Friction Bearing Manufacturers Association	耐磨轴承制造商协会

续表

英文缩写或编号	英文全称	中　　文
AFBMA-B3.15	Load Ratings and Fatigue Life for Ball Bearings	滚珠轴承额定负载和疲劳期限
AFBMA-B3.16	Load Ratings and Fatigue Life for Roller Bearings	滚柱轴承额定负载和疲劳期限
AGA	American Gas Association	美国气体协会
AGA-report No.3	The Suggestion of a Certain Type of Orifice Measuring Device and Used in Measurement of Natural Gas	美国气体协会第3号报告：天然气测量中使用一定类型的孔板测量装置之建议
AGCA	Associated General Contractors of America	美国承包商总会
AGI	The American Geosciences Institute	美国地球科学学会
AGMA	American Gear Manufacturers Association	美国齿轮制造商协会
AIA	The American Institute of Architects	美国建筑师学会
AID	American Institute of Decorators	美国室内装饰家学会
AIEE	American Institute of Electrical Engineers	美国电气工程师学会
AISC	American Institute of Steel Construction	美国钢结构学会
AISC	Code of Standard Practice for Steel Buildings and Bridges	钢结构和桥梁标准实务规范
AISC	Manuals of Steel Construction	钢结构手册
AISC	Specification for Structural Steel Buildings	钢结构建筑规范
AISC-S326	Specification for the Design, Fabrication, and Erection of Structural Steel for Buildings	建筑结构钢的设计、加工及安装规范
AISE	Association of Iron and Steel Engineers	钢铁工程师协会
AISI	American Iron and Steel Institute	美国钢铁学会
AITC	American Institute of Timber Construction	美国木结构学会
AMCA	Air Moving and Conditioning Association	空气运动和空调协会
ANS	American Nuclear Society	美国核协会

续表

英文缩写或编号	英文全称	中文
ANSI	American National Standards Institute	美国国家标准学会
ANSI	Schedules of Preferred Ratings and Rated Required Capabilities for AC High Voltage Circuit Breakers Rated on a Symmetrical Current Basis	以对称电流为基准的交流高压断路器的优化额定值及有关性能一览表
ANSI/ASCE 7-93	Minimum Design Loads for Building and Other Structures	建筑物及其他构筑物的最小设计荷载
ANSI-B16.11	Forged Steel Fittings, Socket-Welded and Threaded	锻钢承插、焊接和螺纹接头
ANSI-B16.20	Ring Joint Gaskets and Grooves for Steel Pipe Flanges	钢管法兰之环接密封垫和槽
ANSI-B16.36	Steel Orifice Flanges	钢制孔板法兰
ANSI-B16.5	Steel Pipe Flanges and Flanged Fittings	钢管法兰和法兰的配件
ANSI-B31.1	Power Piping	动力管道标准
ANSI-B31.5	Code for Piping: Refrigeration Piping	动力管道规程:制冷管道
ANSI-B40.1	Gages-Pressure Indicating Dial Type-Element	压力刻度盘型表-元件
ANSI-C29.1	Test Methods for Electrical Power Insulators, including Addendum C29.2a	电瓷试验方法(电力工业绝缘子试验方法),包括附录 C29.2a
ANSI-C33.10	Safety Standard for Fuseholders	熔断器安全标准
ANSI-C33.38	Safety Standard for Panelboards	配电盘的安全标准
ANSI-C33.65	Electrical Cabinets and Boxes	电气屏和箱(接线盒)
ANSI-C33.77	Attachment Plug and Receptacles	附属插头和插座
ANSI-C37.072	Requirements for Transient Recovery Voltage for AC High Voltage Circuit Breakers Rated on a Symmetrical Current Basis	交流高压断路器以对称电流为基础的暂态恢复电压的要求
ANSI-C37.13	Low Voltage AC Power Circuit Breakers Used in Enclosures	用于外壳内的低压交流断路器

续表

英文缩写或编号	英文全称	中文
ANSI-C37.16	Preferred Ratings, Related Requirements and Application Recommendation for Low Voltage Circuit Breaker and AC Power Circuit Protectors	低压断路器和交流电力回路保护的推荐额定值,有关要求和使用建议
ANSI-C37.20	Switchgear Assemblies, including Metal-Enclosed Bus	开关装置组合,包括金属封闭母线
ANSI-C37.23	Guide for Calculating Losses in Isolated-Phase Bus	离相封闭母线损耗计算导则
ANSI-C37.24	Guide for Evaluating the Effects of Solar Radiation on Outdoor Metal-Clad (General Information Regarding the Effects of Solar Radiation on Outdoor Equipment)	户外金属外壳开关装置太阳辐射影响计算导则(关于户外设备太阳辐射影响的一般说明)
ANSI-C37.30	Standard Definitions and Requirements for High Voltage Air Switches, Insulators and Bus Supports	高压空气开关、绝缘子及母线支架的标准定义及要求
ANSI-C37.32	Schedule of Preferred Ratings, Manufacturing Specification, and Application Guide for High-Voltage Air Switches, Bus Supports, and Switch Accessories	高压空气开关、母线支持结构、开关附件的适用导则及制造厂的规范、推荐的额定值估算一览表
ANSI-C37.35	Guide for the Application, Installation, Operation, and Maintenance of High-Voltage Air Disconnecting and Load Interrupter Switches	高压空气隔离开关和负荷断流器的应用、安装、运行和维护导则
ANSI-C37.90	Relays and Relay Systems Associated with Electric Power Apparatus	电气设备相关的继电器及继电系统
ANSI-C37.90A	Guide for Surge Withstand Capability Tests	冲击承受能力试验指南
ANSI-C39.1	Requirements for Electrical Recording Instruments	电气记录仪表的要求
ANSI-C39.2	Direct Acting Electrical Indicating Instruments	直接作用式电气指示仪表
ANSI-C39.3	Shock Testing for Electric Indicating Instruments	电气指示仪表的冲击试验

续表

英文缩写或编号	英文全称	中文
ANSI-C39.5	Safety Requirements for Electrical and Electronic Measuring and Controlling Instruments	电子、电气测量和控制仪表的安全要求
ANSI-C39.6	Digital Measuring Instruments	数字测量仪表
ANSI-C50.10	General Requirements for the Synchronous Machines	同步电机的一般要求
ANSI-C57.12.00	General Requirements for Liquid-Immersed Distribution, Power and Regulating Transformers	液浸式配电、电力及调压变压器的一般要求
ANSI-C57.12.10	Requirements for Transformers, 230 000 Volts and Below, 833/985 through 8333/10417 kVA, Single-Phase, 750/862 Through 60000/80 000/100000 kVA, Three-Phase	230kV 及以下单相,容量 833/985～8333/10417kVA;三相,容量 750/862 至 60000/80000/100000 kVA 变压器的技术要求
ANSI-C57.12.30	Requirements for Three-Phase Load Tap Changing Transformers, 230000 Volts and Below, 3750/4687 Through 60000/80000/100000kVA	230kV 及以下,容量 3750/4687～60000/80000/100000 kVA 三相带负荷调压变压器的技术要求
ANSI-C57.12.70	Terminal Markings and Connections for Distribution and Power Transformers	配电和电力变压器的端子标志和连接
ANSI-C57.12.80	Terminology (IEC76), including Supplement C57.12.80a	术语(IEC76),包括补遗的 C57.12.80a
ANSI-C57.12.90	Test Code for Distribution, Power and Regulating Transformers, including supplement C57.12.90a	配电、电力和调压变压器的试验标准,包括补遗的 C57.12.90a
ANSI-C57.13	Requirements for Instrument Transformers	仪用变压器的要求
ANSI-C57.92	Guide for Loading Oil-Immersed Distribution and Power Transformers	油浸配电、电力变压器带负荷导则
ANSI-C62.1	Lightning Arresters for Alternating-Current Power Circuits	用于交流电力系统的避雷器
ANSI-C62.2	Guide for Application of Valve Type Lightning Arresters for Alternating-Current System	用于交流电系统的阀型避雷器的应用导则
ANSI-C76.1	Requirements and Test Code for Outdoor Apparatus Bushings	户外电气设备套管的要求及其试验规程

续表

英文缩写或编号	英文全称	中文
ANSI-C80.1	Rigid Steel Conduits, Zinc Coated	镀锌的硬钢导管
ANSI-C80.3	Specifications for Electrical Metallic Tubing, Zinc Coating	电金属管和镀锌层规范
ANSI-C80.4	Fittings for Rigid Metal Conduit and Electric Metallic Tubing	硬金属管和金属管附件
ANSI-C89.2	Dry-Type Transformers for General Applications	通用干式变压器
ANSI-MC96.1	Temperature Measuring Thermocouples	温度测量热电偶
ANSI-N101.1	Efficiency Testing of Air Cleaning Systems Containing Devices for Removal of Particles	含除尘装置的空气净化系统的效率测试
ANSI-Y32.14	Specification for Steam Turbines	汽机规范
API	American Petroleum Institute	美国石油学会
AREA	American Railway Engineering Association	美国铁路工程学会
ASA	American Standards Association	美国标准协会
ASCE	American Society of Civil Engineers	美国土木工程师协会
ASCE Code 52	Guide for Design and Strength of Overhead Transmission Lines	架空输电线路设计及强度导则
ASCII	American Standard Code for Information Interchange	美国信息交换标准码
ASHRAE	American Society of Heating, Refrigerating and Air Conditioning Engineers	美国采暖、制冷及空调工程师学会
ASME	American Society of Mechanical Engineers	美国机械工程师学会
ASME	Boiler and Pressure Vessel Code, Section IX Welding and Brazing Qualification	锅炉和压力容器规程,第IX节:焊接和铜焊限制条件
ASME	Section VIII Boiler and Pressure Vessel Code	锅炉与压力容器规范第VIII节
ASME TDP-1	Recommended Practice for the Prevention of Water Damage to the Steam Turbine used for Electric Power Generation	发电厂蒸汽汽轮机防进水保护的推荐操作规程

续表

英文缩写或编号	英文全称	中文
ASME-PTC10	Fluid Meter, Their Theory & Application	流量计的理论及应用
ASME-PTC19.2	Pressure Measurement	压力测量
ASME-PTC19.3	Temperature Measurement	温度测量
ASME-PTC19.5	Interim Supplement on Instrument and Apparatus Application. Part II of Fluid Meters (Sixth Edition)	关于仪表及仪器装置应用的临时补充,流量表的第Ⅱ部分(第六版)
ASNT	American Society for Nondestructive Testing	美国无损检测学会
ASTM	American Society for Testing and Materials	美国材料及试验学会
ASTM-A105	Specification for Forgings, Carbon Steel, Piping Components	锻件、碳钢管部件的技术要求
ASTM-A106	Specification for Seamless Carbon Steel Pipe for High Temperature Service	高温无缝钢管的技术要求
ASTM-A123	Specification for Zinc Coatings on Products Fabricated from Rolled, Pressed and Forged Steel Shapes, Plates, Bars and Stripes	辊轧、压制及锻造型钢、钢板、钢辊及钢带成品的镀锌规范
ASTM-A164	Electrodeposited Coating of Zinc on Steel	钢的锌电镀层
ASTM-A182	Specification for Forged or Rolled Alloy-Steel Pipe Flanges, Forged Fittings, and Valves and Parts for High Temperature Services	用于高温的锻制或轧制合金钢管法兰、锻制接头、阀门及部件的技术要求
ASTM-A217	Specification for Alloy Steel Castings for Pressure Containing Stainless Steel Tubing for General Services	一般用途的合金钢铸件含压力不锈钢管的技术要求
ASTM-A269	Specification for Seamless and Welded Austenitic Stainless Steel Tubing for General Services	一般用途的无缝和焊接奥氏体不锈钢管的技术要求
ASTM-A307	Standard Specification for Carbon Steel Externally and Internally Threaded Standard Fasteners	用于碳钢标准内外螺纹紧固件的技术要求

续表

英文缩写或编号	英文全称	中文
ASTM-A334	Electrical and Mechanical Properties of Magnetic Materials	磁性材料的电气和机械特性
ASTM-A36	Standard Specification for Structural Steel	结构钢标准规范
ASTM-A376	Specification for Seamless Austenitic Steel Pipe for High Temperature Central Station Services	用于高温中心电站的无缝奥氏体钢管的技术要求
ASTM-A386	Specifications for Zinc Coating (Hot-Dip) on Assembled Steel Products	组装钢件的镀锌层(热镀)规范
ASTM-A525	Specification for Zinc Coated (Galvanized) Iron or Steel Sheet, Coils and Cut Lengths	镀锌层(电镀锌)的铁或钢板、卷带和切割长度的规范
ASTM-B-43	Specification for Seamless Red Brass Pipe, Standard Sizes	无缝紫铜管的标准尺寸和技术要求
ASTM-B432	Copper and Copper Alloy Clad Steel Plate	铜和铜的合金复合钢板
ASTM-B-6	Galvanizing of Structural Steel Assemblies	结构钢组件的镀锌
ASTM-B-75	Specification for Seamless Copper Tube	无缝铜管的技术要求
ASTM-D1533	Test Method for Water in Insulating Liquids (Carl Fischer Methods)	绝缘液体中水分的试验方法(卡尔费斯切尔法)
ASTM-D3487	Specification for Mineral Insulating Oil Used in Electrical Apparatus	用于电气设备的无机绝缘油规范
ASTM-D923	Method of Sampling Electrical Insulating Liquids	电气绝缘液体的取样方法
ASTM-D924	Test Method for Power Factor and Dielectric Constant of Electrical Insulating Liquids	电气绝缘液体功率因数和介电常数的试验方法
AWC	American or Brown Sharpe Gauge for Non-ferrous Sheet and Wire	美国或布朗-夏普有色金属板材、线材标准
AWS	American Welding Society	美国焊接学会
AWS-A2.4	Symbols for Welding, and Nondestructive Testing, including Brazing, DOD Adopted	焊接及无损探伤试验符号,包括采用DOD黄铜焊

续表

英文缩写或编号	英文全称	中文
AWS-D1.1	Structural Welding Code for Steel	钢结构焊接规范
AWWA	American Water Works Association	美国给水工程协会
BRI	Building Research Institute	建筑研究院
CABRA	Copper and Brass Research Association	美国紫铜和黄铜研究学会
CDA	Copper Development Association	美国铜开发学会
CGA	Compressed Gas Association	压缩气体协会
CHy	Commission for Hydrometeorology	水文气象委员会
CIRED	Center for International Research of Environment and Development	国际环境与发展研究中心
CMAA	Crane Manufacturers Association of America	美国起重机制造商协会
CRD	Chief of Research and Development	研究开发总署
CRSI	Concrete Reinforcing Steel Institute	美国混凝土钢筋学会
ECC	European Committee for Concrete	欧洲混凝土委员会
ECCS	European Convention for Constructional Steelworks	欧洲钢结构会议
EIA	Electronic Industries Association	电子工业协会
EIA-RS-232-C	Interface Between Data Terminal Equipment and Data Communication Equipment Using Series Byte Data for Performing Data Exchange	数据终端设备与使用串行二进制数据进行数据交换的数据通信设备之间的接口
EPA	Environmental Protection Agency	环境保护局
EPRI	Electric Power Research Institute	电力研究院
ESC	European Seismological Commission	欧洲地震委员会
FM	Factory Mutual Testing Laboratories	工厂通用实验室
GSA	The Geological Society of America	美国地质学会
HEI	Heat Exchange Institute	热交换学会
HI	Hydraulic Institute	水利学会
HIS	Hydraulic Institute Standards	液压标准学会
HMI	Hoists Manufacture's Institute	起吊机具制造商协会

续表

英文缩写或编号	英文全称	中文
IABSE	International Association for Bridge and Structural Engineering	国际桥梁和结构工程协会
IAEA	International Atomic Energy Agency	国际原子能机构
IAEE	International Association for Earthquake Engineering	国际地震工程协会
IAEG	International Association of Engineering Geology	国际工程地质学会
IAH	International Association of Hydrology	国际水文学学会
IAHR	International Association for Hydraulic Research	国际水力学研究协会
IAM	International Association of Meteorology	国际气象学会
IAS	International Association of Seismology	国际地震学协会
IASS	International Association for Shell Structures	国际壳体结构协会
IAWPR	International Association on Water Pollution Research	国际水污染研究协会
ICEA	Insulated Cable Engineers Association	绝缘电缆工程师协会
ICEA-P-54-440	Standard Publication	标准出版物
ICEA-S-66-524	Cross-linked-thermosetting-polyethylene-insulated Wiring and Cable for the Transmission and Distribution of Electrical	用于输配电工程的交联热固性聚乙烯绝缘电线和电缆
ICEA-S-68-516	Ethylene-Propylene-Rubber Insulated	用于输配电工程的乙丙烯橡胶绝缘电线和电缆
ICG	International Commission on Groundwater	国际地下水委员会
ICI	International Commission on Illumination	国际照明委员会
ICSW	International Commission on Surface Water	国际地表水委员会
ICWQ	International Commission on Water Quality	国际水质委员会

续表

英文缩写或编号	英文全称	中 文
ICWRS	International Commission on Water Resources System	国际水资源系统委员会
IDSA	Industrial Designers Society of America	美国工业设计师学会
IEC	International Electrotechnical Commission	国际电工委员会
IEC-113	Diagrams, Charts and Tables	图表
IEC-117	Recommended Graphical Symbols	推荐的图例符号
IEC-129	Alternating Current Disconnector and Earthing Switches	交流电隔离开关和接地开关
IEC-146	Semiconductor Convertors	半导体换流器
IEC-146,146A,146.2	High Voltage Switches	高压开关
IEC-157-1	Low-voltage Switchgear & Controlling gear	低电压开关设备和控制设备
IEC-168	Tests on Indoor and Outdoor Post Insulators and Post Insulator Units for Systems with Nominal Voltages Greater than 1000V	系统标称电压大于1kV的户内户外柱型绝缘子及绝缘子组件的试验
IEC-183	Guide to the Selection of High-Voltage Cables	高压电缆选择导则
IEC-185	Current Transformers	电流互感器
IEC-186	Voltage Transformers	电压互感器
IEC-228	Conductors of Insulated Cables	绝缘电缆导体
IEC-230	Impulse Tests on Cables and Their Accessories	电缆及其附件的冲击试验
IEC-233	Test on Hollow Insulators for Use in Electrical Equipment	电气设备用的空心绝缘子的试验
IEC-245	Rubber Insulated Cables of Rated Voltage up to and Including 450/750V	450/750V及以上额定电压的橡皮绝缘电缆
IEC-255	Electrical Relays	继电器
IEC-267	Guide to the Testing of Circuit Breaker with Respect to Out-of-phase Switching	断路器失步操作试验导则

续表

英文缩写或编号	英文全称	中　文
IEC-269	Low-voltage Fuses	低压熔断器
IEC-273	Dimensions of Indoor and Outdoor Post Insulators and Post Insulator Units for Systems with Nominal Voltages Greater than 1000V	系统标称电压大于1000V的户内户外柱型绝缘子和绝缘子组件的尺寸
IEC-28	International Standard of Resistance for Copper	铜的电阻国际标准
IEC-282	High Voltage Fuses	高压熔断器
IEC-287	Calculation of the Continuous Current Rating of Cables (100% load factor)	电缆的持续额定电流值的计算(100%负荷系数)
IEC-292	Low Voltage Motor Starters	低压电动机启动器
IEC-296	Specification for Unused Mineral Insulating Oils for Transformers and Switchgear	变压器及断路器用的新无机绝缘油规范
IEC-309	Plugs, Socket Outlets and Couplers for Industrial Purposes	工业用插头、插座出口和接头
IEC-332	Tests on Electric Cables under Fire Conditions	电缆着火条件试验
IEC-337	Control Switches (Low Voltage Switching Devices for Control and Auxiliary Circuits Including Contactor Relays)	控制开关(控制及辅助回路的低压开断电路,包括接触继电器)
IEC-34	Rotating Electrical Machines	旋转电动机械
IEC-376	SF_6 Gas Specifications (include Supplement 376a and 376b)	六氟化硫气体规范(包括补遗的376a和376b)
IEC-44	Instrumentation Transformers	仪用互感器
IEC-445	Identification of Apparatus Terminals and General Rules for a Uniform System of Terminal Marking, Using an Alphanumeric Notation	设备端子的符号和采用字母数字符的端子标记统一系统的通用规则
IEC-446	Identification of Insulated and Bare Conductors by Colors	绝缘及裸露导体的色彩标识
IEC-50	International Electrotechnical Vocabulary	国际电力技术用词表

续表

英文缩写或编号	英文全称	中　文
IEC-502	Extruded Solid Dielectric Insulated Power Cables for Rated Voltages from 1kV up to 30 kV	额定电压1~30kV挤压固体电介质绝缘动力电缆
IEC-51	Recommendation for Direct Acting Indicating Electrical Measuring and Their Accessories	直接动作电测指示仪表及其附件的推荐
IEC-529	Classification of Degrees of Protection Provided by Enclosures	外壳防护等级的分类
IEC-540	Tests Methods for Insulation and Sheaths of Electric Cable and Cords (Elastomeric and Thermoplastic Compounds)	电缆和电线的绝缘和保护套的试验方法(橡胶和热塑性塑料化合物)
IEC-56	High-Voltage Alternating-Current Circuit Breakers	高压交流断路器
IEC-59	Current Ratings	电流额定值
IEC-60	High-Voltage Test Techniques	高压试验技术
IEC-605	Equipment Reliability Testing	设备可靠性测试
IEC-652	Loading Tests on Overhead Line Tower	架空输电线路铁塔荷载试验
IEC-71	Insulation Co-ordination	绝缘配合
IEC-72	Dimensions and Output Ratings for Rotating Electrical Machines. Frame Numbers 56 to 400 and Flange Numbers FF 55 to FF 1080 and FT 55 to FT 1080	旋转电动机械的尺寸和输出额定值机架号56~400和法兰号FF55~FF1080及FT55~FT1080
IEC-726	Dry Type Power Transformers	干式电力变压器
IEC-73	Colors of Indicator Lights and Push-buttons	指示灯和按钮的颜色
IEC-76	Power Transformers	电力变压器
IEC-79	Electrical Apparatus for Explosive Gas Atmospheres	有爆炸性气体场所的电气设备
IEC-826	Loading and Strength of Overhead Transmission Lines	架空输电线路的荷载及强度
IEC-99	Non-Linear Resistor Type Surge Arresters for Alternating-Current System	用于交流电系统的非线性电阻型避雷器

续表

英文缩写或编号	英文全称	中文
IEC-TC37, WG4	Gapless Metal-Oxide Surge Arresters for Alternating-Current System	用于交流电系统的无间隙金属氧化物避雷器
IEEE	Institute of Electrical and Electronics Engineers	电气与电子工程师协会
IEEE 112	Test Procedure for Polyphase Induction Motors and Generators	多相感应电动机和发电机的试验步骤
IEEE 113	Test Code for Direct Current Machines	直流设备的试验规程
IEEE 114	Test Procedure for Single Phase Induction Motors	单相感应电动机的试验步骤
IEEE 119	Recommended Practice for General Principles of Temperature Measurements as Applied to Electrical Apparatus	电气设备测温通则推荐实用作法
IEEE 142	Recommended Practice for Grounding of Industrial and Commercial Power System	工业和商业电力系统接地推荐实用作法
IEEE 18	Standard for Shunt Power Capacitors	并联电力电容器标准
IEEE 21	General Requirements and Test Procedure for Outdoor Apparatus Bushing	户外设备套管试验程序一般要求
IEEE 32	Requirements, Techniques, Test Procedures for Neutral Grounding Devices	中性点接地装置要求、技术和试验步骤
IEEE 4	Standard Techniques for High Voltage Testing	高压试验标准技术
IEEE 472	Guide for Surge Withstanding Capability	冲击耐受能力试验导则
IEEE 48	IEEE Standard Test Procedures and Requirements for High-Voltage Alternating Current Cable Termination	交流高压电缆终端装置IEEE标准试验工序和要求
IEEE 488	Digital Interface for Programmable Metering	可编程仪表的数字接口
IEEE 587	Guide for Surge Voltage in Low-Voltage AC Power Circuits	低压交流电回路冲击电压导则

续表

英文缩写或编号	英文全称	中文
IEEE 691—2001	Guide for Transmission Structure Foundation Design and Testing	输电线路结构基础设计与检验导则
IEEE 80	Guide for Safety in Alternating Current Substation Grounding	变电站交流接地安全指南
IEEE 82	Test Procedure for Impulse Voltage Tests on Insulated Conductors	绝缘导体冲击电压试验程序
IEEE-ATS	Analysis of Seismic Effects on Transmission Structures	输电线路结构的地震影响分析
IEEE-DAGTT	Design and Analysis of Guyed Transmission Towers by Computer	微机辅助输电线路拉线铁塔的设计与分析
IEEE-TSF	Transmission line Structure Foundation for Uplift-Compression Loading	适应上拨下压荷载的输电线路结构基础
IEEE-TTF	Transmission Tower Foundation	输电线路铁塔基础
IES	Illuminating Engineering Society	美国照明工程学会
IFCE	International Federation of Consulting Engineers	国际顾问工程师协会
IFLA	International Federation of Landscape Architects	国际园林设计师协会
IFP	International Federation of Prestressing	国际预应力协会
IGCI	Industrial Gas Cleaning Institute	工业废气洁净学会
IHB	International Hydrographic Bureau	国际水文局
IHP	International Hydrological Program	国际水文计划组织
IIR	International Institute of Refrigeration	国际制冷学会
IISEE	International Institute of Seismology and Earthquake Engineering	国际地震学与地震工程研究所
IJCTB	International Joint Committee for Tall Buildings	国际高层建筑联合委员会
IMF	International Monetary Fund	国际货币基金组织
IPCEA	Insulated Power Cable Engineers Association	绝缘电力电缆工程师协会
ISA	The International Society of Automation	国际自动化学会

续表

英文缩写或编号	英文全称	中文
ISA-BP31.1	Specifications, Installation, and Calibration of Turbine Flowmeters	涡轮流量计的规范、安装和校验
ISA-RP12.1	Electrical Instruments in Hazardous Atmosphere	危险环境下使用的电气仪表
ISA-RP16.1.2.3	Terminology, Dimension and Safety Practices for Indicating 2, 3 Variable Area Meter (Rotameter)	指示用二、三变量范围测量仪表的术语、尺寸及其安全应用(转子流量计)
ISA-RP16.4	Nomenclature and Terminology for Extension Type Variable Area Meters (Rotameters)	扩展型变量范围仪表的名词及术语(转子流量计)
ISA-RP16.5	Installation, Operation and Maintenance Instructions for Glass Tube Variable Area Meters	玻璃管式变量范围仪表的安装、运行和维护指导
ISA-RP16.6	Methods and Equipment for Calibration of Variable Area Meters	变量范围仪表的校验方法和设备
ISA-RP3.2	Flange Mounted Share Edged Orifice Plates Flow Measurement	法兰安装式孔板的流量测量
ISA-RP42.1	Nomenclature for Instrument, Tubing-Fitting (Threaded)	仪表管道接头(带螺纹的)的专用术语
ISA-RP55.1	Hardware Testing of Digital Process Computers	数字计算机的硬件试验
ISA-RP7.1	Pneumatic Control Circuit Pressure Test	气动控制回路压力试验
ISA-S12.11	Electrical Instruments in Hazardous Dust Locations	危险粉尘环境中的电气仪表
ISA-S20	Specification Forms for Process Measurement and Control Instruments, Primary Elements and Control Valves	用于过程测量和控制仪表的一次元件和控制阀门的规范格式
ISA-S37.1	Electrical Transducer Nomenclature and Terminology	电动变送器的专用术语
ISA-S37.12	Specification and Tests for Potentiometric Displacement Transducers	电位测定位移传感器的规范和试验
ISA-S37.3	Specifications and Tests for Strain Pressure Transducers	应变式压力变送器的规范与试验

续表

英文缩写或编号	英文全称	中文
ISA-S37.6	Specification and Test of Potentiometric Pressure Transducers	电位测定压力变送器的规范和试验
ISA-S37.8	Specifications and Tests for Strain Gage Force Transducers	应变式测力传感器的规范和试验
ISA-S5.1	Instrumentation Symbols and Identification	仪表符号与标志
ISA-S5.3	Binary-Logic Diagrams for Process Operation	过程控制中的二进制逻辑图
ISA-S5.4	Instrument Loop Diagram	仪表回路图
ISA-S50.1	Compatibility of Analog Signals for Electronic Industrial Process Instruments	电子工业过程仪表模拟信号的兼容性
ISA-S51.1	Process Instrumentation Terminology	过程仪表术语
ISC	International Seismological Center	国际地震中心
ISO	The International Organization for Standardization	国际标准化组织
ISSMFE	International Society for Soil Mechanics and Foundation Engineering	国际土力学与基础工程学会
ISSS	International Society of Soil Science	国际土壤学会
ITA	International Tunneling Association	国际隧道学会
ITT	International Telephone and Telegraph Corporation	国际电话与电报公司
IUGS	International Union of Geological Science	国际地质科学协会
IWRA	International Water Resources Association	国际水资源协会
IWSA	International Water Supply Association	国际供水协会
LIA	Lead Industries Association, Inc USA	美国铅工业协会股份公司
MBMA	Metal Building Manufacturers Association	金属建材制造者协会
MEPC	Maritime Environment Protection Committee	海洋环境保护委员会

续表

英文缩写或编号	英文全称	中文
MSS	Manufacturers Standardization Society of The Valve and Fitting Industry	阀门及其配件工业制造商标准化协会
NACE	National Association of Corrosion Engineers	全国防腐蚀工程师协会
NAERC	North American Electric Reliability Council	北美电气可靠性委员会
NBS	National Bureau of Standards	美国国家标准局
NCMA	National Concrete Masonry Association	国家混凝土圬工协会
NEBB	National Environmental Balancing Bureau	国家环境平衡局
NEC	National Electrical Code	全国电气规程
NEIC	National Earthquake Information Center	国家地震情报中心
NEMA	National Electrical Manufacturers Association	全国电气制造商协会
NEMA-107	Motor Method of Radio Interference Test for High Voltage Device	高压电气设备的无线电干扰的电机测量方法
NEMA-AB1	Mold Case Circuit Breakers	塑壳断路器
NEMA-CC1	Electric Power Connectors	电力连接器
NEMA-ICS	Industrial Control and Systems	工业控制和系统
NEMA-ICS1	General Standard for Industrial Control Systems	工业控制系统用的一般标准
NEMA-ICS2	Standards for Industrial Control Devices, Controllers and assemblies	工业控制装置、控制设备和成套装置的标准
NEMA-ICS4	Terminal Blocks for Industrial Control Equipment	工业控制设备的端子排
NEMA-ICS6	Enclosures for Industrial Controls and Systems	工业控制和系统的外壳
NEMA-MG1	Motor and Generators	电动机和发电机
NEMA-MG2	Safety Standard for Construction and Guide for Selection, Installation and Use	结构安全标准及选择、安装和使用导则

续表

英文缩写或编号	英文全称	中文
NEMA-ML1	Metal Framing (Continuous Slot Metal Channel Systems)	金属框架(连续切口金属槽系统)
NEMA-PB1	Panelboards	配电板
NEMA-PV3	Safety Codes for Semiconductor Power Conversion Equipment	半导体电力转换(逆变)设备的安全规范
NEMA-PV4	Semiconductor Self-Commutated Converters	半导体自动整流转换器
NEMA-SG4	Alternating Current High Voltage Circuit Breaker	交流高压断路器
NEMA-SG5	Power Switchgear Assemblies	动力开关装置
NEMA-ST20	Dry Type Transformer for General Use	通用干式变压器
NEMA-TR1	Transformers, Regulators and Reactors	变压器、调压器和电抗器
NEMA-TR98	Guide for Loading Oil Immersed Power Transformers with 65℃ Average Winding Rise	线圈平均温升65℃的油浸电力变压器带负荷导则
NEMA-VE1	Cable Tray Systems	电缆托架系统
NEMA-WC30	Color Coating of Wires and Cables	导线和电缆的着色规则
NEMA-WC51	Ampacities of Cables in Open-top Cable Trays	开启式电缆托架内电缆载流量
NEMA-WC7	Insulated Wire and Cable for Transmission and Distribution of Electric Energy	输/配电绝缘电线及电缆
NEMA-WC8	Wire and Cable for Transmission and Distribution of Electrical Energy	输/配电电线及电缆
NEPA	National Environmental Policy Act	国家环境政策法
NESC	National Electric Safety Code	全国电气安全规程
NFPA	National Fire Protection Association	全国防火协会
NFPA-10	Standard for Portable Fire Extinguishers	移动式灭火器标准
NFPA-12	Standard on Carbon Dioxide Extinguishing System	CO_2灭火系统标准
NFPA-13	Standard for the Installation of Sprinkler System	水喷淋系统安装标准

续表

英文缩写或编号	英文全称	中文
NFPA-14	Standard for the Installation of Standpipe Hoses System	立管和水龙带系统安装标准
NFPA-15	Water Spray Fixed Systems for Fire Protection	固定式水喷雾系统消防标准
NFPA-1961	Fire Hose	消防水龙带
NFPA-1963	Fire Hose Connections	消防水龙带接头
NFPA—2001	Standard on Clean Agent Fire Extinguishing System	清洁药剂灭火系统标准
NFPA-24	Private Fire Service Mains and Their Appurtenances	独立消防系统主要部件及附件
NFPA-26	Recommended Practices for the Supervision of Valves Controlling Water Supplies for Fire Protection	消防给水控制阀门的监视推荐作法
NFPA-291	Fire Hydrants	消火栓
NFPA-31	Installation of Oil Burning Equipment	燃油设备的安装
NFPA-70	National Electrical Code	国家电气规程
NFPA-72	National Fire Alarm Code	国家火灾报警规程
NFPA-72A	Local Protective Signaling System	就地保护信号系统
NFPA-72D	Dedicated Protective Signaling System	专用保护信号系统
NFPA-72E	Standard on Automatic Fire Detectors	火警自动监测器标准
NFPA-850	Recommended Practice for Fire Protection for Electric Generating Plant and High Voltage Direct Current Converter Station	发电厂及高压直流换流站防火保护的推荐操作规程
NFPA-85D	Explosion Prevention, Multiple Burner Boiler-Furnaces, Fuel Oil-fired	多燃烧器燃油锅炉炉膛防爆
NFPA-85E	Explosion Prevention, Multiple Burner Boiler-Furnaces, Pulverized Coal-fired	多燃烧器煤粉锅炉炉膛防爆
NFPA-85F	Installation and Operation of Pulverized Fuel System	制粉系统的安装和操作运行

续表

英文缩写或编号	英文全称	中文
NFPA-85G	Prevention of Furnace Implosion in Multiple Burner Boiler-Furnaces	多燃烧器炉膛防内爆
NISEE	National Information Service for Earthquake Engineering	国家地震工程情报服务处
NRB	National Resources Board	国家资源局
NRMCA	National Ready Mixed Concrete Association	全国混凝土制成品协会
OECD	Organization for Economic Cooperation and Development	经济合作与发展组织
OEEC	Organization for European Economic Cooperation	欧洲经济合作组织
OSHA	Occupational Safety and Health Administration	美国劳动部职业安全与卫生局
OSHA-Part 1910: Subpart G, Section 1910.95	Occupational Noise Exposure	第1910部分G分部1910.95节:职业噪声接触
OSHA-Subpart O, Section 1910.219	Mechanical Power Transmission Apparatus	第1910部分O分部1910.219节:机械动力传输装置
OSHA-Subpart Q, Section 1910.252	Welding, Cutting and Brazing	第1910部分Q分部1910.252节:焊接、切割和铜焊
PCI	Prestressed Concrete Institute	预应力混凝土研究所
PFI	Pipe Fabricators Institute	管道制造商学会
PWA	Public Works Administration	公用(事业管理)局
RILEM	International Union of Testing and Research Laboratories for Materials and Structures	国际建筑材料及结构试验及研究实验所联合会
RMA	American Rubber Manufacturers Association	美国橡胶制造商协会
SAE	Society of Automotive Engineers	美国汽车工程师学会
SAMA	Scientific Apparatus Manufacturer Association	科学器具制造商协会
SAMA-RC22-11	Functional Diagramming of I/C System	I/C(仪表与控制)系统功能图

续表

英文缩写或编号	英文全称	中文
SCS	Soil Conservation Service	土壤保持局
SDA	States Department of Army	美国军事部
SDI	Steel Deck Institute, USA	美国钢甲板学会
SEH	Structural Engineering Handbook	结构工程手册
SI	International System of Units	国际单位制
SMACNA	Sheet Metal and Air Conditioning Contractors National Association, Inc.	全国薄板金属和空调承包商协会股份公司
SMACNA-APIDC	Acceptable Practice for Industrial Duct Construction	合格的工业管道制造方法
SMACNA-DLAS	Duct Liner Application Standard	管道衬适用标准
SMACNA-HPDCS	High Pressure Duct Construction Standards	高压管道制造标准
SMACNA-LPDC	Low Pressure Duct Construction Standards	低压管道制造标准
SMACNA-ReIDCS	Rectangular Industrial Duct Construction Standards	矩形工业管道制造标准
SMACNA-RoIDCS	Round Industrial Duct Construction Standards	圆形工业管道制造标准
SSA	Seismological Society of America	美国地震学会
SSC	Seismic Safety Commission	地震安全委员会
SSPC	Structural Steel Painting Council	钢结构油漆委员会
STAF	Standard for Testing Air Filters by the Dust Spot Methods	用聚尘法对滤气器进行测试的标准
TEMA	Tubular Exchanger Manufacturers Association	管式交换器制造商协会
TMB	Taylor Model Basin	泰勒模型池
UBC	Uniform Building Code of USA	美国建筑统一规范
UIA	International Union of Architects	国际建筑师联合会
UICB	International Union of Building Centers	国际建筑中心联合会
UIEO	Union of International Engineering Organizations	国际工程机构联合会
UL	Underwriters Laboratories, Inc.	保险商实验室股份公司

续表

英文缩写或编号	英文全称	中文
UL-198C	Standard for High-Interrupting Capacity Fuses, Current-Limiting Type	限流型、高遮断能力熔断器的标准
UL-198D	Standard for Class K Fuses	K级熔断器标准
UL-198E	Standard for Class R Fuses	R级熔断器标准
UL-44	Safety Standard for Rubber Insulated Wires and Cable	橡胶绝缘电线和电缆的安全标准
UL-486A	Wire Connectors and Soldering Lugs for Use with Copper Conductors	用于铜导线的导线连接装置[接线盒等]与焊接片
UL-489	Molded Case Circuit Breakers and Circuit Breaker Enclosures	塑壳式断路器和断路器外壳
UL-508	Electric Industrial Control Equipment	工业电气控制设备
UL-512	Fuseholders	熔断器基座
UL-845	Electric Motor Control Centers	电动机控制中心
UL-94	Test for Flammability of Plastic Materials for Parts for Devices and Appliances	装置、器具中塑料的耐燃试验
UL-Vol. 1	National Standard for Field Measurements and Instrumentation, Testing & Balancing	国家标准第一卷：现场测试和仪器仪表,测试与平衡
UNDP	United Nations Development Program	联合国开发计划署
UNEP	United Nations Environment Program	联合国环境规划署
UNIDO	United Nations Industrial Development Organization	联合国工业开发组织
USGS	United States Geological Survey	美国地质调查局
USWB	United States Weather Bureau	美国气象局
WEI	World Environment Institute	世界环境学会
WERC	World Environment and Resources Council	世界环境和资源委员会
WMO	World Meteorological Organization	世界气象组织
WPC	World Power Conference	世界动力会议
WRC	Water Resources Congress	水资源大会

续表

英文缩写或编号	英文全称	中　文
WRC	Water Resources Council	水资源理事会
WRL	Water Resources Laboratory	水资源实验室
英国标准		
AEI	Associated Electrical Industries	联合电气公司
BCEA	British Central Electricity Authority	英国中央电气管理局
BEAMA	British Electrical and Allied Manufacturer's Association	英国电气联合制造者协会
BESA	British Engineering Standard Association	英国工程标准协会
BICERA	British Internal Combustion Engine Research Association	英国内燃机研究协会
BNF	British Nuclear Forum	英国核论坛
BRS	Building Research Station	建筑研究所
BS	British Standard	英国标准
BS 449	The Use of Structural Steel in Building	建筑结构钢的使用
BS 4871	Welding Standard	焊接标准
BS 4959—1974	Recommendations for Corrosion and Scale Prevention in Engine Cooling Water System	发动机冷却水系统的防腐蚀和防水垢推荐标准
BS 5155—1984	Specification for Butterfly Valves	蝶形阀门规范
BS 5156—1985	Specification for Diaphragm Valves	隔膜阀规范
BS 599—1966	Methods of Testing Pumps	泵的试验方法
BS 729	Galvanizing	镀锌标准
BSI	British Standards Institution	英国标准学会
BSS	British Standard Specification	英国标准规范
BVMA	British Valve Manufactures Association	英国阀门制造者协会
BWRA	British Welding Research Association	英国焊接研究协会
CERL	Central Electricity Research Laboratories, UK	英国中央电力研究所

英文缩写或编号	英文全称	中文
COID	Council of Industrial Design	英国工业设计会议
CP 312 pt. 1—1973	Code of Practice for Plastics Pipework (Thermoplastics Material)	塑料管道工程(热塑材料)的实用规则
EPEA	Electric Power Engineers' Association	电力工程师协会
GSL	Geological Society of London	伦敦地质学会
ICE	Institution of Civil Engineers	土木工程师学会
IEE	Institute of Electrical Engineers	电机工程师学会
ISE	Institution of Structural Engineers	结构工程师学会
ISI	Iron & Steel Institute	钢铁学会
RCA	Reinforced Concrete Association	钢筋混凝土协会
RIBA	Royal Institute of British Architects	英国皇家建筑师学会
RMS	Royal Meteorological Society	皇家气象学会
RS	Royal Society	皇家学会
SIE	Society of Industrial Engineers	工业工程师协会
TPI	Town Planning Institute	英国城市规划学会
德国标准		
DIN 1012	Galvanized Steel Plate	镀锌钢板
DIN 1013	Round Bars	圆钢
DIN 1017	Flat Bars	扁钢
DIN 1026	Channels	槽钢
DIN 1028	Equal Angles	等边角钢
DIN 1029	Unequal Angles	非等边角钢
DIN 17100	Steel of General Structural Purpose: Quality Standard	通用结构钢:质量标准
DIN 267	Bolt	螺栓标准
DIN 8851 Part 1	Welding Execution	焊接执行标准
DIN EN 10 025(1990)	Hot Rolled Steel	热轧钢
DIN EN 287-1	Welding Standard	焊接标准

续表

英文缩写或编号	英文全称	中文
DIN VDE 0120	Planning and Design of Overhead Powerlines with Rated Voltage above 1kV	1kV 以上等级的架空输电线路的规划与设计
KTG	German Nuclear Society	德国核协会
日本标准		
AAJ	Architectural Association of Japan	日本建筑协会
AIJ	Architectural Institute of Japan Standards	日本建筑学会标准
CPAJ	City Planning Association of Japan	日本都市计划协会
CPIJ	City Planning Institute of Japan	日本都市计划学会
JAA	Japan Architects Association	日本建筑家协会
JAIF	Japan Atomic Industrial Forum	日本原子工业论坛
JCS	Japan Cable Manufactures Association Standards	日本电缆制造商协会标准
JEAC	Japan Electric Association Code	日本电气协会规范
JEC	Japanese Electrotechnical Committee Standards	日本电气技术委员会标准
JEM	The Standard of Japan Electrical Manufacturer's Association Standards	日本电气制造商协会标准
JIDA	Japan Industrial Designers Association	日本工业设计家协会
JIS	Japanese Industrial Standards	日本工业标准
JMA	Japanese Meteorological Agency	日本气象厅
JSCE	Japan Society of Civil Engineering	日本土木工程学会
JSEEP	Japan Society of Earthquake Engineering Promotion	日本地震工程促进协会
MITI	Engineering Standards of the Electrical Facilities	电气设备工程设计标准
NEC	Nippon Electric Company	日本电气公司
NEC	Nippon Electro-technical Committee	日本电工委员会

续表

英文缩写或编号	英文全称	中 文
NRCDP	National Research Center for Disaster Prevention	国家防灾研究中心
SSJ	Seismological Society of Japan	日本地震学会
	Manufacturer's Standards	制造商标准
	Recommended Practice of Explosion Protected Electrical Installations in General Industries	在一般工业中的电气设施防爆建议
	Technical Recommendations of the Research Institute of Industrial Safety of Ministry of Labor in Japan	日本劳工部工业安全研究所技术建议
法国标准		
SFEN	French Nuclear Society	法国核学会
STUP	Societe Technique pour I'Utilisation de la Precontrainte	法国预应力混凝土应用技术学会
澳大利亚标准		
ACSE	Association of Consulting Structural Engineers	结构顾问工程师协会
ANA	Australian Nuclear Association	澳大利亚核协会

国内部分(按标准编号列字母顺序排列)

标准编号	英文全称	中 文
CECS 17:1990	Design specification for UPVC pipeline in use for outdoor water supply engineering	室外硬聚氯乙烯给水管道工程设计规程
CECS 18:1990	Construction and acceptance specification for UPVC pipeline in use for outdoor water supply engineering	室外硬聚氯乙烯给水管道工程施工及验收规程
CECS 30:1991	Design standard for architectural reclaimed water system	建筑中水设计规范
CECS 31:2006	Code for design of steel cable tray engineering	钢制电缆桥架工程设计规范

续表

标准编号	英文全称	中文
CECS 40:1992	Specification for quality control of concrete and precast concrete components	混凝土及预制混凝土构件质量控制规程
CECS 41:1992	Specification for design installation and inspection of UPVC pipeline for water supply in building	建筑给水硬聚氯乙烯管道设计与施工验收规程
CECS 51:1993	Specification for design of reinforced concrete continuous beams and frames considering redistribution of internal forces	钢筋混凝土连续梁和框架考虑内力重分布设计规程
CJ/T 3080—1998	Cast iron spigot and socket drain pipes and fittings with flexible antiseismic joint	承插式柔性抗震接口排水铸铁管及管件
CJJ/T 29—1998	Technical specification of PVC-U pipe work for building drainage	建筑排水硬聚氯乙烯管道工程技术规程
CJJ/T 54—1993	Design code for wastewater stabilization pond	污水稳定塘设计规范
DL 5000—2000	Technical code for designing fossil fuel power plants	火力发电厂设计技术规程
DL 5014—1992	Technical regulation for designing of reactive power compensation equipment for 330~500kV substations	330~500kV 变电所无功补偿装置设计技术规定
DL 5022—2012	Technical regulation for design of civil structure of fuel power plants	火力发电厂土建结构设计技术规定
DL 5024—1993	Ground treatment technical regulation of coal fire power station	火力发电厂地基处理技术规定
DL 5027—1993	Specification for fire protection of electric power installations	电力设备典型消防规程
DL 5028—1993	Standard for electric power engineering drawings	电力工程制图标准
DL 5031—1994	The code of erection and acceptance for electric power construction (Piping Section)	电力建设施工及验收技术规范 管道篇

续表

标准编号	英文全称	中　　文
DL 5053—1996	Design code of labor safety and industrial hygiene for fossil fuel power plants	火力发电厂劳动安全和工业卫生设计规程
DL 5073—2000	Specifications for seismic design of hydraulic structures	水工建筑物抗震设计规范
DL 612—1996	Supervision code for boiler and pressure vessel of the power industry	电力工业锅炉压力容器监察规程
DL/T 403—2000	HV vacuum circuit-breaker for rated voltage 12kV to 40.5kV	12~40.5kV 高压真空断路器订货技术条件
DL/T 448—2000	Technical administrative of electric energy metering	电能计量装置技术管理规程
DL/T 478—2001	General specifications for static protection, security and automatic equipment	静态继电保护及安全自动装置通用技术条件
DL/T 486—2000	Specifications for HV AC disconnectors and earthing switches	交流高压隔离开关和接地开关订货技术条件
DL/T 487—2000	The distribution voltage along insulator string on A.C. overhead line with rated voltage of 330kV and 500kV	330kV 及 500kV 交流架空送电线路绝缘子串的分布电压
DL/T 5029—1994	Standard for designing of architectural finishing of fossil fuel power plant	火力发电厂建筑装修设计标准
DL/T 5032—2005	Technical code of general plan transportation design for fossil fuel power plants	火力发电厂总图运输设计技术规程
DL/T 5035—2004	Technical code for heating, ventilation and air conditioning design of fossil fuel power plant	火力发电厂采暖通风与空气调节设计技术规程
DL/T 5041—1995	Technical regulation for design of the inner communication of fuel power plant	火力发电厂厂内部通信设计技术规定
DL/T 5043—1995	Standard for design of electrical laboratory of thermal power plant	火力发电厂电气实验室设计标准
DL/T 5045—1995	Technical specification for designing of ash slag damming of fossil fuel power plants	火力发电厂灰渣筑坝设计技术规定

续表

标准编号	英文全称	中文
DL/T 5046—2006	Technical code for the design of waste water treatment of fossil fuel power plants	火力发电厂废水治理设计技术规程
DL/T 5047—1995	The code of erection and acceptance of electric power construction section of steam boiler set	电力建设施工及验收技术规范 锅炉机组篇
DL/T 5050—2000	Code of exploratory adits, shafts and trenches for water resources and hydropower project	水利水电工程坑探规程
DL/T 5052—1996	Standard for auxiliary ancillary living and welfare building's area for fossil fuel power plant	火力发电厂辅助、附属及生活福利建筑物建筑面积标准
DL/T 5054—1996	Code for design of thermal power plant steam/water piping	火力发电厂汽水管道设计技术规定
DL/T 5056—1996	Technical code of general plan design for substation	变电所总布置设计技术规程
DL/T 5059—1996	Standard for furnishing maintenance device and architectural area of fuel power plant	火力发电厂修配设备及建筑面积配置标准
DL/T 5068—2006	Technical code for designing chemistry of fossil fuel power plants	火力发电厂化学设计技术规程
DL/T 5072—2007	Code for designing insulation and painting of fossil fuel power plant	火力发电厂保温油漆设计规程
DL/T 5094—1999	Specification for design of building of fossil fuel power plants	火力发电厂建筑设计规程
DL/T 5111—2000	Specifications for construction supervision of hydroelectric and water resources projects	水电水利工程施工监理规范
DL/T 5112—2000	Construction specifications for hydraulic roller compacted concrete	水工碾压混凝土施工规范
DL/T 5114—2000	Design guide of construction diversion for hydropower and water conservancy project	水电水利工程施工导流设计导则
DL/T 5115—2008	Technical specifications for joint seal of concrete face rockfill dam	混凝土面板堆石坝接缝止水技术规范
DL/T 5116—2000	Design guide for construction planning of rolled earth-rock dam for hydropower and water conservancy project	水电水利工程碾压式土石坝施工组织设计导则

续表

标准编号	英文全称	中文
DL/T 5118—2000	Plan design guide for the rural electric power network	农村电力网规划设计导则
DL/T 5119—2000	Design code for the unattended miniaturization substation of rural electric network	农村小型化无人值班变电所设计规程
DL/T 5120—2000	DC System design code for small electric power project	小型电力工程直流系统设计规程
DL/T 5122—2000	Technical code of exploration and surveying for 500kV overhead transmission line	500kV架空送电线路勘测技术规程
DL/T 5123—2000	Specifications for engineering acceptance of hydropower station capital construction	水电站基本建设工程验收规程
DL/T 5136—2001	Technical code for designing of electrical secondary wiring in fossil fuel power plants and substations	火力发电厂、变电所二次接线设计技术规程
DL/T 5137—2001	Technical code for designing electrical measuring and energy metering device	电测量及电能计量装置设计技术规程
DL/T 5153—2002	Code for design of fossil fuel power plant electrical auxiliary system	火力发电厂厂用电设计技术规定
DL/T 552—1995	The testing code of dry cooling tower and condenser of thermal power plant	火力发电厂空冷塔及空冷凝汽器试验方法
DL/T 564—1995	Ripple control receiver for load control	音频负荷控制接收机
DL/T 588—1996	The determination of fouling index of water	水质污染指数测定方法
DL/T 589—1996	Directives of thermal instrumentation and control for coal fired boiler in power plant	火力发电厂燃煤电站锅炉的热工检测控制技术导则
DL/T 590—1996	Directives of thermal instrumentation and control for condensation type turbine in power plant	火力发电厂用凝汽汽轮机的热工检测控制技术导则
DL/T 591—2010	Specification of instrumentation and control for turbine-generator in fossil fuel power plant	火力发电厂汽轮发电机的检测与控制技术条件

续表

标准编号	英文全称	中文
DL/T 592—2010	Specification of instrumentation and control for boiler feed water pump in fossil fuel power plant	火力发电厂锅炉给水泵的检测与控制技术条件
DL/T 600—2001	General rules for drafting electric power professional standards	电力行业标准编写基本规定
DL/T 607—1996	Test on water leakage or hydrogen leakage of steam-turbine generator	汽轮发电机漏水、漏氢的检验
DL/T 608—1996	Guide for 200MW grade steam turbine operation	200MW级汽轮机运行导则
DL/T 609—1996	Guide for 300MW grade steam turbine operation	300MW级汽轮机运行导则
DL/T 610—1996	Guide for 200MW grade boiler operation	200MW级锅炉运行导则
DL/T 611—1996	Guide for 300MW grade boiler operation	300MW级锅炉运行导则
DL/T 712—2000	Guideline for the selection of condenser tube materials in power plant	火力发电厂凝汽器管选材导则
DL/T 713—2000	Immunity requirement for protection and control equipment in 500kV substation	500kV变电所保护和控制设备抗扰度要求
DL/T 715—2000	Selection guidelines for the metallic material of fossil-fired power plants	火力发电厂金属材料选用导则
DL/T 716—2000	Guide on selection of diaphragm valves for power plant	电站隔膜阀选用导则
DL/T 719—2000	Telecontrol equipment and systems Part 5: Transmission protocol Section 102: companion standard for the transmission of integrated totals in electric power systems	远动设备及系统 第5部分:传输规约 第102篇:电力系统电能累计量传输配套标准
DL/T 720—2000	General specifications for protection cabinets and panels of electric power systems	电力系统继电保护柜、屏通用技术条件
DL/T 721—2000	Remote terminal unit of distribution automation system	配电网自动化系统远方终端
DL/T 722—2000	Guide to the analysis and the diagnosis of gases dissolved in transformer oil	变压器油中溶解气体分析和判断导则

续表

标准编号	英文全称	中文
DL/T 723—2000	Technical guide for electric power system security and stability control	电力系统安全稳定控制技术导则
DL/T 724—2000	Specification of operation and maintenance of battery DC power supply equipment for electric power system	电力系统用蓄电池直流电源装置运行与维护技术规程
DL/T 725—2000	Specification of current transformer for electrical power for order	电力用电流互感器订货技术条件
DL/T 726—2000	Specification of voltage transformer for electrical power for order	电力用电压互感器订货技术条件
DL/T 727—2000	Guideline of operation and maintenance for current and voltage transformers	互感器运行检修导则
DL/T 729—2000	On-line condition of indoor insulators Electrical part	户内绝缘子运行条件电气部分
DL/T 731—2000	Error calculator used in electrical energy meter calibrating	电能表测量用误差计算器
DL/T 732—2000	Optical-electrical/acquisition used in electrical energy meter calibrating	电能表测量用光电采样器
DL/T 734—2000	Technical guide of welding repair for boiler drum in fossil-power plants	火力发电厂锅炉汽包焊接修复技术导则
DL/T 735—2000	Measurement and evaluation of the dynamic characteristic on stator end windings of the large turbo-generator	大型汽轮发电机定子绕组端部动态特性的测量及评定
DL/T 737—2000	The operated management standard of unattended-substation for the rural electric power network	农网无人值班变电所运行管理规定
DL/T 741—2001	Operating code for overhead transmission line	架空送电线路运行规程
DL/T 742—2001	Technical specifications for plastic of cooling tower	冷却塔塑料部件技术条件
DL/T 744—2001	General specifications for microprocessor-based motor protection equipment	微机型电动机保护装置通用技术条件
DL/T 746—2001	Guide on selection of butterfly valves for power plant	电站蝶阀选用导则

续表

标准编号	英文全称	中文
DL/T 747—2001	Guide to the quality inspection before acceptance of mechanical coal sampling equipment used by power plant	发电用煤机械采制样装置性能验收导则
DL/T 748.1—2001	Guide of maintenance of boiler unit for thermal power station Part 1: General rules	火力发电厂锅炉机组检修导则 第1部分:总则
DL/T 748.2—2001	Guide of maintenance of boiler unit for thermal power station Part 2: The maintenance of boiler proper	火力发电厂锅炉机组检修导则 第2部分:锅炉本体检修
DL/T 748.3—2001	Guide of maintenance of boiler unit for thermal power station Part 3: The maintenance of valve and steam-water system	火力发电厂锅炉机组检修导则 第3部分:阀门与汽水系统检修
DL/T 748.5—2001	Guide of maintenance of boiler unit for thermal power station Part 5: The maintenance of air & flue system	火力发电厂锅炉机组检修导则 第5部分:烟风系统检修
DL/T 748.6—2001	Guide of maintenance of boiler unit for thermal power station Part 6: The maintenance of precipitator	火力发电厂锅炉机组检修导则 第6部分:除尘器检修
DL/T 748.7—2001	Guide of maintenance of boiler unit for thermal power station Part 7: The maintenance of ash and slag handling system	火力发电厂锅炉机组检修导则 第7部分:除灰渣系统检修
DL/T 748.8—2001	Guide of maintenance of boiler unit for thermal power station Part 8: The maintenance of air preheater	火力发电厂锅炉机组检修导则 第8部分:空气预热器检修
DL/T 748.9—2001	Guide of maintenance of boiler unit for thermal power station Part 9: The maintenance of fly ash system	火力发电厂锅炉机组检修导则 第9部分:干输灰系统检修
DL/T 748.10—2001	Guide of maintenance of boiler unit for thermal power station Part 10: The maintenance of fossil-fired PP FGD equipment	火力发电厂锅炉机组检修导则 第10部分:脱硫装置检修
DL/T 749—2001	Code of performance test for ash handling system	除灰系统试验规程
DL/T 750—2001	Regulations for operation and maintenance of rotary air preheater	回转式空气预热器运行维护规程

续表

标准编号	英文全称	中文
DL/T 751—2001	Code for hydraulic turbine generator	水轮发电机运行规程
DL/T 752—2010	The code of the welding on dissimilar steel for power plant	火力发电异种钢焊接技术规程
DL/T 756—2009	Suspension Clamp	悬垂线夹
DL/T 757—2009	Strain clamp	耐张线夹
DL/T 758—2009	Splicing fittings	接续金具
DL/T 765.1—2001	Technical requirements for distribution fittings	架空配电线路金具技术条件
DL/T 770—2001	General specifications for microprocessor-based transformer protection equipment	微机变压器保护装置通用技术条件
DL/T 780—2001	Specification for neutral-to-earth resistance in distribution system	配电系统中性点接地电阻器
DL/T 781—2001	High frequency switching converter module in power system	电力用高频开关整流模块
DL/T 782—2001	Code of start-up & completion acceptance for power transmission & distribution project of 110kV and above	110kV及以上送变电工程启动及竣工验收规程
DL/T 821—2002	The code of radiographic examination of butt welded-joints of pressure steels and tubes	钢制承压管道对接焊接接头射线检验技术规程
DLGJ 105—2000	Procedure of filing technical documents of electric power engineering and design	电力勘测设计科技文件材料立卷归档办法
DLGJ 118—1997	Regulation for content and depth of feasibility study report of fossil fuel power plants	火力发电厂可行性研究报告内容深度规定
DLGJ 129—1996	Technical standard for designing cable bands	电缆扎带设计技术标准
DLGJ 132—1997	Management specification of verification and documentation for computer aided design of electrical engineering	电力工程计算机辅助设计成品校审及归档管理规定
DLGJ 135—1997	Design regulation for 3D modeling of main building in fossil fuel power plants	火力发电厂主厂房三维模型设计规定

续表

标准编号	英文全称	中文
DLGJ 138—1997	Temporary regulation of FGD part for the content and depth of feasibility study report of fossil fuel power plants	火力发电厂可研报告内容深度规定烟气脱硫部分暂行规定
DLGJ 151—2000	Code of content profundity for feasibility study of power system optical communication project	电力系统光缆通信工程可行性研究内容深度规定
DLGJ 152—2000	Code of content profundity for preliminary design of power system optical communication project	电力系统光缆通信工程初步设计内容深度规定
DLGJ 154—2000	Standard for design fire proof measure and Construction acceptance of cable	电缆防火措施设计和施工验收标准
DLGJ 155—2000	Management procedure of video and pictures of electric power engineering and design	电力勘测设计声像档案管理办法
DLGJ 56—1995	Lighting technical code for designing fossil fuel power plants and substations	火力发电厂和变电所照明设计技术规定
DLGJ 86—1996	Classifying coding method for scientific & technological files of electric engineering survey designs	电力工程勘测设计科技档案分类编号办法
GA 128—1996	Low voltage electrical fire simulated test rules of technicality	低压电器火灾模拟实验技术规程
GB 12348—2008	Emission standard for industrial enterprise noise at boundary	工业企业厂界环境噪声排放标准
GB 12476.1—2000	Electrical apparatus for use in the presence of combustible dust Part 1: Electrical apparatus protected by enclosures and surface temperature limitation-Specification for apparatus	可燃性粉尘环境用电气设备 第1部分:用外壳和限制表面温度保护的电气设备 第1节:电气设备的技术要求
GB 12951—2009	Specification of americium-241 alpha sources for ionization smoke fire detectors	离子感烟火灾探测器用Am241α放射源的技术条件
GB 13223—2003	Air pollutants discharge standard for coal-fired power plant	火电厂大气污染物排放标准

续表

标准编号	英文全称	中文
GB 13271—2001	Emission standard of air pollutants for coal-burning, oil-burning, gas-fired boilers	锅炉大气污染物排放标准
GB 150—2011	Steel made pressurized vessels	钢制压力容器
GB 156—2003	Standard voltage	标准电压
GB 15707—1995	Limits of radio interference from AC high voltage overhead power transmission lines	高压交流架空送电线无线电干扰限值
GB 16279—1996	Fire vehicle-engineering approval evaluation program	消防车定型实验规程
GB 16280—2005	Line type heat detectors	线性感温火灾探测器
GB 16282—1996	General technical conditions for 119 fire alarm system	119火灾报警系统通用技术条件
GB 16367—1996	Radiological protection standards for using geothermal water	地热水应用中的放射卫生防护标准
GB 16543—2008	Safely regulations for the explosion precautions of bituminous coal injection into blast furnace	高炉喷吹烟煤系统防爆安全规范
GB 16668—2010	General technical specifications for powder extinguishing system and components	干粉灭火系统及部件通用技术条件
GB 16669—2010	General technical specifications for components of carbon dioxide fire extinguishing systems	二氧化碳灭火系统及部件通用技术条件
GB 16726—1997	Reinforced concrete horizontal beam and longitudinal beam	钢筋混凝土开间梁、进深梁
GB 16728—1997	Prestressed concrete ribbed roof slabs	预应力混凝土肋形屋面板
GB 16749—1997	Bellows expansion joints for pressure vessel	压力容器波形膨胀节
GB 2099—1980	Single phase & three phase plug & socket technical conditions	单相、三相插头插座技术条件
GB 22337—2008	Emission standard for community noise	社会生活环境噪声排放标准
GB 3095—1996	Ambient air quality standard	大气环境质量标准

续表

标准编号	英文全称	中文
GB 3096—1993	Standard of environmental noise of urban area	城市区域环境噪声标准
GB 3445—2005	Indoor fire hydrant	室内消火栓
GB 3838—1988	Environmental quality standard for surface water	地面水环境质量标准
GB 4272—1992	General principles for thermal insulation technique of equipment and pipes	设备及管道保温技术通则
GB 4452—2011	Outdoor fire hydrant	室外消火栓
GB 4717—2005	Fire alarm control units	火灾报警控制器
GB 50003—2011	Code for design of masonry structures	砌体结构设计规范
GB 50005—2003	Code for design of timber structures	木结构设计规范
GB 50007—2011	Code for design of building foundations	建筑地基基础设计规范
GB 50009—2001	Load code for design of building structures (2006 Version)	建筑结构荷载规范(2006年版)
GB 50010—2010	Code for design of concrete structures	混凝土结构设计规范
GB 50011—2010	Code for seismic design of buildings	建筑抗震设计规范
GB 50013—2006	Code for design of outdoor water supply engineering	室外给水设计规范
GB 50014—2006	Code for design of outdoor wastewater engineering	室外排水设计规范
GB 50015—2003	Code for design of building water supply and drainage (2009 Version)	建筑给水排水设计规范(2009年版)
GB 50016—2006	Code of design on building fire protection and prevention	建筑设计防火规范
GB 50017—2003	Code for design of steel structures	钢结构设计规范
GB 50018—2002	Technical code for design of cold-formed thin-wall steel structures	冷弯薄壁型钢结构技术规范
GB 50019—2003	Code for design of heating ventilation and air conditioning	采暖通风和空气调节设计规范

续表

标准编号	英文全称	中文
GB 50021—1994	Code for investigation of geotechnical engineering (2009 Version)	岩土工程勘察规范(2009年版)
GB 50023—2009	Standard for seismic appraisal of buildings	建筑抗震鉴定标准
GB 50025—2004	Code for building construction in collapsible loess zone	湿陷性黄土地区建筑规范
GB 50026—2007	Code for engineering surveying	工程测量规范
GB 50027—2001	Code for hydrogeological investigation of water supply	供水水文地质勘察规范
GB 50028—2006	Code for design of city gas engineering	城镇燃气设计规范
GB 50029—2003	Code for design of compressed-air station	压缩空气站设计规范
GB 50030—2007	Design Code of oxygen plant	氧气站设计规范
GB 50031—1991	Code for design of acetylene station	乙炔站设计规范
GB 50032—2003	Code for seismic design of outdoor water supply, sewerage, gas and heating engineering	室外给水排水和燃气热力工程抗震设计规范
GB 50033—1991	Standard for daylighting design of industrial enterprises	工业企业采光设计标准
GB 50034—1992	Standard for artificial lighting design of industrial enterprises	工业企业照明设计标准
GB 50037—1996	Code for design of ground surface and floor of building	建筑地面设计规范
GB 50039—2010	Code for fire protection design of rural buildings	村镇建筑设计防火规范
GB 50040—1996	Code for design of dynamic machine foundation	动力机器基础设计规范
GB 50041—1992	Code for design of boiler house	锅炉房设计规范
GB 50045—1995	Code for fire protection design of tall buildings	高层民用建筑设计防火规范
GB 50046—1995	Code for corrosion prevention design of industrial buildings	工业建筑防腐蚀设计规范
GB 50049—1994	Code for design of small-size power plant	小型火力发电厂设计规范

续表

标准编号	英文全称	中文
GB 50050—1995	Code for design of industrial recirculating cooling water treatment	工业循环冷却水处理设计规范
GB 50051—2002	Code for design of chimneys	烟囱设计规范
GB 50052—2009	Code for design of electric power supply systems	供配电系统设计规范
GB 50053—1994	Code for design of 10kV & under electric substation	10kV及以下变电所设计规范
GB 50054—2011	Code for design of low voltage electrical installations	低压配电设计规范
GB 50055—1993	Code for design of power distribution of general electrical installations	通用用电设备配电设计规范
GB 50056—1993	Code for design of electrical equipment of electroheat installations	电热设备电力装置设计规范
GB 50057—2000	Design code for protection of structures against lightning	建筑物防雷设计规范
GB 50058—1992	Electrical installations design code for explosive atmospheres and fire hazard	爆炸火灾危险环境电力装置设计规范
GB 50059—1992	Code for design of 35～110kV substation	35～110kV变电所设计规范
GB 50060—2008	Code for design of high voltage electrical installation (3～110kV)	3～110kV高压配电装置设计规范
GB 50067—1997	Code for fire protection design of garage, motor-repair-shop and parking-area	汽车库、修车库、停车场设计防火规范
GB 50068—2001	Unified standard for reliability design of building structures	建筑结构可靠度设计统一标准
GB 50069—2002	Code for design of water supply and drainage engineering structure	给水排水工程结构设计规范
GB 50070—1994	Code for design of mine electrical power installation	矿山电力装置设计规范
GB 50071—2002	Code for design of small-sized hydropower station	小型水力发电站设计规范
GB 50072—2010	Code for design of cold-storage	冷库设计规范
GB 50073—2001	Code for design of industrial clean rooms	洁净厂房设计规范

续表

标准编号	英文全称	中　文
GB 50074—2002	Code for design of oil depot	石油库设计规范
GB 50077—2003	Code for design of reinforced concrete silos	钢筋混凝土筒仓设计规范
GB 50078—2008	Code for construction and acceptance of chimney engineering	烟囱工程施工及验收规范
GB 50084—2001	Code for design of automatic fire sprinkler systems	自动喷水灭火系统设计规范
GB 50085—2007	Technical code for sprinkling irrigation engineering	喷灌工程技术规范
GB 50086—2001	Technical code for shotcrete rock bolts shore	锚杆喷射混凝土支护技术规范
GB 50089—2007	Safety code for design of engineering of civil explosive materials	民用爆破器材工程设计安全规范
GB 50090—2006	Code for design of railway line	铁路线路设计规范
GB 50091—2006	Code for design of railway station and terminal	铁路车站及枢纽设计规范
GB 50092—96	Code for construction and acceptance of asphalt pavement	沥青路面施工及验收规范
GB 50093—2002	Code for construction and acceptance of industrial automatic instrument works	工业自动化仪表工程施工及验收规范
GB 50094—1998	Code for construction and acceptance of spherical tanks	球形储罐施工及验收规范
GB 50096—1999	Code for design of dwelling houses	住宅建筑设计规范
GB 50098—2009	Code for fire protection design of civil air defence works	人民防空工程设计防火规范
GB 50099—2011	Code for design of school	中小学校设计规范
GB 50111—2006	Code for seismic design of railway engineering	铁路工程抗震设计规范
GB 50115—2009	Code for design of industrial television system	工业电视系统工程设计规范
GB 50116—2008	Code for design of automatic fire alarm system	火灾自动报警系统设计规范
GB 50118—2010	Code for sound insulation design of civil buildings	民用建筑隔声设计规范

续表

标准编号	英文全称	中文
GB 50119—2003	Code for utility technical of concrete admixture	混凝土外加剂应用技术规范
GB 50126—2008	Code for construction of industrial equipment and pipeline insulation engineering	工业设备及管道绝热工程施工规范
GB 50127—2007	Technical code for aerial ropeway engineering	架空索道工程技术规范
GB 50128—2005	Code for construction and acceptance of stand cylindrical steel welded oil storage tank	立式圆筒形钢制焊接油罐施工及验收规范
GB 50134—2004	Code for construction and acceptance of civil air defence works	人民防空工程施工及验收规范
GB 50135—2006	Code for design of high-rising structures	高耸结构设计规范
GB 50136—2011	Code for design of disposal of electroplating waste water	电镀废水治理设计规范
GB 50137—2011	Standard for classification of urban land use and planning standards of development land	城市用地分类与规划建设用地标准
GB 50139—2004	Standard for navigable pass of inland waterways	内河通航标准
GB 50140—2005	Code for design of extinguisher disposition in buildings	建筑灭火器配制设计规范
GB 50141—2008	Code for construction and acceptance of constructional structures of water supply and sewerage	给水排水构筑物施工及验收规范
GB 50147—2010	Code for construction and acceptance of high-voltage electric equipment installation engineering	电气装置安装工程 高压电器施工及验收规范
GB 50148—2010	Code for construction and acceptance of power transformers, oil reactor and mutual inductor	电气装置安装工程 电力变压器、油浸电抗器、互感器施工及验收规范
GB 50149—2010	Code for construction and acceptance of busbar installation of electric equipment installation engineering	电气装置安装工程 母线装置施工及验收规范

续表

标准编号	英文全称	中文
GB 50150—2006	Standard for hand-over test of electric equipment-electric equipment installation engineering	电气装置安装工程 电气设备交接试验标准
GB 50151—2010	Code for design of foam extinguishing systems	泡沫灭火系统设计规范
GB 50152—1992	Method for test of concrete structures	混凝土结构试验方法标准
GB 50153—2008	Unified standard for reliability design of engineering structures	工程结构可靠性设计统一标准
GB 50154—1992	Code for design of underground and earth powder and explosive magazines	地下及覆土火药炸药仓库设计安全规范
GB 50155—1992	Terminology of heating, ventilation and air conditioning	采暖通风与空气调节术语标准
GB 50156—2002	Code for design and construction of automobile gasoline and gas filling station (2006 Version)	汽车加油加气站设计与施工规范(2006年版)
GB 50157—1992	Code for design of underground railway	地下铁道设计规范
GB 50158—2010	Unified standard for reliability design of harbor engineering structures	港口工程结构可靠性设计统一标准
GB 50159—1992	Code for measurement of suspended sediment in open channels	河流悬移质泥沙测验规范
GB 50160—2008	Fire Prevention Code of Petrochemical Enterprise Design	石油化工企业设计防火规范
GB 50161—1992	Safety code for design of fireworks and firecrackers plants	烟花爆竹工厂设计安全规范
GB 50162—1992	Standard for road engineering drawing	道路工程制图标准
GB 50163—1992	Code for design of halon 1301 fire extinguishing systems	卤代烷1301灭火系统设计规范
GB 50164—1992	Standard for quality control of concrete	混凝土质量控制标准
GB 50165—1992	Technical code for maintenance and strengthening of ancient timber buildings	古建筑木结构维护与加固技术规范

续表

标准编号	英文全称	中　　文
GB 50166—2007	Code for installation and acceptance of fire alarm system	火灾自动报警系统施工及验收规范
GB 50167—1992	Code for engineering photogrammetry	工程摄影测量规范
GB 50168—2006	Code for construction and acceptance of cable system-Electric equipment installation engineering	电气装置安装工程　电缆线路施工及验收规范
GB 50169—2006	Code for construction and acceptance of grounding connection-Electric equipment installation engineering	电气装置安装工程　接地装置施工及验收规范
GB 50170—2006	Code for construction and acceptance of rotating electrical machines-Electric equipment installation engineering	电气装置安装工程　旋转电机施工及验收规范
GB 50171—1992	Code for construction and acceptance of switchboard outfit complete cubicle and secondary circuit-Electric equipment installation engineering	电气装置安装工程　盘、柜及二次回路结线施工及验收规范
GB 50172—1992	Code for construction and acceptance of battery-Electric equipment installation engineering	电气装置安装工程　蓄电池施工及验收规范
GB 50173—1992	Code for construction and acceptance of 35kV and under overhead power levels-electric equipment installation engineering	电气装置安装工程　35kV及以下架空电力线路施工及验收规范
GB 50174—2008	Code for design of electronic information system room	电子计算机机房设计规范
GB 50175—1993	Code for construction and acceptance of open coal-mine engineering	露天煤矿工程施工及验收规范
GB 50176—1993	Thermal design code for civil building	民用建筑热工设计规范
GB 50177—2005	Design code for hydrogen station	氢气站设计规范
GB 50178—1993	Standard of climate regionalization for architecture	建筑气候区划标准
GB 50179—1993	Code for measurement of fluid flow in open channels	河流流量测验规范

续表

标准编号	英文全称	中文
GB 50180—2002	Standard for planning of urban residential area	城市居住区规划设计规范
GB 50181—1993	Technical code for constructional engineering in flood detention basin	蓄滞洪区建筑工程技术规范
GB 50182—1993	Code for construction and acceptance of elevators-electric equipment of electrical apparatus installation engineering	电气装置安装工程 电梯电气装置施工及验收规范
GB 50183—2004	Code for fire protection design of petroleum and natural gas engineering	原油和天然气工程设计防火规范
GB 50184—2011	Code for acceptance of construction quality of industrial metallic piping engineering	工业金属管道工程施工质量验收规范
GB 50185—2010	Code for acceptance of construction quality of industrial equipment and pipeline insulation engineering	工业设备及管道绝热工程施工质量验收规范
GB 50186—1993	Standard for fundamental terms of port engineering	港口工程基本术语标准
GB 50187—1993	Code for design of general plan for industrial enterprises	工业企业总平面设计规范
GB 50188—2007	Standard for planning of town	镇规划标准
GB 50189—2005	Design standard for energy efficiency of public buildings	公共建筑节能设计标准
GB 50190—1993	Code for design of anti-microvibration of multistory factory floor	多层厂房楼盖抗微震设计规范
GB 50191—1993	Design code for antiseismic of special structures	构筑物抗震设计规范
GB 50192—1993	Code for design of river port engineering	河港工程设计规范
GB 50193—1993	Code for design of carbon dioxide fire extinguishing systems	二氧化碳灭火系统设计规范
GB 50194—1993	Safety code of power supply and consumption for installation construction engineering	建设工程施工现场供用电安全规范

续表

标准编号	英文全称	中文
GB 50195—1993	Code for design of producer gas station	发生炉煤气站设计规范
GB 50196—1993	Code for design of high & medium expansion foam systems	高倍数、中倍数泡沫灭火系统设计规范
GB 50197—2005	Code for design of open pit mine of coal industry	煤炭工业露天矿设计规范
GB 50198—2011	Technical code for project of civil closed circuit monitoring television system	民用闭路电视监视系统工程技术规范
GB 50199—1994	Unified standard for design of reliability of hydraulic engineering structures	水利水电工程结构可靠度设计统一标准
GB 50200—1994	Technical code for regulation of CATV system	有线电视系统工程技术规范
GB 50201—1994	Standard for flood control	防洪标准
GB 50202—2002	Code for construction and acceptance of foundation engineering	地基与基础工程施工及验收规范
GB 50204—2002	Code for acceptance of constructional quality of concrete structures	混凝土结构工程施工及验收规范
GB 50205—2001	Code for acceptance of constructional quality of steel structures	钢结构工程施工及验收规范
GB 50206—2002	Code for construction and acceptance of timber structures	木结构工程施工及验收规范
GB 50207—2002	Code for acceptance of constructional quality of roofing	屋面工程技术规范
GB 50208—2002	Code for construction and acceptance of under ground water proof engineering	地下防水工程施工及验收规范
GB 50209—2010	Code for acceptance of constructional quality of building ground	建筑地面工程施工及验收规范
GB 50210—2001	Code for construction and acceptance of decoration engineering	装饰工程施工及验收规范

续表

标准编号	英文全称	中文
GB 50211—2004	Code for construction and acceptance of brick works for industrial furnaces	工业炉砌筑工程施工及验收规范
GB 50212—2002	Specification for construction and acceptance of anticorrosion engineering of buildings	建筑防腐蚀工程施工及验收规范
GB 50213—2010	Code for construction and acceptance of mining pit engineering	矿山井巷工程施工及验收规范
GB 50214—2001	Technical code for assembly steel formworks	组合钢模板技术规范
GB 50215—2005	Code for mine design of coal industry	煤炭工业矿井设计规范
GB 50216—2001	Unified standard for design of reliability of railway engineering structures	铁路工程结构可靠度设计统一标准
GB 50217—2007	Code for design of cables of electric engineering	电力工程电缆设计规范
GB 50218—1994	Standard for engineering classification of rock masses	工程岩体分级标准
GB 50219—1995	Code of design for water spray extinguishing systems	水喷雾灭火系统设计规范
GB 50220—1995	Code for transport planning on urban road	城市道路交通规划设计规范
GB 50221—2001	Code for inspection and evaluation of structural steel engineering quality	钢结构工程质量检验评定标准
GB 50222—1995	Code for fire prevention in design of interior decoration of buildings	建筑内部装修设计防火规范
GB 50223—2008	Standard for classification of seismic protection of building construction	建筑抗震设防分类标准
GB 50224—1995	Standard for inspection and evaluation of anticorrosive engineering quality of buildings	建筑防腐蚀工程质量检验评定标准
GB 50225—2008	Code for design of civil air defence works	人民防空工程设计规范

续表

标准编号	英文全称	中 文
GB 50226—2007	Code for design of railway passenger station buildings	铁路旅客车站建筑设计规范
GB 50227—2008	Code for design of installation of shunt capacitors	并联电容器装置设计规范
GB 50229—2006	Code for design of fire protection for fossil fuel power plants and substations	火力发电厂与变电所设计防火规范
GB 50231—2009	General code for construction and acceptance of mechanical equipment installation engineering	机械设备安装工程施工及验收通用规范
GB 50235—2010	Code for construction and acceptance of industrial pipeline	工业管道工程施工及验收规范
GB 50236—2011	Code for construction and acceptance of field equipment, industrial pipe welding engineering	现场设备、工业管道焊接工程施工及验收规范
GB 50242—2002	Code for acceptance of construction quality of water supply drainage and heating works	建筑给水排水及采暖工程施工质量验收规范
GB 50243—2002	Code of acceptance for construction quality of ventilation and air conditioning works	通风与空调工程施工质量验收规范
GB 50251—2003	Code for design of gas transmission pipeline engineering	输气管道工程设计规范
GB 50252—2010	Unified standard for quality inspection and assessment of industrial erection engineering	工业安装工程质量检验评定统一标准
GB 50253—2003	Code for design of oil transmission pipeline engineering	输油管道工程设计规范
GB 50254—1996	Code for construction and acceptance of low-voltage apparatus-electric equipment installation engineering	电气装置安装工程 低压电器施工及验收规范
GB 50255—1996	Code for construction and acceptance of power converter equipment-electric equipment installation engineering	电气装置安装工程 电力变流设备施工及验收规范

续表

标准编号	英文全称	中文
GB 50256—1996	Code for construction and acceptance of electric device of crane-electrical equipment installation engineering	电气装置安装工程 起重机电气装置施工及验收规范
GB 50257—1996	Code for construction and acceptance of electric device for explosion atmospheres and fire hazard electrical equipment installation engineering	电气装置安装工程 爆炸和火灾危险环境电气装置施工及验收规范
GB 50260—1996	Code for design of seismic of electrical installations	电力设施抗震设计规范
GB 50261—2005	Code for installation and commissioning of sprinkler systems	自动喷水灭火系统施工及验收规范
GB 50263—2007	Code for installation and acceptance of gas fire-extinguishing systems	气体灭火系统施工及验收规范
GB 50268—2008	Code for construction and acceptance of water supply and sewerage pipeline works	给水排水管道工程施工及验收规范
GB 50289—1998	Code of urban engineering pipeline comprehensive planning	城市工程管线综合规划规范
GB 50301—2001	Standard for quality inspection and assessment of building construction	建筑工程质量检验评定标准
GB 50309—2007	Code for quality inspection and acceptance of industrial furnaces building	工业炉砌筑工程质量验收规范
GB 50310—2002	Code for acceptance of installation quality of lifts, escalators and passenger conveyors	电梯工程施工质量验收规范
GB 7595—1987	Oil quality standard for turbine in operation	汽机运行中油质标准
GB 7596—1987	Oil quality standard for turbine in power plant operation	电厂运行中汽轮机油质量标准
GB 8978—2002	Integrated wastewater discharge standard	污水综合排放标准
GB/T 10002.1—2006	Unplasticized poly(vinyl chloride) (PVC-U) pipes for water supply	给水用硬聚氯乙烯(PVC-U)管材

续表

标准编号	英文全称	中文
GB/T 10002.2—2003	Fittings made of unplasticized poly (vinyl chloride) (PVC-U) for water supply	给水用硬聚氯乙烯(PVC-U)管件
GB/T 10002.3—1996	Unplasticized polyvinyl chloride (PVC-U) pipes for buried drainage and sewerage systems	埋地排污、废水用硬聚氯乙烯(PVC-U)管材
GB/T 11813—2008	Helium leak testing of nuclear fuel rod for PWR	压水堆燃料棒氦质谱检漏
GB/T 12149—2007	Water for industrial circulating cooling system and boiler-Determination of silica	工业循环冷却水和锅炉用水中硅的测定
GB/T 13234—2009	Calculating methods of energy saved for enterprise	企业节能量计算方法
GB/T 13401—2005	Steel plate butt-welding pipe fittings	钢板制对焊管件
GB/T 13402—2010	Large diameter steel pipe flanges	大直径钢制管法兰
GB/T 13403—2008	Gaskets for large-diameter steel pipe flanges	大直径钢制管法兰用垫片
GB/T 15910—2009	Monitoring and testing for energy saving of heat-transmission and distribution system	热力输送系统节能监测
GB/T 15911—1995	Monitoring and testing method for energy saving of electroheat device in industry	工业电热设备节能监测方法
GB/T 15912.1—2009	Monitoring and testing method for energy conservation of refrigerating systems-Part 1: Cold storage systems	制冷机组及其制冷系统节能测试 第1部分:冷库
GB/T 15913—2009	Monitoring and testing for energy saving of fan's unit and distribute tube system	风机机组与管网系统节能监测
GB/T 15914—1995	Monitoring and testing method for energy saving of steam heating equipment	蒸汽加热设备节能监测方法
GB/T 16146—1995	Standards for controlling radon concentration in dwellings	住房内氡浓度控制标准

续表

标准编号	英文全称	中　　文
GB/T 16283—1996	Fundamental terminology of fixed extinguishing systems	固定式灭火系统基本术语
GB/T 16507—1996	Rules for construction of stationary boilers	固定式锅炉建造规程
GB/T 16614—1996	Statistical method of energy balance in enterprises	企业能量平衡统计方法
GB/T 16618—1996	General principles for thermal insulation technique of industrial furnaces	工业炉窑保温技术通则
GB/T 16632—2008	Determination of scale inhibition performance of water treatment agents-Calcium carbonate precipitation method	水处理剂阻垢性能的测定 碳酸钙沉积法
GB/T 16638.1—2008	Aerodynamics-Concepts, quantities and symbols-Part 1: Aerodynamic terms in common use	空气动力学　概念、量和符号　第1部分:空气动力学常用术语
GB/T 16643—1996	Rolling bearings-combined needle roller thrust roller bearings boundary dimensions	滚动轴承滚针和推力圆柱滚子组合轴承外形尺寸
GB/T 16662—2008	Terms of equipment and materials for building water supply and drainage	建筑给排水设备器材术语
GB/T 16664—1996	Monitoring and testing method for energy saving of power supply distribution system of industrial enterprise	企业供配电系统节能监测方法
GB/T 16665—1996	Monitoring and testing method for energy saving of air compressor unit and air distribution system	空气压缩机组及供气系统节能监测方法
GB/T 16666—1996	Monitoring and testing method for energy saving of motor-pump liquid transport system	泵类及液体输送系统节能监测方法
GB/T 16705—1996	Codes for environmental pollution categories	环境污染类别代码
GB/T 16706—1996	Codes for environmental pollution source categories	环境污染源类别代码

续表

标准编号	英文全称	中文
GB/T 16709.1—2010	Vacuum technology Mounting dimensions of pipeline fittings Part 1: Non knife-edge flange type	真空技术 管路配件的装配 第1部分:非刀口法兰型
GB/T 16709.2—2010	Vacuum technology Mounting dimensions of pipeline fittings Part 2: Knife-edge flange type	真空技术 管路配件的装配 第2部分:刀口法兰型
GB/T 16732—1997	Units and symbols of heating ventilation air conditioning and air cleaning equipment in building	建筑采暖通风空调净化设备计量单位及符号
GB/T 16752—2006	Test methods of concrete and reinforced concrete drainage and sewer pipes	混凝土和钢筋混凝土排水管试验方法
GB/T 17626.10—1998	Electromagnetic compatibility-Testing and measurement techniques-Damped oscillatory magnetic field immunity test	电磁兼容 试验和测量技术 阻尼振荡磁场抗扰度试验
GB/T 17626.11—2008	Electromagnetic compatibility-Testing and measurement techniques-Voltage dips, short interruptions and voltage variations immunity tests	电磁兼容 试验和测量技术 电压暂降、短时中断和电压变化的抗扰度试验
GB/T 17626.1—2006	Electromagnetic compatibility-Testing and measurement techniques-Overview of immunity tests	电磁兼容 试验和测量技术 抗扰度试验总论
GB/T 17626.12—1998	Electromagnetic compatibility-Testing and measurement techniques-Oscillatory waves immunity test	电磁兼容 试验和测量技术 振荡波抗扰度试验
GB/T 17626.2—2006	Electromagnetic compatibility-Testing and measurement techniques-Electrostatic discharge immunity test	电磁兼容 试验和测量技术 静电放电抗扰度试验
GB/T 17626.4—2008	Electromagnetic compatibility-Testing and measurement techniques-Electrical fast transient/burst immunity test	电磁兼容 试验和测量技术 电快速瞬变脉冲群抗扰度试验

续表

标准编号	英文全称	中文
GB/T 17626.6—2008	Electromagnetic compatibility-Testing and measurement techniques-Immunity to conducted disturbances induced by radio-frequency fields	电磁兼容 试验和测量技术 射频场感应的传导骚扰抗扰度
GB/T 17626.7—2008	Electromagnetic compatibility-Testing and measurement techniques-General guide on harmonics and interharmonics measurements and instrumentation for power supply system and equipment connected thereto	电磁兼容 试验和测量技术 供电系统及所连设备谐波、谐间波的测量和测量仪器一般使用说明
GB/T 17626.8—2006	Electromagnetic compatibility-Testing and measurement techniques-Power frequency magnetic field immunity test	电磁兼容 试验和测量技术 工频磁场抗扰度试验
GB/T 17678.1—1999	Requirements for optical disk storage, filing and archival management of CAD electronic records Part 1: filing and archival management of CAD electronic records	CAD电子文件光盘存储、归档与档案管理要求 第一部分:电子文件归档与档案管理
GB/T 17678.2—1999	Requirements for optical disk storage, filing and archival management of CAD electronic records Part 2: Information structure in an optical disk	CAD电子文件光盘存储、归档与档案管理要求 第二部分:光盘信息组织结构
GB/T 17825.1~17825.10—1999	Management of CAD documents	文件管理
GB/T 19000—2008	Quality management systems-Fundamentals and vocabulary	质量管理体系 基础和术语
GB/T 19001—2008	Quality management systems-Requirements	质量管理体系 要求
GB/T 19004—2011	Managing for the sustained success of an organization-A quality management approach	追求组织的持续成功 质量管理方法
GB/T 19016—2005	Quality management systems-Guidelines for quality management in projects	质量管理 项目管理质量指南

续表

标准编号	英文全称	中　　文
GB/T 19022.2—2000	Quality assurance for measuring equipment Part 2: Guidelines for control of measurement processes	测量设备的质量保证　第2部分:测量过程控制指南
GB/T 19889.X—2005	Acoustics-Measurement of sound insulation in buildings and of building elements	声学　建筑和建筑构件隔声测量
GB/T 214—2007	Determination of total sulfur in coal	煤中全硫的测定方法
GB/T 22084.1—2008	Secondary cells and batteries containing alkaline or other non-acid electrolytes-Portable sealed rechargeable single cells-Part 1: Nickel-cadmium	含碱性或其他非酸性电解质的蓄电池和蓄电池组—便携式密封单体蓄电池　第1部分:镉镍电池
GB/T 24020—2000	Environmental management-Environmental labels and declarations-General principles	环境管理　环境标志和声明　通用原则
GB/T 24041.7—2000	Environmental management-Life cycle assessment-Goal and scope definition and inventory analysis	环境管理　生命周期评价目的语范围的确定和清单分析
GB/T 24050—2004	Environmental management-Vocabulary	环境管理　术语
GB/T 26218.1—2010	Selection and dimensioning of high-voltage insulators intended for use in polluted conditions-Part 1: Definitions, information and general principles	污秽条件下使用的高压绝缘子的选择和尺寸确定　第1部分:定义、信息和一般原则
GB/T 26218.2—2010	Selection and dimensioning of high-voltage insulators intended for use in polluted conditions-Part 2: Ceramic and glass insulators for a.c. systems	污秽条件下使用的高压绝缘子的选择和尺寸确定　第2部分:交流系统用瓷和玻璃绝缘子
GB/T 26557—2011	Builders hoists for persons and materials with vertically guided cages	吊笼有垂直导向的人货两用施工升降机
GB/T 2900.1—2008	Electrotechnical terminology-Fundamental terms	电工术语基本　术语
GB/T 2900.36—2003	Electrotechnical terminology-Electric-traction	电工术语　电力牵引
GB/T 3715—2007	Terms relating to properties and analysis of coal	煤质及煤分析有关术语

续表

标准编号	英文全称	中文
GB/T 4473—2008	Synthetic testing of high-voltage alternating current circuit breakers	高压交流断路器的合成实验
GB/T 4648—1996	Rolling bearings-metric tapered roller bearings-flanged cups boundary dimensions	滚动轴承圆锥滚子轴承凸缘外圈外形尺寸
GB/T 4718—2006	Vocabulary used in fire alarm equipment	火灾报警设备专业术语
GB/T 4960.5—1996	Glossary of terms: Nuclear science and technology radiation protection and safety of radiation sources	核科学技术术语:辐射防护与辐射源安全
GB/T 4960.7—2010	Glossary of nuclear science and technology terms-Part 7: nuclear materials control and safeguards	核科学技术术语 第7部分:核材料管制与核保障
GB/T 50001—2010	Unified standard for building drawings	房屋建筑制图统一标准
GB/T 50062—2008	Code for design of relaying protection and automatic device of electric power installations	电力装置的继电保护和自动装置设计规范
GB/T 50063—2008	Code for design of electrical measuring instrumentation of electrical installations	电力装置的电气测量仪表装置设计规范
GB/T 50080—2002	Method for test of properties of ordinary concrete mixture	普通混凝土拌和物性能试验方法
GB/T 50081—2002	Method for test of mechanical properties of ordinary concrete	普通混凝土力学性能试验方法
GB/T 50082—2009	Method for test of long-term performance and durability of ordinary concrete	普通混凝土长期性能和耐久性能试验方法
GB/T 50095-98	Terms and symbols used in hydrometry	水文测验术语和符号标准
GB/T 50100—2001	Standard for modular coordination of residential buildings	住宅建筑模数协调标准
GB/T 50102—2003	Code for design of cooling for industrial recirculating water	工业循环水冷却设计规范
GB/T 50103—2001	Standard for general layout drawings	总图制图标准
GB/T 50104—2001	Standard for architectural drawing	建筑制图标准

续表

标准编号	英文全称	中文
GB/T 50105—2010	Standard for structural drawings	建筑结构制图标准
GB/T 50106—2010	Standard for building water supply and drainage drawings	建筑给水排水制图标准
GB/T 50107—2010	Standard for evaluation of concrete compressive strength	混凝土强度检验评定标准
GB/T 50108—2008	Technical code for waterproofing of underground works	地下工程防水技术规范
GB/T 50109—2006	Design code for softening and demineralization of industrial water	工业用水软化除盐设计规范
GB/T 50114—2010	Standard for heating, ventilation and air conditioning drawings	暖通空调制图标准
GB/T 50121—2005	Rating standard of sound insulation in buildings	建筑隔声评价标准
GB/T 50123—1999	Standard for soil test method	土工试验方法标准
GB/T 50125—2010	Standard for basic terms of water and wastewater engineering	给水排水工程基本术语标准
GB/T 50129—2011	Standard for test methods of basic mechanical properties of masonry	砌体基本力学性能试验方法标准
GB/T 50138—2010	Standard for observation of water level	水位观测标准
GB/T 50228—2011	Standard for foundational terminology of engineering survey	工程测量基本术语标准
GB/T 50265—2010	Design code for pumping station	泵站设计规范
GB/T 50269—1997	Code for measurement method of dynamic properties of subsoil	地基动力特性测试规范
GB/T 50311—2007	Code for engineering design of generic cabling system	综合布线系统工程设计规范
GB/T 50312—2007	Code for engineering acceptance of generic cabling system for building and campus	综合布线工程验收规范
GB/T 50314—2006	Standard for design of intelligent building	智能建筑设计标准
GB/T 5476—1996	Methods for pretreating ion exchange resins	离子交换树脂预处理方法

续表

标准编号	英文全称	中文
GB/T 5750.10—2006	Standard examination methods for drinking water-Disinfection by-products parameters	生活饮用水标准检验方法 消毒副产物指标
GB/T 5750.11—2006	Standard examination methods for drinking water-Disinfectants parameters	生活饮用水标准检验方法 消毒剂指标
GB/T 5750.1—2006	Standard examination methods for drinking water-General principles	生活饮用水标准检验方法 总则
GB/T 5750.2—2006	Standard examination methods for drinking water-Collection and preservation of water samples	生活饮用水标准检验方法 水样的采集和保存
GB/T 5750.3—2006	Standard examination methods for drinking water-Water analysis quality control	生活饮用水标准检验方法 水质分析质量控制
GB/T 5750.4—2006	Standard examination methods for drinking water-Organoleptic and physical parameters	生活饮用水标准检验方法 感官性状和物理指标
GB/T 5750.5—2006	Standard examination methods for drinking water-Nonmetal parameters	生活饮用水标准检验方法 无机非金属指标
GB/T 5750.6—2006	Standard examination methods for drinking water-Metal parameters	生活饮用水标准检验方法 金属指标
GB/T 5750.7—2006	Standard examination methods for drinking water-Aggregate organic parameters	生活饮用水标准检验方法 有机物综合指标
GB/T 5750.8—2006	Standard examination methods for drinking water-Organic parameters	生活饮用水标准检验方法 有机物指标
GB/T 5750.9—2006	Standard examination methods for drinking water-Pesticides parameters	生活饮用水标准检验方法 农药指标
GB/T 5777—2008	Seamless steel pipe and tubing methods for ultrasonic testing	无缝钢管超声波探伤检验方法
GB/T 7064—2008	Specific requirements for cylindrical rotor synchronous machines	隐极同步发电机技术要求
GB/T 7920.1—1996	Construction machinery and equipment general-terminology	建筑机械及设备通用术语
GB/T 7920.3—1996	Builder's hoist-terminology	施工升降机术语

续表

标准编号	英文全称	中文
GB/T 8484—2008	Graduation and test method for thermal insulating properties of doors and windows	建筑外门窗保温性能分级及其检测方法
GB/T 8485—2008	The graduation and test for airborne sound insulating properties of windows and doors	建筑用门窗空气声隔声性能分级及其检测方法
GB/Z 19024—2000	Guidelines for managing the economics of quality	质量经济性管理指南
GBJ 101—1987	Standard for modular coordination of building staircases	建筑楼梯模数协调标准
GBJ 110—1987	Code for design of halon-1211 fire extinguishing systems	1211灭火系统设计规范
GBJ 112—1987	Technical code for buildings in swelled ground zone	膨胀土地区建筑技术规范
GBJ 113—1987	Technical code for construction using hydraulic slip formworks	液压滑动模板施工技术规范
GBJ 114—1988	Standard for heating ventilation and air-conditioning drawings	采暖通风与空气调节制图标准
GBJ 117—1988	Standard for earthquake-resistant evaluation of industrial constructional structures	工业构筑物抗震鉴定标准
GBJ 120—1988	Code for design of community antenna television system in industrial enterprises	工业企业共用天线电视系统设计规范
GBJ 12—1987	Code for design of standard track gauge railway in industrial enterprises	工业企业标准轨距铁路设计规范
GBJ 124—1988	Terms of road construction	道路工程术语标准
GBJ 130—1990	Technical code for reinforced concrete lift-slab structures	钢筋混凝土升板结构技术规范
GBJ 131—1990	Standard for quality inspection and assessment of installation works for automatic instruments	自动化仪表安装工程质量检验评定标准
GBJ 132—1990	Basic terms and general symbols used in structural design of building and civil engineering	工程结构设计基本术语和通用符号
GBJ 133—1990	Standard for artificial lighting design of civil buildings	民用建筑照明设计标准

续表

标准编号	英文全称	中　文
GBJ 142—1990	Standard for protective spacing between medium、short wave broadcast transmitting station and electric cable carrier telecommunications systems	中、短波广播发射台与电缆载波通信系统的防护间距标准
GBJ 143—1990	Standard for protective spacing of radio interference from overhead electric lines, substations to television transposer station and retransmitting station	架空电力线路、变电所对电视差转台、转播台无线电干扰防护间距标准
GBJ 144—1990	Standard for evaluation of reliability of industrial factory buildings	工业厂房可靠性鉴定标准
GBJ 145—1990	Standard for classification of soils	土的分类标准
GBJ 146—1990	Technical code for application of flyash concrete	粉煤灰混凝土应用技术规范
GBJ 201—1983	Code for construction and acceptance of earthwork and blasting engineering	土方与爆破工程施工及验收规范
GBJ 203—1983	Code for construction and acceptance of masonry structures	砖石工程施工及验收规范
GBJ 2—1986	Unified standard for modular coordination of buildings	建筑模数协调统一标准
GBJ 22—1987	Code for design of roads in factories and mining areas	厂矿道路设计规范
GBJ 233—1990	Code for construction and acceptance of 110-500kV over-head electrical power transmission line	110～500kV架空电力线路施工及验收规范
GBJ 300—1988	Unified standard for quality inspection and assessment of building constructional erection works	建筑安装工程质量检验评定统一标准
GBJ 302—1988	Standard for quality inspection and assessment of building heating, sanitary and gas engineering	建筑采暖卫生与煤气工程质量检验评定标准
GBJ 303—1988	Standard for quality inspection and assessment of erection works for building electrical installations	建筑电气安装工程质量检验评定标准
GBJ 304—1988	Standard for quality inspection and assessment of ventilation and air-conditioning works	通风与空调工程质量检验评定标准

续表

标准编号	英文全称	中文
GBJ 321—1990	Standard for quality inspection and assessment of precast concrete members	预制混凝土构件质量检验评定标准
GBJ 42—1981	Code for design of telecommunications in industrial enterprises	工业企业通信设计规范
GBJ 43—1982	Code for seismic assessment of outdoor water supply and drainage engineering facilities	室外给水排水工程设施抗震鉴定标准
GBJ 44—1982	Code for seismic assessment of outdoor gas and heating engineering facilities	室外煤气热力工程设施抗震鉴定标准
GBJ 47—1983	Code for measurement of sound-absorbing coefficient in reverberation room	混响室法吸声系数测量规范
GBJ 48—1983	Standard for disposal of hospital waste water	医院污水排放标准
GBJ 54—1983	Code for design of low voltage electrical installations and wiring systems	低压配电装置及线路设计规范
GBJ 61—1983	Code for design of industrial and civil 35kV or under overhead electrical power transmission line	工业与民用35kV及以下架空电力线路设计规范
GBJ 6—1986	Unified standard for modular coordination of factory buildings	厂房建筑模数协调标准
GBJ 64—1983	Code for design of over-voltage protection of industrial and civil electrical installations	工业与民用电力装置的过电压保护设计规范
GBJ 65—1983	Code for design of earthing of industrial and civil electrical installations	工业与民用电力装置的接地设计规范
GBJ 66—1984	Code for construction and acceptance of installation works of refrigeration equipment	制冷设备安装工程施工及验收规范
GBJ 67—1984	Code for fire protection design of garage	汽车库设计防火规范
GBJ 68—1984	Unified standard for design building structures	建筑结构设计统一标准

续表

标准编号	英文全称	中文
GBJ 76—1984	Code for measurement of reverberation time in hall	厅堂混响时间测量规范
GBJ 79—1985	Code for design of telecommunications earthing in industrial enterprises	工业企业通信接地设计规范
GBJ 83—1985	General symbols, SI-units and basic terms used in structural design of buildings	建筑结构设计通用符号、计量单位和基本术语
GBJ 87—1985	Code for design of noise control in industrial enterprises	工业企业噪声控制设计规范
GBJ 88—1985	Code for measurement of sound-absorbing coefficient and specific acoustic impedance of standing-wave meter	驻波管法吸声系数与声阻抗率测量规范
GBJ 97—1987	Code for construction and acceptance of cement concrete pavement	水泥混凝土路面施工及验收规范
GBZ 1—2010	Hygienic standards for the design of industrial enterprises	工业企业设计卫生标准
HJ/T 75—2001	Technical norm for continuous emissions monitoring of flue gas emitted from thermal power plants	火电厂烟气排放连续监测技术规范
HJ/T 76—2007	Specifications and test procedures for continuous emission monitoring systems of flue gas emitted from stationary sources	固定污染源排放烟气连续监测系统技术要求及检测方法
JG 3035—1996	Building curtain walls	建筑幕墙
JG/T 3064—1999	Steel fiber reinforced concrete	钢纤维混凝土
JGJ 107—2003	General technical specification for mechanical splicing of bars	钢筋机械连接通用技术规程
JGJ 109—1996	Specification for taper threaded splicing of rebars	钢筋锥螺纹接头技术规程
JGJ 115—2006	Technical specification for concrete structural element with cold-rolled and twisted bars	冷轧扭钢筋混凝土构件技术规程
JGJ 118—2011	Code for design of soil and foundation of buildings in frozen soil region	冻土地区建筑地基基础设计规范

续表

标准编号	英文全称	中文
JGJ 18—2003	Specification for welding and acceptance of reinforcing steel bars	钢筋焊接及验收规程
JGJ 3—2010	Specification for design and construction of reinforced concrete structure of tall buildings	钢筋混凝土高层建筑结构设计与施工规程
JGJ 55—2011	Specification for mix proportion design of ordinary concrete	普通混凝土配合比设计规程
JGJ 94—2008	Technical code for building pile foundations	建筑桩基技术规范
JGJ/T 13—1994	Aseismatic technical specification for multistorey masonry building with reinforced concrete tie column	设置钢筋混凝土构造柱多层砖房抗震技术规程
JGJ/T 14—2011	Technical specification for concrete small-sized hollow block masonry building	混凝土小型空心砌块建筑技术规程
JGJ/T 98—2010	Specification for mix proportion design of masonry mortar	砌筑砂浆配合比设计规程
NDGJ 5—1988	Technical stipulation for hydraulic design of fossil fuel power plant	火力发电厂水工设计技术规定
SDJ 26—1989	Code for design of conductor and electrical equipment selection	发电厂、变电所电缆选择与敷设设计技术规程
SDJ 5—1985	Code for design of high voltage switchgear	高压配电装置设计技术规程
SL 105—2007	Specifications for anticorrosion of hydraulic steel structure	水工金属结构防腐蚀规范
SL 174—1996	The construction technical specification of concrete cut off wall used for water resources and hydroelectric project	水利水电工程混凝土防渗墙施工技术规范
SL 176—2007	Assessment specification for construction quality of hydraulic and hydroelectric engineering	水利水电工程施工质量评定规程
SL 189—1996	Design guide for rolled earth-rock fill dams in small size water resources and hydroelectric engineering	小型水利水电工程碾压式土石坝设计导则

续表

标准编号	英文全称	中文
TB 10001—2005	Code for design on road-bed of railway	铁路路基设计规范
TJ 231(1)—1978	Vol. 1 General stipulation Code for construction and acceptance of mechanical equipment installation engineering	第一册 通用规定 机械设备安装工程施工及验收规范
TJ 231(2)—1978	Vol. 2 Installation of metal cutting machine tool Code for construction and acceptance of mechanical equipment installation engineering	第二册 金属切削机床安装 机械设备安装工程施工及验收规范
TJ 231(3)—1978	Vol. 3 Installation of mechanical press, hydropneumatic hammer, hydraulic press and casting facilities Code for construction and acceptance of mechanical equipment	第三册 机械压力机、空气锤、液压机、铸造设备安装 机械设备安装工程施工及验收规范
TJ 231(4)—1978	Vol. 4 Installation of cranes, elevators and continuous transportation Code for construction and acceptance of mechanical equipment installation engineering	第四册 超重设备、电梯、连续运输设备安装 机械设备安装工程施工及验收规范
TJ 231(5)—1978	Vol. 5 Installation of air compressor, fan, pump and separator Code for construction and acceptance of mechanical equipment installation engineering	第五册 压缩机、风机、泵及空气分离器设备安装 机械设备安装工程施工及验收规范
TJ 231(6)—1978	Vol. 6 Installation of crushing and grinding machines, industrial boiler, hoists and fixed diesel engines Code for construction and acceptance of mechanical equipment installation engineering	第六册 破碎粉磨设备、工业锅炉、卷扬机及固定式柴油机安装 机械设备安装工程施工及验收规范
TJ 231(7)—1981	Vol. 7 Installation of forging hammer, hot die forging press, horizontal forging press and shearing machine-Code for construction and acceptance of mechanical equipment installation engineering	第七册 锻锤,热模锻压机,平锻机及剪切机安装 机械设备安装工程施工及验收规范
TJ 231—1975	Code for construction and acceptance of mechanical equipment installation engineering	机械设备安装工程施工及验收规范

续表

标准编号	英文全称	中　文
TJ 305—1975	Standard for quality inspection and assessment of building construction (erection works of general machinery equipment)	建筑安装工程质量检验评定标准（通用机械设备安装工程）
TJ 306—1977	Standard for quality inspection and assessment of building construction (container works)	建筑安装工程质量检验评定标准（容器工程）
TJ 32—1978	Code for earthquake-resistant design of outdoor water supply, sewerage, gas and heating engineering	室外给水排水和煤气热力工程抗震设计规范
YB/T 9033—1998	Inspection-acceptance regulation and quality evaluation standard for water supplying projects of hydrogeological investigation and water wells	供水水文地质勘察和供水管井工程检查、验收和质量评定标准
	People's Republic of China The Compulsory Provisions of Engineering Construction Standards Building	中华人民共和国　工程建设标准强制性条文　房屋建筑部分
	People's Republic of China The Compulsory Provisions of Engineering Construction Standards Engineering Construction of Civil Airports	中华人民共和国　工程建设标准强制性条文　民航机场工程部分
	People's Republic of China The Compulsory Provisions of Engineering Construction Standards Industries Building	中华人民共和国　工程建设标准强制性条文　工业建筑部分
	People's Republic of China The Compulsory Provisions of Engineering Construction Standards Urban and Rural Planning	中华人民共和国　工程建设标准强制性条文　城乡规划部分
	People's Republic of China The Compulsory Provisions of Engineering Construction Standards Urban Construction	中华人民共和国　工程建设标准强制性条文　城市建设部分
	People's Republic of China The Compulsory Provisions of Engineering Construction Standards Water Engineering	中华人民共和国　工程建设标准强制性条文　水利工程部分

附录5　发电工程招标投标文件中技术图纸目录(样例)

SN.	Drawing list	图纸名称
1	Site plan	厂址总体规划图
2	Plant general layout	厂区总平面布置图
3	Boiler air and flue gas system P&I diagram (basic scheme)	锅炉烟风系统管道和仪表流程图(基本方案)
4	Pulverized coal preparation system P&I diagram (basic scheme)	制粉系统管道和仪表流程图(基本方案)
5	Fuel oil system P&I diagram	燃油系统管道和仪表流程图
6	Main, reheat and by-pass steam system P&I diagram	主蒸汽、再热蒸汽及汽机旁路系统管道和仪表流程图
7	Extraction steam system P&I diagram	抽汽系统管道和仪表流程图
8	Feed water system P&I diagram	给水系统管道和仪表流程图
9	Condensate system P&I diagram	凝结水系统管道和仪表流程图
10	HP heater drains and vents system P&I diagram	高压加热器疏水及排气系统管道和仪表流程图
11	LP heater drains and vents system P&I diagram	低压加热器疏水及排气系统管道和仪表流程图
12	Auxiliary steam system P&I diagram	辅助蒸汽系统管道和仪表流程图
13	Boiler drains and blowdown system P&I diagram	锅炉疏水及排气系统管道和仪表流程图
14	Plant and instrument compressed air system P&I diagram	厂用和仪用压缩空气系统管道和仪表流程图
15	Open cycle cooling water system P&I diagram	开式循环冷却水系统管道和仪表流程图
16	Closed cycle cooling water system P&I diagram	闭式循环冷却水系统管道和仪表流程图
17	Service water system P&I diagram	工业水系统管道和仪表流程图
18	Air evacuation and miscellaneous condenser connection P&I diagram	抽真空及凝汽器各连接系统管道和仪表流程图
19	Nitrogen seal system P&I diagram	氮气密封系统管道和仪表流程图
20	Boiler air and flue gas system P&I diagram (optional scheme)	锅炉烟风系统管道和仪表流程图(供选择方案)

续表

SN.	Drawing list	图纸名称
21	Pulverized coal preparation system P&I diagram (optional scheme)	制粉系统管道和仪表流程图（供选择方案）
22	Legends and notes	图例符号及附注
23	Ground floor plan of main power building (basic scheme)	主厂房底层平面布置图（基本方案）
24	Operating floor EL. 12.20m and EL. 6.10m floor plans of main power building (basic scheme)	主厂房12.20m运转层和6.10m层平面布置图（基本方案）
25	Floor plans at EL. 19.00m, EL. 26.45m, EL. 33.60m of main power building (basic scheme)	主厂房19.00m层、26.45m层、33.60m层平面布置图（基本方案）
26	Boiler rear area plan	炉后平面布置图
27	General arrangement section of main power building (basic scheme)	主厂房横断面布置图（基本方案）
28	Ground floor plan of main power building (optional scheme)	主厂房底层平面布置图（供选择方案）
29	Operating floor EL. 12.20m and EL. 6.10m floor plans of main power building (optional scheme)	主厂房12.20m运转层和6.10m层平面布置图（供选择方案）
30	Main power building plans at Floor EL. 19.00m, EL. 26.45m, and EL. 35.40m (optional scheme)	主厂房19.00m层、26.45m层、35.40m层平面布置图（供选择方案）
31	General arrangement section of main power building (optional scheme)	主厂房横断面布置图（供选择方案）
32	Bid area plan	招标区域图
33	Fire fighting system P&I diagram	消防系统管道和仪表流程图
34	Circulating water system P&I diagram	循环水系统管道和仪表流程图
35	Chemical feed system P&I diagram	化学加药系统管道仪表流程图
36	Water-steam sampling system	汽水采样系统图
37	Condensate system P&I diagram	凝结水系统管道和仪表流程图
38	Condensate polishing regeneration system P&I diagram	凝结水精处理再生系统管道和仪表流程图
39	Legends	图例符号
40	Main power building and central control building arrangement section (basic scheme)	主厂房和集中控制楼横断面布置图（基本方案）

续表

SN.	Drawing list	图纸名称
41	Main power building elevations (optional scheme)	主厂房立面图(供选择方案)
42	Bottom ash handling system P&I diagram	除渣系统管道和仪表流程图
43	Fly ash handling system P&I diagram	除灰系统管道和仪表流程图
44	Generator circuit diagram	发电机回路图
45	Auxiliary power supply principle diagram	厂用电源原则接线图
46	Emergency power supply principle diagram	保安电源原则接线图
47	Principle diagram of lighting and maintenance power	照明、检修电源原则接线图
48	220V DC system diagram	220V 直流系统图
49	110V DC system diagram	110V 直流系统图
50	AC UPS conceptual diagram	交流不停电电源示意图
51	General layout for area outside of turbine house column Row A	汽机房 A 排外电气设备布置图
52	Scope of supply of electrical primary equipment for each island	各岛主要电气设备供应分界点图
53	Electrical single line diagram	电气主接线图
54	Schematic drawing of cable interface for each island	各岛电缆接口分界示意图
55	Central control building EL. 12.20m floor plan (basic scheme)	集中控制楼 12.20m 层布置图(基本方案)
56	Central control building EL. 6.10m floor plan (basic scheme)	集中控制楼 6.10m 层布置图(基本方案)
57	Central control building EL. 12.20m floor plan (optional scheme)	集中控制楼 12.20m 层布置图(供选择方案)
58	Central control building EL. 6.10m floor plan (optional scheme)	集中控制楼 6.10m 层布置图(供选择方案)
59	Central control building EL. ±0.00m floor plan (optional scheme)	集中控制楼±0.00m 层布置图(供选择方案)

附录6　国家或地区名称与货币名称、代码

（按照 Country/Region 列中的英文字母顺序排序）

序号	Country/Region	国家/地区	货币	国际标准代码	Currency	货币名称
1	Afghanistan	阿富汗	Af	AFN	Afghan Afghani	阿富汗尼
2	Albania	阿尔巴尼亚	L	ALL	Albanian Lek	阿尔巴尼亚列克
3	Algeria	阿尔及利亚	DA	DZD	Algerian Dinar	阿尔及利亚第纳尔
4	American Samoa	美属萨摩亚	$	USD	United States Dollar	美元
5	Andorra	安道尔	€	EUR	Euro	欧元
6	Angola	安哥拉	Kz	AOA	Angolan Kwanza	安哥拉宽扎
7	Anguilla	安圭拉	EC$	XCD	East Caribbean Dollar	东加勒比元
8	Antigua and Barbuda	安提瓜和巴布达	EC$	XCD	East Caribbean Dollar	东加勒比元
9	Argentina	阿根廷	$	ARS	Argentine Peso	阿根廷比索
10	Armenia	亚美尼亚		AMD	Armenian Dram	亚美尼亚德拉姆
11	Aruba	阿鲁巴岛		AWG	Aruban Florin	阿鲁巴岛弗罗林
12	Australia	澳大利亚	$, A$	AUD	Australian Dollar	澳大利亚元
13	Austria	奥地利	€	EUR	Euro	欧元
14	Azerbaijan	阿塞拜疆		AZN	Azerbaijani Manat	阿塞拜疆马纳特
15	Bahamas	巴哈马	B$	BSD	Bahamian Dollar	巴哈马元
16	Bahrain	巴林	BD	BHD	Bahrain Dinar	巴林第纳尔
17	Bangladesh	孟加拉	৳, Tk	BDT	Bangladeshi Taka	孟加拉塔卡
18	Barbados	巴巴多斯	Bds$	BBD	Barbadian Dollar	巴巴多斯元
19	Belarus	白俄罗斯		BYR	Belarusian Ruble	白俄罗斯卢布

续表

序号	Country/Region	国家/地区	货币	国际标准代码	Currency	货币名称
20	Belgium	比利时	€	EUR	Euro	欧元
21	Belize	伯利兹	BZ$	BZD	Belize Dollar	伯利兹元
22	Benin	贝宁	CFA	XOF	West African CFA Franc	西非法郎（非洲金融共同体法郎）
23	Bermuda	百慕大	BD$	BMD	Bermuda Dollar	百慕大元
24	Bhutan	不丹	Nu	BTN	Bhutanese Ngultrum	不丹努尔特鲁姆
25	Bolivia	玻利维亚	b$	BOB	Bolivian Boliviano	玻利维亚诺
26	Bosnia and Herzegovina	波斯尼亚和黑塞哥维那	KM	BAM	Bosnia and Herzegovina Convertible Mark	波黑可兑换马克
27	Botswana	博茨瓦纳	P	BWP	Botswana Pula	博茨瓦纳普拉
28	Brazil	巴西	R$	BRL	Brazilian Real	巴西雷亚尔
29	British Virgin Islands	英属维尔京群岛	$	USD	United States Dollar	美元
30	Brunei	文莱	B$	BND	Brunei Dollar	文莱元
31	Bulgaria	保加利亚		BGN	Bulgarian Lev	保加利亚列弗
32	Burkina Faso	布基纳法索	CFA	XOF	West African CFA Franc	西非法郎（非洲金融共同体法郎）
33	Burma	缅甸	K	MMK	Burmese Kyat	缅元
34	Burundi	布隆迪	FBu	BIF	Burundian Franc	布隆迪法郎
35	Cambodia	柬埔寨		KHR	Cambodian Riel	柬埔寨瑞尔
36	Cameroon	喀麦隆	FCFA	XAF	Central African CFA Franc	中非金融合作法郎
37	Canada	加拿大	C$	CAD	Canadian Dollar	加拿大元
38	Cape Verde	佛得角		CVE	Cape Verde Escudo	佛得角埃斯库多

续表

序号	Country/Region	国家/地区	货币	国际标准代码	Currency	货币名称
39	Cayman Islands	开曼群岛		KYD	Cayman Islands Dollar	开曼群岛元
40	Central African Republic	中非共和国	FCFA	XAF	Central African CFA Franc	中非金融合作法郎
41	Chad	乍得	FCFA	XAF	Central African CFA Franc	中非金融合作法郎
42	Chile	智利	$	CLP	Chilean Peso	智利比索
43	China	中国	RMB	CNY	Renminbi Yuan	人民币元
44	Columbia	哥伦比亚	$	COP	Colombian Peso	哥伦比亚比索
45	Comoros	科摩罗	CF	KMF	Comorian Franc	科摩罗法郎
46	Congo, Democratic Republic of (DRC)	刚果民主共和国	FC	CDF	Congolese Franc	刚果法郎
47	Congo, Republic of	刚果共和国	FCFA	XAF	Central African CFA Franc	中非金融合作法郎
48	Cook Islands	库克群岛	NZ$	NZD	New Zealand Dollar	新西兰元
49	Costa Rica	哥斯达黎加	c	CRC	Costa Rican Colon	哥斯达黎加科朗
50	Côte d'Ivoire, Republic of	科特迪瓦共和国	CFA	XOF	West African CFA Franc	西非法郎（非洲金融共同体法郎）
51	Croatia	克罗地亚	kn	HRK	Croatian Kuna	克罗地亚库纳
52	Cuba	古巴	$ MN	CUP	Cuban Peso	古巴比索
53	Curaçao	库拉索	NAf, Naf	ANG	Netherlands Antillean Guilder	荷属安的列斯盾
54	Cyprus	塞浦路斯	€	EUR	Euro	欧元
55	Czech Republic	捷克共和国	Kč	CZK	Czech Koruna	捷克克朗
56	Denmark	丹麦	kr.	DKK	Danish Krone	丹麦克朗
57	Djibouti	吉布提	Fdj	DJF	Djiboutian Franc	吉布提法郎

续表

序号	Country/Region	国家/地区	货币	国际标准代码	Currency	货币名称
58	Dominica, Commonwealth of	多米尼加联邦	EC$	XCD	East Caribbean Dollar	东加勒比元
59	Dominican Republic	多米尼加共和国	RD$	DOP	Dominican Peso	多米尼加比索
60	East Timor	东帝汶	$	USD	United States Dollar	美元
61	Ecuador	厄瓜多尔	$	USD	United States Dollar	美元
62	Egypt	埃及	E£	EGP	Egyptian Pound	埃及镑
63	El Salvador	萨尔瓦多	$	USD	United States Dollar	美元
64	Equatorial Guinea	赤道几内亚	FCFA	XAF	Central African CFA Franc	中非金融合作法郎
65	Eritrea	厄立特里亚	Nfk	ERN	Eritrean Nakfa	厄立特里亚纳克法
66	Estonia	爱沙尼亚	€	EUR	Euro	欧元
67	Ethiopia	埃塞俄比亚	Br	ETB	Ethiopian Birr	埃塞俄比亚比尔
68	Faroe Islands	法罗群岛	kr.	DKK	Danish Krone	丹麦克朗
69	Fiji	斐济	F$	FJD	Fiji Dollar	斐济元
70	Finland	芬兰	€	EUR	Euro	欧元
71	France	法国	€	EUR	Euro	欧元
72	French Polynesia	法属波利尼西亚	F	XPF	CFP Franc	太平洋金融共同体法郎
73	Gabon	加蓬	FCFA	XAF	Central African CFA Franc	中非金融合作法郎
74	Gambia	冈比亚	D	GMD	Gambian Dalasi	冈比亚达拉西
75	Georgia	格鲁吉亚		GEL	Georgian Lari	格鲁吉亚拉里
76	Germany	德国	€	EUR	Euro	欧元
77	Ghana	加纳	GH¢	GHS	Ghanaian Cedi	加纳塞地
78	Gibraltar	直布罗陀	£	GIP	Gibraltar Pound	直布罗陀镑
79	Greece	希腊	€	EUR	Euro	欧元
80	Greenland	格陵兰岛	kr.	DKK	Danish Krone	丹麦克朗

续表

序号	Country/Region	国家/地区	货币	国际标准代码	Currency	货币名称
81	Grenada	格林纳达	$	XCD	East Caribbean Dollar	东加勒比元
82	Guam	关岛	$	USD	United States Dollar	美元
83	Guatemala	危地马拉	Q	GTQ	Guatemalan Quetzal	危地马拉格查尔
84	Guinea	几内亚	FG	GNF	Guinean Franc	几内亚法郎
85	Guinea-Bissau	几内亚比绍	CFA	XOF	West African CFA Franc	西非法郎（非洲金融共同体法郎）
86	Guyana	圭亚那	G$	GYD	Guyana Dollar	圭亚那元
87	Haiti	海地	G	HTG	Haitian Gourde	海地古德
88	Honduras	洪都拉斯	L	HNL	Honduran Lempira	洪都拉斯伦皮拉
89	Hong Kong	中国香港	HK$	HKD	Hong Kong Dollar	港元
90	Hungary	匈牙利	Ft	HUF	Hungarian Forint	匈牙利福林
91	Iceland	冰岛	kr, Íkr	ISK	Icelandic Króna	冰岛克朗
92	India	印度	Re	INR	Indian Rupee	印度卢比
93	Indonesia	印度尼西亚	Rp	IDR	Indonesian Rupiah	印度尼西亚卢比（通称盾）
94	Iran	伊朗		IRR	Iranian Rial	伊朗里亚尔
95	Iraq	伊拉克		IQD	Iraqi Dinar	伊拉克第纳尔
96	Ireland	爱尔兰	€	EUR	Euro	欧元
97	Israel	以色列	₪	ILS	Israeli New Sheqel	以色列新谢克尔
98	Italy	意大利	€	EUR	Euro	欧元
99	Jamaica	牙买加	J$	JMD	Jamaican Dollar	牙买加元
100	Japan	日本	¥	JPY	Japanese Yen	日元
101	Jordan	约旦	JD	JOD	Jordanian Dinar	约旦第纳尔
102	Kazakhstan	哈萨克斯坦		KZT	Kazakhstani Tenge	哈萨克斯坦坚戈

续表

序号	Country/Region	国家/地区	货币	国际标准代码	Currency	货币名称
103	Kenya	肯尼亚	KSh	KES	Kenyan Shilling	肯尼亚先令
104	Kiribati	基里巴斯	$, A$	AUD	Australian Dollar	澳大利亚元
105	Korea, Democratic People's Republic of	朝鲜民主主义人民共和国	₩	KPW	Won	朝鲜元
106	Korea, Republic of	韩国	₩	KRW	Won	韩元
107	Kosovo	科索沃	€	EUR	Euro	欧元
108	Kuwait	科威特	KD	KWD	Kuwaiti Dinar	科威特第纳尔
109	Kyrgyzstan	吉尔吉斯斯坦		KGS	Kyrgyzstani Som	吉尔吉斯斯坦索姆
110	Laos	老挝	₭, ₭N	LAK	Lao Kip	老挝基普
111	Latvia	拉脱维亚	Ls	LVL	Latvian Lats	拉脱维亚拉特
112	Lebanon	黎巴嫩	£, L£	LBP	Lebanese Pound	黎巴嫩镑
113	Lesotho	莱索托	L	LSL	Lesotho Loti	莱索托洛蒂
114	Liberia	利比里亚	L$	LRD	Liberian Dollar	利比里亚元
115	Libya	利比亚	LD	LYD	Libyan Dinar	利比亚第纳尔
116	Liechtenstein	列支敦士登	SF	CHF	Swiss Franc	瑞士法郎
117	Lithuania	立陶宛	Lt	LTL	Lithuanian Litas	立陶宛立特
118	Luxembourg	卢森堡	€	EUR	Euro	欧元
119	Macao	中国澳门	MOP$	MOP	Macanese Pataca	澳门元
120	Macedonia	马其顿		MKD	Macedonian Denar	马其顿第纳尔
121	Madagascar	马达加斯加	FMG	MGA	Malagasy Ariary	阿里亚里
122	Malawi	马拉维	MK	MWK	Malawian Kwacha	马拉维克瓦查
123	Malaysia	马来西亚	M$	MYR	Malaysian Ringgit	马来西亚林吉特
124	Maldives	马尔代夫	Rf, MRf	MVR	Maldivian Rufiyaa	马尔代夫拉菲亚

续表

序号	Country/Region	国家/地区	货币	国际标准代码	Currency	货币名称
125	Mali	马里	CFA	XOF	West African CFA Franc	西非法郎（非洲金融共同体法郎）
126	Malta	马耳他	€	EUR	Euro	欧元
127	Marshall Islands	马绍尔群岛	$	USD	United States Dollar	美元
128	Mauritania	毛里塔尼亚	UM	MRO	Mauritanian Ouguiya	毛里塔尼亚乌吉亚
129	Mauritius	毛里求斯	Mau Re	MUR	Mauritius Rupee	毛里求斯卢比
130	Mexico	墨西哥	Mex$	MXN	Mexican Peso	墨西哥比索
131	Micronesia	密克罗尼西亚	$	USD	United States Dollar	美元
132	Moldova	摩尔多瓦		MDL	Moldovan Leu	摩尔多瓦列伊
133	Monaco	摩纳哥	€	EUR	Euro	欧元
134	Mongolia	蒙古	₮	MNT	Mongolian Tugrik	蒙古图格里克
135	Montenegro, Republic of	黑山共和国	€	EUR	Euro	欧元
136	Montserrat	蒙特塞拉特岛	EC$	XCD	East Caribbean Dollar	东加勒比元
137	Morocco	摩洛哥		MAD	Moroccan Dirham	摩洛哥迪拉姆
138	Mozambique	莫桑比克	MT	MZN	Mozambican Metical	莫桑比克梅蒂卡尔
139	Namibia	纳米比亚	N$	NAD	Namibian Dollar	纳米比亚元
140	Nauru	瑙鲁	$, A$	AUD	Australian Dollar	澳大利亚元
141	Nepal	尼泊尔	NRe	NPR	Nepalese Rupee	尼泊尔卢比
142	Netherlands	荷兰	€	EUR	Euro	欧元
143	Netherlands Antilles	荷属安的列斯	NAf, Naf	ANG	Netherlands Antillean Gulden	荷属安的列斯盾
144	New Caledonia	新喀里多尼亚	F	XPF	CFP Franc	太平洋金融共同体法郎

续表

序号	Country/Region	国家/地区	货币	国际标准代码	Currency	货币名称
145	New Zealand	新西兰	NZ$	NZD	New Zealand Dollar	新西兰元
146	Nicaragua	尼加拉瓜	C$	NIO	Nicaraguan Córdoba	尼加拉瓜科多巴
147	Niger	尼日尔	CFA	XOF	West African CFA Franc	西非法郎（非洲金融共同体法郎）
148	Nigeria	尼日利亚	₦	NGN	Nigerian Naira	尼日利亚奈拉
149	Northern Mariana Islands	北马里亚那群岛	$	USD	United States Dollar	美元
150	Norway	挪威	NKr	NOK	Norwegian Krone	挪威克郎
151	Oman	阿曼	RO	OMR	Omani Rial	阿曼里亚尔
152	Pakistan	巴基斯坦	PRe	PKR	Pakistani Rupee	巴基斯坦卢比
153	Palau	帕劳	$	USD	United States Dollar	美元
154	Panama	巴拿马	B	PAB	Panamanian Balboa	巴拿马巴波亚
155	Papua New Guinea	巴布亚新几内亚	K	PGK	Papua New Guinean Kina	巴布亚新几内亚基那
156	Paraguay	巴拉圭	c	PYG	Paraguayan Guarani	巴拉圭瓜拉尼
157	Peru	秘鲁	S/.	PEN	Peruvian Nuevo Sol	秘鲁新索尔
158	Philippines	菲律宾	₱	PHP	Philippine Peso	菲律宾比索
159	Poland	波兰	zł	PLN	Polish Zloty	波兰兹罗提
160	Portugal	葡萄牙	€	EUR	Euro	欧元
161	Puerto Rico	波多黎各	$	USD	United States Dollar	美元
162	Qatar	卡塔尔	QR	QAR	Qatar Riyal	卡塔尔里亚尔
163	Romania	罗马尼亚	L	RON	Romanian New Leu	罗马尼亚新列伊
164	Russia	俄罗斯		RUB	Russian Rouble	俄罗斯卢布

续表

序号	Country/Region	国家/地区	货币	国际标准代码	Currency	货币名称
165	Rwanda	卢旺达	RF	RWF	Rwandan Franc	卢旺达法郎
166	Saint Kitts and Nevis	圣克里斯托弗和尼维斯	EC$	XCD	East Caribbean Dollar	东加勒比元
167	Saint Lucia	圣卢西亚	EC$	XCD	East Caribbean Dollar	东加勒比元
168	Saint Vincent and the Grenadines	圣文森特和格林纳丁斯	EC$	XCD	East Caribbean Dollar	东加勒比元
169	Samoa	萨摩亚	WS$	WST	Samoan Tala	萨摩亚塔拉
170	San Marino	圣马力诺	€	EUR	Euro	欧元
171	Sao Tome and Principe	圣多美和普林西比	Db	STD	Dobra	多布拉
172	Saudi Arabia	沙特阿拉伯	SR	SAR	Saudi Riyal	沙特里亚尔
173	Senegal	塞内加尔	CFA	XOF	West African CFA Franc	西非法郎（非洲金融共同体法郎）
174	Serbia	塞尔维亚	RSD	RSD	Serbian Dinar	塞尔维亚第纳尔
175	Seychelles	塞舌尔	SR, SRe	SCR	Seychelles Rupee	塞舌尔卢比
176	Sierra Leone	塞拉利昂	Le	SLL	Sierra Leonean Leone	塞拉利昂利昂
177	Singapore	新加坡	S$	SGD	Singapore Dollar	新加坡元
178	Sint Maarten	圣马丁	NAƒ, Naf	ANG	Netherlands Antillean Guilder	荷属安的列斯盾
179	Slovakia	斯洛伐克	€	EUR	Euro	欧元
180	Slovenia	斯洛文尼亚	€	EUR	Euro	欧元
181	Solomon Islands	所罗门群岛	SI$	SBD	Solomon Islands Dollar	所罗门群岛元
182	Somalia	索马里	Sh. So.	SOS	Somali Shilling	索马里先令
183	Somaliland	索马里兰	SL. Sh.	None	Somaliland Shilling	索马里兰先令
184	South Africa	南非	R	ZAR	South African Rand	南非兰特

续表

序号	Country/Region	国家/地区	货币	国际标准代码	Currency	货币名称
185	South Ossetia	南奥塞梯		RUB	Russian Rouble	俄罗斯卢布
186	South Sudan	南苏丹		SSP	South Sudanese Pound	南苏丹镑
187	Spain	西班牙	€	EUR	Euro	欧元
188	Sri Lanka	斯里兰卡	SL Rs	LKR	Sri Lanka Rupee	斯里兰卡卢比
189	Sudan	苏丹		SDG	Sudanese Pound	苏丹镑
190	Suriname	苏里南	$	SRD	Surinamese Dollar	苏里南元
191	Swaziland	斯威士兰	L, E	SZL	Swazi Lilangeni	斯威士兰里兰吉尼
192	Sweden	瑞典	kr	SEK	Swedish Krona/Kronor	瑞典克朗
193	Switzerland	瑞士	Fr, SFr	CHF	Swiss Franc	瑞士法郎
194	Syria	叙利亚	LS, £S	SYP	Syrian Pound	叙利亚镑
195	Taiwan, Province of China	中国台湾地区	$, NT$	TWD	New Taiwan Dollar	新台币元
196	Tajikistan	塔吉克斯坦		TJS	Tajikistani Somoni	塔吉克斯坦索莫尼
197	Tanzania	坦桑尼亚		TZS	Tanzanian Shilling	坦桑尼亚先令
198	Thailand	泰国	B	THB	Thai Baht	泰铢
199	Togo	多哥	CFA	XOF	West African CFA Franc	西非法郎（非洲金融共同体法郎）
200	Tonga	汤加	T$	TOP	Tongan Pa'anga	汤加潘加
201	Trinidad and Tobago	特立尼达和多巴哥	TT$	TTD	Trinidad and Tobago Dollar	特立尼达和多巴哥元
202	Tunisia	突尼斯	DT	TND	Tunisian Dinar	突尼斯第纳尔
203	Turkey	土耳其	LT	TRY	Turkish Lira	土耳其里拉
204	Turkmenistan	土库曼斯坦	m	TMT	Turkmenistani Manat	土库曼斯坦马纳特

续表

序号	Country/Region	国家/地区	货币	国际标准代码	Currency	货币名称
205	Turks and Caicos Islands	特克斯和凯科斯群岛	$	USD	United States Dollar	美元
206	Tuvalu	图瓦卢	$, A$	AUD	Australian Dollar	澳大利亚元
207	U. S. Virgin Islands	美属维尔京群岛	$	USD	United States Dollar	美元
208	Uganda	乌干达	U Sh	UGX	Ugandan Shilling	乌干达先令
209	Ukraine	乌克兰	₴	UAH	Ukrainian Hryvnia	乌克兰格里夫纳
210	United Arab Emirates	阿拉伯联合酋长国	Dh	AED	United Arab Emirates Dirham	阿联酋迪拉姆
211	United Kingdom of Great Britain and N. Ireland (U. K.)	英国（大不列颠及北爱尔兰联合王国）	£	GBP	Pound Sterling	英镑
212	United States of America (U. S. A.)	美国（美利坚合众国）	$	USD	United States Dollar	美元
213	Uruguay	乌拉圭	$, $U	UYU	Uruguayan Peso	乌拉圭比索
214	Uzbekistan	乌兹别克斯坦		UZS	Uzbekistani Som	乌兹别克斯坦苏姆
215	Vanuatu	瓦努阿图	VT	VUV	Vanuatu Vatu	瓦努阿图瓦图
216	Vatican City	梵蒂冈城国	€	EUR	Euro	欧元
217	Venezuela	委内瑞拉	Bs. F., Bs.	VEF	Venezuelan Bolívar Fuerte	委内瑞拉强势玻利瓦尔
218	Vietnam	越南	D	VND	Vietnamese Đồng	越南盾
219	Wallis and Futuna	瓦利斯和富图纳群岛	F	XPF	CFP Franc	太平洋金融共同体法郎
220	Yemen, Republic of	也门共和国	YRI	YER	Yemeni Riyal (Rial)	也门里亚尔
221	Zambia	赞比亚	ZK	ZMK	Zambian Kwacha	赞比亚克瓦查
222	Zimbabwe	津巴布韦	$	ZWL	Zimbabwe Dollar	津巴布韦元

附录7 中国国家、省（市）政府部门名称汉英对照

中国国家机关
China's State Organs

1. 全国人民代表大会 National People's Congress (NPC)
主席团 Presidium
提案审查委员会 Motions Examination Committee
常务委员会委员长会议 Chairmen's Council
常务委员会 Standing Committee
代表资格审查委员会 Credentials Committee
专门委员会 Special Committee
民族委员会 Ethnic Affairs Committee
法律委员会 Law Committee
内务司法委员会 Committee for Internal and Judicial Affairs
财政经济委员会 Finance and Economy Committee
教育、科学、文化和卫生委员会 Education, Science, Culture and Public Health Committee
外事委员会 Foreign Affairs Committee
华侨委员会 Overseas Chinese Affairs Committee
环境与资源保护委员会 Environment Protection and Resources Conservation Committee
农业与农村委员会 Agriculture and Rural Affairs Committee
办公厅 General Office
秘书局 Bureau of Secretaries
联络局 Liaison Bureau
外事局 Foreign Affairs Bureau
新闻局 Information Bureau
信访局 Letters and Visits Reception Bureau
人事局 Personnel Bureau
机关事务管理局 Government Offices Administration
培训中心 Training Center
信息中心 Information Center
人民大会堂管理局 Administration Bureau for the Great Hall of the People
全国人大常委会工作委员会 Working and Administration Bodies of the Standing Committee
法制工作委员会 Legislative Affairs Commission
预算工作委员会 Budgetary Affairs Commission
香港基本法委员会 Hong Kong Special Administration Region Basic Law Committee
澳门基本法委员会 Macao Special Administration Region Basic Law Committee
中华人民共和国主席 President of the People's Republic of China
国务院 State Council
中央军事委员会 Central Military Commission
最高人民法院 Supreme People's Court
最高人民检察院 Supreme People's Procuratorate
特定问题调查委员会 Commission of Inquiry into Specific Questions

地方各级人民代表大会和地方各级人民政府　Local People's Congress and Local People's Government at Various Levels

民族自治地方的自治机关　Organs of Self-Government of National Autonomous Areas

纺织工业局　Bureau of Textile Industry

民航局　Civil Aviation Administration

2. **中华人民共和国主席**　President of the People's Republic of China

副主席　Vice President

3. **中央军事委员会**　Central Military Commission

主席　Chairman

副主席　Vice Chairman

委员　Members

4. **国务院组成部门**　Ministries and Commissions under the State Council

总理　Premier

副总理　Vice Premier

国务委员　State Councillor

秘书长　Secretary-General

外交部　Ministry of Foreign Affairs

亚洲司　Department of Asian Affairs

非洲司　Department of African Affairs

西亚北非司　Department of West Asian and North African Affairs

西欧司　Department of West European Affairs

东欧中亚司　Department of East European and Central Asian Affairs

美洲大洋洲司　Department of the Affairs of the Americas and Oceania

领事司　Department of Consular Affairs

礼宾司　Protocol Department

国际司　Department of International Organization and Conference

条法司　Department of Treaty and Law

新闻司　Information Department

国防部　Ministry of National Defense

国家发展和改革委员会　National Development and Reform Commission

教育部　Ministry of Education

科学技术部　Ministry of Science and Technology

工业和信息化部　Ministry of Industry and Information Technology

国家民族事物委员会　State Ethnic Affairs Commission

公安部　Ministry of Public Security

国家安全部　Ministry of State Security

监察部　Ministry of Supervision

民政部　Ministry of Civil Affairs

司法部　Ministry of Justice

财政部　Ministry of Finance

人力资源和社会保障部　Ministry of Human Resources and Social Security

国土资源部　Ministry of Land and Resources

环境保护部　Ministry of Environment Protection

住房和城乡建设部　Ministry of Housing and Urban-Rural Development

交通运输部　Ministry of Transport

铁道部　Ministry of Railways

水利部　Ministry of Water Resources

农业部　Ministry of Agriculture

商务部　Ministry of Commerce

文化部　Ministry of Culture

卫生部　Ministry of Public Health

国家卫生和计划生育委员会　National Health and Family Planning Commission

中国人民银行　People's Bank of China

国家审计署　State Auditing Administration
对外贸易经济合作部　Ministry of Foreign Trade and Economic Cooperation
自然资源部　Ministry of Natural Resources
生态环境部　Ministry of Ecological Environment
农业农村部　Ministry of Agriculture and Rural Affairs
应急管理部　Ministry of Emergency Management
交通部　Ministry of Communications
信息产业部　Ministry of Information Industry
统战部　United Front Work Department
公共关系部　Public Relations Department
新闻出版总署　General Administration of press and Publication
干部司　Personnel Department
人事部　Ministry of Personnel
劳动和社会保障部　Ministry of Labour and Social Security
建设部　Ministry of Construction

5. **国务院办事机构**　Administration Offices under the State Council
国务院侨务办公室　Overseas Chinese Affairs Office of the State Council
国务院港澳事务办公室　Hong Kong and Macao Affairs Office of the State Council
国务院法制办公室　Legislative Affairs Office of the State Council
国务院研究室　Research Office of the State Council
国务院经济体制改革办公室　Economic Restructuring office of the State Council

6. **国务院直属特设机构**　Special Organization Directly under the State Council
国务院国有资产监督管理委员会　State-owned Assets Supervision and Administration Commission of the State Council

7. **国务院直属机构**　Organizations Directly under the State Council
中华人民共和国海关总署　General Administration of Customs
国家税务总局　State Administration of Taxation
国家工商行政管理总局　State Administration for Industry and Commerce
国家质量监督检验检疫总局　General Administration of Quality Supervision, Inspection and Quarantine
国家体育总局　General Administration of Sports
国家安全生产监督管理总局　State Administration of Work Safety
国家统计局　National Bureau Statistics
国家林业局　State Forestry Administration
国家知识产权局　State Intellectual Property Office (SIPO)
国家旅游局　National Tourism Administration
国家宗教事务局　State Administration for Religious Affairs
国务院参事室　Counselors' Office of the State Council
国务院机关事务管理局　Government

Offices Administration of the State Council
国家预防腐败局　National Bureau of Corruption Prevention

8. 国务院直属事业单位　Institutions Directly under the State Council
新华通讯社　Xinhua News Agency
中国科学院　Chinese Academy of Sciences
中国社会科学院　Chinese Academy of Social Sciences
中国工程院　Chinese Acadcmy of Engineering
国务院发展研究中心　Development Research Centre of the State Council
国家行政学院　National School of Administration
中国地震局　China Seismological Bureau
中国气象局　China Meteorological Bureau
中国证券监督管理委员会　China Securities Regulatory Commission (CSRC)
中国银行业监督管理委员会　China Banking Regulatory Commission
中国保险监督管理委员会　China Insurance Regulatory Commission
国家电力监管委员会　State Electricity Regulatory Commission
全国社会保障基金理事会　National Council for Social Security Fund
国家自然科学基金委员会　National Natural Science Foundation
国务院台湾事务办公室　Taiwan Affair Office of the State Council
国务院新闻办公室　Information Office of the State Council
国家档案局　State Archives Administration

中国民用航空总局　Greneral Administration of Civil Aviation of China
国家发展计划委员会　State Development Planning Commission
国家经济贸易委员会　State Economic and Trade Commission
体育运动委员会　Physical Culture and Sports Commission

9. 国务院部委管理的国家局　State Bureau Administrated by Ministries or Commissions
国家信访局　State Bureau for Letters and Calls
国家粮食局　State Administration of Grain
国家能源局　National Energy Board
国家国防科技工业局　State Administration of Science, Technology, and Industry for National Defense
国家烟草专卖局　State Tobacco Monopoly Administration
国家外国专家局　State Administration of Foreign Experts Affairs
国家公务员局　State Bureau of Civil Servants
国家海洋局　State Oceanic Administration
国家测绘地理信息局　State Bureau of Surveying and Mapping
中国民用航空局　Civil Aviation Administration of China
国家邮政局　State Post Bureau
国家文物局　State Administration of Cultural Heritage
国家中医药管理局　State Administration of Traditional Chinese Medicine
国家外汇管理局　State Administration

of Foreign Exchange
国家国内贸易局　State Bureau of Internal Trade
国家煤矿安全监察局　State Administration of Coal Mine Safety
国家保密局　National Administration for Protection of State Secrets
国家密码管理局　State Cryptography Administration
国家航天局　China National Space Administration
国家原子能机构　China Atomic Energy Authority
国家语言文字工作委员会　State Language Commission
国家核安全局　National Nuclear Safety Administration
国家铁路局　State Railway Administration
国家新闻出版广电总局　The State Press and Publication Administration
国家版权局　People's Copyright Administration
国家广播电影电视总局　State Administration of Radio, Film and Television
国家工商行政管理局　State Administration for Industry and Commerce (SAIC)
国家环境保护总局　State Environmental Protection Administration (SEPA)
国家药品监督管理局　State Drug Administration (SDA)
国家安全生产监督管理局　State Administration of Work Safety
10. 党中央各部门　Departments and Committees of the Party Central Committee
中国中央委员会　Central Committee of the Communist Party of China
中央政治局　Political Bureau of the Central Committee
中央政治局常务委员会　Standing Committee of the Political Bureau of the Central Committee
中央书记处　Secretariat of the Central Committee
中央整党工作指导委员会　Central Party Consolidation Guidance Commission
中共中央办公厅　General Office of the CPC Central Committee
中央组织部　Organization Department of the CPC Central Committee
中央宣传部　Propaganda Department of the CPC Central Committee
中央统一战线部　United Front Work Department of the CPC Central Committee
中央对外联络部　International Liaison Department of the CPC Central Committee
中央军事委员会　Military Commission of the CPC Central Committee
中央党的建设工作领导小组　Party Building Directorate of the CPC Central Committee
中央宣传思想工作领导小组　Propaganda and Ideological Work
中央政法委员会　Commission of Politics and Law of the CPC Central Committee
中央爱国卫生运动委员会　Central Patriotic Public Health Campaign Committee
中央党校　Party School of the CPC

Central Committee
中央绿化委员会　Central Afforestation Committee
中央文献研究室　Party Literature Research Centre of the CPC Central Committee
中央政策研究室　Policy Research Centre of the CPC Centre Committee
中央党史研究室　Party History Research Centre of the CPC Central Committee
中央历史人物研究会　Society on the Historical Figure of the CPC
中央文献编辑委员会　Editorial Committee on Party Literature of the Central Committee
中央档案馆　Central Archives
中央国家机关工作委员会　Work Committee of Central Government Departments
中共中央直属机关工作委员会　Work Committee of Department under the Central Committee
中共中央纪律检查委员会　Central Commission for Discipline Inspection of the CPC

11. **军事机构**　Military Establishments
中国人民解放军　Chinese People's Liberation Army (PLA)
中央军事委员会　Central Military Commission of the P.R.C
总参谋部　Headquarters of the General Staff
总政治部　General Political Department
总后勤部　General Logistics Department
中国人民解放军军事法院　PLA Military Court
中国人民解放军军事检察院　PLA Military Procuratorate
退役军人事务部　Ministry of Veterans Affairs

香港政府机构
Organs of Hong Kong Special Administration Region

安全局　Security Bureau
保险特派员处　Office of the Commissioner of Insurance, OCI
财政部门　Department of Finance
财政服务局　Financial Servies Bureau
财政局　Finance Bureau
城市服务部　Urban Services Department
城市委员会　Urban Council
创新与技术主管执行委员会　Chief Executive's Commission on Innovation and Technology
大学专项委员会　University Grants Committee (UGC)
登记和选举办公厅　Registration and Electoral Office
地区发展部　Territory Development Department (TDD)
地区服务部　Regional Services Department
地区委员会　Regional Council
电力和机械服务部　Electrical and Mechanical Services Department (EMSD)
电视和娱乐许可部门　Television and Entertainment Licensing Authority (TELA)
调查统计部　Census and Statistics Department (C&SD)
法律援助部　Legal Aid Department

法律援助服务委员会　Legal Aid Services Council (LASC)
反腐独立委员会（廉政公署）　Independent Commission Against Corruption (ICAC)
房屋管理部门　Housing Authority and Housing Department
房屋局　Housing Bureau
港口处　Post Office
工务局　Works Bureau
工业部　Industry Department
公共服务委员会　Public Service Commission
公路部　Highways Department
公司登记　Companies Registry
官方语言机构　Official Languages Agency (OLA)
管理服务部　Management Services Agency (MSA)
广播部门　Broadcasting Authority (BA)
国库　Treasury
国内航空部　Civil Aviation Department (CAD)
海运部　Marine Department
环境保护部　Environment Protection Department (EPD)
环境资讯委员会　Advisory Council on the Environment (ACE)
机场部　Airport Authority (AA)
机会平等委员会　Equal Opportunities Commission (EOC)
计划、环境和土地局　Planning, Environment and Lands Bureau
计划部　Planning Department
家庭事务部门　Home Affairs Department
家庭事务局　Home Affairs Bureau
监察处　Office of the Ombudsman

建设标准委员会　Construction Standards Committee
建设部门　Building Department
建筑服务部　Architectural Services Department
教育部门　Education Department
教育和人力资源局　Education and Manpower Bureau (EMB)
教育委员会　Education Commission (EC)
经济服务局　Economic Services Bureau
就业再培训局　Employees Retraining Board (ERB)
劳动部门　Labour Department
礼宾处　Official Receiver's Office (ORO)
立法委员会　Legislative Council
立宪事务局　Constitutional Affairs Bureau (CAB)
贸易部　Trade Department
贸易发展委员会　Trade Development Council (TDC)
贸易和工业局　Trade and Industry Bureau
民事服务局　Civil Service Bureau (CSB)
民事服务培训和发展研究所　Civil Service Training and Development Institute (CSTDI)
内税部　Inland Revenue Department (IRD)
农渔部　Agriculture and Fisheries Department (AFD)
排水服务部　Drainage Services Department (DSD)
评估部　Rating and Valuation Department (RVD)
商务和服务促进机构　Business and Services Promotion Unit

社会福利部门　Social Welfare Department
审计委员会　Audit Commission
水供应部　Water Supplies Department (WSD)
税务委员会　Customs and Excise Department
司法部　Department of Justice
通信部门　Office of the Telecommunications Authority (OFTA)
土地部　Lands Department
土地登记　Land Reqistry
土木工程部　Civil Engineering Department (CED)
卫生部门　Department of Health
卫生和福利局　Health and Welfare Bureau
吸烟与健康委员会　Council on Smoking and Health (COSH)
香港安全与未来委员会　Securities and Futures Commission of Hong Kong (SFC)
香港财政局　Hong Kong Monetary Authority (HKMA)
香港港口和海洋局　Hong Kong Port and Maritime Board (PMB)
香港广播电视　Radio Television Hong Kong (RTHK)
香港考试部门　Hong Kong Examinations Authority (HKEA)
香港旅游协会　Hong Kong Tourist Association (HKTA)
香港气象台　Hong Kong Observatory
香港生产力委员会　Hong Kong Productivity Council (HKPC)
香港特别行政区政府　Government of the Hong Kong Special Administration Region
香港体育发展局　Hong Kong Sports Development Board (SDB)
消费者委员会　Consumer Council
信息服务部门　Information Services Department (ISD)
信息技术服务部门　Information Technology and Broadcasting Bureau
行政部门　Department of Administration
选举事务委员会　Electoral Affairs Commission (EAC)
学生财政资助局　Student Financial Assistance Agency (SFAA)
医院部门　Hospital Authority (HA)
医院服务部门　Hospital Services Department (HSD)
印刷部　Printing Department
运输部　Transport Department (TD)
运输局　Transport Bureau
政府财产部　Government Property Agency (GPA)
政府供应部　Government Supplies Department (GSD)
政府陆地运输部　Government Land Transport Agency (GLTA)
政府实验室　Government Laboratory
知识产权部　Intellectual Property Department (IPD)
职业安全和保健委员会　Occupational Safety and Health Councail (OSHC)
职业培训委员会　Vocational Training Council (VTC)

地方机构（省级）

××省人民政府　People's Government of ×× Province
广西壮族自治区人民政府　The People's

Government of Guangxi Zhuang Autonomous Region
内蒙古自治区人民政府　The People's Government of Inner Mongolia Autonomous Region
宁夏回族自治区人民政府　The People's Government of Ningxia Hui Autonomous Region
西藏自治区人民政府　The People's Government of Tibet Autonomous Region
新疆维吾尔自治区人民政府　The People's Government of Xinjiang Uygur Autonomous Region
省政府办公厅　Provincial General Office
自治区政府办公厅　Autonomous Regional General Office
财政厅　Finance Department
地方税务局　Local Taxation Bureau
对外贸易经济合作厅　Foreign Trade and Economic Cooperation Department
发展和改革委员会　Development and Reform Commision
工商行政管理局　Industry and Commerce Administration
工业和信息化委员会　Industry & Information Commission
公安厅　Public Security Department
国有资产监督管理委员会　Stateowned Assets Supervision and Administration
环境保护厅　Environmental Protection Bureau
建设厅　Construction Department
交通厅　Communications Department
交通运输厅　Communications Department

教育厅　Education Department
经济和信息化委员会　Economic and Information Commission
科学技术厅　Science and Technology Department
林业厅　Forest Department
旅游局　Tourism Department
农牧厅　Agriculture and Animal Husbandry Department
农业厅　Agriculture Department
农业委员会　Agriculture Commission
侨务办公室　Overseas Chinese Affairs Office
商务厅　Commerce Department
食品药品监督管理局　Food and Drug Administration
水利厅　Water Resources Department
司法厅　Justice Department
外事办公室　Foreign Affairs Office
外事侨务办公室　Foreign Affairs and Overseas Chinese Affairs Office
卫生厅　Health Department
文化厅　Culture Department
新闻办公室　Information Office
新闻出版局（版权局）　Press and Publication Bureau (Copyright Bureau)
信息产业厅　Information Industry Department
质量技术监督局　Quality and Technical Supervision Bureau
住房和城乡建设厅　Department of Housing and Urban-Rural Development

地方机构（直辖市）

××安全生产监督管理局　×× Municipal Administration of Work Safety
××财政局　×× Municipal Bureau of

Finance

××城乡建设和交通委员会 ×× Municipal Urban and Rural Construction and Transportation Commission

××地方税务局 ×× Municipal Bureau of Local Taxation

××发展和改革委员会 ×× Municipal Commission of Development and Reform

××法制办公室 Legal Affairs Office of the People's Government of ×× Municipality

××工商行政管理局 ×× Municipal Administration of Industry and Commerce

××工业发展局 ×× Municipal Bureau of Industry Development

××公安局 ×× Municipal Bureau of Public Security

××公务员局 ×× Administration of Civil Service

××广播电影电视剧 ×× Municipal Bureau of Radio and Television

××规划和国土资源管理局 ×× Municipal Planning, Land and Resources Administration

××规划委员会 ×× Municipal Commission of City Planning

××国土资源局 ×× Municipal Bureau of Land and Resources

××国有资产监督管理委员会 State-Owned Assets Supervision and Administration Commission of the People's Government of ×× Municipality

××环境保护局 ×× Municipal Bureau of Environmental Protection

××监察局 ×× Municipal Bureau of Supervision

××监狱管理局 ×× Administration Bureau of Prison

××交通委员会 ×× Municipal Committee of Transportation

××交通运输和港口管理局 ×× Municipal Transport and Port Authority

××教育委员会 ×× Municipal Commission of Education

××金融服务办公室 ×× Municipal Office of Finance Service

××经济和信息化委员会 ×× Municipal Commission of Economy and Information Technology

××科学技术委员会 ×× Municipal Commission of Science and Technology

××口岸服务办公室 ×× Municipal Office for Port Services

××劳动和社会保障局 ×× Municipal Bureau of Labor and Social Security

××粮食局 ×× Municipal Grain Bureau

××旅游管理局 ×× Municipal Administration of Tourism

××旅游管理委员会 ×× Municipal Tourism Administration Commission

××绿化和市容管理局 ×× Municipal Greening and City Administration

××民防办公室 ×× Municipal Civil Defense Office

××民防局 ×× Municipal Bureau of Civil Air Defense

××民政局 ×× Municipal Bureau of Civil Affairs

××民族和宗教事务委员会 ×× Municipal Commission of Ethnic and Religious Affairs

××民族事务委员会 ×× Municipal Commission of Ethnic Affairs

××农村工作委员会 ×× Municipal Commission of Rural Affairs

××农业局 ×× Municipal Bureau of Agriculture

××农业委员会 ×× Municipal Agriculture of Commission

××侨务办公室 ×× Overseas Chinese Affairs Office of the People's Government of ×× Municipality

××人口和计划生育委员会 ×× Municipal Commission of Population and Family Planning

××人力资源和社会保障局 ×× Municipal Human Resources and Social Security Bureau

××人力资源局 ×× Municipal Bureau of Human Resources

××人民政府发展研究中心 Development Research Center of the ×× Municipal People's Government

××人民政府法制办公室 The Legislative Affairs Office of ×× Municipal People's Government

××人民政府合作交流办公室 Office for Cooperation and Exchange Affairs of the ×× Municipal People's Government

××人民政府新闻办公室 Information Office of ×× Municipality

××商务委员会 ×× Municipal Bureau of Commerce

××社会团体管理局 ×× Administration Bureau of Non-government Organizations (NGOs)

××审计局 ×× Municipal Audit Bureau

××食品药品监督管理局 ×× Municipal Food and Drug Supervision Administration

××市人民政府外事办公室 Foreign Affairs Office of the ×× Municipal People's Government

××市政市容管理委员会 ×× Municipal Commission of City Administration

××水务局(××海洋局) ×× Municipal Water Authority (×× Municipal Ocean Bureau)

××司法局 The ×× Municipal Bureau of Justice

××体育局 ×× Municipal Bureau of Sports

××统计局 The ×× Municipal Bureau of Statistics

××外事办公室 Foreign Affairs Office of the People's Government of ×× Municipality

××卫生局 ×× Municipal Bureau of Health

××文化广播影视管理局 ×× Municipal Culture, Radio Broadcasting, Film and Television Administration

××文化局 ×× Municipal Bureau of Culture

××文物局 ×× Municipal Bureau of Cultural Heritage

××新闻出版局 ×× Municipal Bureau of Press and Publication

××信访办公室 Office of Letters and Calls of ×× Committee of the Communist party of China and People's Government of ×× Municipality

××信息技术办公室 ×× Municipal Office of Information Technology

××研究室　Research Office of the People's Government of ×× Municipality
××药监局　×× Municipal Drug Administration
××园林绿化局　×× Municipal Bureau of Forestry and Parks
××政府办公厅　×× Municipal General Office
××政府机关事务管理局　General Offices Administration of the ×× Municipal People's Government
××知识产权局　×× Municipal Intellectual Property Bureau
××质量技术监督局　×× Municipal Administration of Quality and Technology Supervision
××住房保障和房屋管理局　×× Municipal Housing, Land and Resources Administration
××住房和城乡建设委员会　×× Municipal Commission of Construction

其 他

行政区划　administration division
特别行政区　Special Administation Region
直辖区　municipality
自治区　autonomous region
经济特区　special economic zone (SEZ)
地级市　prefecture-level city
县级市　country-level city
自治州　autonomous prefecture
自治县　autonomous county
盟　league (prefecture)
旗　banner (county)
市直辖区　sub-municipal district
市直辖县　sub-municipal county
市　city
县级市　county
乡镇　town,township
村　village
对外开放城市　city opened to the outside world

附录8 主要参考文献

[1] 杜振华. 新英汉汉英电力工程技术词典. 北京：中国电力出版社，2013.

[2] 中国电力名词审定委员会. 电力名词. 第2版. 北京：科学出版社，2009.

[3] 中国编委会. 英俄汉电力工业大词典. 北京：中国电力出版社，2008.

[4] Black & Veatch. Power plant Engineering. Chapman & Hall, 1996.

[5] A.K. Raja. Power plant Engineering. New AGE International (P) Ltd, Publishers, New Delhi, 2006.

[6] Phillip A. Laplante. Electrical Engineering Dictionary. CRC Press LLC, 2000.

[7] Tony Burton. Wind Energy Handbook. John Wiley & Sons, Ltd, Baffins Lane, Chichester, West Sussex, PO19 1UD, England, 2001.

[8] Daniel N. Lapedes. Dictionary of Scientific and Technical Terms. McGraw-hill, 1974.

[9] Paul Procter. Longman Dictionary of Contemporary English. Longman Group Ltd, 1995.

[10] P. Breeze. The Future of Global Biomass Power Generation. MBA Group Ltd, London, 2004.

[11] A. Goetzberger & V. U. Hoffmann. Photovoltaic Solar Energy Generation. Springer, New York, 2005.

[12] L.L. Grigsby. The Electric Power Engineering Handbook. Boca Raton: CRC Press, 2001.

[13] John Daintith & E.A. Martin. A Dictionary of Science. Oxford University Press, New York, 2005.

[14] V. Nelson. Wind Energy: Renewable Energy and the Environment. CRC Press, Boca Raton, 2009.

[15] D. Pimentel. Biofuels, Solar and Wind as Renewable Energy Systems. Springer, New York, 2008.

[16] V. Quaschning. Understanding Renewable Energy Systems. Earthscan, London, 2005.

[17] 张兴. 太阳能光伏并网发电及其逆变控制. 北京：机械工业出版社，2011.

[18] 刘万琨. 风能与风力发电技术. 北京：化学工业出版社，2007.

[19] 刘德辉，彭志平. 英汉能源辞典. 北京：能源出版社，1989.

[20] 孙岐昆. 英汉新能源技术词典. 上海：上海交通大学出版社，1989.

[21] 王庆一. 能源词典. 第2版. 北京：中国石化出版社，2005.

[22] 霍志臣，张茂，王君一，徐任学. 英汉新能源词典. 北京：电子工业出版社，1989.

[23] 林景尧，于志成. 英汉能源技术词典. 北京：中国劳动出版社，1991.

[24] 梁维燕. 英汉电站工程辞典. 黑龙江：黑龙江人民出版社，1991.
[25] 清华大学外语系《英汉科学技术词典》编写组. 英汉科学技术词典. 北京：国防工业出版社，1989.
[26] 陆谷孙. 英汉大词典. 上海：上海译文出版社，1993.